MESOSCOPIC PHYSICS OF ELE(

Quantum mesoscopic physics covers a whole class of interference effects related propagation of waves in complex and random media. These effects are ubiquitous in physics, from the behavior of electrons in metals and semiconductors to the propagation of electromagnetic waves in suspensions such as colloids, and quantum systems like cold atomic gases. This book is a modern account of the problem of coherent wave propagation in random media.

As a solid introduction to quantum mesoscopic physics, this book provides a unified overview of the basic theoretical tools and methods. It highlights the common aspects of the various optical and electronic phenomena involved. With over 200 figures, and exercises throughout, the book is ideal for graduate students in physics, electrical engineering, optics, acoustics and astrophysics. It presents a large number of experimental results that cover a wide range of phenomena from semiconductors to optics, acoustics, and atomic physics. It will also be an important reference for researchers in this rapidly evolving field.

ERIC AKKERMANS is Professor of Physics in the Department of Physics at the Technion, Israel Institute of Technology, Israel. GILLES MONTAMBAUX is Directeur de Recherche at the CNRS, Laboratoire de Physique des Solides, Université Paris-Sud, France. Their research interests include the theory of condensed matter physics, mesoscopic quantum physics, and coherent effects in the propagation of waves in random media.

Mesoscopic Physics of Electrons and Photons

Eric Akkermans
Technion, Israel Institute of Technology

Gilles Montambaux
CNRS, Université Paris-Sud

CAMBRIDGE UNIVERSITY PRESS
Cambridge, New York, Melbourne, Madrid, Cape Town,
Singapore, São Paulo, Delhi, Tokyo, Mexico City

Cambridge University Press
The Edinburgh Building, Cambridge CB2 8RU, UK

Published in the United States of America by Cambridge University Press, New York

www.cambridge.org
Information on this title: www.cambridge.org/9780521349475

First published 2007
First paperback edition 2011

A catalogue record for this publication is available from the British Library

ISBN 978-0-521-85512-9 Hardback
ISBN 978-0-521-34947-5 Paperback

Contents

Preface

Wave propagation in random media has been the subject of intense activity for more than two decades. This is now an important area of research, whose frontiers are still fuzzy, and which includes a variety of problems such as wave localization (weak and strong), mesoscopic physics, effects of electron–electron interactions in metals, etc. Moreover, since many disorder effects are not truly specific to a given kind of wave, various approaches have been developed independently in condensed matter physics, in optics, in atomic physics and in acoustics.

A large number of monographs or review articles already exist in the literature and they cover in detail various aspects of the field. Our aim is rather to present the basic common features of the effects of disorder on wave propagation and also to provide the non-specialist reader with the tools necessary to enter and practice this field of research.

Our first concern has been to give a description of the basic physical effects using a single formalism independent of the specific nature of the waves (electrons, electromagnetic waves, etc.). To this purpose, we have started with a detailed presentation of "single-particle" average quantities such as the density of states and elastic collision time using the framework of the so-called "Gaussian model" for the two most important examples of waves, namely Schrödinger and scalar Helmholtz wave equations. We have tried, as much as possible, to make precise the very basic notion of multiple scattering by an ensemble of independent effective scatterers whose scattering cross section may be obtained using standard one-particle scattering theory.

Nevertheless, the quantities of physical interest that are accessible experimentally and used to describe wave propagation in the multiple scattering regime depend essentially on the probability of quantum diffusion which describes the propagation of a wave packet. This probability thus plays a central role and Chapter 4 is devoted to its detailed study. We then see emerging notions such as classical (Diffuson) and coherent (Cooperon) contributions to the probability, which provide basic explanations of the observed physical phenomena such as weak localization corrections to electronic transport, negative magnetoresistance in a magnetic field, coherent backscattering of light, as well as universal conductance fluctuations, optical speckles and mesoscopic effects in orbital magnetism.

It thus happens that all these effects result from the behavior of a single quantity, namely the probability of quantum diffusion. However, in spite of the common background shared by optics and electronics of random media, each of these domains has its own specificity which allows us to develop complementary approaches. For instance, the continuous change in the relative phases of electronic wave functions that can be achieved using a magnetic field or a vector potential has no obvious equivalent in optics. On the other hand, it is possible in optics to change directions of incident and outgoing beams and from this angular spectroscopy to trace back correlations between angular channels.

We have made a special effort to try to keep this book accessible to the largest audience, starting at a graduate level in physics with an elementary acquaintance of quantum mechanics as a prerequisite. We have also skipped a number of interesting but perhaps too specialized issues among which are the study of quantum dots, relations between spectral and transport quantities, strong localization and the Anderson metal–insulator transition, electronic ballistic billiards where "quantum complexity" does not result from disorder but instead from the boundary shape, and metal–superconductor interfaces. All these aspects reflect the richness of the field of "quantum mesoscopic physics" to which this book constitutes a first introduction.

A pleasant task in finishing the writing of a book is certainly the compilation of acknowledgments to all those who have helped us at various stages of the elaboration and writing, either through discussions, criticisms and especially encouragement and support: O. Assaf, H. Bouchiat, B. Huard, J. Cayssol, C. Cohen-Tannoudji, N. Dupuis, D. Estève, A. Georges, S. Guéron, M. Kouchnir, R. Maynard, F. Piéchon, H. Pothier, B. Reulet, B. Shapiro, B. van Tiggelen, D. Ullmo, J. Vidal, E. Wolf. We wish to single out the contribution of C. Texier for his endless comments, suggestions, and corrections which have certainly contributed to improve the quality of this book. Dov Levine accepted to help us in translating the book into English. This was a real challenge and we wish to thank him for his patience. We also wish to thank G. Bazalitsky for producing most of the figures with much dedication.

This venture was in many respects a roller-coaster ride and the caring support of Anne-Marie and Tirza was all the more precious.

Throughout this book, we use the (SI) international unit system, except in Chapter 13. The Planck constant $\hbar$ is generally taken equal to unity, in particular throughout Chapter 4. In the chapters where we think it is important to restore it, we have mentioned this at the beginning of the chapter. In order to simplify the notation, we have sometimes partially restored $\hbar$ in a given expression, especially when the correspondence between energy and frequency is straightforward.

To maintain homogeneous and consistent notation throughout a book which covers fields that are usually studied separately is a kind of challenge that, unfortunately, we have not always been able to overcome.

We have chosen not to give an exhaustive list of references, but instead to quote papers either for their obvious pedagogical value or because they discuss a particular point presented for instance as an exercise.

How to use this book

This book is intended to provide self-contained material which will allow the reader to derive the main results. It does not require anything other than an elementary background in general physics and quantum mechanics.

We have chosen to treat in a parallel way similar concepts occurring in the propagation of electrons and light. The important background concepts are given in Chapter 4, where the notion of *probability of quantum diffusion* in random media is developed. This is a central quantity to which all physical quantities described in this book may be related.

This book is not intended to be read linearly. We have structured it into chapters which are supposed to present the main concepts, and appendices which focus on specific aspects or details of calculation. This choice may sometimes appear arbitrary. For example, the Landauer formalism is introduced in an appendix (A7.2), where it is developed for the diffusive regime, which to our knowledge has not been done in the textbook literature. The standard description of weak localization is presented within the Kubo formalism in the core of Chapter 7, while the Landauer picture of weak localization is developed in Appendix A7.2.

We suggest here a guide for lectures. Although we have tried to emphasize analogies between interference effects in the propagation of electrons and light, we propose two outlines, for two introductory courses respectively on the physics of electrons and the physics of light. We believe that, during the course of study, the interested reader will benefit from the analogies developed between the two fields, for example the relations between speckle fluctuations in optics and universal conductance fluctuations in electronics.

Quantum transport in electronics
Main course

1 *Introduction: mesoscopic physics* Provides a unified and general description of interference and multiple scattering effects in disordered systems. Introduces the physical problems and the main quantities of interest, the different length scales such as the mean free path and the phase coherence length, the notions of multiple scattering and disorder average. Relates the physical properties to the probability of returning to the origin in a random medium. Notion of quantum crossing. Analogies between electronics and optics.

2 *Wave equations and models of disorder* Schrödinger equation for electrons in solids and Helmholtz equation for electromagnetic waves. Gaussian, Edwards, Anderson models for disorder.
3 *Perturbation theory* Presents the minimal formalism of Green's functions necessary for the notions developed further in the book. Multiple scattering and weak disorder expansion.
4.1–4.6 *Probability of quantum diffusion* Definition and description of essential concepts and tools used throughout the book. Iterative structure for the quantum probability, solution of a diffusion equation. Diffuson and Cooperon contributions. Formalism developed in real space. May also be useful to look at the reciprocal space formalism developed in Appendix A4.1.
6 *Dephasing* Proposes a general picture for dephasing and describes several mechanisms due to electron coupling to external parameters or degrees of freedom: magnetic field, Aharonov–Bohm flux, spin-orbit coupling and magnetic impurities. May be skipped at the introductory level, except for magnetic field and Aharonov–Bohm effects.
7 *Electronic transport* Deals with calculations of the average conductivity and of the weak localization correction. The latter is related to the probability of return to the origin for a diffusive particle. Applications to various geometries, plane, ring, cylinder, dimensionality effects. Section A7.2 is a comprehensive appendix on the Landauer formalism for diffusive systems.
10 *Spectral properties of disordered metals* Generalities on random matrix theory. Spectral correlation functions for disordered systems. The last part requires knowledge of correlation functions calculated in Appendix A4.4.
11 *Universal conductance fluctuations* Detailed calculation of the conductance fluctuations in the Kubo formalism, using the diagrammatics developed in Chapters 4 and 7. Many physical discussions on the role of various external parameters.
13 *Interactions and diffusion* Important chapter on the role of electron–electron interaction and its interplay with disorder. Density of states anomaly, correction to the conductivity. Important discussions about lifetime of quasiparticles and phase coherence time.
14 *Persistent currents* Can be considered optional. Thermodynamics and orbital magnetism of mesoscopic systems. Problematics of persistent currents, from the very simple one-dimensional description to the effect of disorder and interaction.

Optional

5 *Properties of the diffusion equation* Provides a comprehensive and self-contained account of properties of the diffusion equation. Diffusion in finite systems, boundary conditions, diffusion on graphs.
Miscellaneous Various appendices are beyond an introductory level, or are not necessary in a first course on mesoscopic physics. They either develop technicalities such as Hikami boxes (A4.2), Cooperon in a time dependent magnetic field (A6.3), or important

extensions such as anisotropic collisions developed in A4.3 and their effect on weak localization (A7.4) and universal conductance fluctuations (A11.1). The Landauer formalism for diffusive systems is developed in A7.2 for the average conductance and the weak localization correction and in A11.2 for conductance fluctuations.

Propagation of light in random media
Main course

This course provides a comprehensive introduction to the propagation of light in random media. It describes coherent effects in multiple scattering: coherent backscattering, diffusing wave spectroscopy and angular and time correlations of speckle patterns. Compared to coherent electronic transport, this course emphasizes notions specific to electromagnetic waves such as angular correlations of transmission (or reflection) coefficients in open space geometry, correlation between channels in a wave guide geometry, as well as the effects of the dynamics of scatterers.

1–4 These chapters are common to the two courses. In addition section 4.6 introduces the important formalism of radiative transfer which is developed in Appendix A5.2.

6 *Dephasing* Generalities on the mechanism of dephasing. Application to the polarization of electromagnetic waves, dynamics of the scatterers and dephasing associated with quantum internal degrees of freedom for the case of scattering of photons by cold atoms (the last topic is treated in Appendix A6.5).

8 *Coherent backscattering of light* Physics of the albedo, reflection coefficient of a diffusive medium. Coherent contribution (Cooperon) to the albedo, and its angular dependence. Uses the formalism developed in Chapter 4. Polarization and absorption effects (see also section 6.6). Extensive discussion of experimental results and coherent backscattering in various physical contexts. This chapter relies upon the results of section 5.6.

9 *Diffusing wave spectroscopy* An experimental technique routinely used to probe the dynamics of scatterers. Calculations result from a simple generalization of the formalism of Chapter 4. Interesting conceptually since diffusing wave spectroscopy exhibits the simplest example of decoherence introduced in a controlled way. Also interesting because this method probes the distribution of multiple scattering trajectories (reflection versus transmission experiments). Study of sections 5.6 and 5.7 is recommended.

12 *Correlations of speckle patterns* Analysis of a speckle pattern. Angular correlations of transmission coefficients. Classification and detailed calculation of the successive contributions C_1, C_2 and C_3. Simple description in terms of quantum crossings. Rayleigh law. Use of the Landauer formalism to relate speckle correlations to universal conductance fluctuations.

Optional

5 *Properties of the diffusion equation* Solutions of the diffusion equation in quasi-one-dimensional geometries, useful for calculations developed in Chapters 8, 9 and 12. Important appendix A5.2 on radiative transfer.

Miscellaneous Various appendices are useful reminders for beginners, for example A2.1 on scattering theory and A2.3 on light scattering by individual scatterers (Rayleigh, Rayleigh–Gans, Mie, resonant). Other appendices go beyond a course at the introductory level, either because they develop additional technicalities such as Hikami boxes (A4.2), useful for the reader interested in detailed calculations of Chapter 12, or because they present additional aspects of multiple scattering of light by random media such as spatial correlations of light intensity (A12.1) or anisotropic collisions (A4.3) and their consequences. The Landauer formalism for diffusive systems is used extensively in Chapter 12 on speckle correlations. Appendix A6.5 gives an overview of the technical tools needed to study the specific problem of multiple scattering of photons by cold atoms.

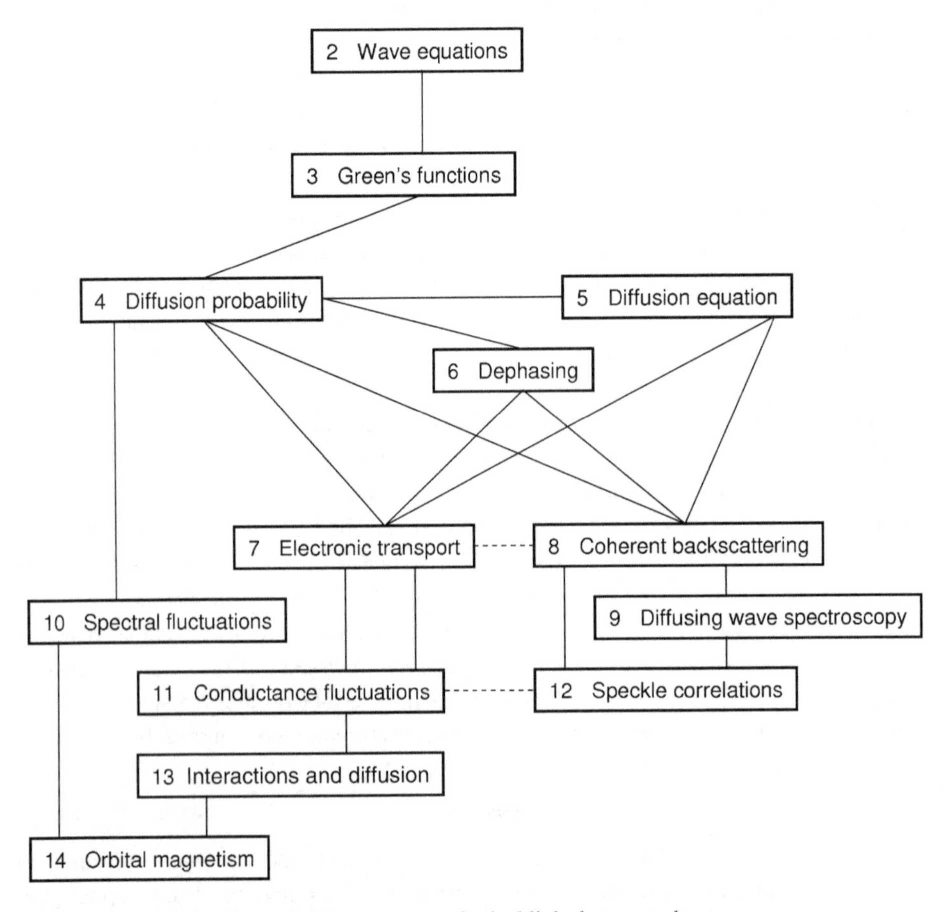

Topics developed in this book. Lines represent logical links between chapters.

1

Introduction: mesoscopic physics

1.1 Interference and disorder

Wave propagation in a random medium is a phenomenon common to many areas of physics. There has been a recent resurgence of interest following the discovery, in both optics and quantum mechanics, of surprising coherent effects in a regime in which disorder was thought to be sufficiently strong to eliminate a priori all interference effects.

To understand the origin of these coherent effects, it may be useful to recall some general facts about interference. Although quite spectacular in quantum mechanics, their description is more intuitive in the context of physical optics. For this reason, we begin with a discussion of interference effects in optics.

Consider a monochromatic wave scattered by an obstacle of some given geometry, e.g., a circular aperture. Figure 1.1 shows the diffraction pattern on a screen placed infinitely far from the obstacle. It exhibits a set of concentric rings, alternately bright and dark, resulting from constructive or destructive interference. According to Huygens' principle, the intensity at a point on the screen may be described by replacing the aperture by an ensemble of virtual coherent point sources and considering the difference in optical paths associated with these sources. In this way, it is possible to associate each interference ring with an integer (the equivalent of a quantum number in quantum mechanics).

Let us consider the robustness of this diffraction pattern. If we illuminate the obstacle by an incoherent source for which the length of the emitted wave trains is sufficiently short that the different virtual sources are out of phase, then the interference pattern on the screen will disappear and the screen will be uniformly illuminated. Contrast this with the following situation: employ a coherent light source and rapidly move the obstacle in its plane in a random fashion. Here too, the interference fringes are replaced by uniform illumination. In this case, it is the persistence of the observer's retina that averages over many different displaced diffraction patterns. This example illustrates two ways in which the diffraction pattern can disappear. In the former case, the disappearance is associated with a random distribution of the lengths of wave trains emanating from the source, while in the latter case, it is the result of an *average* over an ensemble of spatially distributed virtual sources. This example shows how interference effects may vanish upon averaging.

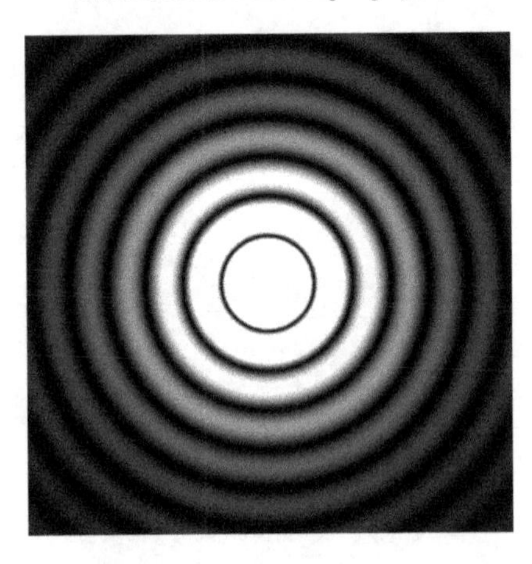

Figure 1.1 Diffraction pattern at infinity for a circular aperture.

Let us now turn to the diffraction of a coherent source by an obstacle of arbitrary type. For instance, suppose that the obstacle is a dielectric material whose refractive index fluctuates in space on a scale comparable to the wavelength of the light. The resulting scattering pattern on a screen placed at infinity, consists of a random distribution of bright and dark areas, as seen in Figure 1.2; this is called a *speckle* pattern.[1] Each speckle associated with the scattering represents a *fingerprint* of the random obstacle, and is specific to it. However, in contrast to the case of scattering by a sufficiently symmetric obstacle (such as a simple circular aperture), it is impossible to identify an order in the speckle pattern, and thus we cannot describe it with a deterministic sequence of integer numbers. This impossibility is one of the characteristics of what are termed complex media.

In this last experiment, for a thin enough obstacle, a wave scatters only once in the random medium before it emerges on its way to the screen at infinity (see Figure 1.3(a)). This regime is called *single scattering*. Consider now the opposite limit of an optically thick medium (also called a turbid medium), in which the wave scatters many times before leaving (Figure 1.3(b)). We thus speak of *multiple scattering*. The intensity at a point on the screen is obtained from the sum of the complex amplitudes of the waves arriving at that point. The phase associated with each amplitude is proportional to the path length of the multiply scattered wave divided by its wavelength λ. The path lengths are randomly distributed, so one would expect that the associated phases fluctuate and average to zero. Thus, the total intensity would reduce to the sum of the intensities associated with each of the paths.

In other words, if we represent this situation as equivalent to a series of thin obstacles, with each element of the series corresponding to a different and independent realization of the random medium, we might expect that for a sufficiently large number of such thin

[1] These speckles resemble those observed with light emitted by a weakly coherent laser, but they are of a different nature. Here they result from static spatial fluctuations due to the inhomogeneity of the scattering medium.

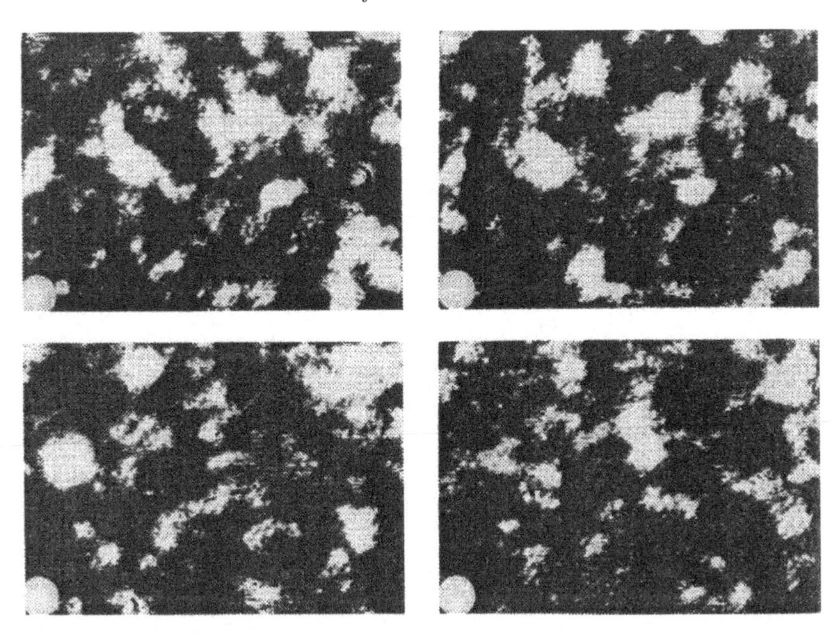

Figure 1.2 Speckle patterns due to scattering through an inhomogeneous medium. Here the medium is optically thick, meaning that the incident radiation undergoes many scatterings before leaving the sample. Each image corresponds to a different realization of the random medium [1].

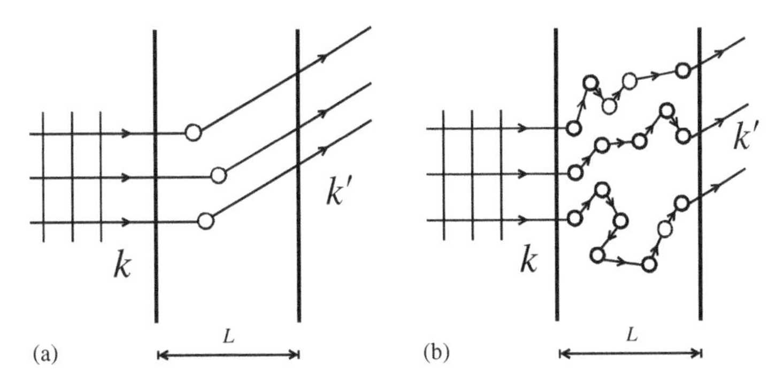

Figure 1.3 Schematic representations of the regimes of (a) single scattering, and (b) multiple scattering.

obstacles, the resulting intensity at a point on the screen would average over the different realizations, causing the speckles to vanish. This point of view corresponds to the classical description, for which the underlying wave nature plays no further role.

Figures 1.2 and 1.4 show that this conclusion is incorrect, and that the speckles survive, *even in the regime of multiple scattering*. If, on the other hand, we perform an ensemble

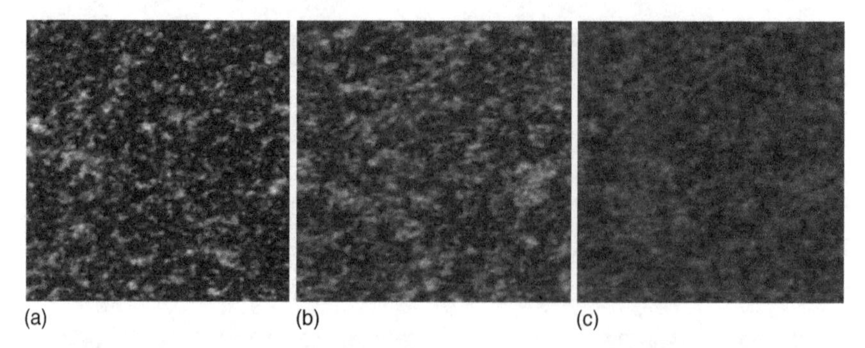

Figure 1.4 Averaging. The first speckle pattern (a) represents a snapshot of a random medium corresponding to a single realization of the disorder. The other two figures (b and c) correspond to an integration over the motion of scatterers, and hence to a self-average. (Figure courtesy of Georg Maret.)

average, the diffraction pattern disappears. This is the case with turbid media such as the atmosphere or suspensions of scatterers in a liquid (milk, for example), where the motion of the scatterers yields an average over different realizations of the random medium, provided we wait long enough. The classical approach, therefore, correctly describes the average characteristics of a turbid medium, such as the transmission coefficient or the diffusion coefficient of the average intensity. It has been employed extensively in problems involving the radiative transfer of waves through the atmosphere or through turbulent media.

This description may be adapted as such to the problem of propagation of electrons in a metal. In this case, the impurities in the metal are analogous to the scatterers in the optically thick medium, and the quantity analogous to the intensity is the electrical conductivity. In principle, of course, it is necessary to use the machinery of quantum mechanics to calculate the electrical conductivity. But since the work of Drude at the beginning of the last century, it has been accepted that transport properties of metals are correctly described by the disorder-averaged conductivity, obtained from a classical description of the degenerate electron gas. However, for a given sample, i.e., for a specific realization of disorder, we may observe interference effects, which only disappear upon averaging [2].

The indisputable success of the classical approach led to the belief that coherent effects would not subsist in a random medium in which a wave undergoes multiple scattering. In the 1980s, however, a series of novel experiments unequivocally proved this view to be false. In order to probe interference effects, we now turn to the Aharonov–Bohm effect, which occurs in the most spectacular of these experiments.

1.2 The Aharonov–Bohm effect

The Young two-slit device surely provides the simplest example of an interference pattern in optics; understanding its analog in the case of electrons is necessary for understanding quantum interference effects. In the Aharonov–Bohm geometry, an infinite solenoid is

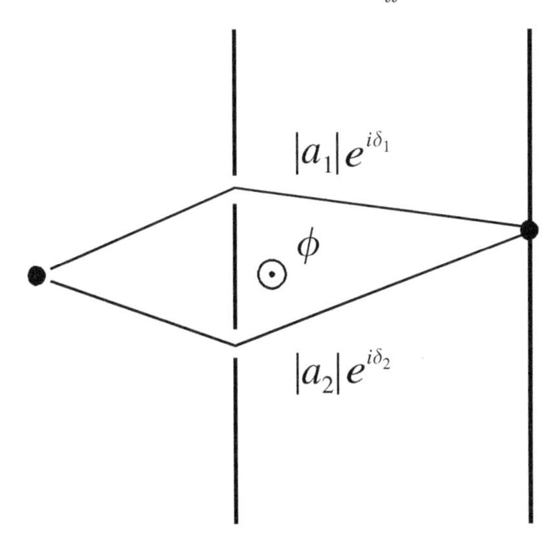

Figure 1.5 Schematic representation of the Aharonov–Bohm effect. A flux tube of flux ϕ is placed behind the two slits.

placed between the slits, such that the paths of the interfering electrons are exterior to it, as indicated in Figure 1.5. The magnetic field outside the solenoid is zero, so that classically it has no effect on the motion of the electrons.

This is not the case in quantum mechanics where, to calculate the intensity, we must sum the complex amplitudes associated with different trajectories. For the two trajectories of Figure 1.5, the amplitudes have the form $a_{1,2} = |a_{1,2}|e^{i\delta_{1,2}}$, where the phases δ_1 and δ_2 are given by ($-e$ is the electron charge):

$$\begin{aligned} \delta_1 &= \delta_1^{(0)} - \frac{e}{\hbar}\int_1 \boldsymbol{A} \cdot d\boldsymbol{l} \\ \delta_2 &= \delta_2^{(0)} - \frac{e}{\hbar}\int_2 \boldsymbol{A} \cdot d\boldsymbol{l}. \end{aligned} \tag{1.1}$$

The integrals are the line integrals of the vector potential $\boldsymbol{A}$ along the two trajectories and $\delta_{1,2}^{(0)}$ are the phases in the absence of magnetic flux. In the presence of a magnetic flux ϕ induced by the solenoid, the intensity $I(\phi)$ is given by

$$\begin{aligned} I(\phi) &= |a_1 + a_2|^2 = |a_1|^2 + |a_2|^2 + 2|a_1 a_2| \cos(\delta_1 - \delta_2) \\ &= I_1 + I_2 + 2\sqrt{I_1 I_2} \cos(\delta_1 - \delta_2). \end{aligned} \tag{1.2}$$

The phase difference $\Delta\delta(\phi) = \delta_1 - \delta_2$ between the two trajectories is now modulated by the magnetic flux ϕ

$$\Delta\delta(\phi) = \delta_1^{(0)} - \delta_2^{(0)} + \frac{e}{\hbar}\oint \boldsymbol{A} \cdot d\boldsymbol{l} = \Delta\delta^{(0)} + 2\pi\frac{\phi}{\phi_0}, \tag{1.3}$$

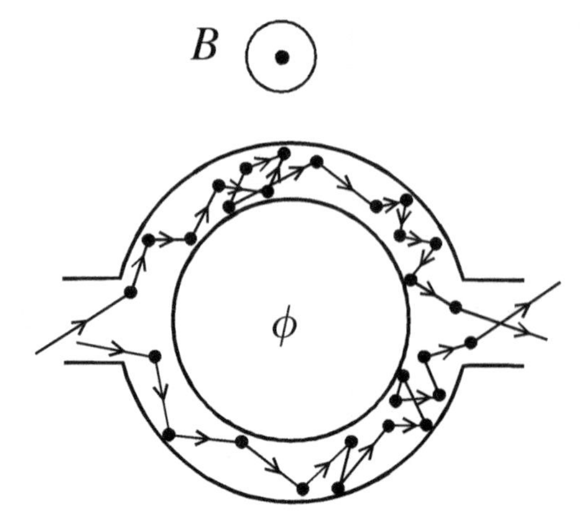

Figure 1.6 Schematic description of the experiment of Webb *et al.* on the Aharonov–Bohm effect in a metal. In this experiment, the applied magnetic field is uniform. ϕ is the flux through the ring.

where $\phi_0 = h/e$ is the quantum of magnetic flux. It is thus possible to vary continuously the state of interference at each point on the screen by changing the magnetic flux ϕ. This is the *Aharonov–Bohm effect* [3]. It is a remarkable probe to study phase coherence in electronic systems [4]. This constitutes an advantage for electronic systems over their optical counterparts.[2]

This effect was observed in the following experiment. A coherent stream of electrons was emitted by an electron microscope and split in two before passing through a toroidal magnet whose magnetic field was confined to the inside of the torus [6]. Thus, the magnetic field was zero along the trajectories of electrons. However, this experiment was performed in vacuum, where the electrons do not undergo any scattering before interfering. In order to demonstrate possible phase coherence in metals, in which the electrons undergo many collisions, R. Webb and his collaborators (1983) measured the resistance of a gold ring [7]. In the setup depicted schematically in Figure 1.6, electrons are constrained to pass through the two halves of the ring, which are analogous to the two Young slits, before being collected at the other end.

The analog of the intensity $I(\phi)$ is the electrical current, or better yet, the conductance $G(\phi)$ measured for different values of the magnetic flux ϕ. The flux is produced by applying a uniform magnetic field, though this does not strictly correspond to the Aharonov–Bohm experiment, since the magnetic field is not zero along the trajectories of electrons. However, the applied field is sufficiently weak that firstly, there is no deflection of the trajectories due to the Lorentz force, and secondly, the dephasing of coherent trajectories due to the magnetic field is negligible in the interior of the ring. Thus, the effect of the magnetic field may be neglected in comparison to that of the flux. Figure 1.7 shows that the magnetoresistance

[2] In a rotating frame, there is an analogous effect, called the Sagnac effect [5].

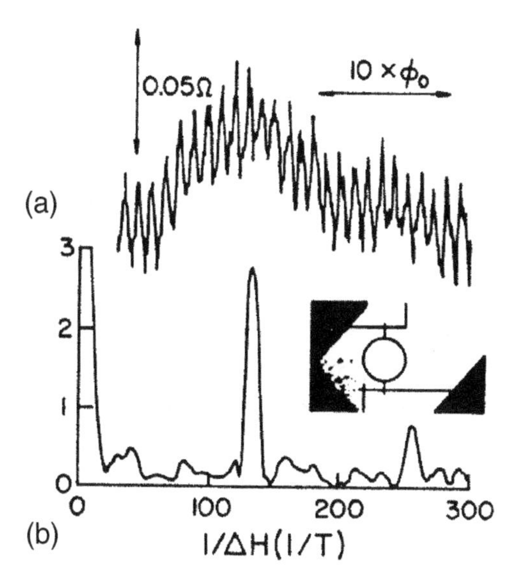

Figure 1.7 (a) Magnetoresistance of a gold ring at low temperature $T = 0.01$ K, (b) Fourier spectrum of the magnetoresistance. The principal contribution is that of the Fourier component at $\phi_0 = h/e$ [7].

of this ring is, to first approximation, a periodic function of the applied flux whose period is the flux quantum $\phi_0 = h/e$. Indeed, since the relative phase of the two trajectories is modulated by the flux, the total current, and therefore the conductance of the ring, are periodic functions of the flux:[3]

$$G(\phi) = G_0 + \delta G \cos\left(\Delta\delta^{(0)} + 2\pi\frac{\phi}{\phi_0}\right). \tag{1.4}$$

This modulation of the conductance as a function of flux results from the existence of *coherent effects* in a medium in which the disorder is strong enough for electrons to be *multiply scattered.* Consequently, the naive argument that phase coherence disappears in this regime is incorrect, and must be reexamined.

1.3 Phase coherence and the effect of disorder

In the aforementioned experiment of Webb *et al.*, the size of the ring was of the order of a micrometer. Now we know that for a macroscopic system, the modulation as a function of magnetic flux disappears. Therefore, there exists a characteristic length such that on

[3] We see in Figure 1.7 that the modulation is not purely periodic. This is due to the fact that the ring is not one dimensional. Moreover, multiple scattering trajectories within the same branch may also be modulated by the magnetic field which penetrates into the ring itself. This is the origin of the low-frequency peak in Figure 1.7(b).

scales greater than this length, there is no longer any phase coherence. This length, called the *phase coherence length* and denoted L_ϕ, plays an essential role in the description of coherent effects in complex systems.

In order to understand better the nature of this length, it is useful to review some notions related to quantum coherence.[4] Consider an ensemble of quantum particles contained in a cubic box of side length L in d dimensions. The possible quantum states are coherent superpositions of wave-functions such that the quantum state of the system is coherent over the whole volume L^d. There are many examples in which quantum coherence extends up to the macroscopic scale: superconductivity, superfluidity, free electron gas at zero temperature, coherent states of the photon field, etc.

For the electron gas at finite temperature, this coherence disappears at the macroscopic scale. It is therefore possible to treat physical phenomena such as electrical or thermal transport, employing an essentially classical approach. The suppression of quantum coherence results from phenomena linked to the existence of incoherent and irreversible processes due to the coupling of electrons to their environment. This environment consists of degrees of freedom with which the electrons interact: thermal excitations of the atomic lattice (phonons), impurities having internal degrees of freedom, interaction with other electrons, etc. This irreversibility is a source of decoherence for the electrons and its description is a difficult problem which we shall consider in Chapters 6 and 13. The phase coherence length L_ϕ generically describes the loss of phase coherence due to irreversible processes. In metals, the phase coherence length is a decreasing function of temperature. In practice, L_ϕ is of the order of a few micrometers for temperatures less than one kelvin.

None of the above considerations are related to the existence of *static* disorder of the type discussed in the two previous sections (e.g., static impurities such as vacancies or substitutional disorder, or variation of the refractive index in optics). Such disorder *does not destroy the phase coherence* and does not introduce any irreversibility. However, the possible symmetries of the quantum system disappear in such a way that it is no longer possible to describe the system with quantum numbers. In consequence, each observable of a random medium depends on the specific distribution of the disordered potential. On average, it is possible to characterize the disorder by means of a characteristic length: the *elastic mean free path* l_e, which represents the average distance travelled by a wave packet between two scattering events with no energy change (see Chapters 3 and 4).

We see, therefore, that the phase coherence length L_ϕ is fundamentally different from the elastic mean free path l_e. For sufficiently low temperatures, these two lengths may differ by several orders of magnitude, so that an electron may propagate in a disordered medium a distance much larger than l_e keeping its phase coherence, so long as the length of its trajectory does not exceed L_ϕ. The loss of coherence, therefore, is not related to the existence of a random potential of any strength, but rather to other types of mechanisms. It may seem surprising that the distinction between the effect of elastic disorder described by l_e and

[4] Most of the notions discussed here use the language of quantum mechanics; however, they have more or less direct analogs in the case of electromagnetic wave propagation.

that associated with irreversible processes of phase relaxation was first demonstrated in the relatively non-trivial case of transport in a metal where the electrons have complex interactions with their environment. However, the same distinction also applies to electromagnetic wave propagation in turbid media in the regime of coherent multiple scattering.

1.4 Average coherence and multiple scattering

If phase coherence leads to interference effects for a specific realization of disorder, it might be thought that these would disappear upon averaging. In the experiment of Webb *et al.* described in section 1.2, the conductance oscillations of period $\phi_0 = h/e$ correspond to a specific ring. If we now average over disorder, that is, over $\Delta\delta^{(0)}$ in relation (1.4), we expect the modulation by the magnetic flux to disappear, and with it all trace of coherent effects. The same kind of experiment was performed in 1981 by Sharvin and Sharvin [8] on a long hollow metallic cylinder threaded by an Aharonov–Bohm flux. A cylinder of height greater than L_ϕ can be interpreted as an ensemble of identical, uncorrelated rings of the type used in Webb's experiment. Thus, this experiment yields an ensemble average. Remarkably, they saw a signal which oscillated with flux but with a periodicity $\phi_0/2$ instead of ϕ_0. How can we understand that coherent effects can subsist *on average*?

The same type of question may be asked in the context of optics. If we average a speckle pattern over different realizations of disorder, does any trace of the phase coherence remain? Here too there was an unexpected result: the reflection coefficient of a wave in a turbid medium (sometimes called its albedo) was found to exhibit an angular dependence that could not be explained by the classical transport theory (Figure 1.8). This effect is known as *coherent backscattering*, and is a signature of phase coherence.

These results show that *even on average, some phase coherence effects remain.* In order to clarify the nature of these effects, let us consider an optically thick random medium. It can be modelled by an ensemble of point scatterers at positions $\boldsymbol{r}_n$ distributed randomly. The validity of this hypothesis for a realistic description of a random medium will be discussed in detail in Chapters 2 and 3. Consider a plane wave emanating from a coherent source (located outside the medium), which propagates in the medium and collides elastically with scatterers, and let us calculate the resulting interference pattern. For this, we study the complex amplitude $A(\boldsymbol{k},\boldsymbol{k}')$ of the wave reemitted in the direction defined by the wave vector $\boldsymbol{k}'$, corresponding to an incident plane wave with wave vector $\boldsymbol{k}$. It may be written, without loss of generality, in the form

$$A(\boldsymbol{k},\boldsymbol{k}') = \sum_{\boldsymbol{r}_1,\boldsymbol{r}_2} f(\boldsymbol{r}_1,\boldsymbol{r}_2)\, e^{i(\boldsymbol{k}\cdot\boldsymbol{r}_1 - \boldsymbol{k}'\cdot\boldsymbol{r}_2)}, \tag{1.5}$$

where $f(\boldsymbol{r}_1,\boldsymbol{r}_2)$ is the complex amplitude corresponding to the propagation between two scattering events located at $\boldsymbol{r}_1$ and $\boldsymbol{r}_2$. This amplitude may be expressed as a sum of the form $\sum_j a_j = \sum_j |a_j| e^{i\delta_j}$, where each path j represents a sequence of scatterings (Figure 1.9)

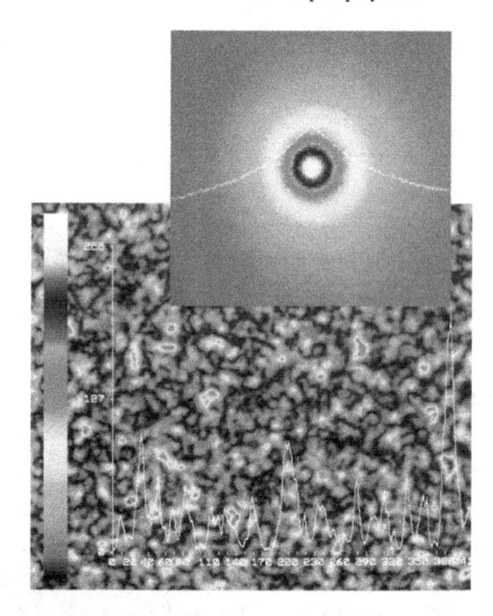

Figure 1.8 Speckle pattern obtained by multiple scattering of light by a sample of polystyrene spheres, as a function of observation angle. The curve in the lower figure represents the intensity fluctuations measured along a given angular direction. The upper figure is obtained by averaging over the positions of the spheres, and the resulting curve gives the angular dependence of the average intensity. (Figure courtesy of G. Maret.)

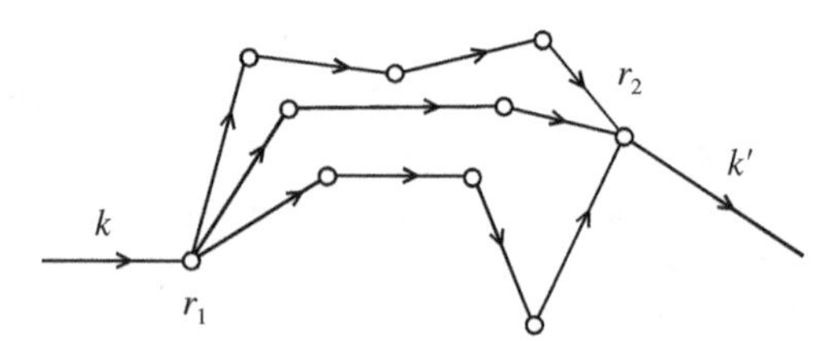

Figure 1.9 Typical trajectories which contribute to the total complex amplitude $f(\boldsymbol{r}_1, \boldsymbol{r}_2)$ of a multiply scattered wave.

joining the points $\boldsymbol{r}_1$ and $\boldsymbol{r}_2$. The associated intensity is given by

$$|A(\boldsymbol{k}, \boldsymbol{k}')|^2 = \sum_{\boldsymbol{r}_1,\boldsymbol{r}_2} \sum_{\boldsymbol{r}_3,\boldsymbol{r}_4} f(\boldsymbol{r}_1, \boldsymbol{r}_2) f^*(\boldsymbol{r}_3, \boldsymbol{r}_4)\, e^{i(\boldsymbol{k}\cdot\boldsymbol{r}_1 - \boldsymbol{k}'\cdot\boldsymbol{r}_2)}\, e^{-i(\boldsymbol{k}\cdot\boldsymbol{r}_3 - \boldsymbol{k}'\cdot\boldsymbol{r}_4)} \tag{1.6}$$

with

$$f(\boldsymbol{r}_1, \boldsymbol{r}_2) f^*(\boldsymbol{r}_3, \boldsymbol{r}_4) = \sum_{jj'} a_j(\boldsymbol{r}_1, \boldsymbol{r}_2)\, a^*_{j'}(\boldsymbol{r}_3, \boldsymbol{r}_4) = \sum_{jj'} |a_j||a_{j'}|\, e^{i(\delta_j - \delta_{j'})}. \tag{1.7}$$

In order to calculate its value averaged over the realizations of the random potential, that is, over the positions of scatterers, it is useful to note that most of the terms in relations (1.6)

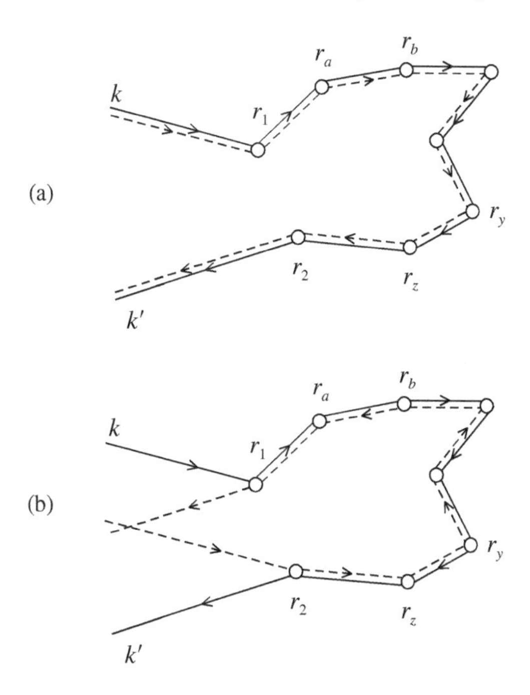

Figure 1.10 Schematic representation of the two types of sequences of multiple scatterings that remain upon averaging. The first (a) corresponds to the classical average intensity. The second (b), for which the two sequences of scattering events are traversed in opposite directions, is the source of the coherent backscattering effect.

and (1.7) average to zero, provided that the phase $\delta_j - \delta_{j'}$, which measures the difference in the lengths of the trajectories of Figure 1.9, is random.

In consequence, the only terms which contribute to the average of $|A(\boldsymbol{k},\boldsymbol{k}')|^2$ are those for which the phases vanish. This can only occur for pairs of *identical* trajectories, those which have the same sequence of scattering events, *either in the same or in opposite directions.*

Such trajectories are schematically represented in Figure 1.10, and correspond to the sequences

$$\boldsymbol{r}_1 \to \boldsymbol{r}_a \to \boldsymbol{r}_b \to \cdots \to \boldsymbol{r}_y \to \boldsymbol{r}_z \to \boldsymbol{r}_2$$

$$\boldsymbol{r}_2 \to \boldsymbol{r}_z \to \boldsymbol{r}_y \to \cdots \to \boldsymbol{r}_b \to \boldsymbol{r}_a \to \boldsymbol{r}_1.$$

The fact that the trajectories are identical imposes on us, in particular, $\boldsymbol{r}_1 = \boldsymbol{r}_3$ and $\boldsymbol{r}_2 = \boldsymbol{r}_4$ for the former process (same direction) and $\boldsymbol{r}_1 = \boldsymbol{r}_4$ and $\boldsymbol{r}_2 = \boldsymbol{r}_3$ for the latter (opposite direction) in relation (1.6). These two processes contribute identically to the intensity provided that the system is invariant under time reversal. Moreover, the second process gives rise, according to (1.6), to an additional dephasing such that the only two non-zero contributions which remain upon averaging are

$$\overline{|A(\boldsymbol{k},\boldsymbol{k}')|^2} = \overline{\sum_{\boldsymbol{r}_1,\boldsymbol{r}_2} |f(\boldsymbol{r}_1,\boldsymbol{r}_2)|^2 \left[1 + e^{i(\boldsymbol{k}+\boldsymbol{k}')\cdot(\boldsymbol{r}_1-\boldsymbol{r}_2)}\right]}, \tag{1.8}$$

where $\overline{\cdots}$ denotes averaging over the realizations of the random potential.

The essence of the present book is a systematic study of the consequences of the existence of these two processes, which survive upon averaging in the course of multiple scattering. The former process is well known. It may be perfectly well understood using a purely classical treatment that does not take into account the existence of an underlying wave equation, since the phases exactly cancel out. In the study of electron transport in metals, this classical analysis is performed in the framework of the Boltzmann equation, while for electromagnetic wave propagation, the equivalent theory, called *radiative transfer*, was developed by Mie and Schwartzchild [9]. Both date from the beginning of the twentieth century.

The second term in relation (1.8) contains a phase factor. This depends on the points $\boldsymbol{r}_1$ and $\boldsymbol{r}_2$, and the sum over these points in the averaging makes this term vanish in general, with two notable exceptions.

- $\boldsymbol{k}+\boldsymbol{k}' \simeq 0$. In the direction exactly opposite to the direction of incidence, the intensity is *twice* the classical value. The classical contribution has no angular dependence on average, and the second term, which depends on $\boldsymbol{k}+\boldsymbol{k}'$, gives an angular dependence to the average intensity reflected by the medium which appears as a peak in the albedo. This phenomenon was observed first in optics and is known as *coherent backscattering*; its study is the object of Chapter 8.
- In the sum (1.8), the terms for which $\boldsymbol{r}_1 = \boldsymbol{r}_2$ are special. They correspond to closed multiple scattering trajectories. Their contributions to the averaged interference term survive even when it is impossible to select the directions $\boldsymbol{k}$ and $\boldsymbol{k}'$. This is the case for metals or semiconductors for which the interference term affects the average transport properties such as the electrical conductivity. This is the origin of the phenomenon of *weak localization*.

1.5 Phase coherence and self-averaging: universal fluctuations

The measurable physical quantities of a disordered quantum system depend on the specific realization of the disorder, at least so long as the characteristic lengths of the system are smaller than the phase coherence length L_ϕ. In the opposite case, that is, for lengths greater than L_ϕ, the phase coherence is lost, and the system becomes classical, i.e., the physical quantities are independent of the specific realization of the disorder. The physics of systems of size less than L_ϕ, called *mesoscopic systems*,[5] is thus particularly interesting because of coherence effects [10, 11]. The physics of mesoscopic systems makes precise the distinction between the complexity due to disorder described by l_e and the decoherence, which depends on L_ϕ:

- disorder (l_e), loss of symmetry and of good quantum numbers (complexity);
- loss of phase coherence (L_ϕ).

[5] The Greek root $\mu\epsilon\sigma o\varsigma$ means intermediate.

Let us now attempt to understand why a disordered quantum system larger than L_ϕ exhibits self-averaging, i.e., why its measurable physical properties are equal to their ensemble averages. If the characteristic size L of a system is much greater than L_ϕ, the system may be decomposed into a collection of $N = (L/L_\phi)^d \gg 1$ statistically independent subsystems, in each of which the quantum coherence is preserved. A physical quantity defined in each subsystem will then take on N random values. The law of large numbers ensures that every macroscopic quantity is equal, with probability one, to its average value. Consequently, every disordered system of size $L \gg L_\phi$ is effectively equivalent to an ensemble average. On the other hand, deviations from self-averaging are observed in systems of sizes smaller than L_ϕ because of the underlying phase coherence. The study of these deviations is one of the main goals of mesoscopic physics. Consider the particularly important example of fluctuations in the electrical conductance of a weakly disordered metal (Chapter 11). In the classical self-averaging limit, for a cubic sample of size L, the relative conductance fluctuations vary as $1/\sqrt{N}$:

$$\frac{\sqrt{\overline{\delta G^2}}}{\overline{G}} \simeq \frac{1}{\sqrt{N}} \simeq \left(\frac{L_\phi}{L}\right)^{d/2} \tag{1.9}$$

where $\delta G = G - \overline{G}$. The average conductance $\overline{G}$ is the classical conductance G_{cl} given by Ohm's law $G_{cl} = \sigma L^{d-2}$ where σ is the electrical conductivity.[6] From relation (1.9), we deduce that $\overline{\delta G^2} \propto L^{d-4}$. For $d \leq 3$, the fluctuations go to zero in the large scale limit, and the system is said to be self-averaging. In contrast, for $L < L_\phi$, it is found experimentally that

$$\sqrt{\overline{\delta G^2}} \simeq \text{constant} \times \frac{e^2}{h}. \tag{1.10}$$

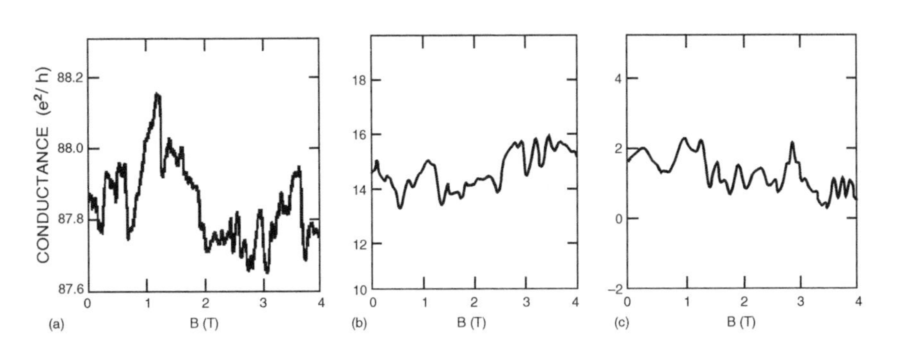

Figure 1.11 Aperiodic variations in the magnetoconductance of three different systems. (a) A gold ring of diameter 0.8 mm, (b) a Si-MOSFET sample, and (c) the result of numerical simulations using the disordered Anderson model (discussed in Chapter 2). The conductance varies by several orders of magnitude from one system to another, but the fluctuations remain of order e^2/h [12].

[6] The expression $G_{cl} = \sigma L^{d-2}$ is a generalization to d dimensions of the standard expression $G_{cl} = \sigma S/L$, for a sample of length L and cross section S.

In the mesoscopic regime, the amplitude of conductance fluctuations is independent of the size L and of the amount of disorder, and one speaks of *universal conductance fluctuations*. The variance of the conductance is the product of a universal quantity e^2/h and a numerical factor which depends solely on the sample geometry. This implies that in the mesoscopic regime, the electrical conductance is no longer a self-averaging quantity. This universality is shown in Figure 1.11 where each plot corresponds to a very different system. One essential characteristic of mesoscopic fluctuations is their *reproducibility*. For a given realization of disorder, the dependence of the fluctuations as a function of an external parameter such as Fermi energy or magnetic field is perfectly reproducible. In this sense, the fluctuations represent, just like speckle patterns in optics, a *fingerprint* of the realization of the disorder, and uniquely characterize it.

1.6 Spectral correlations

We have mentioned the signature of coherent effects on transport properties such as electrical conductance or albedo. For an *isolated* system of finite size, we may wonder about the effect of disorder on the spatial behavior of wave functions and on the correlation of eigen energies. For electromagnetic waves, we are interested in the spectrum of eigenfrequencies. If, for instance, the wave functions are strongly affected by the disorder and are exponentially localized, then the corresponding eigenenergies (or frequencies) may be arbitrarily close to each other since they describe states for which the spatial overlap is exponentially small. These wave functions are uncorrelated, as are their energy levels. If, on the other hand, the wave functions are spatially delocalized over the system and do not exhibit any spatial structure, which corresponds to a regime we may consider as *ergodic*, then the important spatial overlap of the eigenfunctions induces spectral correlations manifested by a "repulsion" of the energy levels. These two extreme situations are very general, and are insensitive to the microscopic details characterizing the disorder. It turns out that the spectral correlations present *universal properties* common to very different physical systems. Consider, for example, the probability $P(s)$ that two neighboring energy levels are separated by s. The two preceding situations are described by two robust limiting cases for the function $P(s)$, corresponding, respectively, to a *Poisson* distribution for the exponentially localized states, and to a *Wigner–Dyson* distribution for the ergodic case. These two distributions, represented in Figure 10.1, describe a wide range of physical problems and divide them, to first approximation, into two classes, corresponding either to integrable systems (Poisson) or to non-integrable (also called *chaotic*) systems (Wigner–Dyson). The latter case may be studied using *random matrix theory* along the general lines discussed in Chapter 10.

Of course, a complex medium exhibits such universal behavior only in limiting cases. From the methods developed in this book, we shall see how to recover certain results of random matrix theory, and to identify corrections to the universal regime. These spectral correlations are extremely sensitive to the loss of phase coherence. They thus depend on L_ϕ and are characteristic of the mesoscopic regime. They are evidenced by the behavior of thermodynamic variables such as magnetization or persistent currents which constitute the orbital response of electrons to an applied magnetic field; this is the object of Chapter 14.

1.7 Classical probability and quantum crossings

Most physical quantities studied in this book are expressed as a function of the product of two complex amplitudes, each being the sum of contributions associated with multiple scattering trajectories:

$$\sum_i a_i^* \sum_j a_j = \sum_{i,j} a_i^* a_j. \tag{1.11}$$

This is the case, for example, of light intensity considered in section 1.4. The combination of amplitudes (1.11) is related to the *probability of quantum diffusion*, whose role is essential in characterizing the physical properties of disordered media. This probability, which describes the evolution of a wave packet between any two points $\boldsymbol{r}$ and $\boldsymbol{r}'$, is written as the product of two complex amplitudes[7] known as propagators or Green's functions. Denoting the average probability by $P(\boldsymbol{r},\boldsymbol{r}')$, we have

$$P(\boldsymbol{r},\boldsymbol{r}') \propto \overline{\sum_{i,j} a_i^*(\boldsymbol{r},\boldsymbol{r}')a_j(\boldsymbol{r},\boldsymbol{r}')}. \tag{1.12}$$

Each amplitude $a_j(\boldsymbol{r},\boldsymbol{r}')$ describes a propagating trajectory j from $\boldsymbol{r}$ to $\boldsymbol{r}'$, and thus $P(\boldsymbol{r},\boldsymbol{r}')$ appears as the sum of contributions of pairs of trajectories, each characterized by an amplitude and a phase. This sum may be decomposed into two contributions, one for which the trajectories i and j are identical, the other for which they are different:

$$P(\boldsymbol{r},\boldsymbol{r}') \propto \overline{\sum_j |a_j(\boldsymbol{r},\boldsymbol{r}')|^2} + \overline{\sum_{i\neq j} a_i^*(\boldsymbol{r},\boldsymbol{r}')a_j(\boldsymbol{r},\boldsymbol{r}')}. \tag{1.13}$$

In the former contribution the phases vanish. In the latter, the dephasing of paired trajectories is large and random, and consequently their contribution vanishes on average.[8] The probability is thus given by a sum of intensities and does not contain any interference term (Figure 1.12):

$$P_{cl}(\boldsymbol{r},\boldsymbol{r}') \propto \overline{\sum_j |a_j(\boldsymbol{r},\boldsymbol{r}')|^2}. \tag{1.14}$$

We shall call this classical term a *Diffuson*. In the weak disorder limit, that is, as long as the wavelength λ is small compared to the elastic mean free path l_e, and for length scales

[7] In this introduction, we do not seek to establish exact expressions for the various physical quantities, but simply to discuss their behavior as a function of multiple scattering amplitudes. We therefore omit time or frequency dependence when it is not essential. More precise definitions are left for Chapters 3 and 4.

[8] We show in the following section that the interference terms of expression (1.13) do not vanish completely, and are at the origin of most of the quantum effects described in this book.

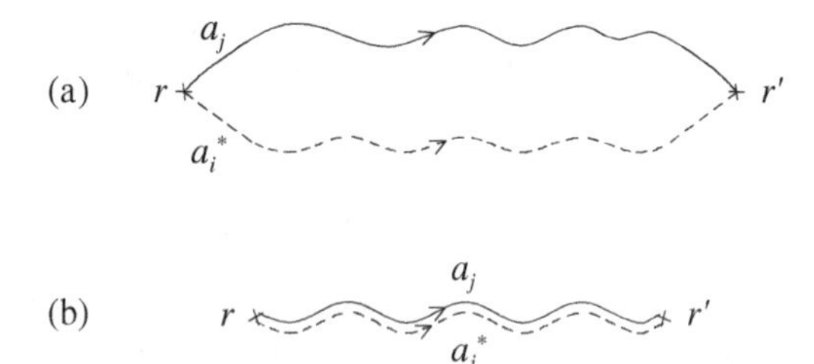

Figure 1.12 By averaging over disorder, the contribution from pairs of different trajectories (a) vanishes, leaving only terms corresponding to identical trajectories to contribute to the average probability (b).

larger than l_e, the Diffuson is well described by the solution of the diffusion equation

$$\left[\frac{\partial}{\partial t} - D\Delta\right]P_{cl}(\boldsymbol{r},\boldsymbol{r}',t) = \delta(\boldsymbol{r}-\boldsymbol{r}')\delta(t), \tag{1.15}$$

where $D = vl_e/d$ is the diffusion constant, v the group velocity of the wave packet, and d the dimension of space.

One quantity of particular importance is the probability of return to the initial point $P_{cl}(\boldsymbol{r},\boldsymbol{r},t)$, as well as its integral over space $Z(t)$. This last quantity is expressed as a function of the eigenfrequencies denoted E_n associated with the diffusion equation (1.15)

$$Z(t) = \int d\boldsymbol{r}P_{cl}(\boldsymbol{r},\boldsymbol{r},t) = \sum_n e^{-E_n t} \tag{1.16}$$

for $t > 0$. For example, for a system[9] of volume Ω, we have

$$Z(t) = \frac{\Omega}{(4\pi Dt)^{d/2}}. \tag{1.17}$$

The dependence as a function of the space dimensionality d plays an essential role, and the physical properties are accordingly more sensitive to the effects of multiple scattering when the dimensionality is small, since the return probability increases with decreasing d.

For a finite system of volume $\Omega = L^d$, boundary conditions may play an important role, since they reflect the nature of the coupling to the environment on which $Z(t)$ depends. This introduces a new characteristic time

$$\tau_D = L^2/D \tag{1.18}$$

called the diffusion time or *Thouless time*. It represents the time to diffuse from one boundary of the sample to the other. If $t \ll \tau_D$, the effect of boundaries is not felt, the diffusion is free and expressed by relation (1.17). If, on the other hand, $t \gg \tau_D$, the entire volume is explored by the random walk, we are in the *ergodic* regime, and $Z(t) \simeq 1$. A characteristic energy is associated with τ_D and called the *Thouless energy* $E_c = \hbar/\tau_D$.

[9] We ignore boundary effects, and thus are dealing with free diffusion.

1.7.1 Quantum crossings

Taking the second contribution of (1.13) to be zero amounts to neglecting all interference effects. In fact, even after averaging over disorder, this contribution is not rigorously zero. There still remain terms describing pairs of distinct trajectories, $i \neq j$, which are sufficiently close that their dephasing is small. As an example, consider the case of Figure 1.13(a), where the two trajectories in a Diffuson follow the same sequence of scatterings but cross, forming a loop with counter propagating trajectories.[10] This notion of crossing is essential because it is at the origin of coherent effects such as weak localization, long range light intensity correlations, or universal conductance fluctuations. As such, it is useful to develop intuition about them. Figure 1.13 shows that one such crossing mixes four complex amplitudes and pairs them in different ways. The crossing, also called a *Hikami box*, is an object which permutes amplitudes [13]. For the induced dephasing to be smaller than 2π, the trajectories must be as close to each other as possible, and the crossing must be localized in space, that is, on a scale of the order of the elastic mean free path l_e. We shall see that the volume associated with a crossing in d dimensions is of order $\lambda^{d-1} l_e$. This may be interpreted by attributing a length vt to a Diffuson – the object built with paired trajectories – propagating during a time t, where v is the group velocity, and a cross section λ^{d-1}, giving a volume $\lambda^{d-1} vt$.

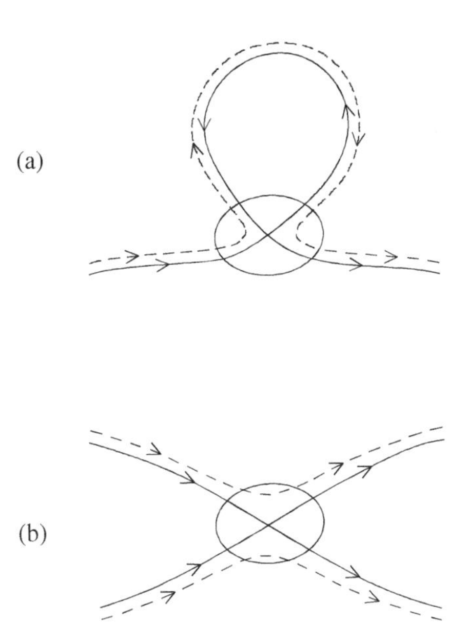

Figure 1.13 (a) The crossing of trajectories contributing to a Diffuson leads to a new pairing of amplitudes. (b) The pairing of four amplitudes resulting from the crossing of two Diffusons.

[10] These quantum crossings, which exchange two amplitudes, should not be confused with the self-crossings of a classical random walk.

To evaluate the importance of quantum effects, let us estimate the probability that two Diffusons will cross, as in Figure 1.13(b). This probability, for a time interval dt, is proportional to the ratio of the Diffuson volume to the system volume $\Omega = L^d$, that is,

$$dp_\times(t) = \frac{\lambda^{d-1} v \, dt}{\Omega} \simeq \frac{1}{g} \frac{dt}{\tau_D}. \tag{1.19}$$

In this expression, we have explicitly indicated the diffusion time $\tau_D = L^2/D$. We have also introduced a dimensionless number g, proportional to the inverse ratio of the two volumes $\lambda^{d-1} v \tau_D / \Omega$. We will show that this number is none other than the classical electrical conductance $g = G_{cl}/(e^2/h)$, in units of the quantum of conductance e^2/h (relation 7.22).

When the disordered medium is coupled to leads, the diffusing waves escape from the system in a time of the order of τ_D which therefore sets the characteristic time for diffusive trajectories. Therefore, the probability for a crossing during the time τ_D is inversely proportional to the dimensionless conductance, namely

$$p_\times(\tau_D) = \int_0^{\tau_D} dp_\times(t) \simeq \frac{1}{g}. \tag{1.20}$$

This parameter allows us to evaluate the importance of quantum corrections to the classical behavior. In the limit of weak disorder $\lambda \ll l_e$, the conductance g is large, so the crossing probability, and hence the effects of coherence, are small.

The quantum crossings, and the dephasing which they induce, introduce a correction to the classical probability (1.14). It is the combination of these crossings, the interference that they describe, and the spatially long range nature of the Diffuson, which allows the propagation of coherent effects over the entire system. These are the effects which lie at the basis of mesoscopic physics. The simple argument developed here straightforwardly implies that the quantum corrections to classical electron transport are of the order $G_{cl} \times 1/g$, that is to say, e^2/h.

In the limit of weak disorder, the crossings are independent of each other. This allows us to write successive corrections to the classical probability as a function of the number of crossings, that is, as a power series in $1/g$.

1.8 Objectives

This book deals with coherent multiple scattering of electronic or electromagnetic waves in disordered media, in the limit where the wavelength[11] $\lambda = 2\pi/k$ is small compared to the elastic mean free path l_e. This is the limit of weak disorder. It is possible to develop a general framework for the description of a large number of physical phenomena, which were effectively predicted, observed, and explained, by employing a small number of rather general ideas. In this section, we briefly outline these ideas, and indicate the relevant chapter in which these phenomena are discussed.

[11] For electrons, λ is the Fermi wavelength.

Weak localization corrections to the conductivity (Chapter 7) and the coherent backscattering peak (Chapter 8)

One particularly important example where the notion of quantum crossing appears is that of electron transport in a weakly disordered conductor. Consider, for example, transport across a sample of size L. The conductance corresponding to the classical probability is the classical or Drude conductance G_{cl}. The quantum correction to the probability leads to a correction to the conductance.

This correction associated with a single crossing is of order $1/g$, but it depends as well on the distribution of loops, that is, on the closed diffusive trajectories (Figure 1.14), whose number is given by the spatial integral (1.16), namely by the integrated probability of returning to the origin $Z(t)$. The probability $p_o(\tau_D)$ of crossing the sample with a single quantum crossing (one loop) is of the form

$$p_o(\tau_D) \sim \frac{1}{g} \int_0^{\tau_D} Z(t)\, \frac{dt}{\tau_D}, \tag{1.21}$$

where $\tau_D = L^2/D$. We obtain the relative correction to the average conductance $\Delta G = G - G_{cl}$ as,

$$\boxed{\frac{\Delta G}{G_{cl}} \sim -p_o(\tau_D)} \tag{1.22}$$

The minus sign in this correction indicates that taking a quantum crossing and therefore a closed loop into account has the effect of reducing the average conductance. This is called the *weak localization* correction.

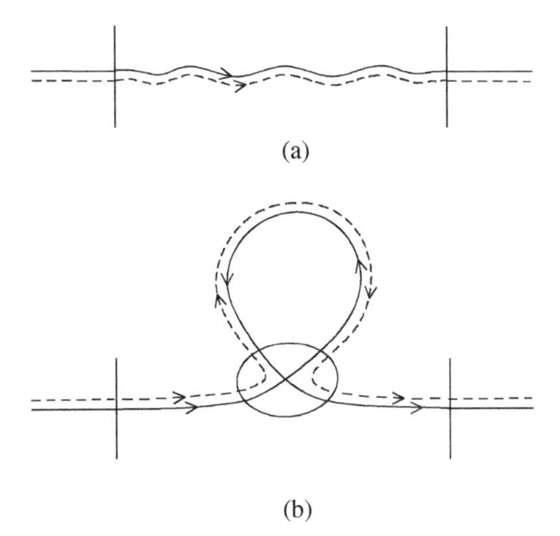

Figure 1.14 The crossing of a Diffuson with itself (b) leads to a quantum correction to the classical Drude conductivity (a).

We note that the two multiple scattering trajectories that form a loop evolve in *opposite* directions. If the system is time reversal invariant, the two amplitudes associated with these trajectories j and j^T are identical $a_{j^T}(\boldsymbol{r},\boldsymbol{r}) = a_j(\boldsymbol{r},\boldsymbol{r})$ so that their product is equal to the product of two amplitudes propagating in the same direction. If there are processes which break this invariance, then the weak localization correction vanishes. This pairing of time-reversed conjugate trajectories is called a *Cooperon*.

This pairing closely resembles that described in the optical counterpart of section 1.4 and Figure 1.10(b), which corresponds to time-reversed multiple scattering amplitudes $a_j(\boldsymbol{r}_1,\boldsymbol{r}_2)$ and $a^*_{j^T}(\boldsymbol{r}_1,\boldsymbol{r}_2)$. For the geometry of a semi-infinite disordered medium, and a plane wave incident along the direction $\boldsymbol{k}$ which emerges along $\boldsymbol{k}'$, the average reflected intensity $I(\boldsymbol{k},\boldsymbol{k}')$ (also called the average *albedo*) depends on the angle between the directions $\boldsymbol{k}$ and $\boldsymbol{k}'$. From (1.8), we have

$$\boxed{I(\boldsymbol{k},\boldsymbol{k}') \propto \int d\boldsymbol{r}\, d\boldsymbol{r}'\, P_{cl}(\boldsymbol{r},\boldsymbol{r}') \left[1 + e^{i(\boldsymbol{k}+\boldsymbol{k}')\cdot(\boldsymbol{r}-\boldsymbol{r}')}\right]} \tag{1.23}$$

We identify $\overline{|f(\boldsymbol{r},\boldsymbol{r}')|^2}$ with the Diffuson $P_{cl}(\boldsymbol{r},\boldsymbol{r}')$ whose endpoints $\boldsymbol{r}$ and $\boldsymbol{r}'$ are taken to be close to the interface between the diffusive medium and the vacuum. The first term in the brackets is the phase independent classical contribution, while the interference term has an angular dependence around the backscattering direction $\boldsymbol{k}' \simeq -\boldsymbol{k}$. The albedo therefore exhibits a peak in this direction, called the *coherent backscattering peak*, whose intensity is twice the classical value.

Correlations of speckle patterns (Chapter 12)

For a given realization of disorder, the intensity distribution of a light wave undergoing multiple scattering is a random distribution of dark and bright spots (Figure 1.2) called a speckle pattern. This interference pattern, arising from the superposition of complex amplitudes, constitutes a "fingerprint" of the specific disorder configuration. In order to characterize a speckle pattern, we can measure the angular distribution of the transmitted (or reflected) intensity using the geometry of a slab of thickness L. In this case, one measures the normalized intensity $\mathcal{T}_{ab}$ transmitted in the direction $\hat{\boldsymbol{s}}_b$ and corresponding to a wave incident in the direction $\hat{\boldsymbol{s}}_a$ (see Figure 12.2). On average, the transmission coefficient $\overline{\mathcal{T}_{ab}}$ depends only slightly on the directions $\hat{\boldsymbol{s}}_a$ and $\hat{\boldsymbol{s}}_b$, and we denote it $\overline{\mathcal{T}}$. The angular correlation of the speckle is defined by

$$C_{aba'b'} = \frac{\overline{\delta\mathcal{T}_{ab}\delta\mathcal{T}_{a'b'}}}{\overline{\mathcal{T}}^2}, \tag{1.24}$$

where $\delta\mathcal{T}_{ab} = \mathcal{T}_{ab} - \overline{\mathcal{T}}$. The fluctuations of the speckle, for a given incidence direction $\hat{\boldsymbol{s}}_a$, are described by the quantity $C_{abab} = \overline{\delta^2\mathcal{T}_{ab}}/\overline{\mathcal{T}}^2$ which happens, as we shall see, to be equal to 1, yielding

$$\overline{\mathcal{T}^2_{ab}} = 2\,\overline{\mathcal{T}}^2. \tag{1.25}$$

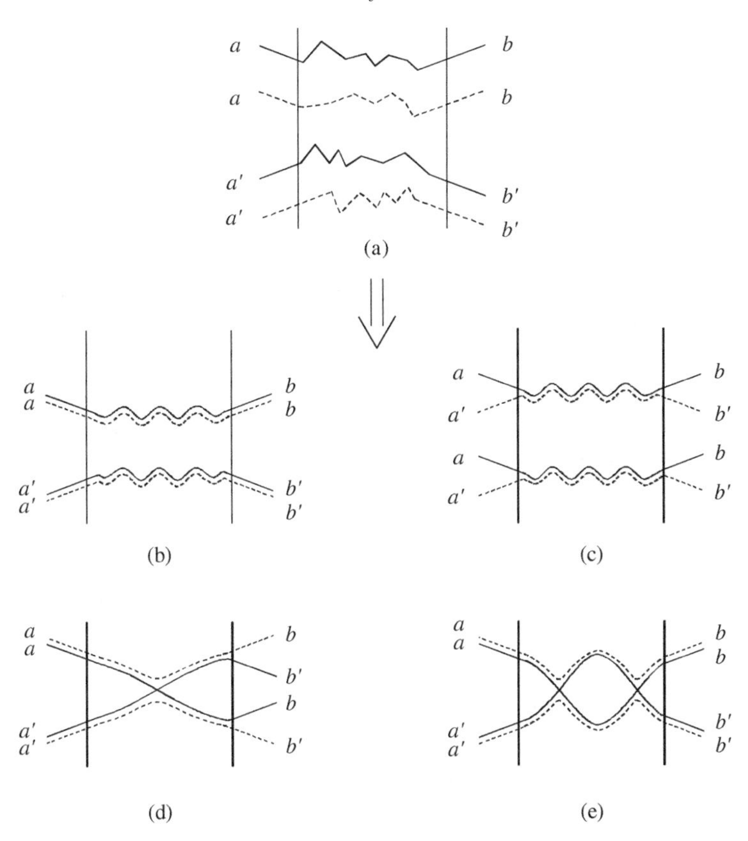

Figure 1.15 (a) The angular correlation function of a speckle pattern is built from the product of four complex amplitudes corresponding to four plane waves incoming along directions $\hat{s}_a$ and $\hat{s}_{a'}$ and emerging along $\hat{s}_b$ and $\hat{s}_{b'}$. The main contributions are obtained by pairing the amplitudes two by two to form Diffusons. This gives rise to contributions (b) and (c). Contribution (c), which corresponds to the correlation function $C^{(1)}_{aba'b'}$ decays exponentially in angle. Contribution (d) contains one quantum crossing, while (e) has two quantum crossings. In this last case, we note that the corresponding correlation function has no angular dependence.

This result, which constitutes the *Rayleigh law*, describes the most "visible" aspect of a speckle pattern, namely, its "granularity," with relative fluctuations of the order of unity.

In contrast to the probability (average conductance or light intensity), a correlation function such as (1.24) is the product of *four complex amplitudes* (Figure 1.15(a)). When averaging over disorder, the only important contributions are obtained by pairing these amplitudes so as to form Diffusons or Cooperons. Neglecting, in a first step, the possibility of a quantum crossing of two Diffusons, there are two possibilities, shown in Figures 1.15(b, c). The first is the product of two average intensities $\overline{\mathcal{T}_{ab}}$ and $\overline{\mathcal{T}_{a'b'}}$. The second gives the principal contribution to the correlation function (1.24), denoted $C^{(1)}_{aba'b'}$. It is non-zero only if $\hat{s}_a - \hat{s}_{a'} = \hat{s}_b - \hat{s}_{b'}$ and it decays exponentially as a function of $k|\hat{s}_a - \hat{s}_{a'}|/L$, that is, over a very small angular range.

It is also possible to pair the amplitudes by interposing one or more quantum crossings. This results in corrections to the angular correlation function in powers of $1/g$. The first one, denoted $C^{(2)}_{aba'b'}$ has a *single crossing*, and is shown in Figure 1.15(d). The presence of a crossing imposes constraints on the pairing of the amplitudes, and thus gives rise to a different angular dependence. We will show in section 12.4.2 that $C^{(2)}_{aba'b'}$ decays as a power of $k|\hat{s}_a - \hat{s}_{a'}|/L$, instead of an exponential decay. It has a weight $1/g \ll 1$ as compared to the term with no crossing.

The contribution $C^{(3)}_{aba'b'}$, with *two crossings* is sketched in Figure 1.15(e). Because of the two-crossing structure, this contribution has no angular dependence, i.e., it yields a uniform background to the correlation function. This result is characteristic of coherent multiple scattering, that is to say, of the combined effect of quantum crossings and of their long range propagation through the Diffusons. Upon averaging the total angular correlation function over all directions of incident and emergent waves, only this last contribution survives, which constitutes the analog for waves of universal conductance fluctuations [14–16].

Universal conductance fluctuations (Chapter 11)

The considerations obtained for electromagnetic waves are easily transposed to the case of electrons in a weakly disordered metal, and in this context lead to conductance fluctuations. In the mesoscopic regime, these fluctuations differ considerably from the classical result: they are universal and of order e^2/h (see section 1.5). This results from the existence of quantum crossings. More precisely, calculation of the fluctuations $\overline{\delta G^2} = \overline{G^2} - \overline{G}^2$ involves the pairing of four complex amplitudes paired into two Diffusons. Moreover, in the framework of the Landauer formalism (Appendix A7.2), the conductance, in units of e^2/h, is related to the transmission coefficient $\mathcal{T}_{ab}$ summed over all incident and emergent directions a and b. Thus, as for the speckle angular correlations, it can be shown that the term with no quantum crossing corresponds to $\overline{G}^2$. The contribution from a single crossing vanishes upon summation over the emergent directions. In contrast, the term with two quantum crossings has no angular dependence (Figure 1.15(e)) and it gives a universal variance $\overline{\delta G^2}$ proportional to $G_{cl}^2/g^2 = (e^2/h)^2$.

We note that, as for the weak localization correction, the variance $\overline{\delta G^2}$ depends on the distribution of *loops*. Here the loops result from *two crossings* (Figure 1.15(e)). For a loop of length vt, the choice of the relative position of the two crossings introduces an additional factor $\lambda^{d-1} vt/\Omega \simeq t/(g\tau_D)$ in the integral (1.21). We thus deduce that

$$\boxed{\frac{\overline{\delta G^2}}{G_{cl}^2} \sim \frac{1}{g^2} \int_0^{\tau_D} Z(t) \frac{t\,dt}{\tau_D^2}} \tag{1.26}$$

This expression resembles the relative correction of weak localization (1.21, 1.22), but the additional factor t has important consequences. The dependence $Z(t) \propto t^{-d/2}$ of the integrated return probability to the origin implies that weak localization correction is universal for $d < 2$, while conductance fluctuations are universal for $d < 4$.

Dephasing (Chapter 6)

The interference effects discussed earlier result from the existence of quantum crossings. They depend on the coherence of the wave-scatterer system, and may be modified in the presence of dephasing processes. Such processes are related to additional degrees of freedom which we may divide into three classes, examples of which are as follows.

- External field: uniform magnetic field, Aharonov–Bohm flux.
- Degrees of freedom of the scattered wave: electron spin and photon polarization.
- Degrees of freedom of the scatterers: magnetic impurities, environment induced by other electrons, motion of scatterers, internal quantum degrees of freedom (atomic Zeeman sublevels).

Let us first consider the case of multiple scattering of electrons, now in the presence of a magnetic field. Full coherence implies that time-reversed trajectories have the same amplitude. This is no longer the case in the presence of a magnetic field, which induces a dephasing between conjugate trajectories:

$$a_{jT}(\boldsymbol{r},\boldsymbol{r}') = a_j(\boldsymbol{r},\boldsymbol{r}')\, e^{i\Phi_j(\boldsymbol{r},\boldsymbol{r}')}. \tag{1.27}$$

Using (1.13) and the discussion on page 20, the correction to the return probability associated with the Cooperon, which we denote P_c, is of the form

$$P_c(\boldsymbol{r},\boldsymbol{r}) \propto \overline{\sum_j |a_j(\boldsymbol{r},\, \boldsymbol{r})|^2\, e^{i\Phi_j(\boldsymbol{r},\boldsymbol{r})}}, \tag{1.28}$$

where $\Phi_j(\boldsymbol{r},\boldsymbol{r})$ is the phase difference accumulated along the closed trajectories. The dephasing due to a magnetic field is

$$\Phi_j(\boldsymbol{r},\boldsymbol{r}') = \frac{2e}{\hbar}\int_{\boldsymbol{r}}^{\boldsymbol{r}'} \boldsymbol{A}\cdot d\boldsymbol{l}, \tag{1.29}$$

where the factor 2 comes from the fact that both of the paired trajectories accumulate the same phase, but with opposite signs, so that their difference adds. The coherent contribution to the return probability is thus affected by this phase factor, and the weak localization correction to the electrical conductance takes the form

$$\boxed{\frac{\Delta G}{G_{cl}} \propto -\int dt\, Z(t) \left\langle e^{i\Phi(t)}\right\rangle} \tag{1.30}$$

where $\left\langle e^{i\Phi(t)}\right\rangle$ is the average phase factor of the ensemble of trajectories of length vt.

The magnetic field thus appears as a way to probe phase coherence. In particular, the Aharonov–Bohm effect gives rise to the spectacular Sharvin–Sharvin effect, in which the average conductance has a contribution which oscillates with period $h/2e$ (section 7.6.2). To evaluate the coherent contribution in the presence of a magnetic field, we must look for solutions of the *covariant diffusion equation* which replaces (1.15)

$$\left[\frac{\partial}{\partial t} - D\left(\nabla_{\boldsymbol{r}} + i\frac{2e}{\hbar}\boldsymbol{A}(\boldsymbol{r})\right)^2\right] P(\boldsymbol{r},\boldsymbol{r}',t) = \delta(\boldsymbol{r}-\boldsymbol{r}')\delta(t). \tag{1.31}$$

The dephasing (1.29) resulting from the application of a magnetic field changes the phase accumulated along a multiple scattering trajectory. In contrast, to describe the coupling to other degrees of freedom, we average locally the relative dephasing between the two complex amplitudes which interfere. This results from our incomplete knowledge of the internal quantum state of the scatterers. The average over the scatterers' degrees of freedom gives rise to an irreversible dephasing, which we describe using a finite phase coherence time τ_ϕ. We show in Chapter 6 that the contributions of the Diffuson and the Cooperon are modified by a phase factor which in general decreases exponentially with time:

$$\left\langle e^{i\Phi(t)}\right\rangle \propto e^{-t/\tau_\phi}. \tag{1.32}$$

Here $\langle \cdots \rangle$ indicates an average over both disorder and these other degrees of freedom. The determination of the phase coherence time requires the evaluation of the average in (1.32). This notion of dephasing extends to any perturbation whose effect is to modify the phase relation between paired multiple scattering trajectories. We present such an example just below.

Dynamics of scatterers – diffusing wave spectroscopy (Chapters 6 and 9)

When properly understood, a source of dephasing is not necessarily a nuisance, but may be used to study the properties of the diffusive medium. Thus, in the case of scattering of electromagnetic waves, it is possible, by measuring the time autocorrelation function of the electromagnetic field, to take advantage of the coherent multiple scattering to deduce information about the dynamics of the scatterers which is characterized by a time scale τ_b. In fact, since the scatterer velocity is usually much smaller than that of the wave, if we send light pulses at different times, 0 and T, we can probe different realizations of the random potential. The paired trajectories thus explore different configurations separated by a time T. This results in a dephasing which depends on the motion of the scatterers during the time interval T. The time correlation function of the electric field E at a point $\boldsymbol{r}$ (with a source at $\boldsymbol{r}_0$) is of the form

$$\langle E(\boldsymbol{r},T)E^*(\boldsymbol{r},0)\rangle \propto \left\langle \sum_j a_j(\boldsymbol{r}_0,\boldsymbol{r},T)a_j^*(\boldsymbol{r}_0,\boldsymbol{r},0)\right\rangle, \tag{1.33}$$

where the average is over both the configurations and the motion of the scatterers. It is given as a function of the classical probability (the Diffuson):

$$\boxed{\langle E(\boldsymbol{r},T)E^*(\boldsymbol{r},0)\rangle \propto \int_0^\infty dt\, P_{cl}(\boldsymbol{r}_0,\boldsymbol{r},t)\, e^{-t/\tau_s}} \tag{1.34}$$

The characteristic time τ_s, which depends on the dynamics of the scatterers, is related to τ_b and to T. This technique which consists in measuring the time correlations of the field or the intensity is known as *diffusing wave spectroscopy*. It gives information about the dynamics of the scatterers. Since the long multiple scattering paths decorrelate very quickly, we may obtain information about the dynamics at very short times. This idea is used in the study of turbid media.

Density of states (Chapter 10)

The preceding examples dealt with the transport of waves or electrons. The case of thermodynamic quantities is more delicate. They are expressed as a function of the density of states, which has the form

$$\rho(\epsilon) \propto \int d\boldsymbol{r} \sum_j a_j(\boldsymbol{r},\boldsymbol{r}). \tag{1.35}$$

When averaging over disorder, the phases vanish, and there is no remnant of phase coherence. On the other hand, quantities which are products of densities of states or thermodynamic potentials involve *pairs of trajectories* and are therefore sensitive to the effects of phase coherence. For example, the *fluctuations of the density of states* are of the form

$$\overline{\rho(\epsilon)\rho(\epsilon')} \propto \int d\boldsymbol{r}\, d\boldsymbol{r}' \overline{\sum_{ij} a_i(\boldsymbol{r},\boldsymbol{r}) a_j^*(\boldsymbol{r}',\boldsymbol{r}')}, \tag{1.36}$$

i.e., they involve paired closed trajectories but with different initial points. In order to keep them paired, we notice that, in the integration over initial points, there appears the lengths $\mathcal{L}_i$ of each multiple scattering closed loops. We thus obtain a structure quite close to the classical probability (1.14):

$$\overline{\rho(\epsilon)\rho(\epsilon')} \propto \int d\boldsymbol{r} \overline{\sum_i \mathcal{L}_i\, |a_i(\boldsymbol{r},\boldsymbol{r})|^2} \tag{1.37}$$

but which contains, besides $P_{cl}(\boldsymbol{r},\boldsymbol{r},t)$, the length $\mathcal{L}_i$ of the trajectories, which is proportional to vt. More precisely, the Fourier transform (with respect to $\epsilon-\epsilon'$) of the correlation function $\overline{\rho(\epsilon)\rho(\epsilon')}$ is proportional not to $Z(t)$ but to $t\,Z(t)$:

$$\boxed{\overline{\rho(\epsilon)\rho(\epsilon')} \xrightarrow{\text{FT}} t\,Z(t)} \tag{1.38}$$

The number of levels $N(E)$ in an energy interval E is the integral of the density of states. A particularly useful quantity for the characterization of the spectral correlations is the variance $\Sigma^2(E) = \overline{N^2} - \overline{N}^2$ of this number of levels given by

$$\Sigma^2(E) = \frac{2}{\pi^2}\int_0^\infty dt \frac{Z(t)}{t}\sin^2\frac{Et}{2}. \tag{1.39}$$

For energies less than the Thouless energy E_c, that is, for times greater than τ_D, we are in the ergodic regime and $Z(t) = 1$. Starting from (1.39), we get

$$\Sigma^2(E) \propto \ln E. \tag{1.40}$$

We recover the behavior of the spectral rigidity described by *random matrix theory*. In the opposite limit, when $E \gg E_c$, that is $t \ll \tau_D$, $Z(t)$ depends on the spatial dimension d via expression (1.17), which leads to the non-universal behavior of the variance $\Sigma^2(E) \propto (E/E_c)^{d/2}$. We thus see the role diffusion plays in spectral properties. It is, in principle, possible to determine the Thouless energy and the diffusion coefficient starting from the spectral correlations. Moreover, these spectral properties enable us to distinguish between a good and a poor conductor.

Fluctuations in thermodynamic properties – orbital magnetism (Chapter 14)

The orbital magnetization of an electron gas is given by the derivative of the total energy with respect to the magnetic field:

$$\mathcal{M} \propto -\frac{\partial}{\partial B}\int_{-\epsilon_F}^0 \epsilon\rho(\epsilon, B)\, d\epsilon. \tag{1.41}$$

In the geometry of a ring threaded by a magnetic field, this magnetic response corresponds to the existence of a *persistent current* circulating along the ring. The fluctuations of the magnetization may be calculated simply from the fluctuations of the density of states. Given the definition (1.41), the variance $\overline{\delta\mathcal{M}^2} = \overline{\mathcal{M}^2} - \overline{\mathcal{M}}^2$ takes the form

$$\overline{\delta\mathcal{M}^2} \propto \frac{\partial}{\partial B}\frac{\partial}{\partial B'}\int_{-\epsilon_F}^0\int_{-\epsilon_F}^0 \epsilon\,\epsilon'\overline{\rho(\epsilon, B)\rho(\epsilon', B')}\,d\epsilon\,d\epsilon'\bigg|_{B'=B} \tag{1.42}$$

which upon Fourier transformation leads to

$$\boxed{\overline{\delta\mathcal{M}^2} \propto \int_0^\infty \frac{\partial^2 Z(t, B)}{\partial B^2}\frac{e^{-t/\tau_\phi}}{t^3}dt} \tag{1.43}$$

where the dependence of $Z(t, B)$ on the magnetic field is obtained by solving equation (1.31). Thus, though the average value of the magnetization is not affected by the phase coherence, its distribution is.

Coulomb interaction (Chapter 13)

Until now, we have ignored the Coulomb interaction between electrons. Taking this into account changes numerous physical properties, particularly in the presence of disorder, since the probability that two electrons interact is increased by the diffusive motion of the electrons. For sufficiently high electron densities, the potential is strongly screened and we may describe the effect of the interactions in the Hartree–Fock approximation. It is enough to add an interaction term of the form

$$\frac{1}{2}\int U(\boldsymbol{r}-\boldsymbol{r}')n(\boldsymbol{r})n(\boldsymbol{r}')\,d\boldsymbol{r}\,d\boldsymbol{r}' \simeq \frac{U}{2}\int n^2(\boldsymbol{r})\,d\boldsymbol{r} \tag{1.44}$$

to the total energy, where $U(\boldsymbol{r}-\boldsymbol{r}') \simeq U\delta(\boldsymbol{r}-\boldsymbol{r}')$ is the screened interaction, and $n(\boldsymbol{r})$ is the local electron density, related to the density of states, so that

$$n(\boldsymbol{r})n(\boldsymbol{r}) \propto \int d\epsilon\, d\epsilon' \sum_{ij} a_i^*(\boldsymbol{r},\boldsymbol{r},\epsilon)a_j(\boldsymbol{r},\boldsymbol{r},\epsilon'). \tag{1.45}$$

The correction to the total energy has the form

$$\delta E_{ee} \propto U \int d\boldsymbol{r}\, d\epsilon_1\, d\epsilon_2\, P_{cl}(\boldsymbol{r},\boldsymbol{r},\epsilon_1-\epsilon_2). \tag{1.46}$$

This correction gives an additional contribution to the average magnetization which reads

$$\boxed{\mathcal{M}_{ee} \propto -U \int_0^\infty \frac{\partial Z(t,B)}{\partial B}\frac{e^{-t/\tau_\phi}}{t^2}dt} \tag{1.47}$$

Density of states anomaly (Chapter 13)

The aforementioned shift in energy (1.46) due to the Coulomb interaction also implies a reduction of the density of states at the Fermi level. Formally, the density of states is expressed as the second derivative of δE_{ee} with respect to the shift ϵ measured relative to the Fermi level which, using (1.46), yields

$$\delta\rho(\epsilon) \propto U \int d\boldsymbol{r}\, P_{cl}(\boldsymbol{r},\boldsymbol{r},\epsilon) \tag{1.48}$$

or

$$\boxed{\delta\rho(\epsilon) \propto U \int_0^\infty Z(t)\cos\epsilon t\, dt} \tag{1.49}$$

This correction, called the *density of states anomaly*, is an important signature of the electron–electron interaction, and it depends on the space dimensionality and the sample geometry via the probability $Z(t)$.

Quasiparticle lifetime (Chapter 13)

The lifetime of a single particle electronic state is limited by electron–electron interaction. Using the Fermi golden rule, it is shown that it is related to the square of a matrix element of the screened Coulomb interaction, that is, a product of four wave functions. The average over disorder introduces the probability of return to the origin. We will show that

$$\boxed{\frac{1}{\tau_{ee}(\epsilon)} \propto \int_0^\infty \frac{Z(t)}{t} \sin^2 \frac{\epsilon t}{2} dt} \tag{1.50}$$

where ϵ is the energy shift measured from the Fermi level. This time can be measured by studying how an electronic current injected at the energy ϵ relaxes to equilibrium. The time $\tau_{ee}(\epsilon)$ diverges as the energy ϵ goes to zero, that is, for a particle at the Fermi level. If it diverges faster than ϵ, then a state at the Fermi level remains well defined, and we may still use the framework of the Landau theory of Fermi liquids in which, to a good approximation, the electronic states may be regarded as weakly interacting quasiparticle states. At finite temperature, even for $\epsilon = 0$, the time τ_{ee} remains finite, and it therefore contributes to the reduction of phase coherence. Its temperature dependence, denoted $\tau_{in}(T)$, involves the spatial dimension through the diffusive motion.

The Coulomb interaction may also be viewed as that of a single electron coupled to a fluctuating longitudinal electromagnetic field which originates from the other electrons. The phase coherence time $\tau_\phi(T)$ which affects the Cooperon results from the dephasing due to this fluctuating electromagnetic field.

Calculational methodology: long and short range correlations, characteristic energies

To complete and summarize this introduction, we give a survey of the characteristic energy scales of the different regimes we have discussed. In the upper part of Figure 1.16, the ergodic regime corresponds to long times over which the diffusing wave uniformly explores all the volume at its disposal. For shorter times, i.e., for higher energies, diffusion is free, with the boundaries having no effect. We will not consider the regime of times shorter than the average collision time $\tau_e = l_e/v$ for which the motion becomes ballistic. On the lower scale, we indicate the limit of validity of the *Diffuson approximation*, for which quantum

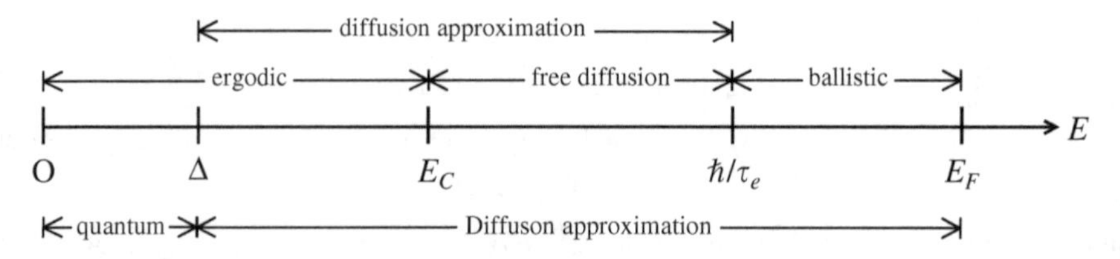

Figure 1.16 Characteristic energy scales defining the different regimes studied in coherent multiple scattering.

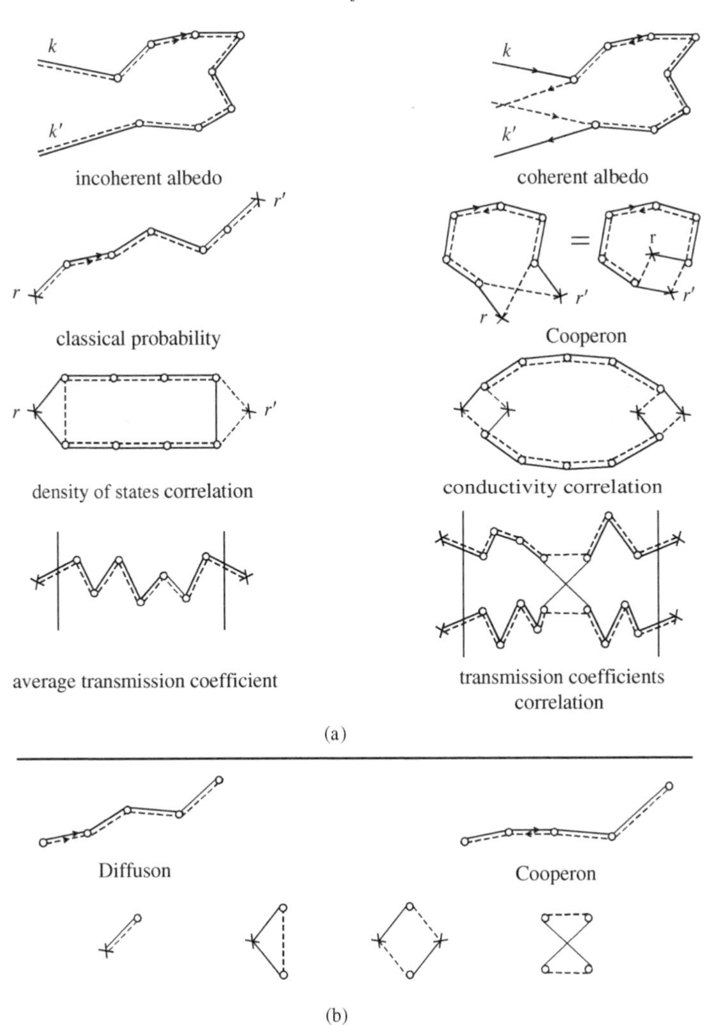

Figure 1.17 (a) Examples of physical quantities whose structure is related to that of the probability $P(\boldsymbol{r}, \boldsymbol{r}')$. (b) The basic building blocks used to calculate these quantities are the classical diffusion probability (the Diffuson), the correction related to phase coherence (the Cooperon), and the short range crossings. The symbol $\circ$ represents a collision and the symbol $\times$ represents an arbitrary point in the medium.

corrections are small. These corrections arise predominantly for energies E smaller than the average level spacing Δ.

The calculation of the physical quantities discussed in the aforementioned examples amounts to evaluating the average of a product of amplitudes associated with multiple scattering trajectories. We have seen how these products may be expressed by means of a pairing of either identical amplitudes or time-reversed amplitudes, the other terms averaging to zero. These pairings can be expressed in terms of the diffusion probability, the solution of a diffusion equation (1.15). This is a long range function whose diagrammatic representation is shown in Figure 1.17. All the quantities of interest involve one or more Diffusons. We have also shown that Diffusons may cross, and we have stressed that these quantum crossings are at the origin of the coherent effects observed in multiple scattering. The crossing is described by a short range function which decays exponentially on a scale of order l_e, and preserves the phase coherence between paired trajectories.

We may thus consider the problem of evaluating different physical quantities as a "construction game" which consists in building a diagrammatic representation which facilitates the calculation. The building blocks in this "game" are the Diffusons, Cooperons, and quantum crossings (Hikami boxes). A few examples are presented in Figure 1.17. We will see that the rules of construction are very precise. This notwithstanding, our hope is that the reader, having read this introduction, will acquire an understanding of the principles which guide the construction of various physical quantities.

2
Wave equations in random media

The general aspects of coherent multiple scattering discussed in Chapter 1 are common to a wide variety of waves which propagate in scattering media. This notwithstanding, each type of wave exhibits its own characteristic behavior. In this chapter, we present several examples of wave equations, and we study two important classes in greater detail: the Schrödinger equation associated with a non-interacting electron gas (weakly disordered metals or semiconductors), and the Helmholtz equation, which describes scalar wave propagation. In addition, we present several models which allow us to describe quantitatively the random nature of the media in which these waves propagate.

2.1 Wave equations

2.1.1 Electrons in a disordered metal

Consider a gas of non-interacting electrons, of charge $-e$ ($e > 0$) and mass m, in the presence of a disordered potential $V(\boldsymbol{r})$. This model describes a metal or a semiconductor. We suppose that the effects of band structure and of electron–electron interactions (Fermi liquid) [17] are taken into account by an effective mass m different from the free electron mass m_0. The states of the system are antisymmetric products of the eigenstates of the single particle Hamiltonian

$$\mathcal{H} = \frac{\boldsymbol{p}^2}{2m} + V(\boldsymbol{r}) \quad \text{with} \quad p = \frac{\hbar}{i}\boldsymbol{\nabla}. \tag{2.1}$$

In the presence of a magnetic field $\boldsymbol{B} = \boldsymbol{\nabla} \times \boldsymbol{A}$, the momentum operator $\boldsymbol{p}$ is replaced by $\boldsymbol{p} + e\boldsymbol{A}$ and the Hamiltonian takes the form

$$\mathcal{H} = -\frac{\hbar^2}{2m}\left(\boldsymbol{\nabla} + \frac{ie}{\hbar}\boldsymbol{A}\right)^2 + V(\boldsymbol{r}). \tag{2.2}$$

In the absence of a potential that explicitly depends on it, the effect of the spin is to introduce a degeneracy factor of 2 for each state. In principle, $V(\boldsymbol{r})$ describes both the confining potential inside the conductor and the scattering potential of the static impurities. We will tacitly assume that the electrons propagate in an infinite medium, so that there is no confining potential.

Exercise 2.1: Hamiltonian in the presence of a magnetic field
Starting from Hamilton's equations,

$$\dot{\boldsymbol{r}} = \frac{\partial \mathcal{H}}{\partial \boldsymbol{p}} \quad \dot{\boldsymbol{p}} = -\frac{\partial \mathcal{H}}{\partial \boldsymbol{r}}, \tag{2.3}$$

verify that in the presence of a magnetic field $\boldsymbol{B} = \nabla \times \boldsymbol{A}$, the Hamiltonian

$$\mathcal{H} = \frac{(\boldsymbol{p} + e\boldsymbol{A})^2}{2m} + V(\boldsymbol{r}) \tag{2.4}$$

leads to Newton's equation:

$$m\ddot{\boldsymbol{r}} = -e\dot{\boldsymbol{r}} \times \boldsymbol{B} - \boldsymbol{\nabla} V(\boldsymbol{r}). \tag{2.5}$$

2.1.2 Electromagnetic wave equation – Helmholtz equation

The case of electromagnetic waves is special, for several reasons. It is probably one of the first examples where the question of the effect of a random medium on phase coherence was posed. In the beginning of the twentieth century, very precise studies were carried out on electromagnetic wave propagation through diffusive media, specifically the atmosphere. From a conceptual viewpoint, this problem stimulated the community working on the theory of probability, who regarded it as a new field for the application of methods developed for the study of Brownian motion [18]. For the atmosphere, the description in terms of a static disordered medium is not appropriate. For many other cases, however, the description in terms of static disorder described by a time-independent potential works well, and it is this case that we consider in the present chapter.[1]

Consider a heterogeneous, non-dissipative, non-magnetic dielectric medium whose dielectric constant $\epsilon(\boldsymbol{r}) = \overline{\epsilon} + \delta\epsilon(\boldsymbol{r})$ is real and positive, and fluctuates about an average value $\overline{\epsilon}$. Suppose that there is no current source in the medium, so that Maxwell's equations for fields oscillating with frequency ω are

$$\begin{aligned} \boldsymbol{\nabla} \times \boldsymbol{E} &= i\omega \boldsymbol{B} \\ \boldsymbol{\nabla} \times \boldsymbol{H} &= -i\omega \boldsymbol{D}, \end{aligned} \tag{2.6}$$

where the fields are related via $\boldsymbol{D} = \epsilon \boldsymbol{E} = [\overline{\epsilon} + \delta\epsilon(\boldsymbol{r})]\boldsymbol{E}$ and $\boldsymbol{B} = \mu_0 \boldsymbol{H}$. From these expressions we see that the form of the wave equation satisfied by the electric

[1] We will reexamine the validity of this hypothesis in Chapters 6 and 9.

field $\boldsymbol{E}$ is

$$-\nabla^2 \boldsymbol{E} + \nabla(\nabla \cdot \boldsymbol{E}) - \frac{\omega^2}{c^2}\frac{\delta\epsilon(\boldsymbol{r})}{\epsilon_0}\boldsymbol{E} = \frac{\overline{\epsilon}}{\epsilon_0}\frac{\omega^2}{c^2}\boldsymbol{E}, \tag{2.7}$$

where $c^2 = 1/(\overline{\epsilon}\mu_0)$. The term $\nabla \cdot \boldsymbol{E}$ represents the polarization charge density, and it has no analog in electronic systems. The term proportional to $\delta\epsilon(\boldsymbol{r})$ gives rise to scattered waves. We will see in section A2.3.1 that it is possible to decouple the effects of polarization and disorder. Leaving aside polarization effects discussed in detail in Chapters 6 and 8, we consider the case of a scalar wave. In this case, the electric field is expressed via a complex function $\psi(\boldsymbol{r})$ which is a solution of the scalar analog of (2.7)

$$\boxed{-\Delta\psi(\boldsymbol{r}) - k_0^2\mu(\boldsymbol{r})\psi(\boldsymbol{r}) = k_0^2\psi(\boldsymbol{r})} \tag{2.8}$$

where $\mu(\boldsymbol{r}) = \delta\epsilon(\boldsymbol{r})/\overline{\epsilon}$ represents the relative fluctuation of the dielectric constant, and where $k_0 = \overline{n}\omega/c \cdot \overline{n}$ is the average optical index $\sqrt{\overline{\epsilon}/\epsilon_0}$. In this form, the wave equation (2.8) has the same structure as the Schrödinger equation, and its solutions depend on the random function $\mu(\boldsymbol{r})$ which plays a role analogous to the disorder potential in the Schrödinger equation and which is responsible for the wave scattering.

It is instructive to discuss the similarities and differences between the Helmholtz equation (2.8) and the Schrödinger equation (2.2). The quantity k_0^2 which plays the role of the energy is always positive, since in a non-dissipative medium the dielectric constant $\epsilon(\boldsymbol{r})$ is real and positive. It is therefore impossible to have a bound state in a negative potential well. We also note that the scattering potential $V(\boldsymbol{r}) = -\, k_0^2\mu(\boldsymbol{r})$ is proportional to the square of the frequency ω. Consequently, in contrast to electronic systems where localization is enhanced as the electron energy decreases, a decrease in the frequency of an electromagnetic wave leads to a reduction of the scattering strength. At high frequency ω, that is, for short wavelengths λ, we obtain the geometrical optics limit, in which interference effects become less relevant.

2.1.3 Other examples of wave equations

The two examples we have discussed are not the only ones which exhibit effects related to coherent multiple scattering. In fact, these effects are common to all wave phenomena (quantum, optical, hydrodynamic, etc.), independent of dispersion relation and space dimension, provided that there are no non-linear effects. Indeed, these may hide disorder effects. Moreover, non-linear equations often have singular solutions (solitons, vortices,...) whose stability is ensured by a topological constraint which is very difficult to destabilize by means of a disordered potential. The competitive role of disorder and non-linearity is poorly understood [19].

Although this book is solely devoted to the description of phase coherence effects and propagation in random media in electronic and optical contexts, we briefly review other wave phenomena for which this description might also be relevant.

Gravity waves

Consider the surface of an incompressible liquid in the gravity field. In equilibrium, the surface is planar. A perturbation to this equilibrium state propagates in the form of a *gravity wave*, whose characteristic frequency depends on the gravity field. The nature of this wave also depends on the depth of the liquid [20] and we distinguish between two regimes by whether the depth is greater than or less than the wavelength. In the former case, which we will not consider here, the propagation is independent of the details of the bottom surface of the vessel, as opposed to the case of small depth, where the structure of the bottom plays the role of a potential. For a periodic modulation of the bottom, Bragg reflections are observed in the transmission coefficient. A disordered bottom gives rise to random but coherent scattering.

In order to determine the gravity wave equation, we begin with the Euler equation for the velocity field $\boldsymbol{v}$ of an incompressible fluid of density ρ in a gravity field $\boldsymbol{g}$

$$\frac{\partial \boldsymbol{v}}{\partial t} + (\boldsymbol{v} \cdot \boldsymbol{\nabla})\, \boldsymbol{v} = -\frac{1}{\rho} \boldsymbol{\nabla} p + \boldsymbol{g}, \tag{2.9}$$

where p is the pressure in the liquid. To simplify the consideration, suppose that the propagation is one dimensional along an axis Ox parallel to the surface, and does not depend on the y-direction. We consider longitudinal waves for which the velocity components are such that $v_x \gg v_z$. Taking the x and z components of the Euler equation (2.9), and neglecting the non-linear terms, we obtain two equations, which, writing $v = v_x$, are

$$\frac{\partial v}{\partial t} = -\frac{1}{\rho} \frac{\partial p}{\partial x} \tag{2.10}$$

and

$$-g = \frac{1}{\rho} \frac{\partial p}{\partial z}. \tag{2.11}$$

To obtain the latter equation, we have neglected v_z. Since the vessel floor is not planar, we denote the equilibrium liquid height by $h_0(x)$ and that of the perturbed liquid by $h(x,t) = h_0(x) + \eta(x,t)$. Integration of the second equation leads to

$$p = p_0 + \rho g (h(x) - z) \tag{2.12}$$

where p_0 is the pressure above the liquid. Inserting this equation in (2.10), we obtain

$$\frac{\partial v}{\partial t} + g \frac{\partial h}{\partial x} = 0. \tag{2.13}$$

To these equations we must add the continuity equation

$$\frac{\partial h}{\partial t} + \frac{\partial}{\partial x}(hv) = 0 \tag{2.14}$$

which expresses that the time dependence of the height at a given point can only arise from the propagation of the perturbation along Ox. Starting from (2.13) and (2.14), and neglecting the non-linear term proportional to ηv, we obtain the wave equation

$$\boxed{\frac{\partial^2 \eta}{\partial t^2} - g\frac{\partial}{\partial x}\left(h_0(x)\frac{\partial \eta}{\partial x}\right) = 0} \qquad (2.15)$$

which describes one-dimensional gravity waves. Their behavior is driven by the constant g and the height of the liquid. For a planar floor $h_0(x) = h_0$, and (2.15) reduces to the standard wave equation with velocity $c = \sqrt{gh_0}$.

The structure of (2.15) is different from that of the Schrödinger equation or the Helmholtz equation. In the present case, the spatial partial derivative introduces the "potential" $h_0(x)$ which resembles a rigidity modulus in the wave equation for a wire under tension. The influence of a random bottom surface on the propagation of gravity waves has been extensively studied experimentally [21]. The observed behavior may be understood qualitatively using the concepts that we will develop here. However, we note that because of the structure of equation (2.15), it is not easy to calculate perturbatively the effects of multiple scattering. On the other hand, in contrast to quantum waves, the absolute values of the amplitude and phase of gravity waves may be measured. It is thus possible to find the analog of an Aharonov–Bohm flux, which is an irrotational vortex, such as is observed in water draining from a bathtub. We may thus observe the structure of phase dislocations in gravity waves, which are experimentally inaccessible in the quantum mechanical context [22].

Acoustic waves

Acoustic waves in compressible fluids (gases or liquids) arise from small amplitude vibrational motion. A perturbation gives rise to a succession of compressions and expansions at each point in the fluid, which are assumed to be adiabatic. In the absence of gravity, by linearizing Euler equation (2.9), we obtain [20]

$$\frac{\partial \boldsymbol{v}}{\partial t} + \frac{1}{\rho_0}\boldsymbol{\nabla} p = 0, \qquad (2.16)$$

where the density ρ of the fluid may be written as the sum of a constant term ρ_0 and a spatially dependent term $\delta\rho(\boldsymbol{r}, t)$. To this we add the linearized continuity equation

$$\frac{\partial \rho}{\partial t} + \rho_0 \boldsymbol{\nabla} \cdot \boldsymbol{v} = 0. \qquad (2.17)$$

This treatment assumes adiabatic evolution of the fluid (that is, there is no heat exchange between the different parts of the fluid), which allows us to relate the pressure and density variations by $\delta p = (\partial p/\partial\rho)_S \delta\rho$. Differentiating equation (2.17) with respect to time we get

$$\frac{\partial^2 p}{\partial t^2} - \left(\frac{\partial p}{\partial \rho}\right)_S \Delta p = 0 \qquad (2.18)$$

which is the equation of sound propagation in the fluid. The velocity c is given by $c^2 = (\partial p/\partial \rho)_S$. For a monochromatic wave, this equation is analogous to the Helmholtz equation (2.8) for scalar electromagnetic waves or to the Schrödinger equation. From a purely theoretical point of view, there is therefore no difference between these waves (acoustic, electromagnetic, and electron waves); all have the same behavior in the presence of disorder. On the other hand, from an experimental point of view, the differences are essential. For example, while it is possible to measure the phase of a sound wave directly, it is difficult to do so for electromagnetic waves and impossible in quantum mechanics. Consequently, the context of acoustic waves has certain advantages in the study of coherence effects and multiple scattering [23]. In addition, for acoustic waves, there is a great variety of random potentials, for which the scattering cross section may be calculated exactly, and varied over many orders of magnitude by exploiting resonance phenomena [24].

It is also possible to extend the previous considerations to waves propagating in elastic media or solids instead of fluids. The main difference thus arises from the vectorial character of the waves with two transversal and one longitudinal components that propagate with different velocities [25,26].

Let us conclude this section by mentioning several other physical examples where multiple scattering plays an important role:

- propagation of third sound on the surface of a superfluid helium film on a rough substrate [27];
- propagation of sound in a "disordered flute" [28];
- propagation of phonons in glasses [29] or excitons in semiconductors [30];
- propagation of seismic waves [31].

2.2 Models of disorder

For electronic systems described by the Hamiltonian (2.2), the sources of the scattering potential $V(\boldsymbol{r})$ are numerous and poorly understood. In a metal, the source may be dislocations, substitutional impurities, vacancies, grain boundaries, etc. In semiconductors, the majority of experiments are performed in a two-dimensional electron gas realized in an inversion layer at the interface between two semiconductors ($GaAs/Al_xGa_{1-x}As$) or between a semiconductor and an insulator (Si–SiO_2 in structures called Si–MOSFET) [32]. In these structures, because of the discontinuity of the band structure parameters, a potential well is created in which perpendicular motion is quantized, and where only the ground state is occupied. The electron dynamics is therefore constrained to the plane of the interface, and it is well described by a quadratic two-dimensional dispersion relation. The considerable interest in these structures is due to the fact that the carrier density may be modified by doping the AlGaAs layer. This doping induces Coulomb scattering centers whose influence on the transport properties may be diminished by moving the doping centers away from the interface. We thus create a two-dimensional electron gas that is almost in the ballistic regime. In this case, the electron dynamics is governed not by scattering from a random potential but only by the shape of the cavity in which the electrons are confined. The study of the transport and thermodynamics of ballistic systems is in itself a vast subject that we will not treat in this book; the interested reader may consult reference [33]. The realization of extremely pure heterojunctions was crucial to the observation of the fractional quantum Hall effect [34].

Static disorder corresponds to the situation in which the electrons scatter elastically, i.e., with no change in energy, and as such it does not describe inelastic scattering processes which are responsible for the loss of phase coherence. In the rest of this chapter, we will only consider the elastic problem for which the complete Hamiltonian is given by (2.2).

Despite its apparent simplicity, there is no exact solution for the diagonalization of the Hamiltonian $\mathcal{H}$, apart from certain special cases in one dimension [35, 36]. In principle, for a given configuration of the disorder, i.e., for a given realization of the random potential $V(\boldsymbol{r})$, the diagonalization of $\mathcal{H}$ is a single-particle problem which admits a complete set of orthonormal eigenfunctions $\psi_n(\boldsymbol{r})$ and eigenvalues ϵ_n. We shall not discuss the possibility of bound states of the potential.

In the case of the Helmholtz equation (2.8), the fluctuations in the dielectric constant described by the function $\mu(\boldsymbol{r})$ may take different forms. The disorder may be a spatially continuous function as in the case of the refractive index for the atmosphere or a turbulent medium. The Gaussian model studied in the following section is well adapted to this type of situation. On the other hand, a problem such as light scattering by a colloidal suspension (for example, an aqueous suspension of submicrometer size dielectric spheres) corresponds to a model of discrete scatterers or localized impurities. In the following sections, we show that these models (continuous or on a lattice) are equivalent in certain limits that will be made precise. In the case where we can employ an effective model of discrete scatterers, these can be characterized by a scattering cross section that can be obtained starting from microscopic models, the simplest of which is that of Rayleigh scattering. This has a great advantage over electronic systems, since it is possible to control physical parameters such as the strength of disorder, the absorption, or the coupling to other degrees of freedom. We will give examples of these in Chapters 6 and 8.

2.2.1 The Gaussian model

Suppose that the disorder potential $V(\boldsymbol{r})$ is a continuous and random function of position. We choose the zero of energy such that the potential is zero on average, $\overline{V(\boldsymbol{r})} = 0$, where $\overline{\cdots}$ denotes the average over realizations of the disorder. In general, a random medium defined by the function $V(\boldsymbol{r})$ is characterized by a normalized probability distribution

$$P[V(\boldsymbol{r})]\,\mathcal{D}V(\boldsymbol{r}) = \frac{1}{\mathcal{Z}} \exp\left[-a \int d\boldsymbol{r}\, F[V(\boldsymbol{r})] \right] \mathcal{D}V(\boldsymbol{r}). \tag{2.19}$$

Remark: generating functional

In order to calculate the various correlation functions, we introduce a generating functional

$$\Phi[g] = \overline{\exp\left[\int d\boldsymbol{r}\, g(\boldsymbol{r})V(\boldsymbol{r})\right]}, \tag{2.20}$$

where $\overline{\cdots} = \int \cdots P[V(\boldsymbol{r})]\mathcal{D}V(\boldsymbol{r})$. The expansion of $\Phi[g]$ in powers of g,

$$\Phi[g] = \sum_{p=0}^{\infty} \frac{1}{p!} \int d\boldsymbol{r}_1 \cdots d\boldsymbol{r}_p\, g(\boldsymbol{r}_1)\cdots g(\boldsymbol{r}_p)\overline{V(\boldsymbol{r}_1)\cdots V(\boldsymbol{r}_p)}, \tag{2.21}$$

facilitates the calculation of correlation functions

$$\overline{V(\boldsymbol{r}_1)V(\boldsymbol{r}_2)\cdots V(\boldsymbol{r}_n)} = \left.\frac{\delta^n \Phi[g]}{\delta g(\boldsymbol{r}_1)\cdots\delta g(\boldsymbol{r}_n)}\right|_{g=0}. \tag{2.22}$$

Moreover, the "connected" correlation functions (also called "cumulants") are defined by

$$\overline{V(\boldsymbol{r}_1)V(\boldsymbol{r}_2)\cdots V(\boldsymbol{r}_n)}^c = \left.\frac{\delta^n \ln\Phi[g]}{\delta g(\boldsymbol{r}_1)\cdots\delta g(\boldsymbol{r}_n)}\right|_{g=0} \tag{2.23}$$

and satisfy

$$\ln\Phi[g] = \sum_{p=1}^{\infty} \frac{1}{p!} \int d\boldsymbol{r}_1 \cdots d\boldsymbol{r}_p\, g(\boldsymbol{r}_1)\cdots g(\boldsymbol{r}_p)\overline{V(\boldsymbol{r}_1)\cdots V(\boldsymbol{r}_p)}^c. \tag{2.24}$$

For example,

$$\begin{aligned}
\overline{V(\boldsymbol{r}_1)}^c &= \overline{V(\boldsymbol{r}_1)}\\
\overline{V(\boldsymbol{r}_1)V(\boldsymbol{r}_2)}^c &= \overline{V(\boldsymbol{r}_1)V(\boldsymbol{r}_2)} - \overline{V(\boldsymbol{r}_1)}\;\overline{V(\boldsymbol{r}_2)}\\
\overline{V(\boldsymbol{r}_1)V(\boldsymbol{r}_2)V(\boldsymbol{r}_3)}^c &= \overline{V(\boldsymbol{r}_1)V(\boldsymbol{r}_2)V(\boldsymbol{r}_3)} - \overline{V(\boldsymbol{r}_1)V(\boldsymbol{r}_2)}\;\overline{V(\boldsymbol{r}_3)}\\
&\quad - \overline{V(\boldsymbol{r}_1)V(\boldsymbol{r}_3)}\;\overline{V(\boldsymbol{r}_2)} - \overline{V(\boldsymbol{r}_2)V(\boldsymbol{r}_3)}\;\overline{V(\boldsymbol{r}_1)}\\
&\quad + 2\overline{V(\boldsymbol{r}_1)}\;\overline{V(\boldsymbol{r}_2)}\;\overline{V(\boldsymbol{r}_3)}\\
&\text{etc.}
\end{aligned} \tag{2.25}$$

One particularly simple model is obtained by assuming that $V(\boldsymbol{r})$ is a Gaussian random potential described by the probability distribution

$$P[V(\boldsymbol{r})]\mathcal{D}V(\boldsymbol{r}) = \frac{1}{\mathcal{Z}}\exp\left[-\frac{1}{2}\int d\boldsymbol{r}\, d\boldsymbol{r}'\, V(\boldsymbol{r})\Delta(\boldsymbol{r}-\boldsymbol{r}')V(\boldsymbol{r}')\right]\mathcal{D}V(\boldsymbol{r}). \tag{2.26}$$

Its generating functional is

$$\Phi[g] = \exp\left[\frac{1}{2}\int d\boldsymbol{r}\, d\boldsymbol{r}'\, g(\boldsymbol{r})B(\boldsymbol{r}-\boldsymbol{r}')g(\boldsymbol{r}')\right], \tag{2.27}$$

where $\Delta(\boldsymbol{r}-\boldsymbol{r}')$ and $B(\boldsymbol{r}-\boldsymbol{r}')$ satisfy

$$\int d\boldsymbol{r}''\, \Delta(\boldsymbol{r}-\boldsymbol{r}'')B(\boldsymbol{r}'-\boldsymbol{r}'') = \delta(\boldsymbol{r}-\boldsymbol{r}'). \tag{2.28}$$

For this model, only the second cumulant is non-zero. The Gaussian model is thus characterized by

$$\boxed{\begin{aligned} \overline{V(\boldsymbol{r})} &= 0 \\ \overline{V(\boldsymbol{r})V(\boldsymbol{r}')} &= B(\boldsymbol{r}-\boldsymbol{r}') \end{aligned}} \tag{2.29}$$

We also assume that $B(\boldsymbol{r}-\boldsymbol{r}')$ only depends on the distance $|\boldsymbol{r}-\boldsymbol{r}'|$, and that it decays with a characteristic length r_c. In this book, we will be particularly interested in the case where the wavelength λ of radiation or scattered electrons is much greater than r_c. In this case, it is a good approximation to take the two-point correlation function to be

$$\boxed{\overline{V(\boldsymbol{r})V(\boldsymbol{r}')} = B\,\delta(\boldsymbol{r}-\boldsymbol{r}')} \tag{2.30}$$

A random potential $V(\boldsymbol{r})$ with this property is called *white noise*. For the Schrödinger equation, the parameter B has dimensions of energy squared times volume. For the Helmholtz equation (2.8), for which $V(\boldsymbol{r}) = -k_0^2\mu(\boldsymbol{r})$, B has dimensions of inverse length, and we define instead

$$k_0^4\, \overline{\mu(\boldsymbol{r})\mu(\boldsymbol{r}')} = B\,\delta(\boldsymbol{r}-\boldsymbol{r}'). \tag{2.31}$$

2.2.2 Localized impurities: the Edwards model

The Gaussian model contains no information about the microscopic nature of disorder. Another model, introduced by Edwards [37], describes the potential $V(\boldsymbol{r})$ as the contribution of N_i identical impurities in a volume Ω, localized at randomly distributed points $\boldsymbol{r}_j$, and characterized by a potential $v(\boldsymbol{r})$. That is,

$$V(\boldsymbol{r}) = \sum_{j=1}^{N_i} v(\boldsymbol{r}-\boldsymbol{r}_j). \tag{2.32}$$

We take the thermodynamic limit $\Omega \to \infty$, while keeping the density $n_i = N_i/\Omega$ constant. The average distance between impurities is $n_i^{-1/d}$ where d is the space dimensionality. We shall consider the case where $v(\boldsymbol{r})$ is a central potential, with a characteristic range r_0.

In order to compare this model with the Gaussian model, we calculate the generating function associated with the potential (2.32) and with a random (Poisson) distribution of impurities such that

$$P(V)\mathcal{D}V = \prod_{j=1}^{N_i} \frac{d\boldsymbol{r}_j}{\Omega} \tag{2.33}$$

and

$$\begin{aligned}\Phi[g] &= \int \prod_{j=1}^{N_i} \frac{d\boldsymbol{r}_j}{\Omega} \exp\Big(\sum_{j=1}^{N_i} F(\boldsymbol{r}_j)\Big) \\ &= \Big(\int \frac{d\boldsymbol{r}}{\Omega} \exp(F(\boldsymbol{r}))\Big)^{N_i},\end{aligned} \tag{2.34}$$

where $F(\boldsymbol{r}) = \int d\boldsymbol{r}'\, g(\boldsymbol{r}')\, v(\boldsymbol{r}-\boldsymbol{r}')$. The functional $\Phi[g]$ may be written in the form:

$$\Phi[g] = \left(1 + \frac{n_i}{N_i} \int d\boldsymbol{r}\, \left(e^{F(\boldsymbol{r})} - 1\right)\right)^{N_i} \tag{2.35}$$

since $n_i = N_i/\Omega$. In the limit of infinite volume Ω and where the number of impurities N_i also tends to infinity, so that the density n_i is constant, $\Phi[g]$ is given by

$$\Phi[g] = \exp\left[n_i \int d\boldsymbol{r}\, \left(e^{F(\boldsymbol{r})} - 1\right)\right]. \tag{2.36}$$

The cumulant expansion is thus

$$\ln \Phi[g] = \sum_{p=1}^{\infty} \frac{n_i}{p!} \int d\boldsymbol{r}\, F^p(\boldsymbol{r}), \tag{2.37}$$

and, using (2.23), the correlation functions are

$$\begin{aligned}\overline{V(\boldsymbol{r}_1)} &= n_i \int d\boldsymbol{r}\, v(\boldsymbol{r}-\boldsymbol{r}_1) \\ \overline{V(\boldsymbol{r}_1)V(\boldsymbol{r}_2)\cdots V(\boldsymbol{r}_n)}^c &= n_i \int d\boldsymbol{r}\, v(\boldsymbol{r}-\boldsymbol{r}_1)\cdots v(\boldsymbol{r}-\boldsymbol{r}_n).\end{aligned} \tag{2.38}$$

We recover the Gaussian model by taking the limit of a high density ($n_i \to \infty$) of weakly scattering impurities ($v(\boldsymbol{r}) \to 0$), such that cumulants of order higher than $n = 2$ vanish. In this case,

$$\overline{V(\boldsymbol{r})V(\boldsymbol{r}')}^c = B(\boldsymbol{r}-\boldsymbol{r}') = n_i \int d\boldsymbol{r}''\, v(\boldsymbol{r}''-\boldsymbol{r})v(\boldsymbol{r}''-\boldsymbol{r}'). \tag{2.39}$$

By shifting the zero of the energy scale, the first-order moment vanishes, and we obtain the Gaussian model defined by (2.30). The Fourier transform $B(\boldsymbol{q})$ of the correlation function $B(\boldsymbol{r}-\boldsymbol{r}')$ is related to that of the potential, $v(\boldsymbol{r})$ and, using (2.39), it is given by:

$$\boxed{B(\boldsymbol{q}) = n_i v(\boldsymbol{q})^2} \tag{2.40}$$

where $v(\boldsymbol{q})$ is the Fourier transform of $v(\boldsymbol{r})$. Note that for a central potential $v(\boldsymbol{r})$, the correlation function $B(\boldsymbol{r}-\boldsymbol{r}')$ only depends on $|\boldsymbol{r}-\boldsymbol{r}'|$, and its Fourier transform $B(\boldsymbol{q})$ only depends on $q = |\boldsymbol{q}|$. Moreover, according to equation (2.39), the range r_c of the correlation function $B(\boldsymbol{r}-\boldsymbol{r}')$ is of the same order of magnitude as the decay length r_0 of the potential $v(\boldsymbol{r})$. We will predominantly use the Gaussian model throughout this book. In particular, we will often consider the case in which the impurity potential is a δ function: $v(\boldsymbol{r}) = v_0 \delta(\boldsymbol{r})$. In this case, the correlation function $B(\boldsymbol{r}-\boldsymbol{r}')$ is

$$B(\boldsymbol{r}-\boldsymbol{r}') = B\,\delta(\boldsymbol{r}-\boldsymbol{r}') = n_i v_0^2\,\delta(\boldsymbol{r}-\boldsymbol{r}'). \tag{2.41}$$

In Appendix A2.1, we recall several basic and important results for scattering from a potential $v(\boldsymbol{r})$.

2.2.3 The Anderson model

In the case of metals, the electrons, rather than being free, are subjected to a strong potential V_{latt} due to the periodic lattice. The Hamiltonian (2.1) becomes $\mathcal{H} = \boldsymbol{p}^2/2m + V(\boldsymbol{r}) + V_{latt}(\boldsymbol{r})$. This case can be treated using the *tight binding* model, for which the eigenstates can be expressed as a linear combination of atomic orbitals

$$|\phi_n\rangle = \sum_i a_{ni}|i\rangle, \tag{2.42}$$

where $|i\rangle$ is an orbital centered on the atom i, that is, an eigenstate of the single atom Hamiltonian. We assume that the overlaps between atomic orbitals are sufficiently small $(\langle i|i'\rangle = \delta_{ii'})$. The Hamiltonian is characterized by its matrix elements written in the $|i\rangle$ basis. For a disordered medium, the diagonal matrix element $\epsilon_i = \langle i|\mathcal{H}|i\rangle$ represents the random site energy corresponding to the atom i, while the off-diagonal element $\langle i|\mathcal{H}|j\rangle = t_{ij}$ describes hopping from site i to site j.

The disorder appears as a random site energy ϵ_i for each site. The Hamiltonian for this seminal model, introduced by Anderson in 1958 [38], is written as

$$\mathcal{H} = -\sum_{i,j} t_{ij}|i\rangle\langle j| + \sum_i \epsilon_i |i\rangle\langle i| \tag{2.43}$$

or, in second quantized form,

$$\mathcal{H} = -\sum_{i,j} t_{ij} a_i{}^{\dagger} a_j + \sum_i \epsilon_i a_i{}^{\dagger} a_i. \tag{2.44}$$

We suppose that the site disorder is uniformly distributed in an interval of width W such that $-W/2 \leq \epsilon_i \leq W/2$, where W is the strength of the disorder. Moreover, we assume that these energies are uncorrelated from site to site. Thus,

$$\overline{\epsilon_i} = 0$$
$$\overline{\epsilon_i \, \epsilon_j} = \frac{W^2}{12} \delta_{ij} \tag{2.45}$$

where δ_{ij} is the Kronecker symbol. One simplification which does not restrict the generality of the model consists in only permitting jumps from a site i to its z nearest neighbors $i + p$ on a cubic lattice. In this approximation, we take $t_{ij} = t\delta_{j,i+p}$ where t is a constant amplitude. The kinetic energy of the electrons depends on their density (i.e., on the Fermi energy). It is proportional to the bandwidth $2zt$. In this description, there is a natural dimensionless parameter W/t which describes the strength of disorder. The case $W = 0$ describes a perfect crystal. The advantage of this model is that it lends itself easily to numerical calculations. It has been used to demonstrate that, in three dimensions, there is a localization transition, the so-called Anderson transition. That is, there is a critical value of the ratio W/t at which the system changes its nature, qualitatively from *diffusive*, where the wave functions are spatially extended, to *localized*, for which the wave functions are spatially localized [39].

In the presence of a magnetic field $\boldsymbol{B} = \nabla \times \boldsymbol{A}$, the Anderson Hamiltonian becomes

$$\mathcal{H} = -\sum_{ij} t_{ij} \, e^{i\theta_{ij}} a_i{}^\dagger a_j + \sum_i \epsilon_i a_i{}^\dagger a_i, \tag{2.46}$$

where θ_{ij} is proportional to the line integral of the vector potential $\boldsymbol{A}$ along the bond (ij):

$$\theta_{ij} = -\frac{e}{\hbar} \int_i^j \boldsymbol{A} \cdot d\boldsymbol{l}. \tag{2.47}$$

Remark: effect of a magnetic field

In the presence of a magnetic field, the form of the Hamiltonian (2.46) is far from obvious. In the absence of disorder, the wave vector is a good quantum number, and the solutions $|\boldsymbol{k}\rangle$ of the tight binding equation

$$\mathcal{H}(\hbar \boldsymbol{k})|\boldsymbol{k}\rangle = \epsilon(\boldsymbol{k})|\boldsymbol{k}\rangle$$

are Bloch waves. In the presence of a magnetic field, the Hamiltonian is replaced by an effective Hamiltonian

$$\mathcal{H}(\hbar \boldsymbol{k} + e\boldsymbol{A}(\boldsymbol{r})),$$

where, by analogy to the case of a free particle, we have replaced $\hbar \boldsymbol{k}$ by $\hbar \boldsymbol{k} + e\boldsymbol{A}$. The justification for this replacement, called the Peierls substitution, is difficult. It has been discussed in particular by Peierls, Kohn, and Wannier [40,41]. It is certainly justified so long as the field is sufficiently weak that the magnetic length l_B, defined by $l_B^2 = \hbar/(eB)$, is large compared to the lattice constant a, which is valid for experimentally accessible fields in metals. This substitution, which derives from the replacement of the translation operator $e^{i\boldsymbol{k}\cdot\boldsymbol{r}}$ by $e^{i\boldsymbol{k}\cdot\boldsymbol{r}-i\frac{e}{\hbar}\int_0^r \boldsymbol{A}\cdot d\boldsymbol{l}}$, leads to the replacement of the second quantized Hamiltonian by the gauge invariant form (2.46).

Appendix A2.1: Theory of elastic collisions and single scattering

The Edwards model relates the characteristics of a Gaussian potential $V(\boldsymbol{r})$ to those of the potential $v(\boldsymbol{r})$ of individual impurities. In this appendix, we review, in broad terms, the theory of scattering of a scalar wave by a *single localized obstacle* described by a potential $v(\boldsymbol{r})$. This theory is the object of detailed studies [42,43]. Having defined some basic quantities, we consider the scattering of electrons by a spherically symmetric potential barrier [43,44]. For scalar waves, the main results apply to different wave equations. We shall consider these in the framework of the Helmholtz equation.

Consider the problem of the scattering of a wave by a *localized* potential $v(\boldsymbol{r})$, that is, a potential which obeys the condition $\lim_{r\to\infty} r v(\boldsymbol{r}) = 0$. We seek solutions to the Helmholtz equation

$$\left(\Delta + k_0^2\right)\psi(\boldsymbol{r}) = v(\boldsymbol{r})\psi(\boldsymbol{r}) \tag{2.48}$$

which satisfy the boundary condition

$$\psi(\boldsymbol{r}) \propto \frac{e^{ik_0 r}}{r} \qquad r \to +\infty, \tag{2.49}$$

corresponding to an asymptotically emergent spherical wave. The solutions of (2.48) are obtained from the associated Green function $G_0(\boldsymbol{r},\boldsymbol{r}',k_0)$,[2] defined by

$$(\Delta_{\boldsymbol{r}} + k_0^2)G_0(\boldsymbol{r},\boldsymbol{r}',k_0) = \delta(\boldsymbol{r}-\boldsymbol{r}'), \tag{2.50}$$

such that the general solution of (2.48) may be written as

$$\psi(\boldsymbol{r}) = \int d\boldsymbol{r}'\, G_0(\boldsymbol{r},\boldsymbol{r}',k_0)v(\boldsymbol{r}')\psi(\boldsymbol{r}'). \tag{2.51}$$

To this solution, we may add the solution of the homogeneous equation

$$\left(\Delta + k_0^2\right)\phi(\boldsymbol{r}) = 0 \tag{2.52}$$

[2] More details on Green's functions are given in section 3.1.

which describes the incident wave, which we choose to be a plane wave $\phi(\boldsymbol{r}) = e^{i\boldsymbol{k}\cdot\boldsymbol{r}}$ with $|\boldsymbol{k}| = k_0$. Using expression (3.48) for the free Green function:

$$G_0(\boldsymbol{r}, \boldsymbol{r}', k_0) = -\frac{1}{4\pi}\frac{e^{ik_0|\boldsymbol{r}-\boldsymbol{r}'|}}{|\boldsymbol{r}-\boldsymbol{r}'|}, \tag{2.53}$$

we obtain

$$\psi(\boldsymbol{r}) = e^{i\boldsymbol{k}\cdot\boldsymbol{r}} - \frac{1}{4\pi}\int d\boldsymbol{r}'\frac{e^{ik_0|\boldsymbol{r}-\boldsymbol{r}'|}}{|\boldsymbol{r}-\boldsymbol{r}'|}v(\boldsymbol{r}')\psi(\boldsymbol{r}'). \tag{2.54}$$

A2.1.1 Asymptotic form of the solutions

In order to study the asymptotic form ($r \to \infty$) of the solution (2.54), we write, for $r \gg r'$ (Figure 2.1)

$$k_0|\boldsymbol{r}-\boldsymbol{r}'| = k_0 r\sqrt{1 + \left(\frac{r'}{r}\right)^2 - 2\frac{\boldsymbol{r}\cdot\boldsymbol{r}'}{r^2}} \simeq k_0 r - \boldsymbol{k}'\cdot\boldsymbol{r}' \tag{2.55}$$

where $\boldsymbol{k}' = k_0\boldsymbol{r}/r$. We thus obtain an approximate form for the free Green function (known as the Fraunhoffer approximation),

$$G_0(\boldsymbol{r}, \boldsymbol{r}', k_0) \simeq -\frac{e^{ik_0 r}}{4\pi r}e^{-i\boldsymbol{k}'\cdot\boldsymbol{r}'}, \tag{2.56}$$

which we substitute into (2.54):

$$\psi(\boldsymbol{r}) \simeq e^{i\boldsymbol{k}\cdot\boldsymbol{r}} - \frac{e^{ik_0 r}}{4\pi r}\int d\boldsymbol{r}'\, e^{-i\boldsymbol{k}'\cdot\boldsymbol{r}'}v(\boldsymbol{r}')\psi(\boldsymbol{r}'). \tag{2.57}$$

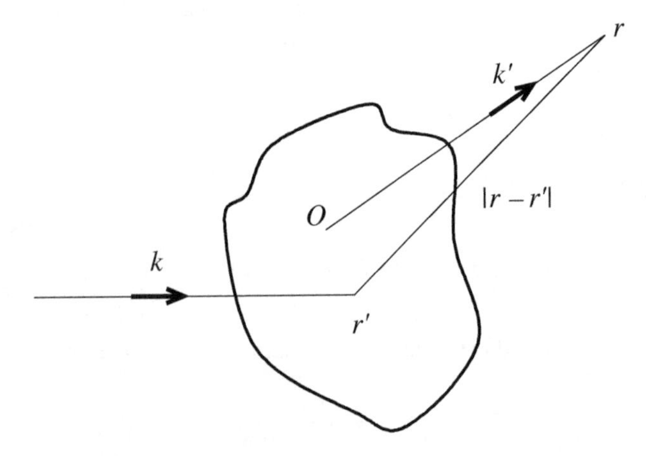

Figure 2.1 Scattered wave after interaction with the potential at point $\boldsymbol{r}'$. In the Fraunhoffer approximation, we have $|\boldsymbol{r}'| \ll |\boldsymbol{r}|$.

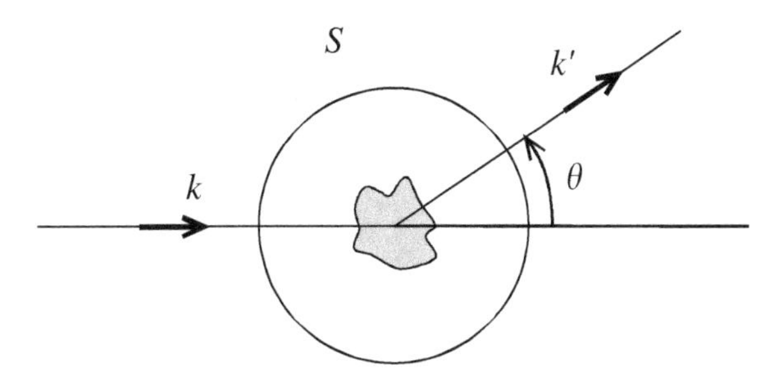

Figure 2.2 Scattering by a localized potential. The total flux $F = F_{inc} + F_{out} + F_s$ crossing the surface S is zero. Since $F_{inc} = \int \boldsymbol{k} \cdot dS = 0$, $F_{out} + F_s$ must vanish. This result constitutes the optical theorem.

Written in the form

$$\psi(\boldsymbol{r}) = e^{i\boldsymbol{k}\cdot\boldsymbol{r}} + \frac{e^{ik_0 r}}{r} f(\boldsymbol{k}, \boldsymbol{k}'), \tag{2.58}$$

this relation defines the *scattering amplitude*

$$\boxed{f(\boldsymbol{k}, \boldsymbol{k}') = -\frac{1}{4\pi} \int d\boldsymbol{r}'\, e^{-i\boldsymbol{k}'\cdot\boldsymbol{r}'} v(\boldsymbol{r}')\psi(\boldsymbol{r}')} \tag{2.59}$$

which has the dimensions of a length. Since the modulus k_0 of the wave vector $\boldsymbol{k}$ is conserved in an elastic collision, the scattering amplitude only depends on $k_0 = |\boldsymbol{k}| = |\boldsymbol{k}'|$ and on the scattering angle θ between the directions $\boldsymbol{k}$ and $\boldsymbol{k}'$ (Figure 2.2), hence $f(\boldsymbol{k}, \boldsymbol{k}') = f(k_0, \theta)$. The expression (2.59) for the scattering amplitude has been defined for the Helmholtz equation. For the Schrödinger equation with a potential $v(\boldsymbol{r})$, it takes the equivalent form:

$$\boxed{f(\boldsymbol{k}, \boldsymbol{k}') = -\frac{m}{2\pi\hbar^2} \int d\boldsymbol{r}'\, e^{-i\boldsymbol{k}'\cdot\boldsymbol{r}'} v(\boldsymbol{r}')\psi(\boldsymbol{r}')} \tag{2.60}$$

In the *low energy limit*, that is, for $k_0 r_0 \ll 1$ where r_0 is the range of the potential $v(\boldsymbol{r})$, the phase factor of equation (2.59) may be taken equal to 1, and the scattering amplitude is independent of $\boldsymbol{k}'$: *the scattering is isotropic*. We thus define the *scattering length* a_s by the limit

$$a_s = -\lim_{k_0 \to 0} f(\boldsymbol{k}, \boldsymbol{k}') \tag{2.61}$$

or equivalently,

$$a_s = \frac{1}{4\pi} \int d\boldsymbol{r}'\, v(\boldsymbol{r}')\psi(\boldsymbol{r}'). \tag{2.62}$$

Exercise 2.2: scattering by a spherical potential
Consider scattering by the central potential $v(r)$ defined as

$$\begin{aligned} v(r) &= U_0 \quad \text{for} \quad r \leq a \\ v(r) &= 0 \quad \text{for} \quad r > a. \end{aligned} \tag{2.63}$$

In the low energy limit, the scattering is isotropic. Calculate the scattering length a_s, defined by (2.61). To that purpose, solve the Helmholtz equation in the limit of zero energy. Introduce the function $u = r\psi$ to reduce the problem to the solution of a simple differential equation, with appropriate continuity conditions at $r = a$. By comparing this solution to the asymptotic form (2.58) for $k_0 \to 0$, show that the scattering length is given by

$$a_s = a - \frac{\tanh Ka}{K} \tag{2.64}$$

where the parameter K is equal to $\sqrt{U_0}$, for the Helmholtz equation. For the Schrödinger equation, the expression for the scattering length is identical, but the parameter K is equal to $\sqrt{2mU_0}/\hbar$.

A2.1.2 Scattering cross section and scattered flux

Starting with a function $\psi(\boldsymbol{r})$ solution of the Helmholtz equation (2.48), we define the flux associated with the current through a surface S by the expression (Figure 2.2):

$$F = \operatorname{Im} \int d\boldsymbol{S} \cdot \psi^*(\boldsymbol{r}) \nabla \psi(\boldsymbol{r}). \tag{2.65}$$

For an incident flux on the potential $v(\boldsymbol{r})$, we measure the emergent flux in a given direction, characterized by an angle θ, in an element of solid angle $d\Omega$. More precisely, we define the *differential scattering cross section* by the ratio between the flux emerging in an element of solid angle $d\Omega$ and the incident flux per unit surface:

$$\sigma(\theta) = \frac{dF_{out}/d\Omega}{dF_{inc}/dS}. \tag{2.66}$$

This relation is quite general, and applies to other types of waves.[3] The normalization of the flux in equation (2.65) contains a prefactor which depends on the specific physical problem we consider. For a quantum particle whose evolution is governed by the Schrödinger equation, the flux is defined through the probability amplitude [45]. The differential scattering cross section thus describes the ratio between the number of particles scattered in a solid angle $d\Omega$ and the incident particle flux per unit surface. In the case of an electromagnetic field, we may use this definition of the scattering cross section where F describes the energy flux $nc\hbar\omega$ of photons, and where n is the photon density [46]. In the classical limit, this is the flux of the Poynting vector.

[3] It also applies to classical particles.

From expression (2.58), we deduce that the incident flux per unit surface equals $dF_{inc}/dS = k_0$. The total flux of the scattered wave is

$$F_{out} = \mathrm{Im} \int dS \cdot f^*(k_0, \theta) \frac{e^{-ik_0 r}}{r} \nabla \left[f(k_0, \theta) \frac{e^{ik_0 r}}{r} \right] \tag{2.67}$$

which, for large distances ($k_0 r \gg 1$), becomes

$$F_{out} = 2\pi k_0 \int_0^{\pi} |f(k_0, \theta)|^2 \sin\theta d\theta \tag{2.68}$$

so that the differential scattering cross section in the direction θ is equal to the square of the modulus of the scattering amplitude:

$$\boxed{\sigma(\theta) = |f(k_0, \theta)|^2} \tag{2.69}$$

We define the total scattering cross section

$$\sigma = 2\pi \int_0^{\pi} \sigma(\theta) \sin\theta d\theta \tag{2.70}$$

which is related by (2.68) to the total scattered flux F_{out} in all directions:

$$F_{out} = k_0 \sigma. \tag{2.71}$$

In the low energy limit, the scattering amplitude has no angular dependence, and the total scattering cross section may be written in terms of the scattering length (2.61):

$$\sigma = 4\pi a_s^2. \tag{2.72}$$

We also define the transport scattering cross section σ^* which measures the flux projected along the direction of the incident wave, which is

$$\sigma^* = 2\pi \int_0^{\pi} \sigma(\theta)(1 - \cos\theta) \sin\theta d\theta. \tag{2.73}$$

The scattering cross sections σ and σ^* are equal if the differential scattering cross section has no angular dependence. The transport scattering cross section σ^* plays an important role in multiple scattering since it is related to the diffusion coefficient (Appendix A4.3) which determines most of the physical properties described in this book.

A2.1.3 Optical theorem Shadow term

Energy conservation imposes an important relation between the total scattering cross section and the scattering amplitude in the forward direction. This relation is known as the optical theorem. Consider the total flux F crossing a spherical surface of radius r. Using the asymptotic form (2.58), we may decompose this flux such that $F = (\boldsymbol{k} \cdot \int d\boldsymbol{S}) + F_{out} + F_s = 0$. The first term, which corresponds to the flux of the incident wave, cancels over the entire surface. The second, given by equation (2.67), is the flux of the scattered wave. The third term F_s describes *the interference between the incident and the scattered waves*. At large distances, it takes the form:

$$F_s = \mathrm{Im} \int d\boldsymbol{S} \cdot \left[i\boldsymbol{k} f^*(k_0, \theta) \frac{e^{i(\boldsymbol{k}\cdot\boldsymbol{r} - k_0 r)}}{r} + e^{-i\boldsymbol{k}\cdot\boldsymbol{r}} \nabla \left(f(k_0, \theta) \frac{e^{ik_0 r}}{r} \right) \right] \tag{2.74}$$

namely,

$$F_s = 2\pi k_0 \int_0^{\pi} \sigma_s(\theta) \sin\theta \, d\theta \tag{2.75}$$

with

$$\sigma_s(\theta) = r(1 + \cos\theta) \, \mathrm{Re} \left[f(k_0, \theta) \, e^{ik_0 r (1 - \cos\theta)} \right]. \tag{2.76}$$

This function oscillates very rapidly as a function of the angle θ. Its contribution to the interference term F_s is only appreciable in an angular window whose size is proportional to $1/\sqrt{k_0 r}$. This contribution, called the "shadow term," represents the diminution due to the scattering of the flux in the direction of the incident wave vector. To calculate this contribution, we integrate by parts after having substituted $x = \cos\theta$, and we obtain

$$F_s = \mathrm{Re} \left[4i\pi f(k_0, 0) - 2i\pi \, e^{ik_0 r} \int_{-1}^{1} dx \, e^{-ik_0 r x} \Big(f(k_0, x) + (1 + x) f'(k_0, x) \Big) \right]. \tag{2.77}$$

The integral in the above expression vanishes in the limit $k_0 r \to \infty$, leaving

$$F_s = -4\pi \, \mathrm{Im} f(k_0, \theta = 0). \tag{2.78}$$

Conservation of total flux implies that the interference term F_s exactly compensates the scattered flux F_{out}, namely $F_{out} + F_s = 0$, yielding, using expressions (2.71) and (2.78):

$$\boxed{\sigma = \frac{4\pi}{k_0} \mathrm{Im} f(k_0, \theta = 0)} \tag{2.79}$$

This result is known as the *optical theorem*, and it is a consequence of the conservation of flux in elastic scattering.

Scattering matrix and optical theorem

In this section, we present a different, more formal derivation of the optical theorem.[4] It applies equally to Helmholtz and Schrödinger equations (here we will use $\hbar^2/2m = 1$). Define first the *scattering operator* or *S-matrix* which relates the incident $|\psi_{in}\rangle$ and emergent $|\psi_{out}\rangle$ states:

$$|\psi_{out}\rangle = S|\psi_{in}\rangle. \tag{2.80}$$

The incident state $|\psi_{in}\rangle = |\boldsymbol{k}\rangle$ is an eigenstate of the free Hamiltonian $\mathcal{H}_0 = -\Delta$, of energy $E = k_0^2$, that is,

$$\mathcal{H}_0|\psi_{in}\rangle = E|\psi_{in}\rangle. \tag{2.81}$$

The emergent state $|\psi_{out}\rangle$ is an eigenstate of the total Hamiltonian $\mathcal{H} = \mathcal{H}_0 + v$ with the same energy, namely

$$\mathcal{H}|\psi_{out}\rangle = E|\psi_{out}\rangle. \tag{2.82}$$

At infinity, the two states $|\psi_{in}\rangle$ and $|\psi_{out}\rangle$ only differ by the scattered wave, whose amplitude tends to zero. As a consequence, the two states are identical up to a phase difference. Subtracting equations (2.81) and (2.82) and using the definition (2.80) of the S-matrix, we obtain

$$vS = (E - \mathcal{H}_0)(S - \mathbb{1}) \tag{2.83}$$

which yields $S = \mathbb{1}$ for $v = 0$. Using the resolvant operator G_0 associated with the free problem and defined by $(E - \mathcal{H}_0)G_0 = \mathbb{1}$, we can rewrite a Dyson equation:[5]

$$\boxed{S = \mathbb{1} + G_0 vS} \tag{2.84}$$

Projecting this equation on an incident state $|\boldsymbol{k}\rangle$ we recover relation (2.54) in the form

$$|\psi_{out}\rangle = |\boldsymbol{k}\rangle + G_0 v|\psi_{out}\rangle. \tag{2.85}$$

Let us now define the *scattering operator* $T = vS$ which satisfies what is known as the Lippman–Schwinger equation:

$$S = \mathbb{1} + G_0 T. \tag{2.86}$$

[4] This section relies on certain results in Chapter 3.

[5] This Dyson equation allows us to obtain the expansion of the S-matrix

$$S = \mathbb{1} + G_0 V + G_0 V G_0 V + \cdots .$$

The Born approximation, which will be studied in greater detail in the following section, consists of only keeping the first two terms in the expansion. This approximation is valid provided the matrix elements of the potential V are small.

Applying this equality to the state $|\boldsymbol{k}\rangle$, we obtain

$$|\psi_{out}\rangle = |\boldsymbol{k}\rangle + G_0 T|\boldsymbol{k}\rangle \tag{2.87}$$

whose projection on $|\boldsymbol{r}\rangle$ yields equation (2.54). Indeed, by using the closure relation, we get

$$\langle \boldsymbol{r}|\psi_{out}\rangle = e^{i\boldsymbol{k}\cdot\boldsymbol{r}} + \int d\boldsymbol{r}' \, \langle \boldsymbol{r}|G_0|\boldsymbol{r}'\rangle\langle \boldsymbol{r}'|T|\boldsymbol{k}\rangle \tag{2.88}$$

and the asymptotic expansion (2.56) of $\langle \boldsymbol{r}|G_0|\boldsymbol{r}'\rangle$ leads to (2.58) with

$$f(\boldsymbol{k},\boldsymbol{k}') = -\frac{1}{4\pi}\langle \boldsymbol{k}'|T|\boldsymbol{k}\rangle. \tag{2.89}$$

To derive the optical theorem, we start with the Lippman–Schwinger equation (2.87). Multiplying by $T^\dagger$ and projecting on the state $|\boldsymbol{k}\rangle$, we obtain for the imaginary part

$$\mathrm{Im}\langle \boldsymbol{k}|T^\dagger|\boldsymbol{k}\rangle = -\mathrm{Im}\langle \boldsymbol{k}|T^\dagger G_0 T|\boldsymbol{k}\rangle, \tag{2.90}$$

since $\mathrm{Im}\langle\psi_{out}|v|\psi_{out}\rangle = 0$. With the aid of (3.52), we have

$$\mathrm{Im}\langle \boldsymbol{k}|T^\dagger|\boldsymbol{k}\rangle = \frac{\pi}{2k_0}\sum_{\boldsymbol{k}'}|\langle \boldsymbol{k}'|T|\boldsymbol{k}\rangle|^2\delta(k'-k_0) \tag{2.91}$$

or

$$-\mathrm{Im}\langle \boldsymbol{k}|T|\boldsymbol{k}\rangle = \frac{k_0}{8\pi}\int_0^\pi |\langle \boldsymbol{k}'|T|\boldsymbol{k}\rangle|^2 \sin\theta d\theta. \tag{2.92}$$

Finally, from (2.70) and (2.89), we deduce that

$$\sigma = \frac{4\pi}{k_0}\mathrm{Im}f(\theta=0) = -\frac{1}{k_0}\mathrm{Im}\langle \boldsymbol{k}|T|\boldsymbol{k}\rangle \tag{2.93}$$

which is the optical theorem (2.79).

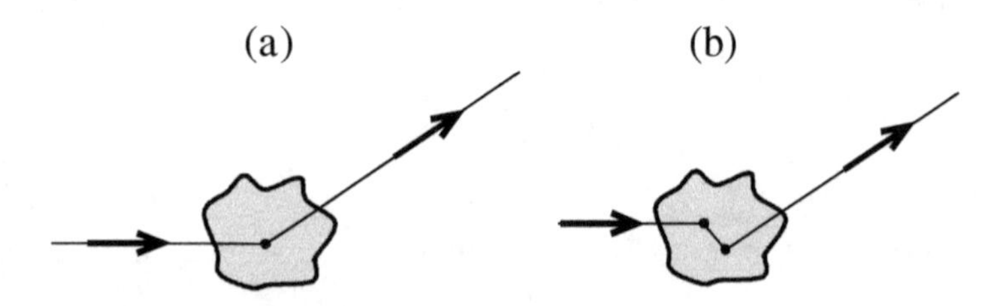

Figure 2.3 (a) In the Born approximation, the wave interacts with the potential only once. (b) Beyond this approximation, the wave is scattered many times by the potential.

A2.1.4 Born approximation

This approximation (Figure 2.3), valid in the limit of a weak potential, consists of retaining only the first two terms in the expansion of the S-matrix derived from equation (2.84), that is, $S = 1 + G_0 v$. In other words, this amounts to assuming that $T = v$.[6] This approximation arises naturally in the framework of the Gaussian model since it is deduced from the Edwards model in the limit of an infinite density of scatterers with a potential $v(\boldsymbol{r})$ whose amplitude tends to zero.

The scattering amplitude is thus simplified considerably, since it suffices to replace in (2.59) the wave function $\psi(\boldsymbol{r}')$ by the plane wave $e^{i\boldsymbol{k}\cdot\boldsymbol{r}'}$ describing the incident wave. This allows us to rewrite the amplitude $f(\boldsymbol{k},\boldsymbol{k}')$ as the Fourier transform of the scattering potential. Thus, the approximate amplitude $f_B(\boldsymbol{k},\boldsymbol{k}')$ is

$$f_B(\boldsymbol{k},\boldsymbol{k}') = -\frac{1}{4\pi}\int d\boldsymbol{r}'\, e^{i(\boldsymbol{k}-\boldsymbol{k}')\cdot\boldsymbol{r}'} v(\boldsymbol{r}') = -\frac{1}{4\pi} v(\boldsymbol{k}-\boldsymbol{k}') \tag{2.94}$$

for the Helmholtz equation. The corresponding differential scattering cross section, defined by (2.69), may be written

$$\sigma_B(k_0,\theta) = \frac{1}{16\pi^2} v^2(\boldsymbol{k}-\boldsymbol{k}'). \tag{2.95}$$

These expressions simplify further in the case where the potential $v(r)$ is central, since the scattering amplitude, and thus the differential scattering cross section, only depend on the modulus $|\boldsymbol{k}-\boldsymbol{k}'|$. Introducing the vector $\boldsymbol{q} = \boldsymbol{k}-\boldsymbol{k}'$ whose norm is $q = 2k_0 \sin(\theta/2)$, where θ is the angle between the directions $\boldsymbol{k}$ and $\boldsymbol{k}'$, we obtain

$$f_B(k_0,\theta) = -\int_0^\infty r dr \frac{\sin qr}{q} v(r). \tag{2.96}$$

This result has an immediate application in the case of the Edwards model discussed in section 2.2.2. Using (2.95), we can relate the correlation function (2.40) of the potential to the differential scattering cross section of a scatterer

$$B(\boldsymbol{k}-\boldsymbol{k}') = B(k_0,\theta) = 16\pi^2 n_i \sigma_B(k_0,\theta), \tag{2.97}$$

where n_i is the density of scatterers. At the Born approximation, the angular average $\langle B(\boldsymbol{k}-\boldsymbol{k}')\rangle$ may be expressed in terms of the total scattering cross section σ_B:

$$\boxed{\langle B(\boldsymbol{k}-\boldsymbol{k}')\rangle = 4\pi n_i \sigma_B} \tag{2.98}$$

For the case of the Schrödinger equation with a potential described by the Edwards model, the expressions for the scattering amplitude and for the correlation function of the Gaussian model may be written:

$$f_B(k_0,\theta) = -\frac{m}{2\pi\hbar^2} v(\boldsymbol{k}-\boldsymbol{k}') \tag{2.99}$$

[6] This notwithstanding, to satisfy the optical theorem in the Born approximation, T must be expanded to the next order $T = V + VG_0V$ in the right hand side of (2.93).

$$\langle B(k_0,\theta)\rangle = \frac{\pi\hbar^4}{m^2} n_i \sigma_B. \tag{2.100}$$

Low energy limit

One particularly simple and useful limit is that of low energy or large wavelength, corresponding to $\lambda \gg r_0$ where r_0 is the range of the potential $v(r)$. In this case, the phase factor in (2.94) vanishes. The scattering amplitude and the differential scattering cross section have no angular dependence; the scattering is *isotropic*:

$$f_B(\mathbf{k},\mathbf{k}') = -\frac{1}{4\pi}\int d\mathbf{r}'\, v(r') \tag{2.101}$$

and

$$\sigma_B(k_0,\theta) = \left[\frac{1}{4\pi}\int d\mathbf{r}' v(r')\right]^2 = \frac{\sigma_B}{4\pi}. \tag{2.102}$$

An example

Consider the scattering by a spherical potential barrier such that $U(r) = U_0$ for $r < a$ and $U(r) = 0$ otherwise (see Exercise 2.2 for the low energy limit). In the Born approximation, the differential scattering cross section, calculated from (2.95), is given by

$$\sigma_B(k_0,\theta) = U_0^2 \frac{(\sin qa - qa\,\cos qa)^2}{q^6}, \tag{2.103}$$

where $q = 2k_0 \sin(\theta/2)$. Its angular dependence is represented in Figure 2.4.

In the low energy limit $k_0 a \to 0$, the scattering becomes isotropic, that is,

$$\sigma_B(k_0,\theta) = \frac{U_0^2 \Omega^2}{16\pi^2} = \frac{\sigma_B}{4\pi} \quad \text{for } k_0 a \to 0, \tag{2.104}$$

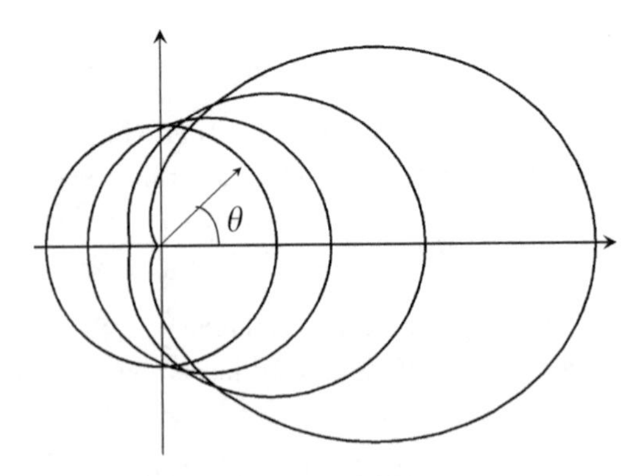

Figure 2.4 Scattering cross section $\sigma_B(k_0,\theta)/\sigma_B$ in the Born approximation, for a spherical potential of radius a and for parameter values $k_0 a = 0.1, 1, 1.5, 2$. As $k_0 a$ increases, the scattering becomes more anisotropic.

Table 2.1 Summary of the results obtained for the opposite of the scattering amplitude $-f(k_0, \theta)$ in the case of a spherical potential of height U_0 and radius a. In the low energy limit, this quantity, independent of the angle θ, is called the scattering length and denoted a_s. The parameter K is such that $K^2 = U_0$ for the Helmholtz equation, and $K^2 = 2mU_0/\hbar^2$ for the Schrödinger equation. We have defined $q = 2k_0 \sin\theta/2$. In the Edwards model, $v_0 \propto K^2 a^3 \to 0$.

	Exact	Born approximation $Ka \ll 1$	Strong potential $Ka \gg 1$
Arbitrary energy		$K^2 \dfrac{\sin qa - qa \cos qa}{q^3}$	
Low energy, $k_0 a \ll 1$	$a_s = a - \dfrac{\tanh Ka}{Ka}$	$a_s = \dfrac{K^2 a^3}{3}$	$a_s = a$

where $\Omega = 4\pi a^3/3$ is the volume of a scatterer. The low energy limit for the case of electrons in a metal ($k_F r_0 \ll 1$ where k_F is the Fermi wave vector) or for light scattering ($k_0 r_0 \ll 1$) imposes a condition on the nature of the scatterer.[7] This limit of isotropic scattering is frequently used to simplify calculations. This notwithstanding, it is important to note that it is very restrictive, and that most of the physical situations encountered correspond to a finite range potential and hence to anisotropic scattering.

The δ function potential

To describe the low energy limit, where the range of the potential is small compared to the wavelength, the scattering potential $v(\boldsymbol{r})$ is often replaced by a δ function potential, $v_0 \delta(\boldsymbol{r})$, obtained by taking the limit where U_0 diverges and a tends to zero, such that the product $v_0 = \int v(\boldsymbol{r}) d\boldsymbol{r} = 4\pi a^3 U_0/3$ remains finite. The total scattering cross section in the Born approximation becomes

$$\sigma_B = \frac{v_0^2}{4\pi}. \tag{2.105}$$

Nevertheless, this limit must be used with care. Indeed, in taking U_0 to infinity, we leave the domain of validity of the Born approximation. In fact, the result of Exercise 2.2 shows that the scattering cross section vanishes (!) at least in the low energy limit, but it is possible to show that this is true for any energy. It is therefore worthwhile understanding the significance

[7] In both cases, the length r_0 must be specified, and it depends on the physics of the scattering process. For the spherical potential, $r_0 = a$.

of the Edwards model (section 2.2.2) applied to the case of δ function impurities:

$$V(r) = \sum_i v_0\, \delta(\boldsymbol{r} - \boldsymbol{r}_i). \tag{2.106}$$

One may well ask whether this choice of potential makes any sense since there is no scattering by a δ function potential. In fact, in this book, we will only consider the case of a Gaussian potential, that is, the limit of the potential (2.106) when $v_0 \propto K^2 a^3 \to 0$, and for which the scattering cross section in the Born approximation is zero. The Gaussian model thus corresponds to an infinite density of scatterers of zero scattering cross section. The model (2.106) is well justified in the limit where $v_0 \to 0$.

Remark

While the δ function potential has a zero scattering cross section, there is another way to define correctly a zero-range potential. This is the pseudo-potential defined as [47]:

$$v(\boldsymbol{r})\psi(\boldsymbol{r}) = v_0\, \delta(\boldsymbol{r}) \frac{\partial}{\partial r} r\psi(\boldsymbol{r}). \tag{2.107}$$

It is easy to show that this expression corresponds to a zero-range potential with a total scattering cross section given by (2.105).

Appendix A2.2: Reciprocity theorem

In order to express time reversal in quantum mechanics, we use the antilinear operator which transforms a state into its complex conjugate. In particular, for a plane wave, we have the transformation $|\boldsymbol{k}\rangle \to |-\boldsymbol{k}\rangle$. When it exists, time reversal invariance is manifested by the following relation for the scattering amplitude f that corresponds to a localized obstacle such as was defined in Appendix A2.1:

$$f(\boldsymbol{k}, \boldsymbol{k}') = f(-\boldsymbol{k}', -\boldsymbol{k}). \tag{2.108}$$

This relation constitutes the *reciprocity theorem* [48]. It expresses the fact that the scattering amplitude is invariant under exchange of the initial and final states, as well as of the directions of propagation. If there are other degrees of freedom present, the preceding relation may be generalized. Thus, if we take, for example, an electromagnetic wave of polarization $\hat{\boldsymbol{\varepsilon}}$, time reversal invariance yields [49]:

$$f(\boldsymbol{k}, \hat{\boldsymbol{\varepsilon}}; \boldsymbol{k}', \hat{\boldsymbol{\varepsilon}}') = f(-\boldsymbol{k}', \hat{\boldsymbol{\varepsilon}}'^*; -\boldsymbol{k}, \hat{\boldsymbol{\varepsilon}}^*) \tag{2.109}$$

where $\hat{\boldsymbol{\varepsilon}}^*$ is the complex conjugate of $\hat{\boldsymbol{\varepsilon}}$. For an electron of spin σ, time reversal flips the spin and the reciprocity theorem becomes [48]

$$f(\boldsymbol{k}, \sigma; \boldsymbol{k}', \sigma') = (-1)^{\sigma - \sigma'} f(-\boldsymbol{k}', -\sigma'; -\boldsymbol{k}, -\sigma). \tag{2.110}$$

These results may be generalized to the case where the scatterers have internal degrees of freedom. For example, in the case of photons scattering from atoms, the fact that the

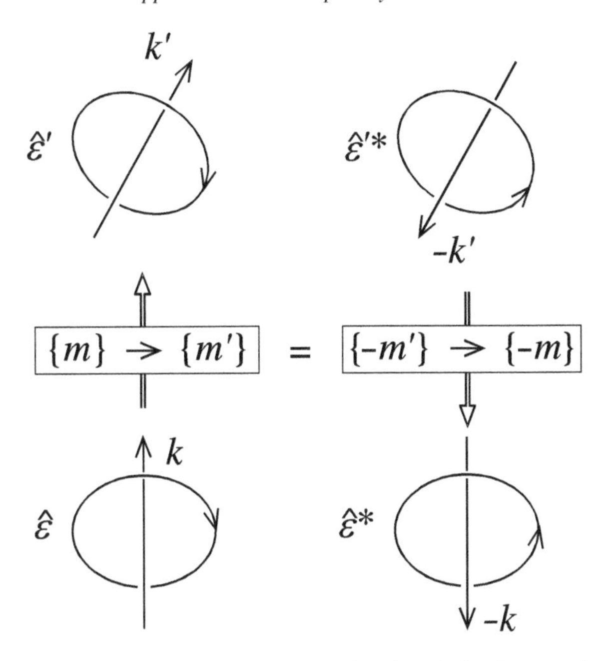

Figure 2.5 Illustration of the reciprocity theorem, equation (2.111), for the scattering of photons $(\boldsymbol{k}, \hat{\boldsymbol{\varepsilon}})$ by atoms, in the presence of an atomic Zeeman degeneracy (see section A2.3.3).

magnetic quantum numbers m of each atomic level may change as a result of scattering (Figure 2.5) must be taken into account. In this case, the reciprocity theorem becomes [48]

$$f(\boldsymbol{k}, \hat{\boldsymbol{\varepsilon}}, \{m\}; \boldsymbol{k}', \hat{\boldsymbol{\varepsilon}}', \{m'\}) = (-1)^{\sum_\alpha (m'_\alpha - m_\alpha)} f(-\boldsymbol{k}', \hat{\boldsymbol{\varepsilon}}'^*, \{-m'\}; -\boldsymbol{k}, \hat{\boldsymbol{\varepsilon}}^*, \{-m\}). \tag{2.111}$$

In a time reversal transformation, it is thus not sufficient to change the direction of the wave vectors and the polarization vectors; the sign of the magnetic quantum numbers must also be changed. We encounter an analogous situation in metals: electrons are scattered by magnetic impurities whose magnetic moments are flipped upon time reversal.

In this book we will usually study sequences of multiple scattering from successive impurities. One could then consider that *these impurities constitute a large scattering object* for which the order of the sequence of scatterings plays the role of an additional degree of freedom. Hence, for a sequence of n scatterings, the reciprocity theorem (2.108) becomes, in the absence of other degrees of freedom,

$$f(\boldsymbol{k}, \boldsymbol{k}', \{1 \to n\}) = f(-\boldsymbol{k}', -\boldsymbol{k}, \{n \to 1\}). \tag{2.112}$$

It is common to denote by T the scattering amplitude associated with a multiple scattering sequence, with the subscript *dir* for the direct sequence $\{1 \to n\}$ and the subscript *rev* for the sequence obtained by time reversal $\{n \to 1\}$. Thus, expression (2.112) becomes

$$T_{dir}(\boldsymbol{k}; \boldsymbol{k}') = T_{rev}(-\boldsymbol{k}'; -\boldsymbol{k}). \tag{2.113}$$

We observe that the direct and time-reversed processes are equal, $T_{dir}(\boldsymbol{k};\boldsymbol{k}') = T_{rev}(\boldsymbol{k};\boldsymbol{k}')$, only in the backscattering direction, that is, when $\boldsymbol{k}' = -\boldsymbol{k}$:

$$T_{dir}(\boldsymbol{k}; -\boldsymbol{k}) = T_{rev}(\boldsymbol{k}; -\boldsymbol{k}). \tag{2.114}$$

Let us note finally that if there is absorption, the notions of reciprocity and time reversal invariance are not identical [50]. In this case, there is no time reversal invariance since absorption is an irreversible process, but the reciprocity is preserved, and is expressed simply by the statement "if you see me, then I see you too," which remains true even through an absorbing fog.

Appendix A2.3: Light scattering

One advantage that light scattering has over electron scattering in metals is that it is often possible to characterize the scatterers and to identify the scattering potential. This allows us to anticipate certain characteristics of multiple scattering, such as the diffusion coefficient, starting from the behavior of single scattering.

We are particularly interested in light scattering from dielectric particles. This is a complicated problem for which an exact solution is only known for a homogeneous sphere of arbitrary radius and index of refraction. This solution was given by G. Mie in 1908 in a study of the scattering properties of light from an aqueous suspension of gold beads. One very useful limit of this problem is that of Rayleigh scattering for which the size of the scatterer is very small compared to the wavelength. This limit corresponds to the Born approximation and will be studied in the following section. The Mie theory is quite formal, but because of its importance, it is treated in detail in several books. We will only give a general overview in section A2.3.2 but provide several references [44, 51, 52] for further reading. Finally, when the wavelength becomes much smaller than the size of the scatterer, we approach the limit of geometrical optics, for which the scattering cross section is constant and of the order of the geometrical cross section πa^2, where a is the radius of the scatterer.

Another important example is that of light scattering by a dipole. In the Born approximation, we recover Rayleigh scattering. It may be resonant, which has the effect of considerably enhancing the cross section.

A2.3.1 Classical Rayleigh scattering

Consider the scattering of a linearly polarized beam by a dielectric medium of volume Ω and typical size a, placed in the vacuum.[8] Its dielectric constant $\epsilon(\boldsymbol{r})$ is of the form $\epsilon(\boldsymbol{r}) = \epsilon_0 + \delta\epsilon(\boldsymbol{r})$. Rayleigh scattering corresponds to the limit where the wavelength is large compared to the size of the scatterer ($k_0 a \ll 1$), or in other words to the low energy limit (see Table 2.1). The potential $U_0 = k_0^2 \delta\epsilon/\epsilon_0$ induced by the scatterer thus corresponds to a small parameter $Ka = k_0 a\sqrt{\delta\epsilon/\epsilon_0}$. Rayleigh scattering may thus be described in the framework of the Born approximation.

[8] The dielectric medium is not necessarily spherical in shape.

From the Maxwell equations (2.6), we derive the form of the wave equation satisfied by the field $\boldsymbol{D}$ [53]

$$\Delta \boldsymbol{D} + k_0^2 \boldsymbol{D} = -\nabla \times \nabla \times \boldsymbol{P}. \tag{2.115}$$

The field $\boldsymbol{P}(\boldsymbol{r}) = \boldsymbol{D}(\boldsymbol{r}) - \epsilon_0 \boldsymbol{E}(\boldsymbol{r}) = \delta\epsilon(\boldsymbol{r})\boldsymbol{E}(\boldsymbol{r})$ thus acts like a source term for the $\boldsymbol{D}$ field. The solution of the wave equation (2.115) may be written, as for (2.51), with the help of the free Green function $G_0(\boldsymbol{r}, \boldsymbol{r}')$ (2.53):

$$\boldsymbol{D}(\boldsymbol{r}) = -\int d^3\boldsymbol{r}'\, G_0(\boldsymbol{r}, \boldsymbol{r}') \nabla' \times \nabla' \times \boldsymbol{P}(\boldsymbol{r}'). \tag{2.116}$$

The solution of this integral equation is difficult since $\boldsymbol{P}$ itself depends on $\boldsymbol{D}$. However, starting from this equation, we may assume, in the limit where $\delta\epsilon$ is small, that $\boldsymbol{P}$ is proportional to the incident field. This approximation, due to Rayleigh, is precisely the Born approximation (section A2.1.4). We denote $\boldsymbol{D} = \boldsymbol{D}_i + \boldsymbol{D}_s$ and $\boldsymbol{E} = \boldsymbol{E}_i + \boldsymbol{E}_s$, where $\boldsymbol{D}_i$ and $\boldsymbol{E}_i$ are the incident fields, and $\boldsymbol{D}_s$ and $\boldsymbol{E}_s$ are the scattered fields. Outside the dielectric medium, $\boldsymbol{D}_i = \epsilon_0 \boldsymbol{E}_i$ and $\boldsymbol{D}_s = \epsilon_0 \boldsymbol{E}_s$. The Born approximation consists in taking $\boldsymbol{P} = \delta\epsilon(\boldsymbol{r})\boldsymbol{E}_i$ inside the dielectric medium, so that starting from (2.116), we obtain the scattered electric field

$$\epsilon_0 \boldsymbol{E}_s = -\int_\Omega d^3\boldsymbol{r}'\, G_0(\boldsymbol{r}, \boldsymbol{r}') \nabla' \times \nabla' \times [\delta\epsilon(\boldsymbol{r}')\boldsymbol{E}_i(\boldsymbol{r}')]. \tag{2.117}$$

The incident electric field is a plane wave of the form $\boldsymbol{E}_i(\boldsymbol{r}) = \boldsymbol{E}_i e^{i\boldsymbol{k}\cdot\boldsymbol{r}}$. Far from the dielectric, the Green function takes the form (2.56) (Fraunhoffer approximation)

$$G_0(\boldsymbol{r}, \boldsymbol{r}') \simeq -\frac{1}{4\pi r} e^{ik_0 r - i\boldsymbol{k}'\cdot\boldsymbol{r}'} \tag{2.118}$$

and the scattered electric field is written as

$$\epsilon_0 \boldsymbol{E}_s(\boldsymbol{r}) = \frac{e^{ik_0 r}}{4\pi r} \int_\Omega d^3\boldsymbol{r}'\, e^{-i\boldsymbol{k}'\cdot\boldsymbol{r}'}\, \nabla' \times \nabla' \times [\delta\epsilon(\boldsymbol{r}')\boldsymbol{E}_i\, e^{i\boldsymbol{k}\cdot\boldsymbol{r}'}]. \tag{2.119}$$

Integrating twice by parts yields

$$\epsilon_0 \boldsymbol{E}_s(\boldsymbol{r}) = -\frac{e^{ik_0 r}}{4\pi r} \int_\Omega d^3\boldsymbol{r}' \delta\epsilon(\boldsymbol{r}')\, e^{i(\boldsymbol{k}-\boldsymbol{k}')\cdot\boldsymbol{r}'} \boldsymbol{k}' \times (\boldsymbol{k}' \times \boldsymbol{E}_i). \tag{2.120}$$

The scattered wave depends on the direction of the polarization of the incident electric field $\hat{\boldsymbol{\varepsilon}}_i = \boldsymbol{E}_i/E_i$. The *vectorial* scattering amplitude $f(\boldsymbol{k}, \boldsymbol{k}', \hat{\boldsymbol{\varepsilon}}_i)$ defined from (2.58) is given by $\boldsymbol{E}_s(\boldsymbol{r}) = E_i f(\boldsymbol{k}, \boldsymbol{k}', \hat{\boldsymbol{\varepsilon}}_i) e^{ik_0 r}/r$, which, from (2.120), leads to

$$f(\boldsymbol{k}, \boldsymbol{k}', \hat{\boldsymbol{\varepsilon}}_i) = -\frac{1}{4\pi} \int_\Omega d^3\boldsymbol{r}' \frac{\delta\epsilon(\boldsymbol{r}')}{\epsilon_0}\, e^{i(\boldsymbol{k}-\boldsymbol{k}')\cdot\boldsymbol{r}'} \boldsymbol{k}' \times (\boldsymbol{k}' \times \hat{\boldsymbol{\varepsilon}}_i). \tag{2.121}$$

Suppose that the medium is homogeneous, namely that $\delta\epsilon(\boldsymbol{r}) = \delta\epsilon = \epsilon - \epsilon_0$ is constant. Moreover, suppose that the wavelength of the radiation is large compared to the size of the dielectric object, such that the exponential may be set equal to one. We then obtain

$$f(\boldsymbol{k},\boldsymbol{k}',\hat{\boldsymbol{\varepsilon}}_i) = -\frac{\alpha_0}{4\pi}\,\boldsymbol{k}' \times (\boldsymbol{k}' \times \hat{\boldsymbol{\varepsilon}}_i) \tag{2.122}$$

where $\alpha_0 = \Omega\delta\epsilon/\epsilon_0$ is the *polarizability*. For the differential scattering cross section, we obtain the expression:

$$\sigma(\boldsymbol{k},\boldsymbol{k}',\hat{\boldsymbol{\varepsilon}}_i) = |f(\boldsymbol{k},\boldsymbol{k}',\hat{\boldsymbol{\varepsilon}}_i)|^2 = \frac{k_0^4\alpha_0^2}{16\pi^2}\left[1 - (\boldsymbol{k}' \cdot \hat{\boldsymbol{\varepsilon}}_i)^2\right]. \tag{2.123}$$

Defining the angle χ between the observation direction $\boldsymbol{k}'$ and the linear polarization of the incident wave $\hat{\boldsymbol{\varepsilon}}_i$, $\cos\chi = \boldsymbol{k}' \cdot \hat{\boldsymbol{\varepsilon}}_i$, we then have

$$\boxed{\sigma(\boldsymbol{k},\boldsymbol{k}',\hat{\boldsymbol{\varepsilon}}_i) = \frac{k_0^4\alpha_0^2}{16\pi^2}\sin^2\chi} \tag{2.124}$$

which is the Rayleigh formula. Aside from the effect of polarization, we recover the result (2.104) for the scattering of a scalar wave by a spherical potential. The k_0^4 dependence is well known as the origin of our blue sky. In the limit of large wavelength $\lambda \gg a$ considered here, the scattering is isotropic: the cross section does not depend on the direction of the incident radiation, but only on its polarization.

This calculation may be generalized to the cases where the incident wave is circularly polarized or unpolarized (natural light). By averaging over the polarizations of the incident beam, we then obtain a relation identical to (2.124) for the differential scattering cross section, upon replacing $\sin^2\chi$ by $\frac{1}{2}(1+\cos^2\theta)$ where θ is the angle between the direction of incidence $\boldsymbol{k}$ and the direction of observation $\boldsymbol{k}'$ (see Figure 2.6 and expression 2.148).

We have not taken account of the fact that in a dielectric, the internal field is not equal to the external field. For the case of a homogeneous sphere, this leads to the substitution of $\boldsymbol{P} = (\epsilon - \epsilon_0)\boldsymbol{E}_i$ by the Clausius–Mosotti equation [54]

$$\boldsymbol{P} = 3\epsilon_0\left(\frac{\epsilon - \epsilon_0}{\epsilon + 2\epsilon_0}\right)\boldsymbol{E}_i \tag{2.125}$$

so that

$$\sigma(\boldsymbol{k},\boldsymbol{k}',\hat{\boldsymbol{\varepsilon}}_i) = k_0^4 a^6\left(\frac{\epsilon - \epsilon_0}{\epsilon + 2\epsilon_0}\right)^2\sin^2\chi, \tag{2.126}$$

where a is the radius of the sphere. Integrating over all emergent directions, we obtain the total scattering cross section σ. It is usual to define the quality factor Q_r as the ratio of this total scattering cross section and the geometric section defined by πa^2. We therefore obtain

$$Q_r(x) = \frac{8}{3}x^4\left(\frac{\epsilon - \epsilon_0}{\epsilon + 2\epsilon_0}\right)^2, \tag{2.127}$$

where the size parameter $x = k_0 a$ is small. The quality factor is thus small, reflecting the fact that Rayleigh scattering is not efficient.

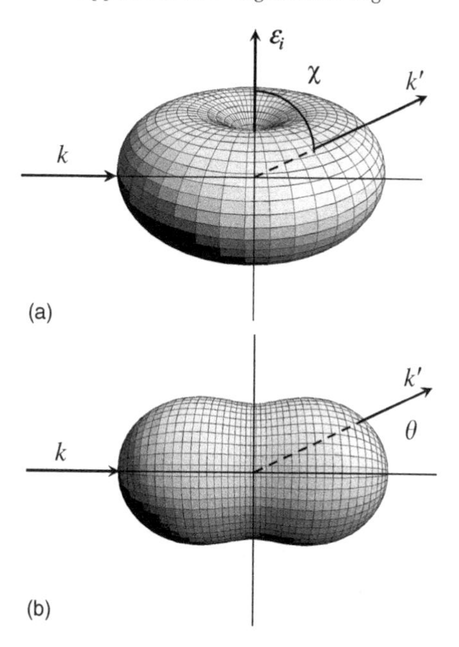

Figure 2.6 Rayleigh scattering. (a) Differential scattering cross section $\sigma(\boldsymbol{k},\boldsymbol{k}',\hat{\boldsymbol{\varepsilon}}_i)$ for a linearly polarized incident beam. (b) Differential scattering cross section for a circularly polarized or unpolarized incident beam (obtained by averaging (a) over the directions of the polarization $\hat{\boldsymbol{\varepsilon}}_i$).

Exercise 2.3: Rayleigh–Gans scattering

If the variation in dielectric constant $\delta\epsilon$ is small, we may be in a regime where the size of the dielectric sphere may no longer be negligible with respect to the wavelength $k_0 a \simeq 1$ but where $Ka = k_0 a\sqrt{\delta\epsilon/\epsilon_0} \ll 1$. We are then no longer in the low energy limit corresponding to Rayleigh scattering, but we are still able to use the Born approximation. This regime is called Rayleigh–Gans scattering.

In this case, upon integrating (2.121) over the volume of the sphere, we find that the scattering amplitude is multiplied by the factor $s(q)$ (see also equation 2.103)

$$s(q) = \frac{3}{2}\frac{\sin qa - qa\cos qa}{q^3a^3}, \tag{2.128}$$

where $q = |\boldsymbol{k} - \boldsymbol{k}'| = 2k_0\sin\theta/2$ and θ is the angle between the incident and emergent directions. We observe that the cross section becomes more and more anisotropic as the wavelength decreases and becomes of the order of the size of the scattering object (Figures 2.4 and 2.7). The scattering amplitude has the form

$$f(\boldsymbol{k},\boldsymbol{k}',\hat{\boldsymbol{\varepsilon}}_i) = -\frac{\alpha_0}{4\pi}s(\boldsymbol{k}-\boldsymbol{k}')\,\boldsymbol{k}'\times(\boldsymbol{k}'\times\hat{\boldsymbol{\varepsilon}}_i) \tag{2.129}$$

with $s(\boldsymbol{k}-\boldsymbol{k}') = s(|\boldsymbol{k}-\boldsymbol{k}'|) = s(q)$. Physically, this approximation is valid for a finite volume of the scatterer, as long as the amplitude and the phase of the wave do not vary too much inside the scatterer.

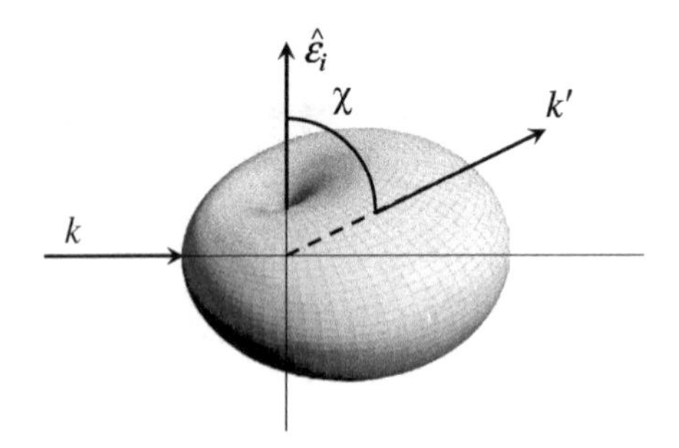

Figure 2.7 Differential scattering cross section $\sigma(\boldsymbol{k},\boldsymbol{k}',\hat{\boldsymbol{\varepsilon}}_i) = |f(\boldsymbol{k},\boldsymbol{k}',\hat{\boldsymbol{\varepsilon}}_i|^2$ in the Rayleigh–Gans regime, for a linearly polarized incident beam and $k_0 a = 1$.

A2.3.2 Mie scattering

The general problem (arbitrary $k_0 a$ and $\delta\epsilon$) of the scattering of an electromagnetic plane wave by a dielectric of finite volume is very complex. In fact, there are very few known exact solutions, and among these, the most important is that of Mie, for a homogeneous spherical volume [51]. The problem consists in solving the wave equations for the incident, emergent, and internal fields, and relating them via the continuity conditions for the electric and magnetic fields at the surface of the sphere. The quality factor (the ratio of the scattering cross section and the geometric section defined by πa^2) is given by

$$Q_{Mie}(x,m) = \frac{2}{x^2}\sum_{n=1}^{\infty}(2n+1)\left(|a_n|^2 + |b_n|^2\right) \tag{2.130}$$

where $m = \sqrt{\epsilon/\epsilon_0}$ is the ratio of the refraction indices in the interior and the exterior of the sphere, and where the amplitudes a_n and b_n are given by

$$a_n = \frac{m\psi_n(mx)\psi_n'(x) - \psi_n(x)\psi_n'(mx)}{m\psi_n(mx)\xi_n'(x) - \xi_n(x)\psi_n'(mx)} \tag{2.131}$$

and

$$b_n = \frac{\psi_n(mx)\psi_n'(x) - m\psi_n(x)\psi_n'(mx)}{\psi_n(mx)\xi_n'(x) - m\xi_n(x)\xi_n'(mx)} \tag{2.132}$$

with $\psi_n(r) = rj_n(r)$ and $\xi_n(r) = rh_n^{(1)}(r)$ and their derivatives. $j_n(r)$ and $h_n^{(1)}$ are the Bessel and Hankel spherical functions of the first type [55]. We see that the quality factor vanishes for $m = 1$ which corresponds to the absence of a scatterer. In Figure 2.8, we depict the quality factor as a function of the size parameter x for a given value of m. We see three distinct regimes. For small values of x, $Q_{Mie} \propto x^4$, we recover Rayleigh scattering. For $x \simeq 1$, that is, when the wavelength is of the order of the size of the scatterer, there is a succession of resonances which correspond to the vanishing of the denominators in the expressions for the amplitudes a_n and b_n. For these values, the scattering is very efficient,

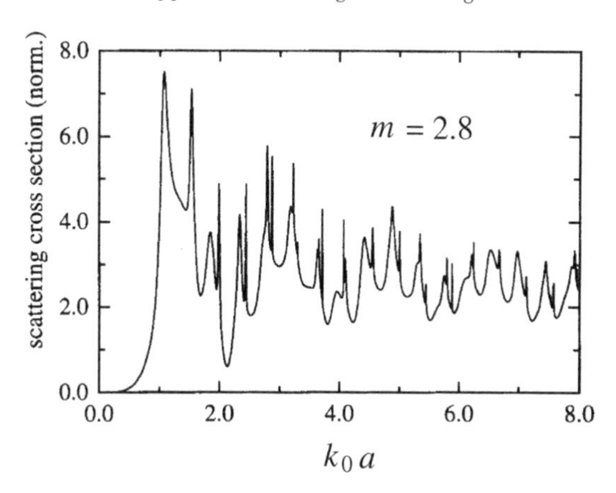

Figure 2.8 Quality factor Q_{Mie} defined as the ratio of the total scattering cross section and the geometric section πa^2, as a function of the size parameter $x = k_0 a$ for the relative refractive index $m = 2.8$. We recover the Rayleigh regime at low frequency, then the Mie resonances, and finally geometric scattering at large values of x [56].

and the scattering cross section is much greater than the geometric section and that of the Rayleigh scattering regime. At resonance, the wave spends a long time in the scatterer and its group velocity is greatly reduced. This result has important consequences in the multiple scattering regime [57] (see remark page 113). Finally, for large values of x (wavelengths much smaller than the size of the scatterer), the total scattering cross section becomes twice the geometric section given by geometrical optics. This factor 2 is known as the extinction paradox and results from diffraction effects at the boundary of the sphere which are not taken into account in geometrical optics in which the geometrical cross section simply represents the shadow term. In contrast to Rayleigh scattering, in the Mie regime, the total cross section has no simple dependence on frequency. This explains, in particular, the gray appearance of a cloudy or polluted sky. Finally, we note that the Mie regime cannot be described by the Gaussian model for which the scatterers have vanishing scattering cross section.

A2.3.3 Atom–photon scattering in the dipole approximation

We have considered the scattering of an electromagnetic wave by a classical object. We now study the case where the scatterer is an atom. The existence of degenerate levels in the atomic spectrum leads to different expressions for the differential scattering cross section which, first of all, acquires a tensorial structure, and second, may become very large at resonance. Moreover the additional internal degrees of freedom arising from the degeneracy of the atomic levels modify the interference in multiple scattering and give rise to a finite phase coherence time (Appendix A6.5).

In order to study the interaction of a monochromatic laser beam of frequency ω with a gas of atoms, we return to the problem of the scattering of a photon of wave vector $\boldsymbol{k}_i$, wavelength $\lambda = 2\pi/k = 2\pi c/\omega$ and polarization $\hat{\boldsymbol{\varepsilon}}_i$ with an atom which we take to be at rest.[9] For visible light, the wavelength λ is much larger than the atomic size (typically by two orders of magnitude). We may thus suppose that the field is constant across the atom, and write the photon–atom interaction in the dipolar approximation. This takes the simple form $-\boldsymbol{d}\cdot\boldsymbol{E}$ where $\boldsymbol{d}$ is the atomic dipole moment operator and $\boldsymbol{E}$ is the electric field of the radiation [58, 59]. The atoms therefore are excellent approximations of point scatterers. Moreover, they are strongly resonant scatterers. At resonance, that is, when the laser frequency ω is adjusted to a resonance frequency ω_0 of the atom, their scattering cross section is very large, of the order of $\lambda^2 \simeq \lambda_0^2$ with $\lambda_0 = 2\pi/\omega_0$.[10] Additionally, since the resonance is very narrow (of the order of several MHz for a resonance frequency of the order of 10^{14} Hz) we may reasonably neglect contributions of the atomic levels that do not correspond to the resonant transition. We may thus model the atom as a degenerate two-level system with spacing $\hbar\omega_0$, each level having a fixed angular momentum. Finally, one more advantage of using atoms to study multiple scattering is that they are all identical (including both the frequency ω_0 and the width of the resonance).

The ground state and the excited state of the atom are characterized by their angular momenta J and J_e, respectively. In the absence of a perturbation, these levels are degenerate, and their magnetic quantum numbers are such that $|m| \leq J$ and $|m_e| \leq J_e$. The dipolar interaction induces transitions between eigenstates of the free Hamiltonian $\mathcal{H}_{at} + \mathcal{H}_{rad}$ describing the atom and the radiation. We distinguish between two types of transition, depending on whether the scattering of the photon does not change the internal atomic state (Rayleigh scattering) or whether the scattering induces a change of magnetic quantum number allowed by the selection rules (Raman scattering). Note that in both cases the scattering is elastic. The possible states of a coupled atom–photon system are of the form $|Jm, 0\rangle$ (0 photon state) or $|Jm, \boldsymbol{k}\hat{\boldsymbol{\varepsilon}}\rangle$ (state with one photon of momentum $\boldsymbol{k}$ and polarization $\hat{\boldsymbol{\varepsilon}}$).[11]

The scattering of a photon by an atom is described by means of the collision operator $\mathcal{T}$ defined in (2.86). To second order in the perturbation expansion, we have $\mathcal{T} = V + VG_0V$ where $V = -\boldsymbol{d}\cdot\boldsymbol{E}$. The scattering process corresponds to the absorption of an incident photon $(\boldsymbol{k}_i, \hat{\boldsymbol{\varepsilon}}_i)$ by the atom, and the emission of a photon $(\boldsymbol{k}', \hat{\boldsymbol{\varepsilon}}')$.[12] It is thus given by the second-order term in V. The associated amplitude $\mathcal{T}_{fi}$ is written in terms of the matrix elements of V obtained from an expansion in the photon modes in a box of volume Ω:

- absorption of a photon $(\boldsymbol{k}_i, \hat{\boldsymbol{\varepsilon}}_i)$

$$\langle J_e m_e, 0|\boldsymbol{d}\cdot\boldsymbol{E}|Jm_i, \boldsymbol{k}_i\hat{\boldsymbol{\varepsilon}}_i\rangle = -i\sqrt{\frac{\hbar\omega}{2\epsilon_0\Omega}}\, e^{i\boldsymbol{k}_i\cdot\boldsymbol{r}} \langle J_e m_e|\boldsymbol{d}\cdot\hat{\boldsymbol{\varepsilon}}_i|Jm_i\rangle; \tag{2.133}$$

[9] This corresponds to a gas of cold atoms in which we may neglect the Doppler effect associated with the atomic velocities.

[10] Recall that the Rayleigh scattering cross section (2.124) is proportional to $\Omega^2\lambda^{-4}$ where Ω is the volume of the scatterer.

[11] The polarization is a unit vector lying in the plane perpendicular to $\boldsymbol{k}$. It may be linear or circular, and will be denoted by the complex vector $\hat{\boldsymbol{\varepsilon}}$.

[12] We consider here a single photon in a laser mode. This description is valid in the limit of a small number of incident photons where we may neglect effects of saturation of the reemitted intensity, as well as optical pumping.

- free propagator G_0 of the atom in its excited state

$$\frac{1}{\hbar(\omega - \omega_0)}; \tag{2.134}$$

- photon emission $(\boldsymbol{k}', \hat{\boldsymbol{\varepsilon}}')$

$$\langle Jm_f, \boldsymbol{k}'\hat{\boldsymbol{\varepsilon}}'|\boldsymbol{d} \cdot \boldsymbol{E}|J_e m_e, 0\rangle = i\sqrt{\frac{\hbar\omega}{2\epsilon_0\Omega}}e^{-i\boldsymbol{k}'\cdot\boldsymbol{r}}\langle Jm_f|\boldsymbol{d} \cdot \hat{\boldsymbol{\varepsilon}}'^*|J_e m_e\rangle. \tag{2.135}$$

Using these expressions, we obtain the non-resonant elastic collision process where the atom is excited by a photon $|\boldsymbol{k}_i\hat{\boldsymbol{\varepsilon}}_i\rangle$ of the laser from the ground state $|Jm_i\rangle$ to an excited state $|J_e m_e\rangle$ and then back to the state $|Jm_f\rangle$. There is another contribution where the atom first emits a photon before being excited. Finally, to lowest order, the matrix element of the collision operator $\mathcal{T}$ for the scattering process $|\Psi_i\rangle = |Jm_i, \boldsymbol{k}_i\hat{\boldsymbol{\varepsilon}}_i\rangle \to |\Psi_f\rangle = |Jm_f, \boldsymbol{k}'\hat{\boldsymbol{\varepsilon}}'\rangle$, is given by [59]:

$$\mathcal{T}_{fi} = \frac{\hbar\omega}{2\epsilon_0\Omega}\sum_{m_e}\left[\frac{\langle Jm_f|\boldsymbol{d} \cdot \hat{\boldsymbol{\varepsilon}}'^*|J_e m_e\rangle\ \langle J_e m_e|\boldsymbol{d} \cdot \hat{\boldsymbol{\varepsilon}}_i|Jm_i\rangle}{\hbar\omega - \hbar\omega_0}\right.$$
$$\left.+\frac{\langle Jm_i|\boldsymbol{d} \cdot \hat{\boldsymbol{\varepsilon}}_i|J_e m_e\rangle\ \langle J_e m_e|\boldsymbol{d} \cdot \hat{\boldsymbol{\varepsilon}}'^*|Jm_f\rangle}{-\hbar\omega - \hbar\omega_0}\right]. \tag{2.136}$$

Non-resonant Rayleigh scattering

For a non-degenerate ground state (for example, if $J = 0$), we recover, in the low frequency limit $\omega \ll \omega_0$, the Rayleigh scattering cross section (2.124) by a classical dipole. To that purpose, we introduce the static polarizability tensor

$$\alpha_{ij} = \frac{1}{\epsilon_0}\sum_{m_e}\frac{\langle Jm|d_i|J_e m_e\rangle\ \langle J_e m_e|d_j|Jm\rangle + \langle Jm|d_j|J_e m_e\rangle\ \langle J_e m_e|d_i|Jm\rangle}{\hbar\omega_0} \tag{2.137}$$

where d_i is the component of the dipole moment operator along the direction i. We assume the tensor to be isotropic, that is $\alpha_{ij} = \alpha_0\delta_{ij}$. We may thus rewrite (2.136) in the form

$$\mathcal{T}_{fi} = -\frac{\hbar\omega}{2\Omega}\,\alpha_0\,\hat{\boldsymbol{\varepsilon}}_i \cdot \hat{\boldsymbol{\varepsilon}}'^*. \tag{2.138}$$

The transition probability per unit time and per unit solid angle is given by the Fermi golden rule:

$$\frac{2\pi}{\hbar}|\mathcal{T}_{fi}|^2\rho = \frac{2\pi}{\hbar}\left(\frac{\hbar\omega\alpha_0}{2\Omega}\right)^2\ |\hat{\boldsymbol{\varepsilon}}_i \cdot \hat{\boldsymbol{\varepsilon}}'^*|^2\ \frac{\Omega}{8\pi^3}\frac{\omega^2}{\hbar c^3}, \tag{2.139}$$

where

$$\rho = \frac{\Omega}{8\pi^3}\frac{\omega^2}{\hbar c^3} \tag{2.140}$$

is the photon density of states per solid angle in a volume Ω.

The differential scattering cross section $\sigma(\boldsymbol{k}_i, \boldsymbol{k}', \hat{\boldsymbol{\varepsilon}}_i)$ for an arbitrary final polarization $\hat{\boldsymbol{\varepsilon}}'$ is obtained by dividing the transition probability (2.139) by the flux of photons c/Ω and summing over the final polarizations $\hat{\boldsymbol{\varepsilon}}'$, namely

$$\sigma(\boldsymbol{k}_i, \boldsymbol{k}', \hat{\boldsymbol{\varepsilon}}_i) = \frac{\alpha_0^2 k^4}{16\pi^2} \sum_{\hat{\boldsymbol{\varepsilon}}' \perp \boldsymbol{k}'} |\hat{\boldsymbol{\varepsilon}}_i \cdot \hat{\boldsymbol{\varepsilon}}'^*|^2 \tag{2.141}$$

where $k = 2\pi/\lambda$. Using the relation

$$\sum_{\hat{\boldsymbol{\varepsilon}} \perp \hat{\boldsymbol{k}}} |\hat{\boldsymbol{\varepsilon}} \cdot \boldsymbol{X}|^2 = \boldsymbol{X} \cdot \boldsymbol{X}^* - (\hat{\boldsymbol{k}} \cdot \boldsymbol{X})(\hat{\boldsymbol{k}} \cdot \boldsymbol{X}^*) \tag{2.142}$$

valid for any complex valued vector $\boldsymbol{X}$, we deduce that

$$\sigma(\boldsymbol{k}_i, \boldsymbol{k}', \hat{\boldsymbol{\varepsilon}}_i) = \frac{\alpha_0^2 k^4}{16\pi^2} \left(1 - |\hat{\boldsymbol{\varepsilon}}_i \cdot \hat{\boldsymbol{k}}'|^2\right), \tag{2.143}$$

where $\hat{\boldsymbol{k}}' = \boldsymbol{k}'/k$. The total scattering cross section σ defined by (2.70) may be written

$$\sigma = \frac{\alpha_0^2 k^4}{6\pi} \tag{2.144}$$

so that we may rewrite the differential scattering cross section (2.141) in the form

$$\boxed{\sigma(\boldsymbol{k}_i, \boldsymbol{k}', \hat{\boldsymbol{\varepsilon}}_i) = \frac{3\sigma}{8\pi} \left(1 - |\hat{\boldsymbol{\varepsilon}}_i \cdot \hat{\boldsymbol{k}}'|^2\right)} \tag{2.145}$$

We thus deduce the angular distribution of Rayleigh scattered polarized light:

- For an incident linear polarization $\hat{\boldsymbol{\varepsilon}}_i$ we have

$$\sigma(\boldsymbol{k}_i, \boldsymbol{k}', \hat{\boldsymbol{\varepsilon}}_i) = \frac{3\sigma}{8\pi} \sin^2 \chi \tag{2.146}$$

  where χ is the angle between $\hat{\boldsymbol{\varepsilon}}_i$ and the wave vector $\boldsymbol{k}'$ of the scattered photon. We thus recover the expression (2.124) for the scattering cross section.
- For a circularly polarized wave, we denote the two polarizations

$$\hat{\boldsymbol{\varepsilon}}_{\pm 1} = \mp \frac{1}{\sqrt{2}} (\hat{\boldsymbol{e}}_x \pm i \hat{\boldsymbol{e}}_y) \tag{2.147}$$

  in a Cartesian basis $(\hat{\boldsymbol{e}}_x, \hat{\boldsymbol{e}}_y, \hat{\boldsymbol{e}}_z)$ where $\hat{\boldsymbol{e}}_z = \hat{\boldsymbol{k}}_i$ for the incident wave. Moreover, we denote by h the *helicity* of the photon, i.e., the projection of its angular momentum along the propagation axis. Thus, a wave of polarization $\hat{\boldsymbol{\varepsilon}}_{+1}$ has positive helicity and one of polarization $\hat{\boldsymbol{\varepsilon}}_{-1}$ has negative helicity. In a mirror projection, the polarization is unchanged, but the wave vector changes sign $\hat{\boldsymbol{k}}_i \to -\hat{\boldsymbol{k}}_i$, so that the helicity is reversed. Denoting the angle between $\boldsymbol{k}_i$ and $\boldsymbol{k}'$ by θ, we have that $|\hat{\boldsymbol{\varepsilon}}_i \cdot \hat{\boldsymbol{k}}'|^2 = \frac{1}{2} \sin^2 \theta$ and, using (2.145), we may write the differential scattering cross section as

$$\sigma(\boldsymbol{k}_i, \boldsymbol{k}') = \frac{3\sigma}{16\pi} (1 + \cos^2 \theta). \tag{2.148}$$

We note that this scattering cross section does not depend on the incident circular polarization. This implies that for unpolarized incident light, the differential scattering cross section is also given by (2.148).

Expression (2.145) corresponds to a sum over final polarizations. We can also analyze an initially polarized light along a given final polarization channel. Thus the linearly polarized light can be analyzed along either the parallel ($l \parallel l$) or perpendicular ($l \perp l$) channels. Similarly, circularly polarized light may be analyzed in channels of parallel ($h \parallel h$) or reversed ($h \perp h$) helicity. The scattering cross section as a function of incident and emitted polarizations is therefore a useful quantity, given by

$$\boxed{\sigma(\boldsymbol{k}_i, \boldsymbol{k}', \hat{\boldsymbol{\varepsilon}}_i, \hat{\boldsymbol{\varepsilon}}') = \frac{3\sigma}{8\pi} |\hat{\boldsymbol{\varepsilon}}_i \cdot \hat{\boldsymbol{\varepsilon}}'^*|^2} \tag{2.149}$$

with the constraints $\boldsymbol{k}_i \perp \hat{\boldsymbol{\varepsilon}}_i$ and $\boldsymbol{k}' \perp \hat{\boldsymbol{\varepsilon}}'$. This expression is the same as (2.141) when omitting the summation over final polarizations. Finally, note that the differential scattering cross section depends on the incident and emitted polarizations, but not on the incident direction $\boldsymbol{k}_i$, and not on the direction of emission $\boldsymbol{k}'$. This implies that the transport scattering cross section (2.73) is equal to the total scattering cross section: $\sigma^* = \sigma$. In this sense,

---

Exercise 2.4 Show that, for non-resonant Rayleigh scattering, the differential scattering cross section in the backscattering direction (2.149) $\boldsymbol{k}' = -\boldsymbol{k}_i$ is non-zero in the channels ($l \parallel l$) and ($h \perp h$) and vanishes in the channels ($l \perp l$) and ($h \parallel h$).

For linear polarization, the scalar product $\hat{\boldsymbol{\varepsilon}}_i \cdot \hat{\boldsymbol{\varepsilon}}'^*$ is zero in the ($l \perp l$) channel.
For circular polarization, the conserved helicity channel ($h \parallel h$) corresponds to emergent polarization $\hat{\boldsymbol{\varepsilon}}' = \hat{\boldsymbol{\varepsilon}}_i^*$. Indeed, the helicity is defined with respect to the direction of propagation, and so the channel ($h \parallel h$) corresponds to a circular polarization opposite to that of $\hat{\boldsymbol{\varepsilon}}_i$. From (2.147) we infer that $\hat{\boldsymbol{\varepsilon}}_i \cdot \hat{\boldsymbol{\varepsilon}}'^* = \hat{\boldsymbol{\varepsilon}}_i \cdot \hat{\boldsymbol{\varepsilon}}_i = 0$. According to (2.139), the transition probability is zero.
On the other hand, in the channel of reversed helicity ($h \perp h$), we have, in backscattering, $\hat{\boldsymbol{\varepsilon}}' = \hat{\boldsymbol{\varepsilon}}_i$ and therefore $\hat{\boldsymbol{\varepsilon}}_i \cdot \hat{\boldsymbol{\varepsilon}}'^* = \hat{\boldsymbol{\varepsilon}}_i \cdot \hat{\boldsymbol{\varepsilon}}_i^* = 1$, see Table 2.2.

Exercise 2.5 Show that for an unpolarized incident wave $\boldsymbol{k}_i$, the light scattered in the direction $\boldsymbol{k}' \perp \boldsymbol{k}_i$ is linearly polarized.

For this, sum expression (2.149) over incident polarizations and use (2.142) to show that the emitted intensity depends on the emergent polarization as $1 - |\hat{\boldsymbol{\varepsilon}}' \cdot \hat{\boldsymbol{k}}_i|^2$. The intensity therefore vanishes if $\hat{\boldsymbol{\varepsilon}}' \parallel \hat{\boldsymbol{k}}_i$, that is, if $\boldsymbol{k}' \perp \boldsymbol{k}_i$. Initially unpolarized light gives rise to completely polarized rays in directions perpendicular to the direction of incidence. This result may be applied to the scattering of sunlight by the atmosphere. It allows navigators to locate the position of the sun even when the sky is clouded over. It is, nevertheless, necessary to have a small portion of blue sky in order to avoid Mie scattering for which these conclusions are incorrect. This fact was probably known to the Vikings, according to archeologists who discovered pieces of cordierite, a naturally occurring polarizer, in tombs.

Table 2.2 Relevant polarization or helicity configurations studied in the backscattering direction, and the corresponding values of the scalar products entering the general expression of the differential cross section (2.167).

		$\hat{\boldsymbol{\varepsilon}}_i \cdot \hat{\boldsymbol{\varepsilon}}'$	$\hat{\boldsymbol{\varepsilon}}_i \cdot \hat{\boldsymbol{\varepsilon}}'^*$
$l \parallel l$	$\hat{\boldsymbol{\varepsilon}}' = \hat{\boldsymbol{\varepsilon}}_i$	1	1
$l \perp l$	$\hat{\boldsymbol{\varepsilon}}' \perp \hat{\boldsymbol{\varepsilon}}_i$	0	0
$h \parallel h$	$\hat{\boldsymbol{\varepsilon}}' = \hat{\boldsymbol{\varepsilon}}_i^*$	1	0
$h \perp h$	$\hat{\boldsymbol{\varepsilon}}' = \hat{\boldsymbol{\varepsilon}}_i$	0	1

Rayleigh scattering is isotropic. This conclusion remains correct for resonant scattering by atoms with degenerate energy levels.

Resonant scattering

We now consider the case of resonant scattering of a photon by an atom. In particular, we evaluate the matrix element $\mathcal{T}_{fi}$ of the collision operator. A first complication comes from the fact that, at resonance, $\omega = \omega_0$ and the amplitude (2.136) diverges. In fact, this divergence is lifted because the excited state J_e has a finite lifetime related to its interaction with the vacuum fluctuations of the electromagnetic field. An atom in the excited state J_e may relax back to the ground state J by spontaneous emission. Upon resumming all the transition amplitudes corresponding to the emission and reabsorption of a photon by an atom in its excited state due to its interaction with vacuum fluctuations [59], the excited states acquire a finite energy width $\hbar\Gamma$, which measures the rate of spontaneous emission.[13] The corresponding scattering amplitude may be obtained from (2.136) and can be put in the form

$$\mathcal{T}_{fi} = \frac{\omega}{2\epsilon_0 \Omega} \frac{1}{\delta + i\Gamma/2} \sum_{m_e} \langle J m_f | \boldsymbol{d} \cdot \hat{\boldsymbol{\varepsilon}}'^* | J_e m_e \rangle \langle J_e m_e | \boldsymbol{d} \cdot \hat{\boldsymbol{\varepsilon}}_i | J m_i \rangle, \tag{2.150}$$

where we have defined $\delta = \omega - \omega_0$. This amplitude involves matrix elements of the operator $\boldsymbol{d} \cdot \hat{\boldsymbol{\varepsilon}}$. In order to evaluate them, we use the standard basis related to the Cartesian basis $(\hat{\boldsymbol{e}}_x, \hat{\boldsymbol{e}}_y, \hat{\boldsymbol{e}}_z)$ by

$$\begin{aligned} \hat{\boldsymbol{e}}_0 &= \hat{\boldsymbol{e}}_z \\ \hat{\boldsymbol{e}}_{\pm 1} &= \mp \frac{1}{\sqrt{2}} (\hat{\boldsymbol{e}}_x \pm i \hat{\boldsymbol{e}}_y) \end{aligned} \tag{2.151}$$

[13] The interaction with vacuum fluctuations also gives rise to a shift in the energy of the excited state (Lamb shift) which we include in our definition of ω_0.

so that any vector $\boldsymbol{X}$ can be decomposed as $\boldsymbol{X} = \sum_{q=0,\pm1}(-1)^q X_q \hat{\boldsymbol{e}}_{-q}$ and

$$\boldsymbol{d} \cdot \hat{\boldsymbol{\varepsilon}} = \sum_q (-1)^q d_q \varepsilon_{-q}. \tag{2.152}$$

The components d_q of the atomic dipole $\boldsymbol{d}$ are operators, and their matrix elements appear in the amplitude $\mathcal{T}_{fi}$. Now, using the Wigner–Eckart theorem, matrix elements $\langle J_e m_e|d_q|Jm\rangle$ of each component d_q are given by [60] (see section 15.5):

$$\langle J_e m_e|d_q|Jm\rangle = \frac{\langle J_e||\boldsymbol{d}||J\rangle}{\sqrt{2J_e+1}}\langle J\,1mq|J_e m_e\rangle, \tag{2.153}$$

where $\langle J_e||\boldsymbol{d}||J\rangle$ is a coefficient independent of m, m_e and q. The term $\langle J\,1mq|J_e m_e\rangle$ is a Clebsch–Gordan coefficient and is real and non-zero whenever $m_e - m = q$ and $J_e = J+1$. The operator $\boldsymbol{d}$ is a vector and the selection rules determine whether the transition is Rayleigh ($q = 0$) or Raman ($q = \pm 1$), as a function of the polarization $\hat{\boldsymbol{\varepsilon}}$ of the photon. We define $d = \langle J_e||\boldsymbol{d}||J\rangle/\sqrt{2J_e+1}$ and a dimensionless operator $\tilde{\boldsymbol{d}} = \boldsymbol{d}/d$. Moreover, it can be shown [59] that the width Γ of the excited state is related to d by

$$\hbar\Gamma = \frac{d^2\omega_0^3}{3\pi\epsilon_0 c^3}. \tag{2.154}$$

Using this relation, and since the dipolar operator only couples states J and J_e, we may rewrite the amplitude (2.150) in the form

$$\mathcal{T}_{fi} = \frac{3\pi\hbar c^3}{\Omega\omega^2}\frac{\Gamma/2}{\delta + i\Gamma/2}\langle Jm_f|(\tilde{\boldsymbol{d}}\cdot\hat{\boldsymbol{\varepsilon}}'^*)(\tilde{\boldsymbol{d}}\cdot\hat{\boldsymbol{\varepsilon}}_i)|Jm_i\rangle. \tag{2.155}$$

The transition rate towards any final Zeeman state is given by the Fermi golden rule

$$\frac{2\pi}{\hbar}\sum_{\hat{\boldsymbol{k}},m_f}|\mathcal{T}_{fi}|^2. \tag{2.156}$$

To compute the differential scattering cross section, we must divide by the incident photon flux c/Ω, and using (2.139) and (2.150), we obtain

$$\frac{d\sigma}{d\Omega_{\boldsymbol{k}'}} = \frac{\Omega}{c}\frac{2\pi}{\hbar}\rho(\omega)\left(\frac{3\pi\hbar c^3}{\Omega\omega^2}\right)^2\frac{\Gamma^2/4}{\delta^2+\Gamma^2/4}\langle Jm_i|(\tilde{\boldsymbol{d}}\cdot\hat{\boldsymbol{\varepsilon}}_i^*)(\tilde{\boldsymbol{d}}\cdot\hat{\boldsymbol{\varepsilon}}')(\tilde{\boldsymbol{d}}\cdot\hat{\boldsymbol{\varepsilon}}'^*)(\tilde{\boldsymbol{d}}\cdot\hat{\boldsymbol{\varepsilon}}_i)|Jm_i\rangle. \tag{2.157}$$

The expression (2.140) for the photon density of states per unit solid angle allows us to write finally

$$\boxed{\frac{d\sigma}{d\Omega_{\boldsymbol{k}'}} = \frac{9}{16\pi^2}\lambda^2\frac{\Gamma^2/4}{\delta^2+\Gamma^2/4}\langle Jm_i|(\tilde{\boldsymbol{d}}\cdot\hat{\boldsymbol{\varepsilon}}_i^*)(\tilde{\boldsymbol{d}}\cdot\hat{\boldsymbol{\varepsilon}}')(\tilde{\boldsymbol{d}}\cdot\hat{\boldsymbol{\varepsilon}}'^*)(\tilde{\boldsymbol{d}}\cdot\hat{\boldsymbol{\varepsilon}}_i)|Jm_i\rangle} \tag{2.158}$$

The total scattering cross section σ is obtained by integrating the differential scattering cross section over the solid angle. Averaging over polarizations of the scattered photons, we get

$$\sigma = \int d\Omega_{k'} \sum_{\hat{\varepsilon}' \perp \hat{k}'} \frac{d\sigma}{d\Omega_{k'}}. \tag{2.159}$$

The sum and the integral only depend on $(\tilde{\boldsymbol{d}} \cdot \hat{\boldsymbol{\varepsilon}}')(\tilde{\boldsymbol{d}} \cdot \hat{\boldsymbol{\varepsilon}}'^*)$ and, using (2.142), give

$$\int d\Omega_{k'} \sum_{\hat{\varepsilon}' \perp k'} (\tilde{\boldsymbol{d}} \cdot \hat{\boldsymbol{\varepsilon}}')(\tilde{\boldsymbol{d}} \cdot \hat{\boldsymbol{\varepsilon}}'^*) = \int d\Omega_{k'} \left(\tilde{\boldsymbol{d}} \cdot \tilde{\boldsymbol{d}} - (\tilde{\boldsymbol{d}} \cdot \hat{\boldsymbol{k}}')(\tilde{\boldsymbol{d}} \cdot \hat{\boldsymbol{k}}')\right) = \frac{8\pi}{3} \tilde{\boldsymbol{d}} \cdot \tilde{\boldsymbol{d}}. \tag{2.160}$$

In equation (2.158), the operator $\tilde{\boldsymbol{d}} \cdot \tilde{\boldsymbol{d}} = \tilde{d}_0 \tilde{d}_0 - \tilde{d}_{+1} \tilde{d}_{-1} - \tilde{d}_{-1} \tilde{d}_{+1}$ acts on the state $(\tilde{\boldsymbol{d}} \cdot \hat{\boldsymbol{\varepsilon}}_i)|Jm_i\rangle$ which is an excited state $|J_e m_e\rangle$. Its action on this state is unity since

$$\begin{aligned} \tilde{\boldsymbol{d}} \cdot \tilde{\boldsymbol{d}} |J_e m_e\rangle = \Big[&\langle J_e m_e | J 1 m_e 0\rangle^2 + \langle J_e m_e | J 1 (m_e - 1) + 1\rangle^2 \\ &+ \langle J_e m_e | J 1 (m_e + 1) - 1\rangle^2 \Big] |J_e m_e\rangle = |J_e m_e\rangle \end{aligned} \tag{2.161}$$

because of the normalization of $|J_e m_e\rangle$. We thus obtain[14]

$$\boxed{\sigma = \frac{3}{2\pi} \lambda^2 \frac{\Gamma^2/4}{\delta^2 + \Gamma^2/4} \langle Jm_i | (\tilde{\boldsymbol{d}} \cdot \hat{\boldsymbol{\varepsilon}}_i^*)(\tilde{\boldsymbol{d}} \cdot \hat{\boldsymbol{\varepsilon}}_i) | Jm_i\rangle} \tag{2.162}$$

This result may also be obtained starting from the optical theorem (2.79). To do this, note that the amplitude $\mathcal{T}_{fi}$ obtained from the Fermi golden rule is proportional to the scattering amplitude $f(\boldsymbol{k}_i m_i, \boldsymbol{k}' m_f)$. The coefficient of proportionality is obtained from the expression (2.157) for the differential scattering cross section and is given by $\sqrt{(\Omega/c)(2\pi/\hbar)\rho(\omega)} = \Omega\omega/2\pi\hbar c^2$. Thus, by defining

$$f(\boldsymbol{k}_i m_i, \boldsymbol{k}' m_f) = -\frac{\Omega\omega}{2\pi\hbar c^2} \mathcal{T}_{fi} \tag{2.163}$$

we get the total scattering cross section (2.162) using the optical theorem (2.79).

Having calculated the scattering cross section, we now average over the statistical distribution of internal Zeeman states $|Jm_i\rangle$ which we assume to be equiprobable. Denoting this average by $\langle \cdots \rangle_{int}$ we have

$$\begin{aligned} \langle (\tilde{\boldsymbol{d}} \cdot \hat{\boldsymbol{\varepsilon}}_i^*)(\tilde{\boldsymbol{d}} \cdot \hat{\boldsymbol{\varepsilon}}_i) \rangle_{int} &= \frac{1}{2J+1} \sum_{m_i, m_e} \langle Jm_i | \tilde{\boldsymbol{d}} \cdot \hat{\boldsymbol{\varepsilon}}_i^* | J_e m_e\rangle \langle J_e m_e | \tilde{\boldsymbol{d}} \cdot \hat{\boldsymbol{\varepsilon}}_i | Jm_i\rangle \\ &= \frac{1}{2J+1} \sum_{m_i, m_e} |\langle Jm_i | \tilde{\boldsymbol{d}} \cdot \hat{\boldsymbol{\varepsilon}}_i | J_e m_e\rangle|^2 = \frac{1}{3} \frac{2J_e + 1}{2J + 1}, \end{aligned} \tag{2.164}$$

[14] Before averaging over the distribution of internal states, there is a preferred direction given by the atomic dipole. The total scattering cross section thus depends on the projection of the incident polarization $\hat{\boldsymbol{\varepsilon}}_i$ along this direction.

where the last expression results from the equality (C15.a) in [60], employed with the identity $\langle J_e || \tilde{\boldsymbol{d}} || J \rangle = \sqrt{2J_e + 1}$. We finally obtain the averaged total scattering cross section:

$$\boxed{\langle \sigma \rangle = A_{JJ_e} \frac{3\lambda^2}{2\pi} \frac{\Gamma^2/4}{\delta^2 + \Gamma^2/4}} \tag{2.165}$$

where $A_{JJ_e} = \frac{1}{3}(2J_e + 1)/(2J + 1)$. We see from this expression, that the averaged total scattering cross section does not depend on the internal structure of the atomic levels (apart from the constant factor A_{JJ_e}). The physical quantities derived from this such as the optical index and the elastic mean free path are also independent of the internal structure of the atomic levels.

In contrast, the differential scattering cross section (2.158) has a richer structure. Averaging (2.158) over the statistical distribution of Zeeman states $|Jm_i\rangle$, we get, using (2.165):

$$\left\langle \frac{d\sigma}{d\Omega_{k'}} \right\rangle_{int} = \frac{3\langle \sigma \rangle}{8\pi A_{JJ_e}} \langle (\tilde{\boldsymbol{d}} \cdot \hat{\boldsymbol{\varepsilon}}_i^*)(\tilde{\boldsymbol{d}} \cdot \hat{\boldsymbol{\varepsilon}}')(\tilde{\boldsymbol{d}} \cdot \hat{\boldsymbol{\varepsilon}}'^*)(\tilde{\boldsymbol{d}} \cdot \hat{\boldsymbol{\varepsilon}}_i) \rangle_{int} \tag{2.166}$$

which is a rank four tensor which clearly depends on the internal structure of the atomic levels. In order to calculate the average $\langle (\tilde{\boldsymbol{d}} \cdot \hat{\boldsymbol{\varepsilon}}_i^*)(\tilde{\boldsymbol{d}} \cdot \hat{\boldsymbol{\varepsilon}}')(\tilde{\boldsymbol{d}} \cdot \hat{\boldsymbol{\varepsilon}}'^*)(\tilde{\boldsymbol{d}} \cdot \hat{\boldsymbol{\varepsilon}}_i) \rangle_{int}$ on the internal states, we must first decompose this tensor into irreducible components [61–63]. Here we just mention the result, and try to justify it. The trace must be invariant under rotation and thus cannot depend on anything other than scalar products between the polarization vectors $(\hat{\boldsymbol{\varepsilon}}_i, \hat{\boldsymbol{\varepsilon}}', \hat{\boldsymbol{\varepsilon}}_i^*, \hat{\boldsymbol{\varepsilon}}'^*)$. More precisely, we may write the differential scattering cross section in the form

$$\left\langle \frac{d\sigma}{d\Omega_{k'}} \right\rangle_{int} = \frac{3\langle \sigma \rangle}{8\pi} \left(w_1 |\hat{\boldsymbol{\varepsilon}}'^* \cdot \hat{\boldsymbol{\varepsilon}}_i|^2 + w_2 |\hat{\boldsymbol{\varepsilon}}' \cdot \hat{\boldsymbol{\varepsilon}}_i|^2 + w_3 \right), \tag{2.167}$$

where the w_i are given in section A6.5.2. The nature of the scattered photon depends not only on the polarization state of the incident photon, but also on the internal Zeeman state of the atom. We will return in greater detail to the role of internal structure on interference effects in multiple scattering in Appendix A6.5. Finally, note that we recover the Rayleigh differential scattering cross section (2.149) for the case of a non-degenerate ground state, where $J = 0$ and $J_e = 1$, for which case we have $(w_1, w_2, w_3) = (1, 0, 0)$.

3

Perturbation theory

In this chapter, we set $\hbar = 1$.

In this and the following chapter, we propose to solve the wave equations (Schrödinger or Helmholtz) in the presence of Gaussian disorder, as introduced in Chapter 2. To this end, we present a method which describes the temporal evolution of a wave packet in random media. This method uses the formalism of Green's functions [64–67] which we employ here to facilitate an iterative expansion in powers of the disorder potential, which is called the multiple scattering expansion. In the limit of weak disorder, which we will define properly, this expansion is expressed in the form of a series of *independent processes*, termed *collision events*, separated by a characteristic time τ_e called the elastic collision time. Associated with this is a characteristic length, the elastic mean free path, defined by $l_e = v\tau_e$ where v is the group velocity of the wave.[1] In order to evaluate the collision time τ_e for the case of the Schrödinger equation, we will use a representation in plane waves which correspond to the eigenstates $|\boldsymbol{k}\rangle$ of the free Hamiltonian. We therefore interpret τ_e as the average lifetime of the states $|\boldsymbol{k}\rangle$. To estimate it in the presence of the potential V to lowest order in perturbation theory, we may use the Fermi golden rule: the lifetime $\tau_{\boldsymbol{k}}$ of a state $|\boldsymbol{k}\rangle$ is given by

$$\frac{1}{\tau_{\boldsymbol{k}}} = 2\pi \sum_{\boldsymbol{k}'} |\langle \boldsymbol{k}|V|\boldsymbol{k}'\rangle|^2 \delta(\epsilon_{\boldsymbol{k}} - \epsilon_{\boldsymbol{k}'}). \tag{3.1}$$

Taking for V the Gaussian disorder potential defined by (2.29) and averaging over disorder, we obtain $\Omega\overline{|\langle \boldsymbol{k}|V|\boldsymbol{k}'\rangle|^2} = B(\boldsymbol{k} - \boldsymbol{k}')$ where $B(\boldsymbol{k} - \boldsymbol{k}')$ is the Fourier transform of the correlation function $\overline{V(\boldsymbol{r})V(\boldsymbol{r}')}$, where $\overline{\cdots}$ denotes the disorder average. Ω is the volume of the system and the wave functions are normalized in this volume. The average lifetime τ_e of a state of energy ϵ is then:[2]

$$\boxed{\frac{1}{\tau_e} = 2\pi\rho_0\gamma_e} \tag{3.2}$$

[1] For degenerate electrons, v is the Fermi velocity v_F. For waves, v is the group velocity, which we will denote by c.
[2] We first show that

$$\frac{1}{\tau_e} = \frac{2\pi}{\Omega}\sum_{\boldsymbol{k}'} B(\boldsymbol{k} - \boldsymbol{k}')\delta(\epsilon_{\boldsymbol{k}} - \epsilon_{\boldsymbol{k}'})$$

and obtain expression (3.2) by passing to the continuum.

where ρ_0 is the density of states per unit volume at the energy considered and where the parameter γ_e which characterizes the disorder is equal to

$$\boxed{\gamma_e = \langle B(\boldsymbol{k} - \boldsymbol{k}') \rangle} \tag{3.3}$$

$\langle \cdots \rangle$ indicates the angular average of $B(\boldsymbol{k} - \boldsymbol{k}')$ with the constraint $\epsilon_{\boldsymbol{k}} = \epsilon_{\boldsymbol{k}'}$. For a white noise as defined by (2.30), we have $\gamma_e = B$. In general, the details of the microscopic nature of the disorder are unknown and they are contained in the phenomenological parameter γ_e. Nevertheless, in the framework of the Edwards model for a density n_i of scatterers of total cross section σ, the relation (2.100) allows us to rewrite the elastic mean free path l_e:[3]

$$\frac{1}{l_e} = n_i \sigma, \tag{3.4}$$

a relation which is also valid for the Helmholtz equation (2.8).[4] In certain cases, the scattering cross section may be calculated by means of a microscopic model. For example, to describe multiple scattering of light by suspensions, we take σ to be the Rayleigh scattering cross section studied in Appendix A2.3. The Gaussian model is thus an *effective* model, which depends on the parameter γ_e that may be related to physical parameters which it may be possible to control experimentally. We will show in Appendix A4.3 that if the potential is of finite range, there is another characteristic time τ^*, called the *transport time*.

Exercise 3.1 Using the Fermi golden rule, show that the elastic lifetime τ_e for the Anderson model (section 2.2.3) with a rectangular distribution of the potential may be put in the form $\frac{1}{\tau_e} \propto W^2/t^2$.

3.1 Green's functions

The Fermi golden rule describes the temporal evolution of the system only for times less than τ_e. To go beyond this regime, we now introduce the Green function and the resolvant operator formalism. We will not seek to describe this formalism in all its generality, but we shall introduce several essential notions [64, 65]. The cases of the Schrödinger equation, which is first order in time, and the Helmholtz equation, which is second order, will be treated separately.

3.1.1 Green's function for the Schrödinger equation

Consider the Schrödinger equation associated with the Hamiltonian (2.1)

$$i\frac{\partial \psi}{\partial t} = \mathcal{H}\psi = (\mathcal{H}_0 + V)\psi, \tag{3.5}$$

[3] For this, we use expression (3.40) for the density of states ρ_0.
[4] For the Helmholtz equation, we will show later, (3.77), that $1/l_e = \gamma_e/(4\pi)$. We thus obtain expression (3.4) by using (2.98).

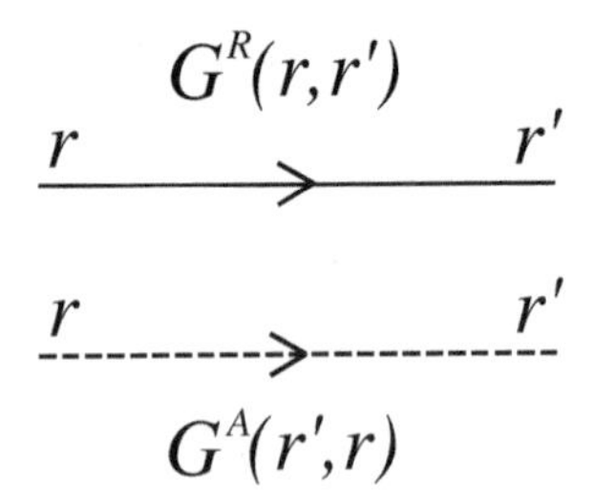

Figure 3.1 Convention employed in this book for the representation of Green's functions. We take $G^A(\boldsymbol{r}',\boldsymbol{r}) = G^R(\boldsymbol{r},\boldsymbol{r}')^*$.

where $\mathcal{H}_0 = -(1/2m)\Delta$ and V is the disorder potential. This describes free spinless electrons. The time evolution of a state $|\psi(t)\rangle$ initially in $|\psi_0(t=0)\rangle$ is described by the unitary evolution operator $U(t)$:

$$|\psi(t)\rangle = U(t)|\psi_0(0)\rangle = e^{-i\mathcal{H}t}|\psi_0(0)\rangle. \tag{3.6}$$

In spatial representation,

$$\psi_f(\boldsymbol{r},t) = \langle \boldsymbol{r}|\psi_f(t)\rangle = \int d\boldsymbol{r}_i \, \langle \boldsymbol{r}|e^{-i\mathcal{H}t}|\boldsymbol{r}_i\rangle \psi_0(\boldsymbol{r}_i,0). \tag{3.7}$$

Following the convention of Figure 3.1, the Green function is defined by[5]

$$G(\boldsymbol{r}_i,\boldsymbol{r},t) = \langle \boldsymbol{r}|e^{-i\mathcal{H}t}|\boldsymbol{r}_i\rangle = \sum_n \phi_n^*(\boldsymbol{r}_i)\phi_n(\boldsymbol{r})\, e^{-i\epsilon_n t} \tag{3.8}$$

where ϵ_n and ϕ_n are respectively the eigenenergies and the normalized eigenstates of $\mathcal{H}$. Thus defined, $G(\boldsymbol{r}_i,\boldsymbol{r},t)$ describes the evolution of a state $|\boldsymbol{r}_i\rangle$ at time $t=0$ to the state $|\boldsymbol{r}\rangle$ at a time t whose sign is not specified. If, on the other hand, we wish to describe the evolution of a state created at $t=0$ to positive or negative times, we are led to define the operators

$$\begin{aligned} \hat{G}^R(t) &= -i\theta(t)\, e^{-i\mathcal{H}t} \\ \hat{G}^A(t) &= i\theta(-t)\, e^{-i\mathcal{H}t} \end{aligned} \tag{3.9}$$

whose spatial representations correspond to retarded G^R and advanced G^A Green's functions

$$\begin{aligned} G^R(\boldsymbol{r}_i,\boldsymbol{r},t) &= -i\theta(t)\, \langle \boldsymbol{r}|e^{-i\mathcal{H}t}|\boldsymbol{r}_i\rangle = -i\theta(t) \sum_n \phi_n^*(\boldsymbol{r}_i)\phi_n(\boldsymbol{r})\, e^{-i\epsilon_n t} \\ G^A(\boldsymbol{r}_i,\boldsymbol{r},t) &= i\theta(-t)\, \langle \boldsymbol{r}|e^{-i\mathcal{H}t}|\boldsymbol{r}_i\rangle = i\theta(-t) \sum_n \phi_n^*(\boldsymbol{r}_i)\, \phi_n(\boldsymbol{r})\, e^{-i\epsilon_n t}. \end{aligned} \tag{3.10}$$

[5] Thus defined, $G^R(\boldsymbol{r}_i,\boldsymbol{r})$ is read from left to right (Figure 3.1). This convention is not the most usual, but we will nevertheless adopt it in this book. It is better adapted to the description of scattering or transport situations, where the source is generally placed on the left and the detector on the right.

The Fourier transforms

$$G^{R,A}(\boldsymbol{r}_i,\boldsymbol{r},\epsilon) = \int_{-\infty}^{\infty} dt\, e^{i\epsilon t}\, G^{R,A}(\boldsymbol{r}_i,\boldsymbol{r},t) \tag{3.11}$$

are written as

$$\boxed{G^{R,A}(\boldsymbol{r}_i,\boldsymbol{r},\epsilon) = \sum_n \frac{\phi_n^*(\boldsymbol{r}_i)\phi_n(\boldsymbol{r})}{\epsilon - \epsilon_n \pm i0}} \tag{3.12}$$

The convergence of the integrals (3.11) at long times necessitates the addition of an imaginary part to the energy $\epsilon \pm i0$, whose sign $\pm$ corresponds to the advanced and retarded parts, respectively. Equation (3.12) allows the formal introduction of operators which are the Fourier transforms of $\hat{G}^{R,A}(t)$:

$$\hat{G}^{R,A}(\epsilon) = \frac{1}{\epsilon - \mathcal{H} \pm i0}. \tag{3.13}$$

Likewise, we define the free particle Green operators associated with the Hamiltonian $\mathcal{H}_0$

$$\hat{G}_0^{R,A}(\epsilon) = \frac{1}{\epsilon - \mathcal{H}_0 \pm i0}. \tag{3.14}$$

The Schrödinger equation (3.5) is therefore expressed as a relation between the operators $\hat{G}$ and $\hat{G}_0$. Multiplying by $\hat{G}_0$ relation (3.13) written in the form

$$(\epsilon - \mathcal{H})\hat{G} = \mathbb{1}, \tag{3.15}$$

and using (3.14) in the form $(\epsilon - \mathcal{H}_0)\hat{G}_0 = \mathbb{1}$, we obtain

$$\hat{G} = \hat{G}_0 + \hat{G}_0 V \hat{G}, \tag{3.16}$$

which lends itself to the iterative expansion we will discuss in section 3.2.

Some properties of Green's functions

We introduced the Green function using the time evolution operator. It also measures the response to a pulse associated with the Schrödinger equation (3.5). Indeed, in the spatial representation, equation (3.13) is written

$$(\epsilon - \mathcal{H} \pm i0)\, G^{R,A}(\boldsymbol{r}_i,\boldsymbol{r},\epsilon) = \delta(\boldsymbol{r} - \boldsymbol{r}_i) \tag{3.17}$$

and relation (3.16) between the operators $\hat{G}$ and $\hat{G}_0$ becomes

$$G(\boldsymbol{r}_i,\boldsymbol{r},\epsilon) = G_0(\boldsymbol{r}_i,\boldsymbol{r},\epsilon) + \int G(\boldsymbol{r}_i,\boldsymbol{r}',\epsilon)V(\boldsymbol{r}')G_0(\boldsymbol{r}',\boldsymbol{r},\epsilon)\, d\boldsymbol{r}'. \tag{3.18}$$

Moreover, the Green function satisfies the following properties:

$$G^A(\boldsymbol{r},\boldsymbol{r}_i,\epsilon) = G^R(\boldsymbol{r}_i,\boldsymbol{r},\epsilon)^* \tag{3.19}$$

and

$$G^A(\boldsymbol{r},\boldsymbol{r}_i,-t) = G^R(\boldsymbol{r}_i,\boldsymbol{r},t)^* \tag{3.20}$$

where * indicates the complex conjugate. We also define the imaginary part of $\hat{G}^R$ by

$$\mathrm{Im}\hat{G}^R = \frac{\hat{G}^R - \hat{G}^A}{2i}. \tag{3.21}$$

Important remark

We use the notation

$$\mathrm{Im}G^R(\boldsymbol{r},\boldsymbol{r}') = \langle \boldsymbol{r}'|\mathrm{Im}\hat{G}^R|\boldsymbol{r}\rangle = \frac{G^R(\boldsymbol{r},\boldsymbol{r}') - G^A(\boldsymbol{r},\boldsymbol{r}')}{2i} \tag{3.22}$$

and not

$$\mathrm{Im}G^R(\boldsymbol{r},\boldsymbol{r}') = \frac{G^R(\boldsymbol{r},\boldsymbol{r}') - G^R(\boldsymbol{r},\boldsymbol{r}')^*}{2i} = \frac{G^R(\boldsymbol{r},\boldsymbol{r}') - G^A(\boldsymbol{r}',\boldsymbol{r})}{2i}. \tag{3.23}$$

Green's function and density of states

All information about the solutions of the Schrödinger equation is contained in the Green function. In particular, it is related to the density of states $\nu(\epsilon)$ defined by

$$\nu(\epsilon) = \sum_n \delta(\epsilon - \epsilon_n). \tag{3.24}$$

We shall also use the density of states per unit volume, denoted $\rho(\epsilon) = \nu(\epsilon)/\Omega$. It is also convenient to define the following.

- The local density of states at a point $\boldsymbol{r}$:

$$\rho_\epsilon(\boldsymbol{r}) = \rho(\boldsymbol{r},\epsilon) = \sum_n |\phi_n(\boldsymbol{r})|^2 \delta(\epsilon - \epsilon_n), \tag{3.25}$$

- The non-local density of states:

$$\rho_\epsilon(\boldsymbol{r},\boldsymbol{r}') = \rho(\boldsymbol{r},\boldsymbol{r}',\epsilon) = \sum_n \phi_n^*(\boldsymbol{r})\phi_n(\boldsymbol{r}')\delta(\epsilon - \epsilon_n). \tag{3.26}$$

Starting from equation (3.12), the expression

$$\frac{1}{x + i0} = pp\frac{1}{x} - i\pi\delta(x) \tag{3.27}$$

enables us to write

$$\rho_\epsilon(\boldsymbol{r},\boldsymbol{r}') = -\frac{1}{\pi}\mathrm{Im}G^R(\boldsymbol{r},\boldsymbol{r}',\epsilon) = -\frac{1}{\pi}\langle \boldsymbol{r}'|\mathrm{Im}\hat{G}^R|\boldsymbol{r}\rangle = \frac{i}{2\pi}\left[G^R(\boldsymbol{r},\boldsymbol{r}',\epsilon) - G^A(\boldsymbol{r},\boldsymbol{r}',\epsilon)\right] \tag{3.28}$$

as well as

$$\rho_\epsilon(\boldsymbol{r}) = -\frac{1}{\pi}\mathrm{Im}G^R(\boldsymbol{r},\boldsymbol{r},\epsilon) \tag{3.29}$$

and to obtain the density of states per unit volume in the form

$$\boxed{\rho(\epsilon) = \frac{\nu(\epsilon)}{\Omega} = -\frac{1}{\pi\Omega}\mathrm{Im}\int d\boldsymbol{r}\, G^R(\boldsymbol{r},\boldsymbol{r},\epsilon)} \tag{3.30}$$

Thus $\rho_\epsilon(\boldsymbol{r}) = \rho_\epsilon(\boldsymbol{r},\boldsymbol{r})$ and $\rho(\epsilon)$ is the spatial average of $\rho_\epsilon(\boldsymbol{r})$.

We can also introduce the Green function in momentum representation

$$G^{R,A}(\boldsymbol{k}_i,\boldsymbol{k},\epsilon) = \langle \boldsymbol{k}|\hat{G}^{R,A}(\epsilon)|\boldsymbol{k}_i\rangle. \tag{3.31}$$

Translation invariance, when it exists, implies that $G(\boldsymbol{r}_i,\boldsymbol{r},\epsilon) = G(\boldsymbol{r}-\boldsymbol{r}_i,\epsilon)$. In this case, $G^{R,A}(\boldsymbol{k}_i,\boldsymbol{k},\epsilon) = G^{R,A}(\boldsymbol{k},\epsilon)\delta_{\boldsymbol{k},\boldsymbol{k}_i}$, with

$$G^{R,A}(\boldsymbol{k},\epsilon) = \int G^{R,A}(\boldsymbol{r},\epsilon)\, e^{-i\boldsymbol{k}\cdot\boldsymbol{r}}\, d\boldsymbol{r}$$

and the density of states per unit volume may be written as

$$\rho(\epsilon) = -\frac{1}{\pi\Omega}\sum_{\boldsymbol{k}}\mathrm{Im}G^R(\boldsymbol{k},\epsilon). \tag{3.32}$$

We see that this expression is analogous to (3.30). The density of states does not depend on the representation under consideration. More generally, we may write

$$\boxed{\nu = -\frac{1}{\pi}\mathrm{Im}\mathrm{Tr}\hat{G}^R} \tag{3.33}$$

Free Green's function

In the absence of disorder, the Hamiltonian $\mathcal{H}_0$ is diagonal in the plane-wave basis, and the Green function is written:

$$G_0^{R,A}(\boldsymbol{k},\epsilon) = \frac{1}{\epsilon - \epsilon(\boldsymbol{k}) \pm i0}, \tag{3.34}$$

where $\epsilon(\boldsymbol{k}) = k^2/2m$. The Green function in spatial representation is the Fourier transform of (3.34). In three dimensions and for an infinite system, we have

$$G_0^{R,A}(\boldsymbol{r}_i, \boldsymbol{r}, \epsilon) = \frac{2m}{(2\pi)^3} \int d\boldsymbol{k}' \, e^{i\boldsymbol{k}'\cdot(\boldsymbol{r}-\boldsymbol{r}_i)} \frac{1}{k^2 - k'^2 \pm i0}. \tag{3.35}$$

Writing $R = |\boldsymbol{r} - \boldsymbol{r}_i|$, the angular integral yields

$$G_0^{R,A}(\boldsymbol{r}_i, \boldsymbol{r}, \epsilon) = \frac{2m}{(2\pi)^2} \int_0^\infty k'^2 \, dk' \frac{1}{ik'R} \frac{e^{ik'R} - e^{-ik'R}}{k^2 - k'^2 \pm i0}. \tag{3.36}$$

We may rewrite the integral over k' as

$$G_0^{R,A}(\boldsymbol{r}_i, \boldsymbol{r}, \epsilon) = -\frac{m}{2i\pi^2 R} \int_{-\infty}^\infty dk' \frac{k' \, e^{ik'R}}{(k' - k \mp i0)(k' + k \pm i0)} \tag{3.37}$$

which may be calculated by the method of residues to give

$$G_0^{R,A}(\boldsymbol{r}_i, \boldsymbol{r}, \epsilon) = -\frac{m}{2\pi} \frac{e^{\pm ikR}}{R} \tag{3.38}$$

with $k = \sqrt{2m\epsilon}$. This result may be obtained directly by noting that the free Green function is a solution of the differential equation

$$\left(\epsilon + \frac{1}{2m}\Delta_{\boldsymbol{r}} \pm i0\right) G_0^{R,A}(\boldsymbol{r}_i, \boldsymbol{r}, \epsilon) = \delta(\boldsymbol{r} - \boldsymbol{r}_i). \tag{3.39}$$

We check that in three dimensions the density of states (without spin), given by equation (3.30), is equal to

$$\boxed{\rho_0(\epsilon) = \frac{mk}{2\pi^2}} \tag{3.40}$$

so that the Green function is of the form

$$\boxed{G_0^{R,A}(\boldsymbol{r}_i, \boldsymbol{r}, \epsilon) = -\pi \rho_0 \frac{e^{\pm ikR}}{kR}} \tag{3.41}$$

Exercise 3.2 For the dispersion relation $\epsilon(\boldsymbol{k}) = \epsilon(|\boldsymbol{k}|)$, verify that the Green function $G_0^{R,A}(\boldsymbol{r}_i, \boldsymbol{r}, \epsilon)$ only depends on $R = |\boldsymbol{r} - \boldsymbol{r}_i|$ and that in $d = 3$, it may be written in the form (3.41).

Exercise 3.3 Show that in $d = 1$, the Green function may be written:

$$G_0^{R,A}(\boldsymbol{r}_i, \boldsymbol{r}, \epsilon) = \mp i \frac{m}{k} e^{\pm ikR} = \mp i\pi\rho_0\, e^{\pm ikR}, \tag{3.42}$$

where $k = |\boldsymbol{k}|$.

Exercise 3.4 Show that in $d = 2$, the Green function may be written:

$$G_0^{R,A}(\boldsymbol{r}_i, \boldsymbol{r}, \epsilon) = \mp i \frac{m}{2} H_0^{(1,2)}(kR), \tag{3.43}$$

where $k = |\boldsymbol{k}|$ and $H_0^{(1,2)}(x) = J_0(x) \pm iY_0(x)$. The functions $J_0(x)$, $Y_0(x)$ and $H_0^{(1,2)}(x)$ are, respectively, Bessel, Neumann, and Hankel functions (also called Bessel functions of the first, second, and third kind) [68]. Calculate the density of states in two dimensions.

Exercise 3.5 Show that in arbitrary dimension d, the density of states $\rho_0(\epsilon)$ for a quadratic dispersion $\epsilon = k^2/2m$ is

$$\rho_0(\epsilon) = \frac{dA_d}{(2\pi)^d} m k^{d-2}, \tag{3.44}$$

where $A_d = \frac{2}{d} \frac{\pi^{d/2}}{\Gamma(d/2)}$ is the volume of the unit sphere. In particular,

$$\rho_0^{1d}(\epsilon) = \frac{m}{\pi k} \qquad \rho_0^{2d}(\epsilon) = \frac{m}{2\pi}. \tag{3.45}$$

3.1.2 Green's function for the Helmholtz equation

As was done in (3.17) for the Schrödinger equation, we define the Green function $G(\boldsymbol{r}_i, \boldsymbol{r}, k_0)$ associated with the Helmholtz equation (2.8) by

$$\left[\Delta_r + k_0^2(1 + \mu(\boldsymbol{r}))\right] G(\boldsymbol{r}_i, \boldsymbol{r}, k_0) = \delta(\boldsymbol{r} - \boldsymbol{r}_i). \tag{3.46}$$

The free Green function $G_0(\boldsymbol{r}_i, \boldsymbol{r}, k_0)$ is thus the solution of the differential equation

$$\left(\Delta_r + k_0^2\right) G_0(\boldsymbol{r}_i, \boldsymbol{r}, k_0) = \delta(\boldsymbol{r} - \boldsymbol{r}_i) \tag{3.47}$$

and is calculated, similar to (3.38), to be

$$\boxed{G_0^{R,A}(\boldsymbol{r}_i, \boldsymbol{r}, k_0) = -\frac{1}{4\pi} \frac{e^{\pm ik_0 R}}{R}} \tag{3.48}$$

in three dimensions, and

$$G_0^{R,A}(\boldsymbol{r}_i, \boldsymbol{r}, k_0) = \mp \frac{i}{2k_0} e^{\pm ik_0 R} \tag{3.49}$$

in one dimension. Finally, note that its Fourier transform is equal to

$$G_0^{R,A}(\boldsymbol{k}, k_0) = \frac{1}{k_0^2 - k^2 \pm i0}. \tag{3.50}$$

As in (3.30), we may relate the density of eigenmodes to the Green function. Since the dispersion relation is linear, $\omega = k_0 c$, the density of modes is given by

$$\nu(\omega) = \sum_{\boldsymbol{k}} \delta(\omega - k_0 c) = \frac{1}{c} \sum_{\boldsymbol{k}} \delta(k - k_0), \tag{3.51}$$

where $k = |\boldsymbol{k}|$. Moreover, the imaginary part of the Green function (3.50) is equal to

$$\mathrm{Im} G_0^R(\boldsymbol{k}, k_0) = -\frac{\pi}{2k_0} \left[\delta(k + k_0) + \delta(k - k_0)\right] \tag{3.52}$$

so that

$$\rho_0(\omega) = -\frac{2k_0}{\pi c \Omega} \sum_{\boldsymbol{k}} \mathrm{Im} G_0^R(\boldsymbol{k}, k_0). \tag{3.53}$$

In the presence of disorder,

$$\rho(\omega) = -\frac{2k_0}{\pi c \Omega} \int d\boldsymbol{r}\, \mathrm{Im} G^R(\boldsymbol{r}, \boldsymbol{r}, k_0) \tag{3.54}$$

or

$$\boxed{\nu = -\frac{2k_0}{\pi c} \mathrm{Im} \mathrm{Tr} \hat{G}^R} \tag{3.55}$$

which is interesting to compare to (3.33) for the Schrödinger equation. The additional prefactor results from the fact that the Schrödinger equation is first order in time, while the Helmholtz equation is second order.

Exercise 3.6 Show that for the free Helmholtz equation, the density of modes in three dimensions is

$$\rho_0(\omega) = \frac{\omega^2}{2\pi^2 c^3}. \tag{3.56}$$

The Green function formalism provides a technical framework for the systematic study of solutions to the Schrödinger equation. It is natural, and indeed indispensable, for the study of electromagnetic wave propagation (the so-called radiative solutions) starting from a distribution of sources $j(\boldsymbol{r})$. This distribution may be a point source described by a

δ function, which corresponds to (3.46). If the wave is emitted by a source distribution $j(\boldsymbol{r})$, the electric field amplitude $\psi(\boldsymbol{r})$ is no longer a solution of (2.8), but of

$$\boxed{\Delta\psi(\boldsymbol{r}) + k_0^2(1+\mu(\boldsymbol{r}))\psi(\boldsymbol{r}) = j(\boldsymbol{r})} \tag{3.57}$$

This differential equation may be reexpressed as an integral equation

$$\psi(\boldsymbol{r}) = \int d\boldsymbol{r}_i \, j(\boldsymbol{r}_i) G(\boldsymbol{r}_i, \boldsymbol{r}, k_0), \tag{3.58}$$

where the Green function $G(\boldsymbol{r}_i, \boldsymbol{r}, k_0)$ is the solution of the differential equation (3.46). Using the free Green function $G_0(\boldsymbol{r}_i, \boldsymbol{r}, k_0)$ which is the solution of (3.47), we obtain another integral equation analogous to (3.18)

$$G(\boldsymbol{r}_i, \boldsymbol{r}, k_0) = G_0(\boldsymbol{r}_i, \boldsymbol{r}, k_0) - k_0^2 \int d\boldsymbol{r}' \, G(\boldsymbol{r}_i, \boldsymbol{r}', k_0)\mu(\boldsymbol{r}')G_0(\boldsymbol{r}', \boldsymbol{r}, k_0), \tag{3.59}$$

so that the solutions of (3.57) are written in a form which allows us to separate the contribution of the source from that of the disorder

$$\psi(\boldsymbol{r}) = \int d\boldsymbol{r}_i j(\boldsymbol{r}_i) G_0(\boldsymbol{r}_i, \boldsymbol{r}, k_0) - k_0^2 \int d\boldsymbol{r}' \, \psi(\boldsymbol{r}')\mu(\boldsymbol{r}')G_0(\boldsymbol{r}', \boldsymbol{r}, k_0). \tag{3.60}$$

This equation is the generalization of (3.59) to the case of an arbitrary distribution of sources.

3.2 Multiple scattering expansion

3.2.1 Dyson equation

We now seek to construct a perturbative expansion starting from equation (3.16) connecting $\hat{G}$ and $\hat{G}_0$. Formally, we may write this expansion as [66, 67]

$$\hat{G} = \hat{G}_0 + \hat{G}_0 V \hat{G}_0 + \hat{G}_0 V \hat{G}_0 V \hat{G}_0 + \cdots. \tag{3.61}$$

In spatial representation this is[6]

$$\begin{aligned} G(\boldsymbol{r}, \boldsymbol{r}') = {} & G_0(\boldsymbol{r}, \boldsymbol{r}') + \int d\boldsymbol{r}_1 G_0(\boldsymbol{r}, \boldsymbol{r}_1) V(\boldsymbol{r}_1) G_0(\boldsymbol{r}_1, \boldsymbol{r}') \\ & + \int d\boldsymbol{r}_1 d\boldsymbol{r}_2 G_0(\boldsymbol{r}, \boldsymbol{r}_1) V(\boldsymbol{r}_1) G_0(\boldsymbol{r}_1, \boldsymbol{r}_2) V(\boldsymbol{r}_2) G_0(\boldsymbol{r}_2, \boldsymbol{r}') \\ & + \int d\boldsymbol{r}_1 d\boldsymbol{r}_2 d\boldsymbol{r}_3 G_0(\boldsymbol{r}, \boldsymbol{r}_1) V(\boldsymbol{r}_1) G_0(\boldsymbol{r}_1, \boldsymbol{r}_2) V(\boldsymbol{r}_2) G_0(\boldsymbol{r}_2, \boldsymbol{r}_3) V(\boldsymbol{r}_3) G_0(\boldsymbol{r}_3, \boldsymbol{r}') \\ & + \cdots. \end{aligned} \tag{3.62}$$

[6] To be concise and when no confusion is possible, the energy (or frequency) dependence of the Green functions is not written explicitly.

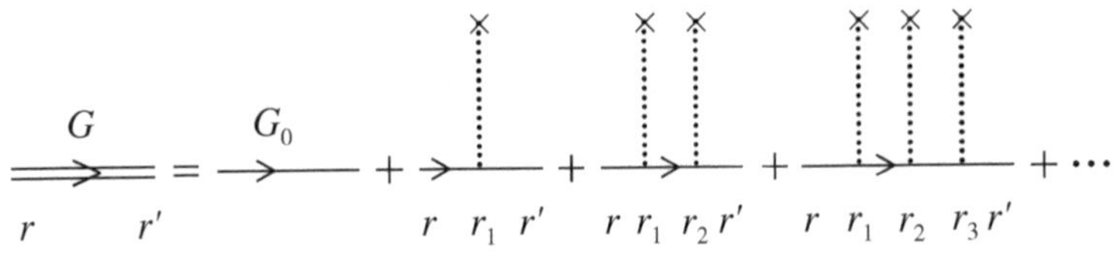

Figure 3.2 Diagrammatic expansion of the Green function before averaging over disorder.

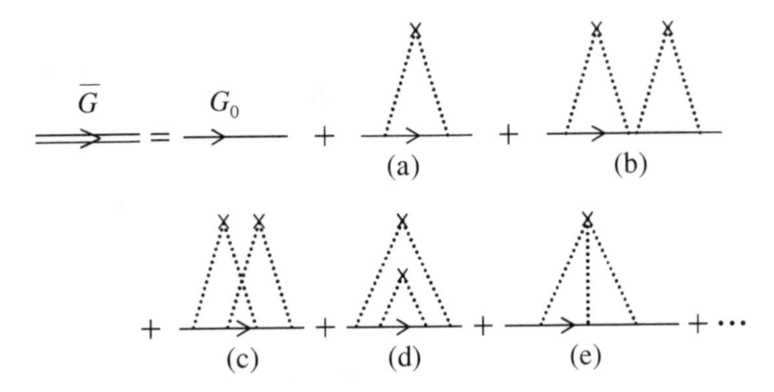

Figure 3.3 Diagrammatic expansion of the average Green function.

This expansion may be expressed pictorially as in Figure 3.2. Upon averaging over disorder and using the particular form of the Gaussian potential (2.29), the expansion of the average Green function $\overline{G}$ simplifies considerably. The term linear in V and all other odd power terms vanish, while the quadratic term yields the correlation function $B(\boldsymbol{r}_1 - \boldsymbol{r}_2)$, leaving us with

$$\overline{G}(\boldsymbol{r},\boldsymbol{r}') = G_0(\boldsymbol{r},\boldsymbol{r}') + \int d\boldsymbol{r}_1\, d\boldsymbol{r}_2\, B(\boldsymbol{r}_1 - \boldsymbol{r}_2) G_0(\boldsymbol{r},\boldsymbol{r}_1) G_0(\boldsymbol{r}_1,\boldsymbol{r}_2) G_0(\boldsymbol{r}_2,\boldsymbol{r}') + \cdots \,. \tag{3.63}$$

This expansion is represented in Figure 3.3, where the pairings of impurity lines represent the different products $\overline{V(\boldsymbol{r}_i)V(\boldsymbol{r}_j)}$. Upon averaging over disorder, translational invariance is recovered, and the Green function depends only on the difference in position: $\overline{G}(\boldsymbol{r},\boldsymbol{r}') = \overline{G}(\boldsymbol{r}-\boldsymbol{r}')$. The Fourier transform of (3.63) is of the form

$$\overline{G}(\boldsymbol{k}) = G_0(\boldsymbol{k}) + \frac{1}{\Omega}\sum_{\boldsymbol{q}} B(\boldsymbol{q}) G_0(\boldsymbol{k}) G_0(\boldsymbol{k}-\boldsymbol{q}) G_0(\boldsymbol{k}) + \cdots \,. \tag{3.64}$$

Upon averaging, we generate all the diagrams corresponding to all possible pairings of interaction lines with the impurities which appear in Figure 3.2. These diagrams are of two types (Figure 3.3). The first are called separable or reducible, that is, they may be separated into two diagrams without cutting an impurity line. All other diagrams are called irreducible. The reducible diagrams may be factorized into a product of irreducible diagrams. This is possible because the integrals over the intermediate wave vectors are independent. For example, diagram (b) of Figure 3.3 may be written

$$\frac{1}{\Omega^2}\sum_{\boldsymbol{q}}\sum_{\boldsymbol{q}'} G_0(\boldsymbol{k})B(\boldsymbol{q})G_0(\boldsymbol{k}-\boldsymbol{q})G_0(\boldsymbol{k})B(\boldsymbol{q}')G_0(\boldsymbol{k}-\boldsymbol{q}')G_0(\boldsymbol{k})$$

$$= G_0(\boldsymbol{k})\left[\sum_{\boldsymbol{q}} \frac{B(\boldsymbol{q})}{\Omega} G_0(\boldsymbol{k}-\boldsymbol{q})G_0(\boldsymbol{k})\right]^2 . \tag{3.65}$$

The sum of the contributions of all the diagrams reduces to the calculation of the geometric series

$$\overline{G}(\boldsymbol{k},\epsilon) = G_0(\boldsymbol{k},\epsilon) + G_0(\boldsymbol{k},\epsilon)\sum_{n=1}^{\infty}\left[\Sigma(\boldsymbol{k},\epsilon)G_0(\boldsymbol{k},\epsilon)\right]^n \tag{3.66}$$

which is a solution of the so-called Dyson equation

$$\overline{G}(\boldsymbol{k},\epsilon) = G_0(\boldsymbol{k},\epsilon) + G_0(\boldsymbol{k},\epsilon)\Sigma(\boldsymbol{k},\epsilon)\overline{G}(\boldsymbol{k},\epsilon). \tag{3.67}$$

The function $\Sigma(\boldsymbol{k},\epsilon)$, called the *self-energy*, is by construction the sum over an infinity of irreducible diagrams (Figure 3.4).

For the Schrödinger equation, the free Green function $G_0(\boldsymbol{k},\epsilon)$ is given by (3.34). The corresponding Dyson equation (3.67) may be written in the form

$$\overline{G}^{R,A}(\boldsymbol{k},\epsilon) = \frac{1}{\epsilon - \epsilon(\boldsymbol{k}) - \Sigma^{R,A}(\boldsymbol{k},\epsilon)}. \tag{3.68}$$

For the Helmholtz equation, the free Green function $G_0(\boldsymbol{k},k_0)$ is given by (3.50) and the average Green function is thus

$$\overline{G}^{R,A}(\boldsymbol{k},k_0) = \frac{1}{k_0^2 - k^2 - \Sigma^{R,A}(\boldsymbol{k},k_0)} \tag{3.69}$$

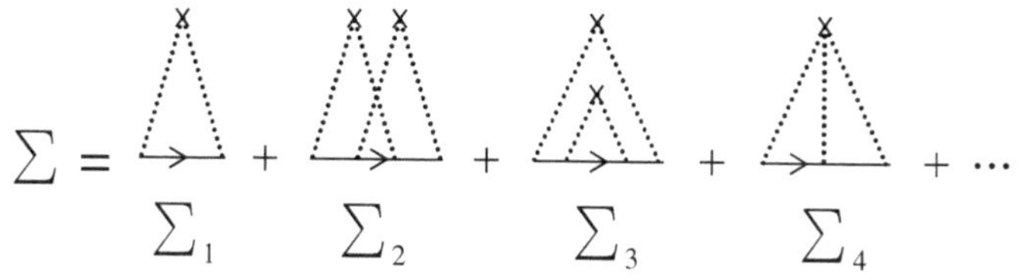

$\Sigma = \Sigma_1 + \Sigma_2 + \Sigma_3 + \Sigma_4 + \cdots$

Figure 3.4 Diagrammatic representation of the self-energy as a sum of irreducible diagrams.

where k_0 is related to the frequency $\omega = k_0 c$ of the wave. The self-energy $\Sigma(\boldsymbol{k}, k_0)$ has, in this case, the dimensions of inverse area.

3.2.2 Self-energy

The calculation of the self-energy is in principle a difficult problem since it contains an infinite number of terms. The first, $\Sigma_1^{R,A}(\boldsymbol{k}, \epsilon)$, may be written as (Figure 3.4)

$$\Sigma_1^{R,A}(\boldsymbol{k}, \epsilon) = \frac{1}{\Omega} \sum_{\boldsymbol{k}'} B(\boldsymbol{k} - \boldsymbol{k}') G_0^{R,A}(\boldsymbol{k}'). \tag{3.70}$$

The real part of the self-energy gives an unimportant shift of the zero of energies (or of frequencies for the Helmholtz equation) and will not be considered further. In the Schrödinger case, equation (3.33) which relates the Green function to the density of states leads to

$$\mathrm{Im}\Sigma_1^R(\boldsymbol{k}, \epsilon) = -\frac{\pi}{\Omega} \sum_{\boldsymbol{k}'} B(\boldsymbol{k} - \boldsymbol{k}') \delta(\epsilon - \epsilon(\boldsymbol{k}')). \tag{3.71}$$

Denoting $\boldsymbol{k} = k\hat{\boldsymbol{s}}$ where $\hat{\boldsymbol{s}}$ is a unit vector, and separating the radial and angular summations, we get

$$\mathrm{Im}\Sigma_1^R(\boldsymbol{k}, \epsilon) = -\pi \rho_0(\epsilon) \, \langle B(k, \hat{\boldsymbol{s}} - \hat{\boldsymbol{s}}') \rangle_{\hat{\boldsymbol{s}}'} \tag{3.72}$$

where we have used the angular average $\langle \cdots \rangle$ of B and the relation $k = \sqrt{2m\epsilon}$. The self-energy defines a characteristic time τ_e called the *elastic collision time* such that

$$\boxed{\frac{1}{2\tau_e} = -\mathrm{Im}\Sigma_1^R(\boldsymbol{k}, \epsilon) = \pi \rho_0(\epsilon) \gamma_e} \quad \textbf{Schrödinger} \tag{3.73}$$

where we denote:

$$\gamma_e = \langle B(\boldsymbol{k} - \boldsymbol{k}') \rangle. \tag{3.74}$$

This time, which has already been obtained using the Fermi golden rule (3.2), is the average lifetime of an eigenstate of wave vector $\boldsymbol{k}$ and energy ϵ. Within the Gaussian model it is independent of the wave vector. The existence of a finite imaginary part for the self-energy implies that the density of states, proportional to the imaginary part of the Green function (relation 3.30), appears as a series of Lorentzians each of which describes an energy eigenstate with a broadening $1/2\tau_e$.

Remark: other scattering mechanisms

We have thus far only considered the contribution of elastic scattering by a random potential due to static impurities. Including, for electronic systems, the spin-orbit coupling leads to a further broadening $1/2\tau_{so}$, and the coupling to magnetic impurities leads to a broadening $1/2\tau_m$. These times τ_{so} and τ_m are defined in section 6.5.2. Taking these scattering mechanisms into account, the self-energy may be written

$$-\mathrm{Im}\Sigma_1^R = \frac{1}{2\tau_e} + \frac{1}{2\tau_{so}} + \frac{1}{2\tau_m} = \frac{1}{2\tau_{tot}} \tag{3.75}$$

which defines the total collision time τ_{tot} in the presence of these new mechanisms (page 213). Any other scattering mechanism may thus be taken into account by an additional contribution to the self-energy. These contributions are additive provided the degrees of freedom are decoupled from each other (Matthiessen rule).

The coupling to dynamical degrees of freedom, which may be either phonons or other electrons, can also be incorporated in a phenomenological way in the single-particle Green function, leading to a further broadening. We will not discuss the electron–phonon interaction in this book (for a discussion, see [69]). The Coulomb interaction between electrons (Chapter 13) also gives rise to a new characteristic time τ_{ee}.

Similarly, in the case of the Helmholtz equation, relations (3.70) and (3.50) lead to

$$\mathrm{Im}\Sigma_1^R(\boldsymbol{k}, k_0) = -\frac{\gamma_e k_0}{4\pi}. \tag{3.76}$$

We define, in this case, the *elastic mean free path* l_e by

$$\boxed{\frac{1}{l_e} = -\frac{1}{k_0}\mathrm{Im}\Sigma_1^R(\boldsymbol{k}, k_0) = \frac{\gamma_e}{4\pi}} \quad \textbf{Helmholtz} \tag{3.77}$$

for $d = 3$. For any dimension, using equations (3.70) and (3.55), we obtain

$$\frac{1}{l_e} = \frac{\pi c \rho_0}{2k_0^2}\gamma_e, \tag{3.78}$$

where ρ_0 is the density of eigenmodes per unit volume.

Table 3.1 summarizes the expressions for the collision time and the mean free path for the Schrödinger and Helmholtz problems.

Table 3.1 Expressions of the collision time τ_e and the elastic mean free path l_e for both Schrödinger and Helmholtz equations.

Schrödinger	Helmholtz
$\frac{1}{2\tau_e} = -\text{Im}\Sigma^R$	$\frac{1}{l_e} = -\frac{1}{k_0}\text{Im}\Sigma^R$
$\text{Im}\Sigma_1^R = -\pi\gamma_e\rho_0$	$\text{Im}\Sigma_1^R = -\frac{\gamma_e k_0}{4\pi}$
$\gamma_e = \frac{1}{2\pi\rho_0\tau_e}$	$\gamma_e = \frac{4\pi}{l_e}$
$\text{Im}\Sigma^R = \gamma_e \text{Im}G^R$	

In the following exercises, we consider a white noise potential (unless mentioned otherwise).

Exercise 3.7 Show that in the spatial representation, the self-energy Σ_1^R may be written

$$\Sigma_1^R(\boldsymbol{r},\boldsymbol{r}',\epsilon) = \gamma_e G_0^R(\boldsymbol{r},\boldsymbol{r}',\epsilon)\delta(\boldsymbol{r}-\boldsymbol{r}') \tag{3.79}$$

so that, from (3.41), equation (3.73) follows immediately.

Exercise 3.8 For the Schrödinger equation, calculate the next correction Σ_2^R in the expansion of the self-energy for $d = 3$ (Figure 3.4).

This term is given by

$$\Sigma_2^R(\boldsymbol{r},\boldsymbol{r}',\epsilon) = \gamma_e^2 \int d\boldsymbol{r}_1\, d\boldsymbol{r}_2\, G_0^R(\boldsymbol{r},\boldsymbol{r}_1)\delta(\boldsymbol{r}-\boldsymbol{r}_2)G_0^R(\boldsymbol{r}_1,\boldsymbol{r}_2)\delta(\boldsymbol{r}_1-\boldsymbol{r}')G_0^R(\boldsymbol{r}_2,\boldsymbol{r}') \tag{3.80}$$

or

$$\Sigma_2^R(\boldsymbol{r},\boldsymbol{r}',\epsilon) = \gamma_e^2\left[G_0^R(\boldsymbol{r},\boldsymbol{r}',\epsilon)\right]^3.$$

Using expression (3.41) for the free Green function in $d = 3$, we have

$$\Sigma_2^R(\boldsymbol{r},\boldsymbol{r}',\epsilon) = -\frac{\pi}{4}\frac{\rho_0}{\tau_e^2}\frac{e^{3ik_0R}}{k_0^3R^3}, \tag{3.81}$$

where $R = |\boldsymbol{r}-\boldsymbol{r}'|$ and where k_0 is the wave vector associated with energy ϵ. The imaginary part $\text{Im}\Sigma_2^R(k,\epsilon)$ of the Fourier transform of $\Sigma_2^R(\boldsymbol{r},\boldsymbol{r}',\epsilon)$ is given by the integral

$$\text{Im}\Sigma_2^R(k,\epsilon) = -\pi^2\frac{\rho_0}{\tau_e^2}\frac{1}{kk_0^3}\int_0^\infty dR\frac{\sin kR \sin 3k_0R}{R^2} = -\frac{\pi^3}{2}\frac{\rho_0}{\tau_e^2k_0^3}\min\left(1,\frac{3k_0}{k}\right).$$

In $d = 3$, the density of states is given by $\rho_0 = \frac{mk_0}{2\pi^2}$ with $\epsilon = \frac{k_0^2}{2m}$. We may thus write

$$\mathrm{Im}\Sigma_2^R(k,\epsilon) = -\frac{1}{2\tau_e}\frac{\pi}{2k_0 l_e}\min\left(1, \frac{3k_0}{k}\right), \tag{3.82}$$

where l_e is the elastic mean free path defined by $l_e = v\tau_e$ and v is the group velocity $v = d\epsilon/dk$. This correction, of order $1/k_0 l_e$, is therefore negligible in $d = 3$.

Exercise 3.9 Generalize the above result to arbitrary dimension d and verify that the correction $\mathrm{Im}\Sigma_2^R(k,\epsilon)$ to the self-energy is proportional to an integral of the form

$$\mathrm{Im}\Sigma_2^R(k,\epsilon) \propto \int_0^\infty \frac{dR}{R^{d-1}}\,. \tag{3.83}$$

The correction is finite for $d = 3$, but diverges in $d = 1$ and is marginal (logarithmic divergence) in $d = 2$. We conclude that for $d \geq 2$, all the higher order terms in the self-energy are negligible compared to $\Sigma_1(k,\epsilon)$ in the limit $k_0 l_e \gg 1$. This is not true for $d = 1$. In this case, there is a systematic method for resumming the diagrams [70].

Exercise 3.10 Show that $\mathrm{Im}\Sigma_3^R$ shown in Figure 3.4 is a correction of order $1/k_0 l_e$ to the self-energy.

It is instructive to give a schematic representation in real space (Figure 3.5) of the sequences of collisions corresponding to Σ_1 and Σ_2 (Figure 3.4). We see that Σ_1 describes a succession of independent collision events while the following terms in the self-energy involve interference terms between successive collisions. The limit $k_0 l_e \gg 1$ corresponds to the regime in which we can neglect these interference effects (see Exercise 3.8). This approximation amounts to assuming that after each collision we may asymptotically reconstruct a free wave. This is the *weak disorder limit* which we treat in this book:

$$\boxed{\textit{weak disorder limit: } k_0 l_e \gg 1} \tag{3.84}$$

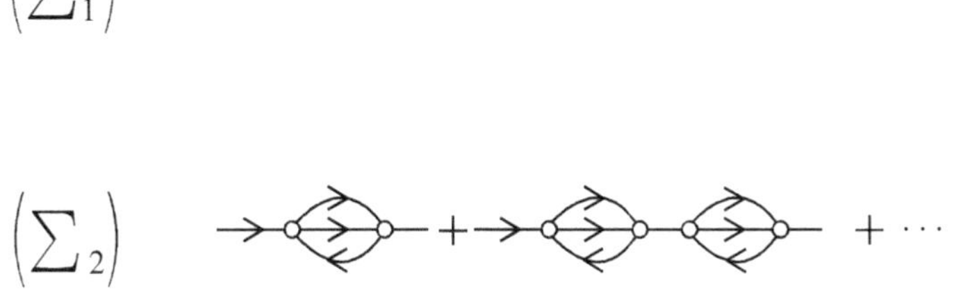

(Σ_1) + + + ···

(Σ_2) + + ···

Figure 3.5 Each term Σ_i which contributes to the self-energy (Figure 3.4) is the basic building block which generates, upon iteration, the sequences of multiple collisions up to a given approximation. The two contributions depicted correspond to iterations generated, respectively, by Σ_1 and Σ_2.

For the Edwards model described by means of equation (2.98) namely,

$$\langle B(\boldsymbol{q})\rangle = 4\pi n_i \sigma, \tag{3.85}$$

where σ is the scattering cross section of an impurity in the Born approximation, we see that the contribution Σ_1, corresponding to independent collisions, is proportional to the density n_i of impurities, and using (3.77), we obtain

$$\boxed{\frac{1}{l_e} = n_i \sigma} \tag{3.86}$$

It is thus possible to relate the mean free path l_e to physical quantities by using, in equation (3.86), the expression for the cross section obtained from the scattering theory. For example, in the case of light scattering by dense suspensions, we use equation (3.86) taking for σ the scattering cross section calculated for a single scatterer (Appendices A2.1 and A2.3). From the dependence of σ as a function of wavelength λ, we infer the behavior of $l_e(\lambda)$.

Remark

We shall regularly employ the terms "impurity" or "scattering center" to describe collision events. The average distance between two successive collisions that separate these "impurities" is the mean free path l_e rather than $n_i^{-1/d}$, which in any case vanishes for the Gaussian model.

Exercise 3.11 Show that for the Edwards model with δ impurities, the diagram corresponding to Σ_4^R (Figure 3.4) is equal to

$$n_i v_0^3 [G_0^R(\boldsymbol{r},\boldsymbol{r}')]^2 \delta(\boldsymbol{r}-\boldsymbol{r}') \tag{3.87}$$

and that it vanishes in the limit of a Gaussian potential for which $v_0 \to 0$ with $n_i v_0^2$ constant.

3.3 Average Green's function and average density of states

By inserting the expression of the self-energy calculated to first order, the average electronic Green function, using (3.68) and (3.73), is given by

$$\overline{G}^{R,A}(\boldsymbol{k},\epsilon) = \frac{1}{\epsilon - \epsilon(\boldsymbol{k}) \pm \frac{i}{2\tau_e}}. \tag{3.88}$$

$\overline{G}^{R,A}(\boldsymbol{r}_i,\boldsymbol{r},\epsilon)$ is the Fourier transform of (3.88). For a quadratic dispersion relation $\epsilon(\boldsymbol{k}) = k^2/2m$, it is

$$\overline{G}^{R,A}(\boldsymbol{r}_i,\boldsymbol{r},\epsilon) = \frac{2m}{(2\pi)^3}\int d\boldsymbol{k}'\, e^{i\boldsymbol{k}'\cdot(\boldsymbol{r}-\boldsymbol{r}_i)} \frac{1}{k_e^2 - k'^2},$$

where we have defined $k_e^2 = k^2 \pm im/\tau_e$. Proceeding in the same manner as for the calculation of the free Green function (3.38), and in the limit $kl_e \gg 1$ such that $k_e = k(1 \pm i/kl_e)^{1/2} \simeq k \pm i/2l_e$, we have

$$\overline{G}^{R,A}(\boldsymbol{r}_i, \boldsymbol{r}, \epsilon) = -\frac{m}{2\pi R} e^{\pm ikR} e^{-R/2l_e}$$
$$= -\pi \rho_0 \frac{e^{\pm ikR}}{kR} e^{-R/2l_e}, \tag{3.89}$$

or equivalently,

$$\boxed{\overline{G}^{R,A}(\boldsymbol{r}_i, \boldsymbol{r}, \epsilon) = G_0^{R,A}(\boldsymbol{r}_i, \boldsymbol{r}, \epsilon)\, e^{-|\boldsymbol{r}-\boldsymbol{r}_i|/2l_e}} \tag{3.90}$$

Similarly, from the expression of the average Green function for waves

$$\overline{G}^{R,A}(\boldsymbol{k}, k_0) = \frac{1}{k_0^2 - k^2 \pm i\frac{k_0}{l_e}}, \tag{3.91}$$

we obtain

$$\overline{G}^{R,A}(\boldsymbol{r}_i, \boldsymbol{r}, k_0) = -\frac{1}{4\pi} \frac{e^{\pm ik_0 R}}{R} e^{-R/2l_e}, \tag{3.92}$$

where $R = |\boldsymbol{r} - \boldsymbol{r}_i|$. Strictly speaking, these last two expressions for the average Green function are correct for an infinite, translation invariant medium. In fact, it is enough that the typical size L of the system is much larger than l_e for these expressions to be valid.

From (3.88) for the average Green function, we deduce that the average density of states is given by

$$\overline{\rho}(\epsilon) = \frac{1}{\Omega} \sum_{\boldsymbol{k}} \frac{1/2\pi\tau_e}{(\epsilon - \epsilon_{\boldsymbol{k}})^2 + (1/2\tau_e)^2} = \frac{1}{2\pi\tau_e} \int_0^\infty \frac{\rho_0(\eta)}{(\eta - \epsilon)^2 + (1/2\tau_e)^2} d\eta. \tag{3.93}$$

For example, in $d = 2$, for which the density of states is constant, $\rho_0(\epsilon) = \rho_0$, we obtain

$$\overline{\rho}(\epsilon) = \rho_0 \left(\frac{1}{2} + \frac{1}{\pi} \arctan 2\epsilon\tau_e \right) \underset{\epsilon\tau_e \to \infty}{\longrightarrow} \rho_0 \left(1 - \frac{1}{2\pi\epsilon\tau_e} \right). \tag{3.94}$$

We thus notice that the average density of states is only slightly affected by disorder, and that the correction is of the order $1/kl_e$ in two dimensions, and $1/(kl_e)^2$ in three dimensions (see Exercise 3.12). Therefore, in the following, we will take $\overline{\rho}(\epsilon) = \rho_0(\epsilon)$.

Exercise 3.12: average electronic density of states in three dimensions
In three dimensions the density of states ρ_0 varies as $\epsilon^{1/2}$. Using (15.66), show that the average density of states may be written:

$$\overline{\rho}(\epsilon) = \frac{m^{3/2}}{\sqrt{2}\pi^2}\left(\epsilon^2 + \frac{1}{4\tau_e^2}\right)^{1/4} \cos\left(\frac{1}{2}\arctan\frac{1}{2\epsilon\tau_e}\right) \underset{\epsilon\tau_e\to\infty}{\longrightarrow} \rho_0(\epsilon)\left(1+\frac{1}{32\epsilon^2\tau_e^2}\right) \tag{3.95}$$

where $\rho_0(\epsilon)$ is given by (3.44).

Exercise 3.13 Show that the time dependence of the Green function $\overline{G}^R(\boldsymbol{r},\boldsymbol{r}',t)$ defined by the Fourier transform of (3.92) may be expressed as

$$\overline{G}^R(\boldsymbol{r},\boldsymbol{r}',t) = -\theta(t)\,\frac{c}{4\pi R}\,\delta(R-ct)\,e^{-t/2\tau_e} \tag{3.96}$$

with $R = |\boldsymbol{r}-\boldsymbol{r}'|$.

Appendix A3.1: Short range correlations

As we shall see in the subsequent chapters, comprehension of the mechanisms underlying the physics of multiple scattering requires a proper description of the propagation from a given source point to the location of the first impurity and then of the sequence of collisions between impurities. It is therefore necessary to have a precise description of the "terminations" of these collision sequences. These are spatially short range functions which involve combinations of average Green's functions. The purpose of this appendix is to review and calculate these different combinations of Green's functions. We first define the short range function

$$g(\boldsymbol{R}) = \begin{cases} -\dfrac{1}{\pi\rho_0}\mathrm{Im}\overline{G}_\epsilon^R(\boldsymbol{r},\boldsymbol{r}') & \textit{Schrödinger} \\[2ex] -\dfrac{4\pi}{k_0}\mathrm{Im}\overline{G}_\epsilon^R(\boldsymbol{r},\boldsymbol{r}') & \textit{Helmholtz} \end{cases} \tag{3.97}$$

with $\boldsymbol{R} = \boldsymbol{r}-\boldsymbol{r}'$. Using equations (3.90, 3.41–3.43) for the average Green functions $\overline{G}_\epsilon^R(\boldsymbol{r},\boldsymbol{r}')$ leads to the expressions

$$\boxed{\begin{array}{ll} g(\boldsymbol{R}) = \dfrac{\sin k_0 R}{k_0 R}\,e^{-R/2l_e} & d=3 \\ g(\boldsymbol{R}) = J_0(k_0 R)\,e^{-R/2l_e} & d=2 \\ g(\boldsymbol{R}) = \cos k_0 R\,e^{-R/2l_e} & d=1 \end{array}} \tag{3.98}$$

with $k_0 = \sqrt{2m\epsilon}$ and $R = |\boldsymbol{R}|$. Moreover, using (3.28), we note that for the Schrödinger equation, the function $g(\boldsymbol{R})$ is related to the average non-local density:

$$\overline{\rho}_\epsilon(\boldsymbol{r},\boldsymbol{r}') = \rho_0(\epsilon)\, g(\boldsymbol{R}). \tag{3.99}$$

Hence, we can show that in the limit $k_0 l_e \gg 1$ we have, for electrons in any dimension,

$$\boxed{\int g^2(\boldsymbol{R})\, d\boldsymbol{R} = \frac{\tau_e}{\pi \rho_0}} \tag{3.100}$$

Exercise 3.14 Calculate the integral $I = \int g^2(\boldsymbol{R}) d\boldsymbol{R}$ and show that, in the limit $k_0 l_e \gg 1$,

$$I = \frac{2\pi l_e}{k_0^2} \ \text{ at } d = 3, \quad I = \frac{2 l_e}{k_0} \ \text{ at } d = 2, \quad I = l_e \ \text{ at } d = 1. \tag{3.101}$$

These expressions are valid for both the Schrödinger and Helmholtz equations.

Show that in the limit $q l_e \to \infty$, the function $a(\boldsymbol{q})$, Fourier transform of $g^2(\boldsymbol{R})$, may be written, for $q < 2k_0$:

$$a(\boldsymbol{q}) = \frac{\pi^2}{k_0^2 q} \ \text{ at } d = 3, \qquad a(\boldsymbol{q}) = \frac{4}{q} \frac{1}{\sqrt{4k_0^2 - q^2}} \ \text{ at } d = 2, \tag{3.102}$$

and that it vanishes for $q > 2k_0$.

The function $g(\boldsymbol{R})$ arises in various combinations of average Green's functions. For example, the integral $f^{1,1}(\boldsymbol{R})$ defined by

$$f^{1,1}(\boldsymbol{R}) = \gamma_e \int \overline{G}^R_\epsilon(\boldsymbol{r},\boldsymbol{r}_1) \overline{G}^A_\epsilon(\boldsymbol{r}_1,\boldsymbol{r}')\, d\boldsymbol{r}_1 \tag{3.103}$$

with $\boldsymbol{R} = \boldsymbol{r} - \boldsymbol{r}'$, can be written

$$\begin{aligned} f^{1,1}(\boldsymbol{R}) &= \frac{\gamma_e}{\Omega} \sum_{\boldsymbol{k}} \overline{G}^R_\epsilon(\boldsymbol{k}) \overline{G}^A_\epsilon(\boldsymbol{k})\, e^{i\boldsymbol{k}\cdot\boldsymbol{R}} \\ &= \frac{\gamma_e}{\Omega} \sum_{\boldsymbol{k}} \frac{e^{i\boldsymbol{k}\cdot\boldsymbol{R}}}{(\epsilon - \epsilon_k)^2 + 1/4\tau_e^2} \end{aligned} \tag{3.104}$$

and thus involves $\text{Im}\overline{G}^R_\epsilon(\boldsymbol{r},\boldsymbol{r}')$. Finally, we obtain[7]

$$f^{1,1}(\boldsymbol{R}) = g(\boldsymbol{R}). \tag{3.105}$$

[7] The result (3.104) is a consequence of the relation

$$\overline{G}^R_\epsilon(\boldsymbol{k}) \overline{G}^A_\epsilon(\boldsymbol{k}) = -2\tau_e \text{Im}\overline{G}^R_\epsilon(\boldsymbol{k}).$$

Table 3.2 Some values of $f^{m,n}$. For waves in $d=3$, replace τ_e by $l_e/(2k_0)$.

$m\backslash n$	1	2	3	4
1	1	$i\tau_e$	$-\tau_e^2$	$-i\tau_e^3$
2	$-i\tau_e$	$2\tau_e^2$	$3i\tau_e^3$	$-4\tau_e^4$
3	$-\tau_e^2$	$-3i\tau_e^3$	$6\tau_e^4$	$10i\tau_e^5$
4	$i\tau_e^3$	$-4\tau_e^4$	$-10i\tau_e^5$	$20\tau_e^6$

More generally, we may introduce the function $f^{m,n}(\boldsymbol{R})$ defined by

$$f^{m,n}(\boldsymbol{R})=\gamma_e\int\prod_{i=1}^{m}d\boldsymbol{r}_i\prod_{j=1}^{n}d\boldsymbol{r}'_j\,\overline{G}_\epsilon^R(\boldsymbol{r},\boldsymbol{r}_1)\cdots\overline{G}_\epsilon^R(\boldsymbol{r}_{m-1},\boldsymbol{r}_m)\overline{G}_\epsilon^A(\boldsymbol{r}_m,\boldsymbol{r}'_1)\cdots\overline{G}_\epsilon^A(\boldsymbol{r}'_n,\boldsymbol{r}')\tag{3.106}$$

i.e., as a product of m retarded Green's functions and n advanced Green's functions. Its value at $\boldsymbol{R}=0$, denoted $f^{m,n}=f^{m,n}(0)$, is given by

$$f^{m,n}=\frac{\gamma_e}{\Omega}\sum_{\boldsymbol{k}}\left[\overline{G}_\epsilon^R(\boldsymbol{k})\right]^m\left[\overline{G}_\epsilon^A(\boldsymbol{k})\right]^n.\tag{3.107}$$

An integration using the residue calculus leads to [71, 72]

$$\boxed{f^{m,n}=i^{n-m}\frac{(n+m-2)!}{(n-1)!(m-1)!}\tau_e^{n+m-2}}\tag{3.108}$$

A similar calculation of $f^{m,n}$ for waves in three dimensions yields

$$f^{m,n}=i^{n-m}\frac{(n+m-2)!}{(n-1)!(m-1)!}\left(\frac{l_e}{2k_0}\right)^{n+m-2}.\tag{3.109}$$

Table 3.2 and Figure 3.6 give expressions for several diagrams as well as values for some $f^{m,n}$.

Exercise 3.15 Calculate the last diagram of Figure 3.6.
Hint: recall that the impurity line corresponds to the quantity $\gamma_e\delta(\boldsymbol{r}-\boldsymbol{r}')$. Then show that this diagram is the product of γ_e and two independent and identical diagrams, each equal to $f^{1,2}(0)/\gamma_e$.

Exercise 3.16 Show that in the limit $k_0 l_e \gg 1$, the sum of a product of two retarded Green's functions is

$$\frac{1}{\Omega}\sum_{\boldsymbol{k}}\overline{G}^R(\boldsymbol{k})\overline{G}^R(\boldsymbol{k})\simeq\frac{\rho_0}{\epsilon},\tag{3.110}$$

to be compared with

$$\frac{1}{\Omega}\sum_{\boldsymbol{k}}\overline{G}^R(\boldsymbol{k})\overline{G}^A(\boldsymbol{k}) = 2\pi\rho_0\tau_e. \tag{3.111}$$

It is therefore negligible, being of order $1/k_0 l_e$. We obtain the same result for the sum of a product of two advanced Green's functions.

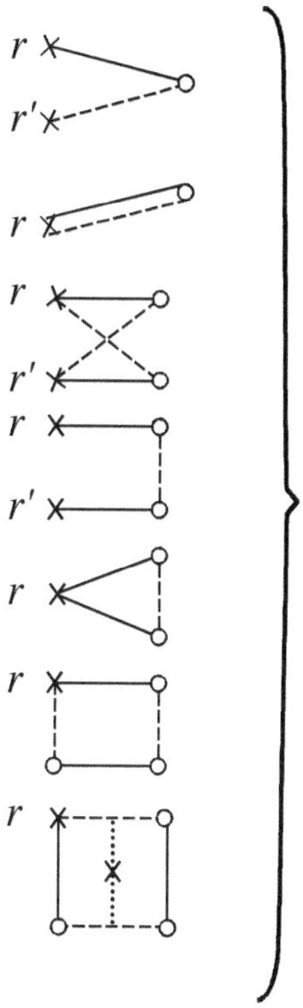

$$= \frac{1}{\gamma_e} \times \begin{cases} f^{1,1}(R) = g(R) \\ f^{1,1}(0) = 1 \\ \dfrac{[f^{1,1}(R)]^2}{\gamma_e} = \dfrac{g^2(R)}{\gamma_e} \\ f^{2,1}(R) = -i\tau_e\, g(R) \\ f^{2,1}(0) = -i\tau_e \\ f^{2,2}(0) = 2\tau_e^2 \\ [f^{2,1}(0)]^2 = -\tau_e^2 \end{cases}$$

Figure 3.6 Different products of Green's functions. The function $f^{m,n}(\boldsymbol{r}-\boldsymbol{r}')$ is independent of the order of the sequence of retarded (solid lines) and advanced (dashed lines) Green's functions. Open circles indicate the points whose positions are integrated over. For waves in $d = 3$, replace τ_e by $l_e/(2k_0)$ and γ_e by $4\pi/l_e$.

4
Probability of quantum diffusion

This chapter contains a description of essential concepts and tools which will be used throughout the book. We take $\hbar = 1$.

The average Green function describes the evolution of a plane wave in a disordered medium, but it does not contain information about the evolution of a wave packet. For optically thick media, or for metals, most physical properties are determined not by the average Green function, but rather by the probability $P(\boldsymbol{r}, \boldsymbol{r}', t)$ that a particle will move from some initial point $\boldsymbol{r}$ to a point $\boldsymbol{r}'$, or possibly return to its initial point.

In this chapter, starting from the Schrödinger equation (or the Helmholtz equation), we establish a general expression describing the quantum probability for the propagation of a particle, i.e., a wave packet, from one point to another. When this probability is averaged over a random potential we can identify, in the weak disorder limit $kl_e \gg 1$, three principal contributions:

- the probability of going from one point to another without any collision;
- the probability of going from one point to another by a *classical* process of multiple scattering;
- the probability of going from one point to another by a *coherent* process of multiple scattering.

We will show that the last two processes, called respectively *Diffuson* and *Cooperon*, satisfy a diffusion equation in certain limits.

4.1 Definition

Starting from the Schrödinger equation, we want to determine the probability of finding a particle of energy ϵ_0 at point $\boldsymbol{r}_2$ at time t, if it was initially at $\boldsymbol{r}_1$ at $t = 0$. To that purpose, we consider the temporal evolution of an initial state denoted by $|\psi_{\boldsymbol{r}_1}\rangle$ and represented by a wave packet of average energy ϵ_0 centered at $\boldsymbol{r}_1$. This wave packet may be decomposed into the orthonormal eigenstates $|\phi_n\rangle$ of the Hamiltonian (2.1). Let us consider a Gaussian wave packet whose width in energy is denoted by σ_ϵ:

$$|\psi_{\boldsymbol{r}_1}\rangle = A \sum_n \langle \phi_n | \psi_{\boldsymbol{r}_1} \rangle \, e^{-(\epsilon_n - \epsilon_0)^2 / 4\sigma_\epsilon^2} |\phi_n\rangle \tag{4.1}$$

such that

$$\psi_{\boldsymbol{r}_1}(\boldsymbol{r}) = \langle \boldsymbol{r}|\psi_{\boldsymbol{r}_1}\rangle = A \sum_n \langle \boldsymbol{r}|\phi_n\rangle\langle\phi_n|\boldsymbol{r}_1\rangle\, e^{-(\epsilon_n-\epsilon_0)^2/4\sigma_\epsilon^2}. \tag{4.2}$$

The normalization coefficient A is chosen so that $A^2 = 1/\sqrt{2\pi}\,\rho_0\sigma_\epsilon$ where ρ_0 is the average density of states.

Remark: normalization

The normalization condition $\int |\psi_{\boldsymbol{r}_1}(\boldsymbol{r})|^2 d\boldsymbol{r} = 1$ of the wave function is written:

$$A^2 \sum_{n,n'} \int \phi_n(\boldsymbol{r})\phi_n^*(\boldsymbol{r}_1)\phi_{n'}^*(\boldsymbol{r})\phi_{n'}(\boldsymbol{r}_1)\, e^{-(\epsilon_n-\epsilon_0)^2/4\sigma_\epsilon^2}\, e^{-(\epsilon_{n'}-\epsilon_0)^2/4\sigma_\epsilon^2}\, d\boldsymbol{r} = 1.$$

Integrating over $\boldsymbol{r}$ and using relation (3.25) leads to

$$A^2 \sum_n |\phi_n(\boldsymbol{r}_1)|^2\, e^{-(\epsilon_n-\epsilon_0)^2/2\sigma_\epsilon^2} = A^2 \int \rho(\boldsymbol{r}_1,\epsilon) e^{-(\epsilon-\epsilon_0)^2/2\sigma_\epsilon^2}\, d\epsilon = 1.$$

We choose to normalize the wave function averaged over disorder. We therefore replace the local density of states $\rho(\boldsymbol{r}_1,\epsilon)$ by its value averaged over disorder $\overline{\rho}(\boldsymbol{r}_1,\epsilon) = \rho_0(\epsilon)$, and for a constant density of states ρ_0 we obtain A.

The evolution of this wave packet from $\boldsymbol{r}_1$ to $\boldsymbol{r}_2$ is described by the matrix element of the evolution operator $e^{-i\mathcal{H}t}$:

$$\begin{aligned}
\langle \boldsymbol{r}_2|^{-i\mathcal{H}t}|\psi_{\boldsymbol{r}_1}\rangle\theta(t) &= \theta(t)\sum_n \langle \boldsymbol{r}_2|\phi_n\rangle\langle\phi_n|\psi_{\boldsymbol{r}_1}\rangle\, e^{-i\epsilon_n t} \\
&= A\theta(t) \sum_n \langle \boldsymbol{r}_2|\phi_n\rangle\langle\phi_n|\boldsymbol{r}_1\rangle\, e^{-(\epsilon_n-\epsilon_0)^2/4\sigma_\epsilon^2} e^{-i\epsilon_n t} \\
&= iA \int \frac{d\epsilon}{2\pi} \sum_n \frac{\langle \boldsymbol{r}_2|\phi_n\rangle\langle\phi_n|\boldsymbol{r}_1\rangle}{\epsilon - \epsilon_n + i0}\, e^{-(\epsilon-\epsilon_0)^2/4\sigma_\epsilon^2} e^{-i\epsilon t} \\
&= iA \int \frac{d\epsilon}{2\pi} G^R(\boldsymbol{r}_1,\boldsymbol{r}_2,\epsilon)\, e^{-(\epsilon-\epsilon_0)^2/4\sigma_\epsilon^2} e^{-i\epsilon t},
\end{aligned} \tag{4.3}$$

where we have used (4.2) and (3.12). At time t, we define the conditional probability as being in the state $|\boldsymbol{r}_2\rangle$ by

$$\begin{aligned}
P(\boldsymbol{r}_1,\boldsymbol{r}_2,t) &= \overline{|\langle \boldsymbol{r}_2|^{-i\mathcal{H}t}|\psi_{\boldsymbol{r}_1}\rangle\theta(t)|^2} \\
&= A^2 \int \frac{d\epsilon}{2\pi}\frac{d\omega}{2\pi} \overline{G^R(\boldsymbol{r}_1,\boldsymbol{r}_2,\epsilon+\omega/2)G^A(\boldsymbol{r}_2,\boldsymbol{r}_1,\epsilon-\omega/2)} \\
&\quad \times e^{-[(\epsilon+\omega/2-\epsilon_0)^2/4\sigma_\epsilon^2 + (\epsilon-\omega/2-\epsilon_0)^2/4\sigma_\epsilon^2]} e^{-i\omega t}.
\end{aligned} \tag{4.4}$$

We suppose now that the product $\overline{G^R(\boldsymbol{r}_1,\boldsymbol{r}_2,\epsilon+\omega/2)G^A(\boldsymbol{r}_2,\boldsymbol{r}_1,\epsilon-\omega/2)}$ depends only slightly on the energy ϵ, which is the case provided the density of states varies slowly around the energy ϵ_0. This product may thus be taken out of the integral. Finally, suppose that the frequency ω is small compared to the width σ_ϵ of the wave packet. In this case,

$$A^2\int d\epsilon\; e^{-[(\epsilon+\omega/2-\epsilon_0)^2/4\sigma_\epsilon^2+(\epsilon-\omega/2-\epsilon_0)^2/4\sigma_\epsilon^2]} \simeq A^2\int d\epsilon\; e^{-(\epsilon-\epsilon_0)^2/2\sigma_\epsilon^2}$$

$$= A^2\sqrt{2\pi}\,\sigma_\epsilon = \frac{1}{\rho_0}. \tag{4.5}$$

We thus obtain

$$P(\boldsymbol{r}_1,\boldsymbol{r}_2,t) = \frac{1}{2\pi}\int_{-\infty}^{\infty} d\omega\, P(\boldsymbol{r}_1,\boldsymbol{r}_2,\omega)\, e^{-i\omega t} \tag{4.6}$$

with

$$P(\boldsymbol{r}_1,\boldsymbol{r}_2,\omega) = \frac{1}{2\pi\rho_0}\overline{G^R(\boldsymbol{r}_1,\boldsymbol{r}_2,\epsilon_0+\omega/2)G^A(\boldsymbol{r}_2,\boldsymbol{r}_1,\epsilon_0-\omega/2)}. \tag{4.7}$$

The probability thus defined[1] describes the evolution of a particle *of given energy* ϵ_0 taken to be a wave packet of width σ_ϵ. This description only makes sense for time scales greater than $1/\sigma_\epsilon$, and thus for frequencies $\omega \ll \sigma_\epsilon$.

It is important to note that the expression (4.7) for the probability is only valid for those systems for which the density of states does not vary too greatly. While (4.4) is general, the considerations which follow it are applicable only if the product $\overline{G^R G^A}$ varies slowly with energy. This is not the case, for example, for a quasicrystal, where the density of states varies strongly over all energy scales. For a disordered system, the density of states may also vary, but the average over disorder smooths it out. We will consider those situations where the density of states $\rho_0(\epsilon)$ may be taken to be constant over the frequency (or energy) interval considered. The average value of the product of two Green's functions is thus independent of ϵ_0 and the *probability of quantum diffusion* may be written in the form[2]

$$\boxed{P(\boldsymbol{r},\boldsymbol{r}',\omega) = \frac{1}{2\pi\rho_0}\overline{G^R_{\epsilon_0}(\boldsymbol{r},\boldsymbol{r}')\,G^A_{\epsilon_0-\omega}(\boldsymbol{r}',\boldsymbol{r})}} \tag{4.9}$$

The Fourier transform $P(\boldsymbol{r},\boldsymbol{r}',t)$ describes the average probability that a particle of energy ϵ_0 evolves from a point $\boldsymbol{r}$ to a point $\boldsymbol{r}'$ in time t.[3] The rest of this chapter will be devoted to studying $P(\boldsymbol{r},\boldsymbol{r}',t)$, that is, to developing a systematic method for the calculation of the average value of the product of two Green's functions $\overline{G^R G^A}$.

[1] Equivalent expressions for the probability have been proposed. For example, see reference [73].

[2] In what follows, we will use the notation

$$G_\epsilon(\boldsymbol{r},\boldsymbol{r}') = G(\boldsymbol{r},\boldsymbol{r}',\epsilon). \tag{4.8}$$

[3] This convention for the arguments of $P(\boldsymbol{r},\boldsymbol{r}')$ corresponds to the choice we made in Chapter 3 (Figure 3.1 and footnote 5, page 72).

The average probability is normalized to unity (see Exercise 4.1). Thus, for all times t we have

$$\int P(\boldsymbol{r},\boldsymbol{r}',t)\,d\boldsymbol{r}' = 1 \tag{4.10}$$

or equivalently

$$\int P(\boldsymbol{r},\boldsymbol{r}',\omega)\,d\boldsymbol{r}' = \frac{i}{\omega}. \tag{4.11}$$

For a system which is translation invariant, $P(\boldsymbol{r},\boldsymbol{r}',\omega)$ depends only on $\boldsymbol{r}-\boldsymbol{r}'$ and its Fourier transform $P(\boldsymbol{q},\omega)$ may be written:

$$P(\boldsymbol{q},\omega) = \frac{1}{2\pi\rho_0\Omega}\sum_{\boldsymbol{k},\boldsymbol{k}'}\overline{G^R_{\epsilon_0}(\boldsymbol{k}_+,\boldsymbol{k}'_+)G^A_{\epsilon_0-\omega}(\boldsymbol{k}'_-,\boldsymbol{k}_-)} \tag{4.12}$$

with $\boldsymbol{k}_\pm = \boldsymbol{k}\pm\boldsymbol{q}/2$.

Exercise 4.1: check the normalization of P

Show that the probability $P(\boldsymbol{r},\boldsymbol{r}',\omega)$ defined by (4.9) is normalized, that is, it satisfies (4.11). To do this, replace the Green functions in (4.9) by their definition to obtain

$$\int P(\boldsymbol{r},\boldsymbol{r}',\omega)\,d\boldsymbol{r}' = \frac{1}{2\pi\rho_0}\int d\boldsymbol{r}'\overline{\sum_{n,m}\frac{\phi_n^*(\boldsymbol{r}')\phi_n(\boldsymbol{r})\phi_m(\boldsymbol{r}')\phi_m^*(\boldsymbol{r})}{(\epsilon_0-\epsilon_n+i0)(\epsilon_0-\epsilon_m-\omega-i0)}}. \tag{4.13}$$

The wave functions are orthonormal, $\int d\boldsymbol{r}'\phi_n^*(\boldsymbol{r}')\phi_m(\boldsymbol{r}') = \delta_{nm}$, so that we get

$$\int P(\boldsymbol{r},\boldsymbol{r}',\omega)\,d\boldsymbol{r}' = \frac{1}{2\pi\rho_0}\int\frac{\overline{\rho}(\boldsymbol{r},\epsilon)}{(\epsilon_0-\epsilon+i0)(\epsilon_0-\epsilon-\omega-i0)}d\epsilon,$$

where $\overline{\rho}(\boldsymbol{r},\epsilon) = \rho_0$. The remaining integral leads to the normalization (4.11).

4.2 Free propagation

It is instructive to begin by studying the probability for a system *in the absence of disorder*. Using the expression (3.41) for the Green function and the definition (4.9), this probability may be written, for $d=3$:

$$P(\boldsymbol{r},\boldsymbol{r}',\omega) = \frac{\pi}{2}\rho_0\frac{e^{i\omega R/v}}{k^2R^2} = \frac{e^{i\omega R/v}}{4\pi R^2 v}, \tag{4.14}$$

where $R=|\boldsymbol{r}'-\boldsymbol{r}|$. We have used the general expression (3.40) for the density of states in $d=3$ and have expanded $k(\epsilon_0)-k(\epsilon_0-\omega)\simeq\omega\partial k/\partial\epsilon = \omega/v$. The time dependence of the probability is obtained by Fourier transforming equation (4.14):

$$P(\boldsymbol{r},\boldsymbol{r}',t) = \frac{\delta(R-vt)}{4\pi R^2}. \tag{4.15}$$

This probability describes the *ballistic* motion of the particle at group velocity v. For a fixed energy, since the modulus v is fixed, the particle moves a distance $R=vt$ in the absence of any collision. This distance being travelled in an arbitrary direction, the probability decreases as $1/R^2$. In general, in d dimensions, it falls off as $1/R^{d-1}$.

4.3 Drude–Boltzmann approximation

In the presence of disorder, in order to calculate the probability (4.9), we must evaluate the average of the product of two Green's functions. The Drude–Boltzmann approximation[4] consists of replacing the average by the product of two average values, that is to say, replacing $P(\boldsymbol{r},\boldsymbol{r}',\omega)$ by

$$\boxed{P_0(\boldsymbol{r},\boldsymbol{r}',\omega) = \frac{1}{2\pi\rho_0}\overline{G}^R_\epsilon(\boldsymbol{r},\boldsymbol{r}')\,\overline{G}^A_{\epsilon-\omega}(\boldsymbol{r}',\boldsymbol{r})} \tag{4.16}$$

In this expression, the disorder appears in the average Green functions which describe the lifetime of an eigenstate of given momentum. Thus $P_0(\boldsymbol{r},\boldsymbol{r}',t)$ is the probability that a particle at $\boldsymbol{r}$ arrives at point $\boldsymbol{r}'$ *without any collision*. Using the expression (3.89) for the average Green function, this probability, in $d=3$, is given by

$$P_0(\boldsymbol{r},\boldsymbol{r}',\omega) = \frac{e^{i\omega R/v - R/l_e}}{4\pi R^2 v} \tag{4.17}$$

or, taking the time Fourier transform

$$\boxed{P_0(\boldsymbol{r},\boldsymbol{r}',t) = \frac{\delta(R-vt)e^{-t/\tau_e}}{4\pi R^2}} \tag{4.18}$$

with $R=|\boldsymbol{r}-\boldsymbol{r}'|$. The probability for a particle not to undergo a collision decreases exponentially with distance. Thus, the total probability that a particle has no collision before time t decreases exponentially with time:

$$\int P_0(\boldsymbol{r},\boldsymbol{r}',t)\,d\boldsymbol{r}' = \theta(t)\,e^{-t/\tau_e} \tag{4.19}$$

and its Fourier transform is

$$\int P_0(\boldsymbol{r},\boldsymbol{r}',\omega)\,d\boldsymbol{r}' = \frac{\tau_e}{1-i\omega\tau_e} \tag{4.20}$$

We show in section 7.2.1 that this integrated probability is proportional to the electrical conductivity in the Drude approximation.

[4] We have chosen this denomination since it is commonly used to name this approximation in the calculation of the conductivity (see also Chapter 7).

4.4 Diffuson or ladder approximation

Equation (4.19) shows that the probability P_0 is not normalized (compare with 4.10). In replacing the average of $G^R G^A$ in the quantum probability (4.9) by the product of the averages, we have omitted certain processes. Indeed, there is another contribution to the probability which describes the *multiple scattering* on the disorder potential. In what follows, we show that, in the weak disorder limit discussed in Chapter 3, this new contribution allows us to normalize the probability anew and under certain conditions it satisfies a classical diffusion equation.

In order to evaluate the probability $P(\boldsymbol{r}, \boldsymbol{r}', \omega)$ given by (4.9), we must consider the disorder average of the product $G^R G^A$ of Green's functions corresponding to all possible multiple scattering sequences such as those represented schematically in Figure 4.1. The Green function $G^R(\boldsymbol{r}, \boldsymbol{r}', \epsilon)$ describes the complex amplitude of a wave packet propagating from $\boldsymbol{r}$ to $\boldsymbol{r}'$ with energy $\epsilon(\boldsymbol{k})$. It has the following structure:[5]

$$G^R(\boldsymbol{r}, \boldsymbol{r}', \epsilon) = \sum_{N=1}^{\infty} \sum_{\boldsymbol{r}_1, \ldots, \boldsymbol{r}_N} |A(\boldsymbol{r}, \boldsymbol{r}', \mathcal{C}_N)| \exp(ik\mathcal{L}_N), \tag{4.21}$$

where $A(\boldsymbol{r}, \boldsymbol{r}', \mathcal{C}_N)$ is the complex amplitude associated with a given sequence of N collisions $\mathcal{C}_N = (\boldsymbol{r}_1, \boldsymbol{r}_2, \ldots, \boldsymbol{r}_N)$ and where the accumulated phase $k\mathcal{L}_N$ measures the length $\mathcal{L}_N$ of the trajectory in units of wavelength λ. The corresponding product $G^R G^A$ is schematically represented in Figure 4.1. The dephasing associated with a product of two trajectories $\mathcal{C}_N$ and $\mathcal{C}_{N'}$ is proportional to the difference in path lengths $\mathcal{L}_N - \mathcal{L}_{N'}$.

The first consequence of averaging over the random potential, within the framework of the Gaussian model (2.29), consists of keeping in $\overline{G^R G^A}$ only those trajectories with an identical ensemble of scattering centers. This results from the short range nature of the Gaussian potential. We therefore obtain configurations of the type depicted in Figure 4.2.

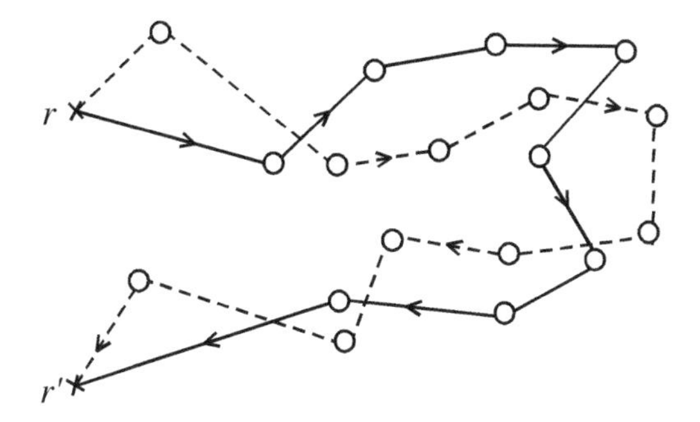

Figure 4.1 Typical trajectories described by the Green functions G^R (solid lines) and G^A (dashed lines). The orientation of the two trajectories follows the convention of Figure 3.1 and represents propagation from $\boldsymbol{r}$ to $\boldsymbol{r}'$.

[5] In Chapter 3, we denoted the scattered wave vector by k_0. Henceforth, we will denote it by k.

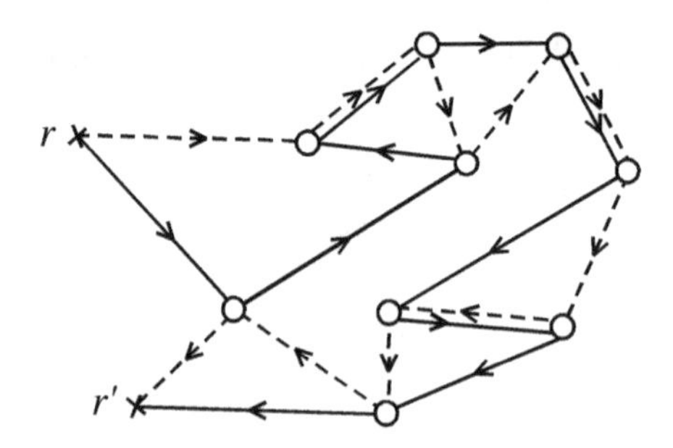

Figure 4.2 Owing to the short range nature of the Gaussian potential, only those trajectories passing through the same scattering centers contribute to $\overline{G^R G^A}$. In this picture, the phase shift between the two trajectories is much larger than 2π.

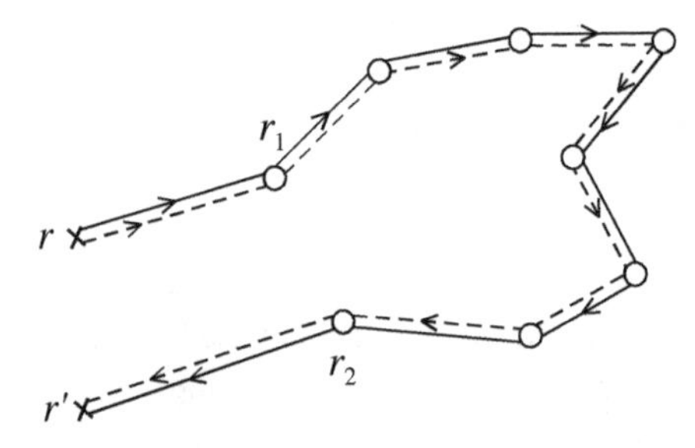

Figure 4.3 Trajectories with identical scattering sequences contribute to the average $\overline{G^R G^A}$.

A second consequence of the averaging consists in replacing the Green functions by their average value. The distance between two collisions is thus of the order of the elastic mean free path l_e. In the weak disorder regime (section 3.2), the collisions are independent and $l_e \gg \lambda$. Consequently, the difference in path lengths between trajectories with non-identical scattering sequences will be at least of order l_e, that is, much greater than the wavelength λ. The resulting dephasing is very large, and these contributions are thus negligible. We therefore retain only those contributions of the type represented in Figure 4.3.

In this approximation, called the *Diffuson approximation*, the expression for the average probability which we denote $P_d(\boldsymbol{r}, \boldsymbol{r}', \omega)$, is the product of three distinct terms. The first describes the propagation from some point $\boldsymbol{r}$ in the medium (which does not necessarily correspond to a scattering event) to the point $\boldsymbol{r}_1$ of the first collision. It is given by

$$\overline{G}^R_\epsilon(\boldsymbol{r}, \boldsymbol{r}_1)\overline{G}^A_{\epsilon-\omega}(\boldsymbol{r}_1, \boldsymbol{r}). \tag{4.22}$$

The second contribution takes into account all possible sequences of collisions between the scatterers $\boldsymbol{r}_1$ and $\boldsymbol{r}_2$. It is characterized by a function denoted $\Gamma_\omega(\boldsymbol{r}_1, \boldsymbol{r}_2)$, which we will call the *structure factor* or *vertex function*. Lastly, the third term describes the propagation from the point $\boldsymbol{r}_2$ of the last scattering event to some termination point $\boldsymbol{r}'$. The resulting

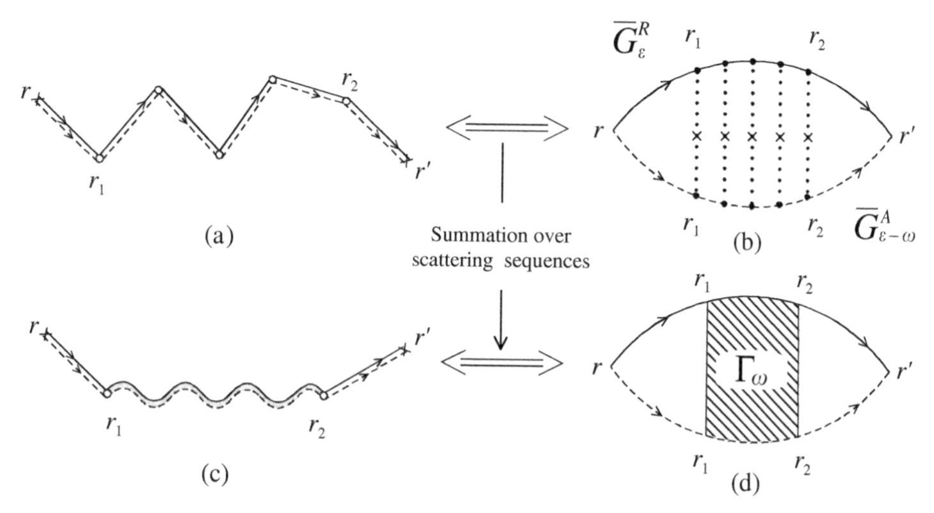

Figure 4.4 Different representations of the probability $P_d(\boldsymbol{r},\boldsymbol{r}',\omega)$. Diagrams (a) and (b) represent trajectories of multiple scattering with five collisions. Diagrams (c) and (d) represent the probability $P_d(\boldsymbol{r},\boldsymbol{r}',\omega)$ obtained by iterating to infinity a series of scattering sequences. The structure factor (or vertex function) Γ_ω is the sum of multiple scattering processes that relate the two scattering events located at points $\boldsymbol{r}_1$ and $\boldsymbol{r}_2$. Diagrams (c) and (d) show that the probability P_d is a product of the form $|G^R|^2\,\Gamma_\omega\,|G^R|^2$. The conventions used to represent the Green functions are those of Figure 3.1.

expression for $P_d(\boldsymbol{r},\boldsymbol{r}',\omega)$ is

$$P_d(\boldsymbol{r},\boldsymbol{r}',\omega) = \frac{1}{2\pi\rho_0}\int \overline{G}^R_\epsilon(\boldsymbol{r},\boldsymbol{r}_1)\overline{G}^A_{\epsilon-\omega}(\boldsymbol{r}_1,\boldsymbol{r})\overline{G}^R_\epsilon(\boldsymbol{r}_2,\boldsymbol{r}')\overline{G}^A_{\epsilon-\omega}(\boldsymbol{r}',\boldsymbol{r}_2)\Gamma_\omega(\boldsymbol{r}_1,\boldsymbol{r}_2)\,d\boldsymbol{r}_1\,d\boldsymbol{r}_2, \tag{4.23}$$

where the integral over the points $\boldsymbol{r}_1$ and $\boldsymbol{r}_2$ is meant to sum over all possible scattering processes. Equation (4.23) is illustrated in Figure 4.4.

In order to evaluate the structure factor Γ_ω, we use the white noise model defined by (2.30), that is, $B(\boldsymbol{r}_1-\boldsymbol{r}_2) = \gamma_e\delta(\boldsymbol{r}_1-\boldsymbol{r}_2)$ with $\gamma_e = 1/2\pi\rho_0\tau_e$. Assuming independent collisions, we construct all possible sequences which contribute to Γ_ω by iterating to infinity a sequence of elementary collision processes of amplitude γ_e. This elementary collision process is sometimes called "elementary vertex." We thus obtain the integral equation[6]

$$\Gamma_\omega(\boldsymbol{r}_1,\boldsymbol{r}_2) = \gamma_e\delta(\boldsymbol{r}_1-\boldsymbol{r}_2) + \gamma_e\int \Gamma_\omega(\boldsymbol{r}_1,\boldsymbol{r}'')\overline{G}^R_\epsilon(\boldsymbol{r}'',\boldsymbol{r}_2)\overline{G}^A_{\epsilon-\omega}(\boldsymbol{r}_2,\boldsymbol{r}'')\,d\boldsymbol{r}'' \tag{4.24}$$

represented diagrammatically in Figure 4.5. The characteristic ladder-like shape of the diagrams entering the iteration has given to the Diffuson approximation the equivalent

[6] An integral equation of this form is called a Bethe–Salpeter equation. In general, the structure factor is a function of four arguments. This is the case if the correlation function $\overline{V(\boldsymbol{r}_1)V(\boldsymbol{r}_2)} = B(\boldsymbol{r}_1-\boldsymbol{r}_2)$ is not a δ function (Appendix A4.3). See also the remark on page 218.

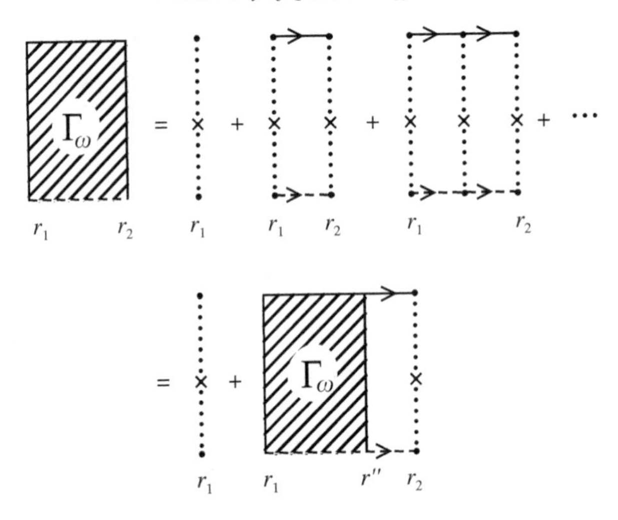

Figure 4.5 Diagrammatic representation of the iterative equation yielding the structure factor, or vertex function, $\Gamma_\omega(\boldsymbol{r}_1, \boldsymbol{r}_2)$. This quantity is called a *Diffuson* or *ladder diagram.*

denomination of *ladder approximation*. In expressions (4.23) and (4.24) the product $\overline{G}^R \overline{G}^A$ appears, which is proportional to the probability P_0 given by (4.16) so that we may rewrite (4.23) as

$$\boxed{P_d(\boldsymbol{r}, \boldsymbol{r}', \omega) = 2\pi\rho_0 \int P_0(\boldsymbol{r}, \boldsymbol{r}_1, \omega)\, \Gamma_\omega(\boldsymbol{r}_1, \boldsymbol{r}_2)\, P_0(\boldsymbol{r}_2, \boldsymbol{r}', \omega)\, d\boldsymbol{r}_1\, d\boldsymbol{r}_2} \qquad (4.25)$$

This result exhibits a very simple structure: the probability for a wave packet to propagate from $\boldsymbol{r}$ to $\boldsymbol{r}'$ depends on the probability of it getting to the first scattering event at $\boldsymbol{r}_1$ with no collision, followed by a process of successive independent collisions, and then finally the probability of it arriving at the final point $\boldsymbol{r}'$ with no further collision. In the literature, this contribution to the probability of quantum diffusion P_d is termed a *Diffuson* or a *ladder*.

The process of multiple scattering is described, for its own part, by the integral equation:

$$\boxed{\Gamma_\omega(\boldsymbol{r}_1, \boldsymbol{r}_2) = \gamma_e \delta(\boldsymbol{r}_1 - \boldsymbol{r}_2) + \frac{1}{\tau_e} \int \Gamma_\omega(\boldsymbol{r}_1, \boldsymbol{r}'') P_0(\boldsymbol{r}'', \boldsymbol{r}_2, \omega)\, d\boldsymbol{r}''} \qquad (4.26)$$

where $\gamma_e = 1/2\pi\rho_0\tau_e$.

Remark

The structure of the probability P_d results from the summation over an infinity of collision sequences. The sequences may be interpreted as a succession of independent events separated by classical ballistic motion. Let us define the probability $w(R)d^3R$ as the probability that two successive collisions are separated by $R = |\boldsymbol{r} - \boldsymbol{r}'|$:

$$w(R) = \frac{\int P_0(\boldsymbol{r},\boldsymbol{r}',t)dt}{\int P_0(\boldsymbol{r},\boldsymbol{r}',t)\,dt\,d\boldsymbol{r}'} = \frac{e^{-R/l_e}}{4\pi R^2 l_e} = \frac{1}{\tau_e} P_0(\boldsymbol{r},\boldsymbol{r}',\omega = 0). \tag{4.27}$$

We may thus explicitly rewrite the probability P_d derived from (4.29) in the form of an infinite sum of independent collision sequences [73]:

$$P_d(\boldsymbol{r},\boldsymbol{r}',t) = \tau_e \sum_{n=1}^{\infty} \int d\boldsymbol{r}_1 \cdots d\boldsymbol{r}_n\, \delta\left(t - \frac{|\boldsymbol{r}-\boldsymbol{r}_1| + \cdots + |\boldsymbol{r}_n - \boldsymbol{r}'|}{v}\right) \times w(\boldsymbol{r},\boldsymbol{r}_1)\cdots w(\boldsymbol{r}_n,\boldsymbol{r}'). \tag{4.28}$$

The probability P_d contains no phase information and may be expressed solely in terms of classical quantities.

Integral equation for the total probability $P(\boldsymbol{r},\boldsymbol{r}',\omega)$

From the integral equation (4.26) for $\Gamma_\omega(\boldsymbol{r},\boldsymbol{r}')$ and the relation (4.25) between $P_d(\boldsymbol{r},\boldsymbol{r}',\omega)$ and $\Gamma_\omega(\boldsymbol{r},\boldsymbol{r}')$, we find for the total probability $P = P_0 + P_d$ the integral equation

$$\boxed{P(\boldsymbol{r},\boldsymbol{r}',\omega) = P_0(\boldsymbol{r},\boldsymbol{r}',\omega) + \frac{1}{\tau_e}\int P(\boldsymbol{r},\boldsymbol{r}'',\omega)P_0(\boldsymbol{r}'',\boldsymbol{r}',\omega)\,d\boldsymbol{r}''} \tag{4.29}$$

Diffuson and reciprocity theorem

We have seen in Appendix A2.2 that, owing to the reciprocity theorem, the multiple scattering amplitudes of a given sequence of collisions and that with the time-reversed sequence are equal (relation 2.113). Let us see how we may apply this principle to the Diffuson approximation. For this, consider Figure 4.5 and time reverse *both* amplitudes associated with multiple scattering sequences. In this way we generate the Diffuson corresponding to propagation between $\boldsymbol{r}_2$ and $\boldsymbol{r}_1$ and, by virtue of the reciprocity theorem, this contribution is identical to that describing propagation between $\boldsymbol{r}_1$ and $\boldsymbol{r}_2$.

There is another possibility offered by the reciprocity theorem. Indeed, since it concerns the amplitude associated with a multiple scattering sequence, we may time reverse *only one* of the two amplitudes of the process shown in Figure 4.5, and obtain an equivalent contribution. In this manner we generate a contribution which is impossible to describe by means of a Diffuson, since the two amplitudes now propagate in opposite directions. We are thus led to the conclusion that the Diffuson approximation does not take into account all the processes allowed by the reciprocity theorem [74]. This fact will be the object of section 4.6.

Diffuson and anisotropic collisions

The relations which we have just established for P_d and for Γ_ω have a simple iterative structure,[7] which is a consequence of the fact that the white noise potential describing an elementary collision process is a δ function. If the potential is of finite range, comparable to the wavelength, collisions are anisotropic and are described by the correlation function $B(\boldsymbol{r}-\boldsymbol{r}')$ (2.29). In this case, we must generalize (4.24) for the structure factor, which now is a function of four points $\Gamma_\omega(\boldsymbol{r}_1,\boldsymbol{r}_2,\boldsymbol{r}_3,\boldsymbol{r}_4)$ instead of two. We may once again establish an expression for Γ_ω and for the probability P. This is the object of Appendices A4.3 and A5.2.

Exercise 4.2 Show that equations (4.28) and (4.29) are equivalent.

Exercise 4.3 Show that the probability to return to the origin after just one collision is [75]

$$P_1(\boldsymbol{r},\boldsymbol{r},t) = \frac{1}{2\pi l_e}(vt)^{1-d}\, e^{-t/\tau_e}.$$

4.5 The Diffuson at the diffusion approximation

The Diffuson, or ladder, approximation corresponding to the weak disorder limit $kl_e \gg 1$ leads to an integral equation for the probability with a simple iterative structure. For certain geometries, it is possible to solve equations (4.25) and (4.26) exactly, and thereby deduce the probability P_d, as is done for an infinite medium in Appendix A5.1. Here, we give an approximate result for P_d valid for long times $t \gg \tau_e$ ($\omega\tau_e \ll 1$), that is to say, after a large number of collisions.[8] This approximation for calculation of the Diffuson is called the *diffusion approximation* and should not be confused with the Diffuson approximation. We now show that in this limit, also called the *diffusive* or *hydrodynamic regime*, the solutions of the integral equations (4.26) and (4.29) are solutions of a classical diffusion equation. Indeed, the spatial variations of $\Gamma_\omega(\boldsymbol{r}_1,\boldsymbol{r}_2)$ are small on the scale of l_e and the integral equation (4.26) simplifies. We may expand $\Gamma_\omega(\boldsymbol{r}_1,\boldsymbol{r}'')$ about $\boldsymbol{r}''=\boldsymbol{r}_2$:

$$\Gamma_\omega(\boldsymbol{r}_1,\boldsymbol{r}'') = \Gamma_\omega(\boldsymbol{r}_1,\boldsymbol{r}_2) + (\boldsymbol{r}''-\boldsymbol{r}_2)\cdot\nabla_{\boldsymbol{r}_2}\Gamma_\omega(\boldsymbol{r}_1,\boldsymbol{r}_2) + \frac{1}{2}[(\boldsymbol{r}''-\boldsymbol{r}_2).\nabla_{\boldsymbol{r}_2}]^2\Gamma_\omega(\boldsymbol{r}_1,\boldsymbol{r}_2) + \cdots . \tag{4.30}$$

[7] This iterative structure is seen more clearly in the Fourier transform (see Appendix A4.3).
[8] In Appendix A5.1, we show that the diffusive regime is quickly attained, for distances of the order of l_e.

The integral of the term linear in the gradient vanishes by symmetry, as do those of the cross terms of the quadratic term, so that the integral (4.26) becomes

$$\Gamma_\omega(\boldsymbol{r}_1,\boldsymbol{r}_2) = \gamma_e \delta(\boldsymbol{r}_1 - \boldsymbol{r}_2) + \Gamma_\omega(\boldsymbol{r}_1,\boldsymbol{r}_2) \int \frac{P_0(\boldsymbol{r}'',\boldsymbol{r}_2,\omega)}{\tau_e} d\boldsymbol{r}''$$

$$+ \frac{1}{2d} \Delta_{\boldsymbol{r}_2} \Gamma_\omega(\boldsymbol{r}_1,\boldsymbol{r}_2) \int \frac{P_0(\boldsymbol{r}'',\boldsymbol{r}_2,\omega)}{\tau_e} (\boldsymbol{r}'' - \boldsymbol{r}_2)^2 \, d\boldsymbol{r}'' + \cdots . \quad (4.31)$$

The two integrals may be easily calculated:

$$\int P_0(\boldsymbol{r}'',\boldsymbol{r}_2,\omega) \, d\boldsymbol{r}'' = \frac{\tau_e}{1 - i\omega\tau_e} \simeq \tau_e(1 + i\omega\tau_e), \quad (4.32)$$

$$\int P_0(\boldsymbol{r}'',\boldsymbol{r}_2,\omega)(\boldsymbol{r}'' - \boldsymbol{r}_2)^2 \, d\boldsymbol{r}'' = 2\frac{l_e^2\tau_e}{(1 - i\omega\tau_e)^2} \simeq 2l_e^2\tau_e. \quad (4.33)$$

The expansion (4.31) may thus be put into the form of a *diffusion equation* for $\Gamma_\omega(\boldsymbol{r}_1,\boldsymbol{r}_2)$:

$$\left(-i\omega - D\Delta_{\boldsymbol{r}_2}\right) \Gamma_\omega(\boldsymbol{r}_1,\boldsymbol{r}_2) = \frac{\gamma_e}{\tau_e} \delta(\boldsymbol{r}_1 - \boldsymbol{r}_2), \quad (4.34)$$

where the diffusion constant D is defined by

$$\boxed{D = \frac{v l_e}{d} = \frac{v^2 \tau_e}{d}} \quad (4.35)$$

d is the space dimension and v is the group velocity. In the diffusive limit of slow spatial variations, Γ_ω varies more slowly than P_0. In (4.25), we may thus take Γ_ω out of the integral, and the probability P_d becomes proportional to the structure factor:

$$P_d(\boldsymbol{r},\boldsymbol{r}',\omega) = 2\pi\rho_0 \, \Gamma_\omega(\boldsymbol{r},\boldsymbol{r}') \int P_0(\boldsymbol{r},\boldsymbol{r}_1,\omega) P_0(\boldsymbol{r}_2,\boldsymbol{r}',\omega) \, d\boldsymbol{r}_1 \, d\boldsymbol{r}_2. \quad (4.36)$$

Using (4.32), we obtain

$$P_d(\boldsymbol{r},\boldsymbol{r}',\omega) = \frac{\tau_e}{\gamma_e} \Gamma_\omega(\boldsymbol{r},\boldsymbol{r}') = 2\pi\rho_0\tau_e^2 \, \Gamma_\omega(\boldsymbol{r},\boldsymbol{r}'), \quad (4.37)$$

so that $P_d(\boldsymbol{r},\boldsymbol{r}',\omega)$ satisfies a diffusion equation as well:[9]

$$\boxed{(-i\omega - D\Delta_{\boldsymbol{r}'}) P_d(\boldsymbol{r},\boldsymbol{r}',\omega) = \delta(\boldsymbol{r} - \boldsymbol{r}')} \quad (4.38)$$

We may now verify that the total probability $P_0 + P_d$ is indeed properly normalized.[10]

[9] Since they are proportional, in practice we will use the same denomination *Diffuson* for Γ_ω and P_d.

[10] This check is simple to perform in Fourier space and is presented in Appendix A4.1.

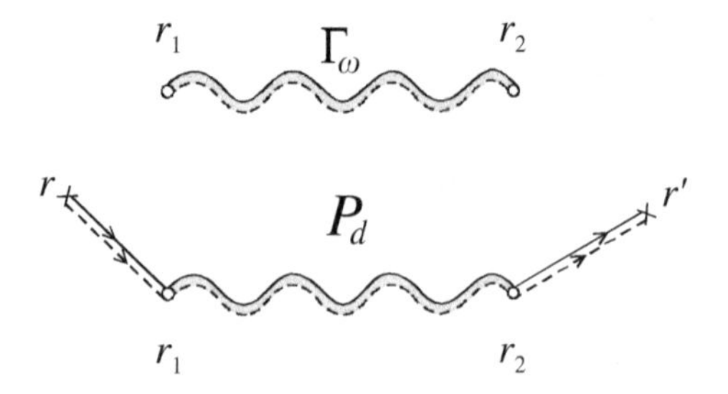

Figure 4.6 The structure factor Γ_ω relates collision events. The probability P_d relates arbitrary points $\boldsymbol{r}$ and $\boldsymbol{r}'$.

Important remark

As indicated in Figure 4.6, the structure factor describes successive collision processes and its arguments correspond to collision events. In contrast, the function P_d describes the propagation between two arbitrary points $\boldsymbol{r}$ and $\boldsymbol{r}'$.

Remark

Chapter 5 is devoted to solutions of the diffusion equation for certain geometries. One particularly useful solution of (4.38) is that corresponding to an infinite three-dimensional medium. For zero frequency, this is:

$$P_d(\boldsymbol{r},\boldsymbol{r}',\omega=0) = \frac{1}{4\pi DR}, \tag{4.39}$$

where $R = |\boldsymbol{r}-\boldsymbol{r}'|$. We will use the Fourier transform of (4.38) as well, which is given by

$$P_d(\boldsymbol{q},\omega) = \frac{1}{-i\omega + Dq^2}. \tag{4.40}$$

This will be studied in section A4.1.2.

In summary, we have shown that, in the limit $kl_e \gg 1$ of weak disorder, the probability of quantum diffusion may be expressed as a classical quantity. This is a consequence of the approximation of independent collisions, which causes all interference effects to average to zero. Then, in the diffusive limit of slow spatial and temporal variations, the probability is the solution of a diffusion equation. This equation (4.38) was established for a translation invariant medium, but its validity, just as for the average Green function, extends to the case of a finite system of characteristic size L, provided $l_e \ll L$.

4.6 Coherent propagation: the Cooperon

The probability $P(\boldsymbol{r},\boldsymbol{r}',t) = P_0(\boldsymbol{r},\boldsymbol{r}',t) + P_d(\boldsymbol{r},\boldsymbol{r}',t)$ calculated in the preceding section is normalized. We may therefore think that we have described all the scattering processes which contribute to the average probability of quantum diffusion. In the limit $kl_e \gg 1$, the latter would just be equal to the classical diffusion probability. Nevertheless, it is not certain that, even in this limit of weak disorder, all possible collision processes have been taken into account, even if the sum of their contributions must cancel. Indeed, the reciprocity theorem (page 101) allows the time reversal of each of the two amplitudes constituting the Diffuson.

Let us reconsider the structure of P_d. It may be thought of as sequences of independent collision events, separated by propagations described by the probability $P_0(\boldsymbol{r}_i, \boldsymbol{r}_{i+1})$, the product of two conjugate Green's functions corresponding to two conjugate amplitudes (Figure 4.7(a)). In consequence, all the phases disappear in the average $\overline{G^R(\boldsymbol{r}, \boldsymbol{r}')G^A(\boldsymbol{r}', \boldsymbol{r})}$ and so P_d is a classical quantity.

Now, there is at least one other contribution which has not been taken into account and which has important physical consequences. To see this, consider the product of two Green's functions describing two trajectories which are identical but for the fact that they are *covered in exactly opposite directions* such as those shown in Figure 4.7(b). The phase factors $k\mathcal{L}_N$ associated with each of the two trajectories (see expression 4.21) are identical *provided the system is time reversal invariant.* In this case, the Green function has the property:

$$G^{R,A}(\boldsymbol{r}, \boldsymbol{r}') = G^{R,A}(\boldsymbol{r}', \boldsymbol{r}). \tag{4.41}$$

A process for which the two trajectories are traversed in opposite directions is allowed by the reciprocity theorem, and thus must contribute equally to the probability. As we see in

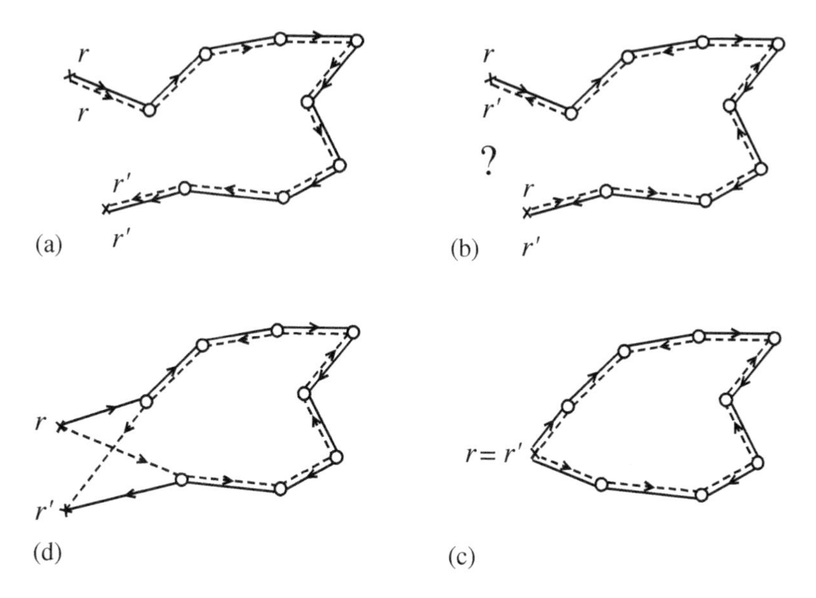

Figure 4.7 (a) Diffuson contribution to the probability. (b) Reversing one of the two trajectories, the points $\boldsymbol{r}$ and $\boldsymbol{r}'$ are exchanged, leading to an impossibility unlike $\boldsymbol{r}$ and $\boldsymbol{r}'$ coincide. In such a case (c), the phases cancel in this new contribution. (d) If $\boldsymbol{r} \neq \boldsymbol{r}'$, there is a mismatch between the two trajectories, leading to a phase shift.

Figures 4.7(b,c), this is not rigorously possible unless $\boldsymbol{r} = \boldsymbol{r}'$. By contrast, for $\boldsymbol{r} \neq \boldsymbol{r}'$, the two trajectories are dephased (Figure 4.7(d)). To what extent may these contribute to the probability?

To evaluate the contribution of this new type of process, we consider the ensemble of diagrams for which the two trajectories correspond to oppositely ordered collision sequences. Figure 4.8 shows several equivalent representations of a sequence of five collisions, an element of an iterative structure resembling that of P_d, which differs only in the reversal of the arguments.

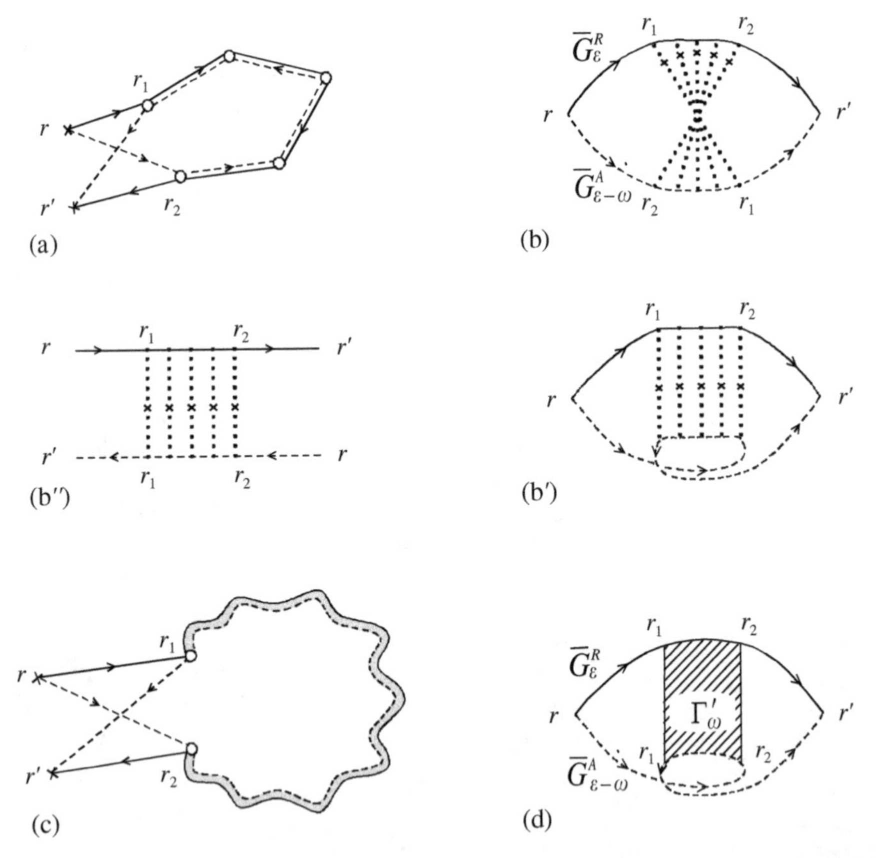

Figure 4.8 (a, b, b′, b′′) Several equivalent representations of a process of five collisions which contribute to X_c. Representation (b) is common in the literature, and is called a Cooperon or *maximally crossed diagram*. Reversing one of the two trajectories (b → b′ → b′′), we see that the Cooperon has a ladder structure very similar to that of the Diffuson. (c,d) Representations of X_c. These two figures should be compared with Figures 4.4(c,d) and demonstrate why the Cooperon is a short range function.

The probability associated with this new process, which we denote X_c, is called a *Cooperon*.[11] It is shown in Figure 4.8(d). Because of the characteristic "fan" structure of the diagrams contributing to the Cooperon, it is also often called a *maximally crossed diagram* in the literature. The Cooperon X_c is given by

$$X_c(\boldsymbol{r},\boldsymbol{r}',\omega) = \frac{1}{2\pi\rho_0}\int \overline{G}^R_\epsilon(\boldsymbol{r},\boldsymbol{r}_1)\overline{G}^R_\epsilon(\boldsymbol{r}_2,\boldsymbol{r}')\overline{G}^A_{\epsilon-\omega}(\boldsymbol{r}',\boldsymbol{r}_1)\overline{G}^A_{\epsilon-\omega}(\boldsymbol{r}_2,\boldsymbol{r})\Gamma'_\omega(\boldsymbol{r}_1,\boldsymbol{r}_2)\, d\boldsymbol{r}_1\, d\boldsymbol{r}_2. \tag{4.42}$$

This contribution to the probability of quantum diffusion has a structure similar to that of P_d given by expression (4.23), with a priori a new structure factor $\Gamma'_\omega(\boldsymbol{r}_1,\boldsymbol{r}_2)$. The latter is a solution of the integral equation[12] illustrated in Figure 4.9:

$$\Gamma'_\omega(\boldsymbol{r}_1,\boldsymbol{r}_2) = \gamma_e\delta(\boldsymbol{r}_1-\boldsymbol{r}_2) + \gamma_e\int \Gamma'_\omega(\boldsymbol{r}_1,\boldsymbol{r}'')\overline{G}^R_\epsilon(\boldsymbol{r}'',\boldsymbol{r}_2)\overline{G}^A_{\epsilon-\omega}(\boldsymbol{r}'',\boldsymbol{r}_2)\, d\boldsymbol{r}'' \tag{4.43}$$

which may usefully be compared with the integral equation (4.24) for $\Gamma_\omega(\boldsymbol{r}_1,\boldsymbol{r}_2)$.

Since the product $\overline{G}^R_\epsilon(\boldsymbol{r},\boldsymbol{r}_2)\overline{G}^A_{\epsilon-\omega}(\boldsymbol{r},\boldsymbol{r}_2)$ is equal to the product $\overline{G}^R_\epsilon(\boldsymbol{r},\boldsymbol{r}_2)\overline{G}^A_{\epsilon-\omega}(\boldsymbol{r}_2,\boldsymbol{r})$, the structure factor $\Gamma'_\omega(\boldsymbol{r}_1,\boldsymbol{r}_2)$ is identical to $\Gamma_\omega(\boldsymbol{r}_1,\boldsymbol{r}_2)$, calculated previously in (4.26) for the Diffuson,[13] a consequence of the reciprocity theorem. Thus, by reversing the direction of

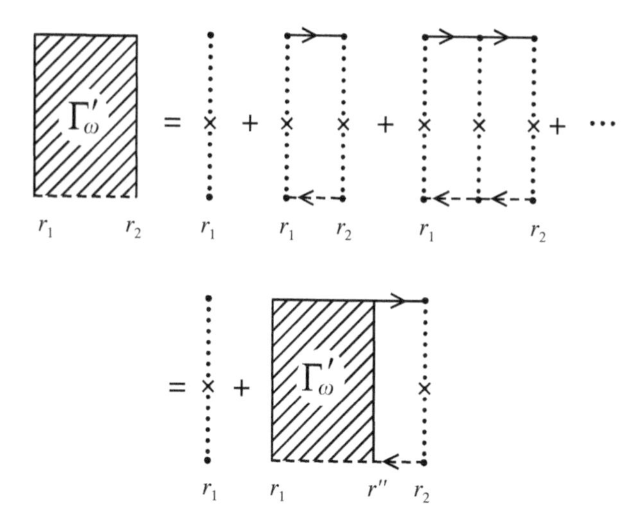

Figure 4.9 Diagrammatic representation of the iterative structure yielding the structure factor Γ'_ω of the Cooperon.

[11] The structure of the diagrams (Figures 4.8 and 4.9) which contribute to the probability X_c resembles the ladder diagrams which arise in the calculation of the superconducting response function studied by L. N. Cooper: they contain two propagators related by the attractive BCS interaction. This is why this contribution to the probability has been called a *Cooperon*.

[12] Employing the conventions of Figure 3.1.

[13] To be rigorous, the first term in the integral equation (4.43) must contain *two* impurities, since the single impurity term has already been taken into account by the Diffuson (see Figures 4.5 and 4.9). This term, which equals $\gamma_e\delta(\boldsymbol{r}_1-\boldsymbol{r}_2)$, does not exist at large distances. We have chosen to incorporate it in the structure factor Γ'_ω so that it coincides with Γ_ω when the system is time reversal invariant. See also footnote 17, page 118.

time in their lower line, the diagrams of Figures 4.5 and 4.9 become identical. However, the product of four Green's functions of (4.42) no longer reduces to the product of two functions P_0 and the resulting dephasing reduces the probability $X_c(\boldsymbol{r},\boldsymbol{r}',\omega)$ as soon as $\boldsymbol{r} \neq \boldsymbol{r}'$. Assuming that the spatial variation is slow, as in section 4.5, (4.42) becomes (neglecting the dependence on frequency ω of the average Green functions)

$$X_c(\boldsymbol{r},\boldsymbol{r}',\omega) = \frac{\Gamma'_\omega(\boldsymbol{r},\boldsymbol{r})}{2\pi\rho_0\gamma_e^2}\left[\gamma_e \int \overline{G}^R_\epsilon(\boldsymbol{r},\boldsymbol{r}_1)\overline{G}^A_\epsilon(\boldsymbol{r}',\boldsymbol{r}_1)\, d\boldsymbol{r}_1\right]^2. \tag{4.44}$$

When $\boldsymbol{r} = \boldsymbol{r}'$, the term in braces is $f^{1,1}(0)$ given in Figure 3.6 and equals 1. Since $\Gamma'_\omega = \Gamma_\omega$, we get

$$X_c(\boldsymbol{r},\boldsymbol{r},\omega) = P_d(\boldsymbol{r},\boldsymbol{r},\omega). \tag{4.45}$$

Because of this additional contribution, *the probability of return to the origin is thus twice as large* as its classical expression $P_d(\boldsymbol{r},\boldsymbol{r},\omega)$. When $\boldsymbol{r} \neq \boldsymbol{r}'$, the term in braces is the short range function $f^{1,1}(\boldsymbol{r}-\boldsymbol{r}')$ which according to (3.105) is equal to $g(\boldsymbol{R})$ defined by (3.97). We thus deduce

$$X_c(\boldsymbol{r},\boldsymbol{r}',\omega) = \frac{\tau_e}{\gamma_e} g^2(\boldsymbol{r}-\boldsymbol{r}')\Gamma'_\omega(\boldsymbol{r},\boldsymbol{r}). \tag{4.46}$$

In three dimensions and in the limit $kl_e \gg 1$, the probability $X_c(\boldsymbol{r},\boldsymbol{r}',\omega)$ is equal to[14]

$$\boxed{X_c(\boldsymbol{r},\boldsymbol{r}',\omega) = X_c(\boldsymbol{r},\boldsymbol{r},\omega)\frac{\sin^2 kR}{k^2R^2}e^{-R/l_e}} \tag{4.47}$$

This contribution is thus negligible for distances $R = |\boldsymbol{r}-\boldsymbol{r}'|$ greater than the mean free path l_e. To recapitulate, the Diffuson and Cooperon contributions are shown in Figure 4.10.

It would appear that, as a result of the Cooperon contribution, the total probability is no longer normalized. We will show in section A4.2.2 that there are additional corrections which indeed restore the normalization.

Important remark: the Cooperon is not the solution of a diffusion equation

The function $X_c(\boldsymbol{r},\boldsymbol{r}',\omega)$ is peaked around $\boldsymbol{r}' \simeq \boldsymbol{r}$. As such it is not the solution of a diffusion equation. On the other hand, Γ'_ω, like Γ_ω, obeys a diffusion equation. We define a probability $P_c(\boldsymbol{r},\boldsymbol{r}',\omega)$ by

$$P_c(\boldsymbol{r},\boldsymbol{r}',\omega) = \frac{\tau_e}{\gamma_e}\Gamma'_\omega(\boldsymbol{r},\boldsymbol{r}') \tag{4.48}$$

which is the analog of P_d and which satisfies a diffusion equation. It is related to $X_c(\boldsymbol{r},\boldsymbol{r}',\omega)$ by

$$X_c(\boldsymbol{r},\boldsymbol{r}',\omega) = P_c(\boldsymbol{r},\boldsymbol{r},\omega)\; g^2(\boldsymbol{r}-\boldsymbol{r}'). \tag{4.49}$$

For convenience, we will employ the same term *Cooperon* for X_c, P_c and Γ'.

[14] See footnote 5, page 97.

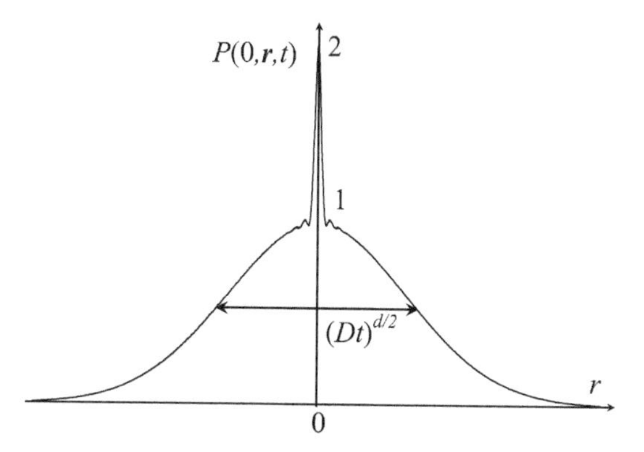

Figure 4.10 Probability $P(0,\boldsymbol{r},t) = P_d + X_c$ as a function of the distance $r = |\boldsymbol{r}|$. The probability is enhanced in a small volume $\lambda^{d-1} l_e$ around the origin where it is doubled. The relative width of the peak is $\lambda^{d-1} l_e/(Dt)^{d/2}$. It is maximal for $t \simeq \tau_e$ and of order $1/(kl_e)^{d-1}$.

The integrated probability associated with the Cooperon is obtained starting from expression (4.47) and is given by (see Exercise 3.14)

$$\int X_c(\boldsymbol{r},\boldsymbol{r}',\omega)\, d\boldsymbol{r}' = X_c(\boldsymbol{r},\boldsymbol{r},\omega)\frac{2\pi\, l_e}{k^2}. \tag{4.50}$$

More generally, this integrated probability is given, in any dimension by

$$\int X_c(\boldsymbol{r},\boldsymbol{r}',\omega)\, d\boldsymbol{r}' = X_c(\boldsymbol{r},\boldsymbol{r},\omega)\int_0^\infty g^2(\boldsymbol{R})\, d\boldsymbol{R} \tag{4.51}$$

which yields, by virtue of (3.100)

$$\boxed{\int X_c(\boldsymbol{r},\boldsymbol{r}',\omega)\, d\boldsymbol{r}' = X_c(\boldsymbol{r},\boldsymbol{r},\omega)\frac{\tau_e}{\pi\rho_0}} \tag{4.52}$$

A good approximation for $X_c(\boldsymbol{r},\boldsymbol{r}',\omega)$ is obtained by replacing (4.47) by

$$X_c(\boldsymbol{r},\boldsymbol{r}',\omega) = \frac{\tau_e}{\pi\rho_0} P_c(\boldsymbol{r},\boldsymbol{r},\omega)\delta(\boldsymbol{r}-\boldsymbol{r}') \tag{4.53}$$

where $P_c(\boldsymbol{r},\boldsymbol{r}')$ obeys a diffusion equation (see the remark above). The Cooperon X_c is a short range contribution that has been obtained by reversing one of two trajectories constituting a multiple scattering sequence, as allowed by the reciprocity theorem. When $\boldsymbol{r} \neq \boldsymbol{r}'$, this contribution is small because of the local dephasing that occurs at those ending points (Figure 4.7(d)). The reciprocity theorem also allows one of the trajectories to be reversed partially so that the corresponding dephasing may occur at any place along the multiple scattering sequence of a Diffuson. The resulting diagram can be viewed as the crossing of a Diffuson with itself (Figure 4.11). It is long range but always smaller than P_d. This kind of diagram will be considered later (sections A4.2.2 and A7.2.3).

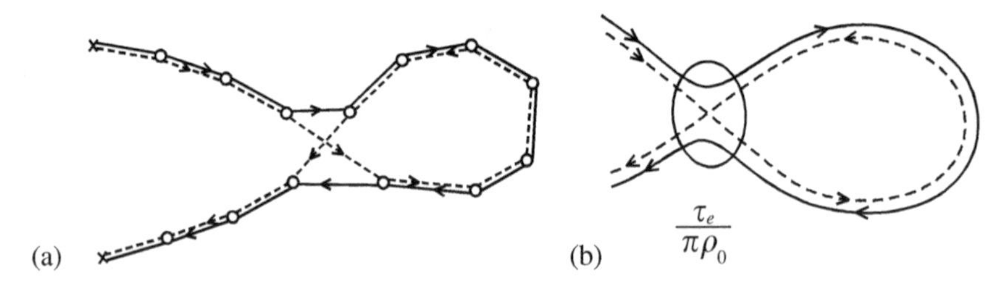

Figure 4.11 (a) The reciprocity theorem allows diagrams where the local dephasing due to a quantum crossing may occur at any place along the multiple scattering sequence. (b) Such a diagram can be seen as resulting from the quantum crossing of a Diffuson with itself. The volume associated with the region of intersection is $\tau_e/(\pi\rho_0) \propto \lambda^{d-1}l_e$.

Once this idea of crossing is understood, it is useful to describe physically the volume $\int_0^\infty g^2(\boldsymbol{R})\,d\boldsymbol{R} = \tau_e/(\pi\rho_0)$ that appears in equation (4.52). It is proportional to $\lambda^{d-1}l_e$ and may be interpreted as follows. Since the Cooperon may be viewed as the result of a quantum crossing of a Diffuson with itself, the spatial volume of the region of intersection is that of a tube of length l_e and cross section λ^{d-1} (see sections 1.7.1 and A4.2.3). It is then appropriate to describe the Diffuson as an object of finite "scattering cross section" λ^{d-1} (Figure 4.11).

4.7 Radiative transfer

Until now, we have been interested in the probability $P(\boldsymbol{r},\boldsymbol{r}',\omega)$ defined in the framework of the Schrödinger equation. In the same way, we may define the probability associated with the Helmholtz equation, which is thus related to the intensity at a given point of electromagnetic radiation regarded as a scalar wave. The Diffuson approximation is frequently used in this context[15] and goes by the name of *radiative transfer* [76].

Consider for example the case of a point source located at a point $\boldsymbol{r}_0$. The electric field ψ_ϵ emanating from this source corresponds to the Green function $G^R_\epsilon(\boldsymbol{r}_0,\boldsymbol{r})$ of the Helmholtz equation (2.8). The average over disorder of $\psi_\epsilon(\boldsymbol{r})$ is given by $\overline{\psi}_\epsilon(\boldsymbol{r}) = \overline{G}_\epsilon(\boldsymbol{r}_0,\boldsymbol{r})$ and is calculated using the methods discussed in Chapter 3. The intensity of the radiation field at a point $\boldsymbol{r}$ is:[16]

$$\begin{aligned} I(\boldsymbol{r}) &= \frac{4\pi}{c}|\psi_\epsilon(\boldsymbol{r})|^2 \\ &= \frac{4\pi}{c}G^R_\epsilon(\boldsymbol{r}_0,\boldsymbol{r})G^A_\epsilon(\boldsymbol{r},\boldsymbol{r}_0). \end{aligned} \tag{4.54}$$

15 The radiative transfer equation is discussed in detail in Appendix A5.2.

16 In this expression, the normalization has been chosen so that the intensity satisfies the Laplace equation with a source of amplitude equal to unity (relation 4.66). This normalization has no real importance since, most of the time, interesting quantities appear as a ration of intensities. See also footnote 3, page 322.

The average intensity

$$\overline{I}(\boldsymbol{r}) = \frac{4\pi}{c} \overline{G_\epsilon^R(\boldsymbol{r}_0,\boldsymbol{r}) G_\epsilon^A(\boldsymbol{r},\boldsymbol{r}_0)} \tag{4.55}$$

has the same structure as the probability P introduced in (4.9) but must be adapted to the case of the Helmholtz equation, so

$$\boxed{P(\boldsymbol{r},\boldsymbol{r}') = \frac{4\pi}{c} \overline{G_\epsilon^R(\boldsymbol{r},\boldsymbol{r}')\, G_\epsilon^A(\boldsymbol{r}',\boldsymbol{r})}} \tag{4.56}$$

where the two Green functions are taken at the same frequency. More generally, we may consider the correlation function of two fields $\psi_\epsilon(\boldsymbol{r})$ and $\psi^*_{\epsilon-\omega}(\boldsymbol{r}')$ at two different energies:

$$\overline{\psi_\epsilon(\boldsymbol{r})\psi^*_{\epsilon-\omega}(\boldsymbol{r}')} = \overline{G_\epsilon^R(\boldsymbol{r}_0,\boldsymbol{r}) G^A_{\epsilon-\omega}(\boldsymbol{r}',\boldsymbol{r}_0)}. \tag{4.57}$$

This correlation function is not directly related to the probability $P(\boldsymbol{r},\boldsymbol{r}',\omega)$, since it involves three points rather than two.

Neglecting correlations between Green's functions, as was done in section 4.3, we obtain the analog of the Drude–Boltzmann contribution (4.17), that is, the intensity of the radiation which has not undergone any collision up to a distance R from the source:

$$I_0(R) = \frac{1}{4\pi R^2 c} e^{-R/l_e}. \tag{4.58}$$

The radiated intensity at distances greater than l_e results from multiple scattering processes. We might thus seek, for the correlation function $\overline{\psi_\epsilon(\boldsymbol{r})\psi^*_{\epsilon-\omega}(\boldsymbol{r}')}$ and for the intensity $\overline{I}(\boldsymbol{r})$, both classical and phase coherent contributions, that is, the Diffuson and the Cooperon. The Diffuson contribution is obtained from Figure 4.12(b) and is written in a manner analogous to (4.23)

$$\overline{\psi_\epsilon(\boldsymbol{r})\psi^*_{\epsilon-\omega}(\boldsymbol{r}')} = \overline{\psi}_\epsilon(\boldsymbol{r})\overline{\psi}^*_{\epsilon-\omega}(\boldsymbol{r}') + \int d\boldsymbol{r}_1 d\boldsymbol{r}_2\, \overline{\psi}_\epsilon(\boldsymbol{r}_1)\overline{\psi}^*_{\epsilon-\omega}(\boldsymbol{r}_1)$$

$$\times\, \Gamma_\omega(\boldsymbol{r}_1,\boldsymbol{r}_2)\overline{G}^R_\epsilon(\boldsymbol{r}_2,\boldsymbol{r})\overline{G}^A_{\epsilon-\omega}(\boldsymbol{r}',\boldsymbol{r}_2) \tag{4.59}$$

which, for $\boldsymbol{r} = \boldsymbol{r}'$ and $\omega = 0$, gives the average radiated intensity $\overline{I}(\boldsymbol{r})$ defined in (4.55)

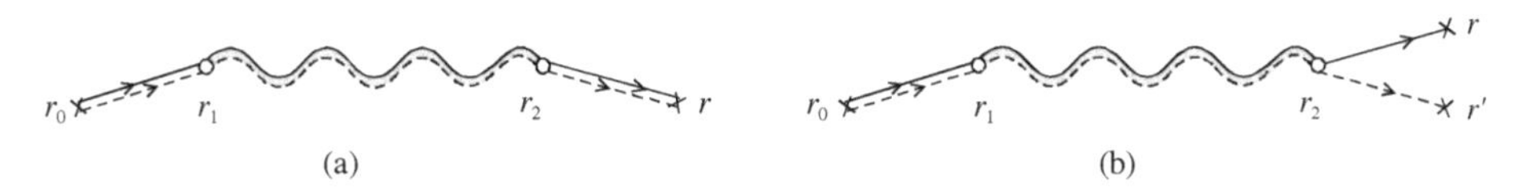

Figure 4.12 (a) Representation of the average intensity $I_d(\boldsymbol{r}) = P_d(\boldsymbol{r}_0,\boldsymbol{r})$, which is a solution of equation (4.60). (b) Representation of the correlation function of the fields.

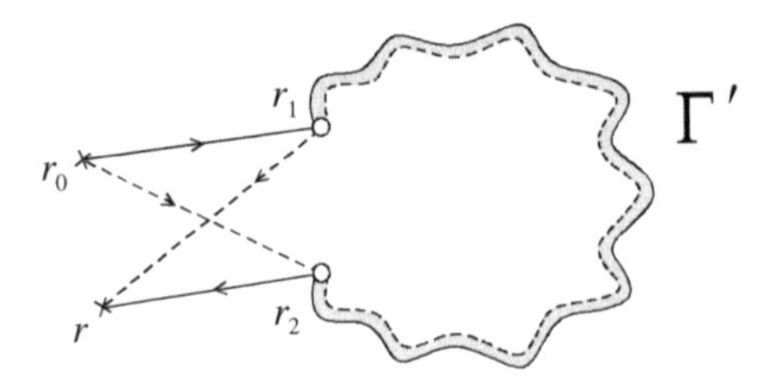

Figure 4.13 Contribution of the Cooperon to the intensity.

$$\begin{aligned}\overline{I}(\boldsymbol{r}) &= I_0(\boldsymbol{r}) + I_d(\boldsymbol{r}) \\ &= I_0(\boldsymbol{r}) + \frac{4\pi}{c}\int d\boldsymbol{r}_1\, d\boldsymbol{r}_2 |\overline{\psi}_\epsilon(\boldsymbol{r}_1)|^2\, \Gamma_{\omega=0}(\boldsymbol{r}_1,\boldsymbol{r}_2)\, |\overline{G}^R_\epsilon(\boldsymbol{r}_2,\boldsymbol{r})|^2,\end{aligned} \tag{4.60}$$

where $I_0(\boldsymbol{r})$ is the Drude–Boltzmann contribution (4.58) and $I_d(\boldsymbol{r})$ is that of the Diffuson. The expression for I_d is analogous to equation (4.25) for the case of the Schrödinger equation. The structure factor $\Gamma_{\omega=0}$ satisfies (4.24). In the same way, the Cooperon contribution is immediately obtained starting from Figure 4.13, which gives for the corresponding average radiated intensity $I_c(\boldsymbol{r})$ a relation analogous to (4.42):

$$I_c(\boldsymbol{r}) = \frac{4\pi}{c}\int d\boldsymbol{r}_1\, d\boldsymbol{r}_2\; \overline{\psi}_\epsilon(\boldsymbol{r}_1)\overline{\psi}^*_\epsilon(\boldsymbol{r}_2)\, \Gamma'_{\omega=0}(\boldsymbol{r}_1,\boldsymbol{r}_2)\, \overline{G}^R_\epsilon(\boldsymbol{r}_2,\boldsymbol{r})\, \overline{G}^A_\epsilon(\boldsymbol{r},\boldsymbol{r}_1) \tag{4.61}$$

where Γ'_ω is given by (4.43).

Remark

The preceding equations correspond to the case of a spatially localized source, represented by a δ function. The expressions (4.60) and (4.61) may be generalized to the case of an arbitrary source described by a function $j(\boldsymbol{r})$. In this case, as we have seen in (3.58), $\overline{\psi}_\epsilon(\boldsymbol{r})$ is not simply a Green's function, but is given by

$$\overline{\psi}_\epsilon(\boldsymbol{r}) = \int d\boldsymbol{r}'\, j(\boldsymbol{r}')\overline{G}_\epsilon(\boldsymbol{r}',\boldsymbol{r}). \tag{4.62}$$

In the diffusive limit, expression (4.60) after integration gives

$$I_d(\boldsymbol{r}) = P_d(\boldsymbol{r}_0,\boldsymbol{r}) = \frac{l_e^2}{4\pi c}\Gamma_{\omega=0}(\boldsymbol{r}_0,\boldsymbol{r}), \tag{4.63}$$

which is the analog of (4.37). Moreover, in this limit, expression (4.24) leads to a diffusion equation, analogous to (4.34)

$$-D\Delta_{\boldsymbol{r}}\Gamma_{\omega=0}(\boldsymbol{r}_0,\boldsymbol{r}) = \frac{\gamma_e}{\tau_e}\delta(\boldsymbol{r}-\boldsymbol{r}_0), \tag{4.64}$$

where the diffusion constant D is given by

$$D = \frac{c l_e}{d}. \tag{4.65}$$

With the aid of (4.63), and since $\gamma_e = 4\pi / l_e$, we obtain a diffusion equation for the intensity as well:

$$-D \Delta I_d(\boldsymbol{r}) = \delta(\boldsymbol{r} - \boldsymbol{r}_0) \tag{4.66}$$

whose solution for free space in $d = 3$ is

$$\boxed{I_d(R) = \frac{1}{4\pi D R}} \tag{4.67}$$

with $R = |\boldsymbol{r} - \boldsymbol{r}_0|$.

Remark: transport velocity
The velocity that shows up in the diffusion coefficient D is, according to the derivation leading to relation (4.35) or equation (4.72), the group velocity of waves propagating in vacuum. It is possible to account for the effect of impurity scattering on the velocity by means of an effective medium approach. There, unlike for the white noise model, the real part of the self-energy is finite and it is related to the refraction index and thus to the group velocity. We have not used this approach because it is model dependent and less easy to handle than the white noise (see e.g. [77]). By applying this approach to resonant scattering of electromagnetic waves, it turns out that the group velocity is not well defined: it is negative at resonance and diverges close to it. This problem was recognized long ago [78] and a better defined velocity v_E has been proposed instead which describes the propagation of energy. Based on the assumption of non-correlated scatterers, it has been proposed to replace the group velocity by v_E in the diffusion coefficient [79] which then happens to be always well defined. It has also been shown subsequently that close to resonant scattering, cooperative effects between scatterers become relevant and lead to a well-defined expression of the group velocity which therefore may still be used in the expression of the diffusion coefficient [80]. In both cases, it is of interest to notice that the velocity of the electromagnetic waves is much lower than its value c in vacuum.

Appendix A4.1: Diffuson and Cooperon in reciprocal space

The Diffuson and the Cooperon are respectively solutions of (4.25) and (4.42). These solutions are easier to obtain after Fourier transform since, owing to their iterative structure, they appear as a geometric series in reciprocal space. The object of this appendix is to demonstrate this structure.

A4.1.1 Collisionless probability $P_0(\boldsymbol{q},\omega)$

First we introduce the quantity $P_0(\boldsymbol{q},\omega)$. This is the Fourier transform of the probability $P_0(\boldsymbol{r},\boldsymbol{r}',t)$ for a particle to propagate between any two points $\boldsymbol{r}$ and $\boldsymbol{r}'$ in a time t, without collision, introduced in section 4.3. The role played by P_0 is very important since it is the building block which, upon iteration, leads to the diffusion process, as seen for example in relation (4.29).

Here we calculate $P_0(\boldsymbol{q},\omega)$ directly in Fourier space. From expression (4.16), we deduce:

$$P_0(\boldsymbol{q},\omega) = \frac{1}{2\pi\rho_0\Omega}\sum_{\boldsymbol{k}} \overline{G}^R_\epsilon(\boldsymbol{k})\,\overline{G}^A_{\epsilon-\omega}(\boldsymbol{k}-\boldsymbol{q}). \tag{4.68}$$

We also define

$$\tilde{P}_0(\boldsymbol{k},\boldsymbol{q},\omega) = \frac{1}{2\pi\rho_0}\overline{G}^R_\epsilon\left(\boldsymbol{k}+\frac{\boldsymbol{q}}{2}\right)\overline{G}^A_{\epsilon-\omega}\left(\boldsymbol{k}-\frac{\boldsymbol{q}}{2}\right) \tag{4.69}$$

such that

$$P_0(\boldsymbol{q},\omega) = \frac{1}{\Omega}\sum_{\boldsymbol{k}} \tilde{P}_0(\boldsymbol{k},\boldsymbol{q},\omega). \tag{4.70}$$

Let us calculate this sum explicitly:

$$P_0(\boldsymbol{q},\omega) = \frac{1}{2\pi\rho_0\Omega}\sum_{\boldsymbol{k}} \frac{1}{\epsilon-\epsilon(\boldsymbol{k})+\frac{i}{2\tau_e}}\,\frac{1}{\epsilon-\omega-\epsilon(\boldsymbol{k}-\boldsymbol{q})-\frac{i}{2\tau_e}}. \tag{4.71}$$

We consider situations where $q \ll k$. We may thus linearize the dispersion relation $\epsilon(\boldsymbol{k}-\boldsymbol{q}) \simeq \epsilon(k) - \boldsymbol{v}\cdot\boldsymbol{q}$ where $\boldsymbol{v} = \nabla_{\boldsymbol{k}}\epsilon$ is the group velocity. Introducing the density of states $\rho_0(\epsilon)$, which we assume has no singularity about ϵ, we get:

$$\begin{aligned} P_0(\boldsymbol{q},\omega) &= \frac{1}{2\pi\rho_0}\int \frac{1}{\epsilon-\eta+\frac{i}{2\tau_e}}\,\frac{1}{\epsilon-\omega-\eta+\boldsymbol{v}\cdot\boldsymbol{q}-\frac{i}{2\tau_e}}\,\rho_0\,d\eta\,d\varpi \\ &= \tau_e\int \frac{d\varpi}{1-i\omega\tau_e+i\boldsymbol{v}\cdot\boldsymbol{q}\tau_e} \end{aligned} \tag{4.72}$$

where $\varpi = (\boldsymbol{k},\boldsymbol{q})$ is the normalized solid angle determined by the two vectors $\boldsymbol{q}$ and $\boldsymbol{k}$. In the diffusive limit $ql_e \ll 1$ and $\omega\tau_e \ll 1$, we expand the denominator, and using the relations

$\int d\varpi\, \boldsymbol{v}\cdot\boldsymbol{q} = 0$ and $\int d\varpi\, (\boldsymbol{v}\cdot\boldsymbol{q})^2 = v^2q^2/d$ (where v is the angular average quadratic velocity and d is the space dimension), we obtain

$$\boxed{P_0(\boldsymbol{q},\omega) = \tau_e(1 + i\omega\tau_e - Dq^2\tau_e + \cdots)} \tag{4.73}$$

where the diffusion constant equals $D = v^2/d\tau_e = vl_e/d$.

Lastly, we define the quantity

$$P_0(\hat{\boldsymbol{s}},\boldsymbol{q},\omega) = \frac{1}{\Omega}\sum_k \tilde{P}_0(\boldsymbol{k},\boldsymbol{q},\omega) \tag{4.74}$$

where the sum is only over the modulus of $\boldsymbol{k}$ and not over its direction, and where $\hat{\boldsymbol{s}} = \boldsymbol{k}/k$. We deduce from expression (4.72) that this function may be formulated as

$$P_0(\hat{\boldsymbol{s}},\boldsymbol{q},\omega) = \tau_e f_\omega(\hat{\boldsymbol{s}},\boldsymbol{q}), \tag{4.75}$$

where $f_\omega(\hat{\boldsymbol{s}},\boldsymbol{q})$ is equal to:

$$f_\omega(\hat{\boldsymbol{s}},\boldsymbol{q}) = \frac{1}{1 - i\omega\tau_e + i\hat{\boldsymbol{s}}\cdot\boldsymbol{q}l_e}. \tag{4.76}$$

Exercise 4.4 Show that in three dimensions,

$$P_0(\boldsymbol{q},\omega) = \frac{1}{qv}\arctan\frac{ql_e}{1 - i\omega\tau_e} \tag{4.77}$$

where $q = |\boldsymbol{q}|$. Show that in two dimensions,

$$P_0(\boldsymbol{r},\boldsymbol{r}',\omega) = \frac{e^{i\omega R/v - R/l_e}}{2\pi Rv} \tag{4.78}$$

or

$$P_0(\boldsymbol{q},\omega) = \frac{\tau_e}{\sqrt{(1 - i\omega\tau_e)^2 + q^2l_e^2}}. \tag{4.79}$$

Check the expansion (4.73) in the diffusive limit.

More generally, calculate the quantities $P_0(\boldsymbol{q},\omega), P_0(\boldsymbol{q},t), P_0(\boldsymbol{r},\omega)$ and $P_0(\boldsymbol{r},t)$, in all space dimensions and check that your results are those displayed on Table 15.1, page 549.

A4.1.2 The Diffuson

The integral equation (4.24) for Γ_ω is difficult to solve in real space and we have only been able to treat it in the limit of slow variations. In reciprocal space, it may be expressed in the form of a geometric series, which leads to a simple expression for the probability P_d

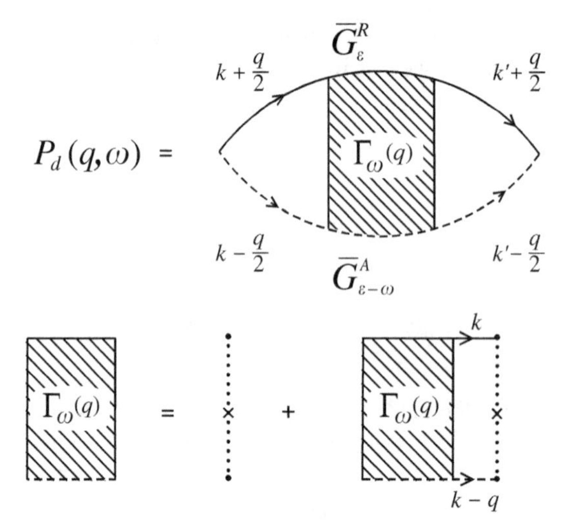

Figure 4.14 Diagrammatic representation of the probability $P_d(\boldsymbol{q},\omega)$ and the structure factor $\Gamma_\omega(\boldsymbol{q})$. The structure calculated here is clearly completely analogous to that obtained in real space in Figures 4.4 and 4.5.

valid beyond the diffusion approximation. In a translation invariant medium, the Fourier transform $\Gamma_\omega(\boldsymbol{q})$ of $\Gamma_\omega(\boldsymbol{r},\boldsymbol{r}')$, which solves (4.24), is given by

$$\Gamma_\omega(\boldsymbol{q}) = \gamma_e + \frac{\gamma_e}{\Omega}\sum_{\boldsymbol{k}} \Gamma_\omega(\boldsymbol{q})\, \overline{G}^R_\epsilon(\boldsymbol{k})\overline{G}^A_{\epsilon-\omega}(\boldsymbol{k}-\boldsymbol{q}). \tag{4.80}$$

This equation has the structure shown in Figure 4.14. In this form, $\Gamma_\omega(\boldsymbol{q})$ factorizes, and may be written

$$\Gamma_\omega(\boldsymbol{q}) = \frac{\gamma_e}{1 - P_0(\boldsymbol{q},\omega)/\tau_e}, \tag{4.81}$$

where $P_0(\boldsymbol{q},\omega)$ is given by (4.68). Moreover, (4.25) becomes

$$P_d(\boldsymbol{q},\omega) = \frac{2\pi\rho_0}{\Omega^2}\sum_{\boldsymbol{k},\boldsymbol{k}'} \tilde{P}_0(\boldsymbol{k},\boldsymbol{q},\omega)\tilde{P}_0(\boldsymbol{k}',\boldsymbol{q},\omega)\; \Gamma_\omega(\boldsymbol{q}) \tag{4.82}$$

where $\tilde{P}_0(\boldsymbol{k},\boldsymbol{q},\omega)$ is defined by (4.69). In this equation, the structure factor does not depend on $\boldsymbol{k}$ or on $\boldsymbol{k}'$. Thus, the sums factorize and give terms proportional to $P_0(\boldsymbol{q},\omega)$ yielding

$$P_d(\boldsymbol{q},\omega) = 2\pi\rho_0 P_0(\boldsymbol{q},\omega)^2\; \Gamma_\omega(\boldsymbol{q}) \tag{4.83}$$

which leads to

$$\boxed{P_d(\boldsymbol{q},\omega) = P_0(\boldsymbol{q},\omega)\frac{P_0(\boldsymbol{q},\omega)/\tau_e}{1 - P_0(\boldsymbol{q},\omega)/\tau_e}} \tag{4.84}$$

where $P_0(\boldsymbol{q}, \omega)$ has been explicitly calculated in two and three dimensions (see Exercise 4.4). The total classical probability $P = P_0 + P_d$ is given by

$$P(\boldsymbol{q}, \omega) = \frac{P_0(\boldsymbol{q}, \omega)}{1 - P_0(\boldsymbol{q}, \omega)/\tau_e}, \tag{4.85}$$

the Fourier transform of (4.29). For a translation invariant system, the expression (4.84) is valid *beyond the diffusion approximation* since no assumptions have been made about the spatial and temporal variations of P_d. We will study this solution as well as the validity of the diffusion approximation in Appendix A5.1.

Remark: normalization of the total probability P

Taking into account that $P_0(\boldsymbol{q} = 0, \omega) = \tau_e/(1 - i\omega\tau_e)$, we check that the probability $P = P_0 + P_d$ is such that

$$P(\boldsymbol{q} = 0, \omega) = \frac{i}{\omega} \tag{4.86}$$

which is the normalization condition (4.11). The normalization of the probability reflects the conservation of particle number (or of energy in the case of waves). This conservation law may also be expressed by the divergence of $P_d(\boldsymbol{q}, \omega = 0)$ for $q = 0$. This diffusion mode is known in the literature as a Goldstone mode.

From (4.81) and (4.84), and using equation (4.73) for $P_0(\boldsymbol{q}, \omega)$, we obtain the following expressions for the structure factor $\Gamma_\omega(\boldsymbol{q})$ and for the probability $P_d(\boldsymbol{q}, \omega)$ in the diffusive limit:

$$\Gamma_\omega(\boldsymbol{q}) = \frac{\gamma_e}{\tau_e} \frac{1}{-i\omega + Dq^2} \tag{4.87}$$

and

$$\boxed{P_d(\boldsymbol{q}, \omega) = \frac{1}{-i\omega + Dq^2}} \tag{4.88}$$

whose Fourier transforms are solutions of the diffusion equations (4.34) and (4.38). We further note that in the diffusive limit $P(\boldsymbol{q}, \omega) \simeq P_d(\boldsymbol{q}, \omega)$ so that the total probability is also a solution of the diffusion equation (4.38).

A4.1.3 The Cooperon

We have seen that the Diffuson $P_d(\boldsymbol{q}, \omega)$ has a simple iterative structure. We may ask whether this is the case for the Cooperon $X_c(\boldsymbol{q}, \omega)$ as well. The Fourier transform of the integral equation (4.42) gives

$$X_c(\boldsymbol{q}, \omega) = \frac{2\pi\rho_0}{\Omega^2} \sum_{\boldsymbol{k},\boldsymbol{k}'} \tilde{P}_0(\boldsymbol{k}, \boldsymbol{q}, \omega) \tilde{P}_0(\boldsymbol{k}', \boldsymbol{q}, \omega) \Gamma'_\omega(\boldsymbol{k} + \boldsymbol{k}'), \tag{4.89}$$

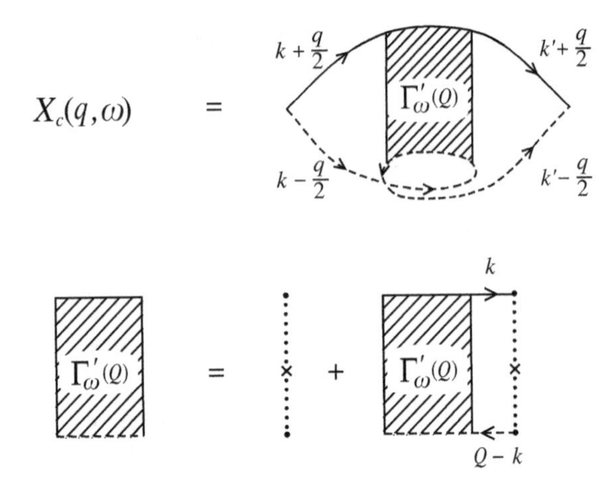

Figure 4.15 Diagrammatic representation of the Cooperon $X_c(\boldsymbol{q},\omega)$ and of the structure factor $\Gamma'_\omega(\boldsymbol{Q}=\boldsymbol{k}+\boldsymbol{k}')$ in reciprocal space. The structure deriving from these diagrams is analogous to that of Figures 4.8 and 4.9.

where $\tilde{P}_0(\boldsymbol{k},\boldsymbol{q},\omega)$ is given by (4.69). It is instructive to compare this expression with (4.82) for the Diffuson. These two quantities are of similar structure; however, for the Diffuson, the wave vector which appears in the structure factor is the *difference* between incoming wave vectors, while for the Cooperon, it is their *sum*.

The structure factor $\Gamma'_\omega(\boldsymbol{Q}=\boldsymbol{k}+\boldsymbol{k}')$ satisfies an integral equation shown schematically in Figure 4.15 and similar to that of the structure factor $\Gamma_\omega(\boldsymbol{q})$. The Fourier transform of (4.26) gives

$$\Gamma'_\omega(\boldsymbol{Q})=\gamma_e+\frac{\gamma_e}{\Omega}\sum_{\boldsymbol{k}}\Gamma'_\omega(\boldsymbol{Q})\,\overline{G}^R_\epsilon(\boldsymbol{k})\overline{G}^A_{\epsilon-\omega}(\boldsymbol{Q}-\boldsymbol{k}) \tag{4.90}$$

with $\overline{G}(\boldsymbol{Q}-\boldsymbol{k})=\overline{G}(\boldsymbol{k}-\boldsymbol{Q})$. The structure factor Γ'_ω of the Cooperon is therefore identical to that (4.80) of the Diffuson[17]

$$\Gamma'_\omega(\boldsymbol{Q})=\frac{\gamma_e}{1-P_0(\boldsymbol{Q},\omega)/\tau_e} \tag{4.92}$$

and in the diffusive limit,

$$\Gamma'_\omega(\boldsymbol{Q})=\frac{\gamma_e}{\tau_e}\frac{1}{-i\omega+DQ^2}. \tag{4.93}$$

[17] To be fully rigorous, since the first term of Γ'_ω contains two impurities, Γ'_ω equals

$$\Gamma'_\omega(\boldsymbol{Q})=\frac{\gamma_e P_0(\boldsymbol{Q},\omega)/\tau_e}{1-P_0(\boldsymbol{Q},\omega)/\tau_e}. \tag{4.91}$$

The expressions (4.91) and (4.92) are identical in the diffusive limit. See also footnote 13, page 107.

As for the Diffuson, calculating $X_c(\boldsymbol{q},\omega)$ is simple in the diffusive limit where we neglect the dependence on $\boldsymbol{q}$, $\boldsymbol{Q}$ and ω in the average Green functions. Using (4.89) and $\boldsymbol{k}\simeq -\boldsymbol{k}'$, we obtain

$$X_c(\boldsymbol{q},\omega) \simeq \frac{1}{2\pi\rho_0\Omega^2}\sum_{\boldsymbol{k}}\left[\overline{G}^R_\epsilon(\boldsymbol{k})\overline{G}^A_\epsilon(\boldsymbol{k})\right]^2\sum_{\boldsymbol{Q}}\Gamma'_\omega(\boldsymbol{Q}). \tag{4.94}$$

The sum over $\boldsymbol{k}$ gives $4\pi\rho_0\tau_e^3$ (see Table 3.2 and relation 3.107). Using (4.93), we obtain

$$\boxed{X_c(\boldsymbol{q},\omega) = \frac{\tau_e}{\pi\rho_0\Omega}\sum_{\boldsymbol{Q}}\frac{1}{-i\omega+DQ^2}} \tag{4.95}$$

A comparison of (4.95) with the probability $P_d(\boldsymbol{q},\omega)$ given by (4.88) suggests two remarks.

- In the diffusion approximation $X_c(\boldsymbol{q},\omega)$ does not depend on $\boldsymbol{q}$. This reflects the fact that the Cooperon is localized in real space. Equation (4.95) thus appears to be the Fourier transform of (4.53).
- In expression (4.82) for the Diffuson, the vectors $\boldsymbol{k}$ and $\boldsymbol{k}'$ both contribute. In other words, a wave packet injected in direction $\boldsymbol{k}$ may exit, after scattering, in an arbitrary direction $\boldsymbol{k}'$. On the other hand, in expression (4.89), the function $\Gamma'_\omega(\boldsymbol{Q}=\boldsymbol{k}+\boldsymbol{k}')$ is peaked around $\boldsymbol{Q}\simeq 0$, and only the vectors $\boldsymbol{k}'$ near to $-\boldsymbol{k}$ contribute to the Cooperon. We may thus associate an angular width $\Delta\theta \simeq Q/k$ of order $1/kl_e$ to the function Γ'_ω. In the limit of weak disorder, although this contribution is small, it is nonetheless directly measurable by means of the optical experiments mentioned in section 1.4, which will be discussed in detail in Chapter 8.

Important remark: $X_c(\boldsymbol{q},\omega)$ and $P_c(\boldsymbol{Q},\omega)$
The Cooperon $X_c(\boldsymbol{r},\boldsymbol{r}',\omega)$ is a function which is peaked about $\boldsymbol{r}'\simeq\boldsymbol{r}$. This is reflected in the fact that $X_c(\boldsymbol{q},\omega)$ is almost independent of $\boldsymbol{q}$ and does not satisfy a diffusion equation. In contrast, the Fourier transform $P_c(\boldsymbol{Q},\omega)$ of $P_c(\boldsymbol{r},\boldsymbol{r}',\omega)$ defined by relation (4.48) is proportional to the structure factor $\Gamma'_\omega(\boldsymbol{Q})$:

$$P_c(\boldsymbol{Q},\omega) = 2\pi\rho_0\tau_e^2\,\Gamma'_\omega(\boldsymbol{Q}) \tag{4.96}$$

and is a solution of a diffusion equation. Moreover, $X_c(\boldsymbol{q},\omega)$ is given by

$$X_c(\boldsymbol{q},\omega) = \frac{1}{\Omega}\sum_{\boldsymbol{Q}} a(\boldsymbol{q}-\boldsymbol{Q})P_c(\boldsymbol{Q},\omega), \tag{4.97}$$

where $a(\boldsymbol{q})$ is the Fourier transform of $g^2(\boldsymbol{r})$. Replacing $g^2(\boldsymbol{r})$ by a δ function in (4.53) amounts to taking $a(\boldsymbol{q})$ constant, leading to (4.95).

Remark: Langer–Neal diagrams

The structure factor $\Gamma'_\omega(\boldsymbol{Q})$ obtained by summing the series represented in Figure 4.15 depends on the momentum $\boldsymbol{Q} = \boldsymbol{k} + \boldsymbol{k}'$. The importance of maximally crossed diagrams in the study of disordered media was first discussed by Langer and Neal [81]. The motivation for their work came from the observation that, in contrast to equilibrium quantities, transport properties of a classical gas do not have a simple analytic expansion in the density n_i of impurities. They addressed the question whether this well-known classical result (for example, for the Lorentz gas [82]) also holds in the quantum case. To this end, they considered the resistivity of a degenerate free electron gas. They noted that there is a class of diagrams, the maximally crossed diagrams, which, for a given order $s \geq 3$ in the expansion, produces terms of the form $n_i^s \ln n_i$ where n_i is the density of scattering centers. These non-analytic contributions are analogous to those of the classical case.

Langer and Neal considered only the contributions of maximally crossed diagrams of a given order, without attempting to sum them. They showed that each term of the geometric series whose sum is (4.92), yields, after summing over $\boldsymbol{k}$ and $\boldsymbol{k}'$, a non-analytic contribution in $\gamma_e^s \ln \gamma_e$. For the series represented in Figure 4.15, it is thus preferable first to sum the crossed diagrams and then to sum over $\boldsymbol{Q} = \boldsymbol{k} + \boldsymbol{k}'$ to obtain the Cooperon contribution to the probability (as well as for the electrical conductivity, as we shall see in detail in section 7.3). This now standard calculation of the Cooperon was subsequently used to treat the anomalous behavior of the diffusivity in a quantum Lorentz gas [82].

Appendix A4.2: Hikami boxes and Diffuson crossings

A4.2.1 Hikami boxes

The notion of quantum crossing of Diffusons or Cooperons is essential for the understanding of interference effects in diffusive systems. These crossings are described by specific combinations of average Green's functions, combinations known in the literature as a *Hikami box* [83, 84].[18] Let us consider the Hikami box, denoted by H, constructed from four average Green's functions, alternately advanced and retarded (Figure 4.16). H is a real function of four arguments, $H(\{\boldsymbol{r}_i\}) = H(\boldsymbol{r}_1, \boldsymbol{r}_2, \boldsymbol{r}_3, \boldsymbol{r}_4)$. It is the sum of three diagrams,

$$H(\{\boldsymbol{r}_i\}) = H^{(A)}(\{\boldsymbol{r}_i\}) + H^{(B)}(\{\boldsymbol{r}_i\}) + H^{(C)}(\{\boldsymbol{r}_i\}) \tag{4.98}$$

which we shall show forms a particular combination. First, note the symmetry

$$H^{(C)}(\boldsymbol{r}_1, \boldsymbol{r}_2, \boldsymbol{r}_3, \boldsymbol{r}_4) = H^{(B)}(\boldsymbol{r}_2, \boldsymbol{r}_3, \boldsymbol{r}_4, \boldsymbol{r}_1). \tag{4.99}$$

[18] This is the currently used term, however they were also introduced in reference [83].

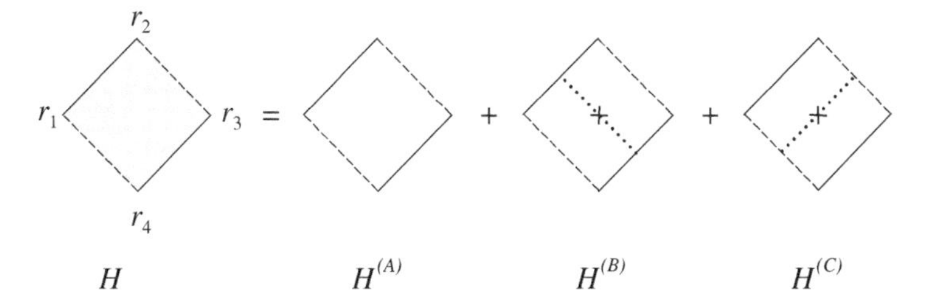

Figure 4.16 Hikami box $H(\boldsymbol{r}_1, \boldsymbol{r}_2, \boldsymbol{r}_3, \boldsymbol{r}_4)$. To obtain the function $H(\boldsymbol{r} - \boldsymbol{r}')$, we integrate over two opposite points. The continuous and the dashed lines are Green's functions. The dotted line is an impurity line.

Integrating over two of the variables $\boldsymbol{r}_i$ (opposite in Figure 4.16), we define the function

$$H(\boldsymbol{r} - \boldsymbol{r}') = \int d\boldsymbol{r}_2\, d\boldsymbol{r}_4\, H(\boldsymbol{r}, \boldsymbol{r}_2, \boldsymbol{r}', \boldsymbol{r}_4) \tag{4.100}$$

and, upon integrating over the third variable, we obtain

$$H = \int d\boldsymbol{r}_2\, d\boldsymbol{r}_3\, d\boldsymbol{r}_4\, H(\boldsymbol{r}, \boldsymbol{r}_2, \boldsymbol{r}_3, \boldsymbol{r}_4). \tag{4.101}$$

Due to translation invariance, the first term depends only on $\boldsymbol{r} - \boldsymbol{r}'$ and the second is constant. In particular,[19] from Table 3.2, we obtain the constants:

$$H^{(A)} = \frac{1}{\gamma_e} f^{2,2}(0) = \begin{cases} 4\pi\rho_0\tau_e^3 & \textit{Schrödinger} \\ \dfrac{l_e^3}{8\pi k^2} & \textit{Helmholtz} \end{cases} \tag{4.102}$$

and

$$H^{(B)} = H^{(C)} = \frac{1}{\gamma_e} \left[f^{1,2}(0) \right]^2 = \begin{cases} -2\pi\rho_0\tau_e^3 & \textit{Schrödinger} \\ -\dfrac{l_e^3}{16\pi k^2} & \textit{Helmholtz.} \end{cases} \tag{4.103}$$

We also note that

$$H = \int H(\boldsymbol{r})\, d\boldsymbol{r} = H^{(A)} + H^{(B)} + H^{(C)} = 0. \tag{4.104}$$

This is a remarkable property. We will show in the following section that it corresponds to the normalization of the probability of quantum diffusion, that is to the conservation of the particle number.

The spatial dependence of $H^{(A)}(\boldsymbol{R})$ is given by

$$H^{(A)}(\boldsymbol{R}) = \frac{1}{\gamma_e^2} |f^{1,1}(\boldsymbol{R})|^2 = \frac{1}{\gamma_e^2} g^2(\boldsymbol{R}). \tag{4.105}$$

[19] The expressions given for waves are valid in three dimensions, and hold equally for electrons. On the other hand, those expressions for electrons which involve the density of states ρ_0 are valid in any dimension. See also footnote 5, page 97.

The functions $H^{(B)}(\boldsymbol{R})$ and $H^{(C)}(\boldsymbol{R})$ are more complicated, and are calculated in Exercise 4.5.

Exercise 4.5 Show that the Fourier transform $H(\boldsymbol{q})$ of $H(\boldsymbol{r})$ satisfies

$$H(\boldsymbol{q}) = \frac{1}{\Omega} H(-\boldsymbol{q}, 0, \boldsymbol{q}, 0) \tag{4.106}$$

where $H(\{\boldsymbol{q}_i\})$ is the Fourier transform of $H(\{\boldsymbol{r}_i\})$. In dimension $d = 3$, show that

$$\begin{aligned} H^{(A)}(\boldsymbol{q}) &= \frac{1}{\gamma_e^2} a(\boldsymbol{q}) \\ H^{(B)}(\boldsymbol{q}) = H^{(C)}(\boldsymbol{q}) &= -\frac{1}{4\gamma_e^3 \tau_e^2} a(\boldsymbol{q})^2, \end{aligned} \tag{4.107}$$

where the function $a(\boldsymbol{q})$ is the Fourier transform of $g^2(\boldsymbol{r})$ given by (3.98). It depends only on $q = |\boldsymbol{q}|$ and is equal to

$$a(q) = \frac{\pi}{k^2 q} \left[\arctan(2k - q)l_e + 2\arctan q l_e - \arctan(2k + q) l_e\right]. \tag{4.108}$$

In particular, for $kl_e \gg 1$ and $ql_e \gg 1$, $a(q) \simeq (\pi^2/k^2 q)\theta(2k - q)$.
Show that for $q \to 0$,

$$a(q) = 2\gamma_e \tau_e^2 \left(1 - \frac{q^2 l_e^2}{3}\right) = \frac{2\pi l_e}{k^2} \left(1 - \frac{q^2 l_e^2}{3}\right). \tag{4.109}$$

Show that the function $H(\boldsymbol{q})$ is given by

$$H(\boldsymbol{q}) = \frac{1}{\gamma_e^2}\left[a(q) - \frac{1}{2\gamma_e \tau_e^2} a(q)^2\right] \simeq \frac{1}{\gamma_e^2}[a(0) - a(q)] \tag{4.110}$$

and that when $q \to 0$, it may be expanded as

$$H(\boldsymbol{q}) = \frac{l_e^5}{24\pi k^2} q^2 + O(q^4) \tag{4.111}$$

such that

$$H(\boldsymbol{q})/H^{(A)}(\boldsymbol{q}) \simeq \frac{1}{3} q^2 l_e^2 = D q^2 \tau_e. \tag{4.112}$$

Show that for $R \to \infty$, we have

$$\begin{aligned} H^{(A)}(\boldsymbol{R}) &\propto \frac{1}{R^2} e^{-R/l_e} \\ H^{(B)}(\boldsymbol{R}) = H^{(C)}(\boldsymbol{R}) &\propto -\frac{1}{R l_e} e^{-R/l_e} \end{aligned} \tag{4.113}$$

and if $R \ll l_e$

$$H^{(B)}(\boldsymbol{r}) \simeq -\frac{1}{\gamma_e^2}\frac{\pi}{16k^2 l_e}\frac{\mathrm{Si}(2kR)}{R}, \tag{4.114}$$

where Si is the sine integral function.
Deduce that

$$H(\boldsymbol{R}=0) \simeq \frac{1}{\gamma_e^2}\left(1-\frac{\pi}{4kl_e}\right). \tag{4.115}$$

Exercise 4.6 Check that the diagram of Figure 4.17 is of order $1/kl_e$ with respect to that of Figure 4.16.

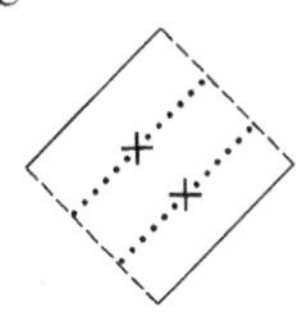

Figure 4.17 Higher order Hikami box.

Due to the two additional Green functions $\overline{G}^A$, this diagram is equal to:

$$H^{(B')} = H^{(B)}\frac{1}{2\pi\rho_0\tau_e}\frac{1}{\Omega}\sum_{\boldsymbol{k}''}\overline{G}^A_\epsilon(\boldsymbol{k}'')\overline{G}^A_\epsilon(\boldsymbol{k}'') \simeq H^{(B)}/kl_e, \tag{4.116}$$

since the sum is carried over the product of two advanced Green's functions (see Exercise 3.15). In this way, deduce a systematic rule for the value of such a diagram as a function of the number n of impurity lines, for $n \geq 2$.

The complete calculation of $H(\{\boldsymbol{r}_i\})$ is difficult. Here we consider only its Fourier transform $H(\{\boldsymbol{q}_i\})$ in the limit $\boldsymbol{q}_i \to 0$. The contribution $H^{(A)}$ of Figure 4.16, detailed in Figure 4.18, is given by

$$H^{(A)}(\{\boldsymbol{q}_i\}) = \delta_{\sum \boldsymbol{q}_i,0}\sum_{\boldsymbol{k}}\overline{G}^R_\epsilon(\boldsymbol{k})\overline{G}^A_\epsilon(\boldsymbol{k}+\boldsymbol{q}_2)\overline{G}^R_\epsilon(\boldsymbol{k}+\boldsymbol{q}_2+\boldsymbol{q}_3)\overline{G}^A_\epsilon(\boldsymbol{k}-\boldsymbol{q}_1) \tag{4.117}$$

where $\delta_{\sum \boldsymbol{q}_i,0}$ is the Kronecker symbol. The contribution $H^{(B)}(\{\boldsymbol{q}_i\})$ contains an additional impurity line and is given by (Figure 4.18)

$$H^{(B)}(\{\boldsymbol{q}_i\}) = \gamma_e\,\delta_{\sum \boldsymbol{q}_i,0}\sum_{\boldsymbol{k}}\overline{G}^R_\epsilon(\boldsymbol{k}-\boldsymbol{q}_4)\overline{G}^A_\epsilon(\boldsymbol{k})\overline{G}^R_\epsilon(\boldsymbol{k}+\boldsymbol{q}_1)$$
$$\times\frac{1}{\Omega}\sum_{\boldsymbol{k}'}\overline{G}^R_\epsilon(\boldsymbol{k}'-\boldsymbol{q}_2)\overline{G}^A_\epsilon(\boldsymbol{k}')\overline{G}^R_\epsilon(\boldsymbol{k}'+\boldsymbol{q}_3) \tag{4.118}$$

and from (4.99), we have $H^{(C)}(\boldsymbol{q}_1,\boldsymbol{q}_2,\boldsymbol{q}_3,\boldsymbol{q}_4) = H^{(B)}(\boldsymbol{q}_2,\boldsymbol{q}_3,\boldsymbol{q}_4,\boldsymbol{q}_1)$. If all the wave vectors $\boldsymbol{q}_i$ are zero, H is zero and we recover (4.104). We therefore expand in small $\boldsymbol{q}_i$, which corresponds to the diffusive regime (section 4.5). Taking equations (3.68) and (3.69) into

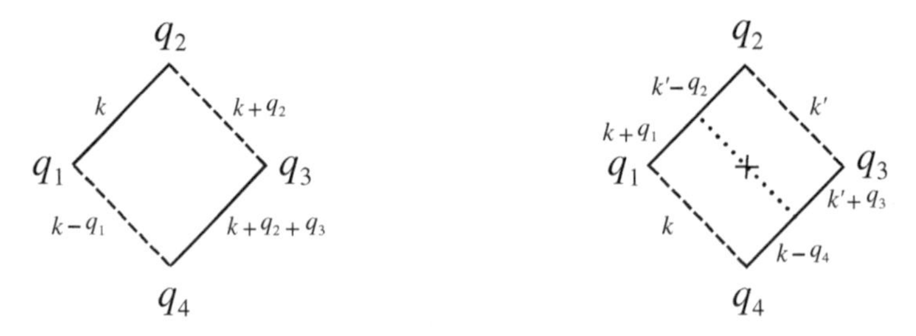

Figure 4.18 Parametrization of Green's functions used in the calculation of $H^{(A)}(\{\boldsymbol{q}_i\})$ and $H^{(B)}(\{\boldsymbol{q}_i\})$.

account, each Green's function is expanded in the following way. For electrons,[20]

$$\overline{G}(\boldsymbol{k}+\boldsymbol{q}) = \overline{G}(\boldsymbol{k}) + \boldsymbol{v}\cdot\boldsymbol{q}\,\overline{G}(\boldsymbol{k})^2 + (\boldsymbol{v}\cdot\boldsymbol{q})^2\,\overline{G}(\boldsymbol{k})^3 \tag{4.119}$$

with $\boldsymbol{v} = \boldsymbol{k}/m$. Inserting this expansion into expression (4.118), the terms linear in $\boldsymbol{q}$ vanish by symmetry after summing over $\boldsymbol{k}$ and $\boldsymbol{k}'$. We are left with

$$H^{(B)}(\{\boldsymbol{q}_i\}) = \frac{\Omega\delta_{\sum \boldsymbol{q}_i,0}}{\gamma_e}\Big[f^{2,1} + \frac{v^2}{3}(q_1^2+q_4^2-\boldsymbol{q}_1\cdot\boldsymbol{q}_4)f^{4,1}\Big] \times \Big[f^{2,1} + \frac{v^2}{3}(q_2^2+q_3^2-\boldsymbol{q}_2\cdot\boldsymbol{q}_3)f^{4,1}\Big] \tag{4.120}$$

where $v = \sqrt{2\epsilon/m}$ is the velocity associated with the energy ϵ. The coefficients $f^{m,n}$ are given by equations (3.108, 3.109) and Table 3.2. Keeping only terms of order q^2, we obtain

$$H^{(B)}(\{\boldsymbol{q}_i\}) = \frac{\tau_e^2}{\gamma_e}\Big[-1 + \frac{l_e^2}{3}\Big(q_1^2+q_2^2+q_3^2+q_4^2-\boldsymbol{q}_1\cdot\boldsymbol{q}_4-\boldsymbol{q}_2\cdot\boldsymbol{q}_3\Big)\Big]\Omega\delta_{\sum \boldsymbol{q}_i,0}. \tag{4.121}$$

Similarly, the diagram $H^{(C)}$ is equal to

$$H^{(C)}(\{\boldsymbol{q}_i\}) = \frac{\tau_e^2}{\gamma_e}\Big[-1 + \frac{l_e^2}{3}\Big(q_1^2+q_2^2+q_3^2+q_4^2-\boldsymbol{q}_1\cdot\boldsymbol{q}_2-\boldsymbol{q}_3\cdot\boldsymbol{q}_4\Big)\Big]\Omega\delta_{\sum \boldsymbol{q}_i,0}. \tag{4.122}$$

Finally, for $H^{(A)}$, equation (4.117) leads to

$$H^{(A)}(\{\boldsymbol{q}_i\}) = 2\frac{\tau_e^2}{\gamma_e}\Big[1 - \frac{l_e^2}{3}\Big(q_1^2+q_2^2+q_3^2+q_4^2+\boldsymbol{q}_1\cdot\boldsymbol{q}_3+\boldsymbol{q}_2\cdot\boldsymbol{q}_4\Big)\Big]\Omega\delta_{\sum \boldsymbol{q}_i,0}. \tag{4.123}$$

Upon summing these three contributions, the constant terms vanish. Using the constraint $\sum_i \boldsymbol{q}_i = 0$ and the relation

$$\Big(\sum_i \boldsymbol{q}_i\Big)^2 = \sum_i q_i^2 + 2\sum_{i<j}\boldsymbol{q}_i\cdot\boldsymbol{q}_j = 0, \tag{4.124}$$

[20] In the expansions (4.119) and (4.126), there is also a term of order $q^2\overline{G}(\boldsymbol{k})^2$ which gives a much weaker contribution to H, of order $1/kl_e$.

the sum $H(\{\boldsymbol{q}_i\})$ of these three contributions yields the small q expansion:

$$H(\{\boldsymbol{q}_i\}) = \Omega h_4 \delta_{\sum \boldsymbol{q}_i,0}\Big[\sum_i q_i^2 + \boldsymbol{q}_1\cdot\boldsymbol{q}_2 + \boldsymbol{q}_2\cdot\boldsymbol{q}_3 + \boldsymbol{q}_3\cdot\boldsymbol{q}_4 + \boldsymbol{q}_4\cdot\boldsymbol{q}_1\Big] \tag{4.125}$$

where $h_4 = l_e^2\tau_e^2/(3\gamma_e)$. The same calculation may be performed for waves[21] and we obtain the same expression (4.125) with

$$h_4 = 2\pi\rho_0 D\tau_e^4 \quad \text{for electrons,} \quad \text{and} \quad h_4 = \frac{l_e^5}{48\pi k^2} \quad \text{for waves} \tag{4.127}$$

We may obtain different expressions equivalent to (4.125) using the constraint $\sum_i \boldsymbol{q}_i = 0$ and relation (4.124).

Exercise 4.7: higher order Hikami boxes

The condition $\sum_i \boldsymbol{q}_i = 0$ and the relation (4.124) allow us to write the expression for Hikami boxes in equivalent forms. Show in particular that

$$H(\boldsymbol{q}_i) = \Omega h_4 \delta_{\sum \boldsymbol{q}_i,0}\Big[-\boldsymbol{q}_1\cdot\boldsymbol{q}_3 - \boldsymbol{q}_2\cdot\boldsymbol{q}_4 + \frac{1}{2}\sum_i q_i^2\Big]. \tag{4.128}$$

Identify the diagrams which arise in the calculation of a *hexagonal* Hikami box (denoted by H_6) and, with the aid of references [84, 85], show that

$$H_6(\boldsymbol{q}_i) = -4\pi\rho_0 D\tau_e^6\, \Omega\delta_{\sum \boldsymbol{q}_i,0}\Big[\sum_i q_i^2 + \boldsymbol{q}_1\cdot\boldsymbol{q}_2 + \boldsymbol{q}_2\cdot\boldsymbol{q}_3 + \boldsymbol{q}_3\cdot\boldsymbol{q}_4 + \boldsymbol{q}_4\cdot\boldsymbol{q}_5 + \boldsymbol{q}_5\cdot\boldsymbol{q}_6 + \boldsymbol{q}_6\cdot\boldsymbol{q}_1\Big]. \tag{4.129}$$

A4.2.2 *Normalization of the probability and renormalization of the diffusion coefficient*

Does the Cooperon contribution to the probability of quantum diffusion modify the normalization of this probability? The doubling of the probability in a volume $\tau_e/(\pi\rho_0)$ about $\boldsymbol{r}' = \boldsymbol{r}$ appears to violate the normalization, the deviation being small, of order $1/(kl_e)^{d-1}$ (see Figure 4.10).

We show here that there are other corrections which preserve the normalization of the probability. Indeed, the diagram representing the Cooperon $X_c(\boldsymbol{r}, \boldsymbol{r}', \omega)$ has the form illustrated in Figure 4.8, or the topologically equivalent form represented in Figure 4.19(a) in Fourier transform. Now, we have seen in section A4.2.1 and in Figure 4.16 that this diagram

[21] For waves, the expansion of Green's functions rewrites

$$\overline{G}(\boldsymbol{k}+\boldsymbol{q}) = \overline{G}(\boldsymbol{k}) + 2\boldsymbol{k}\cdot\boldsymbol{q}\,\overline{G}(\boldsymbol{k})^2 + 4(\boldsymbol{k}\cdot\boldsymbol{q})^2\,\overline{G}(\boldsymbol{k})^3 \tag{4.126}$$

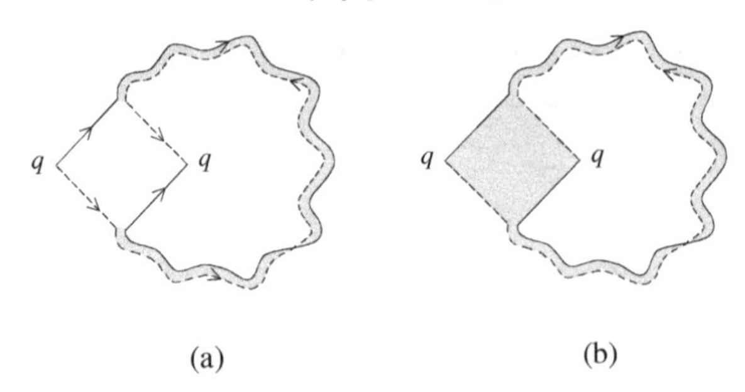

Figure 4.19 Contribution of the Cooperon to the probability. (a) Diagram for the contribution $X_c(\boldsymbol{q},\omega)$. (b) Dressing the Hikami box with impurity lines (Figure 4.16) gives contributions of the same order which must be taken into account. This leads to replacing $X_c(\boldsymbol{q},\omega)$ by $I_c(\boldsymbol{q},\omega) = Dq^2\tau_e X_c(\boldsymbol{q},\omega)$.

is the first of a group of three diagrams defining a Hikami box and giving contributions of the same order. To ensure the normalization of the probability, we must thus consider the contributions of these *three diagrams* (Figure 4.19(b)). The first, calculated in section 4.6, gives

$$X_c(\boldsymbol{r},\boldsymbol{r}',\omega) = \frac{1}{2\pi\rho_0} H^{(A)}(\boldsymbol{r}-\boldsymbol{r}')\Gamma'(\boldsymbol{r},\boldsymbol{r}',\omega) \tag{4.130}$$

where $H^{(A)}(\boldsymbol{R}) = g^2(\boldsymbol{R})/\gamma_e^2$. The other two diagrams involve *negative* quantities (Exercise 4.5 and relation 4.114):

$$H^{(B)}(\boldsymbol{R}) = H^{(C)}(\boldsymbol{R}) \propto -\frac{1}{\gamma_e^2}\frac{1}{kl_e}\frac{\mathrm{Si}(2kR)}{kR}e^{-R/l_e} \tag{4.131}$$

where Si is the sine integral function. The contributions of these two diagrams are negligible when $R \to 0$ and *the probability of return to the origin remains doubled.* On the other hand, (4.104) implies that the integral of the sum of the three contributions compensates exactly, that is

$$H(\boldsymbol{q}=0) = \int H(\boldsymbol{R})\, d\boldsymbol{R} = 0. \tag{4.132}$$

Thus, the Cooperon X_c, which doubles the probability of return to the origin, does not violate the normalization condition of the total probability which is ensured by a small reduction away from the origin (but at distances of order l_e).

We see, therefore, that it is the "dressing" of the Hikami box which restores the normalization of the probability (that is, particle number conservation). This becomes clear in reciprocal space where the normalization of the probability is expressed by condition (4.11),

$$P(\boldsymbol{q}=0,\omega) = \frac{i}{\omega}, \tag{4.133}$$

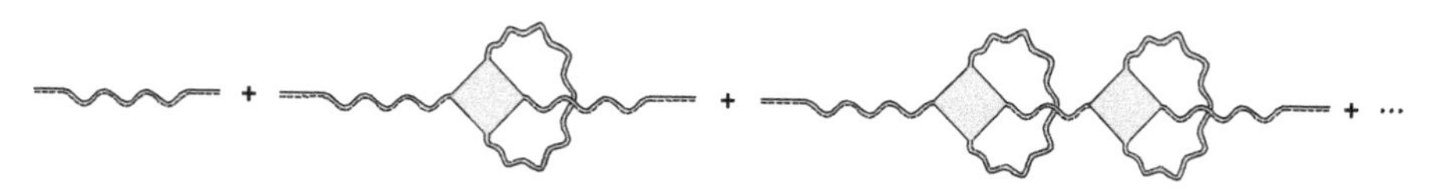

Figure 4.20 Contributions to the total probability $P(\boldsymbol{q},\omega)$. The first term, with a single Diffuson, is $P_d(\boldsymbol{q},\omega)$. The perturbation series illustrated in this figure leads to an expression $P'_d(\boldsymbol{q},\omega)$ for the probability where the diffusion coefficient is modified and given by (4.138).

and any additional contribution must vanish for $\boldsymbol{q}=0$. The diagram in Figure 4.19(a) stands for (4.95). Its contribution doubles the probability of return to the origin but, since it is constant for small $\boldsymbol{q}$, the probability is no longer normalized. However, upon dressing the Hikami box, we replace this diagram by that of Figure 4.19(b) which is equal to

$$I_c(\boldsymbol{q},\omega) = \frac{D\tau_e^2 q^2}{\pi\rho_0\Omega}\sum_Q \frac{1}{-i\omega + DQ^2} = Dq^2\tau_e X_c(\boldsymbol{q},\omega), \tag{4.134}$$

a result obtained using (4.112). We see therefore that dressing a Hikami box multiplies it by $Dq^2\tau_e$. This contribution tends to zero when $q \to 0$ and the probability is properly normalized.

One may now ask whether we should not add other contributions to the classical probability P_d, by inserting more Hikami boxes (and thus Cooperons). The resulting series constituted by the diagrams in Figure 4.20 involves more and more divergent terms since they contain successive powers of the Diffuson. This series may be written

$$P'_d(\boldsymbol{q},\omega) = P_d(\boldsymbol{q},\omega) + P_d^2(\boldsymbol{q},\omega)\frac{I_c(\boldsymbol{q},\omega)}{\tau_e^2} + P_d^3(\boldsymbol{q},\omega)\left[\frac{I_c(\boldsymbol{q},\omega)}{\tau_e^2}\right]^2 + \cdots. \tag{4.135}$$

After summation, we obtain

$$P'_d(\boldsymbol{q},\omega) = \frac{P_d(\boldsymbol{q},\omega)}{1 - \frac{P_d(\boldsymbol{q},\omega)I_c(\boldsymbol{q},\omega)}{\tau_e^2}}. \tag{4.136}$$

We first note that the equality $I_c(\boldsymbol{q}=0,\omega)=0$ implies the normalization of $P'_d(\boldsymbol{q},\omega)$. Moreover, in the limit of small wave vectors, this *renormalized probability* retains a diffusion pole

$$P'_d(\boldsymbol{q},\omega) = \frac{1}{-i\omega + D'q^2} \tag{4.137}$$

but its diffusion coefficient also becomes renormalized:

$$D' = D\left[1 - \frac{X_c(\boldsymbol{q},\omega)}{\tau_e}\right] = D\left[1 - \frac{\Delta}{\pi}\sum_Q \frac{1}{-i\omega + DQ^2}\right]. \tag{4.138}$$

Thus, taking the Cooperon into account to all orders in perturbation theory leads to a reduction in the diffusion coefficient. This phenomenon, called *weak localization*, also gives a similar reduction in the conductivity (see section 7.4).

The renormalization (4.138) of the diffusion coefficient corresponds to the regime of weak disorder $kl_e \gg 1$. We have thus far considered only one class of diagrams, and there is no reason a priori to restrict oneself to this class. For example, the Cooperons of Figure 4.20 may themselves be dressed by Diffusons. Vollhardt and Wölfle [86] were thus led to write a self-consistent equation for the renormalized diffusion coefficient $D(\omega)$ assumed to a be a function of ω only [87]

$$D(\omega) = D - \frac{\Delta}{\pi} \sum_Q \frac{D(\omega)}{-i\omega + D(\omega)Q^2} \cdot \tag{4.139}$$

The solution of this equation permits us to describe the transition to the localized regime for which $kl_e \simeq 1$.

A4.2.3 Crossing of two Diffusons

The Diffuson is a classical contribution to the probability, that is, it does not depend on the phases of the complex amplitudes from which it is constructed. In the diffusive regime, it obeys a diffusion equation.

This notwithstanding, there do appear quantum or wave-like effects associated with these complex amplitudes, owing to the crossing of two Diffusons (or the crossing of a Diffuson with itself). As shown in Figure 4.21, such a crossing is described by a Hikami box. This notion of Diffuson crossing is of particular importance. Indeed, since the Diffusons are classical objects, interference effects can only arise from the existence of such "quantum crossings," which involve Hikami boxes.[22] These combined properties of Diffusons, Cooperons and crossings are at the origin of coherent effects such as weak localization, light intensity correlation functions and conductance fluctuations. It is thus important to develop good intuition about these quantum crossings. Figure 4.21 shows that a crossing mixes the four complex amplitudes of the incident Diffusons and pairs them differently. The two emerging Diffusons are thus formed by two amplitudes deriving respectively from each of the incident Diffusons. Since the Diffuson is a long range object, coherent effects may be propagated over a long distance. A Hikami box appears as an object whose function is to permute the amplitudes.

We are already able to evaluate qualitatively the probability of such a crossing. We have seen in section 4.6 that the volume associated with the crossing of two multiple scattering trajectories is $\lambda^{d-1} l_e$ (Figure 4.11). We may understand this result by considering that a Diffuson propagating for a time t may be thought of as an object of length $\mathcal{L} = vt$ and cross section λ^{d-1}, where v and λ are the group velocity and the wavelength of the corresponding wave. We may thus associate a finite volume $\lambda^{d-1} vt$ with it. The probability for crossing of

[22] Care must be taken not to confuse these "quantum" crossings with the crossings of a classical random walk.

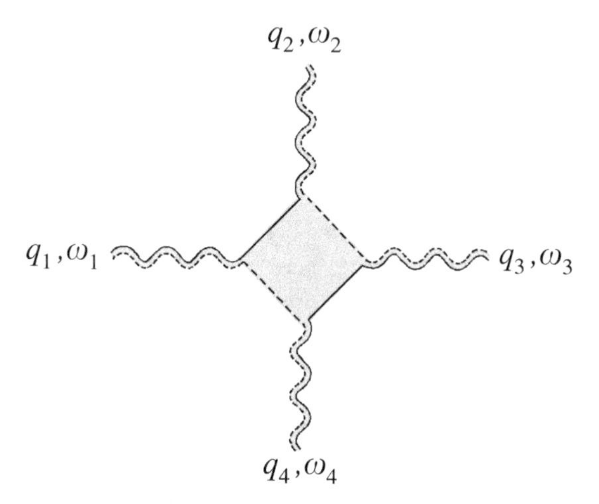

Figure 4.21 The crossing of two Diffusons appears as a set of four Diffusons connected by a Hikami box.

two Diffusons after a time t, in a disordered medium of volume $\Omega = L^d$, is thus proportional to the ratio between the volume of a Diffuson and the volume Ω. Thus, the probability that two Diffusons cross, or that one Diffuson crosses itself, is proportional to[23]

$$p(t) = \frac{\lambda^{d-1} v t}{\Omega} \propto \frac{t}{\rho_0 \hbar \Omega} \tag{4.140}$$

• Consider first an open system, for example a conductor in contact with reservoirs, or in optics, a scattering medium with no reflective boundary. In the diffusive regime, the time required to cross the sample is equal to $\tau_D = L^2/D$ (see section 5.5.1). The probability of having a Diffuson crossing during the time τ_D is thus given by

$$p(\tau_D) = \frac{\tau_D}{\rho_0 \hbar \Omega} = \frac{L^2}{\rho_0 \hbar D \Omega} \simeq \frac{1}{g}. \tag{4.141}$$

This probability is proportional to the inverse of a dimensionless number g, which plays an important role in the description of coherent effects. For the case of electrons in metals, it corresponds to the average dimensionless conductance $g = \overline{G}/(e^2/h)$, where $\overline{G}$ is given by $\overline{G} = 2e^2 D \rho_0 L^{d-2}$ (see Chapter 7).

In the limit of weak disorder $kl_e \gg 1$, g is large, and coherent effects are small. We will study these coherent effects perturbatively in the small parameter $1/g$, which will allow us to characterize systematically the importance of these effects as a function of the number of quantum crossings.

[23] Recall that the density of states in dimension d is proportional to

$$\rho_0 \propto \frac{1}{\hbar v \lambda^{d-1}}.$$

• Consider now an isolated system. The probability $p(t)$ increases linearly and is no longer limited by the time τ_D. The energy scale $\Delta = 1/(\rho_0 \Omega)$ represents, for a finite size system, the average spacing between eigenenergies of the wave equation. We may thus write

$$p(t) \propto \frac{t}{\rho_0 \hbar \Omega} = \frac{\Delta t}{\hbar} \tag{4.142}$$

and introduce the so-called Heisenberg time defined by $\tau_H = h/\Delta$ for which the crossing probability is of order 1. Beyond this time, there are quantum corrections to the probability. An isolated scattering medium can therefore be described classically, that is, in terms of diffusion, only for times less than the Heisenberg time [75]. Similarly, the spectral properties of an isolated system cannot be described classically for energies smaller than Δ (Chapter 10).

Quantitative description of two-Diffuson crossings

The crossing of two Diffusons is described by a Hikami box as calculated in section A4.2.1. Consider the case represented in Figure 4.21 where four Diffusons are connected to a box.[24] The Diffusons may be taken to be at finite frequencies ω_i. In Fourier transform, each Diffuson is, in the diffusive regime, a solution of the diffusion equation (4.88)

$$(-i\omega_i + Dq_i^2)P(\boldsymbol{q}_i, \omega_i) = 1. \tag{4.143}$$

The Hikami box represented in Figure 4.22 connects four Diffusons, each characterized by $\boldsymbol{q}_i$ and ω_i together, with the conditions $\sum_i \boldsymbol{q}_i = 0$ and $\omega_1 + \omega_3 = \omega_2 + \omega_4$.

The calculation of this box is identical to that performed for the zero frequency case in section A4.2.1, but the average Green functions (3.88) now depend on frequency. Thus, for electrons, the Green function

$$\overline{G}^{R,A}_{\epsilon-\omega}(\boldsymbol{k}) = \frac{1}{\epsilon - \omega - \epsilon(\boldsymbol{k}) \pm \frac{i}{2\tau_e}} \tag{4.144}$$

may be expanded as

$$\overline{G}_{\epsilon-\omega}(\boldsymbol{k}+\boldsymbol{q}) = \overline{G}(\boldsymbol{k}) + (\boldsymbol{v}\cdot\boldsymbol{q} + \omega)\ \overline{G}(\boldsymbol{k})^2 + (\boldsymbol{v}\cdot\boldsymbol{q})^2\, \overline{G}(\boldsymbol{k})^3 + \cdots . \tag{4.145}$$

Inserting this expansion into expressions (4.117) and (4.118) for $H^{(A)}$ and $H^{(B)}$, keeping track of the frequencies ω_i carried by the different Green functions, and following the procedure which led to (4.125), we obtain

$$H(\{\boldsymbol{q}_i\}) = \Omega h_4\, \delta_{\sum \boldsymbol{q}_i, 0}\Big[\sum_i \Big(q_i^2 - i\frac{\omega_i}{2D}\Big) + \boldsymbol{q}_1\cdot\boldsymbol{q}_2 + \boldsymbol{q}_2\cdot\boldsymbol{q}_3 + \boldsymbol{q}_3\cdot\boldsymbol{q}_4 + \boldsymbol{q}_4\cdot\boldsymbol{q}_1\Big], \tag{4.146}$$

where the constant h_4 is given by (4.127). Since $\sum_i \boldsymbol{q}_i = 0$, we have

$$\boxed{H(\{\boldsymbol{q}_i\}) = \Omega h_4\, \delta_{\sum \boldsymbol{q}_i, 0}\Big[\frac{1}{2}\sum_i \Big(q_i^2 - i\frac{\omega_i}{D}\Big) - \boldsymbol{q}_1\cdot\boldsymbol{q}_3 - \boldsymbol{q}_2\cdot\boldsymbol{q}_4\Big]} \tag{4.147}$$

[24] This calculation may be immediately generalized to Cooperons.

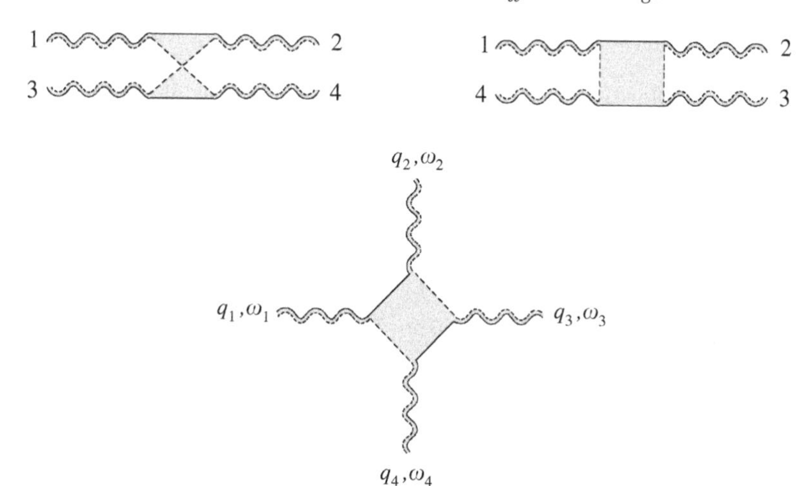

Figure 4.22 Three equivalent representations of a Hikami box describing the crossing of two Diffusons, that is, the change in the pairing of four complex amplitudes.

Taking the diffusion equation (4.143) into account, the first term applied to a Diffuson gives a constant or, in real space, a δ function which is zero away from the sources.[25] The Hikami box relating four Diffusons thus reduces to the form:

$$\boxed{H(\{\boldsymbol{q}_i\}) = \Omega h_4\, \delta_{\sum \boldsymbol{q}_i,0}\Big[-\boldsymbol{q}_1 \cdot \boldsymbol{q}_3 - \boldsymbol{q}_2 \cdot \boldsymbol{q}_4\Big]} \tag{4.148}$$

In practice we will mainly use the spatial representation. Each wave vector is thus a gradient operator $\boldsymbol{q}_i \to -i\boldsymbol{\nabla}_i$. In order to minimize the calculation of these derivatives, it is useful to employ a form for the Hikami box which is less symmetric, but in which only two wave vectors (that is, two gradients) appear. Using the constraint $\sum_i \boldsymbol{q}_i = 0$ and the relation (4.124), we obtain the form

$$H(\{\boldsymbol{q}_i\}) = -2h_4\, \delta_{\sum \boldsymbol{q}_i,0}\, \boldsymbol{q}_1 \cdot \boldsymbol{q}_3 \tag{4.149}$$

or, in Fourier transform,

$$\boxed{H(\{\boldsymbol{r}_i\}) = 2h_4 \int d\boldsymbol{r} \prod_{i=1}^{4} \delta(\boldsymbol{r} - \boldsymbol{r}_i)\boldsymbol{\nabla}_1 \cdot \boldsymbol{\nabla}_3} \tag{4.150}$$

[25] This δ function may lead to divergences which are discussed in references [85, 88, 89]. For example, when the Hikami box connects a Diffuson or a Cooperon making a loop, certain terms of (4.147) produce divergences. The procedure proposed in [88] prevents the appearance of these divergences and leads to the result (4.149, 4.150), adapted to the case where the gradients operate at points far from the sources (that is, on Diffusons whose sources lie outside the diffusive medium). These are called external Diffusons [85].

Appendix A4.3: Anisotropic collisions and transport mean free path

The probability of quantum diffusion was obtained under the assumption that the potential was a Gaussian white noise, that is, with no spatial correlation and such that $\overline{V(\boldsymbol{r})V(\boldsymbol{r}')} = B\delta(\boldsymbol{r} - \boldsymbol{r}')$. In the framework of the Edwards model (section 2.2.2), this implies that the collisions are isotropic. The equivalence between these two points of view is seen by considering δ scatterers, in the limit where the amplitude v_0 of the potential tends to 0 with $n_i v_0^2$ constant. We then have $B = n_i v_0^2$.

We would like to generalize these results to a Gaussian potential whose correlation function has a finite range $\overline{V(\boldsymbol{r})V(\boldsymbol{r}')} = B(\boldsymbol{r} - \boldsymbol{r}')$. For the Edwards model, this corresponds to an impurity potential $v(\boldsymbol{r})$ of finite range. If this range is of the order of the wavelength $\lambda = 2\pi/k$, the scattering becomes anisotropic (see Figure 2.4). The scattering cross section of an impurity calculated in the Born approximation thus depends on the angle θ between the incident and emergent directions (2.95). For the case of the Schrödinger equation and reinserting $\hbar$,

$$\sigma(k,\theta) = \frac{m^2}{\pi\hbar^4} v^2(\boldsymbol{k} - \boldsymbol{k}'). \tag{4.151}$$

The Fourier transform of the correlation function of the potential V also depends on $\boldsymbol{k} - \boldsymbol{k}'$:

$$B(\boldsymbol{k} - \boldsymbol{k}') = n_i v^2(\boldsymbol{k} - \boldsymbol{k}'). \tag{4.152}$$

In Chapter 3, we established a relation between the elastic collision time and the angular average of $B(\boldsymbol{k} - \boldsymbol{k}')$:

$$\frac{1}{\tau_e} = 2\pi\rho_0 \langle B(\boldsymbol{k} - \boldsymbol{k}')\rangle \tag{4.153}$$

together with $1/l_e = n_i\sigma$. The average Green function depends only on the angular average of the scattering potential of an impurity.

What happens to the probability of quantum diffusion when the scattering from impurities becomes anisotropic? Calculation of the structure factor Γ_ω requires taking into account the fact that, in a sequence of successive collisions, each collision must be characterized by incident and emergent directions. As is shown qualitatively in Figure 4.23, we expect a modification of the diffusion process. We will show that there appears a second characteristic time, called the *transport time* and denoted by τ^*, which describes the behavior of Γ_ω and of P_d in multiple scattering. *In contrast to τ_e which is the time separating two elastic collisions, τ^* is the time over which the memory of the incident direction is lost.*

To describe anisotropic collisions, two presentations are possible which employ the equivalence mentioned earlier between a finite-range potential and anisotropic collisions. They correspond respectively to the description of Γ_ω in real and reciprocal space. In the case where the range $B(\boldsymbol{r} - \boldsymbol{r}')$ of the potential is a δ function, the structure factor depends only on the initial and final points of the sequence (Figure 4.24(a)). However, if $B(\boldsymbol{r} - \boldsymbol{r}')$ has a finite range comparable to the wavelength, the structure factor depends on

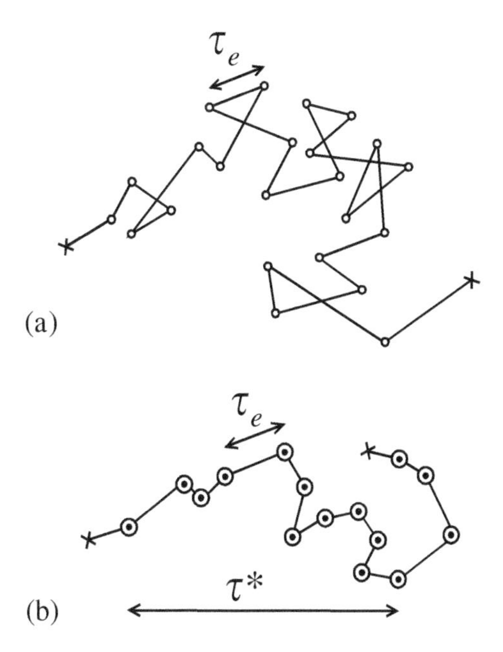

Figure 4.23 Schematic representation of a process of multiple scattering from impurities, when the collisions are either isotropic (a), or anisotropic (b). For isotropic collisions, the times τ_e and τ^* are equal. For anisotropic collisions, more scatterings are needed to lose memory of the incident direction: the time τ^* is greater than τ_e. For colloidal suspensions, their ratio may be of the order of 10^2.

four arguments $\Gamma(\boldsymbol{r}_1, \boldsymbol{r}'_1; \boldsymbol{r}_2, \boldsymbol{r}'_2)$ (Figure 4.24(b)) and satisfies an integral equation of the form (Figure 4.24(c))

$$\Gamma_\omega(\boldsymbol{r}_1, \boldsymbol{r}'_1; \boldsymbol{r}_2, \boldsymbol{r}'_2) = \delta(\boldsymbol{r}_2 - \boldsymbol{r}_1)\delta(\boldsymbol{r}'_2 - \boldsymbol{r}'_1)B(\boldsymbol{r}_1 - \boldsymbol{r}'_1) + \int d\boldsymbol{r}d\boldsymbol{r}'\Gamma_\omega(\boldsymbol{r}_1, \boldsymbol{r}'_1; \boldsymbol{r}, \boldsymbol{r}')\overline{G}^R_\epsilon(\boldsymbol{r}', \boldsymbol{r}'_2)\overline{G}^A_{\epsilon-\omega}(\boldsymbol{r}_2, \boldsymbol{r})B(\boldsymbol{r}_2 - \boldsymbol{r}'_2). \quad (4.154)$$

When $B(\boldsymbol{r} - \boldsymbol{r}') = B\delta(\boldsymbol{r} - \boldsymbol{r}')$, we recover equation (4.24). Moreover the translation invariance implies that Γ_ω depends on only three arguments $\Gamma_\omega(\boldsymbol{r}'_1 - \boldsymbol{r}_1, \boldsymbol{r}'_2 - \boldsymbol{r}_2, \boldsymbol{r}_1 - \boldsymbol{r}_2)$.

An equivalent description of Γ_ω in reciprocal space is obtained by taking the Fourier transform of (4.154). This description is interesting for two reasons. First of all, the use of (4.152) allows the introduction of the function $v^2(\boldsymbol{k} - \boldsymbol{k}')$ which contains information on the anisotropy of the collisions. Moreover, the Fourier transform $\Gamma_\omega(\hat{\boldsymbol{s}}, \hat{\boldsymbol{s}}', \boldsymbol{q})$ takes into account explicitly the functional dependence on the incident and emergent wave vectors, $\hat{\boldsymbol{s}} = \boldsymbol{k}/k$ and $\hat{\boldsymbol{s}}' = \boldsymbol{k}'/k$, respectively. The wave vector $\boldsymbol{q}$ associated with $(\boldsymbol{r}_1 - \boldsymbol{r}_2)$ describes, as in the isotropic case, the long range propagation. Relation (4.80) thus generalizes to

$$\Gamma_\omega(\hat{\boldsymbol{s}}, \hat{\boldsymbol{s}}', \boldsymbol{q}) = B(\hat{\boldsymbol{s}} - \hat{\boldsymbol{s}}') + \frac{1}{\Omega}\sum_{\boldsymbol{k}''}\Gamma_\omega(\hat{\boldsymbol{s}}, \hat{\boldsymbol{s}}'', \boldsymbol{q})\overline{G}^R_\epsilon(\boldsymbol{k}'')\overline{G}^A_{\epsilon-\omega}(\boldsymbol{k}'' - \boldsymbol{q})B(\hat{\boldsymbol{s}}'' - \hat{\boldsymbol{s}}') \quad (4.155)$$

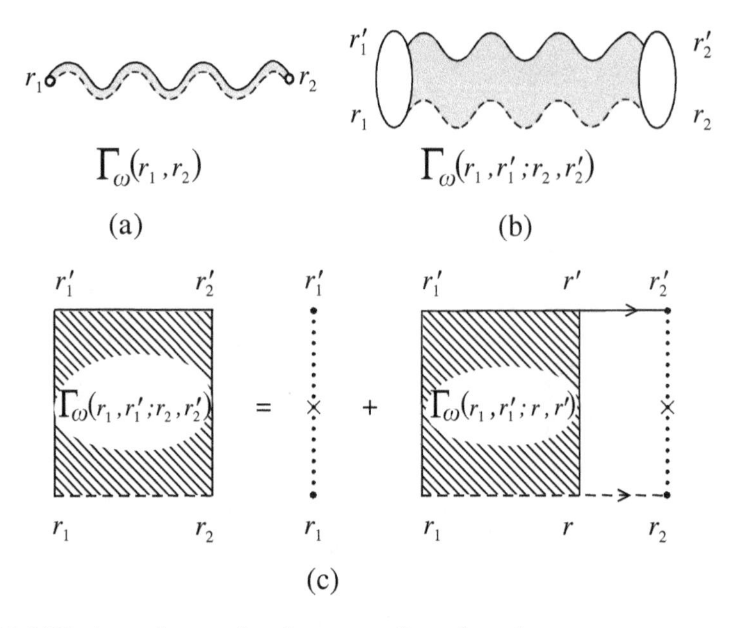

Figure 4.24 (a) For isotropic scattering, the structure factor depends on two arguments $\boldsymbol{r}_1$ and $\boldsymbol{r}_2$. (b) When the correlation function of the potential is of finite range, the structure factor depends on four arguments. (c) Representation of the integral equation for the structure factor $\Gamma_\omega(\boldsymbol{r}_1, \boldsymbol{r}_1'; \boldsymbol{r}_2, \boldsymbol{r}_2')$.

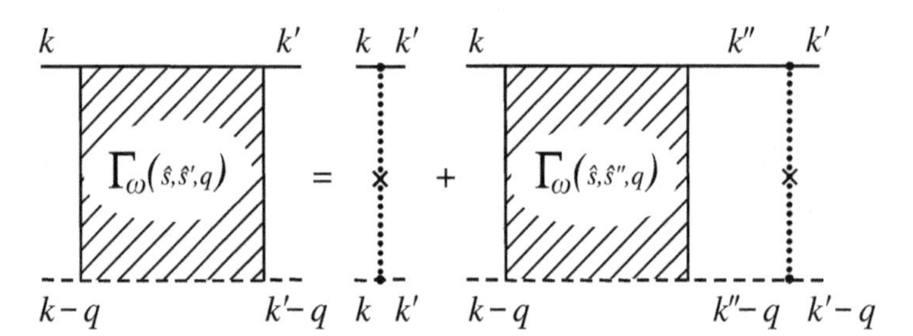

Figure 4.25 Representation of the integral equation for $\Gamma_\omega(\hat{s}, \hat{s}', \boldsymbol{q})$, with $\hat{s} = \hat{\boldsymbol{k}}/k$ and $\hat{s}' = \hat{\boldsymbol{k}}'/k$.

which is diagrammatically represented in Figure 4.25. The structure factor $\Gamma_\omega(\boldsymbol{q})$ is the angular average of $\Gamma_\omega(\hat{s}, \hat{s}', \boldsymbol{q})$:

$$\boxed{\Gamma_\omega(\boldsymbol{q}) = \langle \Gamma_\omega(\hat{s}, \hat{s}', \boldsymbol{q}) \rangle_{\hat{s}, \hat{s}'}} \tag{4.156}$$

For isotropic collisions, $B(\hat{s} - \hat{s}')$ is constant and the structure factor $\Gamma_\omega(\hat{s}, \hat{s}', \boldsymbol{q})$ depends only on $\boldsymbol{q}$. The new difficulty in (4.155) is that Γ_ω no longer factorizes. Consequently, we lose the simple iterative structure (4.80) of the isotropic case and we can no longer solve this equation exactly. We may, however, seek its solution in the diffusive regime, that is, when

q is small. Indeed, the diffusion mode in $1/Dq^2$ (see relation 4.87) is a Goldstone mode which accounts for the conservation of probability. As such, it must survive for anisotropic elastic collisions. However, we shall see that the diffusion coefficient D is affected by the anisotropy of the collisions.

Let us introduce the following notation

$$\begin{aligned} \gamma_e &= \langle B(\hat{s} - \hat{s}')\rangle_{\hat{s}'} = \langle B(\theta)\rangle \\ \gamma_1 &= \langle \hat{s}\cdot\hat{s}' B(\hat{s} - \hat{s}')\rangle_{\hat{s}'} = \langle B(\theta)\cos\theta\rangle, \end{aligned} \tag{4.157}$$

where θ is the angle $(\hat{s}, \hat{s}')$. We note that

$$\langle \hat{s}' B(\hat{s} - \hat{s}')\rangle_{\hat{s}'} = \gamma_1 \hat{s}. \tag{4.158}$$

In the integral equation (4.155), we first rewrite the product of Green's functions, separating the radial $k = |\boldsymbol{k}|$ and angular $\hat{s} = \boldsymbol{k}/k$ dependences of $\boldsymbol{k}$. Performing only the sum over k we obtain[26]

$$\frac{1}{\Omega}\sum_k \overline{G}^R_\epsilon(\boldsymbol{k})\overline{G}^A_{\epsilon-\omega}(\boldsymbol{k}-\boldsymbol{q}) = 2\pi\rho_0\tau_e f_\omega(\hat{s}, \boldsymbol{q}), \tag{4.159}$$

where the function $f_\omega(\hat{s}, \boldsymbol{q})$ defined by (4.76) has the expansion, for $\boldsymbol{q}l_e \to 0$ and $\omega\tau_e \to 0$

$$f_\omega(\hat{s}, \boldsymbol{q}) = 1 + i\omega\tau_e - q^2 l_e^2 (\hat{\boldsymbol{u}}\cdot\hat{s})^2 - iql_e(\hat{\boldsymbol{u}}\cdot\hat{s}) \tag{4.160}$$

with $\hat{\boldsymbol{u}} = \boldsymbol{q}/q$. The integral equation (4.155) becomes

$$\Gamma_\omega(\hat{s}, \hat{s}', \boldsymbol{q}) = B(\hat{s} - \hat{s}') + \frac{1}{\gamma_e}\Big\langle \Gamma_\omega(\hat{s}, \hat{s}'', \boldsymbol{q}) f_\omega(\hat{s}'', \boldsymbol{q}) B(\hat{s}'' - \hat{s}')\Big\rangle_{\hat{s}''}. \tag{4.161}$$

First, we seek to determine $\Gamma_\omega(\hat{s}', \boldsymbol{q}) = \langle \Gamma_\omega(\hat{s}, \hat{s}', \boldsymbol{q})\rangle_{\hat{s}}$ which is the solution of the integral equation

$$\boxed{\Gamma_\omega(\hat{s}', \boldsymbol{q}) = \gamma_e + \frac{1}{\gamma_e}\Big\langle \Gamma_\omega(\hat{s}'', \boldsymbol{q}) f_\omega(\hat{s}'', \boldsymbol{q}) B(\hat{s}'' - \hat{s}')\Big\rangle_{\hat{s}''}} \tag{4.162}$$

The diffusion approximation consists of retaining only the lowest two harmonics ($l = 0$ and $l = 1$) in the expansion of $\Gamma_\omega(\hat{s}', \boldsymbol{q})$ into spherical harmonics:

$$\Gamma_\omega(\hat{s}', \boldsymbol{q}) = \Gamma_\omega(\boldsymbol{q}) + \frac{d}{v}\,\hat{s}'\cdot\boldsymbol{j}_\omega(\boldsymbol{q}), \tag{4.163}$$

[26] Upon averaging this sum over the direction $\hat{s}$, we recover the result (4.73):

$$P_0(\boldsymbol{q}, \omega) = \frac{1}{2\pi\rho_0\Omega}\sum_k \overline{G}^R_\epsilon(\boldsymbol{k})\overline{G}^A_{\epsilon-\omega}(\boldsymbol{k}-\boldsymbol{q}) = \tau_e\langle f_\omega(\hat{s}, \boldsymbol{q})\rangle_{\hat{s}} \simeq \tau_e\left(1 + i\omega\tau_e - Dq^2\tau_e\right).$$

where d is the dimension of space and $\boldsymbol{j}_\omega(\boldsymbol{q})$ is defined as the current density associated with $\Gamma_\omega(\boldsymbol{q}) = \langle \Gamma_\omega(\hat{s}', \boldsymbol{q}) \rangle_{\hat{s}'}$:

$$\boldsymbol{j}_\omega(\boldsymbol{q}) = v \, \langle \hat{\boldsymbol{s}}' \, \Gamma_\omega(\hat{\boldsymbol{s}}', \boldsymbol{q}) \rangle_{\hat{s}'}, \tag{4.164}$$

where v is the group velocity. Inserting (4.163) into (4.162) and separating the harmonics $l = 0$ and $l = 1$, we obtain two equations for the unknowns Γ_ω and $\boldsymbol{j}_\omega$:[27]

$$\begin{aligned} \Gamma_\omega(\boldsymbol{q}) &= \gamma_e + \Gamma_\omega(\boldsymbol{q}) \left(1 + i\omega\tau_e - \frac{q^2 l_e^2}{d} \right) - i l_e \boldsymbol{q} \cdot \boldsymbol{j}_\omega \\ \boldsymbol{j}_\omega(\boldsymbol{q}) &= \frac{\gamma_1}{\gamma_e} \left[\boldsymbol{j}_\omega(\boldsymbol{q}) - i \frac{\boldsymbol{q} l_e}{d} \Gamma_\omega(\boldsymbol{q}) \right] \end{aligned} \tag{4.165}$$

whose solution is

$$\Gamma_\omega(\boldsymbol{q}) = \frac{\gamma_e}{-i\omega\tau_e + \frac{q^2 l_e^2 / d}{1 - \gamma_1/\gamma_e}}. \tag{4.166}$$

This expression introduces a new time scale τ^* defined by

$$\frac{1}{\tau^*} = 2\pi\rho_0(\gamma_e - \gamma_1) \tag{4.167}$$

that is,

$$\boxed{\frac{1}{\tau^*} = 2\pi\rho_0 \Big\langle B(\theta)(1 - \cos\theta) \Big\rangle} \tag{4.168}$$

called the *transport time*. We thus obtain again the characteristic form of a diffusion law

$$\Gamma_\omega(\boldsymbol{q}) = \frac{\gamma_e}{\tau_e} \frac{1}{-i\omega + D^* q^2}, \tag{4.169}$$

where

$$D^* = \frac{1}{d} v^2 \tau^* = \frac{v l^*}{d} \tag{4.170}$$

is the diffusion coefficient and where we have defined the transport mean free path by

$$l^* = v\tau^*. \tag{4.171}$$

[27] To obtain the first expression, we average (4.162) over $\hat{s}'$; for the second, first multiply by $\hat{s}'$ and then average over $\hat{s}'$. Recall that

$$\langle (\hat{s} \cdot \boldsymbol{A}) \, (\hat{s} \cdot \boldsymbol{B}) \rangle_{\hat{s}} = \frac{\boldsymbol{A} \cdot \boldsymbol{B}}{d} \qquad \langle \hat{s} \, (\hat{s} \cdot \boldsymbol{A}) \rangle_{\hat{s}} = \frac{\boldsymbol{A}}{d},$$

for any two vectors $\boldsymbol{A}$ and $\boldsymbol{B}$.

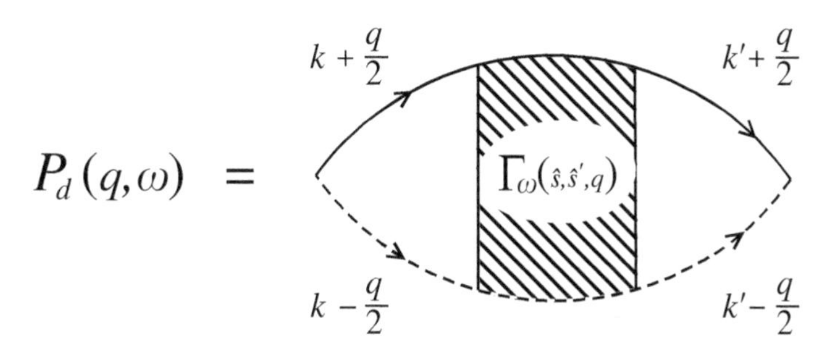

Figure 4.26 Diagrammatic representation of the probability $P_d(\boldsymbol{q},\omega)$ for anisotropic scattering, with $\hat{s}=\hat{\boldsymbol{k}}/k$ and $\hat{s}'=\hat{\boldsymbol{k}}'/k$. This generalizes Figure 4.14 for isotropic scattering.

Just as the elastic mean free path is related to the scattering cross section σ of scatterers (expression 3.86), so is the transport mean free path related to the transport scattering cross section σ^* defined by (2.73):

$$\frac{1}{l^*}=n_i\sigma^*. \tag{4.172}$$

Starting from (4.165), we obtain a *Fick's law*, between $\Gamma_\omega(\boldsymbol{q})$ and its associated current:

$$\boldsymbol{j}_\omega(\boldsymbol{q})=-i\boldsymbol{q}(D^*-D)\Gamma_\omega(\boldsymbol{q}). \tag{4.173}$$

To obtain the behavior of $P_d(\boldsymbol{q},\omega)$ in the diffusive regime, we generalize (4.82) which, because of angular dependences, becomes (Figure 4.26)

$$P_d(\boldsymbol{q},\omega)=\frac{2\pi\rho_0}{\Omega^2}\sum_{\boldsymbol{k},\boldsymbol{k}'}\tilde{P}_0(\boldsymbol{k},\boldsymbol{q},\omega)\tilde{P}_0(\boldsymbol{k}',\boldsymbol{q},\omega)\ \Gamma_\omega(\hat{s},\hat{s}',\boldsymbol{q}). \tag{4.174}$$

Carrying out the integrals over the amplitudes k and k' which factorize, we obtain

$$P_d(\boldsymbol{q},\omega)=2\pi\rho_0\tau_e^2\langle\Gamma_\omega(\hat{s},\hat{s}';\boldsymbol{q})\rangle_{\hat{s},\hat{s}'}=\frac{\tau_e}{\gamma_e}\Gamma_\omega(\boldsymbol{q}). \tag{4.175}$$

The proportionality (4.83) between $P_d(\boldsymbol{q},\omega)$ and $\Gamma_\omega(\boldsymbol{q})$ is thus unchanged, and starting from (4.169), we obtain

$$\boxed{P_d(\boldsymbol{q},\omega)=\frac{1}{-i\omega+D^*q^2}} \tag{4.176}$$

For a finite range potential or, equivalently, for anisotropic scatterers, the Diffuson P_d is characterized by a diffusion constant which depends on the transport time τ^*. In contrast, it is worth noting that the Drude–Boltzmann contribution $P_0(\boldsymbol{q},\omega)$, which gives the probability that a particle undergoes no collision, depends only on the elastic collision time τ_e.

The term *transport time* for τ^* should not cause confusion. This time enters naturally in transport properties such as electrical conductivity (sections 7.2.2 and 7.2.3). We have just seen that it enters also into the probability of quantum diffusion, that is, in the propagation of a wave packet. In fact, the time τ^* enters also into spectral and thermodynamic properties. The elastic collision time τ_e is a property of the single-particle Green function, which can be probed using a physical quantity directly related to this function . Usually, τ_e is deduced from de Haas–van Alphen oscillations, i.e., from the magnetic field dependence of the density of states, that is the imaginary part of the single-particle Green function (section 14.2.2). Measuring the damping of the oscillations in (14.28) gives the elastic collision time τ_e.

Exercise 4.8 Calculate the mean free paths l_e and l^*, for the Edwards model with impurities described by a spherical potential of amplitude U_0 and range a. For this, use the expression for $\sigma(\theta)$ from section A2.1.4:

$$\frac{l^*}{l_e} = \frac{\sigma}{\sigma^*} = \frac{\int_0^\pi \sigma(\theta) \sin\theta d\theta}{\int_0^\pi \sigma(\theta)(1-\cos\theta) \sin\theta d\theta}. \tag{4.177}$$

Show the asymptotic expressions:

$$\frac{l^*}{l_e} \to 1 + \frac{2}{15}(ka)^2 \qquad \text{if } ka \ll 1$$

$$\frac{l^*}{l_e} \to (ka)^2 / \ln[4e^{\gamma-1}ka] \quad \text{if } ka \gg 1 \tag{4.178}$$

where $\gamma = 0.577\cdots$ is the Euler constant.

Appendix A4.4: Correlation of diagonal Green's functions

The probability of quantum diffusion $P(\boldsymbol{r},\boldsymbol{r}',\omega)$ is given by the product of two non-diagonal Green's functions $\overline{G_\epsilon^R(\boldsymbol{r},\boldsymbol{r}')G_{\epsilon-\omega}^A(\boldsymbol{r}',\boldsymbol{r})}$. This quantity plays a major role in the description of a large number of physical phenomena, but it is not the only one. We may also construct, at least formally at this stage, other correlation functions from Green's functions which are not a priori related to P. The field correlation function (4.57) is one example. Another important correlation function defined by

$$K(\boldsymbol{r},\boldsymbol{r}',\omega) = \overline{G_\epsilon^R(\boldsymbol{r},\boldsymbol{r})G_{\epsilon-\omega}^A(\boldsymbol{r}',\boldsymbol{r}')} - \overline{G}_\epsilon^R(\boldsymbol{r},\boldsymbol{r})\overline{G}_{\epsilon-\omega}^A(\boldsymbol{r}',\boldsymbol{r}') \tag{4.179}$$

involves the product of two *diagonal* Green's functions taken at *different* points $\boldsymbol{r}$ and $\boldsymbol{r}'$. We will use this function to study the density of states correlations in Chapter 10. Like the probability of return to the origin $P(\boldsymbol{r},\boldsymbol{r},\omega)$, this correlator describes the product of two *closed* trajectories, whose departure points are *distinct*. Figure 4.27 shows the structure of these two quantities. The correlator $K(\boldsymbol{r},\boldsymbol{r}',\omega)$ contains two contributions represented in Figure 4.27(c and d) and denoted $K^{(1)}$ and $K^{(2)}$.

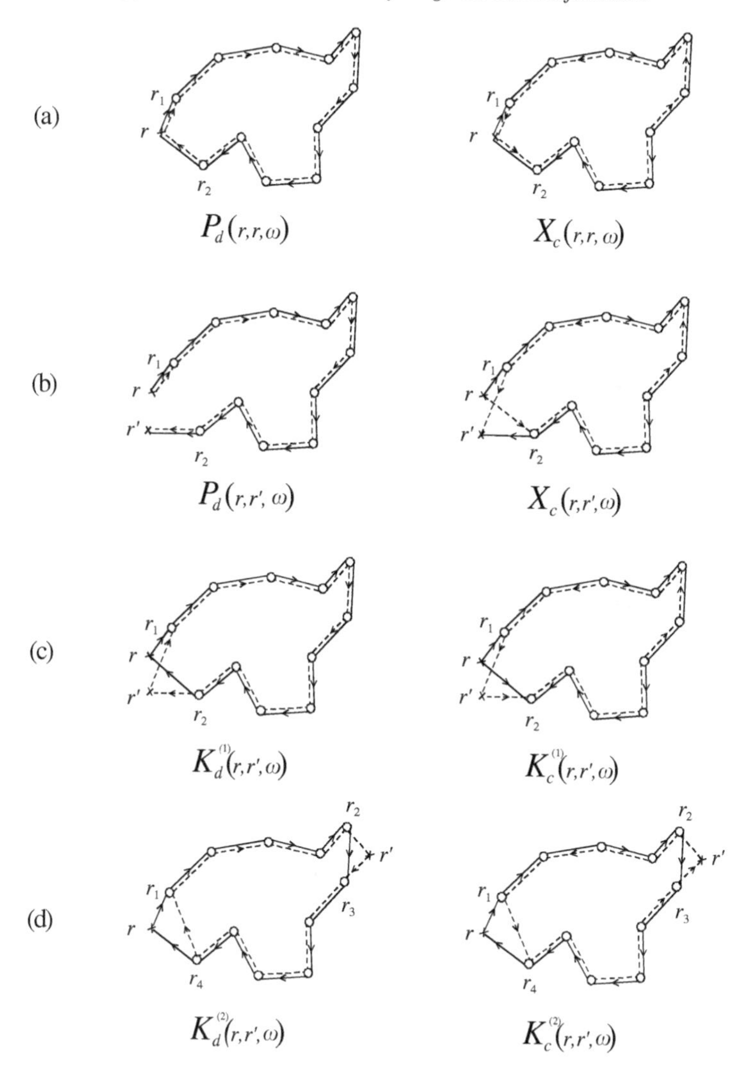

Figure 4.27 Schematic representations of different trajectories corresponding to the correlation functions $P(\boldsymbol{r},\boldsymbol{r},\omega)$, $P(\boldsymbol{r},\boldsymbol{r}',\omega)$, $K^{(1)}(\boldsymbol{r},\boldsymbol{r}',\omega)$ and $K^{(2)}(\boldsymbol{r},\boldsymbol{r}',\omega)$. On the left are the Diffuson contributions, on the right, the Cooperon contributions.

The first contribution contains a Diffuson and is written:

$$K_d^{(1)}(\boldsymbol{r},\boldsymbol{r}',\omega) = \int \overline{G}_\epsilon^R(\boldsymbol{r},\boldsymbol{r}_1)\overline{G}_\epsilon^R(\boldsymbol{r}_2,\boldsymbol{r})\overline{G}_{\epsilon-\omega}^A(\boldsymbol{r}_1,\boldsymbol{r}')\overline{G}_{\epsilon-\omega}^A(\boldsymbol{r}',\boldsymbol{r}_2)\Gamma_\omega(\boldsymbol{r}_1,\boldsymbol{r}_2)\,d\boldsymbol{r}_1\,d\boldsymbol{r}_2. \tag{4.180}$$

In the diffusion approximation (section 4.5), it becomes

$$K_d^{(1)}(\boldsymbol{r},\boldsymbol{r}',\omega) = \Gamma_\omega(\boldsymbol{r},\boldsymbol{r}) \left[\int \overline{G}_\epsilon^R(\boldsymbol{r},\boldsymbol{r}_1)\overline{G}_\epsilon^A(\boldsymbol{r}_1,\boldsymbol{r}')d\boldsymbol{r}_1 \right]^2$$
$$= \Gamma_\omega(\boldsymbol{r},\boldsymbol{r}) \frac{1}{\gamma_e^2} g^2(\boldsymbol{R}), \qquad (4.181)$$

since the integral is the function $f^{1,1}(\boldsymbol{r}-\boldsymbol{r}')/\gamma_e$ defined in (3.103), and equal to the short range function $g(\boldsymbol{r}-\boldsymbol{r}')$ given by (3.97). Finally, using (4.37), we have

$$\boxed{K_d^{(1)}(\boldsymbol{r},\boldsymbol{r}',\omega) = 2\pi\rho_0 g^2(\boldsymbol{r}-\boldsymbol{r}')P_d(\boldsymbol{r},\boldsymbol{r},\omega)} \qquad (4.182)$$

We have thus obtained the result, which will be found to be very useful, that a diagonal correlation function, a priori independent of the probability P_d, may in fact be related to it. Similarly, Figure 4.27(c) shows that there exists a second contribution due to the Cooperon, given by

$$\boxed{K_c^{(1)}(\boldsymbol{r},\boldsymbol{r}',\omega) = 2\pi\rho_0 X_c(\boldsymbol{r},\boldsymbol{r}',\omega)} \qquad (4.183)$$

where $X_c(\boldsymbol{r},\boldsymbol{r}',\omega) = g^2(\boldsymbol{r}-\boldsymbol{r}')P_c(\boldsymbol{r},\boldsymbol{r},\omega)$ is a short range function (4.49). Both contributions $K_d^{(1)}$ and $K_c^{(1)}$ are short range, and vanish for $|\boldsymbol{r}-\boldsymbol{r}'| > l_e$.

Figure 4.27(d) introduces a second contribution $K^{(2)}(\boldsymbol{r},\boldsymbol{r}',\omega)$ which remains important even when $|\boldsymbol{r}-\boldsymbol{r}'|$ becomes greater than l_e. It contains the product of two structure factors Γ_ω and six average Green's functions which vary exponentially. This constrains the points $\boldsymbol{r},\boldsymbol{r}_1,\boldsymbol{r}_4$ on the one hand, and the points $\boldsymbol{r}',\boldsymbol{r}_2,\boldsymbol{r}_3$ on the other, to be distant by not more than l_e. This having been said, writing the correlation function is tedious but straightforward. It contains two terms, the first of which describes scattering trajectories traversed in the same direction (Figure 4.27(d))

$$K_d^{(2)}(\boldsymbol{r},\boldsymbol{r}',\omega) = \int d\boldsymbol{r}_1\, d\boldsymbol{r}_2\, d\boldsymbol{r}_3\, d\boldsymbol{r}_4 \Gamma_\omega(\boldsymbol{r}_1,\boldsymbol{r}_2)\Gamma_\omega(\boldsymbol{r}_3,\boldsymbol{r}_4)$$
$$\times \overline{G}_\epsilon^R(\boldsymbol{r},\boldsymbol{r}_1)\overline{G}_\epsilon^R(\boldsymbol{r}_2,\boldsymbol{r}_3)\overline{G}_\epsilon^R(\boldsymbol{r}_4,\boldsymbol{r})$$
$$\times \overline{G}_{\epsilon-\omega}^A(\boldsymbol{r}',\boldsymbol{r}_2)\overline{G}_{\epsilon-\omega}^A(\boldsymbol{r}_1,\boldsymbol{r}_4)\overline{G}_{\epsilon-\omega}^A(\boldsymbol{r}_3,\boldsymbol{r}'). \qquad (4.184)$$

The second describes scattering trajectories traversed in the opposite direction (Figure 4.27(d))

$$K_c^{(2)}(\boldsymbol{r},\boldsymbol{r}',\omega) = \int d\boldsymbol{r}_1\, d\boldsymbol{r}_2\, d\boldsymbol{r}_3\, d\boldsymbol{r}_4 \Gamma'_\omega(\boldsymbol{r}_1,\boldsymbol{r}_2)\Gamma'_\omega(\boldsymbol{r}_3,\boldsymbol{r}_4)$$
$$\times\, \overline{G}_\epsilon^R(\boldsymbol{r},\boldsymbol{r}_1)\overline{G}_\epsilon^R(\boldsymbol{r}_2,\boldsymbol{r}_3)\overline{G}_\epsilon^R(\boldsymbol{r}_4,\boldsymbol{r})$$
$$\times\, \overline{G}_{\epsilon-\omega}^A(\boldsymbol{r}',\boldsymbol{r}_3)\overline{G}_{\epsilon-\omega}^A(\boldsymbol{r}_4,\boldsymbol{r}_1)\overline{G}_{\epsilon-\omega}^A(\boldsymbol{r}_2,\boldsymbol{r}'). \tag{4.185}$$

In the diffusion approximation, that is, for distances greater than l_e, the integral for $K_d^{(2)}$ decouples, yielding

$$K_d^{(2)}(\boldsymbol{r},\boldsymbol{r}',\omega) = \Gamma_\omega(\boldsymbol{r},\boldsymbol{r}')\Gamma_\omega(\boldsymbol{r}',\boldsymbol{r})$$
$$\times \int \overline{G}_\epsilon^R(\boldsymbol{r},\boldsymbol{r}_1)\overline{G}_{\epsilon-\omega}^A(\boldsymbol{r}_1,\boldsymbol{r}_4)\overline{G}_\epsilon^R(\boldsymbol{r}_4,\boldsymbol{r})\, d\boldsymbol{r}_1\, d\boldsymbol{r}_4$$
$$\times \int \overline{G}_{\epsilon-\omega}^A(\boldsymbol{r}',\boldsymbol{r}_2)\overline{G}_\epsilon^R(\boldsymbol{r}_2,\boldsymbol{r}_3)\overline{G}_{\epsilon-\omega}^A(\boldsymbol{r}_3,\boldsymbol{r}')\, d\boldsymbol{r}_2\, d\boldsymbol{r}_3. \tag{4.186}$$

Introducing the function $f^{2,1}$ defined by expression (3.106), it may be put in the form[28]

$$K_d^{(2)}(\boldsymbol{r},\boldsymbol{r}',\omega) = \frac{1}{\gamma_e^2}|f^{2,1}(0)|^2\Gamma_\omega(\boldsymbol{r},\boldsymbol{r}')\Gamma_\omega(\boldsymbol{r}',\boldsymbol{r}). \tag{4.187}$$

The quantity $f^{2,1}(0)$, given in Table 3.2 and by relation (3.108), is equal to $-i\tau_e$.

Therefore, in the diffusion approximation, the correlation function $K^{(2)}$ appears as the product of four distinct quantities, represented in Figure 4.28: two structure factors $\Gamma_\omega(\boldsymbol{r},\boldsymbol{r}')$ which describe the diffusion between $\boldsymbol{r}$ and $\boldsymbol{r}'$, and two "boxes," which describe how the source points are related to impurities. In the diffusive regime, the structure factor Γ_ω is proportional to the probability P_d (relation 4.37), so that

$$K_d^{(2)}(\boldsymbol{r},\boldsymbol{r}',\omega) = P_d(\boldsymbol{r},\boldsymbol{r}',\omega)P_d(\boldsymbol{r}',\boldsymbol{r},\omega). \tag{4.188}$$

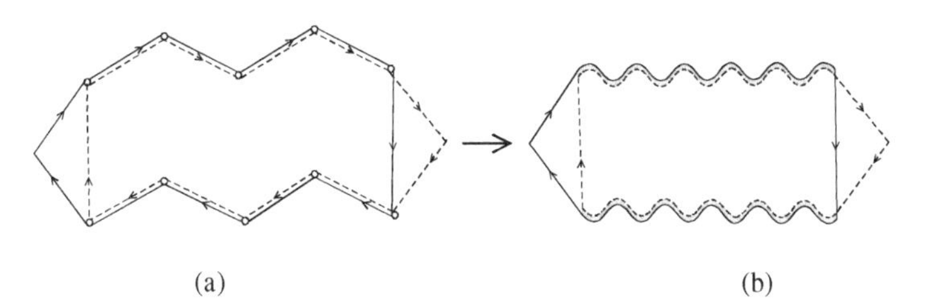

Figure 4.28 Diagrammatic representation (b) of the correlation function $K^{(2)}(\boldsymbol{r},\boldsymbol{r}',\omega)$, obtained by the iteration, *ad infinitum*, of a set of collision sequences, an example of which is shown in (a).

[28] In the limit $\omega\tau_e \ll 1$ we neglect the ω dependence of the average Green functions.

The calculation of the contribution (4.185) associated with the product of two Cooperons is similar, and gives, in the diffusion approximation,

$$K_c^{(2)}(\boldsymbol{r},\boldsymbol{r}',\omega) = \frac{1}{\gamma_e^2}|f^{2,1}(0)|^2\Gamma'_\omega(\boldsymbol{r},\boldsymbol{r}')\Gamma'_\omega(\boldsymbol{r}',\boldsymbol{r}). \tag{4.189}$$

The structure factor Γ'_ω is that of the Cooperon and, like Γ_ω, is a solution of the diffusion equation (4.34). We may rewrite the correlation function $K_c^{(2)}$ with the aid of $P_c(\boldsymbol{r},\boldsymbol{r}',\omega)$ defined in (4.48)

$$K_c^{(2)}(\boldsymbol{r},\boldsymbol{r}',\omega) = P_c(\boldsymbol{r},\boldsymbol{r}',\omega)P_c(\boldsymbol{r}',\boldsymbol{r},\omega). \tag{4.190}$$

Finally, regrouping the contributions (4.182) and (4.188), we write the correlation function K_d as the sum of a short range term and a long range term:

$$\boxed{K_d(\boldsymbol{r},\boldsymbol{r}',\omega) = 2\pi\rho_0\, g^2(\boldsymbol{R})P_d(\boldsymbol{r},\boldsymbol{r},\omega) + P_d(\boldsymbol{r},\boldsymbol{r}',\omega)P_d(\boldsymbol{r}',\boldsymbol{r},\omega)} \tag{4.191}$$

to which we must add the Cooperon contribution

$$\boxed{K_c(\boldsymbol{r},\boldsymbol{r}',\omega) = 2\pi\rho_0\, g^2(\boldsymbol{R})P_c(\boldsymbol{r},\boldsymbol{r},\omega) + P_c(\boldsymbol{r},\boldsymbol{r}',\omega)P_c(\boldsymbol{r}',\boldsymbol{r},\omega)} \tag{4.192}$$

This correlation function appears in the description of spectral correlations in a metal (Chapter 10) and in the study of the effect of electronic interactions on various physical properties (Chapter 13).

Appendix A4.5: Other correlation functions

A4.5.1 Correlations of Green's functions

This appendix contains a number of useful technical results used in the book. At this stage, they may be regarded as an ensemble of exercises intended to improve the reader's understanding of Chapter 4.

- We first consider the correlation function

$$\overline{\mathrm{Im}G^R(\boldsymbol{r}_1,\boldsymbol{r}_2)\mathrm{Im}G^R(\boldsymbol{r}_3,\boldsymbol{r}_4)}\,. \tag{4.193}$$

Using expression (3.22), we see that this product contains terms of the type $\overline{G^R G^A}$ previously studied and additional terms of the form $\overline{G^R G^R}$ and $\overline{G^A G^A}$. To evaluate these additional terms, we may write a disorder expansion of the product $\overline{G^R G^R}$, as displayed in Figure 4.5, where the bottom line now represents a retarded Green's function. From Exercise 3.15, we infer that the successive terms of this expansion decay as powers of $1/(kl_e)$ and may be neglected so that

$$\overline{G^R(\boldsymbol{r}_1,\boldsymbol{r}_2)\,G^R(\boldsymbol{r}_3,\boldsymbol{r}_4)} \simeq \overline{G^R(\boldsymbol{r}_1,\boldsymbol{r}_2)}\;\overline{G^R(\boldsymbol{r}_3,\boldsymbol{r}_4)}. \tag{4.194}$$

Relation (3.22) yields

$$\overline{\mathrm{Im}G^R\ \mathrm{Im}G^R} = \frac{1}{2}\mathrm{Re}\left(\overline{G^R\ G^A} - \overline{G^R\ G^R}\right), \tag{4.195}$$

where, for simplicity, we have not written the arguments of the Green functions. Similarly, we have

$$\overline{\mathrm{Im}G^R}\ \overline{\mathrm{Im}G^R} = \frac{1}{2}\mathrm{Re}\left(\overline{G^R}\ \overline{G^A} - \overline{G^R}\ \overline{G^R}\right). \tag{4.196}$$

Subtracting these two expressions and using (4.194), we prove the following relation, valid in the weak disorder limit $kl_e \gg 1$:

$$\boxed{\overline{\mathrm{Im}G^R\mathrm{Im}G^R}^c \simeq \frac{1}{2}\mathrm{Re}\left(\overline{G^R G^A}^c\right)} \tag{4.197}$$

where the symbol $\overline{\cdots}^c$ denotes the connected parts of these average values.[29] This equation holds for arbitrary arguments of the Green functions, so that we may write in particular

$$\mathrm{Tr}\left(\mathrm{Im}\hat{G}^R\mathrm{Im}\hat{G}^R\right) \simeq \frac{1}{2}\mathrm{Tr}\mathrm{Re}\left(\hat{G}^R\hat{G}^A\right) \tag{4.199}$$

for Green's operators. We have used the result of Exercise 4.9 to eliminate the contributions of the non-connected parts.

Exercise 4.9 Using the spatial representation, check that, for $\omega = 0$,

$$\mathrm{Tr}[\overline{\mathrm{Im}\hat{G}^R}\ \overline{\mathrm{Im}\hat{G}^R}] = \frac{1}{2}\mathrm{Tr}[\overline{\hat{G}^R}\ \overline{\hat{G}^A}] \tag{4.200}$$

in the weak disorder limit $kl_e \gg 1$.

The term on the left is equal to

$$\pi^2\rho_0^2\int_0^\infty \frac{\sin^2 kR}{k^2R^2}e^{-R/l_e}4\pi R^2\,dR = \frac{2\pi^3\rho_0^2 l_e}{k^2}\frac{1}{1+\frac{1}{4k^2l_e^2}} \tag{4.201}$$

while the right hand side is equal to

$$\frac{\pi^2\rho_0^2}{2}\int_0^\infty \frac{1}{k^2R^2}e^{-R/l_e}4\pi R^2\,dR = \frac{2\pi^3\rho_0^2 l_e}{k^2}. \tag{4.202}$$

These two terms are identical in the limit $kl_e \gg 1$. This occurs when the term $\sin^2 kR$ in the integral is replaced by its average $1/2$.

---

[29] We denote the connected part of the average as

$$\overline{A\,B}^c = \overline{A\,B} - \overline{A}\,\overline{B}. \tag{4.198}$$

$\overline{A}\,\overline{B}$ is called the non-connected part.

• Consider now the two correlation functions of the non-local density of states (3.26),

$$\overline{\rho_\epsilon(\boldsymbol{r},\boldsymbol{r}')\rho_{\epsilon-\omega}(\boldsymbol{r}',\boldsymbol{r})} \quad \text{and} \quad \overline{\rho_\epsilon(\boldsymbol{r},\boldsymbol{r}')\rho_{\epsilon-\omega}(\boldsymbol{r}',\boldsymbol{r})}^c \tag{4.203}$$

as well as those of the local density of states (3.25) at different points:

$$\overline{\rho_\epsilon(\boldsymbol{r})\rho_{\epsilon-\omega}(\boldsymbol{r}')} \quad \text{and} \quad \overline{\rho_\epsilon(\boldsymbol{r})\rho_{\epsilon-\omega}(\boldsymbol{r}')}^c. \tag{4.204}$$

Equation (3.28) relates the non-local density of states $\rho_\epsilon(\boldsymbol{r},\boldsymbol{r}')$ to the imaginary part of the Green function. From (4.197), and using the definition (4.9) of the probability $P(\boldsymbol{r},\boldsymbol{r}',\omega)$, we get

$$\overline{\rho_\epsilon(\boldsymbol{r},\boldsymbol{r}')\rho_{\epsilon-\omega}(\boldsymbol{r}',\boldsymbol{r})}^c = \frac{\rho_0}{\pi}\mathrm{Re}[P(\boldsymbol{r},\boldsymbol{r}',\omega) - P_0(\boldsymbol{r},\boldsymbol{r}',\omega)]. \tag{4.205}$$

Recall that this connected part of the probability (i.e., excluding the Drude–Boltzmann contribution) contains two terms, the Diffuson $P_d(\boldsymbol{r},\boldsymbol{r}',\omega)$ and the Cooperon $X_c(\boldsymbol{r},\boldsymbol{r}',\omega)$, which is a short range function given by (4.46) where the function $g(\boldsymbol{R})$ decays exponentially with distance. We thus obtain

$$\boxed{\overline{\rho_\epsilon(\boldsymbol{r},\boldsymbol{r}')\rho_{\epsilon-\omega}(\boldsymbol{r}',\boldsymbol{r})}^c = \frac{\rho_0}{\pi}\mathrm{Re}\left[P_d(\boldsymbol{r},\boldsymbol{r}',\omega) + g^2(\boldsymbol{R})X_c(\boldsymbol{r},\boldsymbol{r},\omega)\right]} \tag{4.206}$$

with $\boldsymbol{R} = \boldsymbol{r} - \boldsymbol{r}'$. Using (3.99), we can deduce the expression for $\overline{\rho_\epsilon(\boldsymbol{r},\boldsymbol{r}')\rho_{\epsilon-\omega}(\boldsymbol{r}',\boldsymbol{r})}$.

• Taking $\boldsymbol{r}' = \boldsymbol{r}$ in (4.206), we thus relate the fluctuations of the local density of states with the real part of the connected part of the quantum probability.

$$\boxed{\overline{\rho_\epsilon(\boldsymbol{r})\rho_{\epsilon-\omega}(\boldsymbol{r})} - \rho_0^2 = \frac{\rho_0}{\pi}\mathrm{Re}\,[P_d(\boldsymbol{r},\boldsymbol{r},\omega) + X_c(\boldsymbol{r},\boldsymbol{r},\omega)]} \tag{4.207}$$

If there is time-reversal invariance, the two terms in brackets are equal.

• From (4.197) and the definition (4.179) of the correlation function $K(\boldsymbol{r},\boldsymbol{r}',\omega)$, the correlation function $\overline{\rho_\epsilon(\boldsymbol{r})\rho_{\epsilon-\omega}(\boldsymbol{r}')}^c$ of the local density of states may be expressed as

$$\overline{\rho_\epsilon(\boldsymbol{r})\rho_{\epsilon-\omega}(\boldsymbol{r}')}^c = \frac{1}{2\pi^2}K(\boldsymbol{r},\boldsymbol{r}',\omega). \tag{4.208}$$

Using (4.191) and (4.192), we find that the Diffuson contribution to this correlation function is

$$\boxed{\overline{\rho_\epsilon(\boldsymbol{r})\rho_{\epsilon-\omega}(\boldsymbol{r}')}^c = \mathrm{Re}\left[\frac{\rho_0}{\pi}g^2(\boldsymbol{R})P_d(\boldsymbol{r},\boldsymbol{r},\omega) + \frac{1}{2\pi^2}P_d(\boldsymbol{r},\boldsymbol{r}',\omega)P_d(\boldsymbol{r}',\boldsymbol{r},\omega)\right]} \tag{4.209}$$

which has both a short and a long range part. We also need to add the Cooperon contribution

$$\mathrm{Re}\left[\frac{\rho_0}{\pi}g^2(\boldsymbol{R})P_c(\boldsymbol{r},\boldsymbol{r}',\omega) + \frac{1}{2\pi^2}P_c(\boldsymbol{r},\boldsymbol{r}',\omega)P_c(\boldsymbol{r}',\boldsymbol{r},\omega)\right] \tag{4.210}$$

where P_c is defined by (4.48).

Exercise 4.10 Show that:

$$\overline{\rho_\epsilon(\boldsymbol{r}_1,\boldsymbol{r}_1')\rho_{\epsilon-\omega}(\boldsymbol{r}_2',\boldsymbol{r}_2)}^c = \frac{\rho_0}{\pi}\mathrm{Re}\Big[P_d(\boldsymbol{r}_1,\boldsymbol{r}_1',\omega)g(\boldsymbol{r}_1-\boldsymbol{r}_2)g(\boldsymbol{r}_1'-\boldsymbol{r}_2') + P_c(\boldsymbol{r}_1,\boldsymbol{r}_1',\omega)g(\boldsymbol{r}_1-\boldsymbol{r}_2')g(\boldsymbol{r}_1'-\boldsymbol{r}_2)\Big], \tag{4.211}$$

where we have only taken into account the contributions that involve only one Diffuson or one Cooperon. Thus recover (4.206) and (4.209, 4.210).

A4.5.2 A Ward identity

Ward identities are aimed to express conservation laws in a systematic way [90]. Technically, they appear as relations between one-particle and two-particle Green's functions, of which the probability of quantum diffusion is an example. The optical theorem (section A2.1.3) is a well-known example of a Ward identity, which relates the scattering amplitude (a one-particle Green's function) to the total scattering cross section (a two-particle Green's function).

We consider here a very simple example of such an identity which relates the product of two (retarded or advanced) Green's functions to the derivative of a single Green's function with respect to energy:

$$\sum_{\boldsymbol{k}_1} G_\epsilon^R(\boldsymbol{k},\boldsymbol{k}_1)G_\epsilon^R(\boldsymbol{k}_1,\boldsymbol{k}') = -\frac{\partial}{\partial\epsilon}G_\epsilon^R(\boldsymbol{k},\boldsymbol{k}'). \tag{4.212}$$

To demonstrate this relation, we use the momentum representation (3.31) of Green's functions:

$$\sum_{\boldsymbol{k}_1} G_\epsilon^R(\boldsymbol{k},\boldsymbol{k}_1)G_\epsilon^R(\boldsymbol{k}_1,\boldsymbol{k}') = \sum_{n,m}\sum_{\boldsymbol{k}_1}\frac{\phi_n^*(\boldsymbol{k})\phi_n(\boldsymbol{k}_1)\phi_m^*(\boldsymbol{k}_1)\phi_m(\boldsymbol{k}')}{(\epsilon-\epsilon_n+i0)(\epsilon-\epsilon_m+i0)}. \tag{4.213}$$

Since $\sum_{\boldsymbol{k}_1}\phi_n(\boldsymbol{k}_1)\phi_m^*(\boldsymbol{k}_1) = \delta_{nm}$, we immediately deduce relation (4.212), which also implies for operators that

$$\boxed{\mathrm{Tr}\left(\hat{G}_\epsilon^R\hat{G}_\epsilon^R\right) = -\frac{\partial}{\partial\epsilon}\mathrm{Tr}\hat{G}_\epsilon^R} \tag{4.214}$$

A4.5.3 Correlations of wave functions

The formalism developed in this chapter provides useful information on the structure of wave functions in disordered systems. We start by considering the wave function autocorrelation $\langle\phi^*(\boldsymbol{r})\phi(\boldsymbol{r}')\rangle_\epsilon$ where $\langle\cdots\rangle_\epsilon$ denotes the average both over disorder configurations and over eigenstates at a given energy ϵ. By definition,

$$\langle \phi^*(\boldsymbol{r})\phi(\boldsymbol{r}')\rangle_\epsilon = \frac{1}{\nu_0}\overline{\sum_n \phi_n^*(\boldsymbol{r})\phi_n(\boldsymbol{r}')\delta(\epsilon-\epsilon_n)} \ . \tag{4.215}$$

This quantity can also be defined from (3.26) as the disorder average of the non-local density of states. Equation (3.28) shows that it is equal to $-1/(\pi\nu_0)\mathrm{Im}\overline{G}^R(\boldsymbol{r},\boldsymbol{r}')$, that is, $g(\boldsymbol{r}-\boldsymbol{r}')/\Omega$, where the short range function $g(\boldsymbol{R})$ is given by (3.98). Therefore the wave function autocorrelation is also short range and depends on the space dimensionality, namely:

$$\Omega\,\langle \phi^*(\boldsymbol{r})\phi(\boldsymbol{r}')\rangle_\epsilon = \begin{cases} \dfrac{\sin k_0 R}{k_0 R}\ e^{-R/2l_e} & d=3 \\ J_0(k_0 R)\ e^{-R/2l_e} & d=2 \\ \cos k_0 R\ e^{-R/2l_e} & d=1 \end{cases} \tag{4.216}$$

where $R=|\boldsymbol{r}-\boldsymbol{r}'|$ and $k_0=\sqrt{2m\epsilon}$ for electrons or $k_0=\epsilon/c$ for electromagnetic waves. It is worth noticing that such behavior for this correlation function is characteristic of quantum systems whose classical limit is chaotic (in this case, there is no disorder, $l_e=\infty$ and the average is performed only over wave functions). M. V. Berry has shown that this corresponds to random wave functions whose structure may be described by a random superposition of plane waves [91]:

$$\phi_n(\boldsymbol{r}) = \frac{1}{\sqrt{\Omega}}\sum_{\boldsymbol{k}} a_n(\boldsymbol{k})\ e^{i\boldsymbol{k}\cdot\boldsymbol{r}}, \tag{4.217}$$

where $|\boldsymbol{k}|=k_0$. For such a random superposition of uncorrelated plane waves, the $a_n(\boldsymbol{k})$ verify $\langle a_n^*(\boldsymbol{k})a_n(\boldsymbol{k}')\rangle = \delta_{\boldsymbol{k}\boldsymbol{k}'}$, so that it follows immediately

$$\Omega\,\langle \phi^*(\boldsymbol{r})\phi(\boldsymbol{r}')\rangle_\epsilon = \sum_{\boldsymbol{k},|\boldsymbol{k}|=k_0} e^{i\boldsymbol{k}\cdot(\boldsymbol{r}-\boldsymbol{r}')}. \tag{4.218}$$

Replacing the sum by an angular integral suitably normalized, we obtain

$$\Omega\,\langle \phi^*(\boldsymbol{r})\phi(\boldsymbol{r}')\rangle_\epsilon = \int_{|\boldsymbol{k}|=k_0} d\Omega_{\boldsymbol{k}}\ e^{i\boldsymbol{k}\cdot(\boldsymbol{r}-\boldsymbol{r}')}, \tag{4.219}$$

which leads to (4.216) with infinite l_e. This result indicates that eigenfunctions in a disordered system share common features with those of chaotic billiards (see also section 10.1.1). The correlation function (4.219) has been obtained numerically for solutions of the Helmholtz equation in chaotic billiards [92].

Relying upon section A4.5.1, we may also obtain interesting results for the correlations of four wave functions [93]. Consider first the correlation of eigenfunctions $|\phi(\boldsymbol{r})|^2$ and $|\phi(\boldsymbol{r}')|^2$ at different energies ϵ and $\epsilon-\omega$. By definition,

$$\langle|\phi_\epsilon(\boldsymbol{r})|^2|\phi_{\epsilon-\omega}(\boldsymbol{r}')|^2\rangle = \frac{1}{\nu_0^2}\overline{\sum_{k,l}|\phi_k(\boldsymbol{r})|^2\delta(\epsilon-\epsilon_k)|\phi_l(\boldsymbol{r}')|^2\delta(\epsilon-\omega-\epsilon_l)}. \tag{4.220}$$

Using (3.25), this quantity can be related to the local density of states $\overline{\rho_\epsilon(\boldsymbol{r})\rho_{\epsilon-\omega}(\boldsymbol{r}')}$ given by (4.209) so that

$$\boxed{\Omega^2\langle|\phi_\epsilon(\boldsymbol{r})|^2|\phi_{\epsilon-\omega}(\boldsymbol{r}')|^2\rangle = 1+\mathrm{Re}\left[g^2(\boldsymbol{R})\Pi_d(\boldsymbol{r},\boldsymbol{r},\omega)+\tfrac{1}{2}\Pi_d^2(\boldsymbol{r},\boldsymbol{r}',\omega)\right]} \tag{4.221}$$

where we have defined $\Pi_d = P_d/(\pi\rho_0)$. To this we must add the Cooperon contribution, obtained from (4.210).

Similarly, we can compute the correlation function

$$\langle\phi_\epsilon^*(\boldsymbol{r})\phi_\epsilon(\boldsymbol{r}')\phi_{\epsilon-\omega}^*(\boldsymbol{r}')\phi_{\epsilon-\omega}(\boldsymbol{r})\rangle = \frac{1}{\rho_0^2}\overline{\sum_{k,l}\phi_k^*(\boldsymbol{r})\phi_k(\boldsymbol{r}')\phi_l^*(\boldsymbol{r}')\phi_l(\boldsymbol{r})\delta(\epsilon-\epsilon_k)\delta(\epsilon-\omega-\epsilon_l)} \tag{4.222}$$

Introducing the correlation function of the non-local density of states $\overline{\rho_\epsilon(\boldsymbol{r},\boldsymbol{r}')\rho_{\epsilon-\omega}(\boldsymbol{r}',\boldsymbol{r})}$ and with the help of (4.206), we obtain

$$\boxed{\Omega^2\langle\phi_\epsilon^*(\boldsymbol{r})\phi_\epsilon(\boldsymbol{r}')\phi_{\epsilon-\omega}^*(\boldsymbol{r}')\phi_{\epsilon-\omega}(\boldsymbol{r})\rangle = g^2(\boldsymbol{R})+\mathrm{Re}\left[\Pi_d(\boldsymbol{r},\boldsymbol{r}',\omega)+g^2(\boldsymbol{R})\Pi_c(\boldsymbol{r},\boldsymbol{r},\omega)\right]} \tag{4.223}$$

where $\Pi_c = P_c/(\pi\rho_0)$.

These quantities exhibit long range correlations, in space and in energy, which follows from the corresponding behavior of the probability of quantum diffusion $P_{d,c}(\boldsymbol{r},\boldsymbol{r}',\omega)$. Within the diffusion approximation, wave functions corresponding to energies separated by ω are correlated spatially over a distance R_ω such that $R_\omega^2 \simeq D/\omega$. Conversely, in a sample of given size L, the wave functions are correlated in a range of energy $\omega \simeq D/L^2$. This behavior has to be contrasted with the short range nature of the correlation of two wave functions given in (4.216). Finally, these expressions have been obtained for energies $\omega \gg \Delta$ and they cannot describe situations where $\omega \simeq \Delta$. In that limit, we need to use the non-linear supersymmetric σ model [93] (see also section 10.1.2).

5

Properties of the diffusion equation

5.1 Introduction

In Chapter 4, we established general expressions for the probability of quantum diffusion $P_d(\boldsymbol{r},\boldsymbol{r}',\omega)$ and for the structure factor $\Gamma_\omega(\boldsymbol{r},\boldsymbol{r}')$. With the aid of these quantities, which are solutions of the integral equations (4.24) and (4.25), we can describe all the physical phenomena studied in this book. It is therefore useful to present these solutions for commonly encountered geometries. Moreover, we have shown that for an infinite medium, in the regime of slow variations, P_d and Γ_ω are solutions of a diffusion equation. In this chapter, we shall study the solutions of this equation for certain geometries. The validity of the diffusion equation will be discussed for the cases of infinite and semi-infinite media in Appendices A5.1 and A5.3.

We shall take particular interest in the Laplace transform $P_\gamma(\boldsymbol{r},\boldsymbol{r}')$ of the probability $P(\boldsymbol{r},\boldsymbol{r}',t)$. This measures the sum of the contributions to the probability from multiple scattering trajectories between $\boldsymbol{r}$ and $\boldsymbol{r}'$ for times less than $1/\gamma$. From this quantity we shall define a characteristic time, the *recurrence time*, which describes the total time spent around an arbitrary point in the medium. It depends on the space dimensionality and on the geometry of the system.

For a finite system, typically a cube of side L, there is a natural characteristic time scale τ_D defined by $L^2 = D\tau_D$. It separates the short time regime, where the role of boundary conditions may be neglected, from the long time regime where these conditions become essential. The inverse of this time defines the so-called *Thouless frequency* $1/\tau_D$ or the *Thouless energy* $E_c = \hbar/\tau_D$.

In this chapter, we shall consider different types of boundary conditions which essentially describe the cases of an isolated system or one coupled to an external environment. The effective space dimensionality may also depend on the nature of the source. For example, in the case of the scattering of electromagnetic waves, depending on the geometry of the incident beam, we may have a problem of effectively one-dimensional diffusion as in the case of a broad beam (that is, one that may be approximated by a plane wave) or a three-dimensional problem for a collimated beam, whose cross section is small compared with the interface of the medium. The case of the plane wave is particularly useful, and we will thus spend considerable time on one-dimensional diffusion.

The reader may perhaps think that dedicating an entire chapter to study of the solutions of the diffusion equation is superfluous, given the vast literature on the subject in both mathematics and physics [94,95]. Our goal here is not to treat the subject exhaustively, but rather to present, in a direct fashion, a certain number of results which will prove useful, and to complement them by relevant references.

5.2 Heat kernel and recurrence time

Let us first consider the situation where the probability of quantum diffusion $P(\boldsymbol{r},\boldsymbol{r}',t)$ is the solution of a diffusion equation. This is an approximation of the integral equations (4.24) and (4.25) which is valid in the limit of slow spatial and temporal variations (section 4.5). A detailed study of the validity of this approximation is presented in Appendices A5.1 and A5.3.

The probability $P(\boldsymbol{r},\boldsymbol{r}',t)$ contains two contributions, the Diffuson and the Cooperon.[1] In the diffusion approximation, each of these contributions is the Green solution of the differential equation which is the temporal Fourier transform of (4.34)

$$\boxed{\left[\frac{\partial}{\partial t} - D\Delta\right] P(\boldsymbol{r},\boldsymbol{r}',t) = \delta(\boldsymbol{r}-\boldsymbol{r}')\delta(t)} \tag{5.1}$$

For the rest of this chapter, $P(\boldsymbol{r},\boldsymbol{r}',t)$ will denote the solution of equation (5.1). We have shown in section 4.5 that within the framework of the diffusion approximation, the structure factor is simply proportional to the probability of quantum diffusion by relation (4.37). We shall thus be interested in the behavior of $P(\boldsymbol{r},\boldsymbol{r}',t)$, knowing that the behavior of the structure factor may be deduced immediately. Equation (5.1) is also known as the *heat equation* and was studied by Fourier in order to describe the diffusive nature of heat propagation.

5.2.1 Heat kernel – probability of return to the origin

In an arbitrary volume Ω, the general solution of equation (5.1) is of the form:

$$P(\boldsymbol{r},\boldsymbol{r}',t) = \theta(t)\sum_n \psi_n^*(\boldsymbol{r})\psi_n(\boldsymbol{r}')\, e^{-E_n t} \tag{5.2}$$

or

$$P(\boldsymbol{r},\boldsymbol{r}',\omega) = \sum_n \frac{\psi_n^*(\boldsymbol{r})\psi_n(\boldsymbol{r}')}{-i\omega + E_n}, \tag{5.3}$$

[1] We disregard here the contribution of P_0. For the Cooperon, we must consider the limit $\boldsymbol{r}=\boldsymbol{r}'$. See section 4.6.

where E_n and ψ_n are the eigenvalues (or eigenfrequencies)[2] and the normalized eigenfunctions (or modes) of the equation:

$$-D\Delta\psi_n(\boldsymbol{r}) = E_n\psi_n(\boldsymbol{r}). \tag{5.4}$$

One important quantity which characterizes the properties of diffusion is the probability of return to the origin, integrated over the point of departure. This function, denoted by $Z(t)$ and called the *heat kernel*, is defined for $t > 0$ by

$$\boxed{Z(t) = \int_\Omega P(\boldsymbol{r},\boldsymbol{r},t)\,d\boldsymbol{r} = \sum_n e^{-E_n t}} \tag{5.5}$$

and thus depends only on the spectrum of eigenvalues.[3] The heat kernel may also be written as the trace of an operator

$$Z(t) = \mathrm{Tr}\, e^{Dt\Delta} \tag{5.6}$$

where Δ is the Laplace operator. The calculation of $Z(t)$ for diffusion in free space is simple, and is recalled in section 5.3. For diffusion in a finite domain, there is not always a simple analytic expression for $Z(t)$. It is, however, possible to obtain asymptotic expansions (Appendix A5.5). In the following, we shall frequently call this quantity the *integrated return probability*. The form (5.4) of the eigenvalue equation is identical to the Schrödinger equation for a free particle of effective mass $m = \hbar/(2D)$. Consequently, $Z(t)$ may also be interpreted as the *partition function* associated with this Schrödinger equation, where the time plays the role of inverse temperature. This observation allows us to transpose certain known solutions of the Schrödinger equation to the case of diffusion.

Exercise 5.1: semi-group structure of the diffusion equation

Show that the solutions of the diffusion equation (5.1) satisfy the so-called semi-group relation:

$$\int dr' P(r,r',t_1)P(r',r'',t_2) = P(r,r'',t_1+t_2) \tag{5.7}$$

for positive times t_1 and t_2.

For this, consider the general solution of (5.2). Using the normalization of the eigenfunctions, $\int dr \psi_n^*(r)\psi_{n'}(r) = \delta_{n\,n'}$, show that

$$\begin{aligned}\int dr' P(r,r',t_1)P(r',r'',t_2) &= \int dr' \sum_{n,n'} \psi_n(r)\psi_n^*(r')\psi_{n'}(r')\psi_{n'}^*(r'')\, e^{-E_n t_1}\, e^{-E_{n'} t_2} \\ &= \sum_n \psi_n(r)\psi_n^*(r'')\, e^{-E_n(t_1+t_2)}\end{aligned} \tag{5.8}$$

which leads to the result.

---

2 The eigenvalues E_n have the dimension of inverse time. Despite this, we shall often use "energy" to designate eigenvalues of the diffusion equation.

3 This is not always strictly exact, as we will see in Appendix A6.1.

5.2.2 Recurrence time

There are several ways to characterize a diffusion process. One of these consists of calculating the time spent by a diffusing particle in the vicinity of a given point. This time depends on space dimensionality, geometry, and on the boundary conditions.

Consider a diffusing particle situated at point $\boldsymbol{r}$ at an initial time $t=0$. Its evolution for later times is given by the probability $P(\boldsymbol{r},\boldsymbol{r}',t)$. In particular, the probability that after a time interval t the particle is located in a given volume $\mathcal{V}$,[4] is equal to

$$\int_{\mathcal{V}} P(\boldsymbol{r},\boldsymbol{r}',t)\,d\boldsymbol{r}'. \tag{5.9}$$

In a given time interval T, the time $\tau(\boldsymbol{r},\mathcal{V},T)$ spent by the diffusing particle in the volume $\mathcal{V}$ is given by the integral

$$\tau(\boldsymbol{r},\mathcal{V},T) = \int_0^T \int_{\mathcal{V}} P(\boldsymbol{r},\boldsymbol{r}',t)\,d\boldsymbol{r}'\,dt \tag{5.10}$$

which depends on the chosen volume $\mathcal{V}$, on the initial departure point $\boldsymbol{r}$, and on T. In the limit $T\to\infty$, the time $\tau(\boldsymbol{r},\mathcal{V},T)$ may have a finite limit, or it may depend asymptotically on T.

Instead of considering a finite time interval T, imagine a diffusing particle with a finite lifetime τ_γ. This lifetime is, for now, introduced by hand; we will discuss its physical significance later. Simply notice that it has no relation to the elastic time τ_e introduced in Chapter 3. The time spent in the volume $\mathcal{V}$ may thus be written

$$\tau(\boldsymbol{r},\mathcal{V},\tau_\gamma) = \int_0^\infty \int_{\mathcal{V}} P(\boldsymbol{r},\boldsymbol{r}',t)\,e^{-t/\tau_\gamma}\,d\boldsymbol{r}'\,dt \tag{5.11}$$

and, introducing the Laplace transform

$$\boxed{P_\gamma(\boldsymbol{r},\boldsymbol{r}') = \int_0^\infty P(\boldsymbol{r},\boldsymbol{r}',t)\,e^{-\gamma t}\,dt} \tag{5.12}$$

we have

$$\tau(\boldsymbol{r},\mathcal{V},\gamma) = \int_{\mathcal{V}} P_\gamma(\boldsymbol{r},\boldsymbol{r}')\,d\boldsymbol{r}' \tag{5.13}$$

where

$$\gamma = \frac{1}{\tau_\gamma} \tag{5.14}$$

is a real number. The Laplace transform $P_\gamma(\boldsymbol{r},\boldsymbol{r}')$ satisfies the diffusion equation (4.38) with $\omega=i\gamma$,

$$\boxed{(\gamma - D\Delta)\,P_\gamma(\boldsymbol{r},\boldsymbol{r}') = \delta(\boldsymbol{r}-\boldsymbol{r}')} \tag{5.15}$$

[4] The point $\boldsymbol{r}$ is not necessarily inside the volume $\mathcal{V}$.

If $\tau(\boldsymbol{r}, \mathcal{V}, \gamma)$ diverges when $\gamma \to 0$, we say that the diffusion process is recurrent. Among all possible choices for the time $\tau(\boldsymbol{r}, \mathcal{V}, \gamma)$, we may consider the time spent in the neighborhood of the point of departure $\boldsymbol{r}$. This neighborhood is defined via the volume $\mathcal{V} = A_d l_e^d$ where the elastic mean free path l_e is the elementary length scale set by the diffusion, and where A_d is the volume of the unit sphere (relation 15.2). The time thus defined is called the *recurrence time* and is directly related to the probability of return to the origin:

$$\tau_R(\boldsymbol{r}, \gamma) = A_d l_e^d \, P_\gamma(\boldsymbol{r}, \boldsymbol{r}). \tag{5.16}$$

The recurrence time is thus the total time spent by the particle around its point of departure. It takes into account only those diffusive trajectories whose length is smaller than $l_e \tau_\gamma / \tau_e$. The recurrence time may possibly diverge when all trajectories are taken into account, that is, when $\tau_\gamma \to \infty$.[5] We shall give an example of free diffusion in dimension $d \leq 2$ in the following section.

To conclude this section, we emphasize once again that the lifetime τ_γ has been introduced in a formal sense as a calculational tool. We shall nonetheless see, in particular in Chapter 6, that this lifetime may have different physical origins related to dephasing between the two trajectories which form the Diffuson or the Cooperon.

We now characterize the nature of the diffusion, that is, its recurrence properties for different space dimensionalities and geometries.

5.3 Free diffusion

The solution of equation (5.1) in free space of dimension d is given by the Gaussian law. To obtain this result, we start from the Fourier transform $P(\boldsymbol{q}, t)$ which satisfies the equation

$$\left(\frac{\partial}{\partial t} + Dq^2\right) P(\boldsymbol{q}, t) = \delta(t) \tag{5.17}$$

whose solution is

$$\boxed{P(\boldsymbol{q}, t) = \theta(t)\, e^{-Dq^2 t}} \tag{5.18}$$

We thus deduce that its Fourier transform

$$P(\boldsymbol{r}, \boldsymbol{r}', t) = \int \frac{d\boldsymbol{q}}{(2\pi)^d} P(\boldsymbol{q}, t)\, e^{i\boldsymbol{q}\cdot(\boldsymbol{r}-\boldsymbol{r}')} \tag{5.19}$$

may be written for $t > 0$:

$$\boxed{P(\boldsymbol{r}, \boldsymbol{r}', t) = \frac{1}{(4\pi D t)^{d/2}}\, e^{-|\boldsymbol{r}-\boldsymbol{r}'|^2/4Dt}} \tag{5.20}$$

[5] For uniform diffusion in a volume Ω, $P(\boldsymbol{r}, \boldsymbol{r}', t)$ is constant and equal to $1/\Omega$. The recurrence time is therefore given by $\tau_\gamma (A_d l_e^d / \Omega)$.

Among all the moments associated with this distribution function, the most common is that giving the typical distance reached via diffusion after a time t

$$\langle R^2(t) \rangle = 2dDt. \tag{5.21}$$

We may also compute the average of the phase factor $e^{i\boldsymbol{q}\cdot\boldsymbol{R}(t)}$:

$$\left\langle e^{i\boldsymbol{q}\cdot\boldsymbol{R}(t)} \right\rangle = e^{-Dq^2 t} \tag{5.22}$$

which is the characteristic function associated with the probability $P(\boldsymbol{r},\boldsymbol{r}',t)$. Finally, the probability of return to the origin after a time t is obtained from (5.20) by taking $\boldsymbol{r}=\boldsymbol{r}'$, that is

$$P(\boldsymbol{r},\boldsymbol{r},t) = \frac{1}{(4\pi Dt)^{d/2}}, \tag{5.23}$$

so that the integrated probability $Z(t)$ defined by (5.5) may be written, for a volume Ω,

$$Z(t) = \frac{\Omega}{(4\pi Dt)^{d/2}}. \tag{5.24}$$

Similarly, we may calculate the temporal Fourier transform in $d=3$,

$$P(\boldsymbol{r},\boldsymbol{r}',\omega) = \frac{1}{4\pi D|\boldsymbol{r}-\boldsymbol{r}'|}\, e^{-(1-i)|\boldsymbol{r}-\boldsymbol{r}'|\sqrt{\omega/2D}} \tag{5.25}$$

and the Fourier transform $Z(\omega)$ of (5.24) is given by relations (15.82–15.84).

Exercise 5.2: Laplace transform in dimension d for free diffusion, and time spent inside a sphere about the origin

• The time integrated probability for a particle to go from the origin to a point $\boldsymbol{r}$ is given by the Laplace transform $P_\gamma(0,\boldsymbol{r})$ (relation 5.12) and depends on space dimension d as

$$P_\gamma^{(3)}(0,\boldsymbol{r}) = \frac{1}{4\pi Dr} e^{-r/L_\gamma}, \quad P_\gamma^{(2)}(0,\boldsymbol{r}) = \frac{1}{2\pi D} K_0(r/L_\gamma), \quad P_\gamma^{(1)}(0,\boldsymbol{r}) = \frac{L_\gamma}{2D} e^{-r/L_\gamma} \tag{5.26}$$

where $r=|\boldsymbol{r}|$ and $L_\gamma = \sqrt{D\tau_\gamma}$ is the associated diffusion length. In three dimensions, this probability converges in the limit $\gamma \to 0$ and we recover the well-known result

$$P(0,\boldsymbol{r},\omega=0) = \frac{1}{4\pi Dr}. \tag{5.27}$$

• Starting from $P_\gamma(0,\boldsymbol{r})$, we obtain from the integral (5.13) the time $\tau_v(\gamma)$ spent in a volume $v = A_d R^d$ about the origin. In the limit $R \ll L_\gamma$, this may be written

$$\tau_v^{(3)}(\gamma) = \frac{R^2}{2D}, \quad \tau_v^{(2)}(\gamma) = \frac{R^2}{2D} \ln b \frac{L_\gamma}{l_e}, \quad \tau_v^{(1)}(\gamma) = \frac{L_\gamma R}{D}, \tag{5.28}$$

where $b \simeq 1.747$. We see that, for $d=3$, the particle spends a finite time about the origin. In contrast, for $d=1$, this time is infinite in the limit of infinite lifetime, and diverges as $\sqrt{\tau_\gamma}$. The case $d=2$ is marginal, with a logarithmic divergence as a function of τ_γ.

We see that the space dimension plays an essential role in the nature of the diffusion. Let us see how the recurrence time $\tau_R(\gamma)$, given by (5.16), depends on it.[6] This may be evaluated starting from the integral[7]

$$\tau_R \simeq A_d l_e^d \int_{\tau_e}^{\infty} \frac{e^{-\gamma t}}{(4\pi D t)^{d/2}} dt. \tag{5.29}$$

We thus obtain

$$\boxed{\begin{array}{ll} d=3 & \dfrac{\tau_R}{\tau_e} \propto 1 \\ d=2 & \dfrac{\tau_R}{\tau_e} \propto \ln \dfrac{L_\gamma}{l_e} \\ d=1 & \dfrac{\tau_R}{\tau_e} \propto \dfrac{L_\gamma}{l_e} \end{array}} \tag{5.30}$$

where the constants of proportionality may be obtained from Exercise 5.2. In dimension $d=3$, τ_R is finite, which means that a diffusing particle, initially at some given point in space, will escape from a small volume l_e^3 after one collision, and will never return. The diffusion is not recurrent. In other words, the phase space is sufficiently large for a diffusing particle to escape to infinity without passing through its point of departure again.

In contrast, for $d \leq 2$, the integral (5.29) diverges for long times and depends on the cutoff τ_γ which suppresses the contribution of long diffusion trajectories. This divergence of τ_R in the limit $\tau_\gamma \to \infty$ may be interpreted by saying that *a diffusing particle passes through its point of departure infinitely many times*. The diffusion is thus recurrent in dimension $d \leq 2$. We may reexpress this result by saying that in the course of a diffusive trajectory made of n collisions, the particle has typically spent a time $\tau_e \sqrt{n}$ $(d = 1)$ (or $\propto \tau_e \ln n$ for $d = 2$) about its departure point. These results constitute Polya's theorem [94] on the recurrence properties of a random walk in free space.

These differences of behavior as a function of dimensionality have essential physical consequences. We shall see many examples such as the quantum correction to the electrical conductivity (Chapter 7), or the correction to the density of states in the vicinity of the Fermi level for an interacting electron gas (section 13.4). Quite generally, it is possible to characterize the nature of the medium in which the waves or electrons propagate, starting from the recurrence property. If the diffusion is not recurrent, we speak of a metal and of extended waves, while in the recurrent case we speak of an insulator and of waves localized in the sense of Anderson. In three dimensions, the classical diffusion is not recurrent, but quantum effects in a strong disorder lead to a transition towards a phase where quantum diffusion becomes recurrent.

The notion of recurrence which we have been discussing is defined from the probability of return to a point (the origin). We may generalize this notion to the case of the probability

[6] For an infinite medium, the recurrence time does not depend on the point of departure.
[7] This may also be derived from the equations (5.28) of the preceding exercise by taking a volume of radius l_e.

of return to a line or to a plane (a hyperplane). For example, the coherent albedo is related to the probability of return to a plane (section 8.4).

5.4 Diffusion in a periodic box

Consider now the solution of the diffusion equation in a box in dimension d, of side lengths L_i $(i=1,\ldots,d)$ and volume Ω with periodic boundary conditions (the volume Ω is thus a torus). The eigenfunctions of equation (5.4) are normalized plane waves $\psi(\boldsymbol{r})=(1/\sqrt{\Omega})e^{i\boldsymbol{q}\cdot\boldsymbol{r}}$, and the eigenvalues are equal to Dq^2, where the values of $\boldsymbol{q}$ are quantized: $q_i=2\pi n_i/L_i$ with $n_i=0,\pm1,\pm2,\pm3,\ldots$. From (5.2), it follows that

$$P(\boldsymbol{r},\boldsymbol{r}',t) = \frac{1}{\Omega}\theta(t)\sum_{\boldsymbol{q}} e^{-Dq^2t}e^{i\boldsymbol{q}\cdot(\boldsymbol{r}'-\boldsymbol{r})}. \tag{5.31}$$

The ensemble of eigenvalues constitutes a rectangular network Γ^* in d dimensions. On this network we define vectors $\boldsymbol{x}$ of coordinates $x_i=2\pi n_i\sqrt{D}/L_i$. We thus derive the expression for the heat kernel

$$Z(t) = \sum_{\boldsymbol{x}\in\Gamma^*} e^{-|\boldsymbol{x}|^2t}. \tag{5.32}$$

Γ^* may be understood as the reciprocal lattice of a real lattice Γ which is the ensemble of points $\boldsymbol{y}$ of coordinates $y_i = m_iL_i/\sqrt{D}$. Applying the Poisson summation formula (15.106), we get

$$Z(t) = \sum_{\boldsymbol{x}\in\Gamma^*} e^{-|\boldsymbol{x}|^2t} = \frac{\Omega}{(4\pi Dt)^{d/2}}\sum_{\boldsymbol{y}\in\Gamma} e^{-|\boldsymbol{y}|^2/4t}. \tag{5.33}$$

The expansion of $Z(t)$ as a function of the integers m_i may be interpreted as the probability of return to the origin after having performed m_i turns around the ring defined by the periodic boundary condition along the coordinate i.[8] The integers m_i are called *winding numbers* or homotopy numbers and they allow a classification of all possible diffusion trajectories on the torus [96]. These are topological numbers, which means that they do not depend on the exact geometrical form of the torus, but only on its topological nature, or more precisely, on the number of its holes, here equal to one. In consequence, the probability of return to the origin, the recurrence time, as well as all the spectral quantities, are identical for all systems having the topology of a torus. We show in Appendix A5.4 how $Z(t)$ depends on the topology of the space in which the diffusion equation is solved.

[8] We can verify that the integers n_i thus defined describe the eigenvalues of angular momentum associated with each of the rings of length L_i.

5.5 Diffusion in finite systems

5.5.1 Diffusion time and Thouless energy

Consider a system of finite size, for example a cube of edge length L, and diffusing particles which are initially inside this volume. For sufficiently small times, the particles diffuse as in an infinite medium. We may define a typical characteristic time after which a diffusing particle begins to "feel" the boundaries: the probability of reaching the boundaries is no longer negligible. For times longer than this, the probability of finding a particle at a point becomes spatially uniform, so that the diffusing particle explores the entire volume at its disposal in an ergodic fashion. It is customary to define this time τ_D using expression (5.21) derived for an infinite medium and taking $\langle R^2(t)\rangle = L^2$ as the typical size, so that

$$\tau_D = \frac{L^2}{D}. \tag{5.34}$$

τ_D is variously called the diffusion time, the ergodic time, or the *Thouless time*.[9] The inverse of this time defines a characteristic frequency, or a characteristic energy in the case of electronic systems, which is called the *Thouless energy* or *Thouless frequency*.[10]

$$\boxed{E_c = \frac{\hbar}{\tau_D} = \frac{\hbar D}{L^2}} \tag{5.35}$$

This energy plays a fundamental role in the description of the physical properties of weakly disordered media.

5.5.2 Boundary conditions for the diffusion equation

We now wish to explore the effects of the boundaries on diffusion. To this end, we must specify the boundary conditions, which are essentially of two types.

- **Neumann conditions** The current associated with the diffusion probability vanishes at the boundaries:

$$\boldsymbol{n} \cdot \nabla_{\boldsymbol{r}'} P(\boldsymbol{r}, \boldsymbol{r}', t)|_{\boldsymbol{r}' \in \partial\Omega} = 0, \tag{5.36}$$

  where $\boldsymbol{n}$ is a unit vector normal to the boundary $\partial\Omega$. This condition describes an isolated system. For the case of electrons, this means that the electrons cannot exit the sample. For the case of waves, it describes the case of *reflecting walls*.
- **Dirichlet conditions** The probability vanishes at the boundaries:

$$P(\boldsymbol{r}, \boldsymbol{r}', t)|_{\boldsymbol{r}' \in \partial\Omega} = 0. \tag{5.37}$$

[9] This expression for τ_D is that commonly used in the literature. It is worth noting that it does not involve the space dimension d which enters through (5.21).

[10] In the chapters in which $\hbar = 1$, we shall use the same notation for both Thouless energy and frequency: $E_c = D/L^2$.

For electronic systems, this describes a system coupled to a "reservoir," in such a way that a particle which leaves the system never returns (connected system). For waves in a diffusive medium, it means that a wave which impinges on the boundary leaves the medium and never returns. This is the case of an *absorbing wall*.

Remarks

- Note that the physical sense of the Neumann and Dirichlet boundary conditions here are opposite to that for the Schrödinger equation, where an isolated system is constrained by an infinite potential barrier, which corresponds to the Dirichlet condition.
- There is no a priori reason why the boundary conditions of the diffusion equation can be deduced simply from those of the initial integral equation (4.25) for the probability of quantum diffusion $P(\boldsymbol{r},\boldsymbol{r}',t)$. We shall see that the solution of this equation in the diffusion approximation leads to mixed boundary conditions (Appendix A5.3):

$$P(\boldsymbol{r},\boldsymbol{r}',t) - z_0\, \boldsymbol{n} \cdot \nabla_{\boldsymbol{r}'} P(\boldsymbol{r},\boldsymbol{r}',t)|_{\boldsymbol{r}'\in\partial\Omega} = 0, \tag{5.38}$$

where z_0 is a length related to the elastic mean free path l_e.

5.5.3 Finite volume and "zero mode"

Consider a finite system of volume Ω. The spectrum of the Laplacian (5.4) is discrete, and is given by the sequence of eigenvalues $\{E_n\}$. The integrated probability of return to the origin is:

$$Z(t) = \sum_n e^{-E_n t}. \tag{5.39}$$

For an isolated system, the Neumann conditions (5.36) imply that the lowest mode of the diffusion equation has zero energy: $E_0 = 0$ (see footnote 2, page 150). When the time t tends to infinity, the dominant contribution to the probability (5.39) comes from the lowest mode. Since it is $E_0 = 0$, the return probability tends to 1. This result is a simple expression of the fact that in a system with reflecting walls, a particle cannot leave the volume Ω. In a finite system, this means that it must return to its departure point infinitely many times. This is the *ergodic regime*. The diffusion is thus recurrent, and if the particle has a finite lifetime, the recurrence time increases as τ_γ. The behavior of the return probability for $t \to \infty$ is controlled uniquely by τ_γ, and is given by

$$Z(t) = \frac{1}{\Omega} e^{-t/\tau_\gamma} \quad \text{(Neumann)}. \tag{5.40}$$

For an open system with Dirichlet conditions (5.37), the energy of the lowest mode is strictly greater than that of a system with Neumann conditions. The mode $E_0 = 0$ is excluded, and this "ground state" energy is finite and proportional to the Thouless energy E_c. This gap determines the long time behavior of the probability of return to the origin. This reflects the fact that there is a finite absorption probability at the boundaries, so that the diffusion is not recurrent. We have that

$$Z(t) = e^{-t/\tau_D - t/\tau_\gamma} \quad \text{(Dirichlet)} \tag{5.41}$$

and for $\tau_\gamma \to \infty$, the long trajectories are naturally cut off after a time of the order of the Thouless time τ_D. The recurrence time τ_R is expressed similarly to (5.30), but where the cutoff time τ_γ is now replaced by τ_D, that is, where L_γ is replaced by the typical system size $L = \Omega^{-1/d}$.

5.5.4 *Diffusion in an anisotropic domain*

Let us consider the case of diffusion in an anisotropic domain,[11] that is, for which the dimensions $L_x > L_y > L_z$ are very different. The behavior of the solutions of the diffusion equation for a particle placed initially at the center of the domain depends on the time scale considered. We can define three characteristic times $\tau_D^{(x)} > \tau_D^{(y)} > \tau_D^{(z)}$ and, therefore, three Thouless energies $E_c^{(x)} < E_c^{(y)} < E_c^{(z)}$. There are four regimes.

- $t < \tau_D^{(z)}$: the diffusion is three dimensional and the probability of return to the origin is close to that in the corresponding infinite medium. Consequently, using (5.24), we have $Z(t) \simeq \Omega/(4\pi Dt)^{3/2}$ where $\Omega = L_x L_y L_z$ is the volume of the domain.
- $\tau_D^{(z)} < t < \tau_D^{(y)}$: for these intermediate times the diffusion is two dimensional and we have $Z(t) \simeq L_x L_y / 4\pi Dt$.
- $\tau_D^{(y)} < t < \tau_D^{(x)}$: one-dimensional diffusion, for which $Z(t) \simeq L_x/(4\pi Dt)^{1/2}$.
- $t > \tau_D^{(x)}$: uniform probability. We are in the ergodic regime defined in section 5.5.1. We speak of the zero-dimensional limit.

Figure 5.1 summarizes the different situations for the case of a two-dimensional system.

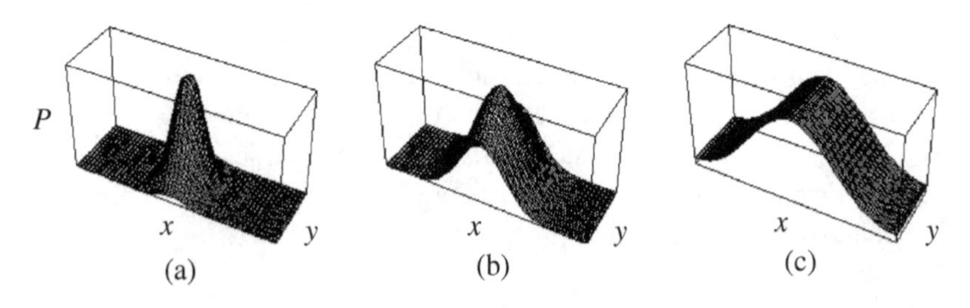

Figure 5.1 Solution of the diffusion equation in a two-dimensional anisotropic domain. (a) For short times $t \ll \tau_D^{(y)}$ the diffusion is isotropic. (b) For a time comparable to $\tau_D^{(y)}$ the diffusion is influenced by the boundaries perpendicular to the direction y, and it finally becomes one dimensional (c) for $t \gg \tau_D^{(y)}$.

[11] We assume a diffusive motion in each of the three directions, that is, $L_i \gg l_e$, with an isotropic diffusion constant which reflects the isotropy on the microscopic scale.

5.6 One-dimensional diffusion

In this section, we study the solutions of the diffusion equation in the commonly encountered case of one-dimensional geometries. For example, a disordered metal may often be described as a $1d$ wire, if its geometry is sufficiently anisotropic and if we are interested in times which are large compared to the transverse Thouless time (see the preceding section and Figure 5.1(c)).[12] Another effectively one-dimensional situation is encountered in optics, where the incident source emits a plane wave directed along an axis Oz, which illuminates a semi-infinite diffusive medium or a slab of thickness L and infinite cross section. In this case, owing to translation invariance in the plane xOy, the probability $P(\boldsymbol{r}, \boldsymbol{r}', t)$ has the form $P(\boldsymbol{\rho}, z, z', t)$ where $\boldsymbol{\rho} = (\boldsymbol{r} - \boldsymbol{r}')_\perp$ is the projection of the vector $(\boldsymbol{r} - \boldsymbol{r}')$ in the plane xOy. The diffusion equation in the medium is thus separable, and the general solution is the product of free diffusion in the plane xOy described by (5.20) and the one-dimensional solution of the diffusion equation along the Oz axis, that is

$$P(\boldsymbol{\rho}, z, z', t) = \frac{e^{-\rho^2/4Dt}}{4\pi Dt} P(z, z', t). \tag{5.42}$$

Conversely, we have

$$P(z, z', t) = \int d^2\boldsymbol{\rho}\, P(\boldsymbol{\rho}, z, z', t). \tag{5.43}$$

We also define the two-dimensional Fourier transform $P(\boldsymbol{k}_\perp, z, z', t)$:

$$P(\boldsymbol{k}_\perp, z, z', t) = \int d^2\boldsymbol{\rho}\, e^{i\boldsymbol{k}_\perp \cdot \boldsymbol{\rho}} P(\boldsymbol{\rho}, z, z', t). \tag{5.44}$$

Its Laplace transform $P_\gamma(\boldsymbol{k}_\perp, z, z') = \int P(\boldsymbol{k}_\perp, z, z', t)\, e^{-\gamma t} dt$ satisfies a differential equation obtained from (5.15):

$$\left(\gamma + Dk_\perp^2 - D\frac{\partial^2}{\partial z^2}\right) P_\gamma(\boldsymbol{k}_\perp, z, z') = \delta(z - z'). \tag{5.45}$$

We shall make frequent use of the following quantities and notation:

$$\begin{array}{ccccc} P(\boldsymbol{r}, \boldsymbol{r}', t) & \stackrel{\text{Laplace}}{\longrightarrow} & P_\gamma(\boldsymbol{r}, \boldsymbol{r}') & \stackrel{\gamma=0}{\longrightarrow} & P(\boldsymbol{r}, \boldsymbol{r}') \\ P(\boldsymbol{\rho}, z, z', t) & \longrightarrow & P_\gamma(\boldsymbol{\rho}, z, z') & \longrightarrow & P(\boldsymbol{\rho}, z, z') \\ P(\boldsymbol{k}_\perp, z, z', t) & \longrightarrow & P_\gamma(\boldsymbol{k}_\perp, z, z') & \longrightarrow & P(\boldsymbol{k}_\perp, z, z') \end{array} \tag{5.46}$$

[12] The diffusion is one dimensional, but the sequence of microscopic multiple scatterings remains three dimensional.

We also note the following correspondences:

$$\begin{aligned} P_\gamma(z,z') &\longleftrightarrow P(\boldsymbol{k}_\perp, z, z') \\ \gamma &\longleftrightarrow Dk_\perp^2 \\ \frac{1}{L_\gamma} &\longleftrightarrow k_\perp \end{aligned} \tag{5.47}$$

$$P_\gamma(k_\perp, z, z') \longleftrightarrow P\left(\sqrt{k_\perp^2 + \frac{1}{L_\gamma^2}}, z, z'\right). \tag{5.48}$$

We next consider solutions to the one-dimensional diffusion equation for the different boundary conditions defined earlier. In particular, we discuss diffusion in a closed loop (a ring), in a wire connected to reservoirs, and in an isolated wire.

5.6.1 The ring: periodic boundary conditions

Consider a ring of perimeter L. The spectrum of the modes may be computed directly from the results of section 5.4. The eigenmodes are $q = 2n\pi/L$ with $n = 0, \pm 1, \pm 2, \pm 3, \ldots$ where n is the angular momentum. The probability $P(z, z', t)$ may be written:

$$P(z,z',t) = \frac{1}{L}\sum_n e^{-4\pi^2 E_c n^2 t}\, e^{2in\frac{\pi}{L}(z-z')}, \tag{5.49}$$

where z is the coordinate of any point along the ring and where $E_c = D/L^2$ is the Thouless energy. Owing to the translation invariance along the ring, the probability depends only on the distance $|z-z'|$ and the probability of return to the origin $P(z,z,t)$ does not depend on the initial point. The function $P_\gamma(z,z')$ is obtained by Laplace transformation and summation over the modes in expression (5.49),[13] or by direct solution of the diffusion equation (5.15):

$$P_\gamma(z,z') = \frac{L_\gamma}{2D}\frac{\cosh(L-2|z-z'|)/2L_\gamma}{\sinh L/2L_\gamma}. \tag{5.50}$$

The recurrence time $\tau_R(\gamma) = 2l_e P_\gamma(z,z)$ may thus be written:

$$\tau_R(\gamma) = \tau_e \frac{L_\gamma}{l_e}\coth\frac{L}{2L_\gamma}. \tag{5.51}$$

In the limit $L_\gamma \ll L$, the trajectories do not travel around the ring, and the result is thus the same as for an infinite wire (5.30). In the inverse limit $L_\gamma \gg L$, the particle may diffuse across the ring many times. The recurrence time becomes of order $2\tau_\gamma l_e/L$, corresponding to uniform diffusion in the ring. This behavior is the expression of a zero mode, that is, the fact that the particle cannot exit the ring, and $Z(t)$ tends to 1 when t goes to infinity.

13 We use the formula (15.65).

In applying the considerations of section 5.4, we may identify the ensemble dual to the angular momentum as that of winding numbers around the ring by using the Poisson transformation (15.106). We thus deduce that the probability $P(z, z', t)$ is given by

$$P(z, z', t) = \frac{1}{\sqrt{4\pi Dt}} \sum_{m=-\infty}^{\infty} e^{-(z-z'+mL)^2/4Dt}. \tag{5.52}$$

The term m in this series represents the probability of going from z to z' having travelled around the ring m times. The description of $P(z, z, t)$ in terms of winding number provides a classification of the possible trajectories. The winding number is a topological number which is independent of the precise form of the ring.

There is an interesting duality between the expansion (5.49) in modes and the expansion (5.52) in winding numbers. Keeping only the zero mode (probability independent of time for $t \gg \tau_D$) in the relation (5.49) corresponds to a continuous sum over all the winding numbers. Conversely, when $t \ll \tau_D$, the winding number $m = 0$ corresponds to a continuous sum over all the modes.

5.6.2 Absorbing boundaries: connected wire

A wire of length L between two contacts is described by Dirichlet boundary conditions since a particle or wave which diffuses close to the boundary is absorbed by a contact. The solutions of equation (5.4) compatible with the constraint (5.37) are $\psi(z) = \sqrt{2/L} \sin qz$, where z is the coordinate along the wire and the eigenmodes are $q = n\pi/L$ with $n = 1, 2, 3, \ldots$. The probability is thus given by

$$P(z, z', t) = \frac{2}{L} \sum_{n>0} e^{-\pi^2 E_c n^2 t} \sin n\pi \frac{z}{L} \sin n\pi \frac{z'}{L}. \tag{5.53}$$

The probability depends on the positions z and z' and not simply on the distance between them. The partition function $Z(t)$ is obtained from (5.53)

$$Z(t) = \sum_{n>0} e^{-\pi^2 E_c n^2 t} \tag{5.54}$$

and tends to 0 when $t \to \infty$, which is indicative of the absence of a zero mode. Using (15.65) or directly solving the diffusion equation (5.15), we may show that the Laplace transform $P_\gamma(z, z')$ defined by (5.12) equals

$$\boxed{P_\gamma(z, z') = \frac{L_\gamma}{D} \frac{(\sinh z_m/L_\gamma)(\sinh(L - z_M)/L_\gamma)}{\sinh L/L_\gamma}} \tag{5.55}$$

or

$$P_\gamma(z, z') = \frac{L_\gamma}{2D} \frac{\cosh(L - z_-)/L_\gamma - \cosh(L - z_+)/L_\gamma}{\sinh L/L_\gamma} \tag{5.56}$$

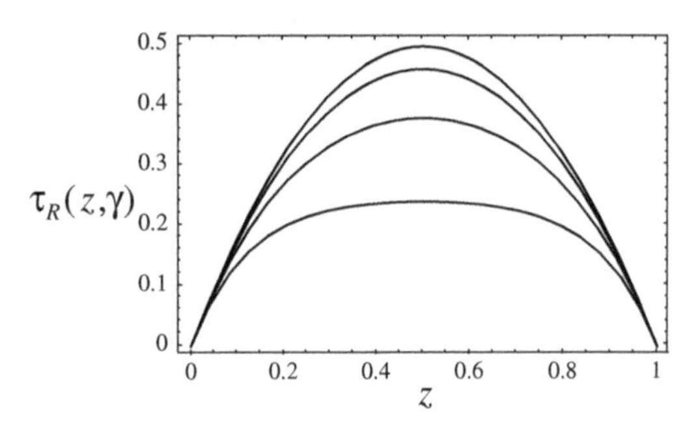

Figure 5.2 Recurrence time $\tau_R(z,\gamma)$ corresponding to absorbing boundary conditions (Dirichlet) for $L/L_\gamma = 0, 1, 2, 4$. This time is finite when $L_\gamma = \infty$. It reaches its maximum value, $L/2v$, in the center, and vanishes at the boundaries. $\tau_R(z,\gamma)$ decreases with decreasing L_γ.

with

$$
\begin{aligned}
z_+ &= z + z' \\
z_- &= |z - z'| \\
z_m &= \min(z, z') = \frac{1}{2}\left(z + z' - |z - z'|\right) \\
z_M &= \max(z, z') = \frac{1}{2}\left(z + z' + |z - z'|\right).
\end{aligned}
\tag{5.57}
$$

$P_\gamma(z, z')$ converges in the limit $\tau_\gamma \to \infty$:

$$
\boxed{P_{\gamma=0}(z, z') = \frac{z_m}{D}\left(1 - \frac{z_M}{L}\right)}
\tag{5.58}
$$

This convergence is due to the fact that the absorbing boundaries eliminate long trajectories. The recurrence time $\tau_R(z,\gamma) = 2l_e P_\gamma(z,z)$ depends on the position z along the wire. As Figure 5.2 shows, it is zero on the boundaries since the particles are absorbed immediately. It is maximal at the center $z = L/2$, where, for $\gamma = 0$, we have $\tau_R(L/2) = \tau_e\ L/2l_e$. By comparing with (5.30), we see that when $L_\gamma \to \infty$, it is the system size that plays the role of a cutoff length and the time spent in the wire in the limit $\tau_\gamma \to \infty$ converges, being of the order of τ_D.

5.6.3 Reflecting boundaries: isolated wire

Electrons cannot exit an isolated metallic wire, so the probability current vanishes at the boundaries. This corresponds to the case of Neumann boundary conditions (5.36). For a plane wave solution of the Helmholtz equation, this describes a perfectly reflecting boundary. In this case, we obtain a spectrum of modes identical to that obtained in the preceding section, but now the value $n = 0$, which corresponds to a constant solution, is

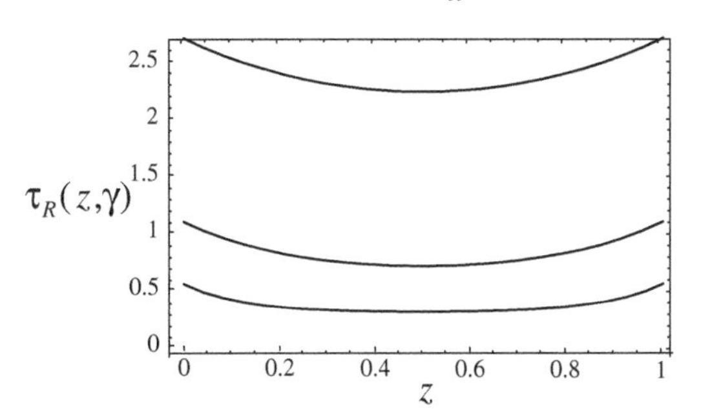

Figure 5.3 Recurrence time $\tau_R(z,\gamma)$ corresponding to Neumann conditions for $L/L_\gamma = 1, 2, 4$. The recurrence time grows with L_γ and diverges when $L_\gamma \to \infty$. It is maximal at the boundaries.

allowed, since it is the current which vanishes at the ends of the wire rather than the eigenfunctions. The allowed values of the wave vector are $q = n\pi/L$ with $n = 0, 1, 2, 3, \ldots$, where L is the length of the wire. The eigenfunctions, however, are different, and are given by $\psi_n(r) = \sqrt{2/L}\cos qz$, so that the probability is

$$P(z,z',t) = \frac{1}{L} + \frac{2}{L}\sum_{n>0} e^{-\pi^2 E_c n^2 t} \cos n\pi\frac{z}{L} \cos n\pi\frac{z'}{L}. \tag{5.59}$$

The partition function $Z(t)$ is given by

$$Z(t) = 1 + \sum_{n>0} e^{-\pi^2 E_c n^2 t} \tag{5.60}$$

which differs from expression (5.54) for absorbing boundary conditions by a constant term corresponding to the zero mode. In contrast to the case of absorbing boundaries, the recurrence time now diverges due to multiple reflections with the boundaries. The function $P_\gamma(z,z')$ is obtained either by Laplace transforming (5.59) or by directly solving the diffusion equation (5.15)

$$\begin{aligned} P_\gamma(z,z') &= \frac{L_\gamma}{D}\frac{(\cosh z_m/L_\gamma)(\cosh(L-z_M)/L_\gamma)}{\sinh L/L_\gamma} \\ &= \frac{L_\gamma}{2D}\frac{\cosh(L-z_-)/L_\gamma + \cosh(L-z_+)/L_\gamma}{\sinh L/L_\gamma}, \end{aligned} \tag{5.61}$$

where z_m, z_M, z_+ and z_- were defined in (5.57). The recurrence time is plotted in Figure 5.3. It is maximal at the boundaries since reflections increase the return probability. In the limit $L \gg L_\gamma$, the trajectories are cut off by L_γ. We verify that in the center of the wire, far from the ends, the boundary conditions do not play a significant role, and we regain the result (5.30) for an infinite wire:

$$\tau_R(L/2) \to \frac{\tau_e}{2}\frac{L_\gamma}{l_e}.$$

Exercise 5.3: average time spent in a wire
Following relation (5.13), we may define the average time spent in a wire of length L by

$$T = \frac{1}{L}\int_0^L P_\gamma(z,z')\,dz\,dz'. \tag{5.62}$$

If the wire is connected to absorbing boundaries, integration of (5.55) leads to

$$T = \frac{\tau_D}{\alpha^3}\left(\alpha - 2\tanh\frac{\alpha}{2}\right) \tag{5.63}$$

with $\alpha = L/L_\gamma$. In the limit $L \ll L_\gamma$, this time converges to $\tau_D/12$, while in the limit $L_\gamma \ll L$, it tends to τ_γ.
In contrast, for a wire connected to reflecting boundaries, show that the time spent in the wire is τ_γ, which simply expresses the fact that the particle never leaves the wire.

Exercise 5.4 Show that in an isolated wire of length $L \gg L_\gamma$, the value of the recurrence time at the boundary is twice that at the center.

5.6.4 Semi-infinite wire

The case of a semi-infinite domain is of special interest since it describes the scattering of a plane wave in a semi-infinite medium, which will be studied in Chapters 8 and 9. It may be deduced simply from the two preceding sections by taking the limit $L \to \infty$. For example, for an absorbing boundary, the probability $P(z,z',t)$ is calculated from (5.53), yielding

$$P(z,z',t) = \frac{2}{\pi}\int_0^\infty e^{-Dq^2 t}\,\sin qz\,\sin qz'\,dq \tag{5.64}$$

which may be put into the form

$$P(z,z',t) = \frac{1}{\sqrt{4\pi Dt}}\left[e^{-(z-z')^2/4Dt} - e^{-(z+z')^2/4Dt}\right]. \tag{5.65}$$

The Laplace transform

$$P_\gamma(z,z') = \frac{L_\gamma}{2D}\left[e^{-\frac{|z-z'|}{L_\gamma}} - e^{-\frac{(z+z')}{L_\gamma}}\right] \tag{5.66}$$

has a finite limit for $L_\gamma \to \infty$, being

$$P(z,z') = \frac{z_m}{D} = \frac{1}{2D}\left(z + z' - |z-z'|\right). \tag{5.67}$$

In similar fashion, starting from (5.59), we obtain, for reflecting boundaries

$$P(z,z',t) = \frac{2}{\pi}\int_0^\infty e^{-Dq^2 t}\,\cos qz\,\cos qz'\,dq \tag{5.68}$$

which may be reexpressed as

$$P(z,z',t) = \frac{1}{\sqrt{4\pi Dt}}\left[e^{-(z-z')^2/4Dt} + e^{-(z+z')^2/4Dt}\right]. \tag{5.69}$$

5.7 The image method

The expressions (5.65) and (5.69) for the case of a semi-infinite wire may be interpreted simply with the aid of the image method employed in electrostatics to describe solutions to the Poisson equation with boundaries. For a reflecting boundary, expression (5.69) may be interpreted as the superposition of the contributions of two charges, one placed at z, and its image placed at $-z$. For the case of an absorbing wall where the probability vanishes at $z'=0$, the image must have a negative "charge," as seen from equation (5.65).

Similarly, in a semi-infinite three-dimensional medium with an absorbing boundary at $z=0$, a source located at $\boldsymbol{r}=(\boldsymbol{r}_\perp, z)$ has an image situated at $\boldsymbol{r}^*=(\boldsymbol{r}_\perp,-z)$. The probability $P(\boldsymbol{r},\boldsymbol{r}')$ given by (5.27) is thus equal to

$$P(\boldsymbol{r},\boldsymbol{r}') = \frac{1}{4\pi D}\left(\frac{1}{|\boldsymbol{r}-\boldsymbol{r}'|} - \frac{1}{|\boldsymbol{r}^*-\boldsymbol{r}'|}\right). \tag{5.70}$$

The one-dimensional probability integrated over time is obtained from (5.43) and is given by

$$P(z,z') = \int d^2\boldsymbol{\rho}\, P(\boldsymbol{\rho},z,z'). \tag{5.71}$$

This may be rewritten:

$$\begin{aligned} P(z,z') &= \frac{1}{4\pi D}\int_S d^2\boldsymbol{\rho}\left(\frac{1}{\sqrt{\rho^2+(z-z')^2}} - \frac{1}{\sqrt{\rho^2+(z+z')^2}}\right) \\ &= \frac{1}{2D}\left(z+z'-|z-z'|\right) = \frac{z_m}{D} \end{aligned} \tag{5.72}$$

where $z_m = \min(z,z')$. We thus recover expression (5.55).

The case of a one-dimensional ring of length L may also be described by the image method. An infinity of images is placed at positions $z+mL$, which enables us to interpret the expansion (5.52) of the probability in winding numbers m.

Finally, for an open wire of finite length, a Poisson transformation of expressions (5.53) and (5.59) enables us to write the probability in the form

$$P(z,z',t) = \frac{1}{\sqrt{4\pi Dt}}\sum_{m=-\infty}^{\infty}\left[e^{-(z-z'+2mL)^2/4Dt} \pm e^{-(z+z'+2mL)^2/4Dt}\right]. \tag{5.73}$$

The $+$ sign corresponds to reflecting boundary conditions and the $-$ sign corresponds to absorbing walls. The resulting form may be simply interpreted as the diffusion from an infinity of positive image charges situated on the abscissa at the points $z+2mL$ and images located on the abscissa at points $-z+2mL$. Depending on the boundary conditions chosen, this second set of images may be of positive or negative charges.

Appendix A5.1: Validity of the diffusion approximation in an infinite medium

The probability of quantum diffusion and the structure factor are, in the Diffuson approximation and *in the limit of slow spatial and temporal variations*, solutions of a diffusion equation (4.34, 4.38). Beyond this limit, we have been able to calculate exactly the Fourier transform of the integral equation (4.29). Thus, for the total probability $P(\boldsymbol{q},\omega)=P_d+P_0$, we obtained (4.85)

$$P(\boldsymbol{q},\omega)=\frac{P_0(\boldsymbol{q},\omega)}{1-P_0(\boldsymbol{q},\omega)/\tau_e}=\tau_e\frac{\arctan ql_e}{ql_e-\arctan ql_e}, \tag{5.74}$$

where $P_0(\boldsymbol{q},\omega)$ is given, for $d=3$, by (4.77). The spatial behavior of $P(\boldsymbol{r},\omega=0)$ is given by the inverse Fourier transform

$$P(\boldsymbol{r},\omega=0)=\frac{1}{4i\pi^2vr}\int_{-\infty}^{+\infty}dq\,e^{iqr}\frac{\arctan ql_e}{1-\frac{\arctan ql_e}{ql_e}} \tag{5.75}$$

with $l_e=v\tau_e$. This is the exact solution of the integral equation (4.29) in $d=3$. We want to compare this exact solution with the result $P(\boldsymbol{r},\omega=0)=1/(4\pi Dr)$, obtained in the diffusion approximation. We calculate this integral by residue calculus. To this end, we employ the representation: $\arctan x=(1/2i)\ln[(1+ix)/(1-ix)]$. The singularities of the integrand arise, on the one hand, from the roots of the equation $\arctan ql_e=ql_e$ and, on the other hand, from two cuts along the imaginary axis going from $q=i/l_e$ to $q=i\infty$ and from $q=-i/l_e$ to $q=-i\infty$. We may thus evaluate the integral (5.75) by using the contour shown in Figure 5.4 and subsequently taking the limit $Q\to\infty$. To calculate the contribution of the pole at zero, we expand the integrand about $q=0$, that is $\arctan ql_e/(1-\arctan ql_e/ql_e)\simeq 3/ql_e$. The

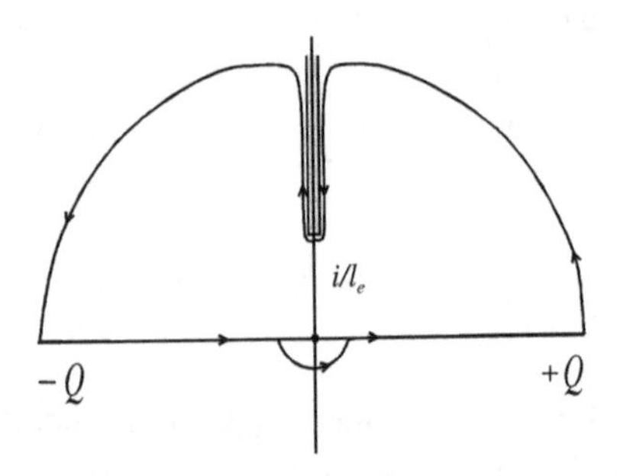

Figure 5.4 The contour in the complex plane used in the calculation of the integral (5.75).

contribution of this pole to the integral is $3i\pi/l_e$. The contribution from the cut is given by the integral

$$\int_1^{+\infty} e^{-rx/l_e} f(x), \tag{5.76}$$

where we have written $x = -iql_e$ and where the function $f(x)$ is given by

$$f(x) = \left[1 + \frac{1}{x}\ln\frac{x-1}{x+1} + \frac{1}{4x^2}\left(\pi^2 + \ln^2\frac{x-1}{x+1}\right)\right]^{-1}. \tag{5.77}$$

Finally, by the Jordan lemma, the contribution to the integral from the two quarters of the circle $q = Qe^{i\theta}$ tends to zero as $Q \to \infty$. What remains is

$$P(r, \omega = 0) = \frac{1}{4\pi Dr}\left(1 + \frac{1}{3}\int_1^{+\infty} e^{-rx/l_e} f(x)\right). \tag{5.78}$$

The first term corresponds to the solution of the diffusion equation (4.39). The second term, which gives corrections to the diffusion approximation, decreases exponentially with the elastic mean free path l_e. Consequently, for distances $r \geq l_e$, the solutions of the diffusion equation constitute an excellent approximation to the total probability $P(r, \omega = 0)$ in an infinite medium.

Similarly, we may deduce the behavior of the probability $P_d = P - P_0$ where $P_0(r, \omega = 0)$ is given by (4.17). The result is

$$\begin{aligned} P_d(r, \omega = 0) &= P(r, \omega = 0) - \frac{1}{4\pi v r^2} e^{-r/l_e} \\ &= \frac{1}{4\pi Dr}\left(1 + \alpha(r/l_e)\right), \end{aligned} \tag{5.79}$$

where the function $\alpha(y)$ is defined by

$$\alpha(y) = \frac{1}{3}\left[\int_1^{+\infty} e^{-yx} f(x)\, dx - \frac{e^{-y}}{y}\right] \tag{5.80}$$

and where $f(x)$ is defined by (5.77). This function, shown in Figure 5.5, decreases exponentially with the elastic mean free path l_e. We thus conclude that the probability $P_d(r, \omega = 0)$ is also a solution of a diffusion equation up to exponentially small corrections. This relative correction is very weak, having the value 0.029 for $r = l_e$.

We may thus conclude that the expression $P_d(r, \omega = 0) = 1/(4\pi Dr)$ for an infinite medium is an excellent approximation. This justifies the use of the diffusion equation (4.38) for distances greater than l_e, and for time scales greater than τ_e. In the presence of boundaries, there are almost no cases for which an exact solution is possible. We shall study the case of a semi-infinite medium with no source (the so-called Milne problem) in Appendix A5.3.

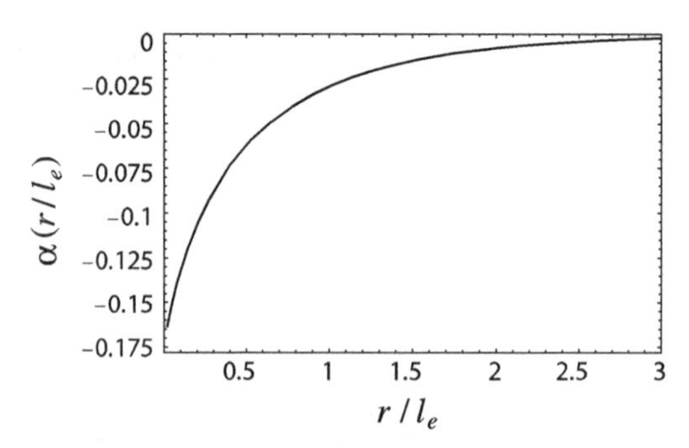

Figure 5.5 Behavior of the function $\alpha(r/l_e)$ giving the relative amplitude of the correction term to the diffusion approximation for the probability P_d. $\alpha(0) = \pi^2/12 - 1 \simeq -0.177$.

Appendix A5.2: Radiative transfer equation

At the approximation of the Diffuson, the probability of quantum diffusion P is a solution of the integral equation (4.29) which, in the limit of slow spatial variations, satisfies a diffusion equation. Here we present a different but equivalent formulation of this integral equation, which, for the case of electromagnetic wave propagation, is called the *radiative transfer equation*. We obtain this equation in the general case of anisotropic collisions.

A5.2.1 Total intensity

In Appendix A4.3, we showed that in the case of anisotropic collisions, it is necessary to generalize the notion of structure factor and to define a function $\Gamma(\hat{s}, \hat{s}', \boldsymbol{q})$ which solves equation (4.155). Similarly, for the probability, starting from relation (4.12) we define, for the solutions of the Helmholtz equation, the function[14]

$$P(\hat{s}, \hat{s}', \boldsymbol{q}) = \frac{4\pi}{c\Omega} \sum_{k,k'} \overline{G^R\left(k\hat{s} + \frac{\boldsymbol{q}}{2}, k'\hat{s}' + \frac{\boldsymbol{q}}{2}\right) G^A\left(k'\hat{s}' - \frac{\boldsymbol{q}}{2}, k\hat{s} - \frac{\boldsymbol{q}}{2}\right)} \tag{5.81}$$

where the sum depends only on the moduli of $\boldsymbol{k}$ and $\boldsymbol{k}'$. Following (4.12), the probability $P(\boldsymbol{q})$ may be expressed as

$$P(\boldsymbol{q}) = \langle P(\hat{s}, \hat{s}', \boldsymbol{q}) \rangle_{\hat{s}, \hat{s}'}. \tag{5.82}$$

We may decompose $P(\hat{s}, \hat{s}', \boldsymbol{q})$ into the sum $P(\hat{s}, \hat{s}', \boldsymbol{q}) = P_0(\hat{s}, \boldsymbol{q})\delta_{\hat{s},\hat{s}'} + P_d(\hat{s}, \hat{s}', \boldsymbol{q})$ where $P_0(\hat{s}, \boldsymbol{q})$ is defined by (4.75).

For isotropic collisions, the relation between P_d and Γ is given by (4.82). Since in the anisotropic case, Γ depends on the directions $\hat{s}$ and $\hat{s}'$, the sums over $\boldsymbol{k}$ and $\boldsymbol{k}'$ cannot be

[14] In what follows, we consider the limit of zero frequency $\omega = 0$. The generalization to finite frequency is immediate. We denote $\Gamma = \Gamma_{\omega=0}$.

factorized, and (4.83) becomes

$$P_d(\hat{s},\hat{s}',q) = \frac{\tau_e}{\gamma_e} f(\hat{s},q) f(\hat{s}',q)\,\Gamma(\hat{s},\hat{s}',q), \tag{5.83}$$

where the function $f(\hat{s},q)$ is given by (4.76). The total probability is thus given by

$$P(\hat{s},\hat{s}',q) = \tau_e f(\hat{s},q)\left[\delta_{\hat{s},\hat{s}'} + \frac{f(\hat{s}',q)}{\gamma_e}\,\Gamma(\hat{s},\hat{s}',q)\right]. \tag{5.84}$$

From the integral equation (4.161) for $\Gamma(\hat{s},\hat{s}',q)$, we may obtain the equation obeyed by $P(\hat{s},\hat{s}',q)$:

$$\boxed{P(\hat{s},\hat{s}',q) = \tau_e f(\hat{s},q)\left[\delta_{\hat{s},\hat{s}'} + \frac{1}{\gamma_e\tau_e}\big\langle P(\hat{s}'',\hat{s}',q) B(\hat{s}''-\hat{s})\big\rangle_{\hat{s}''}\right]} \tag{5.85}$$

Let us now define

$$P(\hat{s},q) = \big\langle P(\hat{s},\hat{s}',q)\big\rangle_{\hat{s}'} \tag{5.86}$$

whose Fourier transform $P(\hat{s},r)$ appears often in the literature under the name *specific intensity* $I(\hat{s},r)$ [97] and is denoted by $I(\hat{s},r)$. The intensity[15] $I(r) = \langle I(\hat{s},r)\rangle_{\hat{s}}$ is none other than the probability $P(r_0,r)$ for a source situated at r_0. We also define the current $J(r)$ called the "energy flux" (or the photon flux) by

$$J(r) = c\big\langle \hat{s} I(\hat{s},r)\big\rangle_{\hat{s}}. \tag{5.87}$$

The study of the specific intensity is the starting point for transport theory. Transport theory has proven to be a fruitful approach to the study of electromagnetic waves propagating through the atmosphere [97–100] or through certain turbid media. The development of this theory was motivated by the fact that it is generally difficult to obtain solutions to the Maxwell equations in a scattering medium, except in very dilute cases.

To obtain the equation obeyed by the specific intensity $I(\hat{s},r)$, it is sufficient to take the angular average of (5.85), that is

$$P(\hat{s},q) = \tau_e f(\hat{s},q)\left[1 + \frac{1}{\gamma_e\tau_e}\big\langle P(\hat{s}',q) B(\hat{s}'-\hat{s})\big\rangle_{\hat{s}'}\right]. \tag{5.88}$$

Using (4.76) and defining

$$p(\hat{s}-\hat{s}') = B(\hat{s}-\hat{s}')/\gamma_e, \tag{5.89}$$

such that $\langle p(\hat{s}-\hat{s}')\rangle = 1$, we obtain, for the zero-frequency specific intensity, the equation

$$(1 + iq\cdot\hat{s} l_e)P(\hat{s},q) = \tau_e + \big\langle P(\hat{s}',q)\, p(\hat{s}-\hat{s}')\big\rangle_{\hat{s}'} \tag{5.90}$$

[15] We use here $I(r)$ to denote the total intensity averaged over disorder and defined by (4.55). This is the sum of two contributions (cf. equation 4.60), $I(r) = I_0(r) + I_d(r)$, that of Drude–Boltzmann and that of the Diffuson.

whose Fourier transform is

$$\boxed{\hat{s} \cdot \nabla I(\hat{s}, \boldsymbol{r}) = -\frac{1}{l_e} I(\hat{s}, \boldsymbol{r}) + \frac{1}{l_e} \langle I(\hat{s}', \boldsymbol{r})\, p(\hat{s} - \hat{s}') \rangle_{\hat{s}'} + \frac{\delta(\boldsymbol{r})}{c}} \tag{5.91}$$

This expression is called the *radiative transfer equation* [97]. Let us emphasize again that this approach is based upon the same assumptions, essentially $kl_e \gg 1$, as the Diffuson approximation. The two approaches are equivalent. The radiative transfer equation describes the variation of the specific intensity of a wave in a given direction $\hat{s}$ as resulting, on the one hand, from the attenuation in this same direction due to elastic scattering and, on the other hand, to the scattering in other directions. The latter is controlled by the function $p(\hat{s} - \hat{s}')$ which is assumed to depend only on the angle between the directions $\hat{s}$ and $\hat{s}'$. The source term here is a $\delta(\boldsymbol{r})$ function, but it may be replaced by an arbitrary source which we shall denote $c\,\varepsilon(\hat{s}, \boldsymbol{r})$.

A5.2.2 Diffuse intensity

The radiative transfer equation provides a tool to describe the evolution of the intensity of a multiply scattered wave with anisotropic collisions. It is possible to simplify this equation and to solve it in the diffusion approximation. For this, we follow the treatment of reference [97]. We separate the total intensity I into two parts, that of Drude–Boltzmann I_0 and that associated with multiple scattering trajectories I_d, which is also called the *diffuse intensity* and which corresponds to the Diffuson. This separation has already been presented in section 4.7. It is of interest since it allows us to determine boundary conditions and thus to solve the radiative transfer equation for $I(\hat{s}, \boldsymbol{r})$. There are no simple boundary conditions for the total intensity because of the Drude–Boltzmann term. However, we are able to find boundary conditions for the multiple scattering contribution I_d.

The intensity I_0 corresponds to a wave emitted by a light source (we further consider the special case of an external source) and it falls off exponentially with decay length l_e. This ballistic component $I_0(\hat{s}, \boldsymbol{r})$ solves the differential equation:

$$\hat{s} \cdot \nabla I_0(\hat{s}, \boldsymbol{r}) = -\frac{1}{l_e} I_0(\hat{s}, \boldsymbol{r}). \tag{5.92}$$

From this relation and the radiative transfer equation (5.91) for $I(\hat{s}, \boldsymbol{r})$, we deduce that $I_d(\hat{s}, \boldsymbol{r})$ is a solution of

$$\hat{s} \cdot \nabla I_d(\hat{s}, \boldsymbol{r}) = -\frac{1}{l_e} I_d(\hat{s}, \boldsymbol{r}) + \frac{1}{l_e} \langle I_d(\hat{s}', \boldsymbol{r})\, p(\hat{s} - \hat{s}') \rangle_{\hat{s}'} + \varepsilon(\hat{s}, \boldsymbol{r}) + \varepsilon_0(\hat{s}, \boldsymbol{r}) \tag{5.93}$$

which has two source terms. The first, denoted $\varepsilon(\hat{s}, \boldsymbol{r})$, is related to a source located in the interior of the disordered medium, and the second, $\varepsilon_0(\hat{s}, \boldsymbol{r})$, arises from the incident intensity. This second term is related to the Drude–Boltzmann component $I_0(\hat{s}, \boldsymbol{r})$:

$$\varepsilon_0(\hat{s}, \boldsymbol{r}) = \frac{1}{l_e} \langle I_0(\hat{s}', \boldsymbol{r})\, p(\hat{s} - \hat{s}') \rangle_{\hat{s}'} \tag{5.94}$$

such that

$$\langle \varepsilon_0(\hat{s}, \boldsymbol{r}) \rangle_{\hat{s}} = \frac{I_0(\boldsymbol{r})}{l_e} \qquad \text{and} \qquad c \, \langle \hat{s} \; \varepsilon_0(\hat{s}, \boldsymbol{r}) \rangle_{\hat{s}} = \left(\frac{1}{l_e} - \frac{1}{l^*} \right) \boldsymbol{J}_0(\boldsymbol{r}), \tag{5.95}$$

where $I_0(\boldsymbol{r}) = \langle I_0(\hat{s}, \boldsymbol{r}) \rangle_{\hat{s}}$ and $\boldsymbol{J}_0(\boldsymbol{r}) = c \, \langle \hat{s} \cdot I_0(\hat{s}, \boldsymbol{r}) \rangle_{\hat{s}}$ are, respectively, the intensity and the current characterizing the Drude–Boltzmann term $I_0(\hat{s}, \boldsymbol{r})$. Here we have introduced the transport mean free path l^* (see also Appendix A4.3 and equation 4.168):

$$\frac{l_e}{l^*} = 1 - \langle p(\hat{s} - \hat{s}') \hat{s} \cdot \hat{s}' \rangle_{\hat{s}'} \tag{5.96}$$

or

$$\boxed{\frac{l_e}{l^*} = 1 - \langle p(\theta) \cos \theta \rangle} \tag{5.97}$$

The intensity $I_0(\boldsymbol{r})$ and the current $\boldsymbol{J}_0(\boldsymbol{r})$ are related by:

$$\text{div} \boldsymbol{J}_0(\boldsymbol{r}) = -\frac{c}{l_e} I_0(\boldsymbol{r}) \tag{5.98}$$

which is derived from (5.92).

In the diffusion approximation, we may proceed along the lines of the calculation of $\Gamma_\omega(\hat{s}, \boldsymbol{q})$ presented in Appendix A4.3. We suppose that the specific intensity $I_d(\hat{s}, \boldsymbol{r})$ is weakly anisotropic and may be expanded as (see also equation 4.163)

$$I_d(\hat{s}, \boldsymbol{r}) = I_d(\boldsymbol{r}) + \frac{3}{c} \boldsymbol{J}_d(\boldsymbol{r}) \cdot \hat{s} \tag{5.99}$$

where $\boldsymbol{J}_d(\boldsymbol{r})$ is the current associated with $I_d(\hat{s}, \boldsymbol{r})$:

$$\boldsymbol{J}_d(\boldsymbol{r}) = c \, \langle \hat{s} \; I_d(\boldsymbol{r}, \hat{s}) \rangle_{\hat{s}}. \tag{5.100}$$

Inserting the expansion (5.99) in (5.93) and using the relation

$$\langle p(\hat{s} - \hat{s}') \hat{s}' \rangle_{\hat{s}'} = \hat{s} \, \langle p(\hat{s} - \hat{s}') \hat{s} \cdot \hat{s}' \rangle_{\hat{s}'}, \tag{5.101}$$

we obtain

$$\hat{s} \cdot \nabla I_d(\boldsymbol{r}) + \frac{3}{c} \hat{s} \cdot \nabla [\boldsymbol{J}_d(\boldsymbol{r}) \cdot \hat{s}] = -\frac{3}{c l^*} \boldsymbol{J}_d(\boldsymbol{r}) \cdot \hat{s} + \varepsilon(\hat{s}, \boldsymbol{r}) + \varepsilon_0(\hat{s}, \boldsymbol{r}). \tag{5.102}$$

This expression is the radiative transfer equation for the specific intensity $I_d(\hat{s}, \boldsymbol{r})$ *in the diffusion approximation*. Averaging over all angles leads to[16]

$$\text{div} \boldsymbol{J}_d(\boldsymbol{r}) = c \langle \varepsilon(\hat{s}, \boldsymbol{r}) \rangle_{\hat{s}} + \frac{c}{l_e} I_0(\boldsymbol{r}) \tag{5.103}$$

[16] We use the expressions

$$\langle \hat{s} (\hat{s} \cdot \boldsymbol{A}) \rangle_{\hat{s}} = \frac{\boldsymbol{A}}{3} \qquad \langle \hat{s} \cdot [\hat{s} \cdot \nabla (\boldsymbol{A} \cdot \hat{s})] \rangle_{\hat{s}} = 0$$

valid for an arbitrary vector $\boldsymbol{A}$.

which expresses flux conservation. Multiplying both sides of (5.102) by $\hat{\boldsymbol{s}}$, taking the angular average and using (5.95), we obtain

$$\boldsymbol{J}_d(\boldsymbol{r}) = -D^* \nabla I_d(\boldsymbol{r}) + \left(\frac{l^*}{l_e} - 1\right) \boldsymbol{J}_0(\boldsymbol{r}) + l^* c \langle \hat{\boldsymbol{s}}\ \varepsilon(\hat{\boldsymbol{s}}, \boldsymbol{r}) \rangle_{\hat{s}} \tag{5.104}$$

where $D^* = cl^*/3$. This expression is a *Fick's law* with a source. Finally, combining (5.103) and (5.104), and employing (5.98), we deduce the following diffusion equation:[17]

$$\boxed{\Delta I_d(\boldsymbol{r}) = -\frac{3}{l_e^2} I_0(\boldsymbol{r}) - \frac{3}{l^*} \langle \varepsilon(\hat{\boldsymbol{s}}, \boldsymbol{r}) - l^* \hat{\boldsymbol{s}} \cdot \nabla \varepsilon(\hat{\boldsymbol{s}}, \boldsymbol{r}) \rangle_{\hat{s}}} \tag{5.105}$$

which generalizes equation (4.66) to the case of sources which are non-point-like and anisotropic. Finally, by inserting (5.104) in (5.99), we obtain a useful angular expansion for the specific scattered intensity $I_d(\hat{\boldsymbol{s}}, \boldsymbol{r})$:

$$I_d(\hat{\boldsymbol{s}}, \boldsymbol{r}) = I_d(\boldsymbol{r}) - l^* \hat{\boldsymbol{s}} \cdot \nabla I_d(\boldsymbol{r}) + \frac{3}{c} \left(\frac{l^*}{l_e} - 1\right) \hat{\boldsymbol{s}} \cdot \boldsymbol{J}_0(\boldsymbol{r}) + 3 l^* c \langle \hat{\boldsymbol{s}}\ \varepsilon(\hat{\boldsymbol{s}}, \boldsymbol{r}) \rangle_{\hat{s}} \tag{5.106}$$

A5.2.3 Boundary conditions

To solve equation (5.105) we must prescribe boundary conditions for the diffuse intensity $I_d(\boldsymbol{r})$. Since the diffusive behavior occurs only in the interior of the disordered medium, the diffuse intensity entering the medium must vanish:

$$I_d(\hat{\boldsymbol{s}}, \boldsymbol{r}) = 0 \qquad \text{for any incoming } \hat{\boldsymbol{s}} \tag{5.107}$$

for each point $\boldsymbol{r}$ on the interface. In the diffusion approximation (5.99), this condition cannot be satisfied exactly. We replace it by the approximate condition which imposes that the incident diffuse flux,

$$J_{d,z}^+(\boldsymbol{r}) = c \langle s_z I_d(\hat{\boldsymbol{s}}, \boldsymbol{r}) \rangle_{\hat{s}_+}, \tag{5.108}$$

vanishes for each point $\boldsymbol{r}$ on the interface. Here s_z is the component normal to the interface, and the angular average $\langle \cdots \rangle_{\hat{s}_+}$ is taken over the half-space $s_z > 0$.

Let us now evaluate this incident diffuse flux for the geometry of Figure 5.6. Starting with equation (5.106) for the diffuse specific intensity $I_d(\hat{\boldsymbol{s}}, \boldsymbol{r})$, we can determine the flux which traverses the interface $z = 0$. For a semi-infinite medium, the different quantities depend only on the space coordinate z. When the medium is illuminated by a plane wave at normal incidence, the Drude–Boltzmann specific intensity is given by

$$I_0(\hat{\boldsymbol{s}}, \boldsymbol{r}) = I_0 \delta(\hat{\boldsymbol{s}} - \hat{\boldsymbol{z}}) e^{-z/l_e} \tag{5.109}$$

[17] For a point source, $\langle \varepsilon(\hat{\boldsymbol{s}}, \boldsymbol{r}) - l^* \hat{\boldsymbol{s}} \cdot \nabla \varepsilon(\hat{\boldsymbol{s}}, \boldsymbol{r}) \rangle_{\hat{s}} = \delta(\boldsymbol{r})/c$, and we thus recover expression (4.66).

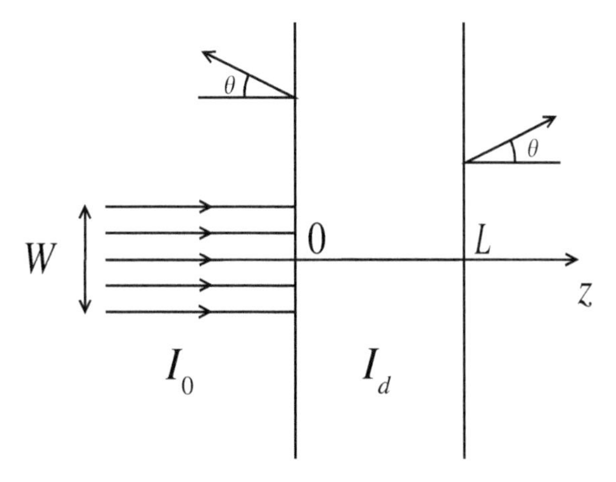

Figure 5.6 Diffusive medium in the geometry of a slab of thickness L, illuminated by a plane wave (possibly collimated). Reflection and transmission coefficients in direction θ are measured. The intensity I_d is non-zero only inside the slab.

and satisfies[18]

$$I_0(\boldsymbol{r}) = \frac{I_0}{4\pi} e^{-z/l_e} \qquad \text{and} \qquad \boldsymbol{J}_0(\boldsymbol{r}) = \frac{cI_0}{4\pi} \hat{z}\, e^{-z/l_e}. \tag{5.110}$$

From (5.106), we can calculate the diffuse flux $J^+_{d,z}(z)$:

$$J^+_{d,z}(z) = \langle s_z \rangle I_d(z) - l^* \langle s_z^2 \rangle \frac{\partial I_d(z)}{\partial z} + \frac{3}{4\pi} \left(\frac{l^*}{l_e} - 1 \right) \langle s_z^2 \rangle I_0\, e^{-z/l_e}, \tag{5.111}$$

where the angular average, performed over the half-space $s_z > 0$, gives $\langle s_z \rangle = 1/2$ and $\langle s_z^2 \rangle = 1/3$, so that

$$J^+_{d,z}(z) = \frac{I_d(z)}{2} - \frac{l^*}{3} \frac{\partial I_d(z)}{\partial z} + \frac{1}{4\pi} \left(\frac{l^*}{l_e} - 1 \right) I_0\, e^{-z/l_e}. \tag{5.112}$$

Let us first consider the case where there is no external source, that is, $I_0 = 0$. The condition for the vanishing of incoming diffuse flux, $J^+_{d,z}(0) = 0$, may be written as

$$\boxed{ I_d - \frac{2}{3} l^* \frac{\partial I_d}{\partial z} \bigg|_{z=0} = 0 } \tag{5.113}$$

We will take this as boundary condition for the case of a semi-infinite medium with no external sources. The behavior of $I_d(z)$ in the vicinity of the interface is thus linear, and is

[18] See equation (4.58) which corresponds to a point source. The present result is for a plane-wave source.

given by

$$I_d(z) \simeq I_d(0)\left(1+\frac{z}{z_0}\right). \tag{5.114}$$

This intensity vanishes at the point of coordinate $z=-z_0$ with $z_0=2l^*/3$. Since this follows from (5.108), this boundary condition holds approximately. We may compare this result with the exact solution of the Milne problem (5.140), obtained for a semi-infinite medium and isotropic collisions, for which we obtain $z_0 \simeq 0.71044\ l_e$ in the absence of a source (Appendix A5.3). This small difference justifies our using the diffusion approximation, which is particularly interesting since it allows us to treat the case of anisotropic collisions and of various geometries as well, with the aid of the condition (5.113). This is not possible using the solution of the Milne problem. We now consider some examples.

Exercise 5.5: boundary conditions and internal reflections

Because of the difference in refractive index between the exterior and the diffusive medium ($n>1$), there are reflections on the internal wall of the interface such that the entering diffuse flux does not vanish completely, but rather satisfies

$$J^+_{d,z}(0) = R\, J^-_{d,z}(0),$$

where $J^-_{d,z}(0)$ is the diffuse flux leaving the medium and where R is a coefficient of internal reflection appropriately averaged over reflection angles [101]. Show that the boundary condition (5.113) may then be written

$$I_d - z_0 \frac{\partial I_d}{\partial z}\bigg|_{z=0} = 0 \tag{5.115}$$

with

$$z_0 = \frac{2}{3} l^* \frac{1+R}{1-R}. \tag{5.116}$$

Exercise 5.6: boundary conditions in d dimensions

Show that, in any dimension, and in the absence of a source, the intensity varies linearly in the vicinity of the interface (relation 5.114), with

$$\frac{z_0}{l^*} = \frac{\langle s_z^2\rangle_+}{\langle s_z\rangle_+} = \frac{\int_0^{\pi/2} \sin^{d-2}\theta \cos^2\theta \, d\theta}{\int_0^{\pi/2} \sin^{d-2}\theta \cos\theta \, d\theta} = \frac{A_d}{2A_{d-1}}. \tag{5.117}$$

The angular averages are performed over the half-space $z>0$. A_d is the volume of the d-dimensional unit sphere (15.2). In two dimensions, $\langle s_z\rangle = 2/\pi$ and $\langle s_z^2\rangle = 1/2$, yielding $z_0=\pi/4$.

A5.2.4 Slab illuminated by an extended source

Consider a slab of thickness L illuminated by a plane wave of normal incidence (Figure 5.6). According to (5.112), the condition for vanishing of the incoming diffuse flux corresponds to

$$I_d(0) - \frac{2l^*}{3} \frac{\partial I_d}{\partial z}\bigg|_0 + \frac{1}{2\pi}\left(\frac{l^*}{l_e} - 1\right) I_0 = 0. \tag{5.118}$$

This same condition, but for the incoming diffuse flux originating in the region $z > L$, is $J^-_{d,z}(z = L) = 0$, that is

$$I_d(L) + \frac{2l^*}{3} \frac{\partial I_d}{\partial z}\bigg|_L - \frac{1}{2\pi}\left(\frac{l^*}{l_e} - 1\right) I_0\, e^{-L/l_e} = 0. \tag{5.119}$$

For $L \gg l_e$, the last term is negligible. The solution of the diffusion equation (5.105) in this one-dimensional geometry is straightforward and is given by

$$I_d(z) = \frac{5}{4\pi} \frac{L + z_0 - z}{L + 2z_0} - \frac{3}{4\pi}\, e^{-z/l_e} \tag{5.120}$$

where $z_0 = 2l^*/3$. The diffuse flux $\boldsymbol{J}_d(\boldsymbol{r})$ is obtained from equation (5.104). In the limit $L \gg l_e$, we have

$$\boldsymbol{J}_d(z) = -\frac{cI_0}{4\pi} \hat{z}\, e^{-z/l_e}. \tag{5.121}$$

In particular, $\boldsymbol{J}_d(z = 0) = -\,cI_0/4\pi\hat{z}$. The diffuse flux is thus equal and opposite to the incident flux $\boldsymbol{J}_0(z = 0)$. Finally, using (5.106) and (5.120), we obtain the intensity at $z = 0$ and the backscattered specific intensity in direction $-\hat{z}$:

$$I_d(z = 0) = \frac{I_0}{2\pi} \qquad \text{and} \qquad I_d(-\hat{z}, 0) = \frac{5}{4\pi} I_0. \tag{5.122}$$

Transmission

Using the specific intensity at $z = L$, we obtain the transmission coefficient in a direction θ, defined by

$$\mathcal{T}(\theta) = s_z \frac{I_d(\hat{s}, L)}{I_0}. \tag{5.123}$$

Inserting (5.120) in (5.106), we obtain $I_d(\hat{s}, L)$ and, using $z_0 = 2l^*/3$, we find

$$\mathcal{T}(\theta) = \frac{5}{4\pi} \frac{l^*}{L + 2z_0} \mu \left(\frac{z_0}{l^*} + \mu\right) \tag{5.124}$$

where $\mu = \cos\theta$. The total transmission coefficient $\mathcal{T} = 2\pi \int_0^{\pi/2} \mathcal{T}(\theta) \sin\theta d\theta$ is equal to

$$\mathcal{T} = \frac{5}{3} \frac{l^*}{L + 2z_0}. \tag{5.125}$$

Reflection

Using the specific intensity at $z=0$, we obtain the reflection coefficient in a direction θ

$$\mathcal{R}(\theta) = s_z \frac{I_d(\hat{s},0)}{I_0}. \tag{5.126}$$

Inserting (5.120) in (5.106), we obtain $I_d(\hat{s},0)$ and, using the value $z_0 = 2l^*/3$, we obtain the reflection coefficient in direction θ

$$\mathcal{R}(\theta) = \frac{3}{4\pi}\mu\left(\frac{z_0}{l^*}+\mu\right) - \frac{5}{4\pi}\frac{l^*}{L+2z_0}\mu\left(\frac{z_0}{l^*}+\mu\right). \tag{5.127}$$

In the limit of a semi-infinite medium, $L\to\infty$, the last term vanishes. The total reflection coefficient $\mathcal{R}=2\pi\int_0^{\pi/2}\mathcal{R}(\theta)\sin\theta d\theta$ is then equal to 1. For finite L, we check that $\mathcal{R}+\mathcal{T}=1$.

A5.2.5 Semi-infinite medium illuminated by a collimated beam

We now consider the case of a semi-infinite medium illuminated by a collimated beam of width W, characterized by the intensity

$$I_0(\hat{s},\boldsymbol{r}) = F_0(\boldsymbol{\rho})\delta(\hat{s}-\hat{z})\, e^{-z/l_e}, \tag{5.128}$$

where $\boldsymbol{\rho}$ is a two-dimensional vector perpendicular to the Oz axis. Let us consider a Gaussian profile $F_0(\boldsymbol{\rho}) = I_0 e^{-\rho^2/W^2}$. The diffuse intensity $I_d(\boldsymbol{r})$ is the solution of equation (5.105) with the boundary condition

$$I_d(\boldsymbol{r}) - \frac{2l^*}{3}\frac{\partial I_d(\boldsymbol{r})}{\partial z}\bigg|_{z=0} + \frac{1}{2\pi}\left(\frac{l^*}{l_e}-1\right)F_0(\boldsymbol{\rho}) = 0 \tag{5.129}$$

at the interface $\boldsymbol{r}=(\boldsymbol{\rho},z=0)$. The solution to this problem for the case of a slab is detailed in reference [102]. Taking the limit $L\to\infty$, we obtain, for a semi-infinite medium,

$$I_d^W(\boldsymbol{\rho},z=0) = \frac{I_0}{4\pi}\frac{l^*}{l_e}\int_0^\infty d\lambda\, J_0(\lambda\rho)W^2\, e^{-\lambda^2W^2/4}\,\frac{\lambda}{1+\frac{2}{3}l^*\lambda}\left(\frac{1}{1+\lambda l_e}-\eta\right), \tag{5.130}$$

where $\eta = 1-l_e/l^*$ and where J_0 is the zeroth order Bessel function. The specific intensity backscattered in direction $-\hat{z}$ is related to the intensity on the interface:

$$I_d^W(-\hat{z},\boldsymbol{\rho},z=0) = \frac{5}{2}I_d^W(\boldsymbol{\rho},z=0). \tag{5.131}$$

For the case of a uniformly illuminated medium ($W\to\infty$), we recover the results (5.122) for a plane wave. For the case of an infinitely thin (and normalized) beam, $F_0(\boldsymbol{\rho}) = I_0\delta(\boldsymbol{\rho})$, we obtain

$$I_d^\delta(\boldsymbol{\rho},z=0) = \frac{I_0}{4\pi^2}\frac{l^*}{l_e}\int_0^\infty d\lambda\, J_0(\lambda\rho)\,\frac{\lambda}{1+\frac{2}{3}l^*\lambda}\left(\frac{1}{1+\lambda l_e}-\eta\right) \tag{5.132}$$

and the specific intensity at an arbitrary point on the interface is given by [102, 103] $I_d^\delta(-\hat{z},\boldsymbol{\rho},z=0)=\frac{5}{2}I_d^\delta(\boldsymbol{\rho},z=0)$. It is interesting to note how the solution for the case of an extended plane wave ($W\to\infty$) is related to that of the diffuse intensity resulting for δ function beam:

$$I_d^{W\to\infty}(-\hat{z})=\frac{5}{2}\int d^2\boldsymbol{\rho}\, I_d^\delta(\boldsymbol{\rho},z=0)=\frac{5}{4\pi}I_0. \tag{5.133}$$

Appendix A5.3: Multiple scattering in a finite medium

For an infinite medium and a point source, the solution of the diffusion equation is an excellent approximation to the probability P_d at the Diffuson approximation in the limit of large distances $r \gg l_e$ (Appendix A5.1). Moreover, for a finite size medium, it is clear that the diffusion approximation is justified sufficiently far from the boundaries. There is, however, an additional difficulty related to the choice of boundary conditions. Replacing the integral equation (4.29) by a diffusion equation, it is not clear that the boundary condition of the latter may be derived simply from those of the initial equation. In order to see to what extent these two problems are related, we shall study a case (one of very few) for which it is possible to obtain an exact solution to the integral equation (4.29); this is the case of a semi-infinite medium with no source (Milne problem).

This solution only exists for the case of isotropic collisions, that is, when $l^*=l_e$. For anisotropic collisions, the integral equations (4.155) and (4.174), for the structure factor Γ and for P, have no simple solution. We must then employ an approximate method which starts directly from the diffusion equation rather than the integral equation. This is the radiative transfer approximation presented in Appendix A5.2.

A5.3.1 Multiple scattering in a half-space: the Milne problem

We begin from the integral equation (4.29) for the total probability. With the aid of expression (4.17) for P_0, this equation may be written as

$$P(\boldsymbol{r}_1,\boldsymbol{r}_2)=\frac{\tau_e}{4\pi l_e|\boldsymbol{r}_1-\boldsymbol{r}_2|^2}e^{-|\boldsymbol{r}_1-\boldsymbol{r}_2|/l_e}+\frac{1}{4\pi l_e}\int d\boldsymbol{r}'\frac{e^{-|\boldsymbol{r}_1-\boldsymbol{r}'|/l_e}}{|\boldsymbol{r}_1-\boldsymbol{r}'|^2}P(\boldsymbol{r}',\boldsymbol{r}_2). \tag{5.134}$$

The first term is negligible when the points $\boldsymbol{r}_1$ and $\boldsymbol{r}_2$ are separated by more than l_e. We thus neglect it, taking point $\boldsymbol{r}_2$ to be at infinity in the interior of the medium. For the geometry of a semi-infinite medium defined by the half-space $z\geq 0$, the probability $P(\boldsymbol{r}_1,\boldsymbol{r}_2)=P(z_1,z_2)$ does not depend on the projection $\boldsymbol{\rho}$ of the vector $(\boldsymbol{r}_1-\boldsymbol{r}_2)$ in the plane $z=0$. Thus, placing z_2 at infinity, equation (5.134) becomes

$$P(z_1)=\frac{1}{4\pi l_e}\int_0^\infty dz'\,P(z')\int d^2\boldsymbol{\rho}\,\frac{e^{-|\boldsymbol{r}_1-\boldsymbol{r}'|/l_e}}{|\boldsymbol{r}_1-\boldsymbol{r}'|^2}. \tag{5.135}$$

The integral in the plane $z=0$ may be calculated:

$$\int d^2\rho \, \frac{e^{-\frac{1}{l_e}\sqrt{\rho^2+(z_1-z')^2}}}{\rho^2+(z_1-z')^2} = 2\pi \int_1^\infty \frac{dt}{t} e^{-t|z_1-z'|/l_e}. \tag{5.136}$$

Expressing lengths in units of l_e, we obtain an integral equation for $P(z)$:

$$\boxed{P(z) = \frac{1}{2}\int_0^\infty dz' P(z') E_1(|z-z'|)} \tag{5.137}$$

where the function E_n is defined by

$$E_n(z) = \int_1^\infty dt \frac{e^{-zt}}{t^n}. \tag{5.138}$$

The problem of solving equation (5.137) is known as the Milne problem. The Wiener–Hopf method gives the exact solution at the price of calculating a difficult inverse Laplace transform [104]. Instead of presenting this method, we use a variational approach [104]. To this end, we first show that sufficiently far from the interface (located at $z=0$), the probability $P(z)$ is a linear function. Indeed, for $z\to\infty$, $P(z)$ varies smoothly, and may thus be expanded in the form

$$P(z') = P(z) + (z'-z)P'(z) + \frac{1}{2}(z'-z)^2 P''(z) + \cdots, \tag{5.139}$$

where P' and P'' are the first and second derivatives of P. Inserting this expansion in equation (5.137) and using the integrals (15.56), we see that the probability is a solution of the equation $P''(z)=0$. Far from the interface, it has the asymptotic behavior:

$$\boxed{P(z) \xrightarrow[z\to\infty]{} z + z_0} \tag{5.140}$$

where z_0 is a constant. The extrapolation of this behavior in the neighborhood of the interface shows that the probability P does not vanish at $z=0$, but on the exterior of the medium at a distance $-z_0$ from it (in units of l_e). The exact value of z_0, as calculated by the Wiener–Hopf method [104], is equal to[19]

$$z_0 = 0.71044609\cdots. \tag{5.141}$$

Here we calculate the length z_0 using a variational method, which also allows the calculation of $P(z)$ for all z. To that purpose, let us define the function $Q(z) = P(z) - z$, such that $Q(\infty) = z_0$. Inserting this into (5.137), we obtain an integral equation for $Q(z)$:[20]

$$Q(z) = \frac{1}{2}\int_0^\infty dz' \; Q(z') E_1(|z-z'|) + \frac{1}{2}E_3(z) \tag{5.142}$$

[19] z_0 is given by the integral

$$z_0 = \frac{6}{\pi^2} + \frac{1}{\pi}\int_0^{\pi/2} dx \left(\frac{3}{x^2} - \frac{1}{1 - x\cot x}\right).$$

[20] Integrals useful for the calculations in this appendix may be found in the formulary on page 582.

as well as the relation

$$z_0 = \frac{3}{2}\int_0^\infty dz Q(z)E_3(z) + \frac{3}{8}. \tag{5.143}$$

This latter equation may be obtained from an exact calculation [104] which results from the following theorem. The solution to the integral equation

$$Q(z) = \frac{1}{2}\int_0^\infty dz' Q(z')\left[E_1(|z-z'|) - E_1(z+z')\right] + S(z), \tag{5.144}$$

where the function $S(z)$ is such that $Q(z)$ is bounded at infinity, satisfies

$$Q(\infty) = 3\int_0^\infty dz\, zS(z). \tag{5.145}$$

The function $S(z)$ corresponding to equation (5.142) is given by

$$S(z) = \frac{1}{2}E_3(z) + \frac{1}{2}\int_0^\infty dz' Q(z')E_1(z+z') \tag{5.146}$$

and, therefore,

$$z_0 = Q(\infty) = \frac{3}{2}\int_0^\infty dz\, zE_3(z) + \frac{3}{2}\int_0^\infty dz' Q(z')\int_0^\infty dz\, zE_1(z+z'). \tag{5.147}$$

The evaluation of the integrals is immediate, and leads to (5.143).

We now solve the integral equation (5.142) approximately, using a variational method. To this end, we introduce the functional:

$$\mathcal{F}(\hat{Q}) = \frac{\int_0^\infty dz\, \hat{Q}(z)\left[\hat{Q}(z) - \frac{1}{2}\int_0^\infty dz'\, \hat{Q}(z')E_1(|z-z'|)\right]}{\left[\int_0^\infty dz\, \hat{Q}(z)E_3(z)\right]^2}. \tag{5.148}$$

It is simple to check, by performing a variation about $\hat{Q}(z)$, that the functional $\mathcal{F}(\hat{Q})$ is minimal for the desired solution $Q(z)$. Moreover, using (5.142) and (5.143), it may be shown that this solution is related to z_0 through

$$\mathcal{F}_{min} = \left(2\int_0^\infty dz\, Q(z)E_3(z)\right)^{-1} = \left(\frac{4}{3}z_0 - \frac{1}{2}\right)^{-1}. \tag{5.149}$$

To determine z_0, we consider several trial functions. The simplest is $\hat{Q}(z) = \text{constant}$. Inserting this function into (5.148) and using the integrals given in (15.52), we obtain $\mathcal{F}_{min} = 9/4 = 2.25$. Using (5.149), we arrive at the variational estimate $z_0 = 17/24 \simeq 0.7083$. To obtain a better approximation for $Q(z)$, and thus for $P(z)$, we consider a higher order approximation obtained by inserting $Q = z_0$ in the right hand side of (5.142), that is,

$$Q(z) = z_0 - \frac{1}{2}z_0E_2(z) + \frac{1}{2}E_3(z). \tag{5.150}$$

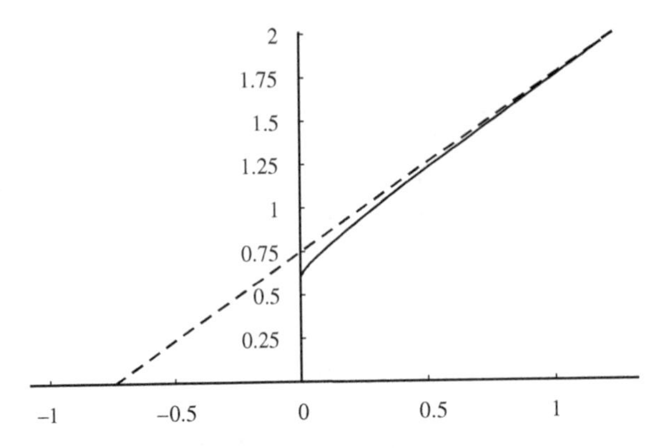

Figure 5.7 Spatial dependence of the probability $P(z)$ for a semi-infinite sourceless medium. The solid curve corresponds to (5.153), whereas the dashed line corresponds to (5.140).

This suggests

$$\hat{Q}(z) = z_0 \left[1 - \lambda E_2(z) + \mu E_3(z)\right] \tag{5.151}$$

as a trial function. Using the integrals (15.57), we find

$$\mathcal{F}(\hat{Q}) = \frac{\frac{1}{4} + \frac{2}{3}\lambda(\ln 2 - 1) + \frac{1}{8}\mu - \frac{1}{3}\lambda\mu(1 - \ln 2) + \lambda^2(\frac{1}{2} - \frac{\pi^2}{24}) + \mu^2(\frac{\pi^2}{36} - \frac{2}{9}\ln 2 - \frac{13}{144})}{(\frac{1}{3} - \frac{1}{8}\lambda + \frac{1}{5}\mu(2\ln 2 - 1))^2}. \tag{5.152}$$

Minimizing with respect to λ and μ, we obtain $\lambda = 0.342895$ and $\mu = 0.315870$. Putting these values into $\mathcal{F}(\hat{Q})$, we obtain $\mathcal{F}_{min} = 2.23584$ and, from (5.149), we have that $z_0 = 0.710445$. Finally, $P(z)$ is calculated from $Q(z)$:

$$P(z) = z + 0.710445 - 0.243608 E_2(z) + 0.244408 E_3(z) \tag{5.153}$$

This function is shown in Figure 5.7.

The diffusion approximation given by (5.140) thus remains excellent for the semi-infinite medium, since the correction to the linear dependence, given by the functions $E_n(z)$, decays exponentially on the scale of the elastic mean free path l_e.

This solution of the integral equation (4.29) for P in a semi-infinite medium exists only for the case of isotropic collisions. For anisotropic collisions, that is, when $l^* \neq l_e$, it does not reduce to the Milne problem. We may, however, use the radiative transfer equation which leads to the estimate $z_0 = (2/3)l^*$ (Appendix A5.2).

A5.3.2 Diffusion in a finite medium

Starting with the radiative transfer equation (Appendix A5.2), we have shown that the probability is well described as the solution of a diffusion equation, but the boundary condition is not Dirichlet. The condition is that the diffuse flux must vanish outside the disordered medium. For the semi-infinite medium geometry, using (5.113) and the relation (4.63) between I_d and the probability, we find that in the vicinity of the interface at $z=0$, P obeys the boundary condition

$$P(z,z') - z_0 \left.\frac{\partial P(z,z')}{\partial z}\right|_{z=0} = 0. \tag{5.154}$$

Since P varies linearly over a distance of the order z_0, we can replace this boundary condition with the condition that the probability P vanishes outside the medium at a distance z_0 given by

$$P(z,z')|_{z=-z_0} = 0. \tag{5.155}$$

From the radiative transfer equation approach, we calculate $z_0 = 2/3\, l^*$ (see equation 5.113), while the treatment based on the exact solution to the Milne problem yields $z_0 \simeq 0.710\, l_e$ (but this is only valid for isotropic collisions).

The conclusion drawn from these two approaches is that in order to calculate P in a scattering medium with boundaries, we may consider the solutions of the diffusion equation (5.15) with Dirichlet boundary conditions with the constraint that the probability vanishes not at the interface but at a distance z_0 outside the medium. This treatment allows us to obtain quantitative expressions which are easily interpreted, in a relatively simple and economical fashion, instead of looking for solutions of the radiated transfer equation that are usually difficult to obtain. We now illustrate this approach in two particularly useful examples.

Semi-infinite medium

The solution obtained for the diffusion equation in half-space corresponds to the case of a source located at $z' \to \infty$. If z and z' are both situated at a finite distance from the interface, the probability $P(z,z')$ is calculated with the aid of equation (5.67) corresponding to Dirichlet boundary conditions (absorbing boundary), by replacing z by $z+z_0$ and z' by $z'+z_0$. We thus obtain

$$P(z,z') = \frac{z_m + z_0}{D} = \frac{1}{2D}\,(z_+ + 2z_0 - z_-)\,, \tag{5.156}$$

where $z_m = \min(z,z')$, $z_+ = z+z'$ and $z_- = |z-z'|$. Similarly, in the case where there is a finite dephasing length L_γ, we replace the Laplace transform $P_\gamma(z,z')$ of (5.66) by

$$P_\gamma(z,z') = \frac{L_\gamma}{2D}\left[e^{-|z-z'|/L_\gamma} - e^{-(z+z'+2z_0)/L_\gamma}\right], \tag{5.157}$$

with the limit $L_\gamma \to \infty$ recovering equation (5.156). This result may also be obtained using the image method (see section 5.7) taking the shifted plane $z=-z_0$ as the plane of symmetry.

Slab of finite width

In the case of a slab of size $L \gg l_e$ contained between two infinite planes, we assume, once again, that we can take the solution corresponding to Dirichlet boundary conditions (5.58) by replacing z by $z+z_0$, z' by $z'+z_0$, and the thickness L by $L+2z_0$, such that[21]

$$P(z,z') = \frac{z_m + z_0}{D}\left(1 - \frac{z_M + z_0}{L+2z_0}\right) \tag{5.158}$$

vanishes when z or z' is equal to $-z_0$ or $L+z_0$. By the same substitution, for the case where the dephasing length L_γ is finite, equation (5.55) becomes[22]

$$P_\gamma(z,z') = \frac{L_\gamma}{D}\frac{(\sinh(z_m+z_0)/L_\gamma)(\sinh(L+z_0-z_M)/L_\gamma)}{\sinh(L+2z_0)/L_\gamma}. \tag{5.159}$$

These solutions will prove useful in the study of the albedo, diffusing wave spectroscopy, and correlations in speckle patterns (Chapters 8, 9 and 12).

Appendix A5.4: Spectral determinant

In section 5.2.1, we defined the integrated probability of return to the origin $Z(t)$ given by (5.5) and wrote it in terms of the eigenvalue spectrum E_n of the diffusion equation. Certain physical properties can be expressed in terms of the Laplace transform of $Z(t)$:

$$\int_0^\infty Z(t)e^{-\gamma t}\,dt, \tag{5.160}$$

where the positive number γ is related to the maximum length of multiple scattering paths, and was introduced in section 5.2.2. By definition of $Z(t)$, the preceding integral may be expressed formally as

$$\sum_n \frac{1}{\gamma + E_n} \tag{5.161}$$

and may be calculated from the determinant

$$\boxed{S(\gamma) = \det(-D\,\Delta + \gamma) = \prod_n (E_n + \gamma)} \tag{5.162}$$

in the form

$$\int_0^\infty Z(t)\,e^{-\gamma t}\,dt = \sum_n \frac{1}{\gamma + E_n} = \frac{\partial}{\partial \gamma}\ln S(\gamma). \tag{5.163}$$

[21] z_m and z_M are defined in equation (5.57).
[22] The condition (5.154) leads to a solution which is slightly different, but very close in the limit where z_0 is small compared to L.

The determinant $S(\gamma)$ is called the *spectral determinant* associated with the diffusion equation. Many physical properties may be calculated from $S(\gamma)$ or its derivatives.

The function $S(\gamma)$ is defined formally by the product of eigenvalues of the diffusion equation. This product is infinite, so its definition is formal. To give an interpretation for $S(\gamma)$, we employ an auxiliary function, called the ζ function, associated with the spectral determinant and defined by

$$\zeta(s,\gamma) = \sum_n \frac{1}{\lambda_n^s} \tag{5.164}$$

where $\lambda_n = E_n + \gamma$. This function is well defined for all values of s for which the series converges. Using the integral

$$\frac{1}{\lambda^s} = \frac{1}{\Gamma(s)} \int_0^\infty dt\, t^{s-1}\, e^{-t\lambda}, \tag{5.165}$$

we may write

$$\zeta(s,\gamma) = \frac{1}{\Gamma(s)} \int_0^\infty dt\, t^{s-1} Z(t)\, e^{-\gamma t}. \tag{5.166}$$

This representation shows that $\zeta(s,\gamma)$ is the Mellin transform of $Z(t)\, e^{-\gamma t}$. The $\zeta(s,\gamma)$ function thus defined is convergent, and its analytic continuation in the complex plane defines a meromorphic function in s which is analytic at $s=0$. We use this analyticity and the identity:

$$\left.\frac{d}{ds}\lambda_n^{-s}\right|_{s=0} = -\ln\lambda_n \tag{5.167}$$

to express the spectral determinant as

$$\ln S(\gamma) = -\left.\frac{d}{ds}\zeta(s,\gamma)\right|_{s=0} \tag{5.168}$$

which is well defined.

Spectral determinant and density of states

The spectral determinant $S(\gamma)$ has interesting properties, and in particular is useful for obtaining the spectrum of eigenvalues of the diffusion equation as well as the density of states $\rho(E) = \sum_n \delta(E - E_n)$. From (5.163) we obtain

$$\rho(E) = -\frac{1}{\pi} \lim_{\eta\to 0^+} \mathrm{Im} \frac{d}{d\gamma} \ln S(\gamma) \tag{5.169}$$

where now γ is complex, and is given by $\gamma = -E + i\eta$.

Example: diffusion in the infinite plane

We now calculate $S(\gamma)$ for the case of diffusion in the plane. The integrated probability of return to the origin is given by the equation (5.24)

$$Z(t) = \frac{1}{4\pi Dt} \tag{5.170}$$

and the associated function $\zeta(s,\gamma)$ given by (5.166) may be written

$$\zeta(s,\gamma) = \frac{1}{4\pi D\Gamma(s)} \int_0^\infty dt\, t^{s-1} \frac{1}{t} e^{-\gamma t} \tag{5.171}$$

which, upon integration, yields

$$\zeta(s,\gamma) = \frac{1}{4\pi D} \frac{\gamma^{1-s}}{s-1}. \tag{5.172}$$

The expression for the spectral determinant may be calculated with the aid of (5.168), and is

$$\ln S(\gamma) = \frac{1}{4\pi D} \gamma(1 - \ln\gamma). \tag{5.173}$$

The density of states may be obtained from (5.169) and we recover the expression (3.44), for $d = 2$, with the replacement of D by $1/2m$.

Example: topological invariant

The probability $Z(t)$ may include an additive constant C of topological origin. Appendix A5.5 gives several examples. In this case, the function $\zeta(s,\gamma)$ and the spectral determinant contain a term of the form:

$$\zeta(s,\gamma) = \frac{C}{\gamma^s} \qquad \ln S(\gamma) = C \ln\gamma. \tag{5.174}$$

From (5.169), we see that there is a singularity in the density of diffusion modes:

$$\rho(E) = C\delta(E). \tag{5.175}$$

Appendix A5.5: Diffusion in a domain of arbitrary shape – Weyl expansion

This chapter deals with the calculation of the integrated probability of return to the origin $Z(t)$ associated with the diffusion equation for some simple geometries – an infinite medium, periodic boundary conditions, one-dimensional diffusion, etc. However, the cases in which $Z(t)$ may be calculated exactly are the exception, not the rule. For a system of arbitrary shape, we do not know generally how to evaluate analytically either the eigenvalues E_n or the heat kernel $Z(t)$.[23]

Nevertheless, in the limit of times $t \ll \tau_D$, where τ_D is the diffusion time (5.34), we may obtain an asymptotic expansion of $Z(t)$. Consider the example of a planar domain with Dirichlet boundary conditions. A particle which begins diffusing from an initial position far from the boundaries does not feel their existence, and to the first approximation, we expect that $Z(t) \simeq S/4\pi Dt$ where S is the surface of the domain (see expression 5.24). We may then ask what are the corrections to this expression due to the boundaries. Looking for this asymptotic expansion, also called the Weyl expansion, is a celebrated problem of contemporary mathematical physics, popularized by M. Kac [105] under the heading "Can we hear the shape of a drum?" In other words, is it possible to reconstruct, starting from the eigenvalue spectrum E_n and the asymptotic expansion of $Z(t)$, the geometrical characteristics of the domain, and in particular, the shape of its contour? This question was posed initially in 1910 by H. A. Lorentz. He asked why the Jeans radiation law (that is, at high frequency) for a finite black body does not depend on the shape of the enclosure containing the radiation, but only on its volume [105, 106]. In fact, the next term in the Weyl expansion depends on the length L of the contour, and it is thus possible to "hear" the length of the contour. To evaluate this term, we approximate the boundary closest to the diffusing particle by a plane tangent to it. We thus obtain a correction term equal to $-(L/4)/\sqrt{4\pi Dt}$ (Exercise 5.7).

Exercise 5.7: diffusion in a semi-infinite plane

With the aid of expression (5.65) generalized to $d=2$, show that the probability of return to the origin in the vicinity of the boundary may be written

$$P(\boldsymbol{r},\boldsymbol{r},t) = \frac{1}{4\pi Dt}\left[1 - e^{-z^2/Dt}\right], \tag{5.176}$$

where z is the distance from the point $\boldsymbol{r}$ to the boundary. Show that the heat kernel $Z(t) = \int_S P(\boldsymbol{r},\boldsymbol{r},t)d\boldsymbol{r}$ is equal to

$$Z(t) = \frac{S}{4\pi Dt} - \frac{L/4}{\sqrt{4\pi Dt}}, \tag{5.177}$$

where S is the surface area of the plane and L is the boundary length. These relations are correct for short times, such that $Dt \ll \min(S, L^2)$, which justifies the consideration of a semi-infinite plane.

---

[23] The same problem arises for the Schrödinger and Helmholtz equations. This statement is the starting point for the study of the so-called complex or chaotic systems in quantum mechanics (in this context, see Chapter 10 and in particular the discussion in section 10.1).

To conclude this somewhat historical introduction, we note that the answer to the question posed by Kac is negative; it is possible to find isospectral domains (having the same eigenvalue spectrum) with the same area and circumference, but with different shapes. This result, demonstrated in 1985, constitutes the Sunada theorem [107].

It is possible to evaluate systematically the terms in the asymptotic expansion of $Z(t)$ [108] for a planar domain. Without going into detail, the general idea consists of calculating the Weyl expansion for the case of a disk, which is an integrable problem for which all the coefficients may be computed in terms of Bessel functions. The case of a domain bounded by a continuous, infinitely differentiable curve may be treated by considering the osculating circles to each point on the contour, which leads to the expression of each term in the Weyl expansion as an integral over the contour, in powers of the local curvature $\kappa(l)$, where l is the arc length along the contour. We thus obtain

$$Z(t) \underset{t\to 0}{\longrightarrow} \frac{\Omega}{4\pi Dt} - \frac{L/4}{\sqrt{4\pi Dt}} + C + O(\sqrt{t}). \tag{5.178}$$

The third term in this expansion is *constant*, and deserves special attention. It may be shown that it equals $C = (1/12\pi)\oint \kappa(l)dl$. It is thus independent of t and of the shape of the contour, since the integral of the curvature equals 2π, yielding a value of $1/6$ for this term. This remains unchanged even if the surface is not planar.

An expansion of the type of (5.178) remains valid for an arbitrary manifold S. It may be shown that $C = \chi(S)/6$, where $\chi(S)$ is a topological invariant called the Euler–Poincaré characteristic of the manifold. The Gauss–Bonnet theorem states that

$$\chi(S) = \frac{1}{2\pi}\int\int_S K\, dS + \frac{1}{2\pi}\oint_{\partial S} \kappa\, dl, \tag{5.179}$$

where K is the Gaussian curvature of the surface S and κ is the geodesic curvature of the boundary ∂S.

For the case of a manifold without boundary, for example a sphere or a torus with one or several holes, the boundary term, that is, the geodesic term in (5.179), vanishes in the expression for $Z(t)$, and $\chi(S)$ only depends on the integral of the Gaussian curvature. It can be shown that $\chi(S) = 2(1-h)$ where h is the connectivity of the surface (the number of holes). Thus $\chi = 2$ for diffusion on a sphere, and χ vanishes for a one-hole torus. This invariant may also be obtained by triangulating the manifold and is thus given by the Euler relation $\chi = V - E + F$ where V, E and F are the numbers of vertices, edges, and cells, respectively. One consequence of (5.178) is that we may hear the Euler–Poincaré characteristic of an arbitrary manifold.

From the Weyl expansion, we can only calculate the behavior of $Z(t)$ for short times, while we often need to know this function for all times. There are, nevertheless, situations where this limit suffices. One example concerns the behavior of high-energy spectral correlations (Chapter 10) [109].

(a) 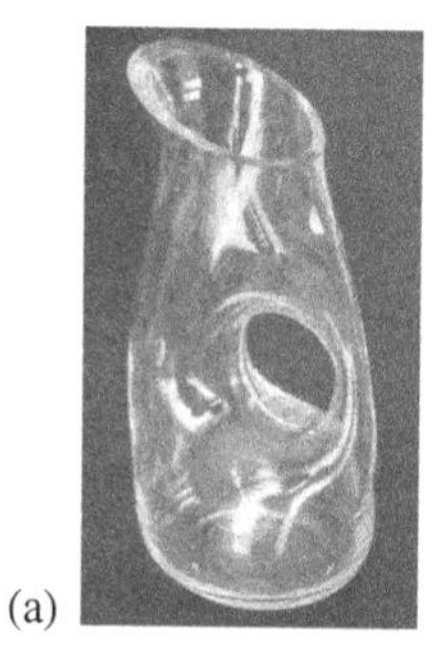 (b) 

Figure 5.8 (a) Carafe (photo T. Akkermans). (b) Descent device used for abseiling (photo G. Montambaux).

Exercise 5.8: Weyl expansion for a sphere
The spectrum of the diffusion equation on a sphere of radius R is given by $E_l = Dl(l+1)/R^2$, these states being $2l+1$ degenerate. Using the Euler–MacLaurin summation formula

$$\sum_{k=0}^{n} f(k) = \int_0^n f(k)\, dk + \frac{1}{2}(f(0) - f(n)) + \frac{1}{12}(f'(n) - f'(0)) - \frac{1}{720}(f'''(n) - f'''(0)) + \cdots \tag{5.180}$$

for the function $f(l) = (2l+1)\, e^{-tE_l}$, show that the heat kernel

$$Z(t) = \sum_{l=0}^{\infty} (2l+1)\, e^{-t\, E_l} \tag{5.181}$$

has the following Weyl expansion:

$$Z(t) = \frac{R^2}{Dt} + \frac{1}{3} - \frac{1}{60}\frac{Dt}{R^2} + \cdots . \tag{5.182}$$

Exercise 5.9 Show that the Euler–Poincaré characteristic $\chi(S)$ of the carafe shown in Figure 5.8(a) is equal to -1, and that the characteristic of the descent device shown in Figure 5.8(b) is equal to -2.

Appendix A5.6: Diffusion on graphs

A5.6.1 Spectral determinant on a graph

Until now, we have solved the diffusion equation for simple geometries. We now consider the case of networks (or graphs) composed of bonds which are one-dimensional diffusive wires. The spectral determinant (5.162) on a finite graph may be simply expressed as a

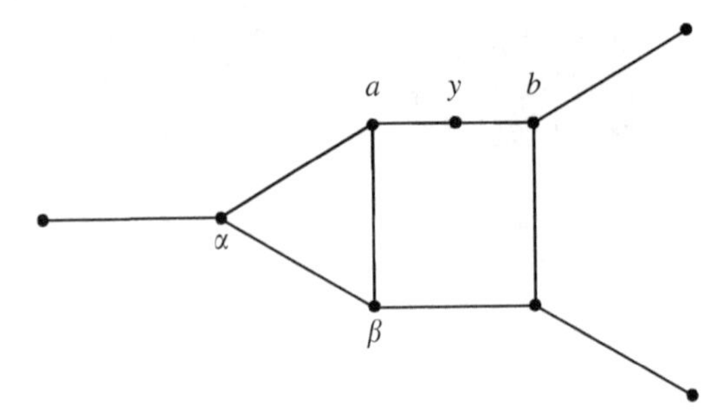

Figure 5.9 Example of a graph: $(\alpha\beta)$ is an arbitrary bond and (ab) is the bond to which the source point y belongs.

function of its geometrical characteristics, that is, the bond lengths and the topology of the connections. For a graph made of E one-dimensional wires and V vertices (see Figure 5.9), we have[24]

$$\boxed{S(\gamma) = \gamma^{\frac{V-E}{2}} \prod_{(\alpha\beta)} \sinh\left(\frac{l_{\alpha\beta}}{L_\gamma}\right) \det M} \tag{5.183}$$

where $l_{\alpha\beta}$ is the length of the bond $(\alpha\beta)$ and $L_\gamma = \sqrt{D/\gamma}$. M is a $V \times V$ square matrix whose diagonal elements are [110–113]

$$M_{\alpha\alpha} = \sum_{\beta}^{m_\alpha} \coth\left(\frac{l_{\alpha\beta}}{L_\gamma}\right). \tag{5.184}$$

The sum is taken over the m_α neighbor vertices of the vertex α. The off-diagonal elements are

$$M_{\alpha\beta} = -\frac{1}{\sinh(l_{\alpha\beta}/L_\gamma)} \tag{5.185}$$

if the vertices α and β are connected by a bond, and are zero otherwise.

To establish this result, we first calculate the probability of return to the origin $P(y, y)$ from any point y of the graph. Then, by integrating over the entire graph, we obtain the spectral determinant from[25]

$$\int dy\, P_\gamma(y, y) = \frac{\partial}{\partial\gamma} \ln S(\gamma). \tag{5.186}$$

24 The prefactor $\gamma^{(V-E)/2}$ is of topological origin. It corresponds to a constant additive term $(V-E)/2$ in the probability of return to the origin $Z(t)$ for diffusion on a $1d$ manifold. On a $2d$ manifold, this constant additive term is $(V-E+F)/6$, where F is the number of cells (see Appendix A5.5).

25 Reference [113] presents an alternative method for calculating the spectral determinant on a graph, based on a path integral formulation.

Writing $D = 1$ into equation (5.15) to simplify the expressions, we begin by solving

$$(\gamma - \Delta_x) P_\gamma(x, y) = \delta(x - y) \tag{5.187}$$

along each bond denoted $(\alpha\beta)$. For a source placed at some point y on the graph, we obtain

$$P_\gamma(x, y) = \frac{P_\gamma(\alpha, y) \sinh \sqrt{\gamma}(l - x) + P_\gamma(\beta, y) \sinh \sqrt{\gamma} x}{\sinh \sqrt{\gamma} l_{\alpha\beta}}, \tag{5.188}$$

where x is the coordinate of a point running along the bond $(\alpha\beta)$ of length $l_{\alpha\beta}$. $P_\gamma(\alpha, y)$ and $P_\gamma(\beta, y)$ are the values taken at the ends of the bond. Current conservation at a vertex α gives

$$-\sum_\beta \partial_{x_{\alpha\beta}} P_\gamma(x_{\alpha\beta} = 0, y) = \delta_{\alpha\, y}, \tag{5.189}$$

where the sum is over the vertices β which are neighbors of α. This leads to a set of linear equations

$$P_\gamma(\alpha, y) \sum_\beta \coth \eta_{\alpha\beta} - \sum_\beta \frac{P_\gamma(\beta, y)}{\sinh \eta_{\alpha\beta}} = \frac{\delta_{\alpha\, y}}{\sqrt{\gamma}} \tag{5.190}$$

with $\eta_{\alpha\beta} = \sqrt{\gamma} l_{\alpha\beta}$. This is a set of $(V + 1)$ linear equations in $(V + 1)$ variables $P_\gamma(\alpha, y)$ where α is either a vertex of the graph or the source y:

$$M_y \begin{pmatrix} P_\gamma(\alpha_1, y) \\ \vdots \\ P_\gamma(y, y) \\ \vdots \\ P_\gamma(\alpha_V, y) \end{pmatrix} = \begin{pmatrix} 0 \\ \vdots \\ 1/\sqrt{\gamma} \\ \vdots \\ 0 \end{pmatrix}. \tag{5.191}$$

The square $(V + 1) \times (V + 1)$ matrix M_y is defined by equations (5.184, 5.185).

We can now calculate $P_\gamma(y, y)$. First of all, using equation (5.190), $P_\gamma(y, y)$ may be written as a function of $P_\gamma(a, y)$ and $P_\gamma(b, y)$ where a and b are the vertices at the endpoints of the bond containing the point y (Figure 5.9):

$$P_\gamma(y, y) \left(\coth \eta_{ay} + \coth \eta_{yb}\right) - \frac{P_\gamma(a, y)}{\sinh \eta_{ay}} - \frac{P_\gamma(b, y)}{\sinh \eta_{by}} = \frac{1}{\sqrt{\gamma}} \tag{5.192}$$

that is,

$$P_\gamma(y, y) = \frac{1}{\sinh \eta_{ab}} \left(\frac{\sinh \eta_{ay} \sinh \eta_{by}}{\sqrt{\gamma}} + P_\gamma(a, y) \sinh \eta_{by} + P_\gamma(b, y) \sinh \eta_{ay} \right). \tag{5.193}$$

Eliminating $P_\gamma(y,y)$ from equation (5.190), we obtain $P_\gamma(a,y)$:

$$P_\gamma(a,y)\sum_\beta \coth\eta_{a\beta} - \sum_\beta P_\gamma(\beta,y)\frac{1}{\sinh\eta_{a\beta}} = \frac{1}{\sqrt{\gamma}}\frac{\sinh\eta_{yb}}{\sinh\eta_{ab}} \tag{5.194}$$

and a similar expression for $P_\gamma(b,\beta)$. The $(V-2)$ remaining equations are unchanged. We are thus left with a set of V linear equations for the variables $P_\gamma(\alpha,y)$:

$$M\begin{pmatrix} P_\gamma(\alpha_1,y) \\ \vdots \\ P_\gamma(a,y) \\ \vdots \\ P_\gamma(b,y) \\ \vdots \\ P_\gamma(\alpha_{V-2},y) \end{pmatrix} = \frac{1}{\sqrt{\gamma}\sinh\eta_{ab}}\begin{pmatrix} 0 \\ \vdots \\ \sinh\eta_{yb} \\ \vdots \\ \sinh\eta_{ay} \\ \vdots \\ 0 \end{pmatrix}. \tag{5.195}$$

The matrix M is defined in (5.184, 5.185) and the term $P_\gamma(y,y)$ related to the source point is no longer present. Inverting the matrix, we obtain $P_\gamma(a,y)$ and $P_\gamma(b,y)$:

$$P_\gamma(a,y) = \frac{1}{\sqrt{\gamma}\sinh\eta_{ab}}\left(T_{aa}\sinh\eta_{yb} + T_{ab}\sinh\eta_{ay}\right) \tag{5.196}$$

where $T = M^{-1}$. Inserting the expressions for $P_\gamma(a,\beta)$ and $P_\gamma(b,\beta)$ in equation (5.193), we finally obtain the return probability $P_\gamma(y,y)$ to any point on the graph:

$$P_\gamma(y,y) = \frac{1}{\sqrt{\gamma}}\left\{\frac{\sinh\eta_{ay}\sinh\eta_{yb}}{\sinh\eta_{ab}} + \frac{1}{\sinh^2\eta_{ab}}\left[T_{aa}\sinh^2\eta_{by} + T_{bb}\sinh^2\eta_{ay} + 2T_{ba}\sinh\eta_{ay}\sinh\eta_{yb}\right]\right\}. \tag{5.197}$$

Spatial integration of $P_\gamma(y,y)$ along the bond (ab) gives

$$\int_a^b dy\, P_\gamma(y,y) = \frac{1}{2\gamma}\left\{\eta_{ab}\coth\eta_{ab} - 1 + (T_{aa}+T_{bb})\left(1+2\gamma\frac{\partial}{\partial\gamma}\right)\coth\eta_{ab} - 2T_{ba}\left(1+2\gamma\frac{\partial}{\partial\gamma}\right)\frac{1}{\sinh\eta_{ab}}\right\} \tag{5.198}$$

where we have used the identities

$$\frac{-\eta}{\sinh^2\eta} = 2\gamma\frac{\partial}{\partial\gamma}\coth\eta \tag{5.199}$$

$$-\eta\frac{\cosh\eta}{\sinh^2\eta} = 2\gamma\frac{\partial}{\partial\gamma}\frac{1}{\sinh\eta}. \tag{5.200}$$

Summing over all the bonds of the graph, and using the additional identities

$$\sum_{(ab)}\left((T_{aa}+T_{bb})\coth\eta_{ab} - 2\frac{T_{ba}}{\sinh\eta_{ab}}\right) = \mathrm{tr}(TM) = V, \tag{5.201}$$

$$\sum_{(ab)}\left((T_{aa}+T_{bb})\frac{\partial}{\partial\gamma}\coth\eta_{ab} - 2T_{ba}\frac{\partial}{\partial\gamma}\frac{1}{\sinh\eta_{ab}}\right) = \mathrm{tr}(T\frac{\partial}{\partial\gamma}M), \tag{5.202}$$

$$\eta\coth\eta = 2\gamma\frac{\partial}{\partial\gamma}\ln\sinh\eta, \tag{5.203}$$

we find that the integral $\int_0^\infty dt\, e^{-\gamma t}Z(t) = \int_{\mathrm{graph}} dy\, P_\gamma(y,y)$ given by equation (5.161) simplifies considerably, and may be put in the form

$$\frac{\partial}{\partial\gamma}\sum_{(ab)}\ln\sinh\eta_{ab} + \frac{V-E}{2\gamma} + \mathrm{tr}M^{-1}\frac{\partial}{\partial\gamma}M. \tag{5.204}$$

Using the property $\mathrm{tr}M^{-1}\frac{\partial}{\partial\gamma}M = \frac{\partial}{\partial\gamma}\ln\det M$, we recover the expression (5.163) for the spectral determinant:

$$\boxed{\int_0^\infty dt\, e^{-\gamma t}Z(t) = \frac{\partial}{\partial\gamma}\ln S(\gamma)} \tag{5.205}$$

where $S(\gamma)$ is given by equation (5.183).

A5.6.2 Examples

We now present expressions for the spectral determinant for several simple geometries [114]. In each case, it suffices to compute the determinant of the matrix M defined by (5.184, 5.185), given by the product of its eigenvalues.

- Isolated wire of length L:

$$S(\gamma) = \frac{L}{L_\gamma}\sinh\frac{L}{L_\gamma}. \tag{5.206}$$

- Wire of length L connected to reservoirs:

$$S(\gamma) = \frac{L_\gamma}{L}\sinh\frac{L}{L_\gamma}. \tag{5.207}$$

- Ring of perimeter L pierced by an Aharonov–Bohm flux ($\theta = 4\pi\phi/\phi_0$):

$$S(\gamma) = 2\left(\cosh\frac{L}{L_\gamma} - \cos\theta\right). \tag{5.208}$$

- Ring of perimeter L with an attached free arm of length b:

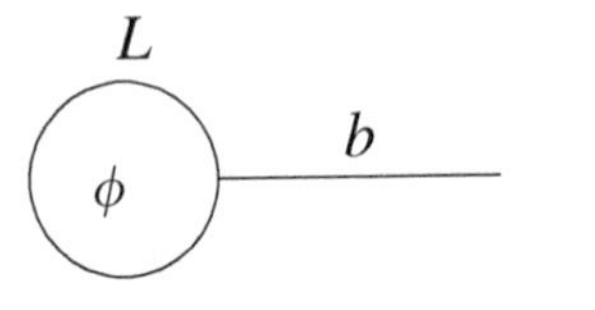

$$S(\gamma) = \sinh\frac{b}{L_\gamma}\sinh\frac{L}{L_\gamma} + 2\cosh\frac{b}{L_\gamma}\left(\cosh\frac{L}{L_\gamma} - \cos\theta\right). \tag{5.209}$$

- Ring of perimeter L with an attached arm of length b which is connected to a reservoir:

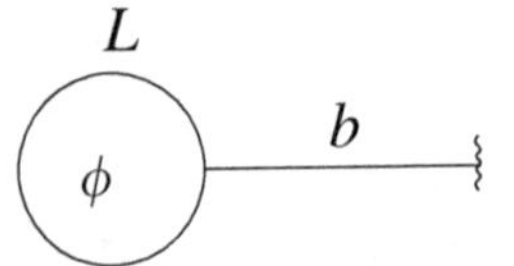

$$S(\gamma) = \frac{L_\gamma}{L}\left(\cosh\frac{b}{L_\gamma}\sinh\frac{L}{L_\gamma} + 2\sinh\frac{b}{L_\gamma}\left(\cosh\frac{L}{L_\gamma} - \cos\theta\right)\right). \tag{5.210}$$

- Ring of perimeter $L = Nl$ with N attached arms of length b (isolated ensemble):

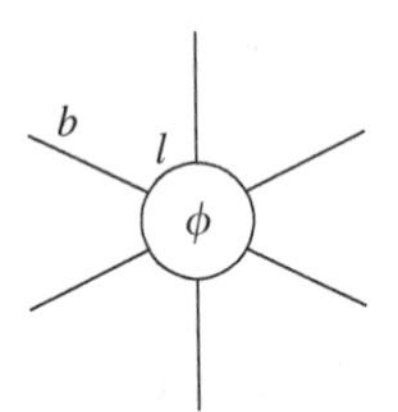

$$S(\gamma) = \prod_{k=0}^{N-1}\left\{\sinh\frac{b}{L_\gamma}\sinh\frac{l}{L_\gamma} + 2\cosh\frac{b}{L_\gamma}\left(\cosh\frac{L}{L_\gamma} - \cos\left[\frac{1}{N}(\theta + 2\pi k)\right]\right)\right\}. \tag{5.211}$$

- Chain of N rings of perimeter $L = 4a$, connected by wires of length b (periodic boundary conditions):

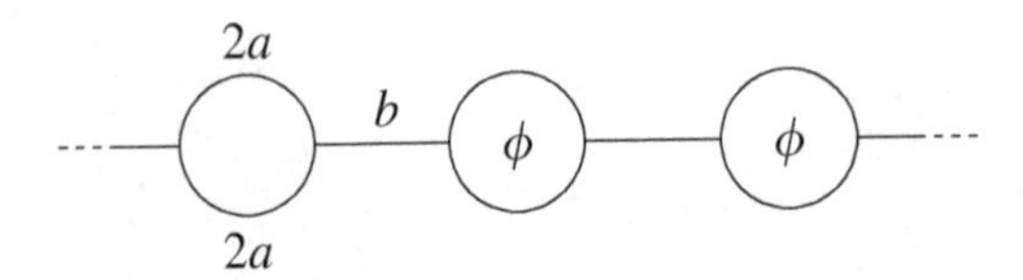

$$S(\gamma) = \left(\frac{L_\gamma}{2a}\right)^N \prod_{k=1}^{N} \left\{ \sinh\left(\frac{b}{L_\gamma}\right) \left[4\cosh^2 \frac{2a}{L_\gamma} + \sinh^2 \frac{2a}{L_\gamma} - 4\cos^2\left(\frac{\theta}{2}\right) \right] \right.$$
$$\left. + 4\sinh\left(\frac{2a}{L_\gamma}\right) \left[\cosh \frac{2a}{L_\gamma} \cosh \frac{b}{L_\gamma} - \cos\left(\frac{\theta}{2}\right) \cos \frac{2k\pi}{N} \right] \right\}. \qquad (5.212)$$

- Ladder of N squares of perimeter $4a$ (periodic boundary conditions):

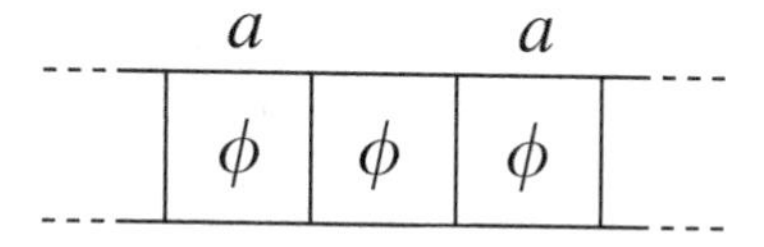

$$S(\gamma) = \prod_{k=1}^{N} \left\{ \frac{L_\gamma}{a} \sinh\left(\frac{a}{L_\gamma}\right) \right.$$
$$\left. \times \left[\left(3\cosh \frac{a}{L_\gamma} - 2\cos \frac{\theta}{2} \cos \frac{2k\pi}{N} \right)^2 - \left(1 + 4\sin^2 \frac{\theta}{2} \sin^2 \frac{2k\pi}{N} \right) \right] \right\}. \qquad (5.213)$$

- Square lattice. We consider a mesh made of N^2 square tiles of side a, in the presence of a perpendicular field, in the limit $N \to \infty$. The spectral determinant is related to solutions of the tight binding problem on a square lattice in a magnetic field, popularized by Hofstadter [115]. For a flux ϕ per plaquette such that $2\phi/\phi_0 = p/q$ where $q \to \infty$, we have the simple form [114]:

$$S(\gamma) = \frac{N^2}{q} \prod_{i=1}^{q} \left[\frac{L_\gamma}{a} \sinh \frac{a}{L_\gamma} \left(4\cosh \frac{a}{L_\gamma} - \epsilon_i \right) \right]^{N/q} \qquad (5.214)$$

where the ϵ_i are the q eigenvalues of the Harper equation:

$$\epsilon \psi_m = \psi_{m-1} + \psi_{m+1} + 2\cos\left(2\pi \frac{p}{q} m\right) \psi_m. \qquad (5.215)$$

A5.6.3 Thermodynamics, transport and spectral determinant

A certain number of physical quantities are related to the probability of return to the origin $Z(t)$ or, more precisely, to its Laplace transform, which is itself related to the spectral determinant by equation (5.205). The result (5.183) obtained for the spectral determinant on an arbitrary graph allows us to calculate these physical quantities directly on networks of diffusive wires. For example, thermodynamic quantities such as the fluctuation in magnetic moment $\overline{\delta \mathcal{M}^2} = \overline{\mathcal{M}^2} - \overline{\mathcal{M}}^2$ or the average magnetization $\mathcal{M}_{ee}$ may be calculated from

$Z(t)$. Thanks to (5.205), the expressions for these quantities (14.35 and 14.51) may be rewritten as a function of the spectral determinant:

$$\overline{\delta\mathcal{M}^2} = \frac{\hbar^2}{2\pi^2}\int_\gamma^\infty d\gamma_1(\gamma-\gamma_1)\frac{\partial^2}{\partial B^2}\ln S(\gamma_1) \tag{5.216}$$

$$\mathcal{M}_{ee} = \frac{F\hbar}{2\pi}\int_\gamma^\infty d\gamma_1\frac{\partial}{\partial B}\ln S(\gamma_1). \tag{5.217}$$

With the aid of (5.183), we may calculate the magnetization of an arbitrary network of diffusive wires [112] from these relations quite simply.

The case of transport properties is more delicate. It is equally possible to relate the weak localization correction or conductance fluctuations to the probability of return to the origin (equations 7.53 and 11.37), and to rewrite them in terms of the spectral determinant:

$$\Delta\sigma = -\frac{se^2D}{\pi\hbar\Omega}\frac{\partial}{\partial\gamma}\ln S(\gamma) \tag{5.218}$$

$$\overline{\delta\sigma^2} = -3\frac{s^2e^4D^2}{\beta\pi^2\hbar^2\Omega^2}\frac{\partial^2}{\partial\gamma^2}\ln S(\gamma). \tag{5.219}$$

It is nonetheless worth being careful: these expressions are not a priori applicable for networks. Indeed, the conductivity is a local function which depends on position in the network. To obtain the conductance between two contacts in the network, we must be able to combine the contributions of different wires, according to the Kirchhoff laws [116]. In this case, the weak localization correction or conductance fluctuations cannot be simply expressed as a function of the spectral determinant. Nonetheless, equations (5.218, 5.219) are useful for simple geometries such as a wire or ring, or for regular infinite networks such as the square lattice where each wire plays the same role. For the wire and ring, for example, we easily recover the results (7.61) and (7.90), with the help of expressions (5.207) and (5.208).

6
Dephasing

6.1 Dephasing and multiple scattering

6.1.1 Generalities

The Diffuson and the Cooperon involve the product of two sequences of multiple scattering passing through the same scatterers in either the same (Diffuson) or the inverse order (Cooperon). These two sequences of multiple scattering are described by two amplitudes which are complex conjugates of each other. For the Diffuson, the phase vanishes, yielding a classical object. For the Cooperon, the two amplitudes correspond to time-reversed sequences. If the system is time reversal invariant, the structure factors associated with the Diffuson and the Cooperon are equal.

There are certain mechanisms which can modify differently the two paired multiple scattering amplitudes and which can result in a dephasing $\Delta\phi(t)$ between the two amplitudes constituting the Diffuson or the Cooperon. In this chapter, we will show that dephasing affects the average probability of quantum diffusion by a global factor, so that upon summation over all multiple scattering sequences, we obtain expressions of the form

$$P(\boldsymbol{r},\boldsymbol{r}',t)\left\langle e^{i\Delta\phi(t)}\right\rangle, \tag{6.1}$$

where $P(\boldsymbol{r},\boldsymbol{r}',t)$ is the probability of quantum diffusion defined in Chapter 4. The relative phase $\Delta\phi(t)$ is a random variable whose distribution depends on the phenomenon at the origin of the dephasing. We denote the average over this distribution by $\langle\cdots\rangle$. In general, it decays exponentially in time[1]

$$\left\langle e^{i\Delta\phi(t)}\right\rangle = e^{-t/\tau_\gamma}, \tag{6.2}$$

which provides a physical interpretation of the cutoff time τ_γ (corresponding to the length $L_\gamma = \sqrt{D\tau_\gamma}$) introduced phenomenologically in section 5.2.2. Many properties may be expressed in terms of the time-integrated probability and are, therefore, direct measures of

[1] This behavior is not necessarily exponential. More generally, it may be a function $f(t/\tau_\gamma)$, as seen for example in section 13.7.3 and in Exercise 13.13.

the Laplace transform (5.12)

$$\int_0^\infty dt\, P(\boldsymbol{r},\boldsymbol{r}',t)\langle e^{i\Delta\phi(t)}\rangle = \int_0^\infty dt\, P(\boldsymbol{r},\boldsymbol{r}',t)e^{-t/\tau_\gamma}. \tag{6.3}$$

The dephasing processes may affect the Diffuson as well as the Cooperon. This may appear paradoxical, since the Diffuson associated with the probability $P_d(\boldsymbol{r},\boldsymbol{r}',t)$ is a classical quantity related to the conservation of particle number or of energy (see the discussion on the normalization of the probability, page 231). In this definition of the Diffuson, the two complex amplitudes which enter it correspond to the same realization of the disorder potential. We may, however, imagine the pairing of two amplitudes corresponding to *distinct realizations of disorder*, or to *different external parameters*. In such cases, we obtain a quantity which retains the structure of a Diffuson, but which may exhibit dephasing. This type of Diffuson no longer corresponds to a Goldstone mode describing particle number or energy conservation.

6.1.2 Mechanisms for dephasing: introduction

In order to clarify these ideas, consider the structure factor of the Diffuson (or the Cooperon) which appears in the integral equation (4.26)[2]

$$\Gamma(\boldsymbol{r},\boldsymbol{r}') = \gamma_e\delta(\boldsymbol{r}-\boldsymbol{r}') + \int \Gamma(\boldsymbol{r},\boldsymbol{r}'')w_0(\boldsymbol{r}'',\boldsymbol{r}')\, d\boldsymbol{r}'', \tag{6.4}$$

where $w_0(\boldsymbol{r}'',\boldsymbol{r}') = P_0(\boldsymbol{r}'',\boldsymbol{r}')/\tau_e$. Upon Fourier transforming, this equation takes the simple form

$$\Gamma(\boldsymbol{q}) = \gamma_e + \Gamma(\boldsymbol{q})w_0(\boldsymbol{q}), \tag{6.5}$$

where $w_0(q) = P_0(\boldsymbol{q},\omega=0)/\tau_e$ (see Appendix A4.1). The structure factor thus has the form

$$\boxed{\Gamma(\boldsymbol{q}) = \frac{\gamma_e}{1-w_0(\boldsymbol{q})}} \tag{6.6}$$

In the limit of slow variations ($ql_e \ll 1$), the integral equation (6.4) is equivalent to a diffusion equation. In this limit, the expansion (4.73):

$$w_0(\boldsymbol{q}) \simeq 1 - Dq^2\tau_e \tag{6.7}$$

leads to

$$\Gamma(\boldsymbol{q}) = \frac{\gamma_e/\tau_e}{Dq^2}. \tag{6.8}$$

This diffusion pole is characteristic of a Goldstone mode. It is intimately linked to the structure of the elementary term $w_0(\boldsymbol{r}_i,\boldsymbol{r}_{i+1})$ which depends on the quantum (or wave)

[2] To keep our notation concise, we shall not write the dependence on frequency ω explicitly.

process involved in the scattering from the impurities. The presence of additional degrees of freedom affects this elementary term, and leads to a modification of the structure factor. We shall meet various different situations.

- The elementary step w_0 is affected by an additional dephasing:

$$w(\boldsymbol{r}_i,\boldsymbol{r}_{i+1}) = w_0(\boldsymbol{r}_i,\boldsymbol{r}_{i+1})\, e^{i\Delta\varphi(\boldsymbol{r}_i,\boldsymbol{r}_{i+1})}. \tag{6.9}$$

The integral equation (6.4) thus becomes

$$\Gamma(\boldsymbol{r},\boldsymbol{r}') = \gamma_e \delta(\boldsymbol{r}-\boldsymbol{r}') + \int \Gamma(\boldsymbol{r},\boldsymbol{r}'')w_0(\boldsymbol{r}'',\boldsymbol{r}')\, e^{i\Delta\varphi(\boldsymbol{r}'',\boldsymbol{r}')}\, d\boldsymbol{r}''. \tag{6.10}$$

If the dephasing $\Delta\varphi(\boldsymbol{r}'',\boldsymbol{r}')$ is a simple and well-defined function, it suffices to solve the integral equation (6.10) in the diffusion approximation. This is the case, for example, for an electron gas in a uniform magnetic field. The dephasing given by the integral of the vector potential vanishes for the Diffuson but remains finite for the Cooperon (section 6.3).

- The dephasing may derive from the existence of degrees of freedom not controlled by the experimentalist. This is the case for multiple scattering from particles in motion. We may then define a new Diffuson by pairing two sequences of multiple scatterings corresponding to scatterers whose positions are measured at different times, 0 and T, that is to say, corresponding to *different realizations of disorder*. The positions of scatterers is not known in general, so we must average the dephasing between the two paired amplitudes over the positions of the scatterers. We shall show (section 6.7) that we can thus define a new elementary step of the form

$$\langle w(\boldsymbol{r}_i,\boldsymbol{r}_{i+1})\rangle_T = b(T) w_0(\boldsymbol{r}_i,\boldsymbol{r}_{i+1}) \tag{6.11}$$

where the parameter $b = \langle e^{i\Delta\phi}\rangle < 1$ is the average of the dephasing associated with the elementary step, and depends on the motion of the scatterers during the time interval T. In equation (6.10), the integral is simply multiplied by b and the Fourier transform of the structure factor satisfies

$$\Gamma(\boldsymbol{q}) = b\gamma_e + \Gamma(\boldsymbol{q}) b w_0(\boldsymbol{q}), \tag{6.12}$$

or

$$\Gamma(\boldsymbol{q}) = \frac{b\gamma_e}{1 - b w_0(\boldsymbol{q})}. \tag{6.13}$$

The parameter $b < 1$ modifies the diffusion pole, which, with the aid of (6.7), becomes

$$\Gamma(\boldsymbol{q}) = \frac{\gamma_e/\tau_e}{1/\tau_\gamma + Dq^2}. \tag{6.14}$$

The average of the dephasing associated with uncontrolled degrees of freedom thus yields a characteristic time of the form

$$\tau_\gamma = \tau_e \frac{b}{1-b}. \tag{6.15}$$

The pole of $\Gamma(\boldsymbol{q})$ describes the attenuation of the diffusion mode. The probability is obtained from the structure factor. By reinserting the frequency dependence and taking the Fourier transform, this results in the expression

$$P(\boldsymbol{r},\boldsymbol{r}',t)\,e^{-t/\tau_\gamma} \tag{6.16}$$

for the probability. The time τ_γ thus defined expresses the loss of phase coherence.

- An additional source of dephasing may arise when the diffusing wave is not scalar, but has additional degrees of freedom such as spin for electrons or polarization for electromagnetic waves. If the scattering potential depends on these degrees of freedom, there will be a dephasing, and thus a finite characteristic time τ_γ. This is the case of spin-orbit scattering, of scattering from magnetic impurities for electrons, and of the rotation of the polarization in Rayleigh scattering or the scattering of photons from atoms with internal degrees of freedom.

  The coupling to these new degrees of freedom may, as in the preceding example, be described in full generality by a modification of the elementary scattering process, also called *elementary vertex*, which then has a tensorial form $\gamma_e b_{\alpha\beta,\gamma\delta}$ where the indices describe either the spin of the electron or the polarization of the wave. More precisely, the tensor $b_{\alpha\beta,\gamma\delta}$ describes the scattering from two initial spin states (α,β) to two final states (γ,δ). The iteration equation for the structure factor now has a tensorial form which generalizes (6.5):

$$\Gamma_{\alpha\beta,\gamma\delta}(\boldsymbol{q}) = \gamma_e b_{\alpha\beta,\gamma\delta} + \sum_{\mu,\nu} \Gamma_{\alpha\beta,\mu\nu}(\boldsymbol{q}) b_{\mu\nu,\gamma\delta} w(\boldsymbol{q}). \tag{6.17}$$

  The diagonalization of this equation involves the eigensubspaces associated with total spin (or its equivalent for the polarization) of the two paired states. In each of the subspaces, singlet or triplet, the diffusion equation may be put in the form (6.12), and we thus obtain a Diffuson (or a Cooperon) characterized by total spin $J = 0$ or $J = 1$:

$$\boxed{\Gamma_J(\boldsymbol{q}) = \frac{\gamma_e/\tau_e}{1/\tau_J + Dq^2}} \tag{6.18}$$

  with a characteristic time of the form

$$\boxed{\tau_J = \tau_e \frac{b_J}{1 - b_J}} \tag{6.19}$$

  where b_J describes, in each subspace, the contribution of the elementary scattering process.

  We thus see how the spin-orbit and the coupling to magnetic degrees of freedom may affect the Diffuson or the Cooperon, and introduce a dephasing. We shall consider these examples in detail, and will treat, in an analogous manner, the effect of polarization on multiple scattering of light by classical dipoles or by atoms possessing a quantum internal structure.

- Finally, there are situations where the dephasing may not be expressed simply in terms of a local interaction vertex. This is, for example, the case for the Cooperon in a time-dependent field. We return to this point in Appendix 6.3.

6.1.3 The Goldstone mode

The classical probability is normalized (see remark page 117). This normalization takes account of the conservation of particle number and is expressed by the divergence of the structure factor (6.8) at low frequency and small wave vector. The introduction of new degrees of freedom (spin or polarization) does not modify this normalization. Indeed, the two paired sequences of multiple scattering of the spin (or polarization) remain *identical*. There is no dephasing and the Diffuson is a Goldstone mode.

More generally, we may construct a Diffuson by pairing two amplitudes corresponding to *independent* rotations of the electron spin (or wave polarization). The resulting Diffuson acquires a pole of the form (6.14) and a finite dephasing time. We are thus led to define two different objects.

- The first is the Diffuson introduced in Chapter 4, that is to say, the Goldstone mode reflecting energy or particle number conservation. This is a classical object and is thus insensitive to dephasing. This scalar mode always exists, even with additional degrees of freedom.[3]
- The second also has the structure of a Diffuson, that is, it results from the pairing of two amplitudes propagating in the same direction but associated with either different realizations of disorder, or different configurations of spin or polarization, or different magnetic fields. In all cases, this kind of Diffuson may have a dephasing and a corresponding characteristic decay time.

In what follows, we shall use these two Diffusons, which we nonetheless choose to denote by the same name. The "phase-sensitive" Diffuson plays an important role in the study of fluctuations (see Chapters 11 and 12).

For the Cooperon, it is clear that there is always a finite phase coherence time since, by construction, we cannot pair two time-reversed amplitudes with the same configurations of spin or polarization.

6.2 Magnetic field and the Cooperon

In order to derive the expression for the Cooperon X_c in section 4.6, we explicitly used the property of time reversal invariance. Owing to this invariance, the structure factor Γ'_ω associated with the Cooperon is identical to that (Γ_ω) associated with the Diffuson P_d.

This is no longer the case when this invariance is broken. One important situation for which there is no time reversal invariance is that of a charged particle in a magnetic field $\boldsymbol{B} = \nabla \times \boldsymbol{A}$.

[3] Except in the presence of absorption, in which case it is attenuated since energy and/or particle number are not conserved.

Remark: magnetic field and the reciprocity theorem

In the presence of a magnetic field B, the generalization of the reciprocity theorem (2.113) is

$$T_{dir}(\boldsymbol{k},\boldsymbol{k}',B) = T_{rev}(-\boldsymbol{k}',-\boldsymbol{k},-B), \tag{6.20}$$

since time reversal changes the sign of the magnetic field. In the backscattering direction $\boldsymbol{k}' = -\boldsymbol{k}$, this condition becomes

$$T_{dir}(\boldsymbol{k},-\boldsymbol{k},B) = T_{rev}(\boldsymbol{k},-\boldsymbol{k},-B) \tag{6.21}$$

such that the direct $T_{dir}(\boldsymbol{k},-\boldsymbol{k},B)$ and inverse $T_{rev}(\boldsymbol{k},-\boldsymbol{k},B)$ processes are different, except in zero field. We say that the magnetic field breaks time reversal invariance (see also [117]).

The Hamiltonian for an electron in a random potential and in a magnetic field may be written as (taking $e > 0$)

$$\mathcal{H} = -\frac{\hbar^2}{2m}\left(\boldsymbol{\nabla} + \frac{ie}{\hbar}\boldsymbol{A}\right)^2 + V(\boldsymbol{r}). \tag{6.22}$$

Classically, the effect of the magnetic field, as expressed by the Lorentz force, is to curve the trajectory of the charged particles. The associated radius of curvature is the cyclotron radius which, for electrons, is $r_c = mv/eB$ where v is the velocity. The length r_c thus appears as an additional scale which must be compared to the elastic mean free path l_e. We shall consider here the case of a weak magnetic field, such that the curvature of the electronic trajectories between two elastic collisions is negligible. In other words, we shall assume that $l_e \ll r_c$, or $\omega_c \tau_e \ll 1$ where $\omega_c = eB/m$ is the cyclotron frequency.

In quantum mechanics, neglecting the curvature of the trajectories does not imply that the magnetic field has no effect. The Aharonov–Bohm effect discussed in Chapter 1 illustrates this point. In the limit of the *eikonal approximation* of slow variation of the vector potential $\boldsymbol{A}(\boldsymbol{r})$, the only effect of the magnetic field is to modify the phase of the wave functions (and thus of the Green function) [118, 119]. In the presence of the field, the average Green function, denoted by $\overline{G}_\epsilon^{R,A}(\boldsymbol{r},\boldsymbol{r}',B)$, may be written as

$$\overline{G}_\epsilon^{R,A}(\boldsymbol{r},\boldsymbol{r}',B) = \overline{G}_\epsilon^{R,A}(\boldsymbol{r},\boldsymbol{r}')\, e^{i\varphi(\boldsymbol{r},\boldsymbol{r}')} \tag{6.23}$$

with

$$\varphi(\boldsymbol{r},\boldsymbol{r}') = -\frac{e}{\hbar}\int_{\boldsymbol{r}}^{\boldsymbol{r}'} \boldsymbol{A}\cdot d\boldsymbol{l}, \tag{6.24}$$

where $\overline{G}_\epsilon^{R,A}(\boldsymbol{r},\boldsymbol{r}') = \overline{G}_\epsilon^{R,A}(\boldsymbol{r},\boldsymbol{r}',B=0)$ is the Green function in zero field. The circulation of the vector potential is calculated along the segment $(\boldsymbol{r},\boldsymbol{r}')$. In this approximation, the vector potential changes appreciably only over distances much larger than the mean free path l_e which controls the decay of the average Green function [119].

In the presence of a magnetic field, the average Green function is no longer translation invariant and the real space formalism developed in sections 4.4 and 4.6 is more appropriate.

Let us first study the case of the Diffuson P_d. For this calculation, we only need combinations of the type $\overline{G}^R_\epsilon(\boldsymbol{r},\boldsymbol{r}',B)\overline{G}^A_{\epsilon-\omega}(\boldsymbol{r}',\boldsymbol{r},B)$ which appear in equations (4.23) and (4.24). In the eikonal approximation, this product remains independent of the field since the phases compensate:

$$\overline{G}^R_\epsilon(\boldsymbol{r},\boldsymbol{r}',B)\overline{G}^A_{\epsilon-\omega}(\boldsymbol{r}',\boldsymbol{r},B) = \overline{G}^R_\epsilon(\boldsymbol{r},\boldsymbol{r}')\overline{G}^A_{\epsilon-\omega}(\boldsymbol{r}',\boldsymbol{r}). \tag{6.25}$$

Thus, in this limit, where we neglect the curvature of the electronic trajectories, the contributions of Drude–Boltzmann P_0 and of the classical diffusion are independent of field.[4]

On the other hand, for the Cooperon, the elementary combination of average Green functions which appear in (4.43) is $\overline{G}^R_\epsilon(\boldsymbol{r},\boldsymbol{r}',B)\overline{G}^A_{\epsilon-\omega}(\boldsymbol{r},\boldsymbol{r}',B)$. In the presence of a field, this product is no longer equal to $\overline{G}^R_\epsilon(\boldsymbol{r},\boldsymbol{r}')\overline{G}^A_{\epsilon-\omega}(\boldsymbol{r}',\boldsymbol{r})$, but rather to

$$\begin{aligned}
\overline{G}^R_\epsilon(\boldsymbol{r},\boldsymbol{r}',B)\overline{G}^A_{\epsilon-\omega}(\boldsymbol{r},\boldsymbol{r}',B) &= \overline{G}^R_\epsilon(\boldsymbol{r},\boldsymbol{r}')\overline{G}^A_{\epsilon-\omega}(\boldsymbol{r},\boldsymbol{r}')\, e^{2i\varphi(\boldsymbol{r},\boldsymbol{r}')} \\
&= \overline{G}^R_\epsilon(\boldsymbol{r},\boldsymbol{r}')\overline{G}^A_{\epsilon-\omega}(\boldsymbol{r}',\boldsymbol{r})\, e^{2i\varphi(\boldsymbol{r},\boldsymbol{r}')} \\
&= 2\pi\rho_0 P_0(\boldsymbol{r},\boldsymbol{r}',\omega)\, e^{2i\varphi(\boldsymbol{r},\boldsymbol{r}')}.
\end{aligned} \tag{6.26}$$

Thus, there is a phase factor which *doubles* that appearing in the average Green function. This derives from the fact that each of the two electronic trajectories joining the points $\boldsymbol{r}$ and $\boldsymbol{r}'$ gives a phase factor of the same sign. This factor 2 is essential for understanding the magnetic field effects on electronic transport properties (see, for example, section 7.6.3). The integral equation (4.43) for the structure factor Γ'_ω associated with the Cooperon now depends on the phase, and becomes

$$\Gamma'_\omega(\boldsymbol{r}_1,\boldsymbol{r}_2) = \gamma_e\delta(\boldsymbol{r}_1-\boldsymbol{r}_2) + \frac{1}{\tau_e}\int \Gamma'_\omega(\boldsymbol{r}_1,\boldsymbol{r}'')P_0(\boldsymbol{r}'',\boldsymbol{r}_2,\omega)\, e^{2i\varphi(\boldsymbol{r}'',\boldsymbol{r}_2)}\, d\boldsymbol{r}''. \tag{6.27}$$

It is thus different from Γ_ω given by (4.26). In the diffusion approximation, that is, for slow spatial variations, we may linearize this equation. Using

$$\nabla_{\boldsymbol{r}}\left[\Gamma'_\omega(\boldsymbol{r}_1,\boldsymbol{r})\, e^{2i\varphi(\boldsymbol{r},\boldsymbol{r}_2)}\right] = e^{2i\varphi(\boldsymbol{r},\boldsymbol{r}_2)}\left[\nabla_{\boldsymbol{r}} + 2i\frac{e}{\hbar}\boldsymbol{A}\right]\Gamma'_\omega(\boldsymbol{r}_1,\boldsymbol{r}), \tag{6.28}$$

a calculation identical to that of section 4.5 leads to the equation:

$$\boxed{\left(-i\omega - D\left[\nabla_{\boldsymbol{r}_2} + i\frac{2e}{\hbar}\boldsymbol{A}(\boldsymbol{r}_2)\right]^2\right)\Gamma'_\omega(\boldsymbol{r}_1,\boldsymbol{r}_2,\omega) = \frac{\gamma_e}{\tau_e}\delta(\boldsymbol{r}_1-\boldsymbol{r}_2)} \tag{6.29}$$

The structure factor Γ'_ω satisfies a *covariant diffusion equation*, in which the magnetic field contribution is accounted for by the substitution

$$\boxed{\nabla \rightarrow \nabla + i\frac{2e}{\hbar}\boldsymbol{A}} \tag{6.30}$$

[4] It is, nonetheless, important to note that there are particular situations (see remark page 203, and sections 11.4.4 and 14.2.3) for which the Diffuson may depend on the applied field.

This substitution is analogous to the covariant substitution $\nabla + ie\mathbf{A}/\hbar$ for the gradient in the Schrödinger equation (2.2), but with an effective charge equal to $(-2e)$ instead of $(-e)$. The covariant form of the differential equation (6.29) is also similar[5] to that obtained when writing the Ginzburg–Landau equations for a superconductor [121], and is a direct consequence of gauge invariance.

We must now determine the Cooperon X_c. In the presence of a magnetic field, equation (4.44) becomes

$$X_c(\mathbf{r},\mathbf{r}',\omega) \simeq \frac{\Gamma'_\omega(\mathbf{r},\mathbf{r})}{2\pi\rho_0}\left[\int \overline{G}^R_\epsilon(\mathbf{r},\mathbf{r}_1)\overline{G}^A_\epsilon(\mathbf{r}',\mathbf{r}_1)\, e^{i\varphi(\mathbf{r},\mathbf{r}_1)+i\varphi(\mathbf{r}',\mathbf{r}_1)}\, d\mathbf{r}_1\right]^2. \tag{6.31}$$

The average Green functions decay exponentially with length l_e. We may thus neglect the dependence of the integral on the field, and we obtain

$$X_c(\mathbf{r},\mathbf{r}',\omega) = \frac{\tau_e}{\gamma_e}\Gamma'_\omega(\mathbf{r},\mathbf{r})g^2(\mathbf{r}-\mathbf{r}') \tag{6.32}$$

where the function $g^2(\mathbf{R})$ defined by (3.98) is short range, and $g^2(0) = 1$. Equations (6.29) and (6.32) allow us to calculate the Cooperon contribution to the probability of return to the origin in the presence of a magnetic field.

Important remark
The Cooperon X_c is not a solution of a diffusion equation (since it vanishes for $|\mathbf{r}-\mathbf{r}'| > l_e$). In section 4.6, we defined the quantity P_c by

$$P_c(\mathbf{r},\mathbf{r}',\omega) = \frac{\tau_e}{\gamma_e}\Gamma'_\omega(\mathbf{r},\mathbf{r}'). \tag{6.33}$$

Following (6.29), P_c obeys the covariant diffusion equation

$$\left(-i\omega - D\left[\nabla_{\mathbf{r}'} + i\frac{2e}{\hbar}\mathbf{A}(\mathbf{r}')\right]^2\right)P_c(\mathbf{r},\mathbf{r}',\omega) = \delta(\mathbf{r}-\mathbf{r}'), \tag{6.34}$$

whose solution, for $\mathbf{r}' = \mathbf{r}$ leads to $X_c(\mathbf{r},\mathbf{r},\omega) = P_c(\mathbf{r},\mathbf{r},\omega)$.

We can see that the covariant diffusion equation (6.29, 6.34) has the same structure as a Schrödinger equation, using the following substitutions:

$$\begin{aligned}\frac{\hbar}{2m} &\to D\\ e &\to 2e.\end{aligned} \tag{6.35}$$

To describe the effects of magnetic field on the probability of return to the origin $P_c(\mathbf{r},\mathbf{r},t)$, we shall mainly be interested in solving this equation for two-dimensional geometries

[5] This similarity is also exploited in [120].

and for different magnetic field configurations. Thus, we now study successively the case of a uniform magnetic field, that of a ring or a cylinder, and in Appendix A6.1, that of a line of magnetic flux crossing a plane. We will calculate the return probability $P_c(\boldsymbol{r},\boldsymbol{r},t)$ and the heat kernel $Z_c(t)$ given by (5.5) where $\{E_n\}$ are the eigenvalues of equation (6.34).

Remark

In the introduction to this chapter, we saw that the Diffuson could be dephased when the two amplitudes which compose it correspond to different realizations of disorder. In similar fashion, there are situations where these amplitudes correspond to *different* magnetic fields B and B'. In this case, combinations of the type $\overline{G}^R_\epsilon(\boldsymbol{r},\boldsymbol{r}',B)\overline{G}^A_{\epsilon-\omega}(\boldsymbol{r}',\boldsymbol{r},B')$ which constitute the Diffuson, contain a phase factor related to the *difference* in phases associated with each amplitude

$$\overline{G}^R_\epsilon(\boldsymbol{r},\boldsymbol{r}',B)\overline{G}^A_{\epsilon-\omega}(\boldsymbol{r}',\boldsymbol{r},B') = \overline{G}^R_\epsilon(\boldsymbol{r},\boldsymbol{r}')\overline{G}^A_{\epsilon-\omega}(\boldsymbol{r}',\boldsymbol{r})\, e^{i[\varphi(\boldsymbol{r},\boldsymbol{r}')-\varphi'(\boldsymbol{r},\boldsymbol{r}')]} \tag{6.36}$$

where $\varphi(\boldsymbol{r},\boldsymbol{r}') = -(e/\hbar)\int_{\boldsymbol{r}}^{\boldsymbol{r}'} \boldsymbol{A}\cdot d\boldsymbol{l}$ and $\varphi'(\boldsymbol{r},\boldsymbol{r}') = -(e/\hbar)\int_{\boldsymbol{r}}^{\boldsymbol{r}'} \boldsymbol{A}'\cdot d\boldsymbol{l}$. In contrast, combinations of the form $\overline{G}^R_\epsilon(\boldsymbol{r},\boldsymbol{r}',B)\overline{G}^A_{\epsilon-\omega}(\boldsymbol{r},\boldsymbol{r}',B')$ entering the Cooperon contain a phase factor which is the *sum* of the phases φ and φ':

$$\overline{G}^R_\epsilon(\boldsymbol{r},\boldsymbol{r}',B)\overline{G}^A_{\epsilon-\omega}(\boldsymbol{r},\boldsymbol{r}',B') = \overline{G}^R_\epsilon(\boldsymbol{r},\boldsymbol{r}')\overline{G}^A_{\epsilon-\omega}(\boldsymbol{r},\boldsymbol{r}')\, e^{i[\varphi(\boldsymbol{r},\boldsymbol{r}')+\varphi'(\boldsymbol{r},\boldsymbol{r}')]}. \tag{6.37}$$

P_c and P_d are thus solutions of the differential equation

$$\left(-i\omega - D\left[\nabla_{\boldsymbol{r}'} + i\frac{e}{\hbar}[\boldsymbol{A}(\boldsymbol{r}') \pm \boldsymbol{A}'(\boldsymbol{r}')]\right]^2\right) P_{c,d}(\boldsymbol{r},\boldsymbol{r}',\omega) = \delta(\boldsymbol{r}-\boldsymbol{r}'). \tag{6.38}$$

The Diffuson depends on the magnetic fields B and B' except if they are equal, in which case we recover the differential equation (6.34) for the Cooperon.

6.3 Probability of return to the origin in a uniform magnetic field

Consider the diffusion of an electron in an infinite two-dimensional plane with a perpendicular magnetic field B. To see the effect of the field on the Cooperon, we must solve the covariant diffusion equation (6.34). Using the substitution (6.30), we see that its eigenvalues are solutions of the Schrödinger equation for a free particle of mass $m = \hbar/2D$ and charge $-2e$ in a uniform field B. We thus obtain the eigenvalues E_n of the diffusion equation, and their degeneracy g_n, from those of the Schrödinger equation (Landau levels):

$$E_n = \left(n+\frac{1}{2}\right)\frac{\hbar e B}{m} \;\rightarrow\; E_n = \left(n+\frac{1}{2}\right)\frac{4eDB}{\hbar}$$
$$\text{energy} \;\rightarrow\; \text{frequency} \tag{6.39}$$

where n is an integer.[6] These levels are infinitely degenerate for an infinite plane, with a degeneracy per unit area S:

$$g_n = \frac{eB}{h}S \rightarrow g_n = \frac{2eB}{h}S. \tag{6.40}$$

The integrated probability of return to the origin $Z_c(t)$ associated with the Cooperon may be obtained from the well-known partition function (5.5) for a free particle of mass $m = \hbar/2D$ and charge $(-2e)$ in a magnetic field, the time playing the role of inverse temperature. This results in

$$\boxed{Z_c(t, B) = \frac{BS/\phi_0}{\sinh(4\pi BDt/\phi_0)}} \tag{6.41}$$

where $\phi_0 = h/e$ is the flux quantum. Z_c is dimensionless, and may be written as a function of the flux $4\pi BDt$ crossing the area $\pi \langle R^2(t) \rangle$ generated by a Brownian trajectory during the time t. In the limit $B \rightarrow 0$, we recover the result (5.23) for free diffusion in an infinite plane: $S/(4\pi Dt)$. The long time behavior of $Z_c(t, B)$ has the form

$$Z_c(t, B) \simeq 2\frac{BS}{\phi_0}e^{-t/\tau_B} \tag{6.42}$$

where the characteristic time τ_B is given by

$$\boxed{\tau_B = \phi_0/4\pi BD} \tag{6.43}$$

To this time, we may associate the magnetic length $L_B = \sqrt{D\tau_B} = \sqrt{\hbar/2eB}$. It plays the role of the cutoff length L_γ introduced in section 5.2.2. In contrast to the case of free diffusion in a plane, the recurrence time, that is, the time spent around the origin, does not diverge with system size, and is given by

$$\tau_R = \frac{\pi l_e^2}{S} \int_{\tau_e}^{\infty} Z_c(t, B)\, dt \simeq \tau_e \ln \frac{\tau_B}{\tau_e}. \tag{6.44}$$

The result (6.41) may be interpreted in the following way: $Z_c(t, B)$ measures the probability of return to the origin after a time t, integrated over all space. It results from the sum of contributions from closed Brownian trajectories, each one having a phase factor $e^{-4i\pi B\mathcal{A}/\phi_0}$ related to the flux $B\mathcal{A}$ through the algebraic area $\mathcal{A}$ bounded by the trajectory. We may thus rewrite $Z_c(t, B)$ as a Fourier transform

$$Z_c(t, B) = Z(t) \int_{-\infty}^{+\infty} \mathcal{P}(\mathcal{A}, t) \cos\left(\frac{4\pi}{\phi_0}B\mathcal{A}\right) d\mathcal{A} \tag{6.45}$$

[6] Note that despite our using the same notation, E_n, we are considering an energy in the case of the Schrödinger equation and a frequency in the case of the diffusion equation.

or

$$Z_c(t,B) = Z(t)\left\langle \exp\left(-i\frac{4\pi}{\phi_0}B\mathcal{A}\right)\right\rangle, \tag{6.46}$$

where $\mathcal{P}(\mathcal{A},t)$ is the probability distribution (normalized to 1) of the algebraic area $\mathcal{A}$ contained inside a closed Brownian path of duration t. This demonstrates that the dephasing due to a uniform magnetic field has the form stated in (6.1). The inverse Fourier transform of (6.45)

$$\mathcal{P}(\mathcal{A},t) = \frac{2}{\phi_0}\int_{-\infty}^{+\infty}\frac{Z_c(t,B)}{Z(t)}\cos\left(\frac{4\pi}{\phi_0}B\mathcal{A}\right)dB$$

leads to

$$\boxed{\mathcal{P}(\mathcal{A},t) = \frac{\pi}{4Dt}\frac{1}{\cosh^2\frac{\pi\mathcal{A}}{2Dt}}} \tag{6.47}$$

This distribution is known as the Lévy law of algebraic areas [122].

Exercise 6.1 Verify that for the Lévy law of algebraic areas (6.47), we have $\langle\mathcal{A}\rangle = 0$ and that the typical area $\sqrt{\langle\mathcal{A}^2\rangle}$ varies linearly with time:

$$\sqrt{\langle\mathcal{A}^2\rangle} = \frac{1}{\sqrt{3}}Dt.$$

The total probability of return to the origin in a uniform field is the sum of the Diffuson and the Cooperon contributions. It is, therefore, given by

$$Z(t,B) = S\left[\frac{1}{4\pi Dt} + \frac{B/\phi_0}{\sinh(4\pi BDt/\phi_0)}\right]. \tag{6.48}$$

In zero field, this probability doubles the classical probability. For a field such that $BDt \simeq \phi_0$, the Cooperon contribution becomes negligible. This may be interpreted simply: after a time t, the typical distance associated with a Brownian walk is $\langle R(t)^2\rangle = 4Dt$ and the typical surface enclosed is $\mathcal{A}(t) \propto \pi\langle R(t)^2\rangle \propto Dt$. We thus see that the Cooperon contribution becomes negligible as soon as there is a flux quantum ϕ_0 or more through the area $\mathcal{A}(t)$.

6.4 Probability of return to the origin for an Aharonov–Bohm flux

The case of a uniform magnetic field is relevant to a large variety of physical situations, in particular for transport in disordered metallic films (section 7.5). Another important case is a field configuration that may be described by an Aharonov–Bohm flux. We shall limit ourselves here to ring and cylinder geometries. The case of an infinite plane pierced by a flux line is discussed in Appendix A6.1.

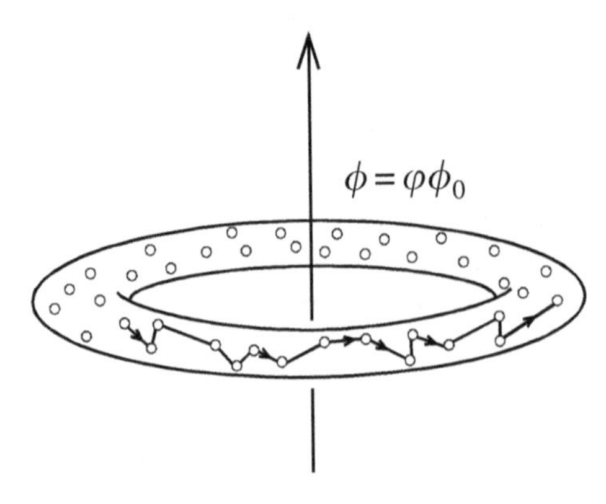

Figure 6.1 A ring threaded by an Aharonov–Bohm flux ϕ. The dimensionless flux is $\varphi = \phi/\phi_0$.

6.4.1 The ring

Consider the diffusion of an electron in a torus of perimeter $L = 2\pi R$ and cross section πa^2 (with $L \gg a$) threaded by an Aharonov–Bohm flux ϕ (Figure 6.1). We assume that $a \gg l_e$ such that the diffusion is three dimensional with a coefficient $D = v_F l_e/3$. For times short compared to the transverse Thouless time a^2/D associated with the transverse dimension a, the diffusion is three dimensional and isotropic. For longer times, however, it becomes one dimensional and anisotropic (see section 5.5.4). We consider this limit here. The vector potential in this geometry is azimuthal and, assuming that $a \ll L$, it does not vary inside the ring, and may be written as

$$A = \frac{\phi}{L}\hat{e}_\theta \tag{6.49}$$

where ϕ is the magnetic flux through the ring and $\hat{e}_\theta$ is a unit vector in the azimuthal direction. Since the vector potential is constant, the magnetic field vanishes. There is thus no Lorentz force and the eikonal approximation becomes exact. The eigenmodes of the covariant equation (6.34) are given by the solutions of:[7]

$$-D(\nabla + 2i\pi\varphi)^2\psi_n(x) = E_n\psi_n(x) \tag{6.50}$$

with the continuity condition

$$\psi_n(x + L) = \psi_n(x) \tag{6.51}$$

that is,

$$E_n = D\frac{4\pi^2}{L^2}(n - 2\varphi)^2 = 4\pi^2 E_c(n - 2\varphi)^2, \tag{6.52}$$

[7] Recall that, even though the diffusion is one dimensional, the motion of the electrons is three dimensional, since the wavelength λ_F is much smaller than a.

where $\varphi = \phi/\phi_0$. The integer $n \in \mathbb{Z}$ is the angular momentum (see section 5.6.1) and $E_c = D/L^2$ is the Thouless frequency (5.35). The heat kernel for the Cooperon is thus given by

$$Z_c(t,\phi) = \sum_{n=-\infty}^{+\infty} e^{-4\pi^2 E_c t (n-2\varphi)^2}. \tag{6.53}$$

Using the Poisson transformation (15.106) we may express Z_c as a function of the number of turns m around the ring, the so-called winding number

$$\boxed{Z_c(t,\phi) = \frac{L}{\sqrt{4\pi D t}} \sum_{m=-\infty}^{+\infty} e^{-m^2 L^2/4Dt} \cos 4\pi m\varphi} \tag{6.54}$$

Each harmonic

$$P(m,t) = \frac{L}{\sqrt{4\pi D t}} e^{-m^2 L^2/4Dt} \tag{6.55}$$

in this expansion represents the probability of return to the origin after m windings around the ring (section 5.6.1). It is thus possible to put Z_c in the form (6.1):

$$Z_c(t,\phi) = Z(t)\langle e^{im\varphi}\rangle. \tag{6.56}$$

This formulation is analogous to the representation (6.46) of the Cooperon for a uniform magnetic field as a function of the distribution of algebraic areas $\mathcal{P}(\mathcal{A},t)$. The quantity which is here coupled to the magnetic flux is the winding number. This result may also be obtained from the functional representation of diffusion (section 6.2.2) using the constraint of fixed winding number in the diffusion law. The typical number of turns after a time t is

$$\langle m^2(t)\rangle = \frac{2Dt}{L^2} = 2\frac{t}{\tau_D}, \tag{6.57}$$

where $\tau_D = L^2/D$ is the Thouless time. The total probability of return to the origin, including both Diffuson and Cooperon contributions, is thus

$$Z(t,\phi) = \frac{L}{\sqrt{4\pi D t}} \sum_{m=-\infty}^{+\infty} e^{-m^2 L^2/4Dt}\left(1 + \cos 4\pi m\varphi\right) \tag{6.58}$$

and has the following two limiting behaviors.

- $t \ll \tau_D$: since τ_D is the typical time for making a full turn around the ring, only the $m = 0$ term has a significant contribution to the sum. The probability of making one or more windings is negligible. The flux dependence is therefore also negligible, and

$$Z(t,\phi) \simeq 2\frac{L}{\sqrt{4\pi D t}}. \tag{6.59}$$

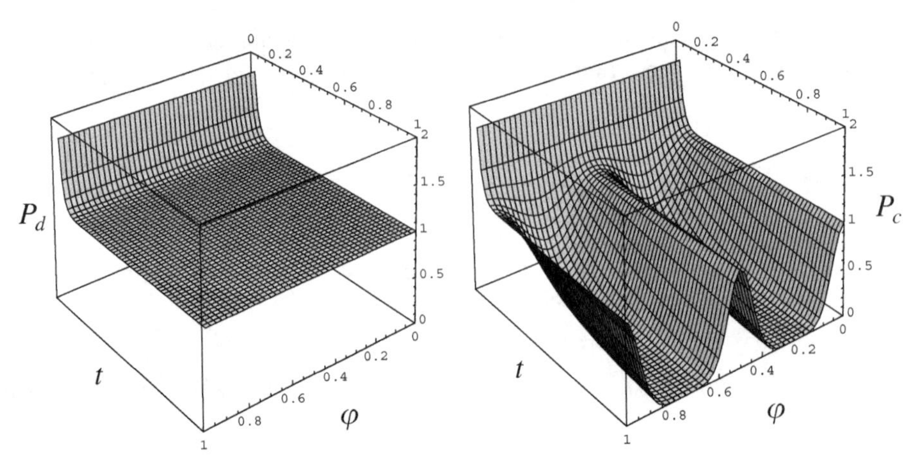

Figure 6.2 Contributions of the Diffuson $P_d(t,\varphi)$ and the Cooperon $P_c(t,\varphi)$ to the probability of return to the origin in a ring, with $\varphi = \phi/\phi_0$. The Diffuson is independent of flux. The Cooperon has a small flux dependence when $t \ll \tau_D$, and for $t \gg \tau_D$ it oscillates with period $\phi_0/2$. It is negligible for $\sqrt{t/\tau_D} < \varphi < \frac{1}{2} - \sqrt{t/\tau_D}$.

- $t \gg \tau_D$: the sum over m may be replaced by the Gaussian integral

$$Z(t,\phi) = \frac{L}{\sqrt{4\pi Dt}} \int_{-\infty}^{+\infty} dm\, e^{-m^2L^2/4Dt}\Big(1 + \cos 4\pi m\varphi\Big). \tag{6.60}$$

For small flux, the return probability becomes

$$Z(t,\phi) \simeq 1 + e^{-16\pi^2\varphi^2 t/\tau_D}. \tag{6.61}$$

In this case, the winding number is large. The only contribution is the zero mode ($n = 0$) which, for the Cooperon, picks up a finite value in the presence of magnetic flux, as seen in expression (6.53).

This duality between modes of diffusion and winding numbers was discussed in section 5.6.1. These limiting behaviors are shown in Figure 6.2.

6.4.2 The cylinder

Another important case is that of a cylinder of perimeter L, thickness a, and height L_z threaded by a flux line (Figure 6.3). Assume that $a \ll L, L_z$, so that the diffusion may be considered two dimensional, and so that the probability has no radial dependence. Moreover, since $a \ll L$, the vector potential has only one azimuthal component in the ring:

$$\mathbf{A} = \frac{\phi}{L}\hat{\mathbf{e}}_\theta. \tag{6.62}$$

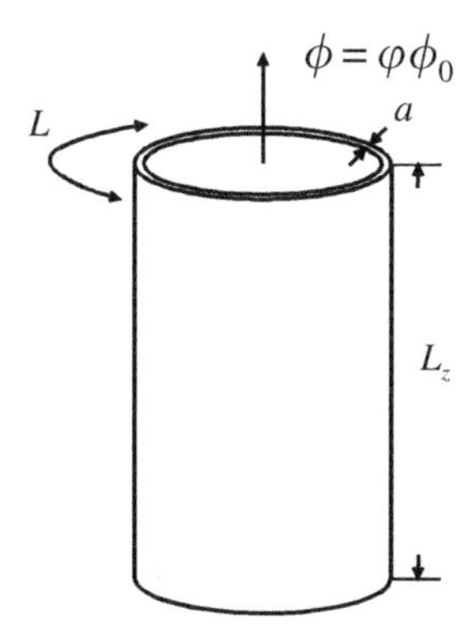

Figure 6.3 A cylinder threaded by an Aharonov–Bohm flux ϕ.

In the x-direction, that is, along the ring, periodic boundary conditions imply that $q_x = (2\pi/L)(n - 2\varphi)$ with $n \in \mathbb{Z}$, as for the ring. In the two other directions y and z, the modes are quantized by the Neumann boundary conditions (hard boundary, section 5.5.2): $q_z = (\pi n_z/L_z)$ where $n_z \in \mathbb{N}$. Moreover, only the mode $q_y = 0$ contributes in the radial direction y, since the diffusion there is uniform.

The eigenvalues of the diffusion equation (6.34) are thus given by

$$E(n_x, n_z) = \frac{4\pi^2 D}{L^2}(n - 2\varphi)^2 + \pi^2 D \frac{n_z^2}{L_z^2} \tag{6.63}$$

and the associated partition function may be factorized:

$$Z_c(t, \phi) = \sum_{n_z \geq 0} e^{-\frac{\pi^2 D n_z^2}{L_z^2} t} \sum_{n=-\infty}^{+\infty} e^{-\frac{4\pi^2 D t}{L^2}(n-2\varphi)^2}. \tag{6.64}$$

We distinguish between two particular geometries.

- Short cylinder: $L_z \ll L_x$. For $t > L_z^2/D$, diffusion is uniform along z, the first sum is dominated by the $n_z = 0$ mode and we recover the case of the ring (6.54).
- Long cylinder: $L_z \gg L_x$. For $t < L_z^2/D$, diffusion in the direction z is free, and we have:

$$Z_c(t, \phi) = \frac{L L_z}{4\pi D t} \sum_{m=-\infty}^{+\infty} e^{-m^2 L^2/4Dt} \cos 4\pi m\varphi. \tag{6.65}$$

The effect of an applied magnetic field is to introduce a global and reversible dephasing for the Cooperon. This allows us to study the properties of the Cooperon in a controlled manner, with the aid of an external parameter.

6.5 Spin-orbit coupling and magnetic impurities

Until now, we have assumed that the electronic Hamiltonian is independent of spin. In fact, an electron with velocity $\dot{\boldsymbol{r}}$ moving in a potential $v(\boldsymbol{r})$ feels an electrostatic field given by $-e\boldsymbol{E} = -\nabla v$. In the rest frame of the electron, there is a magnetic field $\boldsymbol{B} = -\dot{\boldsymbol{r}} \times \boldsymbol{E}/c^2$. The interaction of the spin with the magnetic field is of the form $-\mu_B \boldsymbol{B} \cdot \vec{\sigma}$, and leads to an additional term in the Hamiltonian (2.1):[8]

$$\mathcal{H}_{so} = \frac{\hbar}{4mc^2} \vec{\sigma}.(\nabla v(\boldsymbol{r}) \times \dot{\boldsymbol{r}})\cdot \tag{6.66}$$

The components of the operator $\vec{\sigma}$ are the Pauli matrices σ_x, σ_y and σ_z, which are recalled in (15.40). The importance of this *spin-orbit* coupling depends on the atoms under consideration, and grows quadratically with their atomic number [124].

In addition, in a metal, the electrons may also interact with localized magnetic moments. This local interaction is described by a Hamiltonian of the form

$$\mathcal{H}_m = -J\delta(\boldsymbol{r})\vec{S} \cdot \vec{\sigma}, \tag{6.67}$$

where $\vec{S}$ is the spin of an impurity.

We now study the form of the Diffuson and the Cooperon in the presence of spin-orbit interaction and magnetic impurities. To do this, we must first modify the elastic scattering time τ_e so as to take into account the scattering potentials related to the new couplings. We then study the structure factors associated with the Diffuson and the Cooperon. Each is a solution of an integral equation whose diagonalization explicitly involves the symmetry properties of spin 1/2. Our discussion is based on reference [125]. Another, more qualitative, method [126] is given in Appendix A6.4.

6.5.1 Transition amplitude and effective interaction potential

The spin-orbit interaction provides an additional scattering potential for electrons. The associate scattering cross section is computed from the matrix elements of the spin-orbit Hamiltonian (6.66) evaluated for the states $|\boldsymbol{k}\alpha\rangle$ corresponding to an electron of wave vector $\boldsymbol{k}$ and spin α:

$$\langle \boldsymbol{k}\alpha|\mathcal{H}_{so}|\boldsymbol{k}'\beta\rangle = i\frac{\hbar^2}{4m^2c^2}\vec{\sigma}_{\alpha\beta} \cdot (\boldsymbol{k} \times \boldsymbol{k}')v(\boldsymbol{k}-\boldsymbol{k}') \tag{6.68}$$

where $v(\boldsymbol{k}-\boldsymbol{k}')$, the Fourier transform of the potential $v(\boldsymbol{r})$, is taken constant. We rewrite these matrix elements and define v_{so} so that

$$\langle \boldsymbol{k}\alpha|\mathcal{H}_{so}|\boldsymbol{k}'\beta\rangle = iv_{so}\,\vec{\sigma}_{\alpha\beta} \cdot (\hat{\boldsymbol{k}} \times \hat{\boldsymbol{k}}') \tag{6.69}$$

with $\hat{\boldsymbol{k}} = \boldsymbol{k}/k$ and $\vec{\sigma}_{\alpha\beta} = \langle\alpha|\vec{\sigma}|\beta\rangle$.

[8] This simple argument omits a factor 1/2 originating from Thomas precession [123].

Similarly, the interaction of the electronic spin with a magnetic impurity is described by the matrix element[9]

$$\langle \boldsymbol{k}\alpha|\mathcal{H}_m|\boldsymbol{k}'\beta\rangle = -J\vec{S}\cdot\vec{\sigma}_{\alpha\beta} \tag{6.71}$$

which does not depend on the wave vectors since the corresponding interaction is local. Regrouping the matrix elements, we have

$$\boxed{v_{\alpha\beta}(\boldsymbol{k},\boldsymbol{k}') = v_0\delta_{\alpha\beta} + iv_{so}(\hat{\boldsymbol{k}}\times\hat{\boldsymbol{k}}')\cdot\vec{\sigma}_{\alpha\beta} - J\vec{S}\cdot\vec{\sigma}_{\alpha\beta}} \tag{6.72}$$

with the property $v^*_{\alpha\beta}(\boldsymbol{k},\boldsymbol{k}') = v_{\beta\alpha}(\boldsymbol{k}',\boldsymbol{k})$. This potential is the generalization of the scalar scattering potential denoted v_0 in the context of the Edwards model (section 2.2.2). Here, we recover the limit of a Gaussian model by taking the limit of a high density n_i of weakly scattering impurities.[10] This Gaussian model is now entirely characterized by the elementary vertex, product of transition amplitudes:

$$B_{\alpha\beta,\delta\gamma}(\boldsymbol{k},\boldsymbol{k}') = n_i v_{\alpha\gamma}(\boldsymbol{k},\boldsymbol{k}')v^*_{\beta\delta}(\boldsymbol{k},\boldsymbol{k}') \tag{6.73}$$

which is a generalization of equation (2.40). In the next sections, we calculate the structure factors of the Diffuson and of the Cooperon by iterating this elementary vertex.

Remark: reciprocity theorem

Spin-orbit coupling

In the presence of spin-orbit coupling, the reciprocity theorem, discussed in (2.110) for the case of simple scattering, may be written, for an electron of spin σ [127]:

$$T_{dir}(\boldsymbol{k},\sigma;\boldsymbol{k}',\sigma') = (-1)^{\sigma-\sigma'}T_{rev}(-\boldsymbol{k}',-\sigma';-\boldsymbol{k},-\sigma). \tag{6.74}$$

This relation implies that the amplitudes associated with the direct process $T_{dir}(\boldsymbol{k},\sigma;\boldsymbol{k}',\sigma')$ and inverse process $T_{rev}(\boldsymbol{k},\sigma;\boldsymbol{k}',\sigma')$ are equal (up to sign) if

$$\boldsymbol{k}' = -\boldsymbol{k} \qquad \text{and} \qquad \sigma' = -\sigma. \tag{6.75}$$

The spin-orbit coupling thus preserves time reversal invariance, however the spin must change sign between two time-reversed trajectories.

[9] In this expression, the spin $\vec{S}$ is assumed to be a classical quantity. For a quantum spin, we must specify the states (written $|a\rangle$ and $|b\rangle$), so that the matrix element is

$$\langle \boldsymbol{k}\alpha a|\mathcal{H}_m|\boldsymbol{k}'\beta b\rangle = -J\langle a|\vec{S}|b\rangle\langle\alpha|\vec{\sigma}|\beta\rangle = -J\vec{S}_{ab}\cdot\vec{\sigma}_{\alpha\beta}. \tag{6.70}$$

[10] There is no reason that the densities of static impurities, spin-orbit coupling impurities and magnetic impurities should be the same. This notwithstanding, we shall employ the same notation n_i for each type. This is not of great importance, since we use the Gaussian model, for which the density n_i goes to infinity and the potential to zero.

Magnetic impurities

The spins $\{S_j\}$ constitute internal degrees of freedom of the scatterers and it is now the ensemble (electron + scatterers) which must obey the reciprocity theorem. For an electron of spin σ, this theorem, discussed in (2.110) for single scattering, may be written [127]:

$$T_{dir}(\boldsymbol{k},\sigma,\{S_j\};\boldsymbol{k}',\sigma',\{S_j\}') = (-1)^{\sigma-\sigma'}(-1)^{\sum_j(S_j-S_j')}T_{rev}(-\boldsymbol{k}',-\sigma',\{-S_j'\};-\boldsymbol{k},-\sigma,\{-S_j\}). \tag{6.76}$$

The amplitudes associated with direct and inverse processes are equal if

$$T_{dir}(\boldsymbol{k},\sigma,\{S_j\};\boldsymbol{k}',\sigma',\{S_j'\}) = T_{rev}(\boldsymbol{k},\sigma,\{S_j\};\boldsymbol{k}',\sigma',\{S_j'\}) \tag{6.77}$$

that is, when

$$\boldsymbol{k}' = -\boldsymbol{k} \qquad \sigma' = -\sigma \qquad S_j' = -S_j. \tag{6.78}$$

Thus, we must change the signs of *all the spins*. This condition cannot generally be met for a given configuration of scatterers. The amplitudes associated with the two processes cannot be identical: the coupling to magnetic impurities breaks time reversal invariance.

6.5.2 Total scattering time

In the presence of magnetic impurities and spin-orbit scattering, the collision time τ_e is modified. The new time, denoted τ_{tot}, is still related to the imaginary part of the self-energy given by the diagram of Figure 6.4. Using the results of section 3.2.2, in particular relation (3.72), we obtain

$$\frac{1}{2\tau_{tot}} = -\mathrm{Im}\Sigma_1^R = \pi\rho_0\sum_\beta\langle B_{\alpha\alpha,\beta\beta}(\boldsymbol{k},\boldsymbol{k}')\rangle = \pi\rho_0 n_i\sum_\beta\langle|v_{\alpha\beta}(\boldsymbol{k},\boldsymbol{k}')|^2\rangle \tag{6.79}$$

where $\langle\cdots\rangle$ denotes the angular average taken over the directions of both wave vector $\boldsymbol{k}'$ and spin $\vec{S}$ of the magnetic impurities. Using (6.72), the collision time τ_{tot} satisfies

$$\frac{1}{\tau_{tot}} = \frac{1}{\tau_e} + \frac{1}{\tau_{so}} + \frac{1}{\tau_m} \tag{6.80}$$

with[11]

$$\frac{1}{2\pi\rho_0\tau_e} = n_i v_0^2, \qquad \frac{1}{2\pi\rho_0\tau_{so}} = n_i v_{so}^2\langle(\hat{\boldsymbol{k}}\times\hat{\boldsymbol{k}}')^2\rangle_{\hat{k}'}, \qquad \frac{1}{2\pi\rho_0\tau_m} = n_i J^2\langle S^2\rangle. \tag{6.82}$$

[11] For this, we use the relation

$$\langle(s\cdot\mathbf{A})(s\cdot\mathbf{B})\rangle = \frac{\langle s^2\rangle}{3}\mathbf{A}\cdot\mathbf{B} \tag{6.81}$$

for all vectors $\mathbf{A}$ and $\mathbf{B}$. We denote by $\langle\cdots\rangle$ the angular average over s. Moreover, $\vec{\sigma}^2 = 3$. For quantum spins $\vec{S}$ (see footnote 9), the classical average $\langle S^2\rangle = S^2$ must be replaced by the eigenvalue of the operator $\hat{\vec{S}}^2$. Therefore, $\langle S^2\rangle$ is changed into $S(S+1)$.

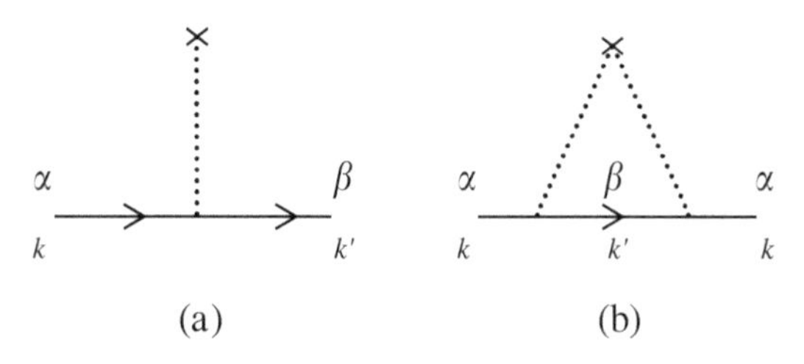

Figure 6.4 (a) Representation of the interaction potential $v_{\alpha\beta}(\boldsymbol{k},\boldsymbol{k}')$. (b) Diagram giving, to lowest order, the self-energy Σ_1^R of an electron in the presence of both spin-orbit interaction and magnetic impurities.

τ_m is the *magnetic scattering time* that represents the average time over which an electron changes the direction of its spin due to the interaction with magnetic impurities. It is the average time between two consecutive scattering events on magnetic impurities. Similarly τ_{so} is the *spin-orbit scattering time* that is, the average time over which an electron changes the direction of its spin due to spin-orbit coupling. The contributions to the inverse of these scattering times are additive: this is the Matthiessen rule (see remarks, pages 83 and 276).

As in the case of a scalar potential (equation 3.3), we introduce the notation

$$\gamma_{tot} = \sum_\beta \langle B_{\alpha\alpha,\beta\beta}(\boldsymbol{k},\boldsymbol{k}')\rangle \tag{6.83}$$

which takes the form

$$\gamma_{tot} = \gamma_e + \gamma_{so} + \gamma_m \tag{6.84}$$

with

$$\boxed{\begin{aligned} \gamma_e &= \frac{1}{2\pi\rho_0\tau_e} &&= n_i v_0^2 \\ \gamma_{so} &= \frac{1}{2\pi\rho_0\tau_{so}} &&= n_i v_{so}^2 \langle|\hat{\boldsymbol{k}}\times\hat{\boldsymbol{k}}'|^2\rangle_{\hat{\boldsymbol{k}}'} \\ \gamma_m &= \frac{1}{2\pi\rho_0\tau_m} &&= n_i J^2\langle S^2\rangle \end{aligned}}$$

Remark: magnetic impurities and the Kondo effect

To describe the coupling of the electronic spin to the magnetic impurities, the Born approximation is insufficient. Kondo showed that, to the next order in perturbation, the coupling constant is modified [128] according to

$$J \longrightarrow J\left(1 - 2J\rho_0 \ln \frac{\epsilon_F}{T}\right), \tag{6.85}$$

where ϵ_F is the Fermi energy. The magnetic scattering τ_m is thus also modified

$$\frac{1}{\tau_m} = 2\pi\rho_0 n_i S(S+1) J^2 \left(1 - 2J\rho_0 \ln \frac{\epsilon_F}{T}\right)^2. \tag{6.86}$$

For an antiferromagnetic coupling ($J < 0$), this leads, for decreasing temperature, to an increase in the electrical resistivity. The logarithmic correction introduces a temperature called the Kondo temperature, which is given by

$$T_K = \epsilon_F \, e^{-1/2|J|\rho_0}. \tag{6.87}$$

In fact, the following terms in the perturbation series are all of the same order, and for temperatures approaching T_K, the perturbation expansion diverges and must be resumed. We will not discuss the different approaches to the solution of this problem [128]. For $T \gtrsim T_K$, this effect may be well described by an effective exchange constant which is temperature dependent [128–130]:

$$J \longrightarrow J(T) = \frac{1}{2\rho_0 \left[\ln^2 \frac{T}{T_K} + \pi^2 S(S+1)\right]^{1/2}} \tag{6.88}$$

which leads to a temperature dependence in the magnetic scattering time $\tau_m(T)$ of the form [131]

$$\frac{1}{\tau_m(T)} = \frac{n_i}{2\pi\rho_0} \frac{\pi^2 S(S+1)}{\ln^2 \frac{T}{T_K} + \pi^2 S(S+1)}. \tag{6.89}$$

Thus, when the temperature decreases, the characteristic time $\tau_m(T)$ first decreases, presents a minimum around the Kondo temperature, and diverges at low temperature, when the magnetic impurities become completely screened. It has been found that the temperature dependence of $\tau_m(T)$ is a universal function of T/T_K:

$$\frac{1}{\tau_m(T)} = \frac{n_i}{\rho_0} f(T/T_K), \tag{6.90}$$

whose expression (6.89) constitutes an approximation above T_K [132].

6.5.3 Structure factor

To build the structure factors of the Diffuson and of the Cooperon, it is important to specify whether or not the paired trajectories belong to the same configuration of magnetic impurities. If the magnetic moment of these impurities is time dependent, we may have to pair trajectories that correspond to magnetic impurity configurations taken at different times 0 and T. Therefore in this section, we will have to specify the dynamics of the magnetic impurities, in order to discuss the possible change of magnetic configuration between times 0 and T (a possible displacement of impurities, that is, a change in the static disorder configuration will be considered in section 6.7). Due to its coupling with conduction electrons, the magnetic moment of impurities relaxes with a characteristic time τ_K.[12] This relaxation can be characterized by a statistical average of the form [134]

$$\langle S(0)S(T)\rangle = \langle S^2\rangle f(T) \qquad \text{with} \qquad f(T) = e^{-T/\tau_K}. \tag{6.91}$$

It may also happen that the magnetic impurities interact via the RKKY interaction and form a so-called "spin-glass" [135] phase where the spins are randomly oriented but their direction is frozen ($\tau_K \to \infty$). We shall therefore consider more specifically two limiting cases and refer to "free magnetic impurities" and "frozen magnetic impurities."

It is thus important to distinguish between situations where the structure factor appears in average quantities (like the average conductance), or in the fluctuation of these quantities (like the conductance fluctuations in Chapter 11). In the first case, the paired trajectories correspond to the *same* disorder and magnetic impurity configurations. This situation falls into the case of "frozen impurities," provided the time τ_K is large compared to the time τ_m [136].

In the second case, the paired trajectories correspond a priori to *different* configurations of magnetic impurities since they correspond to different times. In this case, one should compare the time T between the two measurements and the characteristic time τ_K for relaxation of the magnetic impurities. If the random orientations of the magnetic impurities are frozen during the time T ($\tau_K \gg T$), we can consider that the two trajectories see the same spin configuration of impurities. In the opposite situation for which the dynamics of the magnetic impurities is such that $\tau_K \ll T$, we can consider that the two trajectories experience different spin configurations. Then the paired trajectories which constitute the Diffuson or the Cooperon are not correlated (one may say that they are not connected by magnetic impurity lines). More generally, the pairing between multiple scattering trajectories taken at times 0 and T is weighted by the statistical average (6.91) [137, 138].

Elementary vertex

As in the case of a scalar potential (Chapter 4), the Diffuson and Cooperon are calculated starting from their respective structure factor. In the present case, however, we must specify the states of the spin for each collision (Figure 6.5). The integral equation for the structure factor takes the form represented in Figure 6.6. The elementary vertex now depends on four

[12] The time τ_K, called the electronic Korringa time, is given by $1/\tau_K = 4\pi(\rho_0 J)^2 T$ for temperatures above the Kondo temperature T_K. Below T_K, $1/\tau_K$ saturates to a constant proportional to T_K [133]. We avoid using the common denomination of "spin-flip" time, since it has two different meanings and may lead to confusion as sometimes encountered in the literature. As a result of interaction between conduction electrons and free magnetic impurities, both electrons and impurities relax their spin. Conduction electrons relax the direction of their spin with a time scale τ_m given by (6.81) which depends on the impurity concentration. Magnetic impurities relax their spin over a time scale $1/\tau_K \propto (\rho_0 J)^2 T$ which depends on temperature and not on the impurity concentration.

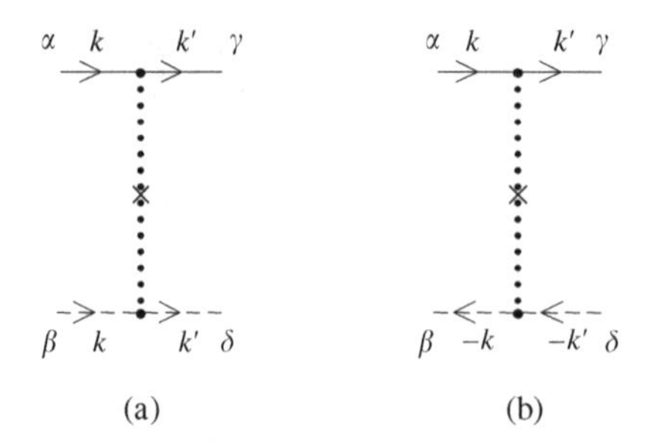

Figure 6.5 Representation of the elementary vertex $\langle B^{(d,c)}_{\alpha\beta,\gamma\delta}(\boldsymbol{k},\boldsymbol{k}')\rangle$ for (a) the Diffuson and (b) the Cooperon.

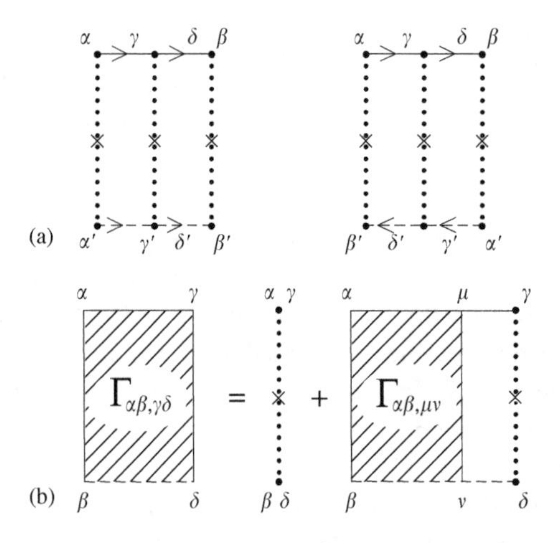

Figure 6.6 (a) Example of pairing of two multiple scattering amplitudes corresponding to different sequences of electronic spins. (b) Integral equation for the structure factor $\Gamma_{\alpha\beta,\gamma\delta}$. The elementary vertex is of the form $\gamma_e b_{\alpha\beta,\gamma\delta}$.

spin states and its strength $\gamma_e = \langle B(\boldsymbol{k},\boldsymbol{k}')\rangle = n_i v_0^2$ in the spinless case is replaced by

$$\langle B^{(d)}_{\alpha\beta,\gamma\delta}(\boldsymbol{k},\boldsymbol{k}')\rangle = n_i \langle v_{\alpha\gamma}(\boldsymbol{k},\boldsymbol{k}') v^*_{\beta\delta}(\boldsymbol{k},\boldsymbol{k}')\rangle = n_i \langle v_{\alpha\gamma}(\boldsymbol{k},\boldsymbol{k}') v_{\delta\beta}(\boldsymbol{k}',\boldsymbol{k})\rangle \tag{6.92}$$

for the Diffuson. It should be noted that the average on the wave vector and on the spin of the impurity is taken at each step of the multiple scattering sequence. The indices $(\alpha\beta)$ and $(\gamma\delta)$ denote, respectively, the electronic spin states before and after the collision (Figure 6.6(b)). For the Cooperon, $\boldsymbol{k}$ and $\boldsymbol{k}'$ are changed into $-\boldsymbol{k}'$ and $-\boldsymbol{k}$ respectively, and the spin β and δ are reversed, giving

$$\langle B^{(c)}_{\alpha\beta,\gamma\delta}(\boldsymbol{k},\boldsymbol{k}')\rangle = n_i \langle v_{\alpha\gamma}(\boldsymbol{k},\boldsymbol{k}') v^*_{\delta\beta}(-\boldsymbol{k}',-\boldsymbol{k})\rangle = n_i \langle v_{\alpha\gamma}(-\boldsymbol{k},-\boldsymbol{k}') v_{\beta\delta}(\boldsymbol{k},\boldsymbol{k}')\rangle. \tag{6.93}$$

Moreover we may have to consider situations where the two paired trajectories correspond to configurations of the magnetic impurities at different times 0 and T. In this case, one should consider the product of two transition amplitudes taken at these different times. Using the matrix elements $v_{\alpha\beta}(\boldsymbol{k},\boldsymbol{k}')$ of the interaction potential (6.72) together with the relation (6.81) gives for the angular and spin impurity average the expression

$$\langle B^{(d,c)}_{\alpha\beta,\gamma\delta}\rangle = \gamma_e b^{(d,c)}_{\alpha\beta,\gamma\delta} \tag{6.94}$$

where we have introduced two dimensionless rank 4 tensors, $b^{(d,c)}_{\alpha\beta,\gamma\delta}$, associated respectively with the Diffuson and the Cooperon[13]

$$\boxed{\begin{aligned} b^{(d)}_{\alpha\beta,\gamma\delta} &= \delta_{\alpha\gamma}\delta_{\beta\delta} + \frac{\gamma_m(T)+\gamma_{so}}{3\gamma_e}\,\vec{\sigma}_{\alpha\gamma}\cdot\vec{\sigma}_{\delta\beta} \\ b^{(c)}_{\alpha\beta,\gamma\delta} &= \delta_{\alpha\gamma}\delta_{\beta\delta} + \frac{\gamma_m(T)-\gamma_{so}}{3\gamma_e}\,\vec{\sigma}_{\alpha\gamma}\cdot\vec{\sigma}_{\beta\delta} \end{aligned}} \tag{6.95}$$

Using (6.91), we have introduced

$$\gamma_m(T) = n_i J^2\langle S(0)S(T)\rangle = \gamma_m f(T) = \frac{1}{2\pi\rho_0\tau_m} f(T). \tag{6.96}$$

The term related to the spin-orbit interaction is *positive* for the Diffuson, and *negative* for the Cooperon. This difference in sign results from time reversing one of the two sequences of the Cooperon. In the basis $|\alpha\beta\rangle = |++\rangle, |+-\rangle, |-+\rangle, |--\rangle$, the two tensors $b^{(d,c)}_{\alpha\beta,\gamma\delta}$ have the matrix form[14]

$$b^{(d)} = \begin{pmatrix} 1 & 0 & 0 & 0 \\ 0 & 1 & 0 & 0 \\ 0 & 0 & 1 & 0 \\ 0 & 0 & 0 & 1 \end{pmatrix} + \frac{\gamma_m(T)+\gamma_{so}}{3\gamma_e}\begin{pmatrix} 1 & 0 & 0 & 2 \\ 0 & -1 & 0 & 0 \\ 0 & 0 & -1 & 0 \\ 2 & 0 & 0 & 1 \end{pmatrix} \tag{6.99}$$

$$b^{(c)} = \begin{pmatrix} 1 & 0 & 0 & 0 \\ 0 & 1 & 0 & 0 \\ 0 & 0 & 1 & 0 \\ 0 & 0 & 0 & 1 \end{pmatrix} + \frac{\gamma_m(T)-\gamma_{so}}{3\gamma_e}\begin{pmatrix} 1 & 0 & 0 & 0 \\ 0 & -1 & 2 & 0 \\ 0 & 2 & -1 & 0 \\ 0 & 0 & 0 & 1 \end{pmatrix}. \tag{6.100}$$

[13] One could also consider a motion of elastic impurities between the times 0 and T, so that the scalar term of the elementary interaction vertex in (6.95) should be changed into $(\gamma_e(T)/\gamma_e)\delta_{\alpha\gamma,\beta\delta}$. The case of mobile impurities is discussed in section 6.7.

[14] Equivalently, they can be written as

$$b^{(d)}_{\alpha\beta,\gamma\delta} = \delta_{\alpha\gamma}\delta_{\beta\delta} + \frac{\gamma_m(T)+\gamma_{so}}{3\gamma_e}\,(2\delta_{\alpha\beta}\delta_{\gamma\delta} - \delta_{\alpha\gamma}\delta_{\beta\delta}), \tag{6.97}$$

$$b^{(c)}_{\alpha\beta,\gamma\delta} = \delta_{\alpha\gamma}\delta_{\beta\delta} + \frac{\gamma_m(T)-\gamma_{so}}{3\gamma_e}\,(2\delta_{\alpha\delta}\delta_{\beta\gamma} - \delta_{\alpha\gamma}\delta_{\beta\delta}). \tag{6.98}$$

Integral equation

The structure factor obtained by iterating the elementary vertex now depends on four spin indices. We denote it $\Gamma_{\alpha\beta,\gamma\delta}(\boldsymbol{q})$ where the indices $(\alpha\beta)$ and $(\gamma\delta)$ refer to the initial and final spin states, respectively. The integral equations for the Diffuson and Cooperon in reciprocal space, (4.80 and 4.90), are generalized as follows (Figure 6.6(b)):[15]

$$\Gamma_{\alpha\beta,\gamma\delta}(\boldsymbol{q}) = \gamma_e b_{\alpha\beta,\gamma\delta} + \frac{\gamma_e}{\Omega} \sum_{\mu,\nu,\boldsymbol{k}} \Gamma_{\alpha\beta,\mu\nu}(\boldsymbol{q})\, \overline{G}^R_\epsilon(\boldsymbol{k})\overline{G}^A_\epsilon(\boldsymbol{k}-\boldsymbol{q})\, b_{\mu\nu,\gamma\delta}. \tag{6.101}$$

Recall that the product of average Green's functions is related to the probability $P_0(\boldsymbol{q})$ via equation (4.68). Therefore, the integral equation, for both Diffuson and Cooperon, takes the form

$$\boxed{\Gamma_{\alpha\beta,\gamma\delta}(\boldsymbol{q}) = \gamma_e b_{\alpha\beta,\gamma\delta} + \sum_{\mu,\nu} \Gamma_{\alpha\beta,\mu\nu}(\boldsymbol{q}) b_{\mu\nu,\gamma\delta} w(\boldsymbol{q})} \tag{6.102}$$

with $w(\boldsymbol{q}) = 2\pi\rho_0\gamma_e P_0(\boldsymbol{q}) = P_0(\boldsymbol{q})/\tau_e$. The probability $P_0(\boldsymbol{q})$ is now defined using average Green's functions characterized by the collision time τ_{tot} given by (6.80). In the diffusive limit, we have $P_0(\boldsymbol{q}) \simeq \tau_{tot}(1 - Dq^2\tau_{tot})$, with a new expression for the diffusion coefficient:

$$D = \frac{1}{d} v_F^2 \tau_{tot}. \tag{6.103}$$

The function $w(\boldsymbol{q})$ varies as

$$w(\boldsymbol{q}) = \frac{1}{\tau_e} P_0(\boldsymbol{q}) \simeq \frac{\tau_{tot}}{\tau_e}(1 - Dq^2\tau_{tot}) = \frac{\gamma_e}{\gamma_{tot}}(1 - Dq^2\tau_{tot}). \tag{6.104}$$

Remark: the Bethe–Salpeter equation

The structure factor associated with the Diffuson (or the Cooperon) is obtained by iterating an elementary vertex $\langle B_{\alpha\beta,\gamma\delta}\rangle$ (6.92) built from the product of two complex scattering amplitudes $v_{\alpha\gamma} v^*_{\beta\delta}$ (or $v_{\alpha\gamma} v^*_{\delta\beta}$ for the Cooperon). This product, which corresponds to four a priori distinct values of the spin, is more general than the differential scattering cross section given by $v_{\alpha\beta} v^*_{\alpha\beta}$. The integral equation (6.102) obtained upon iterating the elementary interaction vertex is known as the Bethe–Salpeter equation (see footnote 6, page 99 and equation 4.154). It is the analog, for the structure factor, of the Dyson equation for the average Green function (Chapter 3). For a scalar potential, that is, with no additional degrees of freedom, the elementary interaction vertex yields the differential scattering cross section.

[15] To keep the notation concise, the calculation here is performed at zero frequency. Moreover, when there is no possible ambiguity, we will omit the subscripts (d) and (c) for the Diffuson and Cooperon. Recall that $G^A_\epsilon(\boldsymbol{q}-\boldsymbol{k}) = G^A_\epsilon(\boldsymbol{k}-\boldsymbol{q})$.

Diagonalization

To solve the integral equation (6.102), it must be diagonalized. To that purpose, we must first diagonalize the tensor b so as to obtain an integral equation in the subspace associated with each eigenvalue b_J of the elementary vertex:

$$\Gamma_J(\boldsymbol{q}) = \gamma_e b_J + \Gamma_J(\boldsymbol{q}) b_J w(\boldsymbol{q}). \tag{6.105}$$

Using (6.104), we calculate the eigenmodes

$$\boxed{\Gamma_J(\boldsymbol{q}) = \frac{\gamma_{tot}/\tau_{tot}}{1/\tau_J + Dq^2}} \tag{6.106}$$

with

$$\boxed{\tau_J = \tau_{tot} \frac{b_J}{\tau_e/\tau_{tot} - b_J}} \tag{6.107}$$

or

$$\gamma_J = \gamma_{tot} \frac{\gamma_{tot}/\gamma_e - b_J}{b_J}. \tag{6.108}$$

Reinserting the dependence on frequency ω, we obtain poles of the form $-i\omega + Dq^2 + 1/\tau_J$. These eigenmodes are thus damped, that is, they decay exponentially with the characteristic time τ_J.

Finally, we obtain the structure factor $\Gamma_{\alpha\beta,\gamma\delta}(\boldsymbol{q})$ from Γ_J and the usual formulae for change of basis. We now study the cases of the Diffuson and Cooperon separately.

6.5.4 The Diffuson

The diagonalization of the matrix (6.99) describing the elementary vertex of interaction is achieved using the singlet-triplet basis, and leads to the two eigenvalues

$$\begin{aligned} b_S^{(d)} &= 1 + \frac{\gamma_m(T) + \gamma_{so}}{\gamma_e} \\ b_T^{(d)} &= 1 - \frac{1}{3}\frac{\gamma_m(T) + \gamma_{so}}{\gamma_e}. \end{aligned} \tag{6.109}$$

The first is non-degenerate, and corresponds to the eigenvector

$$|\tilde{S}\rangle = \frac{1}{\sqrt{2}}(|++\rangle + |--\rangle). \tag{6.110}$$

The second is triply degenerate, and corresponds to the eigenvectors

$$|\tilde{T}_i\rangle = \begin{cases} |+-\rangle \\ |-+\rangle \\ \frac{1}{\sqrt{2}}(|++\rangle - |--\rangle) \end{cases} \tag{6.111}$$

Let us emphasize that the basis of "singlet" and "triplet" states here is unusual. These states correspond not to the addition of two spins, but to the addition of one spin with the complex conjugate of another.[16] Changing the basis, we obtain the following decomposition for $b^{(d)}_{\alpha\beta,\gamma\delta}$:

$$b^{(d)}_{\alpha\beta,\gamma\delta} = b^{(d)}_T\ \delta_{\alpha\gamma}\delta_{\beta\delta} + \frac{1}{2}(b^{(d)}_S - b^{(d)}_T)\delta_{\alpha\beta}\delta_{\gamma\delta}. \tag{6.112}$$

From the eigenvalues of the elementary vertex $b^{(d)}_{\alpha\beta,\gamma\delta}$, we obtain the characteristic relaxation times for the Diffuson and the Cooperon in the singlet and triplet channels. In the limit where the spin-orbit and magnetic impurity effects are weak, that is, for $\tau_{tot} \simeq \tau_e$, we have from (6.108)

$$\frac{1}{\tau_J} \simeq \frac{1}{\tau_e}\left(\frac{\gamma_{tot}}{\gamma_e} - b_J\right) = \frac{1}{\tau_e}\left(1 - b_J + \frac{\gamma_m + \gamma_{so}}{\gamma_e}\right), \tag{6.113}$$

so that, using (6.109, 6.96), we obtain

$$\begin{aligned} \frac{1}{\tau^{(d)}_S} &= \frac{1-f(T)}{\tau_m} \\ \frac{1}{\tau^{(d)}_T} &= \frac{4}{3\tau_{so}} + \frac{1+f(T)/3}{\tau_m} \end{aligned} \tag{6.114}$$

with $f(T)$ given by (6.91). Using (6.106), the two corresponding eigenmodes are given by

$$\Gamma^{(d)}_T(\boldsymbol{q}) = \frac{\gamma_{tot}/\tau_{tot}}{Dq^2 + \frac{4}{3\tau_{so}} + \frac{[1+f(T)/3]}{\tau_m}} \qquad \Gamma^{(d)}_S(\boldsymbol{q}) = \frac{\gamma_{tot}/\tau_{tot}}{Dq^2 + \frac{[1-f(T)]}{\tau_m}}, \tag{6.115}$$

with the two following interesting limits.

- For *frozen impurities*, that is, pairing of trajectories with identical impurity spin configurations ($f(T) = 1$):

$$\Gamma^{(d)}_T(\boldsymbol{q}) = \frac{\gamma_{tot}/\tau_{tot}}{Dq^2 + \frac{4}{3\tau_{so}} + \frac{4}{3\tau_m}} \qquad \Gamma^{(d)}_S(\boldsymbol{q}) = \frac{\gamma_{tot}/\tau_{tot}}{Dq^2}, \tag{6.116}$$

[16] The complex conjugation of a spin eigenstate in the s_z-representation is given by the operator $K = e^{-i\frac{\pi}{2}\sigma_y} = -i\sigma_y$, such that $K|+\rangle = |-\rangle$ and $K|-\rangle = -|+\rangle$ [139], which yields the unusual expressions for the singlet (6.110) and triplet (6.111).

- For *free impurities*, that is, uncorrelated spin configurations ($f(T) = 0$):

$$\Gamma_T^{(d)}(\boldsymbol{q}) = \frac{\gamma_{tot}/\tau_{tot}}{Dq^2 + \frac{4}{3\tau_{so}} + \frac{1}{\tau_m}} \qquad \Gamma_S^{(d)}(\boldsymbol{q}) = \frac{\gamma_{tot}/\tau_{tot}}{Dq^2 + \frac{1}{\tau_m}}. \tag{6.117}$$

Let us stress again that the Diffuson built out of identical trajectories is a Goldstone mode (sections 6.1.3 and 6.5.6). From the above expressions, we deduce the structure factor:

$$\boxed{\Gamma^{(d)}_{\alpha\beta,\gamma\delta} = \Gamma_T^{(d)}\delta_{\alpha\gamma}\delta_{\beta\delta} + \frac{1}{2}(\Gamma_S^{(d)} - \Gamma_T^{(d)})\delta_{\alpha\beta}\delta_{\gamma\delta}} \tag{6.118}$$

$\Gamma^{(d)}_{\alpha\beta,\gamma\delta}(\boldsymbol{q})$ depends on the spin-orbit scattering and on the scattering from magnetic impurities. This result is important since it demonstrates that the Diffuson, which is a priori a classical object, may be dephased in the presence of spin-orbit interaction and magnetic scattering.

6.5.5 The Cooperon

To diagonalize the structure factor associated with the Cooperon, we first seek the eigenvalues of the elementary vertex $b^{(c)}_{\alpha\beta,\gamma\delta}$ that is, we diagonalize the matrix (6.100), in the singlet-triplet basis,

$$|S\rangle = \frac{1}{\sqrt{2}}(|+-\rangle - |-+\rangle)$$
$$|T_i\rangle = \begin{cases} |++\rangle \\ |--\rangle \\ \frac{1}{\sqrt{2}}(|+-\rangle + |-+\rangle). \end{cases} \tag{6.119}$$

The two corresponding eigenvalues are easily calculated:

$$b_S^{(c)} = 1 - \frac{\gamma_m(T) - \gamma_{so}}{\gamma_e}$$
$$b_T^{(c)} = 1 + \frac{1}{3}\frac{\gamma_m(T) - \gamma_{so}}{\gamma_e} \tag{6.120}$$

and, changing the basis,

$$b^{(c)}_{\alpha\beta,\gamma\delta} = \sum_{T_i} b_T\langle\alpha\beta|T_i\rangle\langle T_i|\gamma\delta\rangle + b_S\langle\alpha\beta|S\rangle\langle S|\gamma\delta\rangle, \tag{6.121}$$

we end up with the following decomposition for $b^{(c)}_{\alpha\beta,\gamma\delta}$:

$$b^{(c)}_{\alpha\beta,\gamma\delta} = \frac{1}{2}(b_T^{(c)} + b_S^{(c)})\delta_{\alpha\gamma}\delta_{\beta\delta} + \frac{1}{2}(b_T^{(c)} - b_S^{(c)})\delta_{\alpha\delta}\delta_{\beta\gamma}. \tag{6.122}$$

Remark: change of basis and Clebsch–Gordan coefficients

The coefficients for the change of basis (relation 6.121) are none other than the Clebsch–Gordan coefficients. In general, for particles of spin j (here $j = 1/2$), we diagonalize the diffusion equation according to the eigenstates of total spin J. Thus, there are $2j+1$ possible values of J between $J = 0$ and $J = 2j$. A state J is $(2J+1)$-fold degenerate, so the $J = 0$ state is non-degenerate (singlet), and the $J = 1$ state is threefold degenerate (triplet). The change of basis has the form

$$b_{\alpha\beta,\gamma\delta} = \sum_{J=0}^{2j} \sum_{m=-J}^{J} b_{Jm} C_{\alpha\beta}^{Jm} C_{\gamma\delta}^{Jm}, \tag{6.123}$$

where the coefficients $C_{m_1 m_2}^{Jm} = \langle jjm_1m_2|Jm\rangle$ are the Clebsch–Gordan coefficients. In the present case, b_{Jm} does not depend on m. To describe the multiple scattering of a polarized wave, we must solve the same type of integral equation, but with $j = 1$, which leads to three possible J values: $J = 0$ non-degenerate, $J = 1$ threefold degenerate, and $J = 2$, which is fivefold degenerate (see section 6.6).

The structure factor $\Gamma^{(c)}_{\alpha\beta,\gamma\delta}(\boldsymbol{q})$ has the same structure as the elementary interaction vertex $b^{(c)}_{\alpha\beta,\gamma\delta}$ given by (6.122) and so may be decomposed as

$$\boxed{\Gamma^{(c)}_{\alpha\beta,\gamma\delta} = \frac{1}{2}(\Gamma_T^{(c)} + \Gamma_S^{(c)})\delta_{\alpha\gamma}\delta_{\beta\delta} + \frac{1}{2}(\Gamma_T^{(c)} - \Gamma_S^{(c)})\delta_{\alpha\delta}\delta_{\beta\gamma}} \tag{6.124}$$

The eigenvalues of the elementary interaction vertex $b^{(c)}_{\alpha\beta,\gamma\delta}$ are given by (6.120). In the limit $\tau_{tot} \simeq \tau_e$, and using (6.108), we obtain the relaxation times for the Cooperon in the singlet and triplet channels. They depend on the dynamic of the magnetic impurities. In the limit $\tau_{tot} \simeq \tau_e$, we obtain:

$$\begin{aligned} \frac{1}{\tau_S^{(c)}} &= \frac{1+f(T)}{\tau_m} \\ \frac{1}{\tau_T^{(c)}} &= \frac{4}{3\tau_{so}} + \frac{1-f(T)/3}{\tau_m}. \end{aligned} \tag{6.125}$$

The Cooperon structure factor thus decomposes according to the eigenmodes

$$\Gamma_T^{(c)}(\boldsymbol{Q}) = \frac{\gamma_{tot}/\tau_{tot}}{DQ^2 + \frac{4}{3\tau_{so}} + \frac{[1-f(T)/3]}{\tau_m}} \qquad \Gamma_S^{(c)}(\boldsymbol{Q}) = \frac{\gamma_{tot}/\tau_{tot}}{DQ^2 + \frac{[1+f(T)]}{\tau_m}}, \tag{6.126}$$

with the two interesting limits:

- for *frozen impurities*

$$\Gamma_T^{(c)}(\boldsymbol{Q}) = \frac{\gamma_{tot}/\tau_{tot}}{DQ^2 + \frac{4}{3\tau_{so}} + \frac{2}{3\tau_m}} \qquad \Gamma_S^{(c)}(\boldsymbol{Q}) = \frac{\gamma_{tot}/\tau_{tot}}{DQ^2 + \frac{2}{\tau_m}}, \tag{6.127}$$

- for *free impurities*

$$\Gamma_T^{(c)}(\boldsymbol{Q}) = \frac{\gamma_{tot}/\tau_{tot}}{DQ^2 + \frac{4}{3\tau_{so}} + \frac{1}{\tau_m}} \qquad \Gamma_S^{(c)}(\boldsymbol{Q}) = \frac{\gamma_{tot}/\tau_{tot}}{DQ^2 + \frac{1}{\tau_m}}. \tag{6.128}$$

We note that *the spin-orbit scattering does not affect the singlet channel of the Cooperon* $\Gamma_S^{(c)}(\boldsymbol{Q})$ and that in the absence of magnetic impurities ($\tau_m \to \infty$), we recover the case of the scalar potential (4.93).[17]

6.5.6 The diffusion probability

We have constructed two objects by pairing two amplitudes having *independent* spin configurations. This enabled us to sum independently over the intermediate spin states in the iteration of the elementary scattering process, leading to the simple form of the integral equation (6.102).

The classical contribution P_d to the probability of quantum diffusion is of a particular nature. It is calculated by pairing two multiple scattering amplitudes with *identical* spin configurations at each step in the iteration (Figure 6.7(a)). Consequently, the associated structure factor is a solution of equation (6.102) in a two-dimensional subspace:

$$\Gamma^{(d)}_{\alpha\alpha,\beta\beta}(\boldsymbol{q}) = \gamma_e \, b^{(d)}_{\alpha\alpha,\beta\beta} + \sum_\mu \Gamma^{(d)}_{\alpha\alpha,\mu\mu}(\boldsymbol{q}) \, b^{(d)}_{\mu\mu,\beta\beta} \, w(\boldsymbol{q}) \tag{6.129}$$

with

$$b^{(d)}_{\alpha\alpha,\beta\beta} = \delta_{\alpha\beta} + \frac{\gamma_m + \gamma_{so}}{3\gamma_e} \vec{\sigma}_{\alpha\beta} \cdot \vec{\sigma}_{\beta\alpha} \tag{6.130}$$

which has the matrix representation[18]

$$b^{(d)} = \begin{pmatrix} 1 & 0 \\ 0 & 1 \end{pmatrix} + \frac{\gamma_m + \gamma_{so}}{3\gamma_e} \begin{pmatrix} 1 & 2 \\ 2 & 1 \end{pmatrix}. \tag{6.132}$$

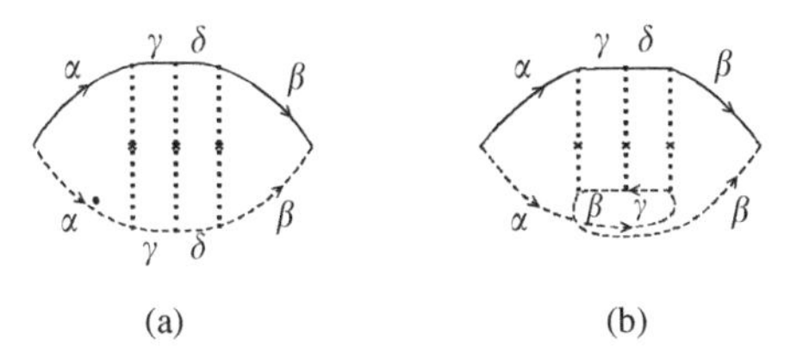

Figure 6.7 (a) Diffuson P_d and (b) Cooperon X_c with a spin-dependent interaction.

[17] The diffusion constant (6.103) is nevertheless slightly modified by the spin-orbit coupling.
[18] Equivalently

$$b^{(d)}_{\alpha\alpha,\beta\beta} = \delta_{\alpha\beta} + \frac{\gamma_m + \gamma_{so}}{3\gamma_e}(2 - \delta_{\alpha\beta}). \tag{6.131}$$

This tensor has two eigenvectors, $\frac{1}{\sqrt{2}}(|++\rangle + |--\rangle)$ and $\frac{1}{\sqrt{2}}(|++\rangle - |--\rangle)$ with eigenvalues $b_S^{(d)}$ and $b_T^{(d)}$ given by (6.109), where $\gamma_m(T)$ is replaced by γ_m. Equation (6.107) shows that the singlet mode is not damped (it is the Goldstone mode[19]) and that the triplet mode has a characteristic relaxation time $\tau_T^{(d)}$ given by (6.116).

The solution of (6.129) is

$$\Gamma^{(d)}_{\alpha\alpha,\beta\beta} = \Gamma_T \delta_{\alpha\beta} + \frac{1}{2}(\Gamma_S - \Gamma_T) \tag{6.133}$$

Thus, if we consider a spin-polarized electron beam, the probability of remaining in the same channel is proportional to

$$\Gamma^{(d)}_{\alpha\alpha,\alpha\alpha} = \frac{1}{2}(\Gamma_S + \Gamma_T) \tag{6.134}$$

and the probability of having an opposite spin polarization ($\beta \neq \alpha$) is

$$\Gamma^{(d)}_{\alpha\alpha,\beta\beta} = \frac{1}{2}(\Gamma_S - \Gamma_T) \tag{6.135}$$

so

$$\sum_\beta \Gamma^{(d)}_{\alpha\alpha,\beta\beta} = \Gamma_S. \tag{6.136}$$

The singlet mode is the Goldstone mode, and we recover particle number conservation.

6.5.7 The Cooperon X_c

In the case of the Cooperon X_c (section 4.6), we pair two time-reversed multiple scattering trajectories. Figure 6.7(b) shows that the two spin sequences are also reversed so that the successive interaction vertices are *correlated*. The structure factor thus involves a succession of interaction vertices which are not independent. In the limit of long sequences, we may neglect this correlation and assume that the two sequences are indeed independent. We may then sum over all intermediate spin configurations, giving a structure factor which obeys the integral equation (6.102). Figure 6.7(b) shows that the Cooperon X_c is proportional to $\sum_\beta \Gamma^{(c)}_{\alpha\beta,\beta\alpha}$. Using (6.124), we get

$$\sum_\alpha \Gamma^{(c)}_{\alpha\beta,\beta\alpha}(\boldsymbol{Q}) = \frac{3}{2}\Gamma_T^{(c)}(\boldsymbol{Q}) - \frac{1}{2}\Gamma_S^{(c)}(\boldsymbol{Q}). \tag{6.137}$$

[19] Note that the Goldstone mode nevertheless depends on spin-orbit scattering and magnetic impurities through the renormalization of the diffusion constant D.

Restoring the frequency dependence in the diffusion poles of the structure factors (6.127), we obtain the generalization of (4.95), that is (for frozen impurities, $\tau_K \gg \tau_m$)

$$X_c^{so+m}(\omega) = \frac{\tau_{tot}}{2\pi\rho_0\Omega}\sum_Q \left(\frac{3}{-i\omega + DQ^2 + \frac{4}{3\tau_{so}} + \frac{2}{3\tau_m}} - \frac{1}{-i\omega + DQ^2 + \frac{2}{\tau_m}}\right). \tag{6.138}$$

By Fourier transforming, we obtain the time dependence of the Cooperon. In the presence of spin-orbit coupling alone, this dependence has the form $X_c^{so}(t) = X_c(t)\langle Q_{so}(t)\rangle$ where $X_c(t)$ is the scalar contribution and where the attenuation factor

$$\langle Q_{so}(t)\rangle = \frac{1}{2}\left(3e^{-4t/3\tau_{so}} - 1\right) \tag{6.139}$$

is an average over the multiple scattering trajectories of the dephasing induced by the spin-orbit coupling. When $t \ll \tau_{so}$, $\langle Q_{so}(t)\rangle = 1$, so that the Cooperon is not modified. For $t \gg \tau_{so}$, the "triplet" term vanishes, but the negative contribution of the singlet is unaffected. Thus, $\langle Q_{so}(t)\rangle$ becomes negative and equal to $-1/2$. The probability $P(0,\boldsymbol{r},t) = P_d(0,\boldsymbol{r},t) + X_c(0,\boldsymbol{r},t)$ is now *reduced* by a factor 2 at $\boldsymbol{r} = 0$, as shown in Figure 6.8.

In the presence of magnetic impurities, the Cooperon has the form $X_c^m(t) = X_c(t)\langle Q_m(t)\rangle$. The attenuation factor, given by

$$\langle Q_m(t)\rangle = \frac{1}{2}\left(3e^{-2t/3\tau_m} - e^{-2t/\tau_m}\right), \tag{6.140}$$

describes the average over multiple scattering trajectories of the dephasing induced by the magnetic impurities. Note that the singlet and triplet contributions are both reduced by the magnetic impurities, such that the Cooperon contribution vanishes at long times.

Finally, the effects of spin-orbit coupling and magnetic impurities add in each channel, singlet and triplet, so that the Cooperon becomes [140] $X_c^{so+m}(t) = X_c(t)\langle Q_{so+m}(t)\rangle$ with

$$\boxed{\langle Q_{so+m}(t)\rangle = \frac{1}{2}\left(3e^{-4t/3\tau_{so}-2t/3\tau_m} - e^{-2t/\tau_m}\right)} \tag{6.141}$$

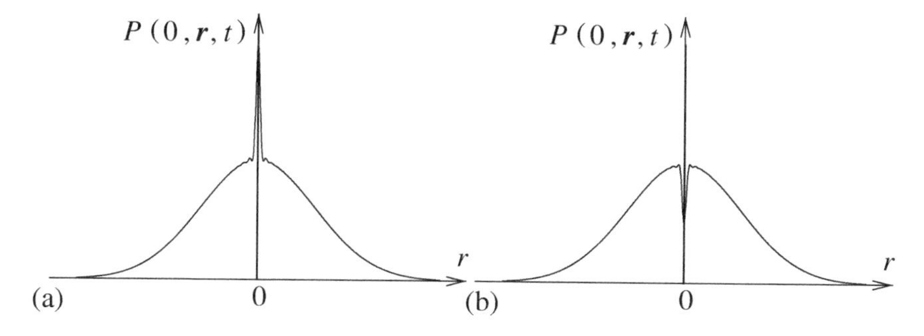

Figure 6.8 Diffusion probability $P(0,\boldsymbol{r},t)$. (a) In the absence of spin-orbit coupling ($1/\tau_{so} = 0$, $\langle Q_{so}(t)\rangle = 1$), the probability of return to the origin is doubled. (b) In the presence of strong spin-orbit coupling ($t \gg \tau_{so}$, $\langle Q_{so}(t)\rangle = -1/2$), it is reduced by a factor 2.

6.6 Polarization of electromagnetic waves

In metals, the existence of an additional degree of freedom, the electron spin, affects the Diffuson and the Cooperon by modifying the relative phase between paired multiple scattering trajectories. For electromagnetic waves or photons, the behavior is qualitatively similar because of the vector nature of the electric field (polarization). This modifies the results obtained for scalar waves in two ways:

- by introducing a dephasing between paired multiple scattering sequences, which thus affects both Diffuson and Cooperon;
- for a given incident polarization direction, it changes the relative weight of the intensity in each polarization channel, for both Diffuson and Cooperon.

We consider the regime of *Rayleigh scattering* for which the wavelength is large compared to the size of the scatterers (Appendix A2.3). This approximation is justified in many situations, in particular for the scattering of photons by cold atoms. Moreover, it provides a rather exhaustive study of the effects of polarization [141–144].

From equation (2.122), we determine the polarization $\boldsymbol{P}'$ of an electric field following a collision as a function of the incident polarization $\boldsymbol{P}$ and the scattered wave vector $\hat{\boldsymbol{k}}' = \boldsymbol{k}'/k'$:

$$\boldsymbol{P}' = -\hat{\boldsymbol{k}}' \times (\hat{\boldsymbol{k}}' \times \boldsymbol{P}). \tag{6.142}$$

This relation is specific to Rayleigh scattering by a dipole (Appendix A2.3) and is not appropriate to an arbitrary scatterer. It may be written in matrix form

$$\boldsymbol{P}' = M(\hat{\boldsymbol{k}}')\boldsymbol{P}, \qquad \text{where} \quad M(\hat{\boldsymbol{k}}) = \begin{pmatrix} 1-\hat{k}_x^2 & -\hat{k}_x\hat{k}_y & -\hat{k}_x\hat{k}_z \\ -\hat{k}_x\hat{k}_y & 1-\hat{k}_y^2 & -\hat{k}_y\hat{k}_z \\ -\hat{k}_x\hat{k}_z & -\hat{k}_y\hat{k}_z & 1-\hat{k}_z^2 \end{pmatrix} \tag{6.143}$$

is a real symmetric matrix, and $\hat{\boldsymbol{k}} = (\hat{k}_x, \hat{k}_y, \hat{k}_z)$. In tensor notation, Rayleigh scattering is thus described by a scattering amplitude of the form

$$\boxed{v_{\alpha\beta}(\boldsymbol{k},\boldsymbol{k}') = v_0 M_{\alpha\beta}(\boldsymbol{k}') = v_0(\delta_{\alpha\beta} - \hat{k}'_\alpha \hat{k}'_\beta)} \tag{6.144}$$

with $v_0 = \alpha_0 k_0^2$, where α_0 is the classical polarizability and k_0 the wave vector (2.122). $v_{\alpha\beta}$ plays the role of an effective amplitude which differs from the scalar potential v_0 of the Edwards model (section 2.2.2). The Gaussian model, the limiting case of the Edwards model for large densities of weak scatterers, is now characterized by the parameter

$$B_{\alpha\beta,\gamma\delta} = n_i v_{\alpha\gamma} v_{\beta\delta} \tag{6.145}$$

which generalizes (2.40). It is interesting to compare this scattering potential to that (6.72) acting on electrons in the presence of magnetic impurities or spin-orbit coupling. In these two cases, the scattering is characterized by a tensor which describes the electron spin or the polarization of the light. For electrons, the non-scalar part is added to the scalar component while it multiplies it in the case of light. For electrons, the tensor is defined in a two-dimensional subspace associated with spin 1/2, while for waves it is defined in a three-dimensional subspace associated with spin 1. The two problems are thus qualitatively similar, but correspond to different symmetries. We shall thus follow a presentation analogous to that used for spin-orbit coupling and magnetic impurities.

Remark: polarization and reciprocity theorem

For an electromagnetic wave of polarization $\hat{\boldsymbol{\varepsilon}}$, time reversal invariance is manifested by the relation (2.109) between complex amplitudes. This is the reciprocity theorem.

For an incident wave of polarization $\hat{\boldsymbol{\varepsilon}}_i$, the amplitudes associated with direct and inverse processes are equal, by virtue of (2.109), the final polarization $\hat{\boldsymbol{\varepsilon}}'$ satisfying $\hat{\boldsymbol{\varepsilon}}' = \hat{\boldsymbol{\varepsilon}}_i^*$.

In the case where the polarization $\hat{\boldsymbol{\varepsilon}}_i$ is linear, this assumes that the emergent light is measured in the same polarization direction ($l \parallel l$). When the polarization $\hat{\boldsymbol{\varepsilon}}_i$ is circular with given helicity, the condition $\hat{\boldsymbol{\varepsilon}}' = \hat{\boldsymbol{\varepsilon}}_i^*$ requires that the emergent light be measured in the channel which conserves helicity ($h \parallel h$).

Exercise 6.2 Show that, since the polarization is transverse ($\boldsymbol{k} \perp \boldsymbol{P}$), the matrix M satisfies

$$M(\hat{\boldsymbol{k}})\boldsymbol{P} = \boldsymbol{P}, \tag{6.146}$$

which reflects the fact that $\boldsymbol{P}$ and $\hat{\boldsymbol{k}}$ are not independent quantities.

6.6.1 Elastic mean free path

The polarization of the electromagnetic wave does not constitute an additional scattering potential, in contrast to the case of spin-orbit scattering (6.72). Nonetheless, the polarization does modify the elastic mean free path. To generalize (3.74), we define

$$\gamma_{pol} = \sum_{\beta} \langle B_{\alpha\alpha,\beta\beta} \rangle . \tag{6.147}$$

With the help of (6.145), it is given by

$$\gamma_{pol} = n_i v_0^2 \sum_{\beta} \langle M_{\alpha\beta} M_{\beta\alpha} \rangle = \frac{2}{3}\gamma_e = \frac{2}{3} n_i v_0^2, \tag{6.148}$$

where $\langle \cdots \rangle$ denotes the angular average over directions $\hat{\boldsymbol{k}}'$ (see 6.152). The new expression for the elastic mean free path, denoted l_{pol}, is thus given by (see relations 3.76 and 3.77)

$$\frac{1}{l_{pol}} = \frac{1}{c\tau_{pol}} = -\frac{1}{k_0} \mathrm{Im}\Sigma_1^R = \frac{\gamma_{pol}}{4\pi} = \frac{\gamma_e}{6\pi} = \frac{2}{3 l_e} \tag{6.149}$$

Taking account of the polarization thus reduces the scattering cross section by a factor 2/3, which is the average of the term $\sin^2 \chi$ in equation (2.124). The final result is

$$\tau_{pol} = \frac{3}{2}\tau_e \qquad l_{pol} = \frac{3}{2}l_e. \tag{6.150}$$

6.6.2 Structure factor

Elementary vertex

To calculate the structure factor of the Diffuson or the Cooperon, we must first specify the initial, (α, β), and final, (γ, δ), polarization components associated with an elementary vertex of the structure factor (Figure 6.6). These states define the tensor which, in Cartesian coordinates, is

$$b_{\alpha\beta,\gamma\delta} = \langle M_{\alpha\gamma} M_{\beta\delta}\rangle = \left\langle (\delta_{\alpha\gamma} - \hat{k}'_\alpha \hat{k}'_\gamma)\,(\delta_{\beta\delta} - \hat{k}'_\beta \hat{k}'_\delta)\right\rangle. \tag{6.151}$$

In contrast to the spin-orbit coupling (6.92 and 6.93), this tensor, which describes the elementary vertex of rotation of the polarization, is the same for the Diffuson and the Cooperon. It is worth noting that the angular average of the elementary vertex of polarization rotation is taken at each step of the multiple scattering sequence. To calculate $b_{\alpha\beta,\gamma\delta}$ explicitly, we use the following angular averages

$$\begin{aligned}
\langle \hat{k}_\alpha \hat{k}_\beta \rangle &= 0 && \text{if} \quad \alpha \neq \beta \\
\langle \hat{k}_\alpha^2 \rangle &= 1/3 && \\
\langle \hat{k}_\alpha \hat{k}_\beta \hat{k}_\gamma \hat{k}_\delta \rangle &= 0 && \text{if} \quad \text{three indices are different} \\
\langle \hat{k}_\alpha^2 \hat{k}_\beta^2 \rangle &= 1/15 && \text{if} \quad \alpha \neq \beta \\
\langle \hat{k}_\alpha^4 \rangle &= 1/5 &&
\end{aligned} \tag{6.152}$$

which may be reexpressed in condensed notation (in Cartesian coordinates):

$$\begin{aligned}
\langle \hat{k}_\alpha \hat{k}_\beta \rangle &= \frac{1}{3}\delta_{\alpha\beta} \\
\langle \hat{k}_\alpha \hat{k}_\beta \hat{k}_\gamma \hat{k}_\delta \rangle &= \frac{1}{15}(\delta_{\alpha\gamma}\delta_{\beta\delta} + \delta_{\alpha\delta}\delta_{\beta\gamma} + \delta_{\alpha\beta}\delta_{\gamma\delta}).
\end{aligned} \tag{6.153}$$

so that

$$\boxed{b_{\alpha\beta,\gamma\delta} = \frac{1}{15}(6\,\delta_{\alpha\gamma}\delta_{\beta\delta} + \delta_{\alpha\delta}\delta_{\beta\gamma} + \delta_{\alpha\beta}\delta_{\gamma\delta})} \tag{6.154}$$

We may represent $b_{\alpha\beta,\gamma\delta}$ in matrix form

$$\frac{1}{15}\begin{pmatrix} 8 & 0 & 0 & 0 & 1 & 0 & 0 & 0 & 1 \\ 0 & 6 & 0 & 1 & 0 & 0 & 0 & 0 & 0 \\ 0 & 0 & 6 & 0 & 0 & 0 & 1 & 0 & 0 \\ 0 & 1 & 0 & 6 & 0 & 0 & 0 & 0 & 0 \\ 1 & 0 & 0 & 0 & 8 & 0 & 0 & 0 & 1 \\ 0 & 0 & 0 & 0 & 0 & 6 & 0 & 1 & 0 \\ 0 & 0 & 1 & 0 & 0 & 0 & 6 & 0 & 0 \\ 0 & 0 & 0 & 0 & 0 & 1 & 0 & 6 & 0 \\ 1 & 0 & 0 & 0 & 1 & 0 & 0 & 0 & 8 \end{pmatrix} \tag{6.155}$$

The structure factor $\Gamma_{\alpha\beta,\gamma\delta}$ iterates the polarization dependence of the elementary scattering process. It solves the integral equation (Figure 6.6):

$$\boxed{\Gamma_{\alpha\beta,\gamma\delta}(\boldsymbol{q}) = \gamma_e b_{\alpha\beta,\gamma\delta} + \sum_{\mu,\nu} \Gamma_{\alpha\beta,\mu\nu}(\boldsymbol{q})\, b_{\mu\nu,\gamma\delta} w(\boldsymbol{q})} \tag{6.156}$$

where the function $w(\boldsymbol{q})$ may be calculated simply from (6.104), by replacing τ_e by τ_{pol}, that is,

$$w(\boldsymbol{q}) \simeq \frac{\tau_{pol}}{\tau_e}(1 - Dq^2\tau_{pol}) = \frac{3}{2}(1 - Dq^2\tau_{pol}) \tag{6.157}$$

with a new diffusion constant

$$D = \frac{1}{3}cl_{pol} = \frac{1}{3}c^2\tau_{pol}. \tag{6.158}$$

To solve this integral equation, we must first diagonalize the tensor $b_{\alpha\beta,\gamma\delta}$, that is, the matrix (6.155). It has three eigenvalues which may be classified according to the values of total spin, denoted $k = 0, 1$ or 2, resulting from the coupling of two polarizations:

$$\begin{aligned} b_{(k=0)} &= 2/3 \\ b_{(k=1)} &= 1/3 \\ b_{(k=2)} &= 7/15. \end{aligned} \tag{6.159}$$

They have degeneracy $2k+1$, that is, $1, 3$ and 5, respectively. The eigenvectors are given by

$$
\begin{aligned}
b_0 = \frac{2}{3} &\longrightarrow \left\{ \frac{1}{\sqrt{3}}(|xx\rangle + |yy\rangle + |zz\rangle) \right. \\
b_1 = \frac{1}{3} &\longrightarrow \left\{ \begin{array}{l} \frac{1}{\sqrt{2}}(|xy\rangle - |yx\rangle) \\ \frac{1}{\sqrt{2}}(|xz\rangle - |zx\rangle) \\ \frac{1}{\sqrt{2}}(|yz\rangle - |zy\rangle) \end{array} \right. \\
b_2 = \frac{7}{15} &\longrightarrow \left\{ \begin{array}{l} \frac{1}{\sqrt{2}}(|xy\rangle + |yx\rangle) \\ \frac{1}{\sqrt{2}}(|xz\rangle + |zx\rangle) \\ \frac{1}{\sqrt{2}}(|yz\rangle + |zy\rangle) \\ \frac{1}{\sqrt{2}}(|xx\rangle - |yy\rangle) \\ \frac{1}{\sqrt{6}}(|xx\rangle + |yy\rangle - 2|zz\rangle). \end{array} \right.
\end{aligned}
\tag{6.160}
$$

Changing the basis (see remark page 222), we get

$$
b_{\alpha\beta,\gamma\delta} = \frac{1}{2}(b_1 + b_2)\delta_{\alpha\gamma}\delta_{\beta\delta} + \frac{1}{2}(-b_1 + b_2)\delta_{\alpha\delta}\delta_{\beta\gamma} + \frac{1}{3}(b_0 - b_2)\delta_{\alpha\beta}\delta_{\gamma\delta}. \tag{6.161}
$$

This tensor is the product of two rank 2 tensors defined in a three-dimensional space associated with the polarization. It is therefore reducible and decomposes into the sum of irreducible components $3 \otimes 3 = 1 \oplus 3 \oplus 5$ containing, respectively, $1, 3$ and 5 independent elements:

$$
b_{\alpha\beta,\gamma\delta} = \sum_{k=0}^{2} b_k \, \mathcal{T}^{(k)}_{\alpha\beta,\gamma\delta}. \tag{6.162}
$$

This decomposition involves three basis tensors: scalar, antisymmetric, and traceless symmetric:

$$
\begin{aligned}
\mathcal{T}^{(0)}_{\alpha\beta,\gamma\delta} &= \frac{1}{3}\delta_{\alpha\beta}\delta_{\gamma\delta} \\
\mathcal{T}^{(1)}_{\alpha\beta,\gamma\delta} &= \frac{1}{2}\left[\delta_{\alpha\gamma}\delta_{\beta\delta} - \delta_{\alpha\delta}\delta_{\beta\gamma}\right] \\
\mathcal{T}^{(2)}_{\alpha\beta,\gamma\delta} &= \frac{1}{2}\left[\delta_{\alpha\gamma}\delta_{\beta\delta} + \delta_{\alpha\delta}\delta_{\beta\gamma}\right] - \frac{1}{3}\delta_{\alpha\beta}\delta_{\gamma\delta}.
\end{aligned}
\tag{6.163}
$$

In each eigensubspace, the diagonalization of the integral equation (6.156) is immediate, as it reduces to the scalar equation

$$\Gamma_k(\boldsymbol{q}) = \gamma_e b_k + \Gamma_k(\boldsymbol{q}) b_k w(\boldsymbol{q}). \tag{6.164}$$

We thus obtain three distinct modes

$$\boxed{\Gamma_k = \frac{\gamma_{pol}/\tau_{pol}}{1/\tau_k + Dq^2}} \tag{6.165}$$

characterized by the relaxation times

$$\tau_k = \tau_{pol} \frac{b_k}{2/3 - b_k}, \tag{6.166}$$

where we have used expression (6.150) for the ratio τ_{pol}/τ_e. From (6.159), we can calculate the expression for the eigenmodes

$$\Gamma_0(\boldsymbol{q}) = \frac{\gamma_{pol}/\tau_{pol}}{Dq^2} \qquad \Gamma_1(\boldsymbol{q}) = \frac{\gamma_{pol}/\tau_{pol}}{Dq^2 + \frac{1}{\tau_{pol}}} \qquad \Gamma_2(\boldsymbol{q}) = \frac{\gamma_{pol}/\tau_{pol}}{Dq^2 + \frac{3}{7\tau_{pol}}}. \tag{6.167}$$

We note that the mode Γ_0 coincides[20] with the scalar mode obtained in the absence of polarization; this is the Goldstone mode reflecting energy conservation. On the other hand, upon making the frequency dependence explicit, we see that the other two modes are characterized by poles of the form $-i\omega + Dq^2 + 1/\tau_k$. Taking the Fourier transform in the time domain, we see that this result implies that their contribution decays exponentially with the characteristic times τ_1 and τ_2 of the order of the average collision time τ_{pol}. The structure factor $\Gamma_{\alpha\beta,\gamma\delta}$ may be written as

$$\Gamma_{\alpha\beta,\gamma\delta} = \sum_{k=0}^{2} \Gamma_k \, \mathcal{T}^{(k)}_{\alpha\beta,\gamma\delta} = \Gamma_0 \mathcal{T}^{(0)} + \Gamma_1 \mathcal{T}^{(1)} + \Gamma_2 \mathcal{T}^{(2)}, \tag{6.168}$$

which, by virtue of (6.163) may be expressed as

$$\boxed{\Gamma_{\alpha\beta,\gamma\delta}(\boldsymbol{q}) = \frac{1}{2}(\Gamma_1 + \Gamma_2)\delta_{\alpha\gamma}\delta_{\beta\delta} + \frac{1}{2}(-\Gamma_1 + \Gamma_2)\delta_{\alpha\delta}\delta_{\beta\gamma} + \frac{1}{3}(\Gamma_0 - \Gamma_2)\delta_{\alpha\beta}\delta_{\gamma\delta}} \tag{6.169}$$

6.6.3 Classical intensity

The scattered intensity is computed by pairing two multiple scattering amplitudes corresponding to the same initial $(\alpha\alpha)$ and final $(\beta\beta)$ polarization states (Figure 6.9). It

[20] Up to the substitution $l_e \rightarrow l_{pol}$.

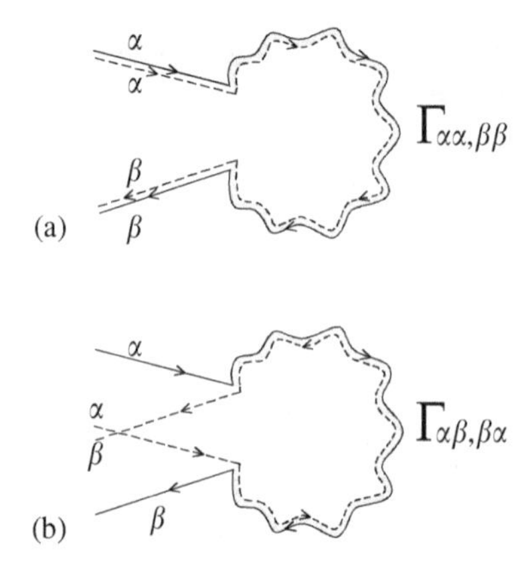

Figure 6.9 The classical intensity and the coherent backscattering contribution involve two different contractions of the tensor associated with the structure factor.

thus results from the iteration of an elementary vertex of the form

$$b_{\alpha\alpha,\beta\beta} = \langle (\delta_{\alpha\beta} - \hat{k}'_\alpha \hat{k}'_\beta)\,(\delta_{\alpha\beta} - \hat{k}'_\alpha \hat{k}'_\beta)\rangle = \frac{1}{15}(7\delta_{\alpha\beta} + 1). \tag{6.170}$$

Although this tensor is of rank 4, to calculate its eigenmodes, it is sufficient to write it in the subspace described by the matrix

$$\frac{1}{15}\begin{pmatrix} 8 & 1 & 1 \\ 1 & 8 & 1 \\ 1 & 1 & 8 \end{pmatrix} \tag{6.171}$$

extracted from (6.155). Its eigenvalues are b_0 and b_2 (doubly degenerate) with corresponding eigenvectors

$$b_0 = \frac{2}{3} \longrightarrow \left\{ \frac{1}{\sqrt{3}}(|xx\rangle + |yy\rangle + |zz\rangle) \right.$$

$$b_2 = \frac{7}{15} \longrightarrow \left\{ \begin{array}{l} \frac{1}{\sqrt{3}}(|xx\rangle + |yy\rangle + |zz\rangle) \\ \\ \frac{1}{\sqrt{2}}(|xx\rangle - |zz\rangle) \\ \\ \frac{1}{\sqrt{6}}(|xx\rangle + |yy\rangle - 2|zz\rangle). \end{array} \right. \tag{6.172}$$

In the parallel polarization channels, that is, when the emergent light is analyzed in the same polarization direction as the incident light, the measured intensity is proportional to

$$\Gamma_{\alpha\alpha,\alpha\alpha} = \frac{1}{3}(\Gamma_0 + 2\Gamma_2). \tag{6.173}$$

In the perpendicular polarization channels ($\alpha \neq \beta$), the measured intensity is proportional to

$$\Gamma_{\alpha\alpha,\beta\beta} = \frac{1}{3}(\Gamma_0 - \Gamma_2). \tag{6.174}$$

The mode Γ_2 is rapidly attenuated, leaving, in these two cases, only the contribution of the scalar mode Γ_0 reduced by one-third in each channel, owing to depolarization of the incident light. In consequence, *the measured intensity is the same in all the polarization channels.*

Note finally that the conservation of probability is expressed by

$$\sum_\beta \Gamma_{\alpha\alpha,\beta\beta} = \Gamma_0. \tag{6.175}$$

6.6.4 Coherent backscattering

In the parallel polarization channels, the structure factor associated with the Cooperon is identical to that of the Diffuson: the only significant contribution comes from the scalar mode Γ_0 reduced by one-third. The polarization does not therefore change the relative contribution of the Diffuson and the Cooperon.

On the other hand, in the perpendicular polarization channels ($\alpha \neq \beta$), the Cooperon contribution to the coherent backscattering is proportional to (Figure 6.9(b))

$$\Gamma_{\alpha\beta,\beta\alpha} = \frac{\Gamma_2 - \Gamma_1}{2}. \tag{6.176}$$

These two modes decay very rapidly, and the Cooperon contribution analyzed in a perpendicular polarization channel vanishes.

Remark: finite size scatterers

The preceding approach may be extended simply to Rayleigh–Gans scattering (see Exercise 2.3) since in this approximation it suffices to replace expression (6.143) by

$$\boldsymbol{P}' = M(\hat{\boldsymbol{k}}')s(\boldsymbol{k}' - \boldsymbol{k})\boldsymbol{P}, \tag{6.177}$$

where $s(\boldsymbol{k}' - \boldsymbol{k})$ is a scalar function which describes the form factor (2.128) of the scatterer. In this case, the scattering becomes anisotropic, and there is another characteristic time which enters: the transport time (Appendix A4.3).

On the other hand, for Mie scattering, this approach is inoperative, since the polarization $\boldsymbol{P}'$ depends on $\boldsymbol{k}$ and $\boldsymbol{k}'$. This notwithstanding, for spherical Mie scatterers, it remains true that in the parallel polarization channels, the polarization does not change the relative contribution of the Diffuson and Cooperon. To see this, note that in the differential scattering cross section associated with a collision event, the components of the polarization parallel and perpendicular to the scattering plane are multiplied by factors which depend only on the angle between $\boldsymbol{k}$ and $\boldsymbol{k}'$, and so are unchanged for the time-reversed sequence.

6.7 Dephasing and motion of scatterers

We have seen in previous examples how the coherent multiple scattering described by the Cooperon is modified by the coupling of a wave or electron to an external field, or by the presence of additional degrees of freedom. We now examine the dephasing induced by the motion of the scatterers.

To this end, consider a medium containing independent random scatterers. In Chapters 3 and 4, we assumed such scatterers to be fixed in space. This assumption is reasonably justified in the case of electronic scattering (see section 11.4.5), but is not generally valid in the case of electromagnetic waves propagating in diffusive media. In this case, the scatterers have a Brownian motion in the surrounding liquid. We may regard the scatterers as fixed if, during the time t over which the wave crosses the diffusive medium, the scatterers move a distance less than the wavelength λ of the light. In other words, the time τ_b, taken by a Brownian scatterer to move a distance of order λ, must be less than t. For electromagnetic waves propagating in suspensions, we have $t = n\tau_e$ where $n \simeq 10^3$ is, in practice, the number of collisions undergone by the wave before leaving the medium. Taking $\tau_e \simeq 6 \times 10^{-14}$s, we have $t \simeq 6 \times 10^{-11}$ s, while τ_b, given in terms of the diffusion coefficient D_b of the scatterers, is of the order of $\tau_b \simeq \frac{\lambda^2}{D_b} \simeq 1$ms. Thus, the time of flight taken by the light to perform a multiple scattering path is negligible compared to the time needed to change the length of this path by λ, meaning that we are completely justified in regarding the scattering centers as static.

On the other hand, the question of the dynamics of the scatterers becomes relevant when we consider the temporal correlation between electric fields $E(T)$ and $E^*(0)$ corresponding to light pulses emitted in a sufficiently long time interval T [145–147]. In this case, the scatterers have each moved a certain distance, and the time correlation function $G_1(T)$ between these fields is a measure of the dephasing between the two trajectories represented in Figure 6.10. In what follows, we evaluate this dephasing and show that the correlation function $G_1(T)$ may be obtained from the expression for the Diffuson established in Chapter 4, modified by a phase term which describes the motion of the scatterers.

6.7.1 General expression for the phase shift

Consider the electric field associated with a scalar electromagnetic wave which is a solution of the Helmholtz equation. The autocorrelation function $G_1(\boldsymbol{r}, T)$ of the electric field is defined by

$$G_1(\boldsymbol{r}, T) = \frac{\langle E(\boldsymbol{r}, T) E^*(\boldsymbol{r}, 0)\rangle}{\langle |E(\boldsymbol{r}, 0)|^2\rangle}, \tag{6.178}$$

where the field $E(\boldsymbol{r}, T)$ is a solution of the Helmholtz equation (3.57) with a source located at $\boldsymbol{r}_0$. The wave trains propagating at times 0 and T are scattered by distinct static configurations of scatterers. The time T thus plays the role of a parameter. The symbol $\langle \cdots \rangle$ denotes an

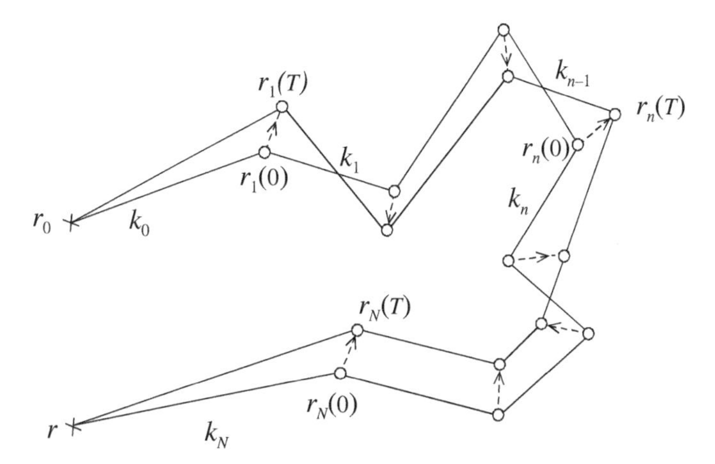

Figure 6.10 When scatterers move, the complex amplitude of a given multiple scattering sequence is modified. In this illustration we assume that we can deform the trajectory continuously.

average taken both over all the possible scattering paths in the medium, and over the random motion of the scatterers, whose character will be discussed later on.

Recall that for $T = 0$, i.e., when the configurations of scatterers are identical, the correlation function $\langle |E(\boldsymbol{r},0)|^2\rangle$ is related to the intensity, that is, to the probability $P(\boldsymbol{r}_0,\boldsymbol{r},\omega=0)$, via relation (4.55)

$$\langle |E(\boldsymbol{r},0)|^2\rangle = \frac{c}{4\pi}P(\boldsymbol{r}_0,\boldsymbol{r},\omega=0) = \frac{c}{4\pi}\int_0^\infty P(\boldsymbol{r}_0,\boldsymbol{r},t)\,dt. \tag{6.179}$$

If the configurations of scatterers at times 0 and T are different, the correlation function has the form

$$\langle E(\boldsymbol{r},T)E^*(\boldsymbol{r},0)\rangle = \frac{c}{4\pi}\int_0^\infty P(\boldsymbol{r}_0,\boldsymbol{r},t)\left\langle e^{i[\phi(t,T)-\phi(t,0)]}\right\rangle dt \tag{6.180}$$

where $\phi(t,T)$ and $\phi(t,0)$ are the phases associated with the configurations of scatterers at times 0 and T. We now determine the dephasing associated with the motion of the scatterers.

In the independent collision approximation (Chapter 3), we may describe the electric field[21] as the superposition of amplitudes associated with all the collision sequences $\mathcal{C}_N = (\boldsymbol{r}_1,\boldsymbol{r}_2,...,\boldsymbol{r}_N)$, that is,

$$E(\boldsymbol{r}_0,\boldsymbol{r},T) = \sum_{N=1}^{\infty}\sum_{\mathcal{C}_N}|A(\boldsymbol{r}_0,\boldsymbol{r},\mathcal{C}_N(T))|\,e^{i\phi_N(T)} \tag{6.181}$$

where the sum is over all possible trajectories, that is, over both all sequence lengths and all possible scatterer positions. The correlation function $\langle E(\boldsymbol{r},T)E^*(\boldsymbol{r},0)\rangle$ may thus be

[21] This is a Green's function such as that given by equation (4.21).

expressed as[22]

$$\langle E(\boldsymbol{r},T)E^*(\boldsymbol{r},0)\rangle = \left\langle \sum_{N,N'}\sum_{\mathcal{C}_N,\mathcal{C}_{N'}} |A(\boldsymbol{r},\mathcal{C}_N(T))A^*(\boldsymbol{r},\mathcal{C}_{N'}(0))|\, e^{i(\phi_N(T)-\phi_{N'}(0))}\right\rangle. \quad (6.182)$$

We assume that the amplitudes A are stationary random variables, namely $A(\boldsymbol{r},\mathcal{C}_N(T)) = A(\boldsymbol{r},\mathcal{C}_N(0))$. Moreover, the contribution of non-diagonal terms $N' \neq N$ cancels upon disorder averaging (section 4.4), so that

$$\langle E(\boldsymbol{r},T)E^*(\boldsymbol{r},0)\rangle = \left\langle \sum_{N,\mathcal{C}_N} |A(\boldsymbol{r},\mathcal{C}_N|^2\, e^{i(\phi_N(T)-\phi_N(0))}\right\rangle. \quad (6.183)$$

To proceed further, we evaluate the phase shift along a collision sequence $\mathcal{C}_N = \{\boldsymbol{r}_1,\boldsymbol{r}_2,...,\boldsymbol{r}_N\}$. For a scalar wave and within the approximation of independent collisions, the phase is given by

$$\phi_N(T) = -\sum_{n=1}^{N}(\boldsymbol{k}_n - \boldsymbol{k}_{n-1})\cdot\boldsymbol{r}_n(T) - \boldsymbol{k}_0\cdot\boldsymbol{r}_0 + \boldsymbol{k}_N\cdot\boldsymbol{r} \quad (6.184)$$

where the wave vectors $\boldsymbol{k}_0$ and $\boldsymbol{k}_N$ relate, respectively, to the incident and emergent waves for this collision sequence. To calculate the phase difference $\Delta\phi_N(T) = \phi_N(T) - \phi_N(0)$, we further assume that the motion of the scatterers is sufficiently slow so that after time T we may neglect the change in direction of the wave vectors (which is second order) so that

$$\Delta\phi_N(T) = -\sum_{n=1}^{N}(\boldsymbol{k}_n - \boldsymbol{k}_{n-1})\cdot\Delta\boldsymbol{r}_n(T) \quad (6.185)$$

with $\Delta\boldsymbol{r}_n(T) = \boldsymbol{r}_n(T) - \boldsymbol{r}_n(0)$. For elastic collisions, the modulus $k = |\boldsymbol{k}|$ is conserved and we write $\boldsymbol{k}_n = k\hat{\boldsymbol{e}}_n$ where $\hat{\boldsymbol{e}}_n$ is a unit vector:

$$\Delta\phi_N(T) = -k\sum_{n=1}^{N}(\hat{\boldsymbol{e}}_n - \hat{\boldsymbol{e}}_{n-1})\cdot\Delta\boldsymbol{r}_n(T). \quad (6.186)$$

At each collision, we may thus associate a phase shift

$$\Delta\varphi(n,T) = -k(\hat{\boldsymbol{e}}_n - \hat{\boldsymbol{e}}_{n-1})\cdot\Delta\boldsymbol{r}_n(T). \quad (6.187)$$

Assuming a decoupling between disorder averages of amplitudes and phases, and for independent collisions, we see that each elementary step (6.11) is now affected by an average phase factor

$$b(T) = \langle e^{i\Delta\varphi(n,T)}\rangle. \quad (6.188)$$

[22] To lighten the notation, we do not specify the source position $\boldsymbol{r}_0$ explicitly.

6.7.2 Dephasing associated with the Brownian motion of the scatterers

To evaluate $b(T)$, we now assume that the motion of the scatterers is Brownian with distribution:[23]

$$\mathcal{P}(\Delta \boldsymbol{r}_n, T) = \left(\frac{1}{4\pi D_b T}\right)^{d/2} e^{-\Delta r_n^2/4D_b T}. \tag{6.189}$$

We first evaluate the average over positions of the scatterers. Since they are independent, we average over each individual scatterer:

$$\begin{aligned}\left\langle e^{i\Delta\varphi(n,T)}\right\rangle_{\Delta \boldsymbol{r}_n(T)} &= \int d\boldsymbol{r}\, e^{-i\boldsymbol{q}_n\cdot \boldsymbol{r}} \left(\frac{1}{4\pi D_b T}\right)^{d/2} e^{-\boldsymbol{r}^2/4D_b T} \\ &= e^{-D_b q_n^2 T},\end{aligned} \tag{6.190}$$

where $\boldsymbol{q}_n = k(\hat{\boldsymbol{e}}_n - \hat{\boldsymbol{e}}_{n-1})$, so that $q_n^2 = 4k^2 \sin^2(\theta_n/2)$ with $\theta_n = (\hat{\boldsymbol{e}}_n, \hat{\boldsymbol{e}}_{n-1})$. It remains to average over the direction of the scattered waves, that is, to calculate

$$b(T) = \left\langle e^{-D_b q^2 T}\right\rangle = \frac{1}{2}\int_0^\pi d\theta\ \sin\theta\, e^{-4D_b k^2 T \sin^2\frac{\theta}{2}}. \tag{6.191}$$

This gives, after integration [146],

$$b(T) = \frac{\tau_b}{T}(1 - e^{-T/\tau_b}). \tag{6.192}$$

The time

$$\boxed{\tau_b = 1/4D_b k^2} \tag{6.193}$$

is the characteristic time for a scatterer to move a distance of the order of λ. The function $b(T)$ is very well approximated by the simpler exponential form

$$b(T) \simeq e^{-T/2\tau_b} \tag{6.194}$$

which we will use in what follows.

Using (6.12) and (6.7), we obtain the structure factor in the diffusive limit and at zero frequency. From equation (4.37), we calculate the expression for the probability $P_d(\boldsymbol{q})$:

$$P_d(\boldsymbol{q}) = \frac{1}{\frac{1-b}{b}\frac{1}{\tau_e} + Dq^2}. \tag{6.195}$$

[23] We shall take care not to confuse the diffusion coefficient D_b of the scatterers with the diffusion coefficient D which describes wave diffusion.

The diffusion pole thus appears with a cutoff time

$$\tau_s(T) = \tau_e \frac{b(T)}{1 - b(T)}. \tag{6.196}$$

We see that the effect of dephasing is to introduce a pole at a finite value of q. In other words, dephasing leads to the disappearance of the zero mode discussed in section 5.5.3. In the limit of short times $T \le \tau_b$, we have $b(T) \simeq 1 - T/2\tau_b$, or

$$\boxed{\tau_s = \tau_e \frac{2\tau_b}{T}} \tag{6.197}$$

Taking the Fourier transform of (6.195) leads finally to

$$\frac{4\pi}{c} \langle E(\boldsymbol{r}, T) E^*(\boldsymbol{r}, 0)\rangle = \frac{1}{4\pi D r} e^{-\frac{r}{l_e}\sqrt{\frac{3T}{2\tau_b}}}. \tag{6.198}$$

For $T = 0$, we recover the intensity (4.67) at point $\boldsymbol{r}$. The correlation function $G_1(\boldsymbol{r}, T)$ defined by (6.178) is given by

$$\boxed{G_1(\boldsymbol{r}, T) = e^{-\frac{r}{l_e}\sqrt{\frac{3T}{2\tau_b}}}} \tag{6.199}$$

The temporal dependence of this correlation function is a stretched exponential, very different from the result (9.11) obtained in the case of single scattering by Brownian scatterers.

We note that the correlation function of the field measures the Laplace transform of the probability. It is of the form

$$\frac{4\pi}{c} \langle E(\boldsymbol{r}, T) E^*(\boldsymbol{r}, 0)\rangle = \int_0^\infty P(\boldsymbol{r}_0, \boldsymbol{r}, t)\, e^{-t/\tau_s}\, dt = P_\gamma(\boldsymbol{r}_0, \boldsymbol{r}) \tag{6.200}$$

with $\gamma = 1/\tau_s$. The exponential factor is indeed the average of a phase factor, appearing in equation (6.180), and the characteristic time $\tau_s = 2\tau_e\tau_b/T$ has been calculated starting from a microscopic description of the impurity motion. Finally, the expression for the dynamic correlation function $G_1(\boldsymbol{r}, T)$ is the ratio

$$G_1(\boldsymbol{r}, T) = \frac{P_\gamma(\boldsymbol{r}_0, \boldsymbol{r})}{P_{\gamma=0}(\boldsymbol{r}_0, \boldsymbol{r})}. \tag{6.201}$$

6.8 Dephasing or decoherence?

From the examples treated in this chapter, it appears that the effect of dephasing on multiple scattering (Cooperon or Diffuson) is to modify the integrated probability of return to the origin $Z(t)$, so that typically

$$Z(t, X) = Z(t) \left\langle e^{i\Delta\phi(t,X)} \right\rangle. \tag{6.202}$$

The phase $\Delta\phi(t, X)$ depends on the physical parameter X at the origin of the dephasing. This results, in general, in an exponential decay in time t for $\langle e^{i\Delta\phi(t,X)}\rangle$ with a characteristic time called the *dephasing time* or the *phase coherence time*.

The dephasing is due to the presence of additional degrees of freedom which may be divided into three classes:

- Degrees of freedom of the diffusing wave: electron spin or photon polarization.
- Degrees of freedom of the scatterers: spin of magnetic impurities, motion of the scatterers, internal quantum degrees of freedom (Appendix A6.5).
- External field: uniform magnetic field, Aharonov–Bohm flux, fluctuating electromagnetic field (Appendix A6.3 and section 13.7.2).

In which situations does one speak of *dephasing* or of *decoherence*? Since these two notions are often confused, we attempt here to give a classification.

We reserve the word decoherence for the case when the dephasing is linked to the notion of *irreversibility*. Thus, a dephasing, such as that associated with a magnetic field, modifies the interference pattern, but causes no decoherence.[24] Similarly, the depolarization of light, or the effect of spin-orbit coupling on the electron spin, are characterized by a reduction of the interference effects, without implying any irreversibility.

In the case where the dephasing arises from the interaction of the wave with degrees of freedom associated with the scatterers, a new notion, linked essentially to the *irreversibility* of the dephasing, comes into play. These degrees of freedom external to the electron or the electromagnetic wave are not controlled by the experimentalist. This lack of information leads us to average the elementary vertex interaction over these degrees of freedom. It is this average which introduces the irreversibility associated with the notion of decoherence. For example, in the case of magnetic impurities, we average the elementary vertex over impurity spin. For cold atoms (Appendix A6.5), we perform a trace over Zeeman sublevels. For Brownian scatterers, we average over positions of the scatterers. Finally, for a fluctuating electromagnetic field (Appendix A6.3), we average over the thermal fluctuations of the field. Each of these averages reflects a lack of information which leads to irreversibility.[25] On the other hand, the dephasing associated with depolarization or with the spin-orbit coupling does not lead to averaging over external degrees of freedom, and thus does not yield irreversibility (in this case, the additional degree of freedom, polarization or spin, is a property of the scattered wave and not of the environment).

We shall use the word decoherence when the environment is described by a statistical average of uncontrolled degrees of freedom. We shall speak, in this case, of a *phase coherence time* which will be denoted τ_ϕ and we shall retain the notation τ_γ for the dephasing time which does not imply decoherence (section 15.9).

[24] The effect of the magnetic field depends greatly on the geometry. For a ring, the field modifies the interference term without reducing it. By contrast, for a plane, the field reduces the interference term due to averaging over the diffusion trajectories. This difference may be described in terms of the *contrast*, defined as the ratio of the quantum and classical contributions. We see in this example that the notion of contrast is independent of that of decoherence.

[25] This implies a change in the state of this environment [148].

Appendix A6.1: Aharonov–Bohm effect in an infinite plane

In section 6.4, we studied the integrated return probability $Z(t, \phi)$ for a ring and a cylinder threaded by an Aharonov–Bohm magnetic flux $\phi = \varphi\phi_0$. These geometries are characterized by a spectrum of eigenfrequencies E_n which are flux dependent.

In this appendix, we calculate $Z(t, \phi)$ for the case of an infinite plane pierced by an Aharonov–Bohm flux line at one point [118]. At first sight, this problem might appear academic, since experimentally relevant geometries are multiply connected and can thus be related to the ring and the cylinder. This notwithstanding, we will give an example in section 7.6.4 for which this case is relevant.

The study of the infinite plane pierced by a flux line is of both pedagogical and methodological interest. This problem arises in different branches of physics (quantum mechanics, field theory, polymer physics ...). The associated function $Z(t)$ has a particular dependence on the flux which derives from topological considerations. To make this point, we could have used a priori the following reasoning. For the infinite plane geometry, the spectrum of eigenmodes of the diffusion equation is a continuum identical to that of the plane without the flux line, since there is no induced force – the magnetic field is zero everywhere except at the excluded point where the line crosses the plane. If the spectra are identical, then the probability $Z(t)$ should be independent of the flux, and given by $S/4\pi Dt$, where S is the surface area. We show here that this result is incorrect.

Consider a magnetic field $B = \phi\delta(\boldsymbol{r})$ concentrated at a single point in the plane[26] and described by the vector potential

$$\boldsymbol{A}(\boldsymbol{r}) = \frac{\phi}{2\pi r}\hat{\boldsymbol{e}}_\theta \tag{6.203}$$

where $\hat{\boldsymbol{e}}_\theta$ is the azimuthal unit vector and ϕ the associated magnetic flux. The Lorentz force on the electrons is thus identically zero.

Just as for the case of the uniform field, we may inquire about the eigenvalues and the eigenfunctions of the equivalent Schrödinger problem. The normalizable solutions of equation (6.34) are

$$\psi_n(q, \boldsymbol{r}) = \frac{1}{\sqrt{2\pi}}\, e^{in\theta}\, J_{|n-2\varphi|}(qr), \tag{6.204}$$

where $n \in \mathbb{Z}$ is the angular momentum, $q = \sqrt{2mE}$ is the wave vector and $J_\alpha(x)$ is a Bessel function of the first kind. The probability $P_\varphi(\boldsymbol{r}_0, \boldsymbol{r}, t)$ in the presence of the flux is given by (5.2) which may be rewritten, by using the eigenfunctions (6.204),

$$P_\varphi(\boldsymbol{r}_0, \boldsymbol{r}, t) = \int_0^\infty \frac{q\,dq}{2\pi} \sum_{n\in\mathbb{Z}} J_{|n-2\varphi|}(qr) J_{|n-2\varphi|}(qr_0)\, e^{in(\theta-\theta_0)}\, e^{-Dq^2 t}. \tag{6.205}$$

26 The δ function is two dimensional.

Integrating over q, we obtain[27]

$$P_\varphi(\boldsymbol{r}_0,\boldsymbol{r},t) = \frac{1}{4\pi Dt} e^{-\frac{r^2+r_0^2}{4Dt}} \sum_{n\in\mathbb{Z}} e^{in(\theta-\theta_0)} I_{|n-2\varphi|}\left(\frac{rr_0}{2Dt}\right), \tag{6.206}$$

where I is a modified Bessel function. The integrated probability of return to the origin $Z(t,\phi)$ is given by

$$\begin{aligned} Z(t,\phi) &= \int d\boldsymbol{r}\, P_\varphi(\boldsymbol{r},\boldsymbol{r},t) \\ &= \int_0^\infty dr \frac{r}{2Dt} e^{-r^2/2Dt} \sum_{n\in\mathbb{Z}} I_{|n-2\varphi|}\left(\frac{r^2}{2Dt}\right). \end{aligned} \tag{6.207}$$

Using the Poisson summation formula (15.106), we may rewrite

$$\sum_{n\in\mathbb{Z}} I_{|n-2\varphi|}\left(\frac{r^2}{2Dt}\right) = \sum_{m\in\mathbb{Z}} e^{-4i\pi m\varphi} \int_{-\infty}^{+\infty} d\nu\, e^{2i\pi m\nu} I_{|\nu|}\left(\frac{r^2}{2Dt}\right) \tag{6.208}$$

where the sum over m now corresponds to a description in terms of winding numbers. To evaluate $Z(t,\phi)$, we calculate the difference in the partition functions:[28]

$$Z(t,\phi) - Z(t,0) = \sum_{m\in\mathbb{Z}} \left(e^{-4i\pi m\varphi} - 1\right) \int_0^\infty dr \frac{r}{2Dt} e^{-r^2/2Dt} \int_{-\infty}^{+\infty} d\nu e^{2i\pi n\nu} I_{|\nu|}\left(\frac{r^2}{2Dt}\right). \tag{6.209}$$

After the change of variables $x = r^2/2Dt$, and using

$$\int_0^\infty dx\, e^{-x} \int_{-\infty}^{+\infty} d\nu e^{2i\pi\nu m} I_{|\nu|}(x) = \frac{1}{2\pi^2 m^2} \tag{6.210}$$

for $m \neq 0$, it remains to evaluate the series[29]

$$Z(t,\phi) - Z(t,0) = \sum_{m\neq 0} \frac{e^{-4i\pi m\varphi} - 1}{4\pi^2 m^2} = -\frac{1}{\pi^2} \sum_{m=1}^{+\infty} \frac{\sin^2(2\pi m\varphi)}{m^2} = -\varphi(1-2\varphi). \tag{6.211}$$

Using (5.24) for $Z(t,0)$, we finally obtain

$$\boxed{Z(t,\phi) = \frac{S}{4\pi Dt} - \varphi(1-2\varphi)} \tag{6.212}$$

for $\varphi \in [0,1]$.

[27] Using (15.69), we can check that $P_{\varphi=0}(\boldsymbol{r}_0,\boldsymbol{r},t)$ is given by its free-space expression (5.20).
[28] Each of the functions diverges by virtue of (5.23) and (5.5).
[29] This Fourier series is a function periodic in φ with period 1, so it suffices to evaluate it for $0 \le \varphi \le 1$.

Spectral determinant and Aharonov–Bohm effect

We stated at the outset of this appendix that the distinction between the functions $Z(t,\phi)$ and $Z(t,0)$ is of topological origin. One way of understanding this is to note that the term associated with the flux is independent of the time t. Indeed, we have seen in Appendix A5.5 that the existence of a constant term in the asymptotic Weyl expansion for $Z(t)$ corresponds to the Euler–Poincaré characteristic of the domain,[30] which is a topological characteristic of the system. This does not result from the existence of a new energy scale, and can thus only reflect the nature of the zero mode. This is a general result which derives from powerful theorems like the index theorem [149] which are beyond the scope of our discussion. Nevertheless, it is instructive to verify this point in the present case. A constant term in the probability $Z(t)$ is reflected in the density of states by a delta function $\delta(E)$ (Appendix A5.5). We thus obtain the density of states of diffusion modes in the presence of a magnetic Aharonov–Bohm flux line

$$\rho_\phi(E) - \rho_0(E) = -\varphi(1-2\varphi)\delta(E). \tag{6.213}$$

This result demonstrates that the two eigenfrequency spectra of the diffusion equation are identical, except for the zero mode.

Appendix A6.2: Functional representation of the diffusion equation

A6.2.1 Functional representation

In section 6.2, we saw the analogy between the diffusion equation and the Schrödinger equation. Here we develop this analogy in order to obtain a functional representation of the diffusion equation.

The Green function G_0 associated with the Schrödinger equation for a free particle of charge q is a solution of the equation

$$\left[-i\hbar\frac{\partial}{\partial t} - \frac{\hbar^2}{2m}\left[\nabla_{r'} - i\frac{q}{\hbar}\boldsymbol{A}(\boldsymbol{r}')\right]^2\right] G_0(\boldsymbol{r},\boldsymbol{r}',t) = \delta(\boldsymbol{r}-\boldsymbol{r}')\delta(t) \tag{6.214}$$

which has the functional integral representation [150]:

$$G_0(\boldsymbol{r},\boldsymbol{r}',t) = \int_{\boldsymbol{r}(0)=\boldsymbol{r}}^{\boldsymbol{r}(t)=\boldsymbol{r}'} \mathcal{D}\{\boldsymbol{r}\}\ \exp\left(\frac{i}{\hbar}\int_0^t \mathcal{L}(\tau)\,d\tau\right) \tag{6.215}$$

where $\mathcal{L}$ is the Lagrangian of a free particle in a magnetic field $\boldsymbol{B} = \nabla\times\boldsymbol{A}$

$$\mathcal{L}(\boldsymbol{r},\dot{\boldsymbol{r}},t) = \frac{1}{2}m\dot{\boldsymbol{r}}^2 + q\dot{\boldsymbol{r}}\cdot\boldsymbol{A}(\boldsymbol{r}). \tag{6.216}$$

[30] For $1d$ diffusion on a finite graph, this constant equals $(V-E)/2$; see Appendix A5.6 for notation.

Similarly, for the diffusion equation, the probability $P_c(\boldsymbol{r},\boldsymbol{r}',t)$ associated with the Cooperon is a solution of the equation (relations 6.29 and 6.34):

$$\left[\frac{\partial}{\partial t} - D\left(\nabla_{\boldsymbol{r}'} + i\frac{2e}{\hbar}\boldsymbol{A}(\boldsymbol{r}')\right)^2\right] P_c(\boldsymbol{r},\boldsymbol{r}',t) = \delta(\boldsymbol{r}-\boldsymbol{r}')\delta(t) \tag{6.217}$$

where the electron charge is denoted $-e$. This solution may be expressed as a functional integral [150]:

$$P_c(\boldsymbol{r},\boldsymbol{r}',t) = \int_{\boldsymbol{r}(0)=\boldsymbol{r}}^{\boldsymbol{r}(t)=\boldsymbol{r}'} \mathcal{D}\{\boldsymbol{r}\} \exp\left(-\int_0^t \mathcal{L}(\tau)\, d\tau\right) \tag{6.218}$$

with[31]

$$\mathcal{L}(\boldsymbol{r},\dot{\boldsymbol{r}},t) = \frac{\dot{\boldsymbol{r}}^2}{4D} + i\frac{2e}{\hbar}\dot{\boldsymbol{r}}\cdot\boldsymbol{A}(\boldsymbol{r}). \tag{6.219}$$

For a time-independent magnetic field $\boldsymbol{B} = \nabla \times \boldsymbol{A}$, we may use (6.218) to express the heat kernel (5.5), yielding

$$Z_c(t,B) = \oint \mathcal{D}\{\boldsymbol{r}\} \exp\left(-\int_0^t \mathcal{L}_0(\tau)\, d\tau\right) \exp\left(-i\frac{2e}{\hbar}\int_0^t \dot{\boldsymbol{r}}\cdot\boldsymbol{A}(\boldsymbol{r})\, d\tau\right) \tag{6.220}$$

where $\oint \mathcal{D}\{\boldsymbol{r}\} = \int d\boldsymbol{r} \int_{\boldsymbol{r}(0)=\boldsymbol{r}}^{\boldsymbol{r}(t)=\boldsymbol{r}} \mathcal{D}\{\boldsymbol{r}\}$ designates the integral over all closed paths and $\mathcal{L}_0 = \dot{\boldsymbol{r}}^2/4D$ is the free Lagrangian. The function $Z_c(t,B)$ may be written in the form:

$$Z_c(t,B) = Z(t)\left\langle \exp\left(-i\frac{4\pi}{\phi_0}\int_0^t \dot{\boldsymbol{r}}\cdot\boldsymbol{A}(\boldsymbol{r})\, d\tau\right)\right\rangle \tag{6.221}$$

where $Z(t)$ is the integrated probability in zero field, and where $\langle\cdots\rangle$ denotes the average over all closed Brownian paths:

$$\langle\cdots\rangle = \frac{1}{Z(t)}\oint \mathcal{D}\{\boldsymbol{r}\} \cdots \exp\left(-\int_0^t \mathcal{L}_0(\tau)\, d\tau\right). \tag{6.222}$$

In the case of a uniform field, by using the symmetric gauge $\boldsymbol{A}(\boldsymbol{r}) = \frac{1}{2}\boldsymbol{B}\times\boldsymbol{r}$, we can write $Z_c(t,B)$ in the form

$$\begin{aligned} Z_c(t,B) &= Z(t)\left\langle \exp\left(-i\frac{2\pi}{\phi_0}B\int_0^t \boldsymbol{r}\times d\boldsymbol{r}\right)\right\rangle \\ &= Z(t)\left\langle \exp\left(-i\frac{4\pi}{\phi_0}B\mathcal{A}(t)\right)\right\rangle \end{aligned} \tag{6.223}$$

where $\mathcal{A}(t) = \frac{1}{2}\int_0^t \boldsymbol{r}\times d\boldsymbol{r}$ is the algebraic area swept out by a closed path during time t.

[31] Note that this Lagrangian has the dimension of inverse time.

A6.2.2 Brownian motion and magnetic field

Expression (6.223) is an example of a characteristic function of diffusion law in the presence of a constraint. More generally, for an expression of the form (6.3), the phase $\phi(t,X)$ depends on a dephasing mechanism which results from a constraint imposed by topological or geometrical characteristics. Thus, (6.223) allows us to express the distribution of algebraic areas bounded by Brownian trajectories. Similarly, the heat kernel for the diffusion around a one-dimensional ring gives the distribution of winding numbers associated with a Brownian trajectory on a closed circuit.

The constrained diffusion laws have been studied extensively in the literature, and yield a great variety of behaviors (Lévy law of algebraic areas [122, 151], Spitzer's law [152] for windings, etc.).

Lévy law of algebraic areas and uniform field

We wish to calculate the distribution $\mathcal{P}(\mathcal{A},t)$ of the algebraic area $\mathcal{A}$ enclosed in the interior of a closed planar Brownian path during a time t (Figure 6.11). This probability distribution is given by[32]

$$\mathcal{P}(\mathcal{A},t) = \langle \delta[\mathcal{A} - \mathcal{A}(t)] \rangle = \frac{1}{2\pi} \int e^{ib\mathcal{A}} \langle e^{-ib\mathcal{A}(t)} \rangle db, \tag{6.224}$$

which relates $\mathcal{P}(\mathcal{A},t)$ to the probability of return to the origin in a uniform magnetic field $B = b\phi_0/4\pi$. Indeed, from (6.223), we have

$$\langle e^{-ib\mathcal{A}(t)} \rangle = \frac{Z_c(t, b\phi_0/4\pi)}{Z(t)}. \tag{6.225}$$

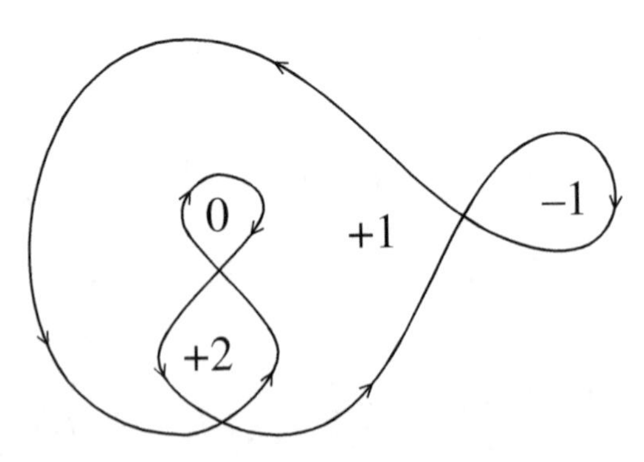

Figure 6.11 Representation of a closed Brownian curve and its associated algebraic area. Each sector is characterized by an integer, its winding number. The sector *0* outside the curve has infinite area.

[32] Conversely

$$\langle e^{-ib\mathcal{A}(t)} \rangle = \int \mathcal{P}(\mathcal{A},t)\, e^{-ib\mathcal{A}}\, d\mathcal{A}.$$

Using relation (6.41) we obtain

$$\langle e^{-ib\mathcal{A}(t)}\rangle = \frac{bDt}{\sinh(bDt)} \tag{6.226}$$

and the Fourier transform (6.224) leads to

$$\boxed{\mathcal{P}(\mathcal{A},t) = \frac{\pi}{4Dt}\frac{1}{\cosh^2\frac{\pi\mathcal{A}}{2Dt}}} \tag{6.227}$$

We thus obtain the Lévy area distribution law for the interior of a Brownian trajectory in time t.

Distribution of winding numbers in a ring

Similarly, the distribution of winding numbers w in a ring is given by

$$\mathcal{P}(w,t) = \big\langle\delta[w - w(t)]\big\rangle = \frac{1}{2\pi}\int e^{ibw}\langle e^{-ibw(t)}\rangle\, db \tag{6.228}$$

which is obtained from the expression of the heat kernel $Z_c(t,\phi)$ for the case of an Aharonov–Bohm flux. From equation (6.223), we have $4\pi\phi/\phi_0$ and

$$\langle e^{-ibw(t)}\rangle = \frac{Z_c(t, b\phi_0/4\pi)}{Z(t)} \tag{6.229}$$

and from (6.54) we obtain the distribution of winding numbers:

$$\mathcal{P}(w,t) = \sum_m \frac{L}{\sqrt{4\pi Dt}}\, e^{-\frac{m^2L^2}{4Dt}}\,\delta(w-m). \tag{6.230}$$

Aharonov–Bohm flux in the plane: the Edwards problem

Another example for which it is possible to relate a diffusion law with a constraint to an expression for the probability of return to the origin is that of an Aharonov–Bohm flux line piercing an infinite plane. We wish to determine the distribution of windings about a point O in an infinite plane (Figure 6.12). We denote by $\mathcal{P}(\theta,t)$ the probability distribution for the angle θ swept out by a Brownian particle in time t with respect to the point O. Here, as opposed to the two preceding examples, there is no translational invariance, since there is one point that has been singled out. Consequently, the distribution depends on both departure and arrival points $\boldsymbol{r}_0$ and $\boldsymbol{r}$. We shall only consider closed paths, for which $\boldsymbol{r} = \boldsymbol{r}_0$.[33] By definition, the distribution $\mathcal{P}(\theta,t)$ is

$$\mathcal{P}(\theta,t) = \big\langle\delta[\theta - \theta(t)]\big\rangle = \frac{1}{2\pi}\int e^{i\varphi\theta}\,\langle e^{-i\varphi\theta(t)}\rangle\, d\varphi \tag{6.231}$$

[33] We may easily generalize this calculation to the case where the endpoints of the Brownian trajectory are different.

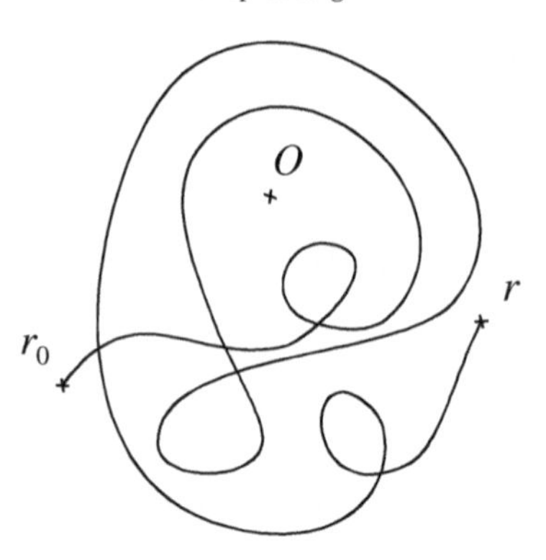

Figure 6.12 Windings of a Brownian trajectory with endpoints $\boldsymbol{r}_0$ and $\boldsymbol{r}$ about the point O. Here we only consider the case where $\boldsymbol{r} = \boldsymbol{r}_0$.

where $\theta(t) = \int_0^t d\tau d\theta/d\tau$ is the angle swept out by the diffusing particle during the time interval $[0, t]$. The average value $\langle e^{-i\varphi\theta(t)}\rangle$ is taken over the ensemble of closed Brownian trajectories. This is the characteristic function associated with the additional constraint: to have a closed trajectory sweeping out an angle $\theta(t)$. This characteristic function is precisely that which appears in the probability $P(t, \varphi)$ to have a closed trajectory in the presence of Aharonov–Bohm flux $\phi = \varphi\phi_0$. Indeed, we may write the curvilinear integral of the vector potential in the form $\int \boldsymbol{A} \cdot d\boldsymbol{l} = \frac{\phi}{2\pi}\int_0^t \dot{\theta}(\tau) d\tau$. By virtue of relation (6.218), the probability $P(t, \varphi)$ may be written in functional representation:

$$
\begin{aligned}
P(t, \varphi) &= \int_{\boldsymbol{r}(0)=\boldsymbol{r}_0}^{\boldsymbol{r}(t)=\boldsymbol{r}_0} \mathcal{D}\{\boldsymbol{r}\} \exp\left(-\int_0^t \frac{\dot{\boldsymbol{r}}^2}{4D} d\tau - i\varphi\theta(t)\right) \\
&= P(t, \varphi = 0)\langle e^{-i\varphi\theta(t)}\rangle.
\end{aligned} \tag{6.232}
$$

Using expressions (6.231) and (6.232), we calculate the distribution $\mathcal{P}(\theta, t)$ of winding angles as a function of the probability of return to the origin $P(t, \varphi)$ in the presence of an Aharonov–Bohm flux $\phi = b\phi_0/2$:

$$
\mathcal{P}(\theta, t) = \frac{1}{2\pi}\int e^{ib\theta} \frac{P(t, \varphi)}{P(t, \varphi = 0)} db. \tag{6.233}
$$

Inserting expression (6.206) for $P(t, \varphi) = P_\varphi(\boldsymbol{r}_0, \boldsymbol{r}_0, t)$, we obtain the distribution law for winding angles about the origin:

$$
\mathcal{P}(\theta, t) = \frac{1}{2\pi}\int db\, e^{ib\theta} \sum_n I_{|n-b|}\left(\frac{r^2}{2Dt}\right) e^{-r^2/2Dt}. \tag{6.234}
$$

Integrating over all angles, we eliminate the angular constraint and we retrieve the probability of having a closed trajectory:

$$\int_{-\infty}^{+\infty} d\theta\, \mathcal{P}(\theta,t) = P(\boldsymbol{r},\boldsymbol{r},t). \tag{6.235}$$

This duality between a constrained diffusion law and the probability associated with the covariant diffusion equation (that is, in the presence of a magnetic field) was first proposed by Edwards [151, 153] in the framework of self-avoiding polymers.

Appendix A6.3: The Cooperon in a time-dependent field

We now consider the behavior of the Cooperon in a time-dependent electromagnetic field [154] characterized by the potentials $\boldsymbol{A}(\boldsymbol{r},t)$ and $V(\boldsymbol{r},t)$. The two multiple scattering sequences corresponding to opposite directions of propagation now see different potentials and therefore pick up different phases. We want to determine the behavior of the Cooperon in this case.

Figure 6.13 indicates the difference between the Diffuson and Cooperon structures. For the Diffuson, the two sequences "see" the same potential. For the Cooperon, if one of the sequences sees the potentials $\boldsymbol{A}(t)$ and $V(t)$, the second sees the potentials $\boldsymbol{A}(\bar{t})$ and $V(\bar{t})$

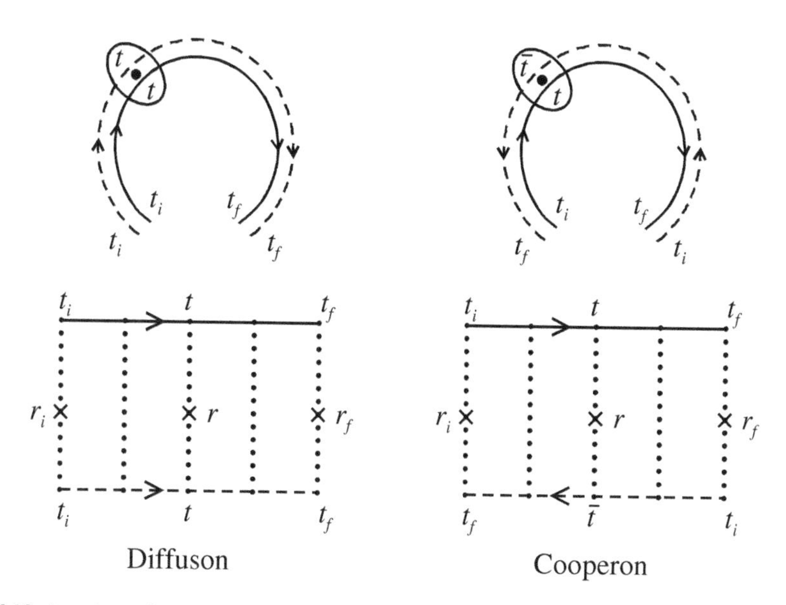

Figure 6.13 At point $\boldsymbol{r}$, the two trajectories of a Diffuson P_d "see" the same external potential $\boldsymbol{A}(t)$. In contrast, the two trajectories of a Cooperon P_c "see" potentials corresponding to different times t and $\bar{t} = t_i + t_f - t$. The figure shows the example of a multiple scattering trajectory with five collisions.

corresponding to time $\bar{t} = t_i + t_f - t$, where t_i and t_f are respectively the initial and final times of the sequence. The two temporal sequences,

$$\begin{array}{ccccccccc} t_i & \rightarrow & \cdots & \rightarrow & t & \rightarrow & \cdots & \rightarrow & t_f \\ t_f & \rightarrow & \cdots & \rightarrow & \bar{t} = t_i + t_f - t & \rightarrow & \cdots & \rightarrow & t_i, \end{array}$$

are related by time reversal (note that $\bar{t_f} = t_i$ and $\bar{t_i} = t_f$). We seek here the evolution equation for the Cooperon $P_c(\boldsymbol{r}_i, \boldsymbol{r}_f, t_i, t_f)$. For this, it is instructive to compare the structures of the Diffuson and the Cooperon in a time-dependent field.

For the Diffuson, we rewrite the integral equation (4.24) for the structure factor Γ in the form (Figure 6.14):

$$\Gamma(\boldsymbol{r}_i, \boldsymbol{r}_f, t_i, t_f) = \gamma_e \delta(\boldsymbol{r}_i - \boldsymbol{r}_f)\delta(t_i - t_f) + \int \Gamma(\boldsymbol{r}_i, \boldsymbol{r}, t_i, t) K(\boldsymbol{r}, \boldsymbol{r}_f, t, t_f)\, d\boldsymbol{r}\, dt$$

where the kernel K of the integral equation is given by

$$K(\boldsymbol{r}, \boldsymbol{r}_f, t, t_f) = \gamma_e \overline{G}^R(\boldsymbol{r}, \boldsymbol{r}_f, t, t_f)\overline{G}^A(\boldsymbol{r}_f, \boldsymbol{r}, t_f, t) \tag{6.236}$$

that is, from (4.16, 4.18):

$$K(\boldsymbol{r}, \boldsymbol{r}_f, t, t_f) = \frac{1}{\tau_e} P_0(\boldsymbol{r}, \boldsymbol{r}_f, t, t_f) = \frac{\delta[R - v(t_f - t)]\, e^{-(t_f - t)/\tau_e}}{4\pi R^2 \tau_e} \tag{6.237}$$

where $R = |\boldsymbol{r} - \boldsymbol{r}_f|$. In the presence of slowly varying external potentials $\boldsymbol{A}(\boldsymbol{r}, t)$ and $V(\boldsymbol{r}, t)$, the average Green function $\overline{G}^R(\boldsymbol{r}, \boldsymbol{r}_f, t, t_f)$ acquires an additional phase $\phi(\boldsymbol{r}, \boldsymbol{r}_f, t, t_f)$ given

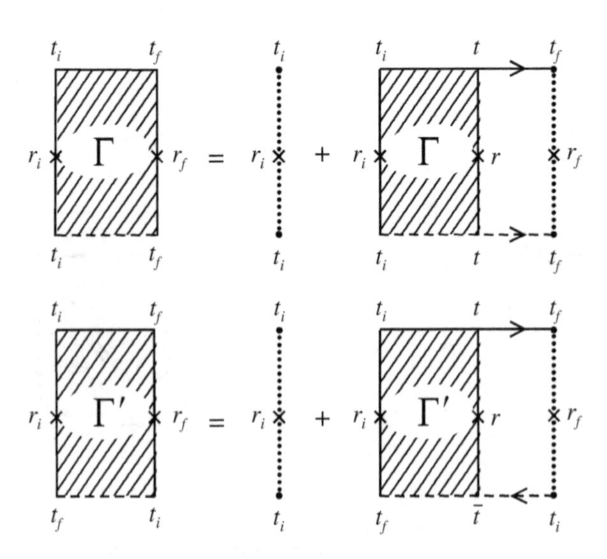

Figure 6.14 Integral equations for the Diffuson $\Gamma(\boldsymbol{r}_i, \boldsymbol{r}_f, t_i, t_f)$ and the Cooperon $\Gamma'(\boldsymbol{r}_i, \boldsymbol{r}_f, t_i, t_f)$. The time $\bar{t}$ is equal to $t_i + t_f - t$.

by [118, 119][34]

$$\phi(\boldsymbol{r},\boldsymbol{r}_f,t,t_f) = -\frac{e}{\hbar}\int_t^{t_f} [\dot{\boldsymbol{r}}(\tau)\cdot\boldsymbol{A}(\boldsymbol{r}(\tau),\tau) - V(\boldsymbol{r}(\tau),\tau)]\,d\tau. \tag{6.238}$$

The Green function $\overline{G}^A(\boldsymbol{r}_f,\boldsymbol{r},t_f,t)$ acquires the opposite phase $\phi(\boldsymbol{r}_f,\boldsymbol{r},t_f,t) = -\phi(\boldsymbol{r},\boldsymbol{r}_f,t,t_f)$ such that the two phases cancel in the integral equation and the Diffuson is not affected by the potentials. This is due to the fact that the two associated sequences correspond to fields taken at the same time. The random phase shifts of each of the sequences thus compensate.

The Cooperon is obtained by reversing the temporal sequence of one of the two sequences $t \to \bar{t} = t_i + t_f - t$. To characterize its temporal evolution, we must introduce the pair of times $\tilde{\boldsymbol{t}} = (t,\bar{t})$ associated with the two sequences which constitute it. We first consider the integral equation for the structure factor Γ'. Its iterative structure must take explicit account of the pair $\tilde{\boldsymbol{t}} = (t,\bar{t})$ associated with the scattering from a single impurity. The integral equation, illustrated in Figure 6.14, takes the form:[35]

$$\Gamma'(\boldsymbol{r}_i,\boldsymbol{r}_f,\tilde{\boldsymbol{t}}_i,\tilde{\boldsymbol{t}}_f) = \gamma_e\delta(\boldsymbol{r}_i-\boldsymbol{r}_f)\delta(t_i-t_f) + \int \Gamma'(\boldsymbol{r}_i,\boldsymbol{r},\tilde{\boldsymbol{t}}_i,\tilde{\boldsymbol{t}})K'(\boldsymbol{r},\boldsymbol{r}_f,\tilde{\boldsymbol{t}},\tilde{\boldsymbol{t}}_f)\,d\boldsymbol{r}\,dt$$

whose kernel K' is given by

$$K'(\boldsymbol{r},\boldsymbol{r}_f,\tilde{\boldsymbol{t}},\tilde{\boldsymbol{t}}_f) = \gamma_e\overline{G}^R(\boldsymbol{r},\boldsymbol{r}_f,t,t_f)\overline{G}^A(\boldsymbol{r}_f,\boldsymbol{r},\bar{t}_f,\bar{t}). \tag{6.239}$$

In the presence of external potentials $\boldsymbol{A}(\boldsymbol{r},t)$ and $V(\boldsymbol{r},t)$, this kernel picks up the phase

$$\boxed{\Phi(\boldsymbol{r},\boldsymbol{r}_f,\tilde{\boldsymbol{t}},\tilde{\boldsymbol{t}}_f) = \phi(\boldsymbol{r},\boldsymbol{r}_f,t,t_f) - \phi(\boldsymbol{r},\boldsymbol{r}_f,\bar{t},\bar{t}_f) \qquad \textit{Cooperon}} \tag{6.240}$$

with $\tilde{\boldsymbol{t}} = (t,\bar{t})$. For the direct sequence, the phase is given by

$$\phi(\boldsymbol{r},\boldsymbol{r}_f,t,t_f) = \frac{e}{\hbar}\int_t^{t_f} [V(\boldsymbol{r}(\tau),\tau) - \dot{\boldsymbol{r}}(\tau)\cdot\boldsymbol{A}(\boldsymbol{r}(\tau),\tau)]\,d\tau \tag{6.241}$$

while for the time-reversed sequence, the direction of time is reversed as is the velocity $\dot{\boldsymbol{r}}(\bar{\tau}) = -\dot{\boldsymbol{r}}(\tau)$:

$$\phi(\boldsymbol{r},\boldsymbol{r}_f,\bar{t},\bar{t}_f) = \frac{e}{\hbar}\int_t^{t_f} [V(\boldsymbol{r}(\tau),\bar{\tau}) + \dot{\boldsymbol{r}}(\tau)\cdot\boldsymbol{A}(\boldsymbol{r}(\tau),\bar{\tau})]\,d\tau, \tag{6.242}$$

so that

$$\Phi(\boldsymbol{r},\boldsymbol{r}_f,\tilde{\boldsymbol{t}},\tilde{\boldsymbol{t}}_f) = \frac{e}{\hbar}\int_t^{t_f} \Big([V(\boldsymbol{r}(\tau),\tau) - V(\boldsymbol{r}(\tau),\bar{\tau})] - \dot{\boldsymbol{r}}(\tau)\cdot[\boldsymbol{A}(\boldsymbol{r}(\tau),\tau) + \boldsymbol{A}(\boldsymbol{r}(\tau),\bar{\tau})]\Big)\,d\tau. \tag{6.243}$$

[34] The potentials must be slowly varying in order to satisfy the eikonal approximation for which only the phase of the Green functions is modified.

[35] To be rigorous, the first term in the integral equation is the two-impurity term, since the single impurity term is already included in the Diffuson. The arbitrary choice of including the single impurity term here does not affect the non-local iterative structure of Γ'. See also footnotes 13 and 17, pages 107 and 118.

The fact that the phase $\Phi(\boldsymbol{r},\boldsymbol{r}_f,\tilde{t},\tilde{t}_f)$ depends not only on t and t_f, but also on the time t_i is a consequence of the *non-local temporal structure* of the Cooperon in a time-dependent field. It is not possible to express the Cooperon in terms of an equation that is local in time, as was done for the Diffuson. Nonetheless, for time-independent fields, the term related to the scalar potential vanishes, while that of the vector potential gives a factor 2 in accord with the results of section 6.2. The phase is in this case a function of the difference $t - t_f$.

By analogy to the time-independent problem, in the limit of slow variations of Γ' and the phase Φ, we may expand $\Gamma'(\boldsymbol{r}_i,\boldsymbol{r},t_i,t)e^{i\Phi(\boldsymbol{r},\boldsymbol{r}_f,t,t_f)}$ for $(\boldsymbol{r},t)$ in the vicinity of $(\boldsymbol{r}_f,t_f)$. This expansion leads to a differential equation for Γ'. Then, using (4.48), we obtain a differential equation for $P_c(\boldsymbol{r}_i,\boldsymbol{r}_f,t_i,t_f)$:

$$\left(\frac{\partial}{\partial t_f} - D\left[\nabla_{\boldsymbol{r}_f} + i\frac{e}{\hbar}\left(\boldsymbol{A}(\boldsymbol{r}_f,t_i) + \boldsymbol{A}(\boldsymbol{r}_f,t_f)\right)\right]^2 - i\frac{e}{\hbar}[V(\boldsymbol{r}_f,t_f) - V(\boldsymbol{r}_f,t_i)]\right) P_c(\boldsymbol{r}_i,\boldsymbol{r}_f,t_i,t_f) = \delta(\boldsymbol{r}_f - \boldsymbol{r}_i)\delta(t_f - t_i). \tag{6.244}$$

In this case, the probability retains the form (6.218):[36]

$$P_c(\boldsymbol{r}_i,\boldsymbol{r}_f,t_i,t_f) = \int_{\boldsymbol{r}(t_i)=\boldsymbol{r}_i}^{\boldsymbol{r}(t_f)=\boldsymbol{r}_f} \mathcal{D}\{\boldsymbol{r}\}\, \exp\left(-\int_{t_i}^{t_f} \mathcal{L}(\tau)\, d\tau\right). \tag{6.245}$$

However, the Lagrangian $\mathcal{L}(\boldsymbol{r},\dot{\boldsymbol{r}},\tau)$ is now given by [126]:

$$\mathcal{L}(\boldsymbol{r}(\tau),\dot{\boldsymbol{r}}(\tau),\tau) = \frac{\dot{\boldsymbol{r}}^2}{4D} + i\frac{e}{\hbar}\dot{\boldsymbol{r}}\cdot[\boldsymbol{A}(\boldsymbol{r}(\tau),\tau) + \boldsymbol{A}(\boldsymbol{r}(\tau),\bar{\tau})] - i\frac{e}{\hbar}\left[V(\boldsymbol{r}(\tau),\tau) - V(\boldsymbol{r}(\tau),\bar{\tau})\right]. \tag{6.246}$$

The probability is thus of the form

$$\begin{aligned} P_c(\boldsymbol{r}_i,\boldsymbol{r}_f,t_i,t_f) &= \int \mathcal{D}\{\boldsymbol{r}\}\exp\left(-\int_{t_i}^{t_f} \frac{\dot{\boldsymbol{r}}^2(\tau)}{4D}d\tau\right)\left\langle e^{i\Phi(t_i,t_f)}\right\rangle \\ &= P_c^{(0)}(\boldsymbol{r}_i,\boldsymbol{r}_f,t_i,t_f)\left\langle e^{i\Phi(t_i,t_f)}\right\rangle_C \end{aligned} \tag{6.247}$$

where $\left\langle e^{i\Phi(t_i,t_f)}\right\rangle_C$ is the average of the phase factor over the diffusive trajectories. For potentials $\boldsymbol{A}(\boldsymbol{r})$ and $V(\boldsymbol{r})$ independent of time, we recover the probability $P_c(\boldsymbol{r},\boldsymbol{r}',t_i,t_f)$ given by (6.218, 6.219).

[36] Recall that the solution of the differential equation

$$\left[\frac{\partial}{\partial t} - D\left(\nabla_{\boldsymbol{r}'} - i\boldsymbol{a}\right)^2 + U(\boldsymbol{r}')\right] F(\boldsymbol{r},\boldsymbol{r}',0,t) = \delta(\boldsymbol{r}-\boldsymbol{r}')\,\delta(t)$$

may be expressed as a functional integral

$$F(\boldsymbol{r},\boldsymbol{r}',0,t) = \int \mathcal{D}\{\boldsymbol{r}\}\exp\left(-\int_0^t \left[\frac{\dot{\boldsymbol{r}}^2(\tau)}{4D} - i\dot{\boldsymbol{r}}\cdot\boldsymbol{a}(\boldsymbol{r}) + U(\boldsymbol{r})\right]d\tau\right).$$

Appendix A6.4: Spin-orbit coupling and magnetic impurities, a heuristic point of view

In section 6.5, we calculated the effect of spin-orbit scattering and magnetic impurities by iterating the tensor $\gamma_e b_{\alpha\beta,\gamma\delta}$ which describes an elementary spin-dependent scattering process. Here we give a somewhat different, less systematic, derivation which has the advantage of elegantly treating the dephasing inherent to the spin rotation due to spin-orbit or magnetic impurities; we base this discussion on references [126, 155].

A6.4.1 Spin-orbit coupling

Consider the evolution of an initial spin state $|s_0\rangle$. We denote by $|s_n\rangle$ the state of the spin after the nth collision. The rotation of the spin due to spin-orbit coupling is described by the rotation operator R_n

$$|s_{n+1}\rangle = R_n|s_n\rangle \tag{6.248}$$

where $R_n = e^{-i\frac{\alpha}{2}\hat{\boldsymbol{u}}\cdot\vec{\sigma}}$. The angle α describes the rotation about an axis defined by the unit vector $\hat{\boldsymbol{u}}$.

After a sequence of N collisions $\mathcal{C}_N = \{\boldsymbol{r}_1, \boldsymbol{r}_2, \ldots, \boldsymbol{r}_N\}$, the spin at time $t = N\tau_e$ is given by

$$|s_t\rangle = \prod_{n=1}^{N} R_n|s_0\rangle = \prod_{n=1}^{N} e^{-i\frac{\alpha}{2}\hat{\boldsymbol{u}}_n.\vec{\sigma}_n}|s_0\rangle. \tag{6.249}$$

In the continuum limit, we may write the state of the spin at time t in the form

$$|s_t\rangle = R_t|s_0\rangle \tag{6.250}$$

where the operator R_t which describes the rotation of the spin during the sequence of collisions is given by[37]

$$R_t = T\exp\left(-i\int_0^t dt'\,\boldsymbol{b}_{t'}\cdot\vec{\sigma}_{t'}\right), \tag{6.251}$$

where the non-commutativity of the Pauli matrices necessitates use of the time-ordered product T. The vector $\boldsymbol{b}_t$ is proportional to the amplitude of the spin-orbit coupling at time t [126]:

$$\boldsymbol{b}_t = \frac{1}{4mc^2}(\nabla v(\boldsymbol{r}_t) \times \dot{\boldsymbol{r}}_t).$$

[37] It is not obvious how to write the integral of a spin operator; this is discussed in [126].

To describe the propagation of an electron which remains in the spin state $|s_0\rangle$, we must consider the matrix element

$$\langle s_0|R_t|s_0\rangle = \langle s_0|T \exp\left(-i\int_0^t dt'\, \boldsymbol{b}_{t'} \cdot \vec{\sigma}_{t'}\right)|s_0\rangle.$$

The average over rotation directions, which we assume to be independent random variables with a Gaussian distribution, is

$$\langle s_0|\langle R_t\rangle|s_0\rangle = \langle s_0|\exp\left(-\frac{1}{2}\left\langle\left[\int_0^t dt'\, \boldsymbol{b}_{t'} \cdot \vec{\sigma}_{t'}\right]^2\right\rangle\right)|s_0\rangle. \qquad (6.252)$$

The argument of the exponential involves the correlation function $\langle b_{\alpha t} b_{\beta t'}\rangle$. Assuming the spin rotations at each collision are uncorrelated, we have

$$\langle b_{t\alpha} b_{t'\beta}\rangle = a_{so}\, \delta_{\alpha\beta}\delta(t-t'), \qquad (6.253)$$

where the parameter a_{so} gives the amplitude of the spin-orbit coupling. Thus, from (6.252), we have

$$\langle s_0|\langle R_t\rangle|s_0\rangle = \langle s_0|e^{-\frac{1}{2}a_{so}\vec{\sigma}^2 t}|s_0\rangle = e^{-\frac{3}{2}a_{so}t} \qquad (6.254)$$

since for spin 1/2, $\vec{\sigma}^2$ has eigenvalue 3. This expression, which weights the probability amplitude of a sequence of duration t, is of the form $e^{-t/2\tau_{so}}$ (that is to say, the average single-particle Green function $\overline{G}^R(\boldsymbol{r}, \boldsymbol{r}', t)$ is multiplied by $e^{-t/2\tau_{so}}$ (see remark on page 83). This defines the spin-orbit collision time τ_{so}

$$\frac{1}{\tau_{so}} = 3\, a_{so}. \qquad (6.255)$$

We now consider *two conjugate time-reversed sequences*. The evolution in one of the two directions is weighted by the factor $\langle s_f|R_t|s_0\rangle$ where $|s_f\rangle$ is the spin of the final state. The evolution in the conjugate direction contains the factor $\langle s_f|R_{-t}|s_0\rangle^* = \langle s_0|R^{\dagger}_{-t}|s_f\rangle$. The contribution of the product of conjugate time-reversed amplitudes to the Cooperon $P_c(\boldsymbol{r}, \boldsymbol{r}', t)$ is thus weighted by the average $\langle Q_{so}(t)\rangle$ over trajectories of the term [126]

$$Q_{so}(t) = \sum_{s_f=\pm} \langle s_0|R^{\dagger}_{-t}|s_f\rangle\langle s_f|R_t|s_0\rangle. \qquad (6.256)$$

The spin-orbit coupling $\boldsymbol{b}_t$ is proportional to the electron velocity and thus changes sign under time reversal, namely,

$$\boldsymbol{b}_{-t} = -\boldsymbol{b}_t. \qquad (6.257)$$

This result is in accordance with the reciprocity theorem which states that the spin must change sign between the two paired sequences (see remark page 211). The change in sign of the coupling $\boldsymbol{b}_t$ leads to

$$R^{\dagger}_{-t} = R_t. \qquad (6.258)$$

Thus, $Q_{so}(t)$ becomes

$$Q_{so}(t) = \sum_{s_f=\pm} \langle s_0 | T \exp\left(-i\int_0^t dt'\, \boldsymbol{b}_{t'} \cdot \vec{\sigma}_{t'}\right) | s_f \rangle \langle s_f | T \exp\left(-i\int_0^t dt'\, \boldsymbol{b}_{t'} \cdot \vec{\sigma}_{t'}\right) | s_0 \rangle .$$

The product of matrix elements can be replaced by a single matrix element written in the product space, so that

$$Q_{so}(t) = \sum_{s_f=\pm} \langle s_f^a s_0^b | T \exp\left(-i\int_0^t dt'\, \boldsymbol{b}_{t'} \cdot (\vec{\sigma}_{t'}^a + \vec{\sigma}_{t'}^b)\right) | s_0^a s_f^b \rangle ,$$

where $|s_0^a s_f^b\rangle$ denotes the tensorial product $|s_0^a\rangle \otimes |s_f^b\rangle$. We assume $\boldsymbol{b}_t$ to be Gaussian distributed so that the average $\langle Q_{so}(t)\rangle$ may be written as

$$\langle Q_{so}(t)\rangle = \sum_{s_f=\pm} \langle s_f^a s_0^b | \exp\left(-\frac{1}{2}\left\langle\left[\int_0^t dt'\, \boldsymbol{b}_{t'} \cdot (\vec{\sigma}_{t'}^a + \vec{\sigma}_{t'}^b)\right]^2\right\rangle\right) | s_0^a s_f^b \rangle .$$

The total spin $(\vec{\sigma}_t^a + \vec{\sigma}_t^b)^2$ is conserved and time independent. Then, making use of (6.253) for the correlator $\langle b_{t\alpha} b_{t'\beta}\rangle$, the time integral is immediate and leads to

$$\langle Q_{so}(t)\rangle = \sum_{s_f=\pm} \langle s_f^a s_0^b | \exp\left(-\frac{1}{2} a_{so}\, t\, (\vec{\sigma}^a + \vec{\sigma}^b)^2\right) | s_0^a s_f^b \rangle \qquad (6.259)$$

which needs to be compared to (6.254) which sets the spin-orbit scattering time. To this purpose, we introduce the singlet state $|S\rangle$ and the three triplet states $|T_\alpha\rangle$, which are eigenstates of the total spin $\vec{\sigma}^a + \vec{\sigma}^b$. Matrix elements of $(\vec{\sigma}^a + \vec{\sigma}^b)^2$ written in this basis are given by

$$\begin{aligned} \langle S | (\vec{\sigma}^a + \vec{\sigma}^b)^2 | S \rangle &= 0 \\ \langle T_\alpha | (\vec{\sigma}^a + \vec{\sigma}^b)^2 | T_\alpha \rangle &= 8. \end{aligned} \qquad (6.260)$$

Inserting the closure relation $|S\rangle\langle S| + \sum_\alpha |T_\alpha\rangle\langle T_\alpha| = 1$, and noting that

$$\begin{aligned} \sum_{s_f,\alpha} \langle s_0 s_f | T_\alpha \rangle \langle T_\alpha | s_f s_0 \rangle &= \tfrac{3}{2} \\ \sum_{s_f} \langle s_0 s_f | S \rangle \langle S | s_f s_0 \rangle &= -\tfrac{1}{2}, \end{aligned} \qquad (6.261)$$

we finally obtain the relation (6.139)

$$\langle Q_{so}(t)\rangle = \frac{1}{2}(3\, e^{-4t/3\tau_{so}} - 1). \qquad (6.262)$$

The singlet state is unaffected by spin-orbit coupling. The probability of return to the origin $P_c(\boldsymbol{r}, \boldsymbol{r}, t)$ is weighted by the factor $\langle Q_{so}(t)\rangle$. We notice that there is no dephasing unless the two spins $\vec{\sigma}^a$ and $\vec{\sigma}^b$ are in a triplet state.

A6.4.2 Magnetic impurities

Consider the evolution of the electron spin along a given multiple scattering trajectory:

$$|s_t\rangle = R_t|s_0\rangle \tag{6.263}$$

where the operator R_t now takes the form

$$R_t = T\exp\left(-i\int_0^t dt'\,\boldsymbol{h}_{t'}\cdot\vec{\sigma}_{t'}\right). \tag{6.264}$$

The magnetic field that the electron spin feels while moving along its trajectory is $\boldsymbol{h}_t = J\sum_j \delta(\boldsymbol{r}-\boldsymbol{r}_j)\boldsymbol{S}_j$. For a given spin state $|s_0\rangle$, electron propagation is weighted by the matrix element $\langle s_0|R_t|s_0\rangle$. By averaging both upon scattering sequences and spin configurations $\boldsymbol{S}_j$ assumed to be independent random variables with Gaussian distribution, we obtain

$$\langle s_0|\langle R_t\rangle|s_0\rangle = \langle s_0|\exp\left(-\frac{1}{2}\left\langle\left[\int_0^t dt'\,\boldsymbol{h}_{t'}\cdot\vec{\sigma}_{t'}\right]^2\right\rangle\right)|s_0\rangle.$$

The correlation function $\langle h_{\alpha t}h_{\beta t'}\rangle$ that appears in the argument of the exponential is of the form

$$\langle h_{t\alpha}h_{t'\beta}\rangle = a_m\,\delta_{\alpha\beta}\delta(t-t') \tag{6.265}$$

where a_m is the strength of the coupling with impurities. Assuming (6.265) amounts to considering that angles induced after each scattering of the electronic spin by a magnetic impurity are uncorrelated. Since $\vec{\sigma}^2 = 3$, we obtain:

$$\langle s_0|\langle R_t\rangle|s_0\rangle = \langle s_0|e^{-\frac{1}{2}a_m\vec{\sigma}^2 t}|s_0\rangle = e^{-\frac{3}{2}a_m t} \tag{6.266}$$

which is of the form $e^{-t/2\tau_m}$. This expression defines the magnetic scattering time τ_m

$$\frac{1}{\tau_m} = 3\,a_m. \tag{6.267}$$

As for spin-orbit coupling, we see that in the presence of magnetic impurities the average Green function is also modified and involves, in addition to the elastic mean free time τ_e, the relaxation time τ_m.

In order to describe how the Cooperon is modified as a result of a coupling to magnetic impurities, we consider two sequences of multiple scattering that are time reversed one from the other. The evolution along one of the trajectories is weighted by the factor $\langle s_f|R_t|s_0\rangle$ whereas the evolution along the second trajectory involves $\langle s_f|R_{-t}|s_0\rangle^* = \langle s_0|R^\dagger_{-t}|s_f\rangle$ where $|s_f\rangle$ is the final spin state. As a whole, the contribution to the Cooperon (i.e., a product of two trajectories) is weighted by the term [126]

$$Q_m(t) = \sum_{s_f=\pm}\langle s_0|R^\dagger_{-t}|s_f\rangle\langle s_f|R_t|s_0\rangle.$$

The field $\boldsymbol{h}_t$ is independent of the electron velocity and therefore it does not change sign under time reversal:

$$\boldsymbol{h}_{-t} = \boldsymbol{h}_t \tag{6.268}$$

which states that reciprocity is broken since

$$R_{-t}^{\dagger} = R_t^* \tag{6.269}$$

(see remark on page 212 and compare with 6.258).

From that point on, the calculation of $Q_m(t)$ is analogous to the case of spin-orbit coupling and $Q_m(t)$ can be rewritten as

$$Q_m(t) = \sum_{s_f=\pm} \langle s_f^a s_0^b | T \exp\left(-i \int_0^t dt' \, \boldsymbol{h}_{t'} \cdot (\vec{\sigma}_{t'}^a - \vec{\sigma}_{t'}^b)\right) | s_0^a s_f^b \rangle .$$

Assuming a Gaussian distribution of $\boldsymbol{h}_t$ we have

$$\langle Q_m(t) \rangle = \sum_{s_f=\pm} \langle s_f^a s_0^b | \exp\left(-\frac{1}{2}\left\langle \left[\int_0^t dt' \, \boldsymbol{h}_{t'} \cdot (\vec{\sigma}_{t'}^a - \vec{\sigma}_{t'}^b)^2\right]\right\rangle\right) | s_0^a s_f^b \rangle .$$

The quantity $(\vec{\sigma}_t^a - \vec{\sigma}_t^b)^2$ is conserved so that using (6.265) for the correlation function $\langle h_{\alpha t} h_{\beta t} \rangle$, the time integral becomes straightforward and we obtain

$$\langle Q_m(t) \rangle = \sum_{s_f=\pm} \langle s_f^a s_0^b | \exp\left(-\frac{1}{2} a_m \, t \, (\vec{\sigma}^a - \vec{\sigma}^b)^2\right) | s_0^a s_f^b \rangle .$$

This expression has to be compared to (6.266) which defines the magnetic scattering time. Using the identity

$$(\vec{\sigma}^a - \vec{\sigma}^b)^2 = 12 - (\vec{\sigma}^a + \vec{\sigma}^b)^2$$

and inserting the closure relation $|S\rangle\langle S| + \sum_\alpha |T_\alpha\rangle\langle T_\alpha| = 1$ where $|S\rangle$ is the singlet state and $|T_\alpha\rangle$ the three triplet states, we obtain immediately from (6.260)

$$\begin{aligned} \langle S | (\vec{\sigma}^a - \vec{\sigma}^b)^2 | S \rangle &= 12 \\ \langle T_\alpha | (\vec{\sigma}^a - \vec{\sigma}^b)^2 | T_\alpha \rangle &= 4. \end{aligned} \tag{6.270}$$

Finally, making use of relations (6.261), we recover relation (6.140):

$$\langle Q_m(t) \rangle = \frac{1}{2}(3 \, e^{-2t/3\tau_m} - e^{-2t/\tau_m}). \tag{6.271}$$

This result is clearly different from the one obtained for spin-orbit scattering in the preceding section since here both singlet and triplet contributions are modified by scattering on magnetic impurities.

Appendix A6.5: Decoherence in multiple scattering of light by cold atoms

For the specific example of Rayleigh scattering of polarized electromagnetic waves, we have seen that multiple scattering induces two distinct features. First, it leads to a depolarization of the incident beam which equally affects the Diffuson and the Cooperon. Second, an additional phase shift appears in the Cooperon that reduces its contribution compared to that of the Diffuson. This reduction is described by finite phase coherence times and is given by expression (6.176).

We study now the case of a polarized light multiply scattered by atoms which have an internal structure due to Zeeman degeneracy of their energy levels. In section A2.3.3, we considered the resonant scattering of a photon by an atom modelled by a degenerate two-level system. The purpose of this appendix is to show that the averaging over internal atomic degrees of freedom leads to the existence of several phase coherence times [156–159]. These characteristic times affect both Diffuson and Cooperon and they thus play an important role in the mesoscopic physics of multiple scattering of photons by cold atomic gases.[38]

A6.5.1 Scattering amplitude and atomic collision time

We first consider the scattering amplitude associated with a photon initially in a state $\boldsymbol{k}$ of polarization $\hat{\boldsymbol{\varepsilon}}$ towards a final state $\boldsymbol{k}'$ of polarization $\hat{\boldsymbol{\varepsilon}}'$. The ground and excited states of the atom are characterized by their respective angular momenta J and J_e. The two levels are assumed to be degenerate and their magnetic quantum numbers are such that $|m| \leq J$ and $|m_e| \leq J_e$. The dipolar interaction $-\boldsymbol{d} \cdot \boldsymbol{E}$, where $\boldsymbol{E}$ is the electric incident field and $\boldsymbol{d}$ is the dipole operator, induces transitions between those states. The scattering amplitude $f(\hat{\boldsymbol{\varepsilon}}\, m_i, \hat{\boldsymbol{\varepsilon}}'\, m_f)$ for near resonant ($\omega \simeq \omega_0$) elastic scattering at energy ω is given by (2.163) and (2.150):

$$f(\hat{\boldsymbol{\varepsilon}}\, m_i, \hat{\boldsymbol{\varepsilon}}' m_f) = -\frac{\omega_0^2}{4\pi\epsilon_0 \hbar c^2} \frac{1}{\delta + i\Gamma/2} \sum_{m_e} \langle J m_f | \boldsymbol{d} \cdot \hat{\boldsymbol{\varepsilon}}'^* | J_e m_e \rangle \langle J_e m_e | \boldsymbol{d} \cdot \hat{\boldsymbol{\varepsilon}} | J m_i \rangle, \quad (6.272)$$

where $\delta = \omega - \omega_0$ is the frequency detuning. Using relation (2.154), we rewrite the scattering amplitude as

$$f(\hat{\boldsymbol{\varepsilon}}\, m_i, \hat{\boldsymbol{\varepsilon}}' m_f) = -\frac{3\lambda}{4\pi} \frac{\Gamma/2}{\delta + i\Gamma/2} \sum_{m_e} \langle J m_f | \tilde{\boldsymbol{d}} \cdot \hat{\boldsymbol{\varepsilon}}'^* | J_e m_e \rangle \langle J_e m_e | \tilde{\boldsymbol{d}} \cdot \hat{\boldsymbol{\varepsilon}} | J m_i \rangle, \quad (6.273)$$

where $\tilde{\boldsymbol{d}} = \boldsymbol{d}/d$ is the dimensionless dipole operator and $d = \langle J_e || \boldsymbol{d} || J \rangle / \sqrt{2J_e + 1}$ is defined on page 67, in terms of the reduced matrix element $\langle J_e || \boldsymbol{d} || J \rangle$. $\lambda = 2\pi c/\omega_0$ is the wavelength of the resonant transition. The scattering amplitude depends on $\boldsymbol{k}$ and $\boldsymbol{k}'$ through polarizations since $\boldsymbol{k} \cdot \hat{\boldsymbol{\varepsilon}} = \boldsymbol{k}' \cdot \hat{\boldsymbol{\varepsilon}}' = 0$. Using (2.101), we define the effective interaction

[38] A gas of cold atoms, for which the Doppler effect due to the velocity of atoms is negligible, corresponds to the regime that we wish to study.

Figure 6.15 Schematic representation of (a) the amplitude $v(\hat{\boldsymbol{\varepsilon}}\, m_i, \hat{\boldsymbol{\varepsilon}}' m_f)$ and (b) the tensor $v_{\alpha\beta}$ as a scattering amplitude of a photon from an initial polarization state $\hat{\boldsymbol{e}}_\alpha$ (denoted α) onto the final polarization state $\hat{\boldsymbol{e}}_\beta$ (denoted β). $|Jm_i\rangle$ and $|Jm_f\rangle$ are respectively the initial and final states of the atom.

potential $v = -4\pi f$:

$$v(\hat{\boldsymbol{\varepsilon}}\, m_i, \hat{\boldsymbol{\varepsilon}}' m_f) = v_0 \sum_{m_e} \langle Jm_f | \tilde{\boldsymbol{d}} \cdot \hat{\boldsymbol{\varepsilon}}'^* | J_e m_e \rangle \langle J_e m_e | \tilde{\boldsymbol{d}} \cdot \hat{\boldsymbol{\varepsilon}} | Jm_i \rangle, \tag{6.274}$$

represented in Figure 6.15(a). This potential generalizes the scalar scattering potential in the Edwards model (section 2.2.2). Here v_0 is given by

$$v_0 = 3\lambda \frac{\Gamma/2}{\delta + i\Gamma/2}. \tag{6.275}$$

Atomic collision time and elastic mean free path

In the presence of internal atomic degrees of freedom, the elastic collision time τ_e defined for a scalar wave is changed and we define a new time τ_{at}. The corresponding elastic mean free path $l_{at} = c\tau_{at}$ can be computed directly using its definition (3.4) in terms of the total scattering cross section which is given by (2.165), namely,

$$\frac{1}{l_{at}} = \frac{1}{c\tau_{at}} = n_i \langle \sigma \rangle = A_{JJ_e} \frac{n_i |v_0|^2}{6\pi}, \tag{6.276}$$

where $A_{JJ_e} = \frac{1}{3}(2J_e + 1)/(2J + 1)$. As in the case of Rayleigh scattering, we introduce the quantity γ_{at} defined by:

$$\gamma_{at} = \frac{4\pi}{l_{at}} = \gamma_e \frac{2}{3} A_{JJ_e} = \gamma_{pol} A_{JJ_e} \tag{6.277}$$

which generalizes the expression for $\gamma_{pol} = \frac{2}{3}\gamma_e$ introduced in (6.148). Here, $\gamma_e = n_i |v_0|^2$. We recover the case of Rayleigh scattering (relation 6.149) by considering $(J, J_e) = (0, 1)$ so that $A_{JJ_e} = 1$. The factor $2/3$ results from averaging over polarizations (see section 6.6.1).

A6.5.2 Elementary atomic vertex

From the scattering amplitude, we can now introduce the elementary atomic vertex for the atom–photon near-resonant scattering. This vertex is obtained by pairing two scattering

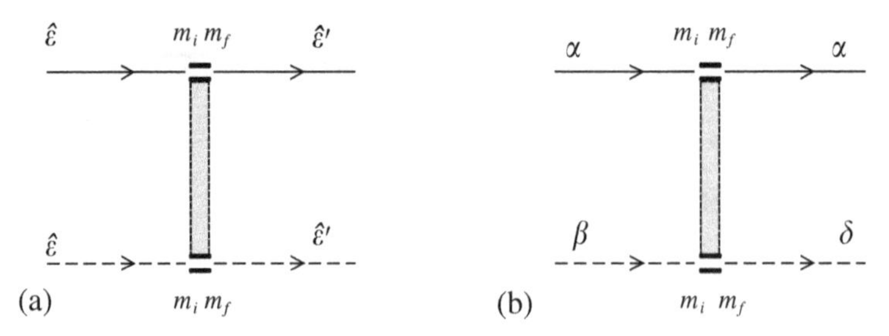

Figure 6.16 (a) Elementary atomic vertex $V_{at}^{(d)}(\hat{\boldsymbol{\varepsilon}}, \hat{\boldsymbol{\varepsilon}}') = V_{at}(\hat{\boldsymbol{\varepsilon}}, \hat{\boldsymbol{\varepsilon}}; \hat{\boldsymbol{\varepsilon}}', \hat{\boldsymbol{\varepsilon}}')$ describing the pairing of two scattering amplitudes of a photon by a single atom as defined in (6.279). (b) Tensor $\mathcal{V}_{\alpha\beta,\gamma\delta}^{(d)}$ defined by equations (6.288, 6.292).

amplitudes. The pairing represented in Figure 6.16(a) defines the elementary atomic vertex $V_{at}(\hat{\boldsymbol{\varepsilon}}, \hat{\boldsymbol{\varepsilon}}; \hat{\boldsymbol{\varepsilon}}', \hat{\boldsymbol{\varepsilon}}')$ which depends on incoming and outgoing polarizations:

$$V_{at}(\hat{\boldsymbol{\varepsilon}}, \hat{\boldsymbol{\varepsilon}}; \hat{\boldsymbol{\varepsilon}}', \hat{\boldsymbol{\varepsilon}}') = n_i \Big\langle v(\hat{\boldsymbol{\varepsilon}} m_i, \hat{\boldsymbol{\varepsilon}}' m_f)\, v^*(\hat{\boldsymbol{\varepsilon}} m_i, \hat{\boldsymbol{\varepsilon}}' m_f) \Big\rangle_{int}, \tag{6.278}$$

where $|Jm_i\rangle$ and $|Jm_f\rangle$ are respectively the initial and final states of the atom. $\langle \cdots \rangle_{int}$ denotes an average over the atomic internal degrees of freedom. For a given total angular momentum J, atoms are generally in a state which is a statistical mixture of Zeeman sublevels. We shall consider here the case where all the Zeeman sublevels have the same statistical weight.[39] Moreover, after scattering, the final state $|Jm_f\rangle$ is not detected, so we must sum over the final states. Thus, the elementary atomic vertex that corresponds to a density n_i of atoms, is given by

$$V_{at}^{(d)}(\hat{\boldsymbol{\varepsilon}}, \hat{\boldsymbol{\varepsilon}}') = V_{at}(\hat{\boldsymbol{\varepsilon}}, \hat{\boldsymbol{\varepsilon}}; \hat{\boldsymbol{\varepsilon}}', \hat{\boldsymbol{\varepsilon}}') = \frac{n_i}{2J+1} \sum_{m_i m_f} v(\hat{\boldsymbol{\varepsilon}} m_i, \hat{\boldsymbol{\varepsilon}}' m_f)\, v^*(\hat{\boldsymbol{\varepsilon}} m_i, \hat{\boldsymbol{\varepsilon}}' m_f), \tag{6.279}$$

where we have used a shorter notation $V_{at}^{(d)}(\hat{\boldsymbol{\varepsilon}}, \hat{\boldsymbol{\varepsilon}}')$ to define the elementary vertex.

Remark: generalized atomic vertex

We limit our study of the photon–atom interaction to the case of the intensity propagation described by the Diffuson and the coherent backscattering described by the Cooperon. In this case, the Diffuson (and the Cooperon) is obtained by pairing trajectories corresponding to the *same* Zeeman initial state (m_i) and to the *same* final state (m_f).

More generally, one may consider pairing trajectories with *different* initial Zeeman states (denoted m_1 and m_2 in Figure 6.17) and different final states (denoted m_3 and m_4), introducing the generalized vertex:

$$n_i \Big\langle v(\hat{\boldsymbol{\varepsilon}}_1 m_1, \hat{\boldsymbol{\varepsilon}}_3 m_3)\, v^*(\hat{\boldsymbol{\varepsilon}}_2 m_2, \hat{\boldsymbol{\varepsilon}}_4 m_4) \Big\rangle_{int}, \tag{6.280}$$

[39] It is nevertheless important to keep in mind that if this is not so, we need to introduce the corresponding density matrix. This may lead to an additional asymmetry between the Diffuson and Cooperon contributions [159].

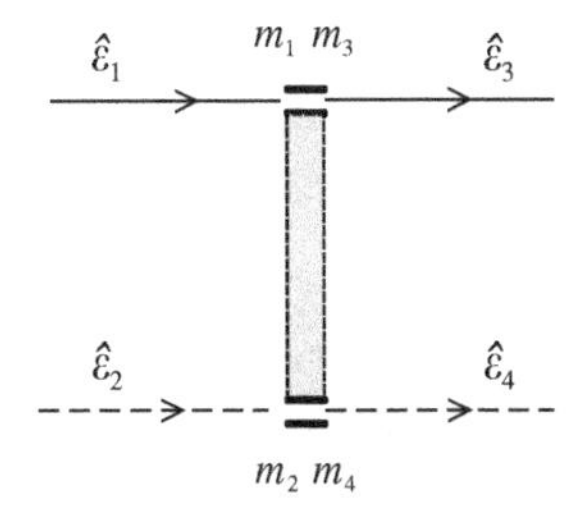

Figure 6.17 Generalized vertex describing the pairing of two scattering amplitudes of a photon by a single atom, as defined in (6.280). Contrary to the vertex (6.278) entering the Diffuson, the four Zeeman states m_j may be different.

where $\langle \cdots \rangle_{int}$ denotes a trace over the internal degrees of freedom $m_j, j \in [1,4]$ with the appropriate density matrix. Indeed, Zeeman quantum numbers change with time as a result of a large variety of processes such as scattering, collisions between atoms, spontaneous emission. Then, Diffusons (or Cooperons) that enter into time correlation functions (see Chapter 9) or into static quantities such as angular correlations of speckle patterns (Chapter 12) are obtained by pairing trajectories that correspond to different Zeeman configurations m_j. The treatment of this generalized vertex and of the corresponding Diffuson is more complicated [159].

In order to construct the Diffuson (and the Cooperon), it will be helpful to expand the polarization in a given basis. In section 6.6 devoted to Rayleigh scattering, a Cartesian basis was used. It turns out that the elementary vertex V_{at} can be calculated very conveniently in the standard (or spherical) basis $\{\hat{\boldsymbol{e}}_q\}$ defined in (2.151).[40]Any vector $\hat{\boldsymbol{X}}$ can be expanded in this standard basis under the form $\hat{\boldsymbol{X}} = \sum_{q=0,\pm 1} X_q \hat{\boldsymbol{e}}_q^* = \sum_{q=0,\pm 1} (-1)^q X_q \hat{\boldsymbol{e}}_{-q}$, so that the scalar product $\tilde{\boldsymbol{d}} \cdot \hat{\boldsymbol{\varepsilon}}$ has the form

$$\tilde{\boldsymbol{d}} \cdot \hat{\boldsymbol{\varepsilon}} = \sum_q (-1)^q \tilde{d}_q \varepsilon_{-q}, \tag{6.283}$$

where $\tilde{d}_q$ and ε_q are the standard components of the reduced dipole operator and the polarization respectively. We can thus write the scattering amplitude in the form

$$\boxed{v(\hat{\boldsymbol{\varepsilon}}\, m_i, \hat{\boldsymbol{\varepsilon}}'m_f) = \sum_{\alpha\beta} \varepsilon_\alpha (\varepsilon'_\beta)^* \, v_{\alpha\beta}} \tag{6.284}$$

40 The vectors of the standard basis verify the relations

$$\hat{\boldsymbol{e}}_q \cdot \hat{\boldsymbol{e}}_{q'}^* = \delta_{q,q'} \qquad \hat{\boldsymbol{e}}_q^* = (-1)^q \hat{\boldsymbol{e}}_{-q} \qquad \hat{\boldsymbol{e}}_q \cdot \hat{\boldsymbol{e}}_{q'} = (-1)^q \delta_{q,-q'}. \tag{6.281}$$

The dipole operator $\boldsymbol{d}$ is Hermitian and its components d_q satisfy

$$d_q = \boldsymbol{d} \cdot \hat{\boldsymbol{e}}_q \qquad d_q^\dagger = (-1)^q d_{-q}. \tag{6.282}$$

where the rank 2 tensor $v_{\alpha\beta}$ is represented in Figure 6.15(b) and is given by

$$v_{\alpha\beta} = v_0(-1)^\alpha \sum_{m_e} \langle Jm_f|\tilde{d}_\beta|J_e m_e\rangle\langle J_e m_e|\tilde{d}_{-\alpha}|Jm_i\rangle. \tag{6.285}$$

Making use of the Wigner–Eckart theorem (15.34), the matrix elements of the dimensionless dipole operator $\tilde{d}_q$ are written in terms of a $3j$-symbol:[41]

$$\langle J_e m_e|\tilde{d}_q|Jm\rangle = (-1)^{J_e - m_e}\sqrt{2J_e+1}\begin{pmatrix} J_e & 1 & J \\ -m_e & q & m \end{pmatrix}. \tag{6.286}$$

This allows us to express the tensor $v_{\alpha\beta}$ as

$$v_{\alpha\beta} = v_0(-1)^j(2J_e+1)\sum_{m_e}\begin{pmatrix} J_e & 1 & J \\ -m_e & -\beta & m_f \end{pmatrix}\begin{pmatrix} J_e & 1 & J \\ -m_e & -\alpha & m_i \end{pmatrix}, \tag{6.287}$$

where $j = J - m_f + J_e - m_e + \alpha$. The summation over the quantum number m_e contains only one term since the $3j$-symbols are non-zero only for $m_e = m_i - \alpha = m_f - \beta$, as a result of conservation of the projection of angular momentum along a given quantization axis.

As for the scattering amplitude in relation (6.284), we can write the vertex V_{at} (6.279) in terms of a rank 4 tensor $\mathcal{V}^{(d)}_{\alpha\beta,\gamma\delta}$ using the standard basis (2.151),

$$V^d_{at}(\hat{\boldsymbol{\varepsilon}}, \hat{\boldsymbol{\varepsilon}}') = V_{at}(\hat{\boldsymbol{\varepsilon}}, \hat{\boldsymbol{\varepsilon}}; \hat{\boldsymbol{\varepsilon}}', \hat{\boldsymbol{\varepsilon}}') = \sum_{\alpha\beta\gamma\delta} \varepsilon_\alpha(\varepsilon_\beta)^*(\varepsilon'_\gamma)^*\varepsilon'_\delta\ \mathcal{V}^{(d)}_{\alpha\beta,\gamma\delta} \tag{6.288}$$

$$\mathcal{V}^{(d)}_{\alpha\beta,\gamma\delta} = \frac{n_i}{2J+1}\sum_{m_i m_f} v_{\alpha\gamma}(m_i, m_f)v^*_{\beta\delta}(m_i, m_f). \tag{6.289}$$

From equation (6.287), we obtain the expression for the tensor $\mathcal{V}^{(d)}_{\alpha\beta,\gamma\delta}$ represented in Figure 6.16(b):

$$\begin{aligned}\mathcal{V}^{(d)}_{\alpha\beta,\gamma\delta} &= n_i|v_0|^2\frac{(2J_e+1)^2}{(2J+1)}(-1)^{\alpha+\beta} \\ &\times \sum_{m_i m_f}\sum_{m_e m'_e}(-1)^{-m_e - m'_e}\begin{pmatrix} J_e & 1 & J \\ -m_e & -\alpha & m_i \end{pmatrix}\begin{pmatrix} J_e & 1 & J \\ -m_e & -\gamma & m_f \end{pmatrix} \\ &\times \begin{pmatrix} J_e & 1 & J \\ -m'_e & -\delta & m_f \end{pmatrix}\begin{pmatrix} J_e & 1 & J \\ -m'_e & -\beta & m_i \end{pmatrix}.\end{aligned} \tag{6.290}$$

[41] In this appendix, we assume that J and J_e are integers. The manipulation of the expressions involving $3j$- and $6j$-symbols is done accordingly.

Notice that the summations over m_e and m'_e reduce to a single term since the $3j$-symbols are non-vanishing only for $m_e = m_i - \alpha = m_f - \gamma$ and $m'_e = m_i - \beta = m_f - \delta$. This implies that

$$m_i - m_f = \alpha - \gamma = \beta - \delta. \tag{6.291}$$

From these constraints, it is straightforward (see Exercise 6.3) to find that the tensor $\mathcal{V}^{(d)}_{\alpha\beta,\gamma\delta}$ contains only terms of the form $\delta_{\alpha\beta}\delta_{\gamma\delta}$, $\delta_{\alpha\gamma}\delta_{\beta\delta}$ or $\delta_{\alpha,-\delta}\delta_{\beta,-\gamma}$, and that it is given by

$$\boxed{\mathcal{V}^{(d)}_{\alpha\beta,\gamma\delta} = n_i|v_0|^2 A_{JJ_e}(w_1\,\delta_{\alpha\gamma}\delta_{\beta\delta} + w_2\,(-1)^{\alpha+\beta}\delta_{\alpha,-\delta}\delta_{\beta,-\gamma} + w_3\,\delta_{\alpha\beta}\delta_{\gamma\delta})} \tag{6.292}$$

where $A_{JJ_e} = \frac{1}{3}(2J_e+1)/(2J+1)$ and the coefficients (w_1, w_2, w_3) are

$$\begin{aligned} w_1 &= \frac{1}{3}(s_0 - s_2) \\ w_2 &= \frac{1}{2}(-s_1 + s_2) \\ w_3 &= \frac{1}{2}(s_1 + s_2), \end{aligned} \tag{6.293}$$

with the definition of the coefficients s_k in terms of a $6j$-symbol:

$$s_k = 3(2J_e+1)\begin{Bmatrix} 1 & 1 & k \\ J & J & J_e \end{Bmatrix}^2. \tag{6.294}$$

These coefficients obey the sum rule $\sum_{k=0}^{2}(2k+1)s_k = 3$, so we have $w_1 + w_2 + 3w_3 = 0$.

Remark

Using another approach that we shall not develop here, it may be shown [157] that the sum entering the expression of the atomic vertex (6.279)

$$V_{at}(\hat{\boldsymbol{\varepsilon}}, \hat{\boldsymbol{\varepsilon}}') = V_{at}(\hat{\boldsymbol{\varepsilon}}, \hat{\boldsymbol{\varepsilon}}; \hat{\boldsymbol{\varepsilon}}', \hat{\boldsymbol{\varepsilon}}') = \frac{n_i|v_0|^2}{2J+1}\sum_{m_i m_f}\sum_{m_e m_{e'}} \langle Jm_f|\tilde{\boldsymbol{d}}\cdot\hat{\boldsymbol{\varepsilon}}'^*|J_e m_e\rangle\langle J_e m_e|\tilde{\boldsymbol{d}}\cdot\hat{\boldsymbol{\varepsilon}}|Jm_i\rangle$$
$$\times\ \langle Jm_f|\tilde{\boldsymbol{d}}\cdot\hat{\boldsymbol{\varepsilon}}'^*|J_e m_{e'}\rangle^*\langle J_e m_{e'}|\tilde{\boldsymbol{d}}\cdot\hat{\boldsymbol{\varepsilon}}|J_i m_i\rangle^* \tag{6.295}$$

is equal to:

$$V^{(d)}_{at}(\hat{\boldsymbol{\varepsilon}}, \hat{\boldsymbol{\varepsilon}}') = n_i|v_0|^2 A_{JJ_e}\left(w_1|\hat{\boldsymbol{\varepsilon}}'\cdot\hat{\boldsymbol{\varepsilon}}^*|^2 + w_2|\hat{\boldsymbol{\varepsilon}}'\cdot\hat{\boldsymbol{\varepsilon}}|^2 + w_3\right), \tag{6.296}$$

which is the expression (2.167) for the differential cross section. Here we obtain this result by inserting (6.292) into (6.288). The total cross section is then obtained by integrating this differential cross section using the prescription (2.159).

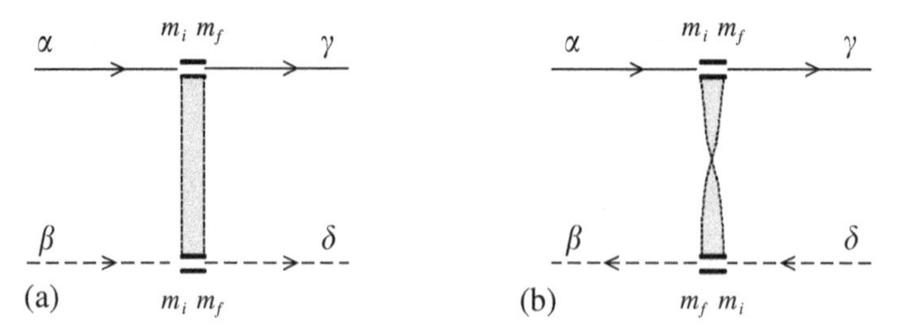

Figure 6.18 Representation of the tensors $\mathcal{V}^{(d,c)}_{\alpha\beta,\gamma\delta}$. They are related through (6.299).

Exercise 6.3 Derive (6.292) from (6.290).
Hint: using the identity (15.37) and the relation (15.38), the quadruple sum in equation (6.290) becomes

$$\sum_{kq}(2k+1)\left\{\begin{matrix} 1 & 1 & k \\ J & J & J_e \end{matrix}\right\}^2 \left(\begin{matrix} 1 & 1 & k \\ -\alpha & \gamma & q \end{matrix}\right)\left(\begin{matrix} 1 & 1 & k \\ -\beta & \delta & q \end{matrix}\right). \tag{6.297}$$

Using relation (15.39) of the formulary, we obtain (6.292).

Remark: elementary atomic vertex of the Cooperon
The elementary atomic vertex for the Cooperon is easily deduced from expression (6.289) obtained for the Diffuson. Here, β and δ are reversed (Figure 6.18(b)), so that the tensor $\mathcal{V}^{(c)}_{\alpha\beta,\gamma\delta}$ for the Cooperon is

$$\mathcal{V}^{(c)}_{\alpha\beta,\gamma\delta} = \frac{n_i}{2J+1}\sum_{m_i m_f} v_{\alpha\gamma}(m_i,m_f)v^*_{\delta\beta}(m_i,m_f), \tag{6.298}$$

instead of (6.289) for the Diffuson. Using (6.290), it is straightforward to check that the correspondence between the two tensors is

$$\mathcal{V}^{(c)}_{\alpha\beta,\gamma\delta} = (-1)^{\beta+\delta}\,\mathcal{V}^{(d)}_{\alpha-\delta,\gamma-\beta}. \tag{6.299}$$

The Cooperon tensor is obtained from the Diffuson tensor by simply exchanging w_2 and w_3 in equation (6.292).

A6.5.3 Structure factor

Integral equation for the Diffuson

The structure factor of either the Diffuson or the Cooperon is obtained from iteration of the elementary vertex. This vertex has a tensorial structure and, likewise, the structure factor acquires a tensorial structure. When written in a standard basis, we denote it by $\Gamma_{\alpha\beta,\gamma\delta}(\boldsymbol{q})$.

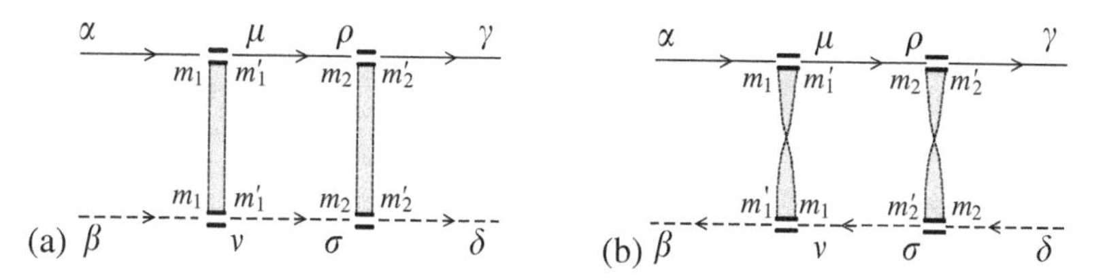

Figure 6.19 Representation of double scattering for (a) the Diffuson and (b) the Cooperon.

In order to understand how to build the iteration properly, we consider first the case of the double scattering Diffuson represented in Figure 6.19(a).

The two incoming amplitudes have the same wave vector but may have different polarization components (α, β). After being scattered by an atom which makes a transition between states m_1 and m_1', the two outgoing amplitudes propagate with another wave vector, random in direction but identical for both, and with two different polarization components (μ, ν). The further propagation of the photon in between two scattering events is characterized by its retarded propagator which is a rank 2 tensor of the form (see relation 3.50)

$$D^R_{\mu\rho}(\boldsymbol{k}, k_0) = M_{\mu\rho} G^R(\boldsymbol{k}, k_0) = \frac{\delta_{\mu,\rho} - (-1)^\rho \hat{k}_\mu \hat{k}_{-\rho}}{k_0 - k^2 + i0}, \tag{6.300}$$

where $\hat{\boldsymbol{k}} = \boldsymbol{k}/k_0$ and the factor

$$M_{\mu\rho} = \delta_{\mu,\rho} - (-1)^\rho \hat{k}_\mu \hat{k}_{-\rho}, \tag{6.301}$$

already defined in (6.144) in Cartesian coordinates, accounts for the transversality condition. The second photon amplitude involves the propagator $D^A_{\nu\sigma}(\boldsymbol{k}, \omega) = M_{\sigma\nu} G^A(\boldsymbol{k}, \omega)$. Since the outgoing wave vector $\boldsymbol{k}$ is random, the propagation of the intensity in between two successive scattering events is given by the average product $\langle D^R_{\mu\nu} D^A_{\rho\sigma} \rangle = w(\boldsymbol{q})\, b_{\mu\rho,\nu\sigma}$, where the quantity $w(\boldsymbol{q}) = P_0(\boldsymbol{q})/\tau_e$ was introduced in Appendix A4.1 and, in the diffusion approximation, is given by (see equation 6.104):

$$w(\boldsymbol{q}) \simeq \frac{\gamma_e}{\gamma_{at}} (1 - Dq^2 \tau_{at}), \tag{6.302}$$

where $D = c l_{at}/3$ is the diffusion coefficient of the light wave in three dimensions. This function describes the scalar part of the intensity propagator. We define the tensor

$$b_{\alpha\beta,\gamma\delta} = \langle M_{\alpha\gamma} M_{\delta\beta} \rangle \tag{6.303}$$

in which $\langle \cdots \rangle$ denotes an average over photon wave vectors. This tensor is identical to the one given in equations (6.151, 6.154) written in Cartesian coordinates, whereas it is written here in a standard basis and it takes the form[42]

$$\boxed{b_{\alpha\beta,\gamma\delta} = \frac{1}{15}(6\,\delta_{\alpha\gamma}\delta_{\beta\delta} + (-1)^{\alpha+\beta}\delta_{\alpha,-\delta}\delta_{\beta,-\gamma} + \delta_{\alpha\beta}\delta_{\gamma\delta})} \tag{6.305}$$

instead of (6.154). Finally, the double scattering contribution to the Diffuson can be written in the form

$$\mathcal{V}^{(d)}_{\alpha\beta,\mu\nu}\, b_{\mu\nu,\rho\sigma}\, \mathcal{V}^{(d)}_{\rho\sigma,\gamma\delta} \tag{6.306}$$

which allows us to define a suitable matrix multiplication law and to derive an iteration equation for the structure factor $\Gamma^{(d)}_{\alpha\beta,\gamma\delta}(\boldsymbol{q})$ of the Diffuson:

$$\Gamma^{(d)}_{\alpha\beta,\gamma\delta}(\boldsymbol{q}) = \gamma_e\, \mathcal{V}^{(d)}_{\alpha\beta,\gamma\delta} + w(\boldsymbol{q}) \sum_{\mu\nu\rho\sigma} \Gamma^{(d)}_{\alpha\beta,\mu\nu}(\boldsymbol{q})\, b^{(d)}_{\mu\nu,\rho\sigma}\, \mathcal{V}^{(d)}_{\rho\sigma,\gamma\delta}. \tag{6.307}$$

Notice that the first term in this iteration involves only the elementary atomic vertex and not the tensor b, since the latter accounts for the photon propagation between two scattering events.[43]

Integral equation for the Cooperon

The iteration equation for the Cooperon is derived in a similar way. The double scattering case, represented in Figure 6.19(b), is deduced from the Diffuson by reversing the lower amplitude. In order to build the iteration law for the Cooperon, we see that we are led to use the elementary atomic vertex $\mathcal{V}^{(c)}_{\alpha\beta,\gamma\delta}$ for the Cooperon defined in (6.299). The iteration equation for the Diffuson and the Cooperon can then be cast in the single form and it is represented in Figure 6.20,

$$\Gamma^{(d,c)}_{\alpha\beta,\gamma\delta}(\boldsymbol{q}) = \gamma_e\, \mathcal{V}^{(d,c)}_{\alpha\beta,\gamma\delta} + w(\boldsymbol{q}) \sum_{\mu\nu\rho\sigma} \Gamma^{(d,c)}_{\alpha\beta,\mu\nu}(\boldsymbol{q})\, b_{\mu\nu,\rho\sigma}\, \mathcal{V}^{(d,c)}_{\rho\sigma,\gamma\delta}. \tag{6.308}$$

Notice that the tensor $b_{\alpha\beta,\gamma\delta}$ remains unchanged while considering the Diffuson or the Cooperon, a property that was used in section 6.6.

Diagonalization

To proceed further and calculate the structure factors $\Gamma^{(d,c)}_{\alpha\beta,\gamma\delta}$, we notice that the kind of multiplication that shows up in the right hand side of the iteration equation (6.308) suggests

[42] In the standard basis, the following angular averages are given by

$$\begin{aligned} (-1)^{\alpha}\langle \hat{k}_\alpha \hat{k}_\beta\rangle &= \tfrac{1}{3}\delta_{\alpha\beta} \\ (-1)^{\alpha+\delta}\langle \hat{k}_{-\alpha}\hat{k}_\beta\hat{k}_\gamma\hat{k}_{-\delta}\rangle &= \tfrac{1}{15}[\delta_{\alpha\gamma}\delta_{\beta\delta} + (-1)^{\alpha+\beta}\delta_{\alpha,-\delta}\delta_{\beta,-\gamma} + \delta_{\alpha\beta}\delta_{\gamma\delta}] \end{aligned} \tag{6.304}$$

instead of (6.153) in a Cartesian basis.

[43] This might nevertheless seem at odds with the iteration equation (6.156) for the structure factor of an electromagnetic wave in the absence of atomic internal degrees of freedom, where b was incorporated in the scattering vertex. However, notice that the vertex $\mathcal{V}^{(d)}_{\alpha\beta,\gamma\delta}$ describes the atom–photon scattering and not the average behavior of polarization as the elementary vertex in the iteration equation (6.156).

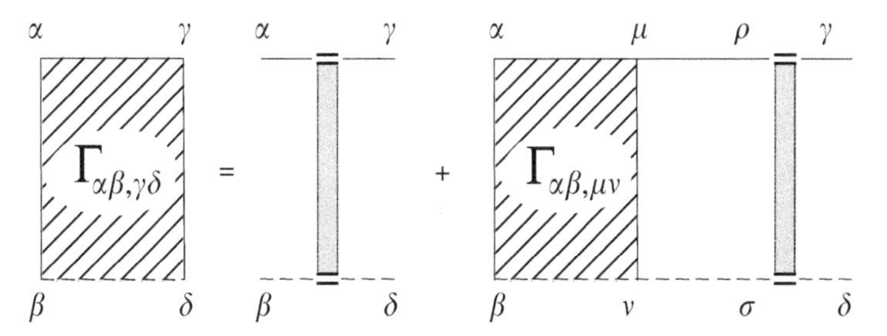

Figure 6.20 Representation of the integral equation (6.308) for the structure factor $\Gamma^{(d,c)}_{\alpha\beta,\gamma\delta}$. The elementary vertex $\mathcal{V}^{(d,c)}_{\alpha\beta,\gamma\delta}$ is represented in Figure 6.18(a) for the Diffuson and Figure 6.18(b) for the Cooperon.

that we can use matrix forms for the tensors Γ, b and $\mathcal{V}$. We then drop indices and treat them as (9×9) matrices. We can rewrite the iteration equation in the symbolic form

$$\boxed{\Gamma(\boldsymbol{q}) = \gamma_e \mathcal{V} + w(\boldsymbol{q})\, \Gamma(\boldsymbol{q})\, b\mathcal{V}} \tag{6.309}$$

As done in section 6.6.2, we can now diagonalize the matrices b and $\mathcal{V}^{(d,c)}$. They have three eigenvalues denoted respectively b_k and $\lambda_k^{(d,c)}$, $k = 0, 1, 2$, with respective degeneracies $1, 3$ and 5. The iteration equation (6.309) shows that Γ depends on two different tensors $\mathcal{V}$ and $b\mathcal{V}$ which share a common set of projectors.[44] (This is not generally the case, see remark on page 268.) The tensors $\mathcal{V}^{(d,c)}_{\alpha\beta,\gamma\delta}$ and $b_{\alpha\beta,\gamma\delta}$ defined respectively in (6.292) and (6.305) can therefore be decomposed using a set of three projectors. These projectors were given in (6.163) in a Cartesian basis. Here they are given by

$$\begin{aligned}
\mathcal{T}^{(0)}_{\alpha\beta,\gamma\delta} &= \frac{1}{3}\delta_{\alpha\beta}\delta_{\gamma\delta} \\
\mathcal{T}^{(1)}_{\alpha\beta,\gamma\delta} &= \frac{1}{2}\left[\delta_{\alpha\gamma}\delta_{\beta\delta} - (-1)^{\alpha+\beta}\delta_{\alpha,-\delta}\delta_{\beta,-\gamma}\right] \\
\mathcal{T}^{(2)}_{\alpha\beta,\gamma\delta} &= \frac{1}{2}\left[\delta_{\alpha\gamma}\delta_{\beta\delta} + (-1)^{\alpha+\beta}\delta_{\alpha,-\delta}\delta_{\beta,-\gamma}\right] - \frac{1}{3}\delta_{\alpha\beta}\delta_{\gamma\delta},
\end{aligned} \tag{6.310}$$

so that

$$\mathcal{V}^{(d,c)}_{\alpha\beta,\gamma\delta} = \sum_{k=0}^{2} \lambda_k^{(d,c)}\, \mathcal{T}^{(k)}_{\alpha\beta,\gamma\delta} \tag{6.311}$$

$$b_{\alpha\beta,\gamma\delta} = \sum_{k=0}^{2} b_k\, \mathcal{T}^{(k)}_{\alpha\beta,\gamma\delta} \tag{6.312}$$

[44] This is due to the rotational invariance of the elementary vertex. For a more general vertex as defined in the remark on page 258, there are in general nine eigenvalues and the spectral decomposition is more complex, see remark on page 268.

with[45]

$$\lambda_0^{(d)} = A_{JJ_e}(w_1 + w_2 + 3w_3) = \frac{A_{JJ_e}}{3}(s_0 + 3s_1 + 5s_2) = A_{JJe}$$
$$\lambda_1^{(d)} = A_{JJ_e}(w_1 - w_2) = \frac{A_{JJ_e}}{6}(2s_0 + 3s_1 - 5s_2) \qquad (6.313)$$
$$\lambda_2^{(d)} = A_{JJ_e}(w_1 + w_2) = \frac{A_{JJ_e}}{6}(2s_0 - 3s_1 + s_2)$$

and

$$\lambda_0^{(c)} = A_{JJ_e}(w_1 + 3w_2 + w_3) = \frac{A_{JJ_e}}{3}(s_0 - 3s_1 + 5s_2) = A_{JJ_e}(1 - 2s_1)$$
$$\lambda_1^{(c)} = A_{JJ_e}(w_1 - w_3) = \frac{A_{JJ_e}}{6}(2s_0 - 3s_1 - 5s_2) \qquad (6.314)$$
$$\lambda_2^{(c)} = A_{JJ_e}(w_1 + w_3) = \frac{A_{JJ_e}}{6}(2s_0 + 3s_1 - s_2).$$

The eigenvalues b_k are those defined in (6.159).

Remark

To summarize, the results obtained here can be simply cast in the following form which makes clear the relation between the s_k and the λ_k:

$$\mathcal{V}^{(d)}_{\alpha\beta,\gamma\delta} = \sum_{k=0}^{2} s_k\, \mathcal{T}^{(k)}_{\alpha\gamma,\beta\delta} = \sum_{k=0}^{2} \lambda_k^{(d)}\, \mathcal{T}^{(k)}_{\alpha\beta,\gamma\delta} \qquad (6.315)$$

$$\mathcal{V}^{(c)}_{\alpha\beta,\gamma\delta} = (-1)^{\beta+\delta} \sum_{k=0}^{2} s_k\, \mathcal{T}^{(k)}_{\alpha\gamma,-\delta-\beta} = \sum_{k=0}^{2} \lambda_k^{(c)}\, \mathcal{T}^{(k)}_{\alpha\beta,\gamma\delta}. \qquad (6.316)$$

Indeed, note that the natural coupling scheme for elementary vertices $\mathcal{V}^{(d,c)}_{\alpha\beta,\gamma\delta}$ is the "vertical" combination $(\alpha\gamma) \leftrightarrow (\beta\delta)$ between the elementary scattering amplitudes $v_{\alpha\gamma}$ and $v_{\beta\delta}$ as it is explicit in relation (6.289). On the other hand, in the construction of the structure factor describing multiple scattering (relations 6.306, 6.307), the intensity vertex has to be chained "horizontally" in the direction $(\alpha\beta) \leftrightarrow (\gamma\delta)$, in order to be iterated [160].

Using the orthogonality property of the projectors $\mathcal{T}^{(k)}$, the product $b\,\mathcal{V}^{(d,c)}$ can be written

$$b\mathcal{V}^{(d,c)} = \sum_{k=0}^{2} b_k\, \lambda_k^{(d,c)}\, \mathcal{T}^{(k)}. \qquad (6.317)$$

[45] The eigenvalues $\lambda_k^{(d,c)}$ can be rewritten in terms of 6j- and 9j-symbols [158] :

$$\lambda_k^{(d)} = 3A_{JJ_e}(2J_e + 1)\begin{Bmatrix} 1 & 1 & k \\ J_e & J_e & J \end{Bmatrix}^2$$

$$\lambda_k^{(c)} = 3A_{JJ_e}(2J_e + 1)\begin{Bmatrix} 1 & J_e & J \\ 1 & J & J_e \\ k & 1 & 1 \end{Bmatrix}.$$

From these results, we conclude that in the presence of a Zeeman degeneracy of the atomic levels, the elementary vertices associated respectively with the Diffuson and the Cooperon no longer coincide. In particular, a Goldstone mode exists for the Diffuson and corresponds to $\lambda_0^{(d)} = A_{JJe}$, whereas it does not exist for the Cooperon since $\lambda_0^{(c)} = A_{JJ_e}(1-2s_1)$. Then, unlike the case of classical Rayleigh scattering, the Cooperon is reduced in all the modes. We recover the classical case by taking $(J, J_e) = (0,1)$, which leads to $s_1 = 0$, giving $\lambda_0^{(c)} = 1$.

The structure factors $\Gamma^{(d,c)}_{\alpha\beta,\gamma\delta}$ can be decomposed into three eigenmodes

$$\Gamma^{(d,c)}_{\alpha\beta,\gamma\delta} = \sum_{k=0}^{2} \Gamma_k^{(d,c)} \mathcal{T}^{(k)}_{\alpha\beta,\gamma\delta} \tag{6.318}$$

where

$$\Gamma_k^{(d,c)} = \frac{\gamma_e \lambda_k^{(d,c)}}{1 - w(\boldsymbol{q}) b_k \lambda_k^{(d,c)}}. \tag{6.319}$$

The function $w(\boldsymbol{q})$ being given by equation (6.302), we obtain

$$\Gamma_k^{(d,c)}(\boldsymbol{q}) = \frac{1}{b_k} \frac{\gamma_{at}/\tau_{at}}{Dq^2 + \frac{1}{\tau_k^{(d,c)}}}, \tag{6.320}$$

with the characteristic decay times given by:

$$\boxed{\tau_k^{(d,c)} = \tau_{at} \frac{b_k \lambda_k^{(d,c)}}{\frac{2}{3} A_{JJ_e} - b_k \lambda_k^{(d,c)}}} \tag{6.321}$$

which generalizes (6.166). For the case of atomic transitions between states such that $J_e = J$ where $J_e = J \pm 1$, it is possible to express these characteristic times $\tau_k^{(d,c)}$ as simple fractions. Their value is given in Table 6.1. In section 8.9.3, we analyze the coherent backscattering peak for light scattered by a gas of cold rubidium atoms, corresponding to the case $(J = 3, J_e = 4)$.

Remark

Expression (6.320) for the eigenmodes $\Gamma_k^{(d,c)}(\boldsymbol{q})$ of the structure factor differs from (6.165) obtained for the case of polarized electromagnetic waves by an extra factor $1/b_k$. The origin of this factor is the difference between the iteration equations (6.307) and (6.156). In the former case, we consider iterating the atom–photon scattering elementary vertex, while in the latter case we write an iteration equation for the randomization of the polarization as described by the tensor $b_{\alpha\beta,\gamma\delta}$. This randomization naturally appears in the iteration (6.307) since, in between two successive scattering events, the polarization is also randomized.

Table 6.1 Values of the depolarization and attenuation times associated with the Diffuson and the Cooperon for different atomic transitions characterized by the quantum numbers J and $J_e = J+1$; the times are given in units of τ_{at}.

	$\tau_0^{(d)}$	$\tau_1^{(d)}$	$\tau_2^{(d)}$	$\tau_0^{(c)}$	$\tau_1^{(c)}$	$\tau_2^{(c)}$
J	∞	$\frac{J+2}{3J+2}$	$\frac{7(J+2)(2J+5)}{3(62J^2+79J+10)}$	$\frac{1}{J(2J+3)}$	$\frac{1}{4J+1}$	$\frac{7(6J^2+12J+5)}{58J^2+66J+15}$
$J=0$	∞	1	7/3	∞	1	7/3
$J=1$	∞	0.6	0.32	0.2	0.2	1.16
$J=2$	∞	0.5	0.20	0.071	0.11	0.98
$J=3$	∞	0.45	0.16	0.037	0.077	0.90
$J=4$	∞	0.43	0.14	0.022	0.059	0.86

Finally, the Diffuson contribution is obtained from the expression (6.311) for the tensors $\mathcal{T}^{(k)}$:

$$\Gamma^{(d)}_{\alpha\beta,\gamma\delta} = \frac{\Gamma_2^{(d)} + \Gamma_1^{(d)}}{2}\delta_{\alpha\gamma}\delta_{\beta\delta} + \frac{\Gamma_2^{(d)} - \Gamma_1^{(d)}}{2}(-1)^{(\alpha+\beta)}\delta_{\alpha-\delta}\delta_{\beta-\gamma} + \frac{\Gamma_0^{(d)} - \Gamma_2^{(d)}}{3}\delta_{\alpha\beta}\delta_{\gamma\delta}, \tag{6.322}$$

which corresponds to

$$\Gamma^{(d)}(\hat{\boldsymbol{\varepsilon}}, \hat{\boldsymbol{\varepsilon}}') = \frac{\Gamma_2^{(d)} + \Gamma_1^{(d)}}{2}|\hat{\boldsymbol{\varepsilon}} \cdot \hat{\boldsymbol{\varepsilon}}'^*|^2 + \frac{\Gamma_2^{(d)} - \Gamma_1^{(d)}}{2}|\hat{\boldsymbol{\varepsilon}} \cdot \hat{\boldsymbol{\varepsilon}}'|^2 + \frac{\Gamma_0^{(d)} - \Gamma_2^{(d)}}{3} \tag{6.323}$$

which is the iteration of the elementary atomic vertex $\mathcal{V}_{at}^{(d)}(\hat{\boldsymbol{\varepsilon}}, \hat{\boldsymbol{\varepsilon}}')$ defined in (6.296) for incoming $\hat{\boldsymbol{\varepsilon}}$ and outgoing $\hat{\boldsymbol{\varepsilon}}'$ polarizations. Similarly, the Cooperon contribution is obtained from (6.299), so that

$$\Gamma^{(c)}(\hat{\boldsymbol{\varepsilon}}, \hat{\boldsymbol{\varepsilon}}') = \frac{\Gamma_2^{(c)} + \Gamma_1^{(c)}}{2}|\hat{\boldsymbol{\varepsilon}} \cdot \hat{\boldsymbol{\varepsilon}}'^*|^2 + \frac{\Gamma_0^{(c)} - \Gamma_2^{(c)}}{3}|\hat{\boldsymbol{\varepsilon}} \cdot \hat{\boldsymbol{\varepsilon}}'|^2 + \frac{\Gamma_2^{(c)} - \Gamma_1^{(c)}}{2}. \tag{6.324}$$

Remark: generalized vertex

We have seen in the remark on page 259 that for quantities like speckle correlations, we have to pair trajectories where atoms have different Zeeman states, so we have to define a generalized vertex where the four internal atomic Zeeman states are different. In contrast to the scalar vertex defined above, this vertex is not necessarily rotation invariant. The nine eigenvalues of this generalized vertex are no longer degenerate, and the spectral decomposition of the vertex involves nine projectors in the most general case [159]. Moreover, the iteration equation (6.309) involves two tensors $\mathcal{V}$ and $b\mathcal{V}$ which now do not share a common set in their spectral decomposition, since b is always scalar and is decomposed using three degenerate modes. To find the decomposition of Γ, we need to work with one set of projectors only. We thus define a new tensor $U(\boldsymbol{q})$ by

$$\Gamma(\boldsymbol{q}) = U(\boldsymbol{q})\mathcal{V}. \tag{6.325}$$

It is straightforward to rewrite the integral equation (6.309) for $\Gamma(\boldsymbol{q})$ into a new equation for $U(\boldsymbol{q})$:

$$U(\boldsymbol{q}) = \gamma_e + w(\boldsymbol{q})\, U(\boldsymbol{q})\, \mathcal{V}b \tag{6.326}$$

which gives U in terms of $\mathcal{V}b$ and the identity tensor. The diagonalization of $\mathcal{V}b$ yields the decomposition of $U(\boldsymbol{q})$ from which $\Gamma(\boldsymbol{q})$ is obtained.

Exercise 6.4 Write the tensor $b_{\alpha\beta,\gamma\delta}$ in a matrix form and compare with (6.155). Check that these two matrices, written respectively in a standard and in a Cartesian basis, have the same eigenvalues. Do the same exercise for the projectors $\mathcal{T}^{(k)}_{\alpha\beta,\gamma\delta}$ written either in a standard (6.311) or in a Cartesian (6.163) basis.

7
Electronic transport

In this chapter, we restore $\hbar$ and we denote the spin degeneracy by s; $e > 0$.

7.1 Introduction

This chapter deals with the study of the electrical conductivity of weakly disordered metals. The general description of transport in metals is a fundamental problem in which the phenomena of coherence and multiple scattering play a key role. The purpose of this chapter is to calculate and study the average electric conductivity. We shall consider in detail the so-called "*weak localization*" quantum correction to classical conductivity, which originates from the pairing of time-reversed trajectories associated with the Cooperon. Higher moments of the conductivity distribution will be studied in Chapter 11.

We consider the diagonal component $\sigma_{xx}(\omega)$ along a direction Ox of the electrical conductivity tensor[1] $\sigma_{\alpha\beta}$, whose derivation is recalled in Appendix A7.1. For a degenerate electron gas at temperature $T \ll T_F = \epsilon_F/k_B$, the real part $\sigma_{xx}(\omega)$ is given by its $T = 0\,\mathrm{K}$ expression:[2]

$$\sigma_{xx}(\omega) = s\frac{\hbar}{\pi\Omega}\,\mathrm{Tr}\left[\hat{j}_x\,\mathrm{Im}\hat{G}^R_{\epsilon_F}\,\hat{j}_x\,\mathrm{Im}\hat{G}^R_{\epsilon_F-\omega}\right], \tag{7.1}$$

where $\hat{j}_x = -e\hat{p}_x/m$ is the current operator in the Ox direction. The factor $s = 2$ accounts for the spin degeneracy. The Green function is given by relation (3.13) and the Hamiltonian $\mathcal{H}$ by (2.1).

Since $\mathrm{Im}\hat{G}^R = (\hat{G}^R - \hat{G}^A)/2i$, expression (7.1) for the conductivity contains the products $\hat{G}^R\hat{G}^A$, $\hat{G}^R\hat{G}^R$ and $\hat{G}^A\hat{G}^A$. In Appendix A4.5 we show that the terms $\hat{G}^R\hat{G}^R$ and $\hat{G}^A\hat{G}^A$ give negligible contributions. The trace of their average is of order $1/k_F l_e$ with respect to the terms $\hat{G}^R\hat{G}^A$. Moreover, using a Ward identity (4.214), it is possible to relate the product $\hat{G}^R\hat{G}^R$ to $\hat{G}^R$. Such a term contributes to neither the interference effects considered in this chapter

[1] The system is supposed to be isotropic, so that $\sigma_{xx} = \sigma_{yy} = \sigma_{zz}$.

[2] This expression for the conductivity is valid for a degenerate electron gas ($T \ll T_F$). For a non-degenerate gas, relation (7.124) should be used. Moreover, taking the zero temperature expression does not exclude a possible temperature dependence included in the Green functions due to coupling with other degrees of freedom (see remark on page 83). Here we keep the notation $\hat{G}_{\epsilon_F-\omega}$ instead of $\hat{G}_{\epsilon_F-\hbar\omega}$.

nor the fluctuations considered in Chapter 11. Finally we retain

$$\boxed{\sigma_{xx}(\omega) = \frac{s\hbar}{2\pi\Omega} \mathrm{Re}\,\mathrm{Tr}\left[\hat{j}_x \hat{G}^R_{\epsilon_F} \hat{j}_x \hat{G}^A_{\epsilon_F-\omega}\right]} \tag{7.2}$$

The current operator is equal to $\hat{j}_x = i(e\hbar/m)(\partial/\partial x)$. We can evaluate the trace in a spatial representation, so that the disordered average conductivity $\overline{\sigma}_{xx}(\omega)$, now written $\sigma(\omega)$, is

$$\sigma(\omega) = \overline{\sigma}_{xx}(\omega) = -s\frac{e^2\hbar^3}{2\pi m^2\Omega}\int d\boldsymbol{r}\, d\boldsymbol{r}'\ \mathrm{Re}\ \overline{\partial_x G^R_\epsilon(\boldsymbol{r},\boldsymbol{r}')\partial_{x'} G^A_{\epsilon-\omega}(\boldsymbol{r}',\boldsymbol{r})} \tag{7.3}$$

whereas in the momentum representation, we obtain[3]

$$\boxed{\sigma(\omega) = s\frac{e^2\hbar^3}{2\pi m^2\Omega}\sum_{\boldsymbol{k},\boldsymbol{k}'} k_x k'_x\ \mathrm{Re}\ \overline{G^R_\epsilon(\boldsymbol{k},\boldsymbol{k}')G^A_{\epsilon-\omega}(\boldsymbol{k}',\boldsymbol{k})}} \tag{7.4}$$

The Green function $G^R_\epsilon(\boldsymbol{k},\boldsymbol{k}')$ is given by (3.31).

Remark

The spatial dependence $\sigma(\boldsymbol{q},\omega)$ of the average conductivity is given by a relation similar to (7.4):

$$\sigma(\boldsymbol{q},\omega) = s\frac{e^2\hbar^3}{2\pi m^2\Omega}\sum_{\boldsymbol{k},\boldsymbol{k}'} k_x k'_x\ \overline{G^R_\epsilon(\boldsymbol{k}_+,\boldsymbol{k}'_+)G^A_{\epsilon-\omega}(\boldsymbol{k}'_-,\boldsymbol{k}_-)}, \tag{7.5}$$

where $\boldsymbol{k}_\pm = \boldsymbol{k} \pm \boldsymbol{q}/2$. This expression resembles the one obtained for the probability $P(\boldsymbol{q},\omega)$ given by (4.12) (we restore $\hbar$ in the Green functions and in the density of states ρ_0):

$$P(\boldsymbol{q},\omega) = \frac{\hbar}{2\pi\rho_0\Omega}\sum_{\boldsymbol{k},\boldsymbol{k}'} \overline{G^R_\epsilon(\boldsymbol{k}_+,\boldsymbol{k}'_+)G^A_{\epsilon-\omega}(\boldsymbol{k}'_-,\boldsymbol{k}_-)}. \tag{7.6}$$

However, because of the factor $k_x k'_x$ in the conductivity, there is no simple relation between $\sigma(\boldsymbol{q},\omega)$ and $P(\boldsymbol{q},\omega)$.

In Chapter 4 we saw that P can be considered as the sum of three terms. The first one, $P_0(\boldsymbol{r},\boldsymbol{r}',\omega)$, represents the probability that a particle reaches point $\boldsymbol{r}'$ without any collision (Drude–Boltzmann). The first two contributions P_0 and P_d are classical and do not account for any phase effect, whereas the third contribution, X_c, describes the Cooperon. Here the procedure is quite similar and we represent (Figure 7.1) the average conductivity as the sum of three terms corresponding respectively to P_0, P_d, and X_c. We calculate these three terms in the following sections.

[3] For simplicity we write $\epsilon = \epsilon_F$ in the argument of the Green functions, keeping in mind that ϵ is the Fermi energy.

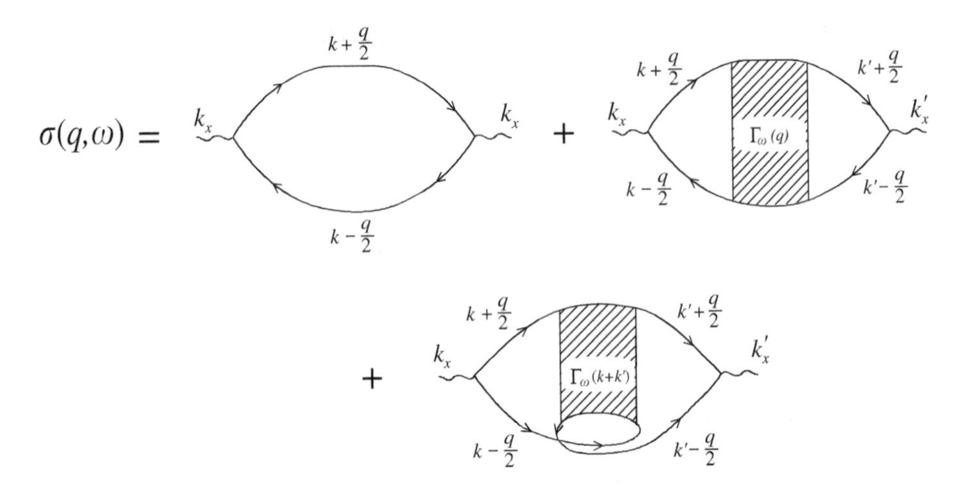

Figure 7.1 Representations of the Drude–Boltzmann, Diffuson, and Cooperon contributions to the average conductivity. In the case of isotropic collisions, the second diagram (the Diffuson) vanishes (section 7.2.2). In the case of anisotropic collisions, the third diagram must be "dressed" by an additional impurity line and by "vertex corrections" (Appendix A7.4).

Remark

Within linear response theory, it is possible to define a non-local conductivity $\sigma_{\alpha\beta}(\boldsymbol{r},\boldsymbol{r}')$, which is the response to the local electric field $\boldsymbol{E}(\boldsymbol{r}')$. The local current is $j_\alpha(\boldsymbol{r}) = \int d\boldsymbol{r}' \sigma_{\alpha\beta}(\boldsymbol{r},\boldsymbol{r}')E_\beta(\boldsymbol{r}')$. This conductivity is written [161]

$$\sigma_{\alpha\beta}(\boldsymbol{r},\boldsymbol{r}') = -s\frac{e^2\hbar^3}{2\pi m^2}\left[\partial_\alpha \mathrm{Im}G_\epsilon^R(\boldsymbol{r},\boldsymbol{r}')\partial'_\beta \mathrm{Im}G_\epsilon^R(\boldsymbol{r}',\boldsymbol{r}) - \mathrm{Im}G_\epsilon^R(\boldsymbol{r},\boldsymbol{r}')\partial_\alpha\partial'_\beta \mathrm{Im}G_\epsilon^R(\boldsymbol{r}',\boldsymbol{r})\right]. \quad (7.7)$$

Contrary to (7.3) where they have been neglected, here it is necessary to retain the products G^RG^R and G^AG^A in order to obtain a correct expression of $\sigma(\boldsymbol{r},\boldsymbol{r}')$ which, in particular, obeys current conservation. Their contribution is negligible after spatial integration (7.8). We check that the conductivity σ defined by (7.1) verifies

$$\sigma = \frac{1}{\Omega}\int\int d\boldsymbol{r}\, d\boldsymbol{r}'\, \sigma(\boldsymbol{r},\boldsymbol{r}'). \quad (7.8)$$

The calculation of the disorder averaged conductivity $\overline{\sigma}_{\alpha\beta}(\boldsymbol{r},\boldsymbol{r}')$ shows that, for isotropic collisions, it becomes [161]

$$\overline{\sigma}_{\alpha\beta}(\boldsymbol{r},\boldsymbol{r}') = \sigma_0\left[\frac{3}{4\pi l_e}\frac{R_\alpha R_\beta}{R^4}e^{-R/l_e} - D\partial_\alpha\partial'_\beta P_d(\boldsymbol{r},\boldsymbol{r}')\right] \quad (7.9)$$

where $\boldsymbol{R} = \boldsymbol{r} - \boldsymbol{r}'$ and $\sigma_0 = ne^2\tau_e/m$. The current conservation $\sum_\alpha \partial_\alpha\sigma_{\alpha\beta}(\boldsymbol{r},\boldsymbol{r}') = \sum_\beta \partial'_\beta\sigma_{\alpha\beta}(\boldsymbol{r},\boldsymbol{r}') = 0$ implies that P_d is a solution of the diffusion equation $-D\Delta P_d(\boldsymbol{r},\boldsymbol{r}') = \overline{\delta}(\boldsymbol{r}-\boldsymbol{r}')$ where $\overline{\delta}$ is a normalized function of range l_e. After integration of $\overline{\sigma}_{\alpha\beta}(\boldsymbol{r},\boldsymbol{r}')$ on the volume (7.8), the second term gives a vanishing contribution, so that we recover the Drude conductivity, $\overline{\sigma}_{\alpha\beta} = \sigma_0\delta_{\alpha\beta}$ (see the following section). Nevertheless, this second term is essential since it ensures current conservation and conservation of the number of particles. The structure of (7.9) is very similar to the structure of the probability $P(\boldsymbol{r},\boldsymbol{r}') = P_0(\boldsymbol{r},\boldsymbol{r}') + P_d(\boldsymbol{r},\boldsymbol{r}')$ for which we have shown in section 4.4 that the Diffuson P_d ensures conservation of the number of particles. Like $P(\boldsymbol{r},\boldsymbol{r}')$, the conductivity $\sigma(\boldsymbol{r},\boldsymbol{r}')$ is a *non-local* quantity, with short range and long range contributions. Its structure reflects the profound relationship between the Kubo and Landauer formalisms (see page 310).

7.2 Incoherent contribution to conductivity

7.2.1 Drude–Boltzmann approximation

This approximation is equivalent to the one we used to obtain the probability P_0. It consists in assuming that

$$\overline{G_\epsilon^R(\boldsymbol{k},\boldsymbol{k}')G_{\epsilon-\omega}^A(\boldsymbol{k}',\boldsymbol{k})} \simeq \overline{G}_\epsilon^R(\boldsymbol{k},\boldsymbol{k}')\overline{G}_{\epsilon-\omega}^A(\boldsymbol{k}',\boldsymbol{k}), \tag{7.10}$$

where the average Green functions are given by (3.88):

$$\overline{G}_\epsilon^{R,A}(\boldsymbol{k},\boldsymbol{k}') = \overline{G}_\epsilon^{R,A}(\boldsymbol{k})\delta_{\boldsymbol{k},\boldsymbol{k}'} = \frac{\delta_{\boldsymbol{k},\boldsymbol{k}'}}{\epsilon - \epsilon_{\boldsymbol{k}} \pm i\frac{\hbar}{2\tau_e}}. \tag{7.11}$$

From this assumption, we obtain a contribution denoted by $\sigma_0(\omega)$ for the conductivity (7.4):

$$\sigma_0(\omega) = s\frac{e^2\hbar^3}{2\pi m^2\Omega}\sum_{\boldsymbol{k}} k_x^2 \,\mathrm{Re}\,\frac{1}{(\epsilon_F - \epsilon_{\boldsymbol{k}} + i\frac{\hbar}{2\tau_e})}\frac{1}{(\epsilon_F - \omega - \epsilon_{\boldsymbol{k}} - i\frac{\hbar}{2\tau_e})}. \tag{7.12}$$

In this expression, we replace the discrete sum by an integral, using

$$\frac{1}{\Omega}\sum_{\boldsymbol{k}} f(\epsilon_{\boldsymbol{k}}) \to \int d\epsilon\, \rho_0(\epsilon)\int d\varpi\, f(\epsilon,\varpi), \tag{7.13}$$

where ϖ is the normalized solid angle ($\int d\varpi = 1$) and where ρ_0 is the density of states per unit volume and per spin direction. The energy dependence of k_x^2 is regular, and we assume that the density of states varies slowly in an energy range $\hbar/\tau_e$ around ϵ_F. The product $\rho_0 k_x^2$ can therefore be extracted from the integral, and the angular average of k_x^2 is then k_F^2/d. Calculating the integral by the method of residues and introducing the diffusion coefficient $D = v_F^2\tau_e/d$, we finally obtain for the zero frequency conductivity $\sigma_0 = \sigma_0(\omega = 0)$

$$\boxed{\sigma_0 = se^2 D\rho_0(\epsilon_F)} \tag{7.14}$$

which is the Einstein relation [162–164]. For free electrons, the density of states can be simply written (for all dimensions d) in terms of the electronic density n:

$$\rho_0(\epsilon) = \frac{n(\epsilon)d}{2s\epsilon}. \tag{7.15}$$

We deduce the well-known expression for the Drude conductivity [165]:

$$\sigma_0 = \frac{ne^2\tau_e}{m}. \tag{7.16}$$

From expression (3.44) for the density of states, we can also write the conductivity in the form

$$\sigma_0 = sA_d\frac{e^2}{h}\left(\frac{k_F}{2\pi}\right)^{d-1} l_e \tag{7.17}$$

where $A_d = \pi^{d/2}/\Gamma(d/2+1)$ is the volume of the unit sphere in dimension d.

In the above Drude approximation, the term $k_x k_x'$ has been extracted from the sum in relation (7.4) and replaced by k_F^2/d. Comparing expressions (7.5) and (7.6), we see that the Drude conductivity is proportional to the corresponding contribution P_0 to classical probability

$$\sigma_0(\omega) = se^2\rho_0\frac{v_F^2}{d}\mathrm{Re}P_0(\boldsymbol{q}=0,\omega) \tag{7.18}$$

or

$$\sigma_0(\omega) = \frac{ne^2}{m}\int d\boldsymbol{r}'\mathrm{Re}P_0(\boldsymbol{r},\boldsymbol{r}',\omega). \tag{7.19}$$

Using (4.20) implies

$$\sigma_0(\omega) = \frac{ne^2}{m}\mathrm{Re}\frac{\tau_e}{1-i\omega\tau_e}. \tag{7.20}$$

Finally the average conductance is given by Ohm's law

$$\overline{G} = \sigma_0 L^{d-2}. \tag{7.21}$$

This relation is a generalization in d dimensions of the usual form $\overline{G} = \sigma_0 S/L$ for a sample of length L and section S. In d dimensions, the average conductance is

$$\overline{G} = sA_d\frac{e^2}{h}\frac{k_F l_e}{2\pi}(k_F L)^{d-2}. \tag{7.22}$$

It is size independent in dimension $d=2$. For a wire of length L and section S, it is written

$$\overline{G} = s\frac{e^2}{h}\frac{k_F^2 l_e S}{3\pi L}. \tag{7.23}$$

Remark: conductance and Thouless energy

Since the Drude conductivity is proportional to the diffusion coefficient D, it can be rewritten in terms of the Thouless energy $E_c = \hbar D/L^2$, defined in section 5.5.1 as the inverse of the diffusion time through a system of finite size L. The density of states per unit volume is $1/(\Omega\Delta)$, where Δ is the average level spacing, so the conductivity σ_0 given by the Einstein formula (7.14) can be rewritten as $\sigma_0 = se^2 D/\Omega\Delta$. The average conductance is given by Ohm's law $\overline{G} = \sigma_0 L^{d-2}$. It is thus equal to

$$\overline{G} = s\frac{e^2}{\hbar}\frac{E_c}{\Delta}. \tag{7.24}$$

This expression displays the quantum of conductance e^2/h, and its inverse h/e^2 is equal to 25.8 kΩ. Dimensionless conductance $g = \overline{G}/(e^2/h)$ then appears as the ratio of two characteristic energies, the Thouless energy and the level spacing:

$$\boxed{g = 2\pi s\frac{E_c}{\Delta}} \tag{7.25}$$

In the literature, the Thouless energy is sometimes defined as $E_c = hD/L^2$, so that sometimes we encounter the expression $g = sE_c/\Delta$. For an isolated system, the conductance can also be interpreted as a measure of the sensitivity of the energy levels to a change in boundary conditions. This point of view has been developed by Thouless [166].

Remark: conductance as a measure of disorder

The limit $k_F l_e \gg 1$ corresponds to the case of a weakly disordered system (see page 85), in which the wave functions are well described by plane waves weakly perturbed by the disorder potential. This limit corresponds to a good conducting metal. Another natural way to characterize the quality of a conductor is to introduce its dimensionless electrical conductance g, that is, the conductance G in units of $e^2/h = 1/(25.8\,\text{k}\Omega)$. A large value $g \gg 1$ corresponds to a good conductor, whereas $g \ll 1$ describes an insulator. What is the correspondence between these two criteria? To answer this question we use expression (7.17) for the Drude conductivity σ_0. From Ohm's law $\overline{G} = \sigma_0 L^{d-2}$, we deduce that

$$g \propto k_F l_e (k_F L)^{d-2}. \tag{7.26}$$

For $d = 2$, g is directly proportional to $k_F l_e$ and the two criteria are equivalent. For $d = 3$, the condition $k_F l_e \gg 1$ implies $g \gg 1$. In the opposite limit $k_F l_e \ll 1$, Ohm's law (7.26) is no longer valid and the pertubative approach does not apply. The electronic states are spatially localized. In this case, the conductance decreases exponentially with size and $g \ll 1$. For $d = 1$, the $1/k_F l_e$ perturbative approach is not valid (Exercise 3.9): the electronic states are localized and g is always small as long as $L > l_e$.

So the conductance g is a good measure of disorder and can be used as a parameter to interpolate between the limit $g \gg 1$ for a conductor and the limit $g \ll 1$ for an insulator [167].

In section A4.2.3, we have seen that the parameter $1/g$ measures the probability of two Diffusons crossing, that is, the importance of quantum corrections to classical transport.

Remark: coupling to other degrees of freedom
The Drude conductivity (7.16) is proportional to the elastic collision time τ_e, and it is directly related to the inverse of the imaginary part of the self-energy (3.73). Up to now, we have only considered the contribution of elastic collisions due to static impurities. It is also possible to take into account coupling to other degrees of freedom (see section 6.5) supposed to be independent of each other (magnetic impurities, spin-orbit coupling, etc.) and to generalize relation (7.16) in the form

$$\sigma_0 = \frac{ne^2 \tau_{tot}}{m}, \tag{7.27}$$

where the effective time τ_{tot} is given by the Matthiessen rule (see remark on page 83)

$$\frac{1}{\tau_{tot}} = \frac{1}{\tau_e} + \frac{1}{\tau_{so}} + \frac{1}{\tau_m}. \tag{7.28}$$

Coupling to other dynamic degrees of freedom (phonons, interaction with other electrons) can also be incorporated. Electron–phonon interaction is not discussed in this book (see for example [163, 165]). The Coulomb interaction between electrons can also be described by a characteristic time. We shall treat this point in detail in section 13.6.
Generally, the smallest time is the elastic collision time $\tau_e \ll \tau_{so}, \tau_m$. Moreover we shall consider a low temperature limit where the interaction with phonons can be neglected, a necessary condition for observing phase coherent effects. In these conditions, the Drude conductivity remains equal to $ne^2\tau_e/m$.

7.2.2 The multiple scattering regime: the Diffuson

The normalization of the probability of quantum diffusion necessitates taking into account, in addition to the Drude–Boltmann term, the contribution of the multiple scattering within the Diffuson approximation (section 4.4). Similarly, let us now consider the contribution of the Diffuson to the conductivity. It is obtained from relation (7.4) where we introduce the Diffuson (4.82) into the average of the Green functions. This contribution, written σ_d, is equal to

$$\sigma_d(\omega) = s\frac{e^2\hbar^3}{2\pi m^2 \Omega^2}\left(\frac{2\pi\rho_0}{\hbar}\right)^2 \sum_{\boldsymbol{k},\boldsymbol{k}'} k_x k'_x \,\mathrm{Re}\, \tilde{P}_0(\boldsymbol{k},\boldsymbol{q},\omega)\Gamma_\omega(\hat{\boldsymbol{s}},\hat{\boldsymbol{s}}',\boldsymbol{q})\tilde{P}_0(\boldsymbol{k}',\boldsymbol{q},\omega) \tag{7.29}$$

with $\tilde{P}_0(\boldsymbol{k},\boldsymbol{q},\omega) = (\hbar/2\pi\rho_0)\overline{G}^R_\epsilon(\boldsymbol{k}_+)\overline{G}^A_{\epsilon-\omega}(\boldsymbol{k}_-)$ and where $\Gamma_\omega(\hat{\boldsymbol{s}},\hat{\boldsymbol{s}}',\boldsymbol{q})$ is the structure factor (4.155) which depends a priori on the vector $\boldsymbol{q}$ and on the incoming and outgoing directions of the diffusive process. This expression is similar to the one (4.82) obtained for $P_d(\boldsymbol{q},\omega)$, but here the product $k_x k'_x$ plays a crucial role.

First we notice that the function $\tilde{P}_0$ is peaked around k_F. Then we separate the sums on the moduli k and k' from the angular integrations. Using (4.75) and rearranging the prefactors, we deduce the contribution of the Diffuson to the conductivity, in the limit $\boldsymbol{q} \to 0$:

$$\sigma_d(\omega) = \sigma_0(\omega) \operatorname{Re} \frac{1}{(1 - i\omega\tau_e)^2} \frac{d}{\gamma_e} \left\langle (\hat{\boldsymbol{s}} \cdot \hat{\boldsymbol{x}})(\hat{\boldsymbol{s}}' \cdot \hat{\boldsymbol{x}})\, \Gamma_\omega(\hat{\boldsymbol{s}}, \hat{\boldsymbol{s}}') \right\rangle_{\hat{s}, \hat{s}'} \tag{7.30}$$

where $\langle \cdots \rangle_{\hat{s}, \hat{s}'}$ is the angular average over the directions $\hat{\boldsymbol{s}}$ and $\hat{\boldsymbol{s}}'$. We write $\Gamma_\omega(\hat{\boldsymbol{s}}, \hat{\boldsymbol{s}}') = \Gamma_\omega(\hat{\boldsymbol{s}}, \hat{\boldsymbol{s}}'; \boldsymbol{q} = 0)$. For more details see Appendix A4.3.

- For *isotropic collisions*, $\Gamma_\omega(\hat{\boldsymbol{s}}, \hat{\boldsymbol{s}}'; \boldsymbol{q})$ does not depend on the incoming and outgoing directions. Therefore the angular average in (7.30) cancels since $\langle \hat{\boldsymbol{s}} \cdot \hat{\boldsymbol{s}}' \rangle_{\hat{s}} = 0$. The contribution of the Diffuson to the conductivity is thus zero: $\sigma_d(\omega) = 0$.
- If *the collisions are anisotropic*, that is, if the potential $v(\boldsymbol{k} - \boldsymbol{k}')$ has an angular structure, then $\Gamma_\omega(\hat{\boldsymbol{s}}, \hat{\boldsymbol{s}}')$ depends on $\hat{\boldsymbol{s}}$ and $\hat{\boldsymbol{s}}'$ and obeys the integral equation (4.161). Now the Diffuson gives a finite contribution. In order to calculate the angular average in (7.30), we introduce the vector [168, 169]:

$$\Lambda_\omega \hat{\boldsymbol{s}} = \left\langle \hat{\boldsymbol{s}}' \Gamma_\omega(\hat{\boldsymbol{s}}, \hat{\boldsymbol{s}}') \right\rangle_{s'} \tag{7.31}$$

which is colinear to $\hat{\boldsymbol{s}}$. Multiplying (4.161) by $\hat{\boldsymbol{s}}'$ and averaging on $\hat{\boldsymbol{s}}'$, we obtain the equation

$$\Lambda_\omega \hat{\boldsymbol{s}} = \gamma_1 \hat{\boldsymbol{s}} + \frac{1}{\gamma_e} f_\omega \Lambda_\omega \left\langle \hat{\boldsymbol{s}}'' B(\hat{\boldsymbol{s}}'' - \hat{\boldsymbol{s}}) \right\rangle_{\hat{s}''}, \tag{7.32}$$

where the function $f_\omega = f_\omega(\hat{\boldsymbol{s}}'', \boldsymbol{q} = 0) = 1/(1 - i\omega\tau_e)$ defined by (4.76) has no angular dependence when $\boldsymbol{q} = 0$. This expression for f_ω is valid for any $\omega\tau_e$, even beyond the diffusive regime. The solution of (7.32) is

$$\Lambda_\omega = \frac{\gamma_1}{1 - f_\omega \gamma_1/\gamma_e}, \tag{7.33}$$

where the quantities γ_1 and γ_e are defined by (4.157). Finally, writing s_x the component of $\hat{\boldsymbol{s}}$ along Ox, we have

$$\left\langle s_x s'_x \operatorname{Re}\Gamma_\omega(\hat{\boldsymbol{s}}, \hat{\boldsymbol{s}}', \boldsymbol{q}) \right\rangle_{\hat{s}, \hat{s}'} = \frac{\Lambda_\omega}{d}. \tag{7.34}$$

We deduce for $\sigma_d(\omega)$

$$\sigma_d(\omega) = \sigma_0 \operatorname{Re} \frac{1}{(1 - i\omega\tau_e)^2} \frac{\gamma_1/\gamma_e}{1 - f_\omega \gamma_1/\gamma_e}, \tag{7.35}$$

so that the total conductivity, denoted $\sigma_{cl}(\omega) = \sigma_0(\omega) + \sigma_d(\omega)$, is

$$\boxed{\sigma_{cl}(\omega) = s e^2 D^* \rho_0 \operatorname{Re} \frac{1}{1 - i\omega\tau^*}} \tag{7.36}$$

The transport time τ^* is given by[4]

$$\boxed{\frac{1}{\tau^*} = 2\pi\rho_0(\gamma_e - \gamma_1) = 2\pi\rho_0 n_i \langle v^2(\theta)(1-\cos\theta)\rangle} \tag{7.37}$$

and the diffusion constant D^* is

$$D^* = \frac{v_F^2\tau^*}{d}. \tag{7.38}$$

We finally obtain an expression for the total conductivity $\sigma_{cl}(\omega)$ which is similar to the Drude conductivity (7.20), except that the elastic collision time has been replaced by the transport time τ^*. The two expressions coincide for isotropic collisions, in which case $\tau^* = \tau_e$.

It is useful to return to the structure of the expressions for $P_d(\boldsymbol{q},\omega)$ and $\sigma_d(\omega)$. Indeed, one could ask why the two quantities are so different, given that the two original expressions (7.5) and (7.6) are quite similar. They only differ by the product $k_x k'_x$ which enters the conductivity. Both quantities contain the structure factor which, in the limit $\boldsymbol{q} \to 0$, has the following angular structure (see relation 4.163):

$$\Gamma_\omega(\hat{s},\hat{s}',\boldsymbol{q}) = \Gamma_\omega(\boldsymbol{q}) + \hat{s}\cdot\hat{s}'\,\Lambda_\omega. \tag{7.39}$$

The calculation of $P_d(\boldsymbol{q},\omega)$ involves the average $\langle\Gamma_\omega(\hat{s},\hat{s}',\boldsymbol{q})\rangle_{\hat{s},\hat{s}'} = \Gamma_\omega(\boldsymbol{q})$ which has a pole in $\omega = 0$, whereas the conductivity $\sigma_d(\omega)$ involves the average $\langle s_x s_{x'}\Gamma_\omega(\hat{s},\hat{s}',\boldsymbol{q})\rangle_{\hat{s},\hat{s}'}$ proportional to Λ_ω and with no pole in $\omega = 0$.

7.2.3 Transport time and vertex renormalization

In this book, most of the transport properties are studied in the simplest case of isotropic collisions. Nevertheless, we can ask ourselves how these properties are modified in the most general case of anisotropic collisions. First we have to replace D by D^* in the diffusion pole. Secondly, *the current operator has to be renormalized.* In order to understand the origin of this renormalization, we consider the conductivity. The calculation of $\sigma_{cl} = \sigma_0 + \sigma_d$ involves the sum of the two first diagrams in Figure 7.1. Based on properties of the Diffuson, we see that this sum has the same structure as the first diagram for σ_0, provided we replace the incoming momentum $\boldsymbol{k} = k\hat{s}$ by

$$k\left[\hat{s} + \frac{1}{\Omega}\sum_{\boldsymbol{k}'}\hat{s}' G^R(\boldsymbol{k}')G^A(\boldsymbol{k}')\Gamma_\omega(\hat{s}',\hat{s})\right]. \tag{7.40}$$

[4] We have already introduced the time τ^* in the calculation of the probability P_d in the presence of anisotropic collisions (relation 4.168).

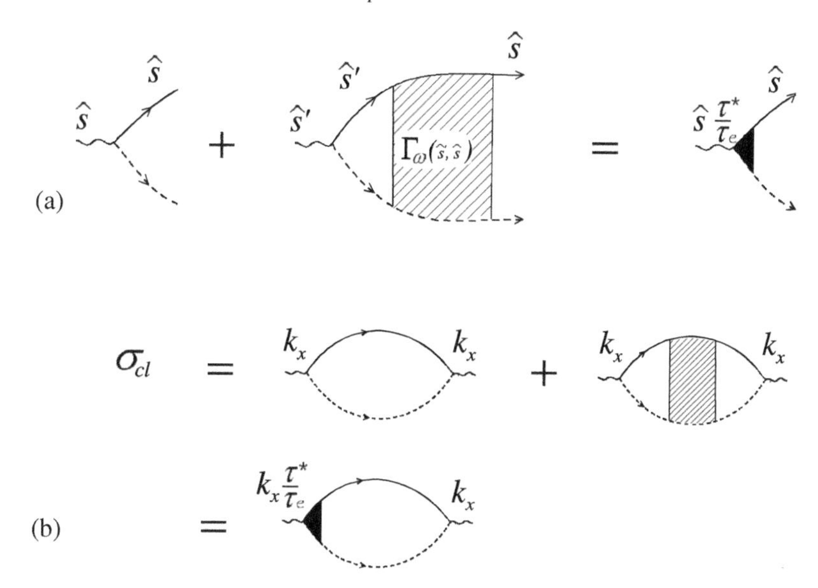

Figure 7.2 (a) Renormalization of the current vertex for anisotropic collisions. (b) Classical conductivity σ_{cl} for anisotropic collisions.

With the help of relations (4.159, 7.31), this amounts to performing the substitution

$$\hat{s} \longrightarrow \hat{s} + \frac{f_\omega}{\gamma_e} \langle \hat{s}' \Gamma_\omega(\hat{s}', \hat{s}) \rangle_{\hat{s}'} = \hat{s} \left(1 + f_\omega \frac{\Lambda_\omega}{\gamma_e} \right) \tag{7.41}$$

represented in Figure 7.2(a). At zero frequency the quantity Λ_0 given by (7.33) is equal to $\Lambda_0 = \gamma_1 \gamma_e / (\gamma_e - \gamma_1)$ so that $(1 + \Lambda_0/\gamma_e) = \gamma_e/(\gamma_e - \gamma_1) = \tau^*/\tau_e$. Thus we must perform the substitution

$$\hat{s} \longrightarrow \hat{s}\, \frac{\tau^*}{\tau_e}, \tag{7.42}$$

where τ^* is the transport time (7.37). So, to obtain the classical conductance σ_{cl}, it is enough to consider the diagram for σ_0 and to replace one of the wave vectors k_x by $k_x(1 + \Lambda_0/\gamma_e) = k_x\tau^*/\tau_e$ (Figure 7.2(b)). The diffusion coefficient $D = \langle v_x^2 \rangle \tau_e$ must therefore be replaced by $D^* = \langle v_x^2 \rangle \tau^*$ [170].

7.3 Cooperon contribution

We have seen in section 4.6 that there is a third contribution to the probability of quantum diffusion, called the Cooperon. It corresponds to the same pairs of trajectories as for the Diffuson, but they are time reversed (Figure 4.7). The Cooperon gives an additional contribution to the conductivity (7.4) which is written

$$\sigma_c(\omega) = s\frac{e^2\hbar^3}{2\pi m^2\Omega^2}\left(\frac{2\pi\rho_0}{\hbar}\right)^2 \sum_{\boldsymbol{k},\boldsymbol{k}'} k_x k'_x \mathrm{Re}\left[\tilde{P}_0(\boldsymbol{k},\boldsymbol{q},\omega)\tilde{P}_0(\boldsymbol{k}',\boldsymbol{q},\omega)\Gamma'_\omega(\boldsymbol{k}+\boldsymbol{k}')\right] \quad (7.43)$$

and which corresponds to the last diagram of Figure 7.1. Let us remark on the similarity between this expression and the one (4.89) obtained for $X_c(\boldsymbol{q},\omega)$. The function $\Gamma'_\omega = \Gamma_\omega$ is given by (4.92) and depends on $\boldsymbol{k}+\boldsymbol{k}'$. *Contrary to the Diffuson case, the Cooperon contribution to conductivity has an angular structure*, even though the interaction potential $v(\boldsymbol{k}-\boldsymbol{k}')$ has not. This contribution is singular and diverges in the limit $\omega \to 0$ and $\boldsymbol{k}+\boldsymbol{k}' = 0$. The main contribution therefore comes from the vicinity of the particular value $\boldsymbol{k}' = -\boldsymbol{k}$, and the product $k_x k'_x$ can be averaged to $-k_F^2/d$. We can write directly $\sigma_c(\omega)$ in terms of the Cooperon contribution $X_c(\boldsymbol{q}=0,\omega)$ to the quantum diffusion probability:[5]

$$\sigma_c(\omega) = -se^2\rho_0\frac{v_F^2}{d}\mathrm{Re}\, X_c(\boldsymbol{q}=\boldsymbol{0},\omega). \quad (7.44)$$

Using expression (4.95) for $X_c(\boldsymbol{q},\omega)$, we finally obtain

$$\boxed{\sigma_c(\omega) = -s\frac{e^2 D}{\pi\hbar\Omega}\sum_{\boldsymbol{Q}}\mathrm{Re}\frac{1}{-i\omega + DQ^2}} \quad (7.45)$$

or, after a Fourier transform and for a translation invariant system,

$$\sigma_c(\omega) = -\frac{ne^2}{m}\int d\boldsymbol{r}'\,\mathrm{Re}X_c(\boldsymbol{r},\boldsymbol{r}',\omega). \quad (7.46)$$

The total conductivity is

$$\sigma(\omega) = \frac{ne^2}{m}\int d\boldsymbol{r}'\,\mathrm{Re}\left[P_0(\boldsymbol{r},\boldsymbol{r}',\omega) - X_c(\boldsymbol{r},\boldsymbol{r}',\omega)\right]. \quad (7.47)$$

Using (4.52), we can also write

$$\sigma_c(\omega) = -s\frac{e^2 D}{\pi\hbar}\mathrm{Re}X_c(\boldsymbol{r},\boldsymbol{r},\omega). \quad (7.48)$$

This correction taken at zero frequency can be expressed in terms of the return probability $X_c(\boldsymbol{r},\boldsymbol{r},t)$ for diffusive motion:

$$\sigma_c(\omega=0) = -s\frac{e^2 D}{\pi\hbar}\int_0^\infty dt\, X_c(\boldsymbol{r},\boldsymbol{r},t), \quad (7.49)$$

[5] We neglect the ω dependence of $\tilde{P}_0(\boldsymbol{k},0,\omega)$ which is not singular.

or, using (5.5),

$$\boxed{\sigma_c(\omega = 0) = -s\frac{e^2 D}{\pi\hbar\Omega}\int_0^\infty dt\, Z_c(t)} \tag{7.50}$$

where $Z_c(t)$ is the contribution of the Cooperon to the integrated return probability. For anisotropic collisions, one has simply to replace τ_e by τ^*, and D by $D^* = v_F^2\tau^*/d$ (see Appendix A7.4).

7.4 The weak localization regime

Relations (7.45) and (7.50) give the quantum correction to the classical conductivity due to the Cooperon. The negative sign of this correction comes from selection of the direction $\boldsymbol{k}' \simeq -\boldsymbol{k}$ for which the coherent contribution is maximum. This reduction in conductivity by interference effects is called the *weak localization* correction. The adjective "weak" means that the correction is small compared to the main Drude contribution by a factor $1/k_F l_e \ll 1$. We shall see in section 7.5.2 that in the presence of spin-orbit coupling, the correction may become positive.

From relation (7.49), we can deduce the relative zero frequency weak localization correction $\Delta\sigma/\sigma_0 = \sigma_c/\sigma_0$ in the integral form:

$$\frac{\Delta\sigma}{\sigma_0} = -\frac{1}{\pi\hbar\rho_0}\int_0^\infty dt\, X_c(\boldsymbol{r},\boldsymbol{r},t) \tag{7.51}$$

or, using (7.50),

$$\frac{\Delta\sigma}{\sigma_0} = -\frac{\Delta}{\pi\hbar}\int_0^\infty dt\, Z_c(t). \tag{7.52}$$

The energy $\Delta = 1/(\Omega\rho_0)$ is the mean level spacing per spin direction (at the Fermi level).

Up to now, we have assumed that the pairs of time-reversed trajectories are phase coherent. This is no longer the case in the presence of dephasing processes which may induce a relative phase shift between these paired trajectories (see Chapter 6). Later on in this chapter, we consider some of them: effect of a magnetic field, scattering by magnetic impurities, spin-orbit scattering. We shall not consider the dephasing due to dynamic degrees of freedom such as electron–phonon coupling and electron–electron interaction which will be postponed to Chapter 13. Depending on the origin of the phase shift, i.e., whether the corresponding physical process is deterministic or not, the characteristic time will be called the "cutoff time" τ_γ or phase coherence time τ_ϕ (see sections 6.8 and 15.9). Finally the correction (7.52) associated with the Cooperon can be written in the form

$$\boxed{\Delta\sigma = -s\frac{e^2 D}{\pi\hbar\Omega}\int_0^\infty dt\, Z_c(t)e^{-t/\tau_\phi}} \tag{7.53}$$

which accounts for the suppression of long trajectories after a time τ_ϕ (or τ_γ).[6]

[6] We shall see in section 13.7.3 that the time dependence of the decoherence induced by electron–electron interactions is not always exponential.

7.4.1 Dimensionality effect

The behavior of the quantum correction to the conductivity is driven by the recurrence time τ_R defined by (5.16), that is, the time spent by a diffusive particle in the neighborhood of its starting point. For a finite system or for a system with a characteristic size much larger than the phase coherence length $L_\phi = \sqrt{D\tau_\phi}$, the behavior of the recurrence time depends crucially on the space dimensionality d, following relations (5.30). We therefore expect the weak localization, which is proportional to the integrated return probability, to depend also on d.

Let us first remark that expression (5.24) is only meaningful in the diffusive regime, that is for $t > \tau_e$. Therefore the possible divergence of the integral (7.53) for short times has to be cut off at time τ_e. We thus rewrite (7.53) in the symmetric form [171]

$$\Delta\sigma = -s\frac{e^2 D}{\pi\hbar\Omega}\int_0^\infty dt\, Z_c(t)(e^{-t/\tau_\phi} - e^{-t/\tau_e}). \tag{7.54}$$

Consider a system whose characteristic size $L = \Omega^{1/d}$ is larger than L_ϕ. For times smaller than τ_ϕ, the diffusion is the same as in an infinite medium, and the return probability is given by (5.24), that is $Z_c(t) = \Omega/(4\pi Dt)^{d/2}$. Using relation (15.85), the integral takes the form, for $d < 4$:[7]

$$\Delta\sigma = -s\frac{e^2}{\pi\hbar}\frac{1}{(4\pi)^{d/2}}\Gamma(1-d/2)[L_\phi^{2-d} - l_e^{2-d}]. \tag{7.55}$$

For the conductivity expressed in terms of the quantum of conductance e^2/h, we deduce the important results:

$$\begin{array}{|lll|}
\hline
\textit{quasi-}1d & \Delta\sigma = & -\, s\dfrac{e^2}{h}\dfrac{1}{S}(L_\phi - l_e) \\
d=2 & \Delta\sigma = & -\, s\dfrac{e^2}{\pi h}\ln\dfrac{L_\phi}{l_e} \\
d=3 & \Delta\sigma = & -\, s\dfrac{e^2}{2\pi h}\left(\dfrac{1}{l_e} - \dfrac{1}{L_\phi}\right) \\
\hline
\end{array} \tag{7.56}$$

There is no diffusive regime in strictly one dimension (see Exercise 3.9). We consider rather the case of a *quasi-one-dimensional* wire with a finite section S. "Quasi-one-dimensional" means that the diffusion is one dimensional in a three-dimensional wire with a section which is large compared to the Fermi wavelength λ_F, so that the number of transverse channels is large (see section A7.2.1): this limit corresponds to a very anisotropic three-dimensional system. We then use relation (7.54) with $\Omega = LS$ and $Z_c(t) = L/\sqrt{4\pi Dt}$.

[7] With the usual definitions of L_ϕ and l_e, namely $L_\phi^2 = D\tau_\phi$ and $l_e^2 = dD\tau_e$, relation (7.55) should exhibit the difference $L_\phi^{2-d} - (l_e/d)^{2-d}$. However, the cutoff is introduced "by hand" and the dependence on l_e is obtained only within a multiplicative constant. In the literature, one usually encounters the more "symmetric" expression (7.55), leading to relations (7.56) and (7.57). The dependence on l_e is only indicative.

The two-dimensional result corresponds to an electron gas strictly confined in a plane. For a quasi-2d system of width a, the conductivity has different dimensions and must be divided by a.

Given the expression (7.17) for σ_0, the relative correction to the conductivity can be rewritten as, for $L_\phi \gg l_e$,

$$\begin{aligned} &\textit{quasi-1d} \qquad \frac{\Delta\sigma}{\sigma_0} = -\frac{3L_\phi}{4Ml_e} \\ &d=2 \qquad \frac{\Delta\sigma}{\sigma_0} = -\frac{2}{\pi k_F l_e}\ln\frac{L_\phi}{l_e} \\ &d=3 \qquad \frac{\Delta\sigma}{\sigma_0} = -\frac{3}{2(k_F l_e)^2}. \end{aligned} \tag{7.57}$$

In the quasi-1d case, the relative correction depends on the number of channels $M = k_F^2 S/4\pi$ given by relation (7.142). In three dimensions, the correction does not depend on L_ϕ.

Exercise 7.1 Show that the conductivity (7.53) can be rewritten in the form

$$\Delta\sigma(\omega) = -s\frac{e^2}{\pi\hbar\Omega}\sum_Q \frac{1}{Q^2 + \frac{1}{L_\phi^2} - i\frac{\omega}{D}}. \tag{7.58}$$

Replacing the sum by an integral,

$$\Delta\sigma(\omega) = -\frac{e^2 d}{\pi\hbar}A_d \int \frac{dQ}{(2\pi)^d}\frac{Q^{d-1}}{Q^2 + \frac{1}{L_\phi^2} - i\frac{\omega}{D}}, \tag{7.59}$$

where A_d is the volume of the unit sphere in dimension d (15.2), show that we recover expressions (7.56) and (7.57) for the static weak localization correction.

When $\omega = 0$ and $L_\phi \to \infty$, the convergence of the integral (7.59) for $Q = 0$ and $Q \to \infty$ depends on the sign of $d-2$. The ultraviolet divergence at $Q \to \infty$ is artificial: the diffusion approximation $Ql_e \ll 1$ describes electronic trajectories of length larger than l_e, and therefore the integral over Q must have $1/l_e$ as an upper cutoff. For $d \leq 2$, the divergence at $Q = 0$ is more serious, and a finite value of L_ϕ is necessary to make the integral converge.

Exercise 7.2 From relation (7.58), show that for a quasi-one-dimensional wire connected to leads and with finite L_ϕ, the weak localization correction to the average conductance, expressed in units of e^2/h, is

$$\Delta g = -\frac{2s}{\pi^2}\sum_{n=1}^{\infty}\frac{1}{n^2 + (x/\pi)^2} \tag{7.60}$$

with $x = L/L_\phi$. This sum is equal to

$$\Delta g = -s\left(\frac{1}{x}\coth x - \frac{1}{x^2}\right) \tag{7.61}$$

and it has the limiting behaviors

$$\begin{aligned} \Delta g &\to -\frac{s}{3} \quad \text{for} \quad L_\phi \gg L \\ \Delta g &\to -s\frac{L_\phi}{L} \quad \text{for} \quad L_\phi \ll L. \end{aligned} \tag{7.62}$$

When should we consider the diffusion in a wire to be one or three dimensional? We have seen in section 5.5.4 that the effective dimensionality depends on the time scale being considered. For a wire of finite width, the diffusion is three dimensional for short times and one dimensional for long times. The weak localization correction, which results from the contribution of all diffusive trajectories, is thus three dimensional for times smaller than the transverse diffusion time (section 5.5.4) and one dimensional for longer times. The variation of the conductivity with L_ϕ is thus given by the three-dimensional result, if L_ϕ is smaller than the transverse dimension of the wire, and by the one-dimensional result in the other case. Consequently, by lowering the temperature, that is, by increasing L_ϕ, in principle it is possible to display the crossover from three-dimensional to one-dimensional behavior [172].

7.4.2 Finite size conductors

For a finite size conductor, electronic transport is characterized by a conductance which is related to the conductivity by Ohm's law $G = \sigma L^{d-2}$. Introducing dimensionless conductances written in units of e^2/h, we obtain from (7.17) and (7.56), the classical (Drude) conductance g and the weak localization correction Δg:

$$\begin{array}{lll} \textit{quasi-1d} & g = s\dfrac{4}{3}\dfrac{Ml_e}{L} & \Delta g = -s\dfrac{L_\phi}{L} \\ d = 2 & g = \dfrac{s}{2}k_F l_e & \Delta g = -\dfrac{s}{\pi}\ln\dfrac{L_\phi}{l_e} \\ d = 3 & g = s\dfrac{k_F^2 L l_e}{3\pi} & \Delta g = -\dfrac{s}{2\pi}\dfrac{L}{l_e} \end{array} \tag{7.63}$$

These results correspond to the case where the phase coherence length L_ϕ is smaller than the system size L. In the opposite case $L < L_\phi$, defined as the *mesoscopic regime*, the weak localization correction is more difficult to calculate because it depends explicitly on the solutions of the diffusion equation and on the expression of $Z(t)$ for a finite size system. An analytical solution exists only for a few specific geometries (Chapter 5). Nevertheless, it is natural to consider that the cutoff length, which is L_ϕ for an infinite system, is now replaced by the smallest length, namely the system size L. Therefore, the weak localization correction in the mesoscopic regime is obtained qualitatively by replacing L_ϕ by L in expressions (7.63). In Exercise 7.2, the quasi-one-dimensional case is solved exactly for all values of the ratio L/L_ϕ. In particular, Δg becomes constant in the mesoscopic regime $L \ll L_\phi$ and varies as L_ϕ/L for $L_\phi \ll L$.

Remark: localization length

The calculation of the weak localization correction is meaningful only if this correction remains small compared to the classical conductance. Relations (7.63) lead naturally to a new characteristic length, called *localization length* and denoted ξ, which corresponds to $\Delta g/g \simeq 1$.

In the quasi-$1d$ case, this length is proportional to the elastic mean free path and to the number of channels:

$$\xi_{1d} \simeq Ml_e. \tag{7.64}$$

In two dimensions, the localization length is exponentially large in the weak disorder limit $k_F l_e \gg 1$:

$$\xi_{2d} \simeq l_e \exp\left(\frac{\pi}{2} k_F l_e\right). \tag{7.65}$$

Therefore, even though $k_F l_e \gg 1$, the perturbative calculation becomes incorrect in the quasi-$1d$ case, when the system size becomes larger than ξ_{1d}. In two dimensions, this length remains very large as long as $k_F l_e \gg 1$. In three dimensions, it is infinite in the weak disorder limit.

The nature of electronic states changes entirely on lengths larger than ξ. The wave functions are no longer extended but rather localized over this length scale. For a sample size larger than ξ, electronic transport is no longer described by Ohm's law. This behavior is difficult to measure experimentally since it results from phase coherence, and L_ϕ must be larger than ξ. In three dimensions, the localization length is finite when $k_F l_e$ becomes smaller or of order unity. There is a phase transition, called the Anderson transition, between a regime where all states are extended and another state where they are localized. We shall not consider the physics of strong localization. For a review of this topic, see [167, 173–175].

7.4.3 Temperature dependence

In a metal, the weak localization correction to the conductivity is small. Experimentally, modification of an external parameter such as temperature, frequency or magnetic field can be used as a tool to measure it. For instance, temperature modifies the phase coherence time $\tau_\phi(T)$ and therefore modifies the weak localization correction. Assuming a power law variation $\tau_\phi \propto T^{-p}$, the decrease in phase coherence length $L_\phi(T) \propto T^{-p/2}$ with temperature leads to a reduction in the conductivity given by (7.56) which depends on space dimensionality d:

$$\begin{array}{lll} d=1 & \Delta\sigma(T) & \propto \; -T^{-p/2} \\ d=2 & \Delta\sigma(T) & \propto \; \ln T \\ d=3 & \Delta\sigma(T) & \propto \; T^{p/2}. \end{array} \tag{7.66}$$

As an example, Figure 7.3 shows the first experimental evidence for a $\ln T$ dependence in metallic films. The exponent p of the temperature depends on the physical mechanisms

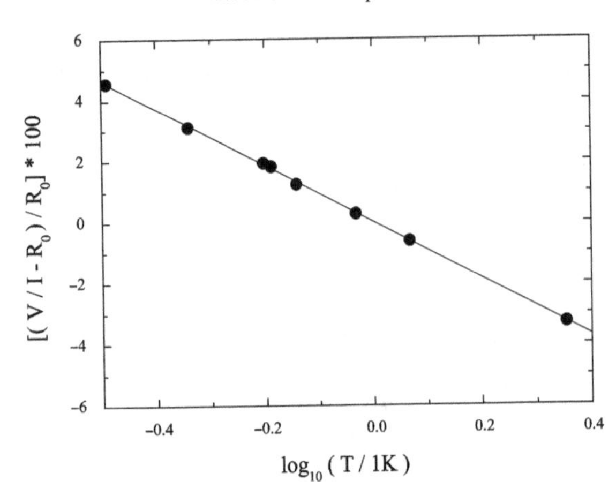

Figure 7.3 Logarithmic temperature dependence of the resistivity in a PdAu film [176].

which limit the phase coherence. The electron–phonon interaction corresponds to $p = 3$ whereas the electron–electron interaction gives $p = d/2$ for $d = 2, 3$ and $p = 2/3$ for $d = 1$ (see section 13.7). These different behaviors have been observed experimentally and are discussed in review papers [172, 174, 177–179].

7.5 Weak localization in a magnetic field

7.5.1 Negative magnetoresistance

Applying a magnetic field provides a very elegant way to probe the weak localization correction. In Chapter 6, we saw how the Cooperon is modified by the field. Since weak localization gives a negative contribution to the conductivity (in the absence of spin-orbit coupling), its suppression by a magnetic field leads to an increase in conductivity, that is a *negative magnetoresistance* which is certainly one of the most spectacular signatures of weak localization.[8]

We consider a magnetic field weak enough to fulfill $\omega_c \tau_e \ll 1$. In this limit we can neglect the bending of electronic trajectories between two elastic collisions (section 6.3).[9]

[8] The electron–electron interaction leads to an additional contribution to the conductivity which varies as $T^{d/2-1}$ for $d = 1, 3$ and as $\ln T$ for $d = 2$. It is magnetic field independent (see section 13.5).

[9] Remember that the Lorentz force leads to a change in the classical longitudinal conductivity which is written

$$\sigma_0(B) = \frac{\sigma_0(0)}{1 + (\omega_c \tau_e)^2},$$

$\omega_c = eB/m$ being the cyclotron frequency. The effects discussed in this section are weak field effects. They correspond to a limit where we can neglect the Lorentz force and the curvature of electronic trajectories. In this limit where $\omega_c \tau_e \ll 1$, σ_0 is assumed to remain constant.

The correction (7.53) to the conductivity due to the Cooperon becomes

$$\Delta\sigma(B) = -s\frac{e^2 D}{\pi\hbar\Omega}\int_0^\infty dt\, Z_c(t,B)e^{-t/\tau_\phi} \tag{7.67}$$

where $Z_c(t,B)$ is the partition function associated with the covariant diffusion equation in a magnetic field.

• For a two-dimensional gas in a uniform magnetic field perpendicular to the film, $Z_c(t,B)$ is given by (6.41)

$$Z_c(t,B) = \frac{BS/\phi_0}{\sinh(4\pi DtB/\phi_0)}. \tag{7.68}$$

For long times it decreases exponentially as e^{-t/τ_B}, where the characteristic time $\tau_B = \phi_0/(4\pi DB)$ acts as a cutoff time for large trajectories involved in the Cooperon. The convergence of the integral (7.67) at large times is thus driven by the smallest of the times τ_B and τ_ϕ, so that the magnetic field has an appreciable effect only if $\tau_B < \tau_\phi$. As in the absence of field, $Z_c(t,B)$ still gives a divergent contribution for $t \to 0$, which can be cured by taking τ_e as a lower cutoff. The weak localization correction becomes

$$\Delta\sigma(B) = -s\frac{e^2 D}{\pi\hbar}\int_0^\infty \frac{B/\phi_0}{\sinh 4\pi BDt/\phi_0}\left(e^{-t/\tau_\phi} - e^{-t/\tau_e}\right) dt. \tag{7.69}$$

It can be expressed as the difference between two digamma functions $\Psi(x)$ (defined in relation 15.49):

$$\Delta\sigma(B) = -s\frac{e^2}{4\pi^2\hbar}\left[\Psi\left(\frac{1}{2}+\frac{\hbar}{4eDB\tau_e}\right) - \Psi\left(\frac{1}{2}+\frac{\hbar}{4eDB\tau_\phi}\right)\right]. \tag{7.70}$$

Let us introduce the characteristic field B_ϕ, defined by $B_\phi = \phi_0/8\pi L_\phi^2$, which corresponds to a flux quantum through an area of the order of L_ϕ^2. The weak localization correction vanishes beyond this field which is of the order of 10^{-3}T for $L_\phi \simeq 1\mu$m. Since $\tau_e \ll \tau_\phi$ and $B \ll \phi_0/4\pi l_e^2$, the above expression can be replaced by

$$\boxed{\Delta\sigma(B) = -s\frac{e^2}{4\pi^2\hbar}\left[\ln\left(\frac{\hbar}{4eDB\tau_e}\right) - \Psi\left(\frac{1}{2}+\frac{\hbar}{4eDB\tau_\phi}\right)\right]} \tag{7.71}$$

In zero field, we recover the $d=2$ expression (7.56). Since σ_0 is field independent, the magnetoconductivity $\sigma(B)-\sigma(0)$ is the difference between the weak localization corrections $\Delta\sigma(B)-\Delta\sigma(0)$. Using the relations (7.56) for $d=2$ and (7.71), the magnetoconductivity becomes

$$\Delta\sigma(B) - \Delta\sigma(0) = -s\frac{e^2}{4\pi^2\hbar}\left[\ln\left(\frac{\hbar}{4eDB\tau_\phi}\right) - \Psi\left(\frac{1}{2}+\frac{\hbar}{4eDB\tau_\phi}\right)\right]. \tag{7.72}$$

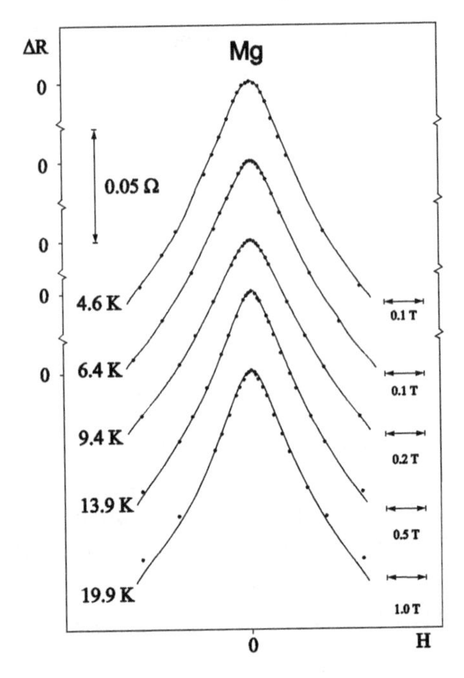

Figure 7.4 Magnetic field dependence of the magnetoresistance of a magnesium film for different temperatures. The points are experimental results and the solid curves correspond to (7.71). The time $\tau_\phi(T)$ is a fitting parameter [179].

The magnetoresistance given by $\Delta R(B) \propto -\Delta\sigma(B)/\sigma_0^2$ is thus negative. In the limit $B \ll B_\phi$, the asymptotic expansion (15.45) leads to

$$\Delta\sigma(B) - \Delta\sigma(0) \simeq \frac{s}{96\pi^2}\frac{e^2}{\hbar}\left(\frac{B}{B_\phi}\right)^2. \tag{7.73}$$

The field dependence of the weak localization correction has been studied extensively, particularly in metallic films, as shown in Figure 7.4 in a well-known example [179]. The fit of the experimental results to the theoretical prediction (7.71) gives the temperature dependence of the phase coherence time $\tau_\phi(T)$. This has become the standard method used to measure the phase coherence length and its temperature dependence [172].

• For a quasi-one-dimensional wire of section $W \times W$ in a uniform field perpendicular to its axis, the field dependence of the weak localization correction is given by (see Exercise 7.3)

$$\Delta\sigma(B) = -s\frac{e^2}{hS}\left(L_\phi^{-2} + \frac{e^2 W^2 B^2}{3\hbar^2}\right)^{-1/2}. \tag{7.74}$$

• For a conductor of size smaller than L_ϕ, in principle one should calculate the return probability $Z_c(t, B)$ in relation (7.67). This calculation is difficult because the eigenvalues

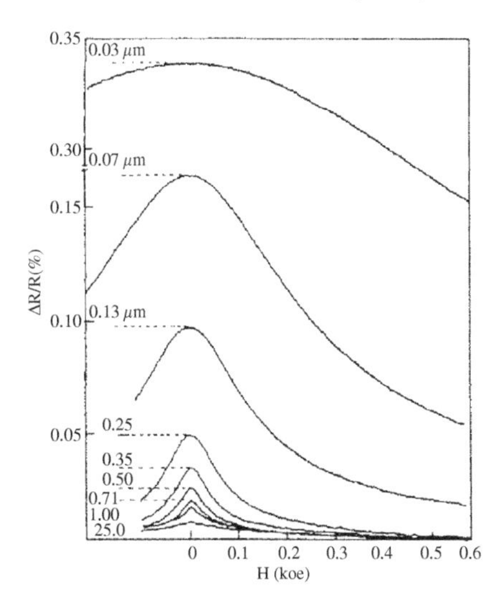

Figure 7.5 Magnetoresistance of lithium films of different widths W [180].

of the diffusion equation depend on the geometry of the sample. Practically, it is sufficient to replace the cutoff length $L_\phi = \sqrt{D\tau_\phi}$ in expressions (7.63) by the system size. If the sample is anisotropic, of length $L < L_\phi$ and width W, the area L_ϕ^2 must be replaced by the area LW. The weak localization correction disappears when there is one flux quantum through the sample. This behavior is clearly observed in Figure 7.5, which displays the magnetoresistance of wires of different widths. We see that the correction decreases as the inverse of W, as predicted by relation (7.57), where the number M of transverse channels is proportional to W for $d = 2$ (relation 7.172). Moreover, the characteristic field B_ϕ corresponds to a flux quantum through the area LW and varies as $1/W$.

Exercise 7.3: magnetoresistance of a quasi-1d wire

Consider a wire of square section $W \times W$, in a magnetic field B perpendicular to its axis Ox. Show that in a weak field, the weak localization correction is still given by the quasi-1d expression (7.56), with the substitution [181]:

$$\frac{1}{L_\phi^2} \longrightarrow \frac{1}{L_\phi^2(B)} = \frac{1}{L_\phi^2} + \frac{e^2B^2W^2}{3\hbar^2} = \frac{1}{L_\phi^2} + \frac{W^2}{12L_B^4}, \tag{7.75}$$

where $L_B = \sqrt{\hbar/2eB}$ is the magnetic length.

Hint: solve the diffusion equation with the magnetic field as a perturbation. The wire being quasi-one-dimensional, this equation is one dimensional (the transverse modes are $n_y = n_z = 0$).

At zero frequency, the eigenvalues of the diffusion equation (6.34) are solutions of

$$-D\left(\partial_x + i\frac{2eA_x}{\hbar}\right)^2 \psi_{n_x} = E_{n_x}\psi_{n_x}. \tag{7.76}$$

Choose the gauge $A_x = By$, $y \in [-y/2, y/2]$. In perturbation, the diffusion modes become

$$DQ_{n_x}^2 \longrightarrow DQ_{n_x}^2 + \frac{4e^2DB^2}{\hbar^2}\langle\psi_{n_x}|y^2|\psi_{n_x}\rangle. \tag{7.77}$$

The eigenstate ψ_{n_x} is independent of y (uniform transverse mode) so that calculation of the matrix element gives the same shift, $e^2B^2W^2/3\hbar^2$, to all eigenvalues. The weak localization correction therefore retains the quasi-1d form (7.56), given the substitution (7.75). We notice that the characteristic field for the magnetoresistance in this geometry corresponds to a flux quantum through an area $L_\phi W$.

7.5.2 Spin-orbit coupling and magnetic impurities

A magnetic field gives rise to a reversible dephasing which modifies the weak localization correction. Similar behavior holds for spin-orbit coupling. Its effect (section 6.5) is to multiply the Cooperon contribution by the factor $\langle Q_{so}(t)\rangle$ given by expression (6.139), so that (7.67) becomes

$$\Delta\sigma(B) = -s\frac{e^2D}{\pi\hbar\Omega}\int_0^\infty dt\, Z_c(t,B)\langle Q_{so}(t)\rangle\left(e^{-t/\tau_\phi} - e^{-t/\tau_e}\right). \tag{7.78}$$

The time dependence of $\langle Q_{so}(t)\rangle$ is set by the time scale τ_{so} which measures the strength of spin-orbit scattering. For a strong spin-orbit coupling, i.e., when $t \gg \tau_{so}$, the factor $\langle Q_{so}(t)\rangle$ changes its sign and tends to $-1/2$. As a consequence, the sign of the weak localization correction is changed [182]. This phenomenon is called *antilocalization*. There is a *decrease* in the resistance when the temperature is lowered. Since the magnetic field destroys the Cooperon contribution, we now expect a *positive* magnetoresistance (Figure 7.6). A straightforward generalization of the calculation leading to (7.70) gives the weak localization correction in a magnetic field and with spin-orbit coupling. We obtain for a two-dimensional electron gas:

$$\Delta\sigma(B) = -s\frac{e^2}{4\pi^2\hbar}\left[\Psi\left(\frac{1}{2}+\frac{B_e}{B}\right) - \frac{3}{2}\Psi\left(\frac{1}{2}+\frac{B_\phi + 4B_{so}/3}{B}\right) + \frac{1}{2}\Psi\left(\frac{1}{2}+\frac{B_\phi}{B}\right)\right] \tag{7.79}$$

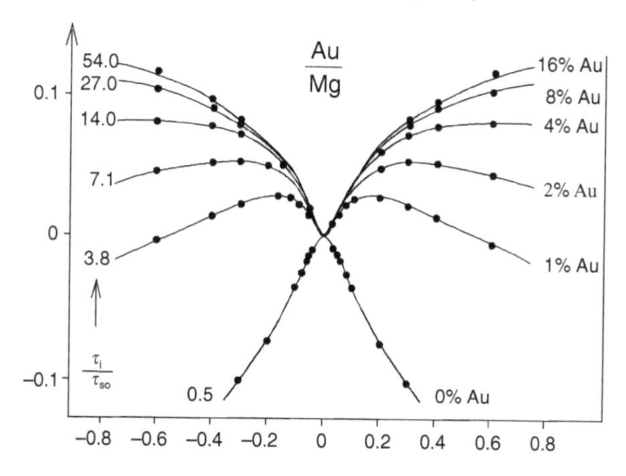

Figure 7.6 Magnetoresistance of a magnesium film with different concentrations of gold impurities. The spin-orbit coupling increases with the gold concentration and the magnetoresistance becomes positive [179].

where we have defined the characteristic fields

$$\boxed{\begin{aligned} B_e &= \frac{\hbar}{4eD\tau_e} \\ B_\phi &= \frac{\hbar}{4eD\tau_\phi} \\ B_{so} &= \frac{\hbar}{4eD\tau_{so}} \end{aligned}} \tag{7.80}$$

In contrast to spin-orbit coupling, scattering of electrons by magnetic impurities leads to an irreversible dephasing and to a finite phase coherence time (see sections 6.8 and 15.9). We have seen in section 6.5.7 that this scattering modifies the Cooperon by the factor $\langle Q_m(t)\rangle$ given by expression (6.140). This term multiplies $Z_c(t, B)$ in expression (7.67). When both spin-orbit coupling and scattering on magnetic impurities are present (relation 6.141), their respective reduction factors multiply *in each channel*, triplet or singlet, so that the global reduction factor becomes[10]

$$\langle Q_{so+m}(t)\rangle = \frac{1}{2}\left(3e^{-4t/3\tau_{so}-2t/3\tau_m} - e^{-2t/\tau_m}\right). \tag{7.81}$$

We deduce for the two-dimensional electron gas:

$$\Delta\sigma(B) = -s\frac{e^2}{4\pi^2\hbar}\left[\Psi\left(\frac{1}{2}+\frac{B_e}{B}\right) - \frac{3}{2}\Psi\left(\frac{1}{2}+\frac{B_1}{B}\right) + \frac{1}{2}\Psi\left(\frac{1}{2}+\frac{B_2}{B}\right)\right] \tag{7.82}$$

[10] See discussion page 215.

where the fields B_1 and B_2 are defined by $B_1 = B_\phi + \frac{4}{3}B_{so} + \frac{2}{3}B_m$ and $B_2 = B_\phi + 2B_m$ with[11]

$$\boxed{B_m = \frac{\hbar}{4eD\tau_m}} \tag{7.83}$$

Similarly, for the quasi-one-dimensional wire, the weak localization correction is given by

$$\Delta\sigma(B) = -s\frac{e^2}{hS}\left[\frac{3}{2}\left(\frac{1}{L_\phi^2(B)} + \frac{4}{3D\tau_{so}} + \frac{2}{3D\tau_m}\right)^{-1/2} - \frac{1}{2}\left(\frac{1}{L_\phi^2(B)} + \frac{2}{D\tau_m}\right)^{-1/2}\right], \tag{7.84}$$

where $L_\phi(B)$ is given by equation (7.75).

Exercise 7.4 Show that the weak localization correction to the conductivity at finite frequency can be written in the form:

$$\Delta\sigma(B) = -s\frac{e^2 D}{2\pi\hbar\Omega}\mathrm{Re}\sum_n\left(\frac{3}{-i\omega + E_n + \frac{1}{\tau_\phi} + \frac{4}{3\tau_{so}} + \frac{2}{3\tau_m}} - \frac{1}{-i\omega + E_n + \frac{1}{\tau_\phi} + \frac{2}{\tau_m}}\right) \tag{7.85}$$

where E_n are the eigenvalues of the diffusion equation. For the different cases studied in this chapter, we have

$$E_n \rightarrow \begin{cases} DQ^2 & \text{no magnetic field} \\ \left(n+\frac{1}{2}\right)\frac{4eDB}{\hbar} & \text{uniform field (infinite plane)} \\ 4\pi^2 E_c(n-2\varphi)^2 & \text{Aharonov–Bohm flux (ring)} \end{cases}$$

with the appropriate degeneracy.

7.6 Magnetoresistance and Aharonov–Bohm flux

An applied uniform magnetic field provides a very sensitive probe for the precise measurement of weak localization corrections. This quantitative description is a great achievement of the theory of coherent effects in weakly disordered metals and it thus constitutes a justification of the covariant diffusion equation for the Cooperon (6.34).

We have already noticed that in quantum mechanics, a charged particle is sensitive to the vector potential, even in the absence of a magnetic field. This is the Aharonov–Bohm effect (section 1.2). Then, from the covariant form of the diffusion equation and its close analogy

[11] Here we denote by B_ϕ the characteristic field related to all physical processes other than magnetic impurities and leading to a finite value of τ_ϕ.

with the Schrödinger equation (section 6.2), we expect the weak localization correction to be sensitive to the presence of an Aharonov–Bohm flux. This will show up through the flux dependence of the return probability Z_c (see section 6.4). The observation of the Aharonov–Bohm effect represents a truly spectacular success of the theory of weakly disordered metals. We first consider the case of a ring.

7.6.1 Ring

We consider a disordered ring of perimeter L and section S threaded by an Aharonov–Bohm flux (section 6.4.1). The flux dependence of the conductivity results from the diffusive trajectories that perform at least one turn around the ring. This is why we are interested in time scales necessarily larger than the transverse diffusion time (section 5.5.4). The diffusion in the ring is thus one dimensional if $L \gg \sqrt{S}$. In the presence of an Aharonov–Bohm flux ϕ, the weak localization correction (7.53) becomes

$$\Delta\sigma(\phi) = -s\frac{e^2 D}{\pi\hbar\Omega}\int_0^\infty dt\, Z_c(t,\phi)\, e^{-t/\tau_\phi}. \tag{7.86}$$

The integrated return probability $Z_c(t,\phi)$ is given by expression (6.54) and its expansion in terms of the winding number gives immediately the harmonics expansion of the correction to the conductivity:

$$\Delta\sigma(\phi) = \sum_{m=-\infty}^{\infty} \Delta\sigma_m \cos(4\pi m\varphi), \tag{7.87}$$

where $\varphi = \phi/\phi_0$ and

$$\Delta\sigma_m = -s\frac{e^2}{\pi\hbar}\frac{L}{S}\sqrt{\frac{E_c}{4\pi}}\int_0^\infty \frac{e^{-m^2/4E_c t - t/\tau_\phi}}{\sqrt{t}} dt. \tag{7.88}$$

With the help of relation (15.72), we have

$$\Delta\sigma = -s\frac{e^2}{h}\frac{L_\phi}{S}\left(1 + 2\sum_{m=1}^{\infty} e^{-mL/L_\phi}\cos 4\pi m\varphi\right) \tag{7.89}$$

which leads to[12]

$$\boxed{\Delta\sigma = -s\frac{e^2}{h}\frac{L_\phi}{S}\frac{\sinh(L/L_\phi)}{\cosh(L/L_\phi) - \cos 4\pi\varphi}} \tag{7.90}$$

The average conductivity of a ring is thus a periodic function of the flux ϕ with period $\phi_0/2 = h/2e$. The amplitude of the oscillations decreases exponentially with the perimeter L of the ring, and becomes negligible when $L \gg L_\phi$.

[12] This expression for $\phi = 0$ has already been obtained when calculating the recurrence time in a ring (5.51).

The periodicity $\phi_0/2$ is a direct consequence of the structure of the Cooperon. Indeed, the covariant diffusion equation (6.34) for the Cooperon describes a fictitious charge $(-2e)$ whose origin is the pairing of time-reversed diffusive trajectories.

It is difficult to observe this effect on a single ring, since it is not self-averaging. However, experiments performed on chains of rings [183] do display this periodicity.

7.6.2 Long cylinder: the Sharvin–Sharvin effect

Consider now the geometry of a cylinder of perimeter L, height L_z, and thickness a, as discussed in section 6.4.2. In this case, $Z_c(t, \phi)$ has a time dependence, different from the case of the ring, which originates from free diffusion along the cylinder axis. By inserting the expression (6.65) for $Z_c(t, \phi)$ obtained in this case into relation (7.53), we obtain an expansion similar to (7.87) for the conductivity, with harmonics given by

$$\Delta\sigma_m = -s\frac{e^2}{\pi\hbar a}\int_0^\infty \frac{e^{-m^2/4E_c t - t/\tau_\phi}}{4\pi t}dt = -s\frac{e^2}{\pi h a}K_0(mL/L_\phi) \tag{7.91}$$

where $K_0(x)$ is a modified Bessel function [184] which decreases exponentially for large argument (15.75). For $m \neq 0$, the integral is given by (15.70). For $m = 0$, it diverges at small times, and we must introduce the elastic time τ_e as a lower cutoff:

$$\int_0^\infty \frac{e^{-t/\tau_\phi} - e^{-t/\tau_e}}{4\pi t}dt = \frac{1}{2\pi}\ln\frac{L_\phi}{l_e}. \tag{7.92}$$

Consequently, the conductance correction ΔG measured along the cylinder height L_z, namely $\Delta G = \Delta\sigma a L/L_z$, is given by

$$\boxed{\Delta G(\phi) = -s\frac{e^2}{\pi h}\frac{L}{L_z}\left[\ln\frac{L_\phi}{l_e} + 2\sum_{m=1}^{+\infty}K_0(mL/L_\phi)\cos 4\pi m\varphi\right]} \tag{7.93}$$

As for the ring, the amplitude of oscillations of period $\phi_0/2$ decreases exponentially with the perimeter L of the cylinder, and becomes negligible when $L \gg L_\phi$.

These magnetoresistance oscillations were predicted by Altshuler, Aronov and Spivak [185] and observed by Sharvin and Sharvin [186]. In Figure 7.7, we see that the amplitude of the oscillations and the average resistance decrease when the field increases. This additional effect may be explained by the penetration of the uniform applied magnetic field inside the metal, so that the conditions of the Aharonov–Bohm effect are not exactly fulfilled. The penetration of the field can be simply accounted for by replacing the length L_ϕ in (7.93), by a field dependent phase coherence length given by (7.75).[13] The agreement between the experimental results and the theory is then excellent.

[13] The result of Exercise 7.3 obtained for a quasi-one-dimensional wire also holds for a cylinder, since the third dimension z plays no role and the perimeter L is much longer than the thickness a.

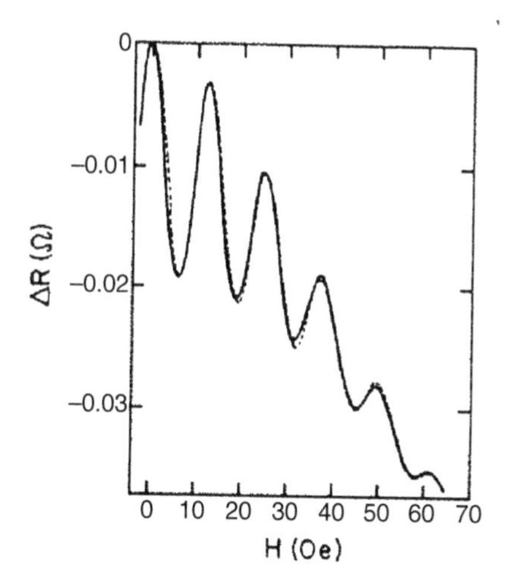

Figure 7.7 Magnetoresistance $\Delta R(B)$ of a lithium film evaporated on a cylindrical quartz filament (continuous line) and comparison with expression (7.93) (dashed line). This fit provides a value of the phase coherence length L_ϕ equal to 2.2 μm at the temperature $T = 1.1$ K (after [186, 187]).

7.6.3 Remark on the Webb and Sharvin–Sharvin experiments: ϕ_0 versus $\phi_0/2$

In the experiment of R. Webb *et al.* [188] described in Chapter 1 (section 1.2), we have seen that the magnetoresistance of a single ring pierced by a magnetic flux ϕ displays oscillations of period ϕ_0. This is the Aharonov–Bohm effect. However, we have just seen that the same type of experiment performed earlier by Sharvin and Sharvin on a long cylinder pierced by a magnetic flux [186] also displays magnetoresistance oscillations, but with a period of $\phi_0/2$. Why does the harmonics ϕ_0 disappear in the second experiment? To answer this question, let us imagine that we have many rings with different disorder configurations. The magnetoresistance of each of them oscillates with a period ϕ_0, but with a random phase that depends on the disorder configuration and vanishes upon averaging over disorder configurations [189].[14] The same argument holds for the case of a cylinder which can be viewed as an ensemble of such rings with random phases. However, the Sharvin–Sharvin experiment shows unambiguously that there exists an interference effect which survives the disorder average. This effect results from the phase coherence between time-reversed multiple scattering trajectories and it corresponds to the flux-dependent part of the weak localization correction [185]. It gives rise to oscillations of period $\phi_0/2$, since the relative phase of time-reversed trajectories that contribute to the Cooperon is $4\pi\phi/\phi_0$. There are alternative ways of performing an ensemble average, for instance by considering the magnetoresistance of a network made of quasi-one-dimensional diffusive wires [191–195].

[14] For a two-point measurement, the phase is restricted to being 0 or π, because of Onsager relations [190].

7.6.4 The Aharonov–Bohm effect in an infinite plane

We now consider the case of a single Aharonov–Bohm magnetic flux line, impenetrable to electrons and perpendicular to an infinite plane. This limit may be useful to describe the situation of a very inhomogeneous magnetic field. The corresponding Aharonov–Bohm vector potential is

$$\boldsymbol{A}(\boldsymbol{r}) = \frac{\phi}{2\pi r}\,\hat{\boldsymbol{e}}_\theta \tag{7.94}$$

so that the magnetic field is confined in a single point, namely

$$\boldsymbol{B}(\boldsymbol{r}) = \phi\,\delta(\boldsymbol{r})\hat{\boldsymbol{e}}_z. \tag{7.95}$$

To evaluate $\Delta\sigma(\phi)$, we must calculate the corresponding partition function $Z(t,\phi)$. This was obtained in Appendix A6.1:

$$Z(t,\phi) - Z(t,0) = -\varphi(1-2\varphi) \tag{7.96}$$

where $\varphi = \phi/\phi_0$ and $\varphi \in [0,1]$. Using relation (7.67), we obtain for the weak localization correction at small flux ϕ

$$\boxed{\Delta\sigma(\phi) - \Delta\sigma(0) = 2s\frac{e^2}{h}\frac{L_\phi^2}{S}\frac{|\phi|}{\phi_0}} \tag{7.97}$$

where S is the area of the sample. The magnetoconductance has a *triangular singularity* for $\phi \to 0$ which contrasts with the quadratic behavior (7.73) obtained for a weak uniform magnetic field.

This singularity has been observed in a metallic film above which a type II superconductor has been evaporated (Figure 7.8). The dilute limit for which the vortex density is small corresponds to an ensemble of independent flux lines and it can be described by relation (7.97). Increasing the temperature, and thus the vortex density, a crossover occurs towards a uniform magnetic field where quadratic behavior in low field is observed as predicted by relation (7.73) [196, 197].

Appendix A7.1: Kubo formulae

A7.1.1 Conductivity and dissipation

Our purpose here is to calculate the static conductivity of a disordered metal. We consider a system at equilibrium which is perturbed by an external voltage (for more details see for instance [164]). In the framework of linear response theory, this voltage induces an electric field $\boldsymbol{E}$ and a current density $\boldsymbol{j}$ which is proportional[15] to it, namely $\boldsymbol{j} = \sigma\boldsymbol{E}$. To calculate

[15] We consider an isotropic medium for which the conductivity tensor is spherical.

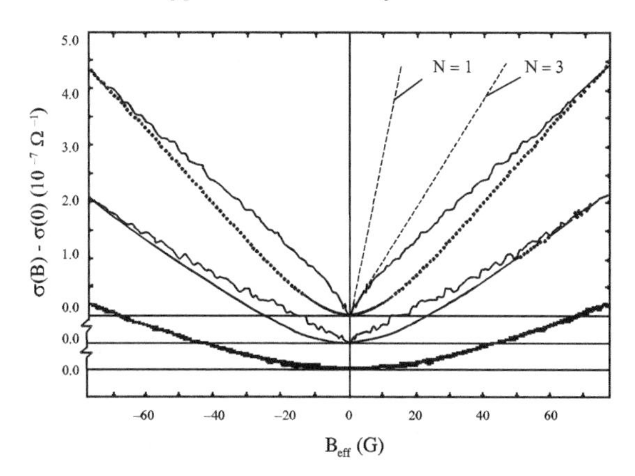

Figure 7.8 Magnetoconductance of a two-dimensional electron gas on top of which a lead superconducting film has been evaporated. The field B_{eff} is created by the vortices induced in the type II superconductor. The different curves correspond to different temperatures. The integer N is the number of flux quanta. We notice the crossover between linear behavior at small field and quadratic behavior at larger field [196]. For more details of this geometry, see [196, 197].

the conductivity σ, we introduce a perturbation term associated to the electric field in the Hamiltonian (2.1) which becomes

$$\mathcal{H} = \frac{p^2}{2m} + V(\boldsymbol{r}) + e\boldsymbol{E} \cdot \boldsymbol{r}. \tag{7.98}$$

However, for a macroscopic conductor, the additional term of this Hamiltonian cannot be considered as a perturbation, even though the electric field is small. It is more convenient to use the equivalent Hamiltonian

$$\mathcal{H} = \frac{[\boldsymbol{p} + e\boldsymbol{A}(t)]^2}{2m} + V(\boldsymbol{r}), \tag{7.99}$$

where the electric field results from the time-dependent vector potential $\boldsymbol{A}(t)$ as

$$\boldsymbol{E} = -\frac{\partial \boldsymbol{A}}{\partial t} \quad \text{and} \quad \nabla \times \boldsymbol{A} = 0. \tag{7.100}$$

The current density is given by $\boldsymbol{j} = \mathrm{Tr}(\rho \hat{\boldsymbol{j}})$ where $\hat{\boldsymbol{j}}$ is the current density operator and $\rho(t)$ is the one-particle density matrix which obeys

$$i\hbar \frac{\partial \rho}{\partial t} = [\mathcal{H}, \rho]. \tag{7.101}$$

We can write ρ in the form $\rho = \rho_0 + \delta\rho$, where ρ_0 is the equilibrium density matrix that corresponds to $\boldsymbol{A} = 0$. A stationary electric field is suitably described by a monochromatic field of frequency ω, $\boldsymbol{E}(\omega) = i\omega \boldsymbol{A}(\omega)$, in the limit $\omega \to 0$. For a macroscopic system, (7.101) leads to an infinitely long transient regime without relaxation towards a stationary regime. One way to cure this problem is to introduce a term of the form $-i\gamma\delta\rho(t)$ into (7.101), leading to an exponential relaxation on the time scale $\hbar/\gamma$. We finally take the limit $\gamma \to 0$, assuming that the end result does not depend on γ.

Another way to interpret the relaxation rate γ is to consider a finite size conductor coupled to the external environment (the reservoirs) constituted by the measurement leads. Electrons can penetrate into the reservoirs, change their phase randomly, and be reinjected incoherently into the sample. A weak coupling between the conductor and the reservoirs can be modelled with a broadening γ of the energy levels of the isolated sample, so that $\hbar/\gamma$ represents the lifetime of an electron in the conductor. Instead of (7.101), we obtain

$$i\hbar\frac{\partial\rho}{\partial t} = [\mathcal{H}, \rho] - i\gamma\left(\rho(t) - \rho_{eq}(t)\right), \tag{7.102}$$

where the density matrix ρ_{eq} describes the total system at thermal equilibrium. We may conclude from this description that an intrinsic irreversible behavior appears as a result of the coupling to reservoirs. But we should keep in mind that the whole system, namely conductor together with reservoirs, remains quantum mechanically coherent, and it is the averaging over the degrees of freedom of the reservoirs that introduces irreversibility.

Remark: conductivity, dissipation and the Feynman chain

The conductivity measures the dissipation by Joule heating in a wire of section S, length L and resistance $R = (1/\sigma)(L/S)$. It may seem paradoxical that the conductivity, proportional to the *elastic* collision time τ_e, corresponds to an irreversible dissipation effect.

The solution to this paradox rests in the thermodynamic limit. To understand this point, let us consider the well-known case [198] of a chain of electrical quadrupoles LC (made of a self-inductance L and a capacitance C) in series connected to a generator at one extremity of the chain. Each quadrupole is a purely reactive element with a purely imaginary impedance, so that it does not absorb energy. However, the equivalent impedance Z of the infinite chain of quadrupoles is easily found to be

$$Z = \sqrt{L/C - \omega^2 L^2/4}$$

which is real at low frequency $\omega < 2/\sqrt{LC}$. This means that the infinite chain dissipates energy! This is a consequence of the infinite limit. In this limit, the circuit can absorb energy and transfer it to infinity. The same consideration applies to electronic disordered systems with elastic collisions, that is, the dissipation occurs in the reservoirs which are considered as infinite systems.

Within the linear response approximation, we retain only the linear terms in $\boldsymbol{A}(t)$ in the expressions of $\delta\rho(t)$, of $\delta\rho_{eq}(t) = \rho_{eq} - \rho_0$, and of the Hamiltonian (7.99). Rewriting $\mathcal{H} = \mathcal{H}_0 + \mathcal{H}_1(t)$ where $\mathcal{H}_0$ is given by (2.1) and

$$\mathcal{H}_1(t) = \frac{e}{2m}(\boldsymbol{p}\cdot\boldsymbol{A} + \boldsymbol{A}\cdot\boldsymbol{p}), \tag{7.103}$$

equation (7.102) can be cast in the form

$$i\hbar\frac{\partial\delta\rho}{\partial t}=[\mathcal{H}_0,\delta\rho(t)]+[\mathcal{H}_1,\rho_0]-i\gamma\left(\delta\rho(t)-\delta\rho_{eq}(t)\right). \tag{7.104}$$

Fourier transforming this equation expressed in a basis $(|\alpha\rangle,\epsilon_\alpha)$ of eigenstates of $\mathcal{H}_0$ leads to

$$\langle\alpha|\delta\rho(\omega)|\beta\rangle=\frac{[f(\epsilon_\alpha)-f(\epsilon_\beta)]\langle\alpha|\mathcal{H}_1(\omega)|\beta\rangle-i\gamma\langle\alpha|\delta\rho_{eq}(\omega)|\beta\rangle}{(\epsilon_\alpha-\epsilon_\beta)-\hbar\omega-i\gamma} \tag{7.105}$$

where $f(\epsilon_\alpha)$ is the occupation factor of the state $|\alpha\rangle$. The stationary condition for $\delta\rho_{eq}$ is expressed by

$$\langle\alpha|\delta\rho_{eq}(\omega)|\beta\rangle=\frac{f(\epsilon_\alpha)-f(\epsilon_\beta)}{\epsilon_\alpha-\epsilon_\beta}\langle\alpha|\mathcal{H}_1(\omega)|\beta\rangle. \tag{7.106}$$

In addition, the current density operator $\hat{\boldsymbol{j}}$ can also be written as the sum of two terms: $\hat{\boldsymbol{j}}=\hat{\boldsymbol{j}}_0+\hat{\boldsymbol{j}}_1$ with

$$\begin{aligned}\hat{\boldsymbol{j}}_0&=-\frac{e}{2m}\left(\hat{n}(\boldsymbol{r})\boldsymbol{p}+\boldsymbol{p}\,\hat{n}(\boldsymbol{r})\right)\\ \hat{\boldsymbol{j}}_1&=-\frac{e^2}{2m}\left(\hat{n}(\boldsymbol{r})\boldsymbol{A}+\boldsymbol{A}\hat{n}(\boldsymbol{r})\right)\end{aligned} \tag{7.107}$$

so that the current $\boldsymbol{j}$ becomes in linear response,

$$\boldsymbol{j}=\mathrm{Tr}\left(\rho(t)\hat{\boldsymbol{j}}\right)\simeq\mathrm{Tr}\left(\rho_0\hat{\boldsymbol{j}}_1\right)+\mathrm{Tr}\left(\delta\rho(t)\hat{\boldsymbol{j}}_0\right). \tag{7.108}$$

The first term corresponds to the diamagnetic current $\mathrm{Tr}\left(\rho_0\hat{\boldsymbol{j}}_1\right)=-ne^2A/m$, where n is the electronic density. With the help of (7.108), (7.105) and (7.106), we obtain

$$j_x(\omega)=-A_x(\omega)\left[\frac{ne^2}{m}+\frac{e^2}{m^2\Omega}\sum_{\alpha\beta}\frac{f(\epsilon_\alpha)-f(\epsilon_\beta)}{\epsilon_\alpha-\epsilon_\beta}\frac{\epsilon_\alpha-\epsilon_\beta-i\gamma}{\epsilon_\alpha-\epsilon_\beta-\hbar\omega-i\gamma}|\langle\alpha|p_x|\beta\rangle|^2\right] \tag{7.109}$$

for an electric field applied along the Ox axis, so that the conductivity is

$$\sigma_{xx}(\omega)=\frac{i}{\omega}\left[\frac{ne^2}{m}+\frac{e^2}{m^2\Omega}\sum_{\alpha\beta}\frac{f(\epsilon_\alpha)-f(\epsilon_\beta)}{\epsilon_\alpha-\epsilon_\beta}\frac{\epsilon_\alpha-\epsilon_\beta-i\gamma}{\epsilon_\alpha-\epsilon_\beta-\hbar\omega-i\gamma}|\langle\alpha|p_x|\beta\rangle|^2\right]. \tag{7.110}$$

In order to simplify this expression, we use the so-called "*f-sum rule*" which states that

$$n+\frac{1}{m\Omega}\sum_{\alpha\beta}\frac{f(\epsilon_\alpha)-f(\epsilon_\beta)}{\epsilon_\alpha-\epsilon_\beta}|\langle\alpha|p_x|\beta\rangle|^2=0. \tag{7.111}$$

To prove this equality, we define the oscillator strength[16]

$$f_{\alpha\beta} = \frac{2}{m}\frac{|\langle\alpha|p_x|\beta\rangle|^2}{\epsilon_\alpha - \epsilon_\beta} \tag{7.112}$$

and use the commutator $[x, \mathcal{H}_0] = i\hbar p_x/m$, so that $\sum_{\beta\neq\alpha} f_{\alpha\beta} = -1$. Therefore,

$$\frac{1}{\Omega}\sum_\alpha f(\epsilon_\alpha)\sum_\beta \frac{|\langle\alpha|p_x|\beta\rangle|^2}{\epsilon_\alpha - \epsilon_\beta} = -\frac{nm}{2}. \tag{7.113}$$

A permutation of the sums over α and β changes the sign of the right hand side term. After subtraction of the two relations, we obtain (7.111). Using this relation in expression (7.110) together with the identity

$$\frac{\epsilon_\alpha - \epsilon_\beta - i\gamma}{\epsilon_\alpha - \epsilon_\beta - \hbar\omega - i\gamma} = 1 + \frac{\hbar\omega}{\epsilon_\alpha - \epsilon_\beta - \hbar\omega - i\gamma}, \tag{7.114}$$

yields the following expression for the conductivity:

$$\sigma_{xx}(\omega) = i\frac{e^2\hbar}{m^2\Omega}\sum_{\alpha\beta}\frac{f(\epsilon_\alpha) - f(\epsilon_\beta)}{\epsilon_\alpha - \epsilon_\beta}\frac{|\langle\alpha|p_x|\beta\rangle|^2}{\epsilon_\alpha - \epsilon_\beta - \hbar\omega - i\gamma}. \tag{7.115}$$

Introducing the matrix element of the current operator $j_{\alpha\beta} = \frac{e}{m}\langle\alpha|p_x|\beta\rangle$ (not to be confused with the current density operator defined by 7.107) and restoring the spin s, we obtain finally

$$\boxed{\sigma_{xx}(\omega) = is\frac{\hbar}{\Omega}\sum_{\alpha\beta}\frac{f(\epsilon_\alpha) - f(\epsilon_\beta)}{\epsilon_\alpha - \epsilon_\beta}\frac{|j_{\alpha\beta}|^2}{\epsilon_\alpha - \epsilon_\beta - \hbar\omega - i\gamma}} \tag{7.116}$$

In particular, the real part is

$$\operatorname{Re}\sigma_{xx}(\omega) = -s\frac{\pi\hbar}{\Omega}\sum_{\alpha\beta}\frac{f(\epsilon_\alpha) - f(\epsilon_\beta)}{\hbar\omega}|j_{\alpha\beta}|^2\delta_\gamma(\epsilon_\alpha - \epsilon_\beta - \hbar\omega), \tag{7.117}$$

where δ_γ is a δ function "broadened" by the coupling to the external environment

$$\delta_\gamma(x) = \frac{\gamma/\pi}{x^2 + \gamma^2}. \tag{7.118}$$

Inserting an additional integral over energy and using (15.59), we rewrite (7.117) as

$$\operatorname{Re}\sigma_{xx}(\omega) = -s\frac{\pi\hbar}{\Omega}\sum_{\alpha\beta}\int d\epsilon\frac{f(\epsilon) - f(\epsilon - \hbar\omega)}{\hbar\omega}|j_{\alpha\beta}|^2\delta_{\gamma/2}(\epsilon - \epsilon_\alpha)\delta_{\gamma/2}(\epsilon - \hbar\omega - \epsilon_\beta) \tag{7.119}$$

[16] This terminology originates from atomic physics.

which can be expressed as a trace:

$$\mathrm{Re}\,\sigma_{xx}(\omega) = -s\frac{\pi\hbar}{\Omega}\int d\epsilon \frac{f(\epsilon)-f(\epsilon-\hbar\omega)}{\hbar\omega}\mathrm{Tr}\left[\hat{j}_x\,\delta_{\gamma/2}(\epsilon-\mathcal{H}_0)\,\hat{j}_x\,\delta_{\gamma/2}(\epsilon-\hbar\omega-\mathcal{H}_0)\right] \tag{7.120}$$

or

$$\mathrm{Re}\,\sigma_{xx}(\omega) = -s\frac{\hbar}{\pi\Omega}\int d\epsilon \frac{f(\epsilon)-f(\epsilon-\hbar\omega)}{\hbar\omega}\mathrm{Tr}\left[\hat{j}_x\,\mathrm{Im}\hat{G}^R_\epsilon\,\hat{j}_x\,\mathrm{Im}\hat{G}^R_{\epsilon-\hbar\omega}\right] \tag{7.121}$$

with

$$\hat{G}^{R,A}_\epsilon = \frac{1}{\epsilon-\hat{\mathcal{H}}_0 \pm i\gamma/2}. \tag{7.122}$$

When the temperature is much lower than the Fermi temperature, we can safely replace this expression by its zero temperature limit:[17]

$$\boxed{\mathrm{Re}\,\sigma_{xx}(\epsilon_F,\omega) = s\frac{\hbar}{\pi\Omega}\mathrm{Tr}\left[\hat{j}_x\,\mathrm{Im}\hat{G}^R_{\epsilon_F}\,\hat{j}_x\,\mathrm{Im}\hat{G}^R_{\epsilon_F-\hbar\omega}\right]} \tag{7.123}$$

which is relation (7.1). At finite temperature this expression generalizes to

$$\sigma_{xx}(T,\omega) = -\int d\epsilon \frac{f(\epsilon)-f(\epsilon-\hbar\omega)}{\hbar\omega}\sigma_{xx}(\epsilon,\omega), \tag{7.124}$$

whose zero frequency limit is

$$\sigma_{xx}(T) = \int d\epsilon\left(-\frac{\partial f}{\partial\epsilon}\right)\sigma_{xx}(\epsilon). \tag{7.125}$$

A7.1.2 Density-density response function

We consider the response of a free electron gas to a scalar potential $V(\boldsymbol{q},\omega)$. In the linear response approximation, this potential induces a variation of the electronic density $\delta n(\boldsymbol{q},\omega)$ of the form

$$\delta n(\boldsymbol{q},\omega) = -\chi_0(\boldsymbol{q},\omega)V(\boldsymbol{q},\omega), \tag{7.126}$$

where $\chi_0(\boldsymbol{q},\omega)$ is called the susceptibility or the density-density response function. Within the linear response theory, it can be shown that this response function is given by

$$\chi_0(\boldsymbol{q},\omega) = -\frac{s}{\Omega}\sum_{\alpha\beta}\frac{f_\alpha - f_\beta}{\epsilon_\alpha-\epsilon_\beta-\hbar\omega-i0^+}|\langle\alpha|e^{i\boldsymbol{q}\cdot\boldsymbol{r}}|\beta\rangle|^2 \tag{7.127}$$

[17] There is another temperature dependence that arises from the coupling to other degrees of freedom, described by the energy broadening $\gamma = \hbar/\tau_\phi(T)$, where the phase coherence time τ_ϕ depends on temperature.

which can be expressed in terms of Green's functions [199]

$$\chi_0(\boldsymbol{q},\omega) = s\int \frac{d\epsilon}{2i\pi}\left[f(\epsilon-\hbar\omega)\Phi_\omega^{RR} - f(\epsilon)\Phi_\omega^{AA} + [f(\epsilon)-f(\epsilon-\hbar\omega)]\,\Phi_\omega^{RA}\right] \quad (7.128)$$

with

$$\Phi_\omega^{RA}(\boldsymbol{q},\omega) = \frac{1}{\Omega}\mathrm{Tr}\left[e^{-i\boldsymbol{q}\cdot\hat{\boldsymbol{r}}}\,\hat{G}_\epsilon^R\,e^{i\boldsymbol{q}\cdot\hat{\boldsymbol{r}}}\,\hat{G}_{\epsilon-\hbar\omega}^A\right] \quad (7.129)$$

and

$$\hat{G}_\epsilon^{R,A} = \frac{1}{\epsilon - \hat{\mathcal{H}}_0 \pm i0^+}. \quad (7.130)$$

The trace is conveniently rewritten in the momentum basis:

$$\Phi_\omega^{RA}(\boldsymbol{q},\omega) = \frac{1}{\Omega}\sum_{\boldsymbol{k},\boldsymbol{k}'} G_\epsilon^R(\boldsymbol{k}_+,\boldsymbol{k}'_+)G_{\epsilon-\hbar\omega}^A(\boldsymbol{k}_-,\boldsymbol{k}'_-) \quad (7.131)$$

with $\boldsymbol{k}_\pm = \boldsymbol{k} \pm \boldsymbol{q}/2$. There are similar expressions for Φ^{RR} and Φ^{AA}. The disorder average $\overline{\Phi}^{RA}$ is related to the probability of quantum diffusion through (4.12)

$$\overline{\Phi}_\omega^{RA}(\boldsymbol{q},\omega) = \frac{2\pi\rho_0}{\hbar}P(\boldsymbol{q},\omega). \quad (7.132)$$

Moreover, in the limit $\boldsymbol{q},\omega \to 0$, we have

$$\mathrm{Im}\overline{\Phi}_\omega^{RR}(\boldsymbol{q},\omega) = -\mathrm{Im}\overline{\Phi}_\omega^{AA}(\boldsymbol{q},\omega) = \pi\frac{\partial}{\partial\epsilon}\rho(\epsilon) \quad (7.133)$$

where $\rho(\epsilon)$ is the density of states (3.30) and $\mathrm{Re}\overline{\Phi}_\omega^{RR}(\boldsymbol{q},\omega) = \mathrm{Re}\overline{\Phi}_\omega^{AA}(\boldsymbol{q},\omega)$. At zero temperature and in the diffusive limit, we finally obtain for $\overline{\chi}_0(\boldsymbol{q},\omega)$[18]

$$\boxed{\overline{\chi}_0(\boldsymbol{q},\omega) = s\rho_0[1 + i\omega P_d(\boldsymbol{q},\omega)] = s\rho_0\frac{Dq^2}{-i\omega + Dq^2}} \quad (7.134)$$

There exists an additional term $(s/2i\pi)\hbar\omega\,\mathrm{Re}\phi^{RR}$ of order $\hbar\omega/\epsilon_F$, which is negligible [199, 200].

Appendix A7.2: Conductance and transmission

A7.2.1 Introduction: Landauer formula

In this chapter, the electronic transport is characterized by the conductivity σ. It is the linear response of the conductor to an electric field. In its most general form, the conductivity is

18 We neglect the contribution of P_0 (section A4.1.1).

a non-local function $\sigma(\boldsymbol{r},\boldsymbol{r}')$, given by (7.7), which relates the current density to the local field

$$\boldsymbol{j}(\boldsymbol{r}) = \int d\boldsymbol{r}' \sigma(\boldsymbol{r},\boldsymbol{r}')\boldsymbol{E}(\boldsymbol{r}'). \tag{7.135}$$

For a conductor invariant under translation and a constant local electric field,[19] this relation reduces to $\boldsymbol{j} = \sigma \boldsymbol{E}$ where the conductivity σ is given by

$$\sigma = \int d\boldsymbol{r}' \, \sigma(\boldsymbol{r},\boldsymbol{r}') = \frac{1}{\Omega} \int d\boldsymbol{r} \, d\boldsymbol{r}' \, \sigma(\boldsymbol{r},\boldsymbol{r}'), \tag{7.136}$$

where Ω is the volume of the conductor. The classical conductivity is an intrinsic property of the system, which is defined in the bulk and depends neither on the geometry of the sample nor on the position of the voltage sources.

On the other hand, an electric transport experiment does not measure the conductivity but rather the conductance G, that is the current $I = GV$ induced by a potential drop. For a sample of length L and section S:

$$I = \int_S \boldsymbol{j}(\boldsymbol{r}) \, d\boldsymbol{\rho} = \frac{V}{L^2} \int d\boldsymbol{r} \, d\boldsymbol{r}' \, \sigma(\boldsymbol{r},\boldsymbol{r}') = \frac{V}{L^2} \Omega \sigma \tag{7.137}$$

which leads to Ohm's law[20]

$$G = \sigma \frac{S}{L}. \tag{7.138}$$

For a sample with any shape, the conductance G depends on the geometry and on the location of the voltage sources.

In this appendix, we present an approach due to Landauer [201, 202] which relates the conductance G to the transmission properties of the disordered conductor, viewed as a quantum potential barrier. For a finite length one-dimensional wire connected to perfect conductors, the Landauer formula,

$$G = s \frac{e^2}{h} T, \tag{7.139}$$

relates the conductance to the transmission coefficient T through the wire.[21] This relation holds for each realization of disorder and not only on average.

[19] The relation (7.135) is valid for any distribution of the electric field which satisfies the boundary conditions given by the potential at the boundaries of the sample [161].

[20] Here we restrict on purpose the discussion to general considerations on the relation between conductivity and conductance. For a more detailed description, see [161].

[21] The conductance we consider here is that of the entire system composed of the barrier itself and the two ideal leads. The conductance of the barrier alone is

$$G = s \frac{e^2}{h} \frac{T}{1-T}, \tag{7.140}$$

so expression (7.139) can be viewed as the equivalent conductance of the barrier in series with a "contact conductance" se^2/h due to the coupling between the sample and the two ideal leads. Both expressions give the same result in the limit of a thick barrier for which $T \ll 1$. For a detailed discussion, see reference [202].

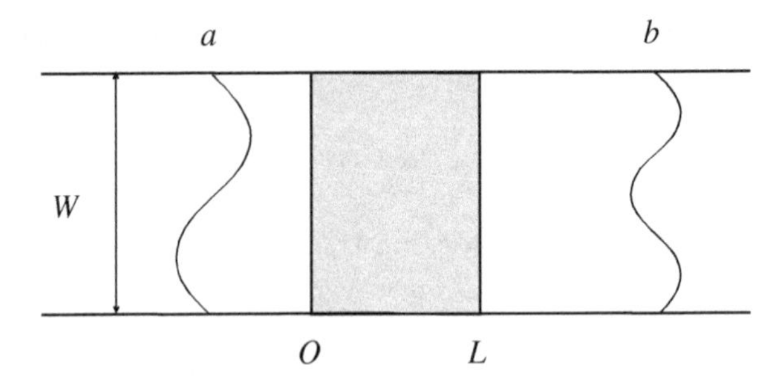

Figure 7.9 Wave guide geometry used in the Landauer formalism. The disordered conductor of length L and section $S = W^{d-1}$ is perfectly connected to ideal leads defined as wave guides propagating incoming, reflected and transmitted plane waves.

This Landauer formula (7.139) can be generalized to higher dimensionalities. To that purpose, we consider a disordered conductor of length L and square section $S = W^{d-1}$, connected to ideal leads (Figure 7.9) viewed as wave guides propagating transmitted and reflected plane waves. In this geometry, the transverse wave vector of the eigenmodes of the wave guide (also called *transverse channels*) are quantized by boundary conditions. We can thus define the transmission coefficient T_{ab} between an incoming channel a and an outgoing channel b. The Landauer formula (7.139) generalizes to

$$\boxed{G = s\frac{e^2}{h}\sum_{a,b} T_{ab}} \tag{7.141}$$

To determine the number of transverse channels, we consider that electrons are injected at the Fermi energy, so that the square of their wave vector is $k_F^2 = k^2 + |\boldsymbol{q}|^2$ where k and $\boldsymbol{q}$ are the longitudinal and transverse components. The transverse modes are quantized in units of $2\pi/W$. In three dimensions, the number of transverse channels is

$$M = \frac{\pi k_F^2}{4\pi^2/W^2} = \frac{k_F^2 S}{4\pi}. \tag{7.142}$$

The average conductance is obtained from (7.139):1

$$\overline{G} = s\frac{e^2}{h}\overline{T} = s\frac{e^2}{h}\sum_{ab}\overline{T_{ab}} \tag{7.143}$$

where T is the transmission coefficient defined by $T = \sum_{ab} T_{ab}$.

It is interesting to rewrite the Drude formula for the average conductivity σ_0 in a form which displays the number of transverse channels. For a slab of length L and square section $S = W^{d-1}$, the average conductance is given by Ohm's law $\overline{G} = \sigma_0 S/L$ so that using the

Einstein relation (7.14) yields

$$\overline{G} = se^2 \rho_0 D \frac{W^{d-1}}{L}. \tag{7.144}$$

In three dimensions, the density of states at the Fermi level is given by (3.44), that is, $\rho_0 = k_F^2/(2\pi^2 \hbar v_F)$. We thus obtain for the corresponding average conductance

$$\overline{G}^{3d} = s\frac{e^2}{h}\frac{k_F^2 l_e}{3\pi}\frac{S}{L} = s\frac{e^2}{h}\frac{4M}{3}\frac{l_e}{L} \tag{7.145}$$

written in terms of the number of transverse channels $M = k_F^2 S/4\pi$. Accordingly, in dimensions $d = 1$ and $d = 2$, we have

$$\overline{G}^{1d} = s\frac{e^2}{h}\frac{2l_e}{L} \quad \text{and} \quad \overline{G}^{2d} = s\frac{e^2}{h}\frac{\pi M}{2}\frac{l_e}{L} \tag{7.146}$$

where, in $d = 2$, the number of transverse channels is

$$M = \frac{2k_F}{2\pi/W} = \frac{k_F W}{\pi}. \tag{7.147}$$

The Landauer formalism is also perfectly suited to describe the propagation of electromagnetic waves in a random scattering medium. In this case, there is no local description equivalent to the Kubo formula, and the medium is only characterized by its transmission and reflection coefficients. The equivalence between Landauer and Kubo formalisms, which is discussed below, clarifies the unity which exists between the a priori quite different fields of electronic transport and light propagation in random media. For the case of optics, we shall not restrict ourselves to a wave guide geometry but we shall rather consider the geometry of a diffusive medium in an open space, where an incident plane wave is either reflected or transmitted as a spherical wave at large distance (Chapters 8, 9 and 12).

The Landauer formalism for wave guide geometry has been generalized by M. Büttiker to account for more complex geometries called "multi-terminal geometries" with several contacts [190].

A7.2.2 From Kubo to Landauer

The purpose of this section is to show the equivalence of the Kubo and Landauer formalisms, and to deduce the relation (7.141) from the Kubo formula for the conductivity. We follow the derivation of reference [203] and consider a disordered conductor of length L and section S connected to infinitely long wave guides of identical section, which represent the reservoirs. A voltage bias is applied between the two reservoirs, which gives rise to an electric field along the Ox direction. From relations (7.137) and (7.3), we end up with the following expression for the real valued zero frequency conductance

$$G = -s\frac{e^2\hbar^3}{2s\pi m^2L^2}\int d\boldsymbol{r}\,d\boldsymbol{r}'\,\partial_x G^R(\boldsymbol{\rho},\boldsymbol{\rho}',x,x')\partial_{x'}G^A(\boldsymbol{\rho}',\boldsymbol{\rho},x',x). \tag{7.148}$$

The Green functions describe here the overall system (disordered conductor plus reservoirs) and are taken at Fermi energy. The spatial coordinates are written as $\boldsymbol{r} = (\boldsymbol{\rho}, x)$ and Fourier transforming with respect to the transverse coordinates gives[22]

$$G^R_{ab}(x,x') = G^R(\boldsymbol{q}_a,\boldsymbol{q}_b,x,x') = \frac{\hbar}{S}\int d\boldsymbol{\rho}d\boldsymbol{\rho}' G^R(\boldsymbol{\rho},\boldsymbol{\rho}',x,x')e^{i(\boldsymbol{q}_a\cdot\boldsymbol{\rho}-\boldsymbol{q}_b\cdot\boldsymbol{\rho}')}, \tag{7.149}$$

with an equivalent expression for G^R_{ab}. In a wave guide geometry, the boundary conditions in the transverse directions quantize the corresponding wave vectors $\boldsymbol{q}_a$ and $\boldsymbol{q}_b$, leading to a finite number of transverse channels (7.142). The energy $\epsilon = (\hbar^2/2m)(k^2 + |\boldsymbol{q}|^2)$ is taken at the Fermi level and k is the longitudinal component. Upon integrating over the transverse coordinates $\boldsymbol{\rho}$ and $\boldsymbol{\rho}'$, and using $\int d\boldsymbol{\rho}\; e^{i\boldsymbol{q}\cdot\boldsymbol{\rho}} = S\delta_{\boldsymbol{q},0}$, we obtain

$$G = -s\frac{e^2\hbar}{2\pi m^2L^2}\int dx\,dx' \sum_{\boldsymbol{q}_a,\boldsymbol{q}_b}\partial_x G^R_{ab}(x,x')\partial_{x'}G^A_{ba}(x',x). \tag{7.150}$$

In this relation, the integrand can be written in terms of the longitudinal current density associated with the Schrödinger equation. This component of the current density is conserved.[23] Therefore the integrand does not depend on x, and can be calculated far from the disordered region, in the wave guides where Green's functions are plane waves whose dependence in x is simply $e^{ik_a x}$. Thus, performing the derivatives ∂_x and $\partial_{x'}$ amounts to multiplying by k_a and k_b. The integral is then straightforward and

$$G = s\frac{e^2}{h}\sum_{\boldsymbol{q}_a,\boldsymbol{q}_b} v_a v_b |G^R_{ab}(x,x')|^2, \tag{7.151}$$

where $v_a = \hbar k_a/m$ is the longitudinal velocity and x and x' are any positions along the incoming and outgoing leads. By choosing $x = 0$ and $x' = L$ and defining

$$T_{ab} = v_a v_b |G^R_{ab}(0,L)|^2, \tag{7.152}$$

[22] The inverse Fourier transform is

$$G^R(\boldsymbol{\rho},\boldsymbol{\rho}',x,x') = \frac{1}{\hbar S}\sum_{\boldsymbol{q}_a,\boldsymbol{q}_b} G^R_{ab}(x,x')\,e^{i(-\boldsymbol{q}_a\cdot\boldsymbol{\rho}+\boldsymbol{q}_b\cdot\boldsymbol{\rho}')}.$$

It corresponds to periodic transverse boundary conditions which are assumed here. More generally one can define the Green function:

$$G^R(\boldsymbol{\rho},\boldsymbol{\rho}',x,x') = \sum_{\boldsymbol{q}_a,\boldsymbol{q}_b}\phi^*_a(\boldsymbol{\rho})G^R_{ab}(x,x')\phi_b(\boldsymbol{\rho}')$$

where $\phi_a(\boldsymbol{\rho})$ is the transverse wave function in the leads for the eigenmode a.

[23] A detailed discussion of this point can be found in [202, 203].

we obtain

$$G = s\frac{e^2}{h}\sum_{q_a,q_b} T_{ab} \tag{7.153}$$

This expression for the conductance can also be written in terms of the transmission matrix t whose elements t_{ab} in the basis of free eigenmodes are

$$t_{ab} = i\sqrt{v_a v_b}G^R_{ab}(0, L), \tag{7.154}$$

so that

$$G = s\frac{e^2}{h}\sum_{q_a,q_b} |t_{ab}|^2, \tag{7.155}$$

or

$$\boxed{G = s\frac{e^2}{h}\text{Tr}\, tt^\dagger} \tag{7.156}$$

Since the trace is independent of the basis, the conductance can be calculated in the eigenbasis of the matrix $tt^\dagger$. This approach is complementary to the method discussed in this book. It is based on statistical properties of the eigenvalues of the matrix $tt^\dagger$ (for a review, see [204]).

The transmission coefficient T_{ab} given by (7.152) may be interpreted as the probability of traversing the disordered medium. Later in this appendix, we calculate the average transmission coefficient within the Diffuson approximation.

The Landauer approach is well adapted to the description of electronic transport and it provides an alternative and equivalent approach to the Kubo formula. It becomes essential if we wish to describe transport in conductors with a complex geometry or in optics where there is no equivalent to the Kubo description. Moreover, in optics, it is possible to measure the contribution T_{ab} of each transmission channel. The Landauer approach allows us to retrieve all the results already obtained within the Kubo formalism, such as the weak localization correction (7.63) to the conductance. This is the subject of the next section.

A7.2.3 Average conductance and transmission

As represented in Figure 7.10(a), the average of the transmission coefficient (7.152) appears as the product of three contributions. The first one describes how an incident wave packet penetrates into the scattering medium. It is the product of two average Green's functions, and it decreases exponentially when entering the sample. The second contribution accounts for multiple scattering in the disordered conductor, and is given by the structure factor Γ. The third term describes the outgoing wave packet. Therefore, by analogy with (4.23), we

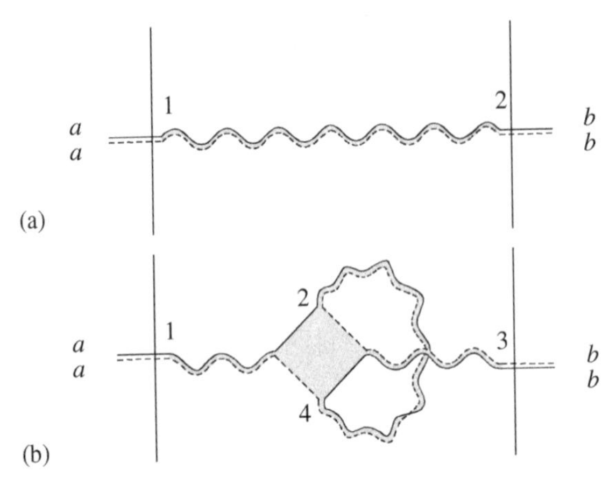

Figure 7.10 Schematic representations of (a) the average transmission coefficient $\overline{T}_{ab}$ and (b) the weak localization correction ΔT_{ab} in the Landauer formalism.

can write

$$\overline{|G^R_{ab}(0,L)|^2} = \frac{1}{S}\int dx_1\, dx_2\, |\overline{G}^R_a(0,x_1)|^2 \Gamma(x_1,x_2) |\overline{G}^R_b(x_2,L)|^2, \tag{7.157}$$

where, in the geometry of Figure 7.9, the average Green function $\overline{G}^R_a(x,x')$ is such that

$$\overline{G}^R_a(x,x') = \int_S d^2\boldsymbol{\rho}\ \overline{G}^R(\boldsymbol{\rho},x,x')\, e^{i\boldsymbol{q}_a\cdot\boldsymbol{\rho}}. \tag{7.158}$$

From (3.42), it can be shown that

$$\overline{G}^R_a(x,x') = -\frac{i}{v_a} e^{ik_a|x-x'|-|x-x'|/2l_e\mu_a}, \tag{7.159}$$

where $k_a = k_F\cos\theta_a$ and

$$\mu_a = \cos\theta_a = \frac{v_a}{v_F}. \tag{7.160}$$

The function $\Gamma(x_1,x_2)$ is the two-dimensional Fourier transform $\Gamma(\boldsymbol{q},x_1,x_2)$ of the structure factor $\Gamma(\boldsymbol{\rho},x_1,x_2)$ taken at $\boldsymbol{q} = 0$, so that the average transmission coefficient between channels a and b is

$$\overline{T}_{ab} = \frac{1}{Sv_a v_b}\int dx_1\, dx_2\, e^{-x_1/l_e\mu_a}\, e^{-|L-x_2|/l_e\mu_b}\, \Gamma(x_1,x_2). \tag{7.161}$$

In the diffusion approximation, the structure factor Γ is proportional to the probability P_d (relation 4.37). For the case of Dirichlet boundary conditions that correspond to a quasi-one-dimensional wire connected to reservoirs, both quantities behave linearly near the interface, as (5.58). Using (7.160), we have

$$\overline{T}_{ab} = \frac{\tau_e^2}{S}\Gamma(l_e\mu_a, L - l_e\mu_b) \tag{7.162}$$

or[24]

$$\overline{T}_{ab} = \frac{1}{2\pi\rho_0 S}P_d(l_e\mu_a, L - l_e\mu_b). \tag{7.163}$$

From relation (5.58), we deduce[25]

$$\overline{T}_{ab} = \frac{1}{2\pi\rho_0 S}\frac{l_e^2}{DL}\mu_a\mu_b \tag{7.164}$$

which can also be expressed with the help of the number M of transverse channels (7.142) as

$$\overline{T}_{ab} = \frac{3}{4}\frac{l_e}{ML}\mu_a\mu_b. \tag{7.165}$$

The conductance in three dimensions is obtained by summing over the channels. Since $|\boldsymbol{q}_a| = k_F \sin\theta_a$, we find

$$\sum_{\boldsymbol{q}_a} f(\boldsymbol{q}_a) = \frac{k_F^2 S}{2\pi}\int_0^{\pi/2} f(\boldsymbol{q}_a)\sin\theta_a\cos\theta_a\, d\theta_a, \tag{7.166}$$

and in particular

$$\sum_{\boldsymbol{q}_a} 1 = M \quad \sum_{\boldsymbol{q}_a}\mu_a = \frac{2}{3}M. \tag{7.167}$$

Finally the average conductance is

$$\overline{G} = s\frac{e^2}{h}\sum_{\boldsymbol{q}_a,\boldsymbol{q}_b}\overline{T}_{ab} = s\frac{e^2}{h}\frac{M}{3}\frac{l_e}{L}. \tag{7.168}$$

This result is four times smaller than the Drude conductance (7.145). This discrepancy comes from the incompatibility between the Dirichlet boundary conditions and the diffusion approximation. This point is discussed in the following section.

Anticipating the solution to this boundary problem in the next section, we have shown that the Drude conductance can be obtained equivalently within Landauer and Kubo formalisms. However, these two approaches seem a priori conceptually different. In Kubo formalism, the conductivity is obtained from the short range Drude–Boltzmann term which is neglected

[24] We use the notation $P_d(x, x') = P_d(q = 0, x, x')$, solution of (5.45) with $\gamma = k_\perp = 0$.

[25] We shall see in the next section how to modify relations (7.164, 7.165, 7.168) in order to implement properly the boundary conditions.

in the present derivation, whereas the contribution of the Diffuson vanishes (see page 277) for isotropic collisions. In the Landauer formalism, the main contribution to the average conductance results from the long range Diffuson, and the short range Drude–Boltzmann term is negligible. Indeed, the two points of view are complementary and equivalent. The conductivity is a *volume quantity* defined by the integral

$$\sigma = \frac{1}{\Omega} \int\int d\boldsymbol{r}\, d\boldsymbol{r}'\, \sigma(\boldsymbol{r},\boldsymbol{r}'), \tag{7.169}$$

whereas the conductance is essentially a *surface quantity* calculated at the boundaries of the conductor which can be written in the form

$$G = \sigma \frac{S}{L} = \int\int d\boldsymbol{S}\, d\boldsymbol{S}'\, \sigma(\boldsymbol{r},\boldsymbol{r}'), \tag{7.170}$$

where the two integrals are taken over the boundaries of the sample. This last expression is equivalent to the Landauer formula: it expresses the conductance as a transmission coefficient between two interfaces [161]. The equivalence between relations (7.169) and (7.170) results from current conservation which leads to the diffusion equation relating the long range and short range contributions (see remark on page 272).

Exercise 7.5: average transmission coefficient in dimension d

Expression (7.164) is valid in any dimension for a density of states ρ_0 of the form

$$\rho_0 = \frac{dA_d}{(2\pi)^d} \frac{k_F^{d-1}}{\hbar v_F} \tag{7.171}$$

where A_d is the volume of the unit sphere (15.2). In d dimensions, the number of channels is

$$M = \frac{A_{d-1}}{(2\pi)^{d-1}} (k_F W)^{d-1} \tag{7.172}$$

so that the following relation holds between density of states and number of transverse channels:

$$\rho_0 W^{d-1} = \frac{dA_d}{A_{d-1}} \frac{M}{hv_F}, \qquad \rho_0 W^2 = \frac{2M}{\pi \hbar v_F} \quad \text{for} \quad d = 3. \tag{7.173}$$

From (7.162), we then obtain the average transmission coefficient in dimension d:

$$\overline{T_{ab}} = \frac{A_{d-1}}{A_d} \frac{l_e}{ML} \mu_a \mu_b. \tag{7.174}$$

This expression must be modified in order to account properly for boundary conditions (see the following section) by means of the substitution $\mu_i \to \mu_i + z_0/l_e$ where $z_0/l_e = 2/3$.

Exercise 7.6: local and non-local contributions to the conductivity in a quasi-one-dimensional wire

The purpose of this exercise is to discuss the correspondence between the Kubo and the Landauer formalisms which is encoded in the expression of the non-local conductivity $\sigma(x,x')$.

Following the procedure explained in the remark on page 272, show that, for a quasi-one-dimensional wire, the classical non-local conductivity $\sigma(x,x')$ is the sum of two terms, the Drude–Boltzmann and the Diffuson:

$$\sigma_{cl}(x,x') = \sigma_0(x,x') + \sigma_d(x,x') \tag{7.175}$$

where, on a scale larger than l_e, the Drude–Boltzmann conductivity is local $\sigma_0(x,x') = \sigma_0\delta(x - x')$ and the non-local contribution of the Diffuson is long range and is given by $\sigma_d(x,x') = -\sigma_0 D\partial_x\partial_{x'}P_d(x,x')$ with $\sigma_0 = ne^2\tau_e/m$.

Check the current conservation expressed as $\partial_x\sigma_{cl}(x,x') = 0$. Show that in the Kubo formalism, the Drude–Boltzmann term gives the conductivity in the form (7.169) and that the contribution of the second term cancels. Show that in the Landauer formalism, the conductance (7.170) results from the Diffuson term.

A7.2.4 Boundary conditions and impedance matching

The expression for the average transmission (7.165) is based on an approximate expression for P_d obtained by assuming Dirichlet boundary conditions, that is, cancellation of the probability outside the diffusive medium. In section 5.2.3, we proposed another possible choice of boundary conditions better suited to describe the transformation of a ballistic wave into a diffusive wave across the interface. This choice is based on the requirement that the incoming diffuse flux must vanish at the interface. In this case, the probability does not vanish exactly at the interface, but at a distance z_0 slightly *outside the diffusive medium*. We must therefore consider the expression (5.158) for $P_d(x,x')$ with $z_0 = 2l_e/3$ in three dimensions. Instead of (7.165), we obtain the modified expression

$$\overline{T_{ab}} = \frac{3l_e}{4M(L+4l_e/3)}\left(\mu_a + \frac{2}{3}\right)\left(\mu_b + \frac{2}{3}\right). \tag{7.176}$$

For a long conductor, $L + 4l_e/3 \simeq L$. Summing over outgoing channels, we obtain the transmission coefficient for a given incoming mode a

$$\overline{T_a} = \sum_{q_b}\overline{T_{ab}} = \frac{l_e}{L}\left(\mu_a + \frac{2}{3}\right). \tag{7.177}$$

By integrating over the incoming channels, we now recover the Drude result (7.145) for the conductance

$$\boxed{\overline{T} = \sum_{q_a, q_b} \overline{T}_{ab} = \frac{4M}{3}\frac{l_e}{L}} \tag{7.178}$$

In order to describe properly the continuity between the diffusive wire and the contacts within the diffusion approximation, we must take into account the finite distance $z_0 = 2l_e/3$. An interesting outcome of (7.178) is that the average conductance may be viewed as the sum of contributions of M channels each giving on average a contribution of order l_e/L.

Exercise 7.7: average transmission coefficient in dimension $d = 2$

Using the results of section A5.2.3 for boundary conditions, show that for the geometry of a two-dimensional wave guide, we have [205]

$$\overline{T}_{ab} = \frac{2}{\pi}\frac{l_e}{ML}\left(\mu_a + \frac{\pi}{4}\right)\left(\mu_b + \frac{\pi}{4}\right), \tag{7.179}$$

since, in $d = 2$, the distance z_0 is given by $z_0/l_e = \pi/4$ (see relation 5.117). The number of transverse channels is $M = k_F W/\pi$. Using the equality

$$\sum_{q_a} f(q_a) = M \int_0^{\pi/2} f(q_a) \cos\theta_a \, d\theta_a, \tag{7.180}$$

show that

$$\sum_{q_a} 1 = M \quad \sum_{q_a} \mu_a = \frac{\pi}{4} M \tag{7.181}$$

and deduce that

$$\overline{T}_a = \sum_{q_b} \overline{T}_{ab} = \frac{l_e}{L}\left(\mu_a + \frac{\pi}{4}\right) \quad \text{and} \quad \overline{T} = \sum_{q_a} \overline{T}_a = \frac{\pi M}{2}\frac{l_e}{L}. \tag{7.182}$$

The Drude result is also recovered in dimension $d = 2$ (7.146).

Exercise 7.8: reflection coefficient

The reflection coefficient R_{ab} between two modes a and b can be defined in a way analogous to the transmission coefficient (7.152), namely

$$R_{ab} = v_a v_b |G^R_{ab}(0,0)|^2. \tag{7.183}$$

Using an expression similar to (7.157), show that [205]

$$\overline{R}_{ab} = \frac{2}{\pi M}\left(\frac{\pi}{4} + \frac{\mu_a \mu_b}{\mu_a + \mu_b}\right) \quad \text{and} \quad \overline{R}_{ab} = \frac{3}{4M}\left(\frac{2}{3} + \frac{\mu_a \mu_b}{\mu_a + \mu_b}\right) \tag{7.184}$$

in dimensions 2 and 3.

Exercise 7.9: current conservation

Energy conservation implies that, for each incoming channel, $\overline{R}_a + \overline{T}_a = 1$. When $l_e \ll L$, the average transmission coefficient becomes negligible so that $\overline{R}_a \simeq 1$. Show that expression (7.184) obtained for $\overline{R}_{ab}$ does not satisfy this condition and that, for the incident mode $\theta_a = 0$, we obtain $\overline{R}_a = \sum_{q_a} \overline{R}_{ab} \simeq 0.773$ in $d = 2$ and $\overline{R}_a \simeq 0.790$ in $d = 3$.

This discrepancy is a consequence of the diffusion approximation, which does not describe the reflection coefficient correctly because of the existence of short trajectories in the vicinity of the interface. In this respect, it is interesting to notice that $\overline{R}_a$ is closer to 1 in dimension 3. This is because the weight of short trajectories is smaller for larger space dimensionalities. Formally, we can check that $\overline{R}_a \to 1$ if $d \to \infty$. In transmission, the diffusion approximation is better justified since all diffusive trajectories that cross the sample are long.

A7.2.5 Weak localization correction in the Landauer formalism

The weak localization correction to the conductivity was calculated in section 7.4 within the Kubo formalism, and we obtained the expressions (7.63, 7.62) for the average conductance. In the framework of the Landauer formalism, it is possible to recover these expressions. To that purpose, we start by calculating the weak localization correction ΔT_{ab} to the average transmission coefficient $\overline{T}_{ab}$. It is represented by Figure 7.10(b), which comprises two Diffusons and one Cooperon forming a loop. By generalization of expression (7.162), we can write the correction ΔT_{ab} in the form

$$\Delta T_{ab} = \frac{\tau_e^2}{S} \int \prod_i dx_i \, H(\{x_i\}) \, \Gamma(l_e \mu_a, x_1) \, \Gamma'(x_2, x_4) \, \Gamma(x_3, L - l_e \mu_b), \tag{7.185}$$

where Γ and Γ' are the respective structure factors of the Diffuson and of the Cooperon. The contribution of the Hikami box $H(\{x_i\})$ is given by relations (4.150, 4.127). Using (4.37) and applying the gradients on the Diffusons, we obtain

$$\Delta T_{ab} = \frac{2D}{(2\pi \rho_0)^2 S} \int_0^L \partial_x P_d(l_e \mu_a, x) \, P_c(x, x) \, \partial_x P_d(x, L - l_e \mu_b) \, dx. \tag{7.186}$$

Given the expression (5.158) for $P_d(x, x')$, the gradients $\partial_x P_d$ are constant terms and in the limit $L \gg l_e$, we have

$$\partial_x P_d(l_e \mu_a, x) = -\frac{l_e \mu_a + z_0}{DL} \qquad \partial_x P_d(x, L - l_e \mu_b) = \frac{l_e \mu_b + z_0}{DL} \tag{7.187}$$

where $z_0 = 2l_e/3$. The relative correction is obtained with the help of (7.176) and (7.173):

$$\frac{\Delta T_{ab}}{\overline{T}_{ab}} = -\frac{1}{\pi \hbar \rho_0 SL} \int P_c(x,x)\, dx. \tag{7.188}$$

We recover the weak localization correction (7.51) calculated within the Kubo formalism. We notice that this relative correction does not depend on the specific transmission mode ($a \to b$) considered. The return probability is given by

$$P_c(x,x) = \frac{x}{D}\left(1 - \frac{x}{L}\right). \tag{7.189}$$

Summing over the modes, we obtain the weak localization correction to the dimensionless conductance:

$$\Delta g = -\frac{s}{3} \tag{7.190}$$

which is the result (7.62) of Exercise 7.2, page 283.

A7.2.6 Landauer formalism for waves
Wave guide geometry

Within the Landauer formalism, we have calculated the average transmission coefficient $\overline{T}_{ab}$ and the conductance of a quasi-one-dimensional conductor in the geometry of a wave guide. This description can also be adapted to the propagation of an electromagnetic wave through a diffusive medium. By analogy with (7.152), that is, for the geometry of a wave guide, the transmission coefficient between two channels a and b is given by

$$\boxed{T_{ab} = k_a k_b |G^R_{ab}(0,L)|^2} \tag{7.191}$$

Equation (7.157) remains unchanged, but the Green function corresponds now to the Helmholtz equation (3.49)

$$\overline{G}^R_a(x,x') = -\frac{i}{2k_a} e^{ik_a|x-x'| - |x-x'|/2l_e\mu_a} \tag{7.192}$$

so that

$$\overline{T}_{ab} = \frac{1}{4k_a k_b S} \int dx_1\, dx_2\, e^{-x_1/l_e\mu_a}\, e^{-|L-x_2|/l_e\mu_b}\, \Gamma(x_1,x_2). \tag{7.193}$$

The calculation is similar to the electronic case and leads to

$$\overline{T}_{ab} = \frac{l_e^2}{4k^2 S} \Gamma(l_e\mu_a, L - l_e\mu_b). \tag{7.194}$$

Using (4.63), we obtain

$$\overline{\mathcal{T}_{ab}} = \frac{\pi c}{k^2 S} P_d(l_e\mu_a, L - l_e\mu_b) \tag{7.195}$$

which is analogous to (7.163), and must be corrected by the replacement of μ_i with $\mu_i+2/3$ to account for proper boundary conditions.

Open space geometry

For waves, we can also consider the geometry of a medium illuminated by a plane wave, which transmits (or reflects) spherical waves at large distance (Chapters 8, 9 and 12). Here we can no longer define transverse channels. Although the scattering behavior inside the disordered medium remains unchanged, the boundary conditions are different. We now define a transmission coefficient $\mathcal{T}_{ab}$ per unit solid angle whose average is[26]

$$\overline{\mathcal{T}_{ab}} = \frac{1}{(4\pi)^2} \int dx_1\, dx_2\, e^{-x_1/l_e\mu_a}\, e^{-|L-x_2|/l_e\mu_b}\, \Gamma(x_1, x_2). \tag{7.196}$$

The incoming and outgoing wave vectors are no longer quantized as in a wave guide geometry. Using relation (4.63) and the extrapolation length $z_0 = \frac{2}{3}l_e$, we obtain

$$\overline{\mathcal{T}_{ab}} = \frac{c}{4\pi}\mu_a\mu_b P_d(l_e\mu_a, L - l_e\mu_b) = \frac{3}{4\pi}\frac{l_e}{L}\mu_a\mu_b\left(\mu_a + \frac{2}{3}\right)\left(\mu_b + \frac{2}{3}\right). \tag{7.197}$$

The angular integration over the half-space gives

$$\overline{\mathcal{T}_a} = \frac{l_e}{L}\mu_a\left(\mu_a + \frac{2}{3}\right) \tag{7.198}$$

and, for the total transmission coefficient, we have

$$\overline{\mathcal{T}} = \frac{4\pi}{3}\frac{l_e}{L} \tag{7.199}$$

which is different from (7.178). The difference between these results and expressions (7.176), (7.177) and (7.178) is not surprising; it expresses the fact that, in a wave guide geometry, there are M discrete channels whereas, for an open space geometry, the incoming and outgoing directions are continuous angular variables defined between 0 and π. We shall return to this description in detail in Chapters 8, 9 and 12.

Appendix A7.3: Real space description of conductivity

Expression (7.47) for the conductivity is given in terms of the probability of quantum diffusion written in real space. This suggests that it is possible to establish this relation

[26] For the reflection coefficient see relation (8.10) and for a complete derivation see section 12.3.

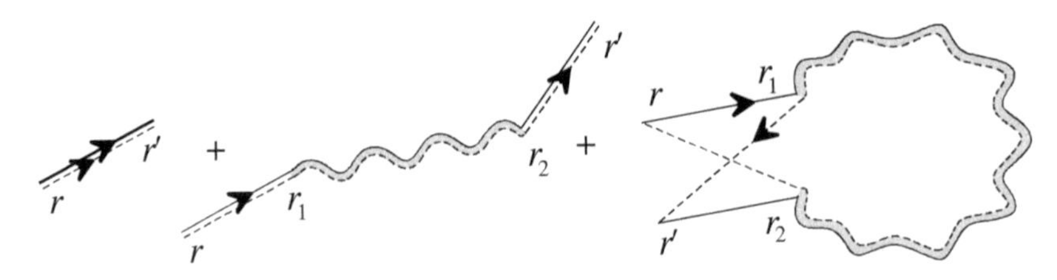

Figure 7.11 Representation of the Drude–Boltzmann, of the Diffuson, and of the Cooperon contributions to conductivity. Here, the arrows represent the incoming (∂_x) and outgoing ($\partial_{x'}$) current operators. It is clear that for the Diffuson the currents have uncorrelated directions, whereas for the Cooperon the currents j_x and $j_{x'}$ are opposite when $\boldsymbol{r} \simeq \boldsymbol{r}'$.

directly, without working in reciprocal space. This is the purpose of this appendix. From its definition (7.3), the average conductivity can be rewritten:

$$\sigma(\omega) = s\frac{e^2\hbar^3}{2\pi m^2\Omega}\int\int d\boldsymbol{r}\, d\boldsymbol{r}'\ \mathrm{Re}\ \overline{\partial_x G^R_\epsilon(\boldsymbol{r},\boldsymbol{r}')\partial_{x'}G^A_{\epsilon-\omega}(\boldsymbol{r}',\boldsymbol{r})} \tag{7.200}$$

where ϵ is the Fermi energy. This expression resembles very much the expression (4.9) for the quantum probability, except for the two spatial derivatives that remind us that $\sigma(\omega)$ is indeed a current-current correlation function. The derivatives ∂_x and $\partial_{x'}$ act respectively on the retarded and advanced Green functions.

The diagrammatic expansion of the conductivity is similar to that for the quantum probability (see relations 4.23 and 4.42 and Figure 7.11). We obtain

$$\begin{aligned}\overline{\partial_x G^R_\epsilon(\boldsymbol{r},\boldsymbol{r}')\partial_{x'}G^A_{\epsilon-\omega}(\boldsymbol{r}',\boldsymbol{r})} &= \partial_x\overline{G}^R_\epsilon(\boldsymbol{r},\boldsymbol{r}')\partial_{x'}\overline{G}^A_{\epsilon-\omega}(\boldsymbol{r}',\boldsymbol{r}) \\ &+ \int \partial_x\overline{G}^R_\epsilon(\boldsymbol{r},\boldsymbol{r}_1)\overline{G}^A_{\epsilon-\omega}(\boldsymbol{r}_1,\boldsymbol{r})\Gamma_\omega(\boldsymbol{r}_1,\boldsymbol{r}_2)\overline{G}^R_\epsilon(\boldsymbol{r}_2,\boldsymbol{r}')\partial_{x'}\overline{G}^A_{\epsilon-\omega}(\boldsymbol{r}',\boldsymbol{r}_2)\, d\boldsymbol{r}_1\, d\boldsymbol{r}_2 \\ &+ \int \partial_x\overline{G}^R_\epsilon(\boldsymbol{r},\boldsymbol{r}_1)\partial_{x'}\overline{G}^A_{\epsilon-\omega}(\boldsymbol{r}',\boldsymbol{r}_1)\Gamma'_\omega(\boldsymbol{r}_1,\boldsymbol{r}_2)\overline{G}^R_\epsilon(\boldsymbol{r}_2,\boldsymbol{r}')\overline{G}^A_{\epsilon-\omega}(\boldsymbol{r}_2,\boldsymbol{r})\, d\boldsymbol{r}_1\, d\boldsymbol{r}_2.\end{aligned} \tag{7.201}$$

The spatial variation of the Green functions given by (3.89) is mainly determined by the phase factor $e^{ik_F|\boldsymbol{r}-\boldsymbol{r}'|}$ and consequently, in the limit $k_F l_e \gg 1$:

$$\partial_x\overline{G}^R_\epsilon(\boldsymbol{r}-\boldsymbol{r}') = ik_F\frac{x-x'}{|\boldsymbol{r}-\boldsymbol{r}'|}\overline{G}^R_\epsilon(\boldsymbol{r},\boldsymbol{r}'). \tag{7.202}$$

Introducing the quantity

$$u_{x,x'} = \frac{x-x'}{|\boldsymbol{r}-\boldsymbol{r}'|} \tag{7.203}$$

allows us to write equation (7.201) as

$$\overline{\partial_x G^R_\epsilon(\boldsymbol{r},\boldsymbol{r}')\partial_{x'} G^A_{\epsilon-\omega}(\boldsymbol{r}',\boldsymbol{r})} = k_F^2 \Big[u^2_{x,x'} \overline{G}^R_\epsilon(\boldsymbol{r},\boldsymbol{r}')\overline{G}^A_{\epsilon-\omega}(\boldsymbol{r}',\boldsymbol{r}) \tag{7.204}$$

$$+ \int u_{x,x_1} u_{x_2,x'} \overline{G}^R_\epsilon(\boldsymbol{r},\boldsymbol{r}_1)\overline{G}^A_{\epsilon-\omega}(\boldsymbol{r}_1,\boldsymbol{r})\Gamma_\omega(\boldsymbol{r}_1,\boldsymbol{r}_2)\overline{G}^R_\epsilon(\boldsymbol{r}_2,\boldsymbol{r}')\overline{G}^A_{\epsilon-\omega}(\boldsymbol{r}',\boldsymbol{r}_2)\, d\boldsymbol{r}_1\, d\boldsymbol{r}_2$$

$$+ \int u_{x,x_1} u_{x_1,x'} \overline{G}^R_\epsilon(\boldsymbol{r},\boldsymbol{r}_1)\overline{G}^A_{\epsilon-\omega}(\boldsymbol{r}',\boldsymbol{r}_1)\Gamma'_\omega(\boldsymbol{r}_1,\boldsymbol{r}_2)\overline{G}^R_\epsilon(\boldsymbol{r}_2,\boldsymbol{r}')\overline{G}^A_{\epsilon-\omega}(\boldsymbol{r}_2,\boldsymbol{r})\, d\boldsymbol{r}_1\, d\boldsymbol{r}_2 \Big]$$

Upon spatial integration, the first term on the right hand side of (7.204) involves the average value:

$$\langle u^2_{x,x'} \rangle = \frac{1}{d}. \tag{7.205}$$

The contribution of the Diffuson (second term) cancels out since the incoming and outgoing currents are uncorrelated. This is specific to the case of isotropic collisions

$$\langle u_{x,x_1} u_{x_2,x'} \rangle = 0. \tag{7.206}$$

Finally, the integral of the Cooperon involves only neighboring points $\boldsymbol{r}$ and $\boldsymbol{r}'$, so that in the last term of (7.204)

$$\langle u_{x,x_1} u_{x_1,x'} \rangle \simeq \langle u_{x,x_1} u_{x_1,x} \rangle = -\frac{1}{d}. \tag{7.207}$$

Altogether the conductivity is written

$$\sigma(\omega) = \frac{ne^2}{m\Omega} \int\!\!\int d\boldsymbol{r}\, d\boldsymbol{r}' \ \mathrm{Re}\ \left[P_0(\boldsymbol{r},\boldsymbol{r}',\omega) - P_c(\boldsymbol{r},\boldsymbol{r}',\omega) \right], \tag{7.208}$$

where we have introduced the electronic density $n = (2s/d)\epsilon\rho_0(\epsilon)$. For a translation invariant system, one of the spatial integrals is straightforward, and we recover the relation (7.47).

Appendix A7.4: Weak localization correction and anisotropic collisions

The weak localization correction has been calculated for the case of isotropic collisions on impurities. Quite remarkably this correction, given by (7.56), does not depend on the diffusion coefficient.[27] We wish now to address the question whether this remains correct for anisotropic collisions, that is, when there are two characteristic times τ_e and τ^*. In this case, the classical conductivity is still given by the Einstein relation provided we replace D by $D^* = v_F^2 \tau^*/d$ (relation 7.36). How then is the modified weak localization correction?

In order to understand the structure of this correction properly, it is useful to rewrite the diagram of the Cooperon shown in Figure 7.1 in the form sketched in Figure 7.12(a).

[27] There are exceptions for low-dimensional systems $d \leq 2$ where a cutoff is needed.

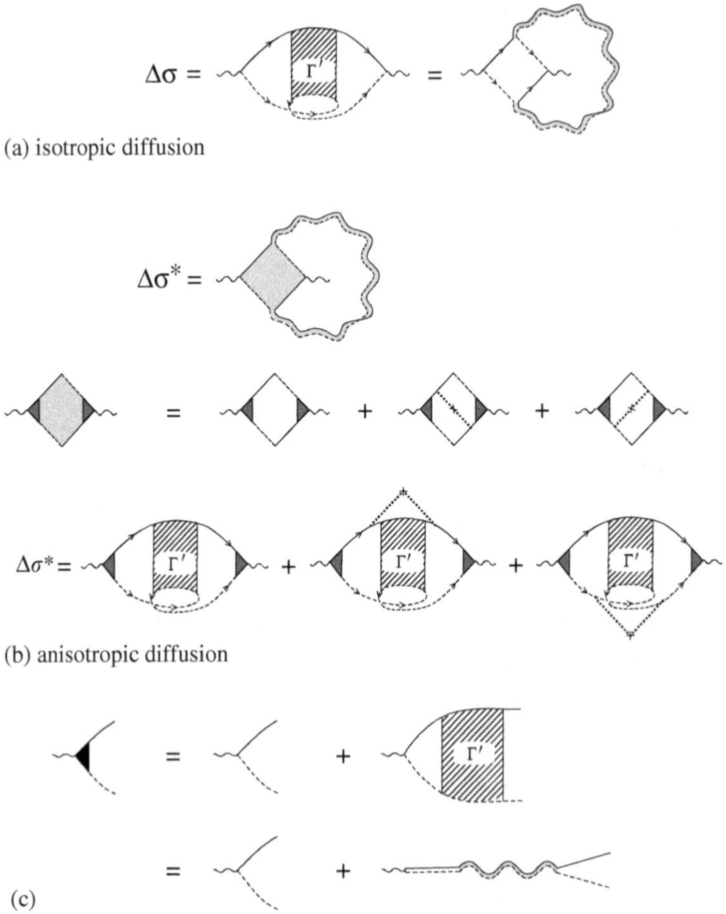

Figure 7.12 Diagrams for the weak localization correction to the conductivity: (a) isotropic collisions, (b) anisotropic collisions, (c) representation of the current vertex dressed by a Diffuson.

A first modification consists in replacing the structure factor Γ by (4.169), that is, $\Gamma^* \propto 1/(-i\omega + D^*q^2)$. On the other hand, as shown in Figure 7.12(b), one must renormalize the current vertices (this point is discussed in section 7.2.3). The Cooperon involves a Hikami box which must be dressed by additional impurity lines in order to account for all contributions of the same order (see Figure 4.16). We check easily that, in the isotropic case, they do not contribute because of current vertices, so that we are left with the bare Hikami box shown in Figure 7.12(a), which is given by

$$\langle s^2 \rangle H^{(A)} = H^{(A)}. \tag{7.209}$$

For the anisotropic case, it must be replaced by the diagram $\tilde{H}^{(A)}$:

$$\tilde{H}^{(A)} = H^{(A)} \left(\frac{\tau^*}{\tau_e}\right)^2 \tag{7.210}$$

because of relation (7.42) which renormalizes the current vertex. On the other hand, the diagrams $\tilde{H}^{(B)}$ and $\tilde{H}^{(C)}$ with one additional impurity line (Figure 7.12(b)) involve the angular average $\langle \hat{s} \cdot \hat{s}' B(\hat{s} - \hat{s}') \rangle = \gamma_1$, so that

$$\tilde{H}^{(B)} = \tilde{H}^{(C)} = H^{(B)} \left(\frac{\tau^*}{\tau_e}\right)^2 \frac{\langle \hat{s} \cdot \hat{s}' B(\hat{s} - \hat{s}') \rangle}{\gamma_e} \tag{7.211}$$

where $H^{(B)} = -H^{(A)}/2$ is the bare Hikami box as defined and calculated in section A4.2.1. Therefore

$$\tilde{H}^{(B)} = \tilde{H}^{(C)} = -\frac{1}{2} H^{(A)} \frac{\gamma_1}{\gamma_e} \left(\frac{\tau^*}{\tau_e}\right)^2 . \tag{7.212}$$

Altogether, the bare box which enters into the calculation of the conductivity for the isotropic case is now replaced by $\tilde{H} = \tilde{H}^{(A)} + \tilde{H}^{(B)} + \tilde{H}^{(C)}$, that is,

$$\tilde{H} = H^{(A)} \left(\frac{\tau^*}{\tau_e}\right)^2 \left(1 - \frac{\gamma_1}{\gamma_e}\right) = H^{(A)} \frac{\tau^*}{\tau_e} . \tag{7.213}$$

Because of the multiplicative factor τ^*/τ_e, the weak localization correction $\Delta\sigma^*$ keeps the form (7.45), provided we perform the substitution $D \to D^*$.

8

Coherent backscattering of light

8.1 Introduction

Phase coherence is at the basis of the interference effects which lead to weak localization in electronics. This phase coherence also has important consequences in optics. Moreover, using an incident laser beam, it is possible in optics to study the angular behavior of both transmitted and reflected waves. This is difficult in electronic devices, where electrons are injected and collected from reservoirs and do not have an accessible angular structure. In this chapter, we study the intensity of the light reflected by a diffusive medium and we show that it has an angular structure that is due to the coherent effects associated with the Cooperon. We also show that it is possible to single out and analyze the contribution of multiple scattering paths as a function of their length. This leads to a kind of "spectroscopy" of diffusive trajectories.

The issue of wave scattering in disordered media has a long history. At the turn of the twentieth century, a purely classical approach to the description of radiative transfer of electromagnetic waves through the atmosphere, based on the Boltzmann equation, had already been proposed by Schuster [206]. This problem was subsequently extended to include the related domains of turbulent media, meteorology and liquids. It was only during the 1980s, however, that the possibility of phase coherent effects in the multiple scattering of waves in random media was raised. The interest surrounding this question is certainly related to new developments in similar questions in the quantum theory of scattering [207–209]. A systematic description of coherent effects emphasizing the role of the Cooperon was initially proposed in references [210] and [211]. These new developments came together with the first experimental results [212–214], giving rise to a large number of works which it would be rather difficult to list comprehensively [215]; we shall quote only a few in this chapter. Nevertheless, we follow here the approach developed in references [216–218].[1]

The variety of phenomena that result from multiple scattering of electromagnetic waves is rather broad. Consequently, the coherent backscattering phenomenon that we study in this chapter has a large range of applications that have been developed only during the last

[1] This choice of references should not be understood as resulting from a formed opinion on other references. Our aim is simply to use those references with notation similar to that used in this book.

few years. We shall not study them all, but rather present some of them towards the end of this chapter.

We consider first the case of a scalar wave, and then we include effects of polarization. We define and study the reflection coefficient (sometimes called the *albedo*) of a semi-infinite diffusive medium in terms of the Diffuson and of the Cooperon. We then extend these results to the case of finite absorption. Finally, we present a rather detailed account of the experimental situation which shows quite spectacularly the success of the present ideas about coherent multiple scattering and its large field of applicability.

8.2 The geometry of the albedo

8.2.1 Definition

The physical situation we aim to describe is as follows. A far-field and point-like source emits a monochromatic light which we assimilate to a plane wave directed towards the interface between vacuum and the diffusive medium. The direction of this plane wave is characterized by the unit vector $\hat{s}_i$. The wave scattered by the medium emerges through the same interface (i.e., in reflection) and is detected far from the interface along the direction $\hat{s}_e$. We are thus interested in the angular dependence of the reflection coefficient, called the *albedo* (it is also sometimes called the *bistatic coefficient* [219, 220]). There are several definitions of this coefficient which are appropriate to different fields of physics such as astrophysics, atomic physics, nuclear physics, etc. For instance, in astronomy, the albedo of a planet is defined as the ratio between the total reflected light flux and the incident flux coming from the Sun. Thus defined, the albedo of the Earth is 35% while that of the Moon is only 6%. For more details, see references [219, 220].

The detector of the outgoing light along the direction $\hat{s}_e$ essentially measures the intensity $I(R\hat{s}_e) \propto E^2$ of the electromagnetic field $E(R\hat{s}_e)$. For a spherical wave detected at a distance R which is large compared to the size of the interface, the energy flux per unit time and per unit solid angle is

$$\frac{dF}{d\Omega} = cR^2 I(R\hat{s}_e)\,, \tag{8.1}$$

where F is the flux of the Poynting vector. The incident flux is given by[2]

$$F_0 = cSI_0\,, \tag{8.2}$$

and we define the albedo $\alpha(\hat{s}_e)$ by the dimensionless ratio

$$\boxed{\alpha(\hat{s}_e) = \frac{1}{F_0}\frac{dF}{d\Omega} = \frac{R^2}{S}\frac{I(R\hat{s}_e)}{I_0}} \tag{8.3}$$

[2] We consider here the situation of normal incidence. In the general case, the incident flux depends on the cosine of the angle between the incident direction and the perpendicular to the surface.

The albedo appears to be a quantity close to a differential cross section (see also the relation 2.66) up to a multiplicative factor related to the shape of the interface of the diffusive medium.

8.2.2 Albedo of a diffusive medium

The albedo characterizes the light scattered by a diffusive medium. To calculate this quantity, we need to evaluate the intensity $I(R\hat{s}_e)$ of the scattered field as defined by relation (4.54).[3] We start by presenting a heuristic derivation that allows us to calculate the Diffuson and Cooperon contributions and to understand the characteristics of the coherent albedo (triangular singularity, algebraic decrease etc.).

We consider a semi-infinite diffusive medium filling the half-space $z \geq 0$. The half-space $z \leq 0$ is a free space that contains the source and the detectors (Figure 8.1). The incident (assumed to be normal to the interface) and emergent beams are respectively characterized by the wave vectors $\boldsymbol{k}_i = k\hat{s}_i$ and $\boldsymbol{k}_e = k\hat{s}_e$ where $\hat{s}_i$ and $\hat{s}_e$ are unit vectors. Since the waves experience elastic scattering, only their direction $\hat{s}$ changes while the amplitude $k = \omega_0/c$ remains constant. Moreover, we also assume that the difference in optical index between the two media is negligible.

In order to calculate the intensity $I(R\hat{s}_e)$, we first consider the case of a scalar wave solution of the Helmholtz equation (2.8). This solution corresponds to an incident plane wave and to a spherical outgoing wave detected at a point $\boldsymbol{R} = R\hat{s}_e$ in the far field, namely at a distance which is very large compared to the size of the interface of the diffusive medium (Figure 8.1). For a given incident direction $\hat{s}_i$, the amplitude $\psi_{\omega_0}(\hat{s}_e)$ of the outgoing wave

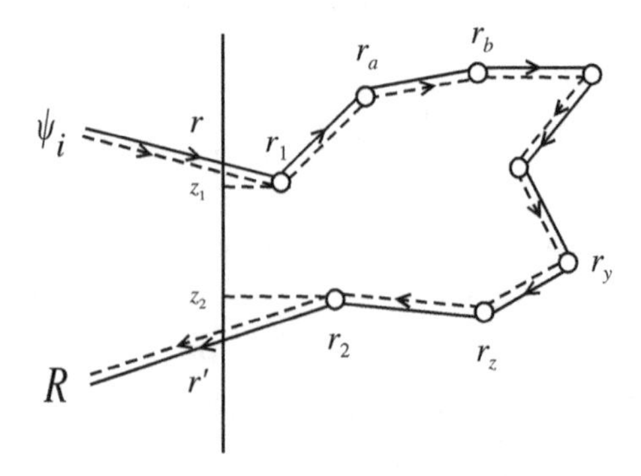

Figure 8.1 Contribution of the Diffuson to the albedo averaged over disorder. We have assumed in the calculation that the incident beam is perpendicular to the interface.

[3] In the definition (4.54) the source of the field is a δ function of unit strength so that the intensity thus defined does not have the dimensions of a light intensity. This is not very important since the albedo appears as the ratio of two such intensities.

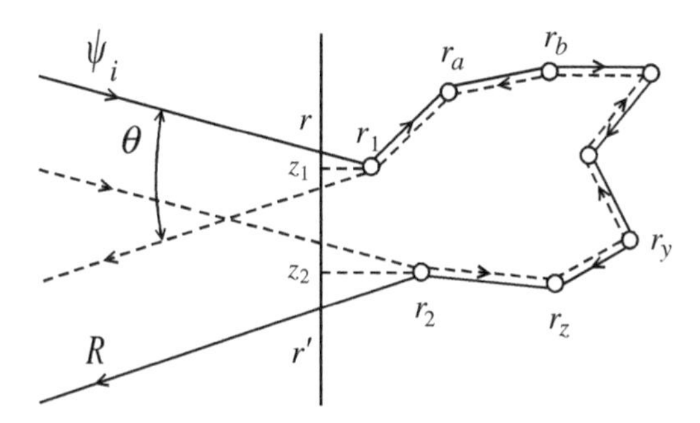

Figure 8.2 Contribution of the Cooperon to the albedo averaged over disorder. We have assumed in the calculation that the incident beam is perpendicular to the interface.

has the following structure

$$\psi_{\omega_0}(\hat{s}_e) \propto \int d\boldsymbol{r}\, d\boldsymbol{r}'\, e^{ik(\hat{s}_i\cdot\boldsymbol{r}-\hat{s}_e\cdot\boldsymbol{r}')}\, G(\boldsymbol{r},\boldsymbol{r}',\omega_0)\,. \tag{8.4}$$

This expression is nothing but the Fourier transform of the Green function (4.21), and it has the form

$$\psi_{\omega_0}(\hat{s}_e) = \int d\boldsymbol{r}\, d\boldsymbol{r}'\, e^{ik(\hat{s}_i\cdot\boldsymbol{r}-\hat{s}_e\cdot\boldsymbol{r}')} \sum_{N=1}^{\infty} \sum_{\boldsymbol{r}_1,\dots,\boldsymbol{r}_N} |A(\boldsymbol{r},\boldsymbol{r}',\mathcal{C}_N)| \exp\left(i\frac{2\pi\mathcal{L}_N}{\lambda}\right) \tag{8.5}$$

and the intensity is $I(R\hat{s}_e) = (4\pi/c)|\psi_{\omega_0}(\hat{s}_e)|^2$. In this expression the points $\boldsymbol{r}$ and $\boldsymbol{r}'$ can be anywhere on the interface, whereas the points $\boldsymbol{r}_i$ correspond to scattering events. $\mathcal{C}_N = (\boldsymbol{r}_1, \boldsymbol{r}_2, \dots, \boldsymbol{r}_N)$ is a sequence of N scattering events and $\mathcal{L}_N/\lambda$ is the total length of the corresponding trajectory measured in units of the wavelength $\lambda = (2\pi/k)$. The field ψ_{ω_0} thus appears as a sum over all the possible multiple scattering sequences in the half-space $z \geq 0$, weighted by the phase factors that account for the two incident and emergent waves in the free half-space $z \leq 0$.

There are three distinct contributions to the albedo α. The first includes all the terms for which the sequences that contribute respectively to ψ_{ω_0} and $\psi^*_{\omega_0}$ are different (as represented in Figure 4.1). The second contribution includes all the terms that correspond to identical scattering sequences (Figure 8.1). The third and last contribution corresponds also to identical scattering sequences propagating in opposite directions (Figure 8.2).

For a given realization of the disorder, i.e., of the position of the scatterers, the first term provides the main contribution α_s to the albedo, but vanishes upon disorder averaging (Figure 1.4). This term corresponds to the fluctuations or speckle patterns that we shall study in Chapter 12. The two remaining contributions denoted α_d and α_c have a finite average value and correspond respectively to the Diffuson and the Cooperon, namely to the incoherent and coherent parts.

8.3 The average albedo

8.3.1 Incoherent albedo: contribution of the Diffuson

As stated before, the only multiple scattering trajectories that still contribute to disorder averaging are those for which the sequences C_N of scattering events that enter the amplitudes ψ_{ω_0} and $\psi^*_{\omega_0}$ are identical. In order to derive an expression for the average albedo, we need to evaluate at the approximation of the Diffuson the reflected average intensity, $I_d(R\hat{s}_e)$ at a point $\boldsymbol{R} = R\hat{s}_e$ at infinity.

In the experimental setup for the albedo measurement, both the source and the detector are placed *outside* the diffusive medium so that we must also include the conversion of the incident plane wave into a diffusing wave and of the diffusing wave back into an outgoing spherical wave. Several approaches are thus possible. The most natural is to proceed as in Chapter 4, namely to describe multiple scattering as a product $\overline{G}\ \overline{G}\ \Gamma\ \overline{G}\ \overline{G}$ (see relation 4.37 or 4.60), where $\overline{G}$ is an average Green's function and where Γ is the structure factor[4] of the Diffuson. We thus obtain an expression analogous to (4.60) which takes the form,

$$I_d(R\hat{s}_e) = \frac{4\pi}{c} \int d\boldsymbol{r}_1\, d\boldsymbol{r}_2\, |\overline{\psi}_i(\boldsymbol{r}_1)|^2 |\Gamma(\boldsymbol{r}_1, \boldsymbol{r}_2)| \overline{G}^R(\boldsymbol{r}_2, \boldsymbol{R})|^2 . \tag{8.6}$$

The structure of this expression is represented in Figure 8.1. $|\overline{\psi}_i(\boldsymbol{r}_1)|^2$ is the average intensity at point $\boldsymbol{r}_1$ that originates directly from the light source, i.e., without intermediate scattering. It can thus be calculated within the Drude–Boltzmann approximation. The term $|\overline{G}^R(\boldsymbol{r}_2, \boldsymbol{R})|^2$ describes, at the same approximation, the wave propagation between the last scattering event in $\boldsymbol{r}_2$ and any point $\boldsymbol{R}$. Finally, $\Gamma(\boldsymbol{r}_1, \boldsymbol{r}_2)$ is the Diffuson structure factor, which obeys equation (4.24).

For the geometry that we are considering, the source term is well described by an incident plane wave so that

$$\overline{\psi}_i(\boldsymbol{r}_1) = \sqrt{\frac{cI_0}{4\pi}}\, e^{-|\boldsymbol{r}_1 - \boldsymbol{r}|/2l_e}\, e^{ik\hat{s}_i \cdot \boldsymbol{r}_1} , \tag{8.7}$$

in which $\boldsymbol{r}$ is the intersection of the incident beam with the interface placed at $z = 0$, $\boldsymbol{r}_1$ is the location of the first scattering event, and $k\hat{s}_i$ is the wave vector of the incident plane wave (Figure 8.1). The average albedo at the Diffuson approximation is obtained from relations (8.3) and (8.6) and it takes the form

$$\alpha_d = \frac{R^2}{S} \int d\boldsymbol{r}_1\, d\boldsymbol{r}_2\, e^{-|\boldsymbol{r}_1 - \boldsymbol{r}|/l_e}\, \Gamma(\boldsymbol{r}_1, \boldsymbol{r}_2) |\overline{G}^R(\boldsymbol{r}_2, \boldsymbol{R})|^2. \tag{8.8}$$

[4] From now on, we shall no longer specify the frequency ω_0 of the waves.

For $|\boldsymbol{R}-\boldsymbol{r}_2| \to \infty$ (Fraunhoffer approximation), the average Green function $\overline{G}^R(\boldsymbol{r}_2,\boldsymbol{R})$ given by relation (3.48) can be expanded as (see 2.56):

$$\begin{aligned}\overline{G}^R(\boldsymbol{r}_2,\boldsymbol{R}) &= e^{-|\boldsymbol{r}'-\boldsymbol{r}_2|/2l_e}\,\frac{e^{ik|\boldsymbol{R}-\boldsymbol{r}_2|}}{4\pi\,|\boldsymbol{R}-\boldsymbol{r}_2|}\\ &\simeq e^{-|\boldsymbol{r}'-\boldsymbol{r}_2|/2l_e}\,e^{-ik\hat{\boldsymbol{s}}_e\cdot\boldsymbol{r}_2}\,\frac{e^{ikR}}{4\pi R}\,. \end{aligned} \tag{8.9}$$

Introducing the projection[5] μ of the unit vector $\hat{\boldsymbol{s}}_e$ on the Oz axis, we obtain $|\boldsymbol{r}_2-\boldsymbol{r}'| = z_2/\mu$ and $|\boldsymbol{r}_1-\boldsymbol{r}| = z_1$ so that the expression of the incoherent albedo associated to the Diffuson finally takes the form

$$\boxed{\alpha_d = \frac{1}{(4\pi)^2 S}\int d\boldsymbol{r}_1\,d\boldsymbol{r}_2\,e^{-z_1/l_e}\,e^{-z_2/\mu l_e}\,\Gamma(\boldsymbol{r}_1,\boldsymbol{r}_2)} \tag{8.10}$$

Moreover, by assuming slow spatial variations, the structure factor $\Gamma(\boldsymbol{r}_1,\boldsymbol{r}_2)$, solution of equation (4.26), is shown to obey the diffusion equation

$$-D\Delta_{\boldsymbol{r}_2}\Gamma(\boldsymbol{r}_1,\boldsymbol{r}_2) = \frac{4\pi c}{l_e^2}\,\delta(\boldsymbol{r}_1-\boldsymbol{r}_2) \tag{8.11}$$

and it is related to the probability $P_d(\boldsymbol{r}_1,\boldsymbol{r}_2)$ through the relation (4.63):

$$P_d(\boldsymbol{r}_1,\boldsymbol{r}_2) = \frac{l_e^2}{4\pi c}\,\Gamma(\boldsymbol{r}_1,\boldsymbol{r}_2)\,. \tag{8.12}$$

Inserting this relation into (8.10), we obtain

$$\boxed{\alpha_d = \frac{c}{4\pi l_e^2}\int_0^\infty dz_1\,dz_2\,e^{-z_1/l_e}\,e^{-z_2/\mu l_e}\,P_d(z_1,z_2)} \tag{8.13}$$

with $P_d(z_1,z_2) = \int_S d^2\boldsymbol{\rho}\,P_d(\boldsymbol{\rho},z_1,z_2)$. For the geometry of a semi-infinite medium, the function $P_d(\boldsymbol{\rho},z_1,z_2)$ depends on the coordinates z_1 and z_2 as well as on the projection $\boldsymbol{\rho}$ of the vector $\boldsymbol{r}_1-\boldsymbol{r}_2$ on the plane $z=0$.

The calculation of the average albedo thus reduces to that of the probability P_d in a semi-infinite medium. We have shown (section A5.2.3) that, for this geometry, P_d is well described by the solution of a diffusion equation, provided we choose as an effective boundary condition that P_d vanishes at the point $-z_0$ with $z_0 = \frac{2}{3}l_e$ (for a discussion of this point see page 142 of reference [215]).[6] The solution of the corresponding stationary diffusion equation is obtained using the image method (section 5.7 and Appendix A5.3).

[5] We assume that the incident beam is perpendicular to the interface.

[6] The exact solution of the Milne problem (Appendix A5.3) gives $z_0 \simeq 0.710\,l_e$. But this value is not consistent with the diffusion approximation. We shall thus consider instead the value $z_0 = 2/3 l_e$ obtained within this approximation.

The images of the points $\boldsymbol{r}_1$ and $\boldsymbol{r}_2$ are determined with respect to the plane $-z_0$, so that the probability $P_d(\boldsymbol{\rho}, z_1, z_2)$ becomes

$$P_d(\boldsymbol{\rho}, z_1, z_2) = \frac{1}{4\pi D}\left(\frac{1}{\sqrt{\rho^2 + (z_1 - z_2)^2}} - \frac{1}{\sqrt{\rho^2 + (z_1 + z_2 + 2z_0)^2}}\right). \tag{8.14}$$

The integration over $\boldsymbol{\rho}$ leads to (relation 5.156)

$$P_d(z_1, z_2) = \frac{1}{2D}\left[(z_1 + z_2 + 2z_0) - |z_1 - z_2|\right] = \frac{z_m + z_0}{D} \tag{8.15}$$

where $z_m = \min(z_1, z_2)$. Using (8.13), we finally obtain for the albedo α_d the expression

$$\boxed{\alpha_d = \frac{3}{4\pi}\mu\left(\frac{z_0}{l_e} + \frac{\mu}{\mu + 1}\right)} \tag{8.16}$$

From this relation, it appears that within the Diffuson approximation the average albedo of an optically thick medium is almost independent of the angle between the incident beam and the direction $\hat{\boldsymbol{s}}_e$ of the outgoing wave (see Figure 8.7).

Remarks

- The previous expression for α_d results from the calculation of the intensity $I_d(\boldsymbol{R})$ at a point outside the diffusive medium. This is an approximation. The theory of radiative transfer (Appendix A5.2) amounts to calculating the specific intensity $I_d(z = 0, \hat{\boldsymbol{s}}_e)$ at the interface along a given outgoing direction $\hat{\boldsymbol{s}}_e$ (5.131). This approach leads to a different expression for the albedo, namely,

$$\alpha_d = \frac{3}{4\pi}\mu\left(\frac{z_0}{l_e} + \mu\right) \tag{8.17}$$

which, in contrast to (8.16), is normalized, $2\pi\int_0^{\pi/2}\alpha_d(\theta)\sin\theta d\theta = 1$. Nevertheless, we have chosen here to present the simplest derivation. It will be useful to get some intuition about the angular dependence of the coherent albedo and of the coherent backscattering cone.

- At first glance, the previous derivation of the average albedo applies to isotropic scatterers only, i.e., for a transport mean free path l^* equal to l_e. For the case of anisotropic scattering, the angular dependence of the structure factor should be taken into account (Appendix A4.3). The albedo α_d becomes

$$\alpha_d = \frac{1}{(4\pi)^2 S}\int d\boldsymbol{r}_1\, d\boldsymbol{r}_2\, e^{-z_1/l_e}\, e^{-z_2/\mu l_e}\,\Gamma(\hat{\boldsymbol{s}}_i, \hat{\boldsymbol{s}}_e, \boldsymbol{r}_1, \boldsymbol{r}_2) \tag{8.18}$$

where $\Gamma(\hat{\boldsymbol{s}}_i, \hat{\boldsymbol{s}}_e, \boldsymbol{r}_1, \boldsymbol{r}_2)$ is the Fourier transform of $\Gamma(\hat{\boldsymbol{s}}_i, \hat{\boldsymbol{s}}_e, \boldsymbol{q})$ defined by (4.155). We shall return to this point in section 8.6.

8.3.2 The coherent albedo: contribution of the Cooperon

We consider now the contribution α_c of the Cooperon to the average albedo as represented in Figure 8.2. As for the incoherent contribution, we describe the conversion of a plane wave into a diffusing wave by means of an average Green's function that decreases exponentially while entering the diffusive medium. The intensity $I_c(R\hat{s}_e)$ is given by the relation (4.61), namely,

$$I_c(R\hat{s}_e) = \frac{4\pi}{c} \int d\boldsymbol{r}_1\, d\boldsymbol{r}_2\, \overline{\psi}_i(\boldsymbol{r}_1)\overline{\psi}_i^*(\boldsymbol{r}_2)\Gamma'(\boldsymbol{r}_1,\boldsymbol{r}_2)\overline{G}^R(\boldsymbol{r}_2,\boldsymbol{R})\overline{G}^A(\boldsymbol{R},\boldsymbol{r}_1) \tag{8.19}$$

where $\Gamma'(\boldsymbol{r}_1,\boldsymbol{r}_2)$ is the Cooperon structure factor (section 4.6). Time reversal invariance implies that $\Gamma' = \Gamma$. Considering as previously an incident plane wave (8.7) and using the Fraunhoffer approximation (8.9) for the Green functions, we obtain

$$\alpha_c = \frac{R^2}{S} \int d\boldsymbol{r}_1\, d\boldsymbol{r}_2\, e^{-z_2/2\mu l_e} \frac{e^{-ik\hat{s}_e\cdot\boldsymbol{r}_2}}{4\pi R} e^{-z_1/2\mu l_e} \frac{e^{ik\hat{s}_e\cdot\boldsymbol{r}_1}}{4\pi R} \Gamma(\boldsymbol{r}_1,\boldsymbol{r}_2) \\ \times\, e^{-z_1/2l_e}\, e^{-z_2/2l_e}\, e^{ik\hat{s}_i\cdot\boldsymbol{r}_1}\, e^{-ik\hat{s}_i\cdot\boldsymbol{r}_2}\,, \tag{8.20}$$

so that

$$\boxed{\alpha_c(\hat{s}_e) = \frac{1}{(4\pi)^2 S} \int d\boldsymbol{r}_1\, d\boldsymbol{r}_2\, e^{-(\frac{\mu+1}{2\mu})\frac{z_1+z_2}{l_e}}\, \Gamma(\boldsymbol{r}_1,\boldsymbol{r}_2)\, e^{ik(\hat{s}_i+\hat{s}_e)\cdot(\boldsymbol{r}_1-\boldsymbol{r}_2)}} \tag{8.21}$$

The phase that appears in this relation leads to an angular dependence of the Cooperon contribution to the albedo. It is also important to notice that the exponential attenuation factors appearing in relations (8.10) and (8.21) are different. Along the *backscattering direction*, defined by the condition $\hat{s}_i + \hat{s}_e = 0$, the phase factor disappears and, since $\mu = 1$, we obtain

$$\alpha_c(\theta = 0) = \alpha_d \tag{8.22}$$

where θ is the angle between the incident and emergent directions $\hat{s}_i$ and $\hat{s}_e$, as represented in Figure 8.2. Therefore the total average albedo $\alpha(\theta) = \alpha_d + \alpha_c(\theta)$ is such that

$$\boxed{\alpha(\theta = 0) = 2\alpha_d} \tag{8.23}$$

This relation should be compared with the doubling of the probability to return to the origin due to the Cooperon contribution as discussed in section 4.6. The corresponding physical phenomenon is usually known as *coherent backscattering*.

Using the relation (4.63) between the structure factor Γ and P_d, (8.21) becomes

$$\alpha_c = \frac{c}{4\pi l_e^2} \int_0^\infty dz_1\, dz_2\, e^{-\left(\frac{\mu+1}{2\mu}\right)\frac{z_1+z_2}{l_e}} \int_S d^2\boldsymbol{\rho}\, P_d(\boldsymbol{\rho}, z_1, z_2)\, e^{i\boldsymbol{k}_\perp\cdot\boldsymbol{\rho}}\,, \tag{8.24}$$

where $\boldsymbol{k}_\perp = (\boldsymbol{k}_i + \boldsymbol{k}_e)_\perp = k(\hat{s}_i + \hat{s}_e)_\perp$ is the projection on the xOy plane of the vector $\boldsymbol{k}_i + \boldsymbol{k}_e$. If the length of $(\hat{s}_i + \hat{s}_e)$ is small enough, we can neglect its projection along the z axis. Using both relation (8.14) and the integral

$$\int_S d^2\boldsymbol{\rho}\, \frac{e^{i\boldsymbol{k}_\perp\cdot\boldsymbol{\rho}}}{\sqrt{\rho^2 + A^2}} = 2\pi \frac{e^{-k_\perp |A|}}{k_\perp}\,, \tag{8.25}$$

with the notation $k_\perp = |\boldsymbol{k}_\perp|$, we obtain

$$\boxed{\alpha_c = \frac{c}{4\pi l_e^2} \int_0^\infty dz_1\, dz_2\, e^{-\left(\frac{\mu+1}{2\mu}\right)\frac{z_1+z_2}{l_e}} P_d(k_\perp, z_1, z_2)} \tag{8.26}$$

with[7]

$$P_d(k_\perp, z_1, z_2) = \frac{1}{2Dk_\perp}\left(e^{-k_\perp|z_1-z_2|} - e^{-k_\perp(z_1+z_2+2z_0)}\right). \tag{8.27}$$

Upon integrating, relation (8.26) becomes

$$\boxed{\alpha_c(k_\perp) = \frac{3}{8\pi} \frac{1}{\left(k_\perp l_e + \frac{\mu+1}{2\mu}\right)^2} \left(\frac{1 - e^{-2k_\perp z_0}}{k_\perp l_e} + \frac{2\mu}{\mu+1}\right)} \tag{8.28}$$

As for the incoherent albedo, the dependence of $\alpha_c(\theta)$ on μ is negligible. In most calculations, we shall thus use the previous expression with $\mu = 1$, namely

$$\alpha_c(\theta) = \frac{3}{8\pi} \frac{1}{(1 + k_\perp l_e)^2} \left(1 + \frac{1 - e^{-2k_\perp z_0}}{k_\perp l_e}\right). \tag{8.29}$$

At small angles, we have $k_\perp \simeq (2\pi/\lambda)|\theta|$. The coherent contribution is non-zero within a cone of angular aperture $\lambda/2\pi l_e$ near the backscattering direction ($k_\perp \to 0$). By expanding (8.29) we obtain

$$\begin{aligned} \alpha_c(\theta) &\simeq \alpha_d \left(1 - 2\frac{(l_e + z_0)^2}{l_e + 2z_0} k_\perp\right) + O\left(k_\perp^2\right) \\ &\simeq \alpha_c(0) - \frac{3}{4\pi} \frac{(l_e + z_0)^2}{l_e} k_\perp + O\left(k_\perp^2\right) \end{aligned} \tag{8.30}$$

and $\alpha_c(0) = \alpha_d$. This result can be cast in the form

$$\alpha_c(\theta) \simeq \alpha_c(0) - \beta k_\perp l_e = \alpha_c(0) - \beta k l_e |\theta| \tag{8.31}$$

where the parameter β is defined by [217]

$$\beta = \frac{3}{4\pi}\left(1 + \frac{z_0}{l_e}\right)^2 = \frac{25}{12\pi} \tag{8.32}$$

[7] Equation (8.27) can also be deduced from relation (5.157) and the correspondence (5.47).

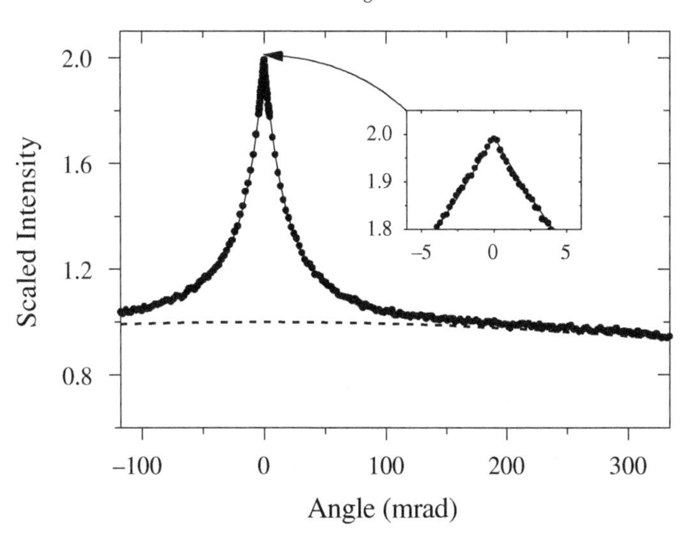

Figure 8.3 Average intensity backscattered as a function of the angle θ, as measured in a powder (solid solution) of ZnO. The albedo at exact backscattering, i.e., for $\theta = 0$, is twice as large as its background value. In the inset, we see the triangular singularity. A more quantitative analysis of this behavior is given in section 8.8 [221].

with $z_0 = \frac{2}{3} l_e$. A *triangular singularity* thus appears in the angular dependence of the albedo (Figure 8.3). This singularity results from the diffusive character of the propagation of the light intensity and, in the next section, we show that it is a measure of the distribution of the lengths of the multiple scattering paths. At large angles, $\alpha_c \to 0$, and only the classical, incoherent, contribution of the Diffuson remains.

The expressions that we have just established for the coherent albedo were obtained using the diffusion approximation. We may question the validity of this approximation for the short multiple scattering trajectories. This point is discussed in section 8.8. An exact solution to the coherent albedo problem has been proposed for a scalar wave and isotropic scattering [222]. This solution is based on a slightly different formulation of the Milne problem (Appendix A5.3). The corresponding solution, obtained for a semi-infinite geometry, uses the Wiener–Hopf method and is inconvenient both for numerical handling and for comparing with experimental results. Moreover, it cannot be extended to more physical situations for which the scattering is anisotropic and the polarization of the waves plays a role. Nevertheless, it accounts for the contribution of the single scattering, an important point to which we shall return in section 8.8.2.

Exercise 8.1 Using relation (8.27), show that the expansion of $P_d(k_\perp, z_1, z_2)$ around $k_\perp \to 0$ is given by

$$P_d(k_\perp, z_1, z_2) = P_d(0, z_1, z_2) - \frac{k_\perp}{D}(z_1 + z_0)(z_2 + z_0) . \tag{8.33}$$

Exercise 8.2: modification of the albedo for a slab of finite width

How is the triangular singularity of the coherent backscattering peak modified when the diffusive medium is a slab of finite width L? Show that in relation (8.30), $k_\perp$ must be replaced by $k_\perp \coth k_\perp (L+2z_0)$.

For a finite width, we still keep the relation (8.26) and we replace (8.27) by (5.159), using the correspondence (5.47), namely

$$P_d(k_\perp, z_1, z_2) = \frac{1}{Dk_\perp} \frac{\sinh k_\perp (z_m + z_0) \sinh k_\perp (L + z_0 - z_M)}{\sinh k_\perp (L + 2z_0)} \tag{8.34}$$

with $z_m = \min(z, z')$ and $z_M = \max(z, z')$. We then expand this expression for small values of the arguments $k_\perp z_m$, $k_\perp z_M$ and $k_\perp z_0$, and we obtain a relation similar to (8.33) where $k_\perp$ has been replaced by $k_\perp \coth k_\perp (L + 2z_0)$. Finally, the integral (8.26) leads to the announced result.

The aim of this exercise is to show that the characteristic cusp in the backscattering direction disappears for values of $k_\perp$ smaller than $1/L$. This singularity thus results from the contribution of long diffusive trajectories. Cutting off these trajectories beyond the length L modifies the cusp for small values of $k_\perp$, i.e., for small angles. This relation between long trajectories and small angles is discussed in more detail in the next section.

More generally, using (8.34) and assuming $z_0 = 0$, show that for a slab of finite width L, the incoherent albedo $\alpha_d(L)$ and the coherent contribution $\alpha_c(k_\perp, L)$ are given by the expressions

$$\alpha_d(L) = \frac{3}{8\pi} \left(1 - e^{-2b}\right) \left(1 - \frac{\tanh(b/2)}{b/2}\right) \tag{8.35}$$

$$\alpha_c(k_\perp, L) = \frac{3}{8\pi} \frac{1 - e^{-2b}}{(1 - k_\perp l_e)^2} \left[1 + \frac{2k_\perp l_e}{(1 + k_\perp l_e)^2} \frac{1 - \cosh\left(b\ (k_\perp l_e + 1)\right)}{\sinh b\ \sinh(b\ k_\perp l_e)}\right],$$

valid only in the diffusive limit, i.e., when the optical depth defined by $b = L/l_e$ becomes much larger than 1. Check that in the backscattering direction we still have the relation $\alpha_c(0, L) = \alpha_d(L)$, but without the cusp.

8.4 Time dependence of the albedo and study of the triangular cusp

Of interest is an alternative derivation of the albedo starting from the time dependence of the diffusion probability $P_d(\boldsymbol{r}, \boldsymbol{r}', t)$. For the geometry of a semi-infinite medium, with the help of relations (5.42) and (5.65), and the vanishing boundary condition at $-z_0$, we obtain

$$P_d(\boldsymbol{r}, \boldsymbol{r}', t) = \frac{e^{-\rho^2/4Dt}}{(4\pi Dt)^{3/2}} \left[e^{-(z-z')^2/4Dt} - e^{-(z+z'+2z_0)^2/4Dt}\right], \tag{8.36}$$

where we have taken for the two-dimensional Fourier transform, the expression $P_d(k_\perp, z, z', t)$,

$$P_d(k_\perp, z, z', t) = \frac{e^{-Dk_\perp^2 t}}{(4\pi Dt)^{1/2}} \left[e^{-(z-z')^2/4Dt} - e^{-(z+z'+2z_0)^2/4Dt}\right]. \tag{8.37}$$

We can now define, at least formally, a time-dependent albedo $\alpha(t) = \alpha_d(t) + \alpha_c(\theta, t)$ by

$$\alpha_d = \int_0^\infty dt\, \alpha_d(t)\,, \quad \alpha_c(\theta) = \int_0^\infty dt\, \alpha_c(\theta, t)\,. \tag{8.38}$$

We consider the small angle limit so that $\mu = 1$ and the angular dependence of α_d is negligible. Noticing that $\alpha_c(0, t) = \alpha_d(t)$ and using (8.26), we obtain for $\alpha_d(t)$ and $\alpha_c(\theta, t)$ the two expressions

$$\alpha_d(t) = \frac{c}{4\pi l_e^2} \int_0^\infty dz\, dz'\, e^{-z/l_e} e^{-z'/l_e} P_d(z, z', t) \tag{8.39}$$

$$\alpha_c(\theta, t) = \frac{c}{4\pi l_e^2} \int_0^\infty dz\, dz'\, e^{-z/l_e} e^{-z'/l_e} P_d(k_\perp, z, z', t)\,. \tag{8.40}$$

The integrals over z and z' have no angular dependence. The only remaining angular dependence comes from the factor $e^{-Dk_\perp^2 t}$ that is nothing but the *Fourier transform of a two-dimensional diffusion process* restricted to the interface.

Because of the exponential factors that appear in the integrals over z and z', these integrals are cut off at a length of the order of the elastic mean free path l_e. Then, in the long time limit ($t \gg \tau_e$), the Gaussian terms in the brackets of (8.37) can be expanded. Upon integrating over z and z', we obtain

$$\alpha_c(\theta, t) = \alpha_d(t)\, e^{-Dk_\perp^2 t} \tag{8.41}$$

with

$$\alpha_d(t) \simeq c(z_0 + l_e)^2 \frac{1}{(4\pi D t)^{3/2}} \tag{8.42}$$

and $k_\perp \simeq 2\pi|\theta|/\lambda = k|\theta|$. This expression allows us to rewrite $\alpha_c(\theta, t)$ in the form

$$\alpha_c(\theta) \propto \int_0^\infty \frac{1}{t^{3/2}}\, e^{-\frac{1}{3}(kl_e\theta)^2 t/\tau_e} \left(1 - e^{-t/\tau_e}\right) dt\,. \tag{8.43}$$

The integrand in this expression accounts for the contribution of diffusive trajectories of total length t. An additional factor $(1 - e^{-t/\tau_e})$ has been introduced in order to cut off the integral at small times.

In the backscattering direction ($k_\perp = 0$), we have $\alpha_c(\theta = 0, t) = \alpha_d(t) \propto (Dt)^{-3/2}$. At a given time t, the coherent backscattered echo is enhanced by a factor $(1 + e^{-Dk_\perp^2 t})$ compared to its incoherent value. This enhancement occurs inside a cone whose angular aperture is $\theta(t) = \lambda/(2\pi\sqrt{Dt})$, so that the smaller the length t of the diffusive paths, the larger is the angular aperture of their contribution to the coherent backscattering cone (Figure 8.4). This implies that in the backscattering direction, *the factor* 2 *between the coherent and incoherent contributions remains at all times.*

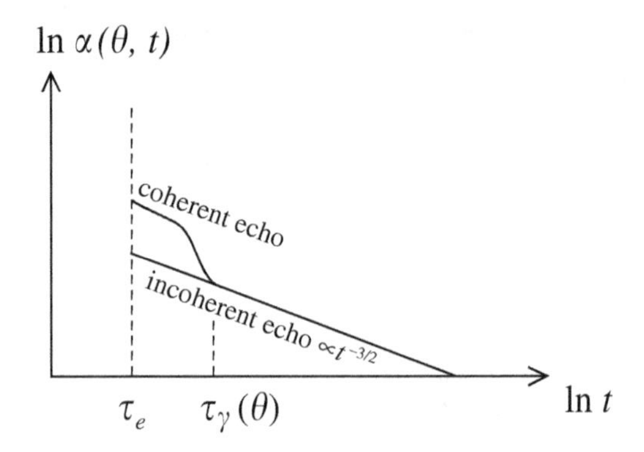

Figure 8.4 Time dependence of the albedo $\alpha(\theta, t)$ at a fixed angle θ in a logarithmic plot. The coherent contribution appears for times shorter than $\tau_\gamma(\theta)$ given by the relation (8.46).

It is of interest that the albedo $\alpha_d(t)$ can be interpreted as the probability of reaching the plane $z = -z_0$ after a time t. This probability varies as $t^{-3/2}$ and does not depend on the space dimensionality, in the sense that the same result remains true for a two-dimensional diffusive process and a one-dimensional interface or more generally in d dimensions with a $(d-1)$-dimensional interface.

Another useful representation of the albedo $\alpha_c(\theta)$ is obtained in the form of the following Laplace transform

$$\alpha_c(\theta) = \int_0^\infty dt\, \alpha_d(t) \left\langle e^{i\boldsymbol{k}_\perp \cdot \boldsymbol{\rho}} \right\rangle \tag{8.44}$$

in which the average is taken using the Gaussian law:

$$\left\langle e^{i\boldsymbol{k}_\perp \cdot \boldsymbol{\rho}} \right\rangle = e^{-Dk_\perp^2 t} . \tag{8.45}$$

The comparison of these two relations with (6.3) allows us to define a dephasing time $\tau_\gamma(\theta) = 1/Dk_\perp^2$, that can also be written as

$$\boxed{\frac{\tau_\gamma(\theta)}{\tau_e} = \frac{3}{(kl_e\theta)^2}} \tag{8.46}$$

The triangular cusp characteristic of the coherent backscattering can then be interpreted as arising from the sum of a series of Gaussian terms weighted by the probability $(Dt)^{-3/2}$. Although each of these terms behaves parabolically near backscattering $\theta \simeq 0$, the integral becomes singular around this value. The angle θ thus appears as a variable conjugate to the length t of the diffusive paths. A given value of θ selects all paths of lengths $t/\tau_e \leq 3/(kl_e\theta)^2$ that contribute to the coherent backscattering peak (Figure 8.4) . In other words, *long trajectories provide the main contribution to the coherent albedo at small angles.* The quantity $L_\gamma = \sqrt{D\tau_\gamma(\theta)} \simeq \lambda/2\pi\theta$ appears to be the characteristic length beyond which diffusive paths no longer contribute to the Cooperon and to the coherent backscattering cone. In that sense, L_γ can be viewed as a dephasing length (see Chapter 6) associated with the controlled and reversible phase shift driven by the angle θ.

8.5 Effect of absorption

The effect of absorption on the albedo can be accounted for by means of an *absorption length* l_a and the phenomenological expression[8]

$$\alpha(\theta, l_a) = \int_0^\infty dt\, \alpha(\theta, t)\, e^{-t/\tau_a} \tag{8.47}$$

where $\tau_a = l_a/c$ is the absorption time. Then, for the coherent albedo we obtain

$$\alpha_c(\theta, l_a) = \int_0^\infty dt\, \alpha_d(t)\, e^{-Dk_\perp^2 t}\, e^{-t/\tau_a} . \tag{8.48}$$

These expressions of coherent albedo in either the presence or the absence of absorption are related by

$$\boxed{\alpha_c(k_\perp, l_a) = \alpha_c\left(\sqrt{k_\perp^2 + k_a^2}, \infty\right)} \tag{8.49}$$

with $k_a^{-1} = \sqrt{D\tau_a} = \sqrt{l_e l_a/3}$. In the backscattering direction, the previous expression yields

$$\alpha_c(0, l_a) = \alpha_c(k_a, \infty) . \tag{8.50}$$

The overall effect of a finite absorption length is to cut off the contributions of diffusive trajectories of length longer than $\sqrt{D\tau_a}$ to both the coherent and incoherent albedos. Based on the analysis presented in the previous section and on the expression (8.43), we expect the coherent albedo to exhibit a parabolic behavior for angles θ such that $\tau_\gamma > \tau_a$, or equivalently $\theta < \lambda/(2\pi\sqrt{l_e l_a})$. This results from the suppression of the diffusive trajectories of length larger than $\sqrt{D\tau_a}$. Such behavior has indeed been observed experimentally (see Figure 8.5 and section 8.8.3).

It is also of interest to compare the effect of a finite absorption length with that of a phase coherence time τ_ϕ as it appeared in the description of coherent transport in metals (section 7.4). Indeed it is important to notice that the absorption (τ_a) has a different effect than dephasing (τ_ϕ) since it affects both coherent and incoherent contributions, so that the factor 2 at backscattering remains unchanged. This is to be contrasted with the case of electrons where a finite τ_ϕ affects the coherent contribution responsible for weak localization by cutting off the trajectories of length longer than $\sqrt{D\tau_\phi}$ but leaves the incoherent contribution unchanged.

[8] Care must be taken to distinguish between the absorption length l_a defined as $l_a = c\tau_a$ which corresponds to a diffusive trajectory of total length τ_a, and the length $L_a = \sqrt{D\tau_a}$ which is the typical distance reached by diffusion in a time τ_a. The latter plays a role analogous to L_γ defined previously. It is usually more convenient to use l_a.

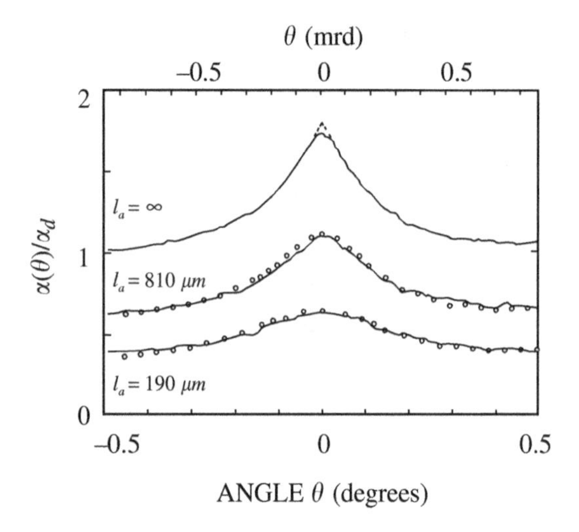

Figure 8.5 Behavior of the albedo in the presence of absorption for different values of the absorption length l_a. See also Figure 8.11 [223].

8.6 Coherent albedo and anisotropic collisions

So far, we have limited ourselves to the case of isotropic scattering. The only relevant characteristic length in the diffusive medium is then the elastic mean free path l_e. But in most physically relevant situations, the scattering is anisotropic and the transport mean free path l^* differs from the elastic mean free path l_e (Appendix A4.3). How is the albedo modified in that case? We still expect the diffusion approximation to describe the long multiple scattering trajectories properly, provided we replace D by D^* (i.e., l_e by l^*), so that the small angle behavior of the coherent albedo α_c remains unchanged. In other words, we expect (8.30) still to hold provided we replace l_e by l^* and z_0 by $\frac{2}{3}l^*$.

To understand better the distinct roles of l_e and l^*, let us start from the definitions (8.6) and (8.19) of the albedo. In the expression for α_d, the mean free path l_e appears in the average Green functions, and in the structure factor Γ. For the case of anisotropic scattering it seems justified to replace the diffusion coefficient D by D^* in the expression of Γ (relation 4.169). On the other hand, the average Green functions, that account respectively for the first and last scattering events, remain functions of l_e and not of l^*. Performing the above mentioned changes leads straightforwardly to a set of expressions for the small angle coherent albedo and for the incoherent one, that depend on the ratio l_e/l^*, in contradiction to the argument presented in the first paragraph and, more importantly, to experimental results.

To grasp the nature of this apparent contradiction, it must be realized that the relations (8.6) and (8.19) need to be generalized in order to account for the angular dependence of the structure factor (relation 8.18). Instead, we now present an alternative description of light transport based on radiative transfer, the details of which may be found in Appendix A5.2. Equation (5.93) for the specific intensity $I_d(\hat{s}, \boldsymbol{r})$ relies on the same approximations as used for the Diffuson P_d. The advantage of the radiative transfer approach is that it allows a more systematic description that includes the geometry of the diffusive medium and the nature of the light sources. For instance, light sources appear explicitly in (5.93) and the nature of the boundaries sets the choice of the boundary conditions. Within the diffusion approximation used thus far for calculation of the albedo, the radiative transfer equation reduces to the diffusion equation (5.105). But an advantage of the radiative transfer description as compared to (4.66) resides in the fact that it contains explicitly the source of the radiation. Another advantage is that it allows the conversion of an incident plane wave into a diffusing wave to be described simply with the boundary condition (5.108) of a vanishing incoming diffuse flux. In the absence of sources, this boundary condition is given by (5.113), and it depends on the extrapolation length $z_0 = \frac{2}{3}l^*$ we have been using previously. If there is a source, this boundary condition takes the form (5.129), which depends on both l^* and l_e.

The calculation of the albedo based on the radiative transfer equation and on this boundary condition is more consistent, although less intuitive, than the calculation that led to relations (8.16) and (8.28). But it is important to emphasize that these two descriptions rely on the diffusion approximation and, as such, they are equivalent. They differ in the fact that they do not treat the contribution of the short multiple scattering trajectories on the same footing.

To proceed further, we again consider the geometry of a semi-infinite diffusive medium whose interface is illuminated by a point-like light source of intensity $I_0\delta(\boldsymbol{\rho})$, in which $\boldsymbol{\rho}$ is a vector contained in the plane $z = 0$ defining the interface. We have shown that the average intensity $I_d(\boldsymbol{r})$ is, at the diffusion approximation, the solution of the equation (5.105) with the boundary condition (5.129). For the geometry of a slab, the solution of this problem is given by (5.132), namely

$$I_d(\boldsymbol{\rho}, z=0) = \frac{I_0}{4\pi^2}\frac{l^*}{l_e}\int_0^\infty d\lambda\, J_0(\lambda\rho)\,\frac{\lambda}{1+\frac{2}{3}l^*\lambda}\left(\frac{1}{1+\lambda l_e} - \eta\right) \tag{8.51}$$

where $\eta = 1 - l_e/l^*$ and J_0 is a Bessel function. The outgoing flux at a point of the interface is given by (5.131), namely $\frac{5}{2}I_d(\boldsymbol{\rho}, z=0)$ [216, 224]. The coherent albedo is simply given by the Fourier transform of this flux with respect to the variable $\boldsymbol{\rho}$:

$$\alpha_c(\theta) = \frac{5}{2I_0}\int_S d^2\boldsymbol{\rho}\, I_d(\boldsymbol{\rho}, z=0)\, e^{i\boldsymbol{k}_\perp\cdot\boldsymbol{\rho}}\,. \tag{8.52}$$

The integral over $\boldsymbol{\rho}$ is easily performed using (15.61). We obtain finally

$$\boxed{\alpha_c(\theta) = \frac{5}{4\pi}\frac{1}{1-\eta}\frac{1}{1+\frac{2}{3}k_\perp l^*}\left(\frac{1}{1+k_\perp l_e} - \eta\right)} \tag{8.53}$$

In the small angle limit, only long diffusive trajectories contribute to α_c, which then happens to depend only on the transport mean free path l^*, as stated previously. In this limit, the expression (8.53) is rewritten as

$$\alpha_c(\theta) \simeq \alpha_d - \beta^* k_\perp l^* = \alpha_d - \beta^* k l^* |\theta| \tag{8.54}$$

with

$$\beta^* = \frac{3}{4\pi}\left(1 + \frac{z_0}{l^*}\right)^2 . \tag{8.55}$$

By inserting $z_0 = 2l^*/3$, we notice that $\beta^* = 25/12\pi$, i.e., that it is independent of the exact nature of the scattering process[9] (see 8.32). The expression (8.53) for α_c does not apply in the large angle limit (it gives an overall negative value to the total albedo) since its range of validity is restricted to the diffusion approximation, i.e., to diffusive trajectories of length longer than l^*.

8.7 The effect of polarization

Up to now, we have considered the Cooperon contribution to the coherent albedo under the assumption that the light is a scalar wave, thus ignoring the effect of polarization. We now reconsider this assumption [226]. We saw in section 6.6 that the effect of polarization is to introduce a phase shift between paired sequences of multiple scattering. Here we discuss this phase shift in the framework of the Rayleigh approximation[10] (sections A2.1.4 and 6.6). We shall consider a linearly polarized wave (denoted by l) or a circularly polarized wave characterized by its helicity h.

8.7.1 Depolarization coefficients

In section 6.6 we studied the evolution of a given polarization state of a light beam under multiple scattering. This evolution is described by the integral equation (6.156). A first and obvious effect of multiple scattering is the depolarization of the incident light. After a time t (i.e., for diffusive trajectories of length ct), the depolarization of a wave can be characterized with the help of two depolarization coefficients $d_\parallel(t)$ and $d_\perp(t)$ which measure the relative intensity analyzed in the backscattering direction, within each polarization channel respectively parallel ($l \parallel l$ or $h \parallel h$) and perpendicular ($l \perp l$ or $h \perp h$) to the incident polarization. These coefficients are defined by

$$d_\parallel(t) = \frac{\Gamma_\parallel^{(d)}(t)}{\Gamma_\parallel^{(d)}(t) + \Gamma_\perp^{(d)}(t)} \qquad d_\perp(t) = \frac{\Gamma_\perp^{(d)}(t)}{\Gamma_\parallel^{(d)}(t) + \Gamma_\perp^{(d)}(t)} , \tag{8.56}$$

9 To the best of our knowledge, there is no full fledged "microscopic" calculation of the albedo for the case of anisotropic scattering. Moreover, it would not be very useful since such an expression of the albedo at large angles would necessarily depend on other parameters so far neglected (e.g. polarization, optical index mismatch, etc.). Nevertheless, it is worth mentioning the result obtained in reference [225] for the coherent albedo in the anisotropic case, that was obtained by combining together a microscopic description and the radiative transfer approach. The absolute slope obtained is $-3/4\pi(z_0/l^* + l_e/l^*)^2$. It depends on l_e whereas it is expected to be universal.

10 The results obtained in this approximation can be readily extended to the case of Rayleigh–Gans scattering (remark on page 233).

where $\Gamma_{\parallel}^{(d)}$ and $\Gamma_{\perp}^{(d)}$ are the amplitudes of the structure factor in the corresponding polarization channels. Clearly, in the long time limit, the depolarization is complete:

$$d_{\parallel}(t) \underset{t\to\infty}{\longrightarrow} \frac{1}{2} \qquad d_{\perp}(t) \underset{t\to\infty}{\longrightarrow} \frac{1}{2} \,. \tag{8.57}$$

This results from the fact that the contribution of the scalar mode Γ_0 is not attenuated and is the same in the different components $\Gamma_{\alpha\alpha,\beta\beta}$ of the structure factor (see equations 6.173, 6.174), so that the contribution of long trajectories ($t \to \infty$) to the intensity is the same in all polarization channels.

Depolarization affects equally the coherent and incoherent contributions to the albedo, i.e., the Diffuson and the Cooperon. This effect must be taken into account to describe properly the multiple scattering of an initially polarized light. For instance, the incoherent albedo is obtained from expression (8.38), which, after inserting the time-dependent depolarization factors, becomes

$$\alpha_d^{\parallel} = \int_0^{\infty} dt\, \alpha_d(t) d_{\parallel}(t) \simeq \frac{1}{2}\alpha_d \tag{8.58}$$

$$\alpha_d^{\perp} = \int_0^{\infty} dt\, \alpha_d(t) d_{\perp}(t) \simeq \frac{1}{2}\alpha_d \,. \tag{8.59}$$

The net effect of depolarization is to reduce the incoherent background described by $\alpha_d^{\perp}$ and $\alpha_d^{\parallel}$ almost in the same proportion. Half of the signal is detected in each of the two polarization channels, parallel and perpendicular. The approximation $d_{\parallel,\perp} \simeq \frac{1}{2}$ is well justified for the almost fully depolarized long trajectories, that is for times larger than the characteristic times τ_1 and τ_2 of the decaying modes (equation 6.166). On the other hand, the weight of short trajectories, still partially polarized, is then underestimated.

8.7.2 *Coherent albedo of a polarized wave*

In addition to depolarization, there is an additional phase shift between multiple scattering sequences propagating in opposite directions. This phase shift thus affects the Cooperon and the coherent backscattering. To see how this happens, we consider an incident polarized beam. The coherent albedo can be analyzed either in the same polarization channel or in the perpendicular one. We have shown in section 6.6.4 that if the light is analyzed in the same polarization channel, the vectorial nature of the light does not play any role, apart from the depolarization effects previously discussed. The corresponding coherent albedo $\alpha_c^{\parallel}$, obtained from (8.43), remains unchanged and it includes only the depolarization (8.57) of the incident light:

$$\alpha_c^{\parallel}(\theta) = \int_0^{\infty} \alpha_d(t) d_{\parallel}(t)\, e^{-Dt\left(\frac{2\pi}{\lambda}\theta\right)^2} (1 - e^{-t/\tau_e})\, dt \,. \tag{8.60}$$

Is this result also true for the perpendicular channel? For an initially polarized light, the coherent albedo analyzed in the perpendicular channel is attenuated by a multiplicative factor $Q_\perp(t)$ defined as the ratio of the Cooperon and the Diffuson contributions to the structure factor:

$$Q_\perp(t) = \frac{\Gamma_\perp^{(c)}(t)}{\Gamma_\perp^{(d)}(t)} \, . \tag{8.61}$$

This ratio vanishes at large times, since the Cooperon in the perpendicular channel involves two rapidly decaying modes $k = 1$ and $k = 2$ (see equations 6.176 and 6.167), whereas $\Gamma_\perp^{(d)}(t)$ is driven by the Goldstone mode Γ_0, see the discussion on page 231. The coherent albedo $\alpha_c^\perp(\theta)$ becomes

$$\alpha_c^\perp(\theta) = \int_0^\infty \alpha_d(t) d_\perp(t) Q_\perp(t)\, e^{-Dt\left(\frac{2\pi}{\lambda}\theta\right)^2} (1 - e^{-t/\tau_e})\, dt \, . \tag{8.62}$$

For a beam analyzed along a polarization parallel to the incident one, we check using (8.58) and (8.60) that $\alpha_c^\parallel(\theta = 0) = \alpha_d^\parallel$. Therefore, as a result of depolarization, both coherent

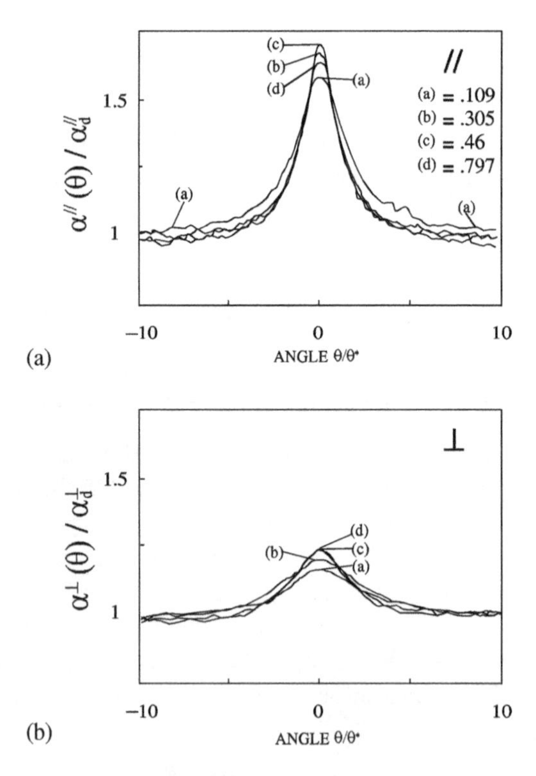

Figure 8.6 Angular behavior of the coherent backscattering peak for a linearly polarized light which is analyzed (a) along the same polarization channel, and (b) along the perpendicular channel. The diffusive medium is made of polystyrene spheres of diameters 0.109 μm, 0.305 μm, 0.460 μm and 0.797 μm. The angles are scaled by the peak widths (θ^*) and the intensities by the multiply scattered (‖) incoherent intensity so that all curves have the same width [223].

and incoherent contributions are reduced by half, but their ratio remains unchanged, namely there is still a factor 2 for the coherent backscattering peak as in the scalar case. On the other hand, for an emergent beam analyzed in a channel perpendicular to the incident one, the coherent albedo is reduced owing to the exponential decay of the Cooperon $\Gamma_{\perp}^{(c)}(t)$ at large time. Consequently, the contribution of long diffusive trajectories is reduced. The ratio between the coherent and incoherent contributions in backscattering ($\theta = 0$) is thus given by

$$r = \frac{\int_0^\infty dt\, \alpha_d(t) d_\perp(t) Q_\perp(t)}{\int_0^\infty dt\, \alpha_d(t) d_\perp(t)} . \tag{8.63}$$

The height of the cone is then reduced. The behaviors described above have been observed experimentally, and are represented in Figure 8.6. We should keep in mind that the previous results apply only for the limiting case of Rayleigh scattering. For bigger scatterers, which is the common case, we deal with anisotropic Mie scattering. The depolarization thus occurs on longer multiple scattering trajectories.

8.8 Experimental results

In the introduction of this chapter, we gave a brief historical overview of the study of coherent backscattering. In the first measurement of an interference effect near the backscattering direction [212], the relative enhancement factor was about 15% instead of the expected 100% predicted theoretically. This partly explains why this effect was not observed beforehand, even by chance. The angular aperture of the cone, given by the ratio λ/l^*, is of the order of a few milliradians only. Its observation thus requires a very sensitive angular resolution which has been achieved only quite recently. Following this first observation and triggered by a number of theoretical predictions, several groups have designed more and more precise experimental setups. The existing experiments can be divided into two main groups: those performed on diffusive media made of liquid suspensions of scatterers [212–214, 223] and those using solid solutions [227–229]. In the first setting, the average results simply from the motion of the scatterers integrated over a long enough time. In the second case, the average is obtained either by collecting the results obtained from different configurations or by rotating a cylindrical sample along its axis, each position providing a different configuration. The equivalence of the results obtained from these different methods can be viewed as a justification of the ergodic hypothesis.

The best angular resolution obtained so far is less than 50 μrad [229]. A good angular resolution is an experimental constraint for the observation of the effects predicted theoretically, namely the factor 2 enhancement of the coherent albedo and the triangular cusp near the backscattering direction. Experimental confirmation of these predictions has paved the road to the quantitative study of the other effects presented in this chapter, i.e., those related to the presence of absorption, the size of the scatterers, the role of polarization, etc. The results that have been obtained demonstrate beyond any doubt that, besides a good understanding of coherent backscattering, we now have at our disposal a tool that allows us to characterize very precisely multiple scattering systems.

8.8.1 The triangular cusp

In the course of our study of coherent albedo, we have assumed that the incoherent albedo α_d has no angular structure. This appears clearly in Figure 8.7 which presents backscattering by teflon. In Figure 8.8, we observe the enhancement by a factor 2 and the cusp characteristics of coherent backscattering. The agreement between these experimental results and expression (8.28) is excellent, for both liquid and solid solutions, and for a broad range of wavelengths and of elastic mean free paths [221, 229]. This agreement nevertheless raises a number of questions, bearing in mind the approximations underlying the derivation of expression (8.28). It has indeed been obtained for a scalar wave and within the diffusion approximation which is well justified only for long trajectories, i.e., for small angles. This approximation underestimates the relative weight of short trajectories as compared to the exact solution [222].

This excellent agreement can be partially explained by the role played by polarization. We have indeed obtained that, for a polarized wave analyzed along the incident polarization channel, there is no phase shift between the two trajectories paired into the Cooperon, thus justifying the use of a scalar wave. Moreover, in contrast to long trajectories, the short trajectories are only partially depolarized, as shown by relations (8.57). As a result of their remaining polarization, the contribution of these short trajectories is partially washed out

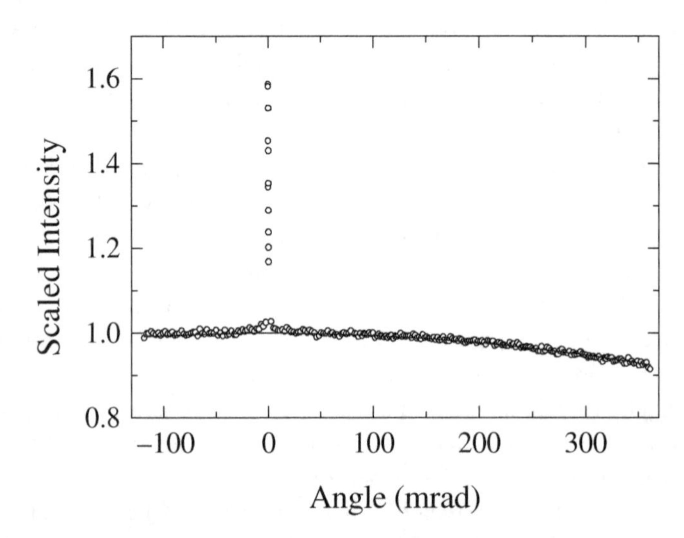

Figure 8.7 Angular dependence of the intensity backscattered by a sample of teflon. The extremely narrow cone results from the large elastic mean free path. The intensity is normalized to one near zero angle. The continuous line corresponds to a fit using expression (8.16) for α_d [221].

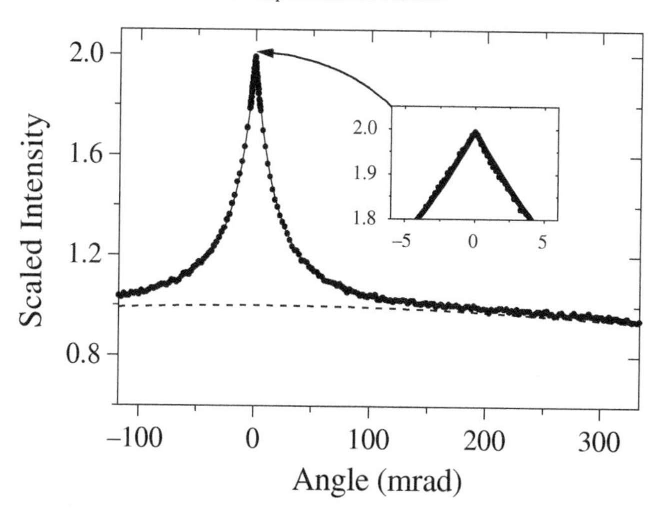

Figure 8.8 Angular dependence of the intensity backscattered by a powder (solid solution) of ZnO. The elastic mean free path is $l^* = 1.9 \pm 0.1$ μm. The normalization is such that the incoherent contribution (dashed line) is equal to one in the backscattering direction ($\theta = 0$). The enhancement factor is $\alpha(\theta = 0) = 1.994 \pm 0.012$. The continuous line corresponds to the expression (8.28) established in the diffusion approximation (weighted by a factor 1/2 that accounts for depolarization). The inset shows the behavior of the coherent albedo near backscattering and the triangular cusp fitted using expression (8.28) [221].

when it is analyzed along the incident polarization channel.[11] The contribution of the short trajectories is not, then, as important as predicted by the exact scalar theory.

8.8.2 Decrease of the height of the cone

Figure 8.9 shows that the width of the coherent backscattering cone decreases when l^* increases as predicted by (8.54). We also notice in the inset of the figure that the height depends on the mean free path. It decreases with l^* and reaches values smaller than the expected factor 2.

A first and immediate source for this discrepancy is single scattering. Let us recall that in our respective calculations of the Diffuson and the Cooperon, we included in the latter the single scattering contribution for the sake of convenience only (footnote 13, page 107). However, it does not contribute to the coherent albedo since it is angle-independent, whereas it does contribute to the incoherent part α_d. Strictly speaking, to evaluate the enhancement factor $\mathcal{A}$, i.e., the total value of the albedo at $\theta = 0$, we must take out of α_c the single

[11] The same argument applies equally to the justification for neglecting the single scattering contribution to α_c, since it remains fully polarized.

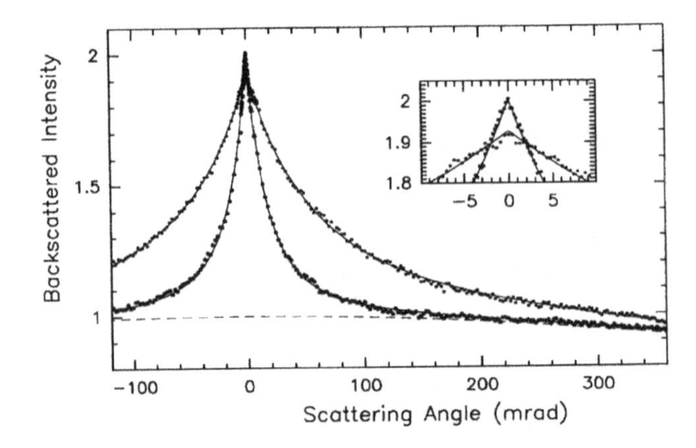

Figure 8.9 Two measurements of the coherent backscattering cone that correspond respectively to a small value of l^* (broad cone) and to a larger value (narrow cone). The latter corresponds to a solid solution ($BaSO_4$) characterized by the parameter $kl^* = 22.6 \pm 1.0$, whereas the broader cone corresponds to a liquid suspension of TiO_2 characterized by $kl^* = 5.8 \pm 1.0$. The solid and dashed line curves correspond respectively to relations (8.28) and (8.16). In the inset, we see clearly the deviation from factor 2 for the smallest value of l^* [229].

scattering contribution α_0. With the help of the equality $\alpha_c = \alpha_d$, we thus obtain the relation

$$\mathcal{A} = \frac{(\alpha_c - \alpha_0) + \alpha_d}{\alpha_d} = 2 - \frac{\alpha_0}{\alpha_d} . \tag{8.64}$$

As stated previously, polarization plays an important role in the behavior of the single scattering contribution to $\mathcal{A}$. For the case of Rayleigh scattering by point-like scatterers, single scattering is described by the differential scattering cross section (2.149) where the unit vectors $\hat{\boldsymbol{\varepsilon}}_i$ and $\hat{\boldsymbol{\varepsilon}}'$ account for the polarization of the incident and emergent fields, respectively. These vectors define four possible polarization channels: for a linearly polarized incident wave, we have two channels $(l \parallel l)$ or $(l \perp l)$, while for a circularly polarized incident wave, we have the two other channels $(h \parallel h)$ and $(h \perp h)$ where h is the helicity defined with respect to the direction of propagation.

For the case of single scattering, $\hat{\boldsymbol{\varepsilon}}_i \cdot \hat{\boldsymbol{\varepsilon}}'^*$ remains finite in backscattering only when $(l \parallel l)$ or $(h \perp h)$ (see Exercise 2.4). The latter channel describes a reflection in a mirror of a circularly polarized wave. It is thus possible, using channels $(l \perp l)$ or $(h \parallel h)$, to get rid of the single scattering contribution in backscattering for the Rayleigh case. Moreover, relation (8.62) indicates that in the channel $(l \perp l)$ there is a non-zero phase shift between the multiple scattering trajectories paired into the Cooperon. This is not the case for the channel $(h \parallel h)$ for which the attenuation factor $\langle Q_{\alpha\alpha} \rangle = \langle Q_\parallel \rangle = 1$ (relation 8.61) [230]. It is thus possible, in this channel $(h \parallel h)$, to obtain an enhancement factor equal to its

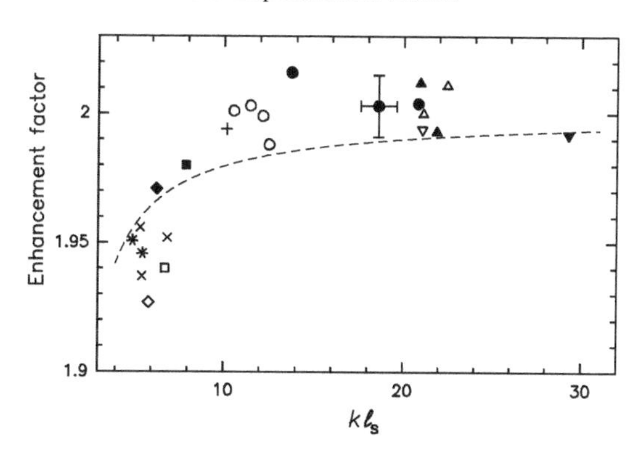

Figure 8.10 Deviation of the amplification factor $\mathcal{A}$ with respect to the value $\mathcal{A} = 2$ plotted as a function of the parameter kl^*. Each point corresponds to a different solid solution. The dashed line results from a calculation of $\mathcal{A}$ which incorporates double scattering on the same scatterer [229].

maximum 2, since both the single scattering and the finite phase shift in the Cooperon are absent.

For anisotropic scattering, i.e., in the Mie regime (section A2.3.2), all polarization channels contribute to backscattering. This leads to a smaller enhancement factor. In most cases, however, the anisotropy of the scattering cross section makes the single scattering contribution negligible in backscattering, thus restoring an enhancement factor close to 2.

Yet another contribution to a reduction of $\mathcal{A}$ exists. It originates in multiple scattering processes in which the wave is scattered more than once by a given scatterer. This contribution is independent of the angle θ, and therefore is added to the incoherent background, thus reducing $\mathcal{A}$. It has been calculated for the case of two scattering events [229] and it modifies the α_0 term in $\mathcal{A}$ by a factor proportional to $1/kl^*$. This behavior is in good agreement with the results shown in Figure 8.10.

8.8.3 The role of absorption

The role of absorption in coherent backscattering does not simply amount to a decrease in the height of the cone as given by (8.49). The introduction of a new characteristic length l_a allows a more quantitative study of the comparative roles of l^* and l_e for the case of anisotropic scattering [223].

For instance, the results given in Figure 8.11 were obtained for a liquid suspension of polystyrene spheres of diameter 0.46 μm and a light of wavelength $\lambda = 0.389\,\mu\text{m}$, a situation that corresponds to very anisotropic scattering.[12] The absorption length l_a measured independently at this concentration of scatterers is of the order of 100 μm. From (8.47), we expect suppression of the contributions of the long trajectories. This is indeed

[12] Calculation of l^* using Mie scattering theory for spheres with this diameter yields $l^* = 21.5\,\mu\text{m}$, whereas $l_e = 4.1\,\mu\text{m}$ (section A2.3.2).

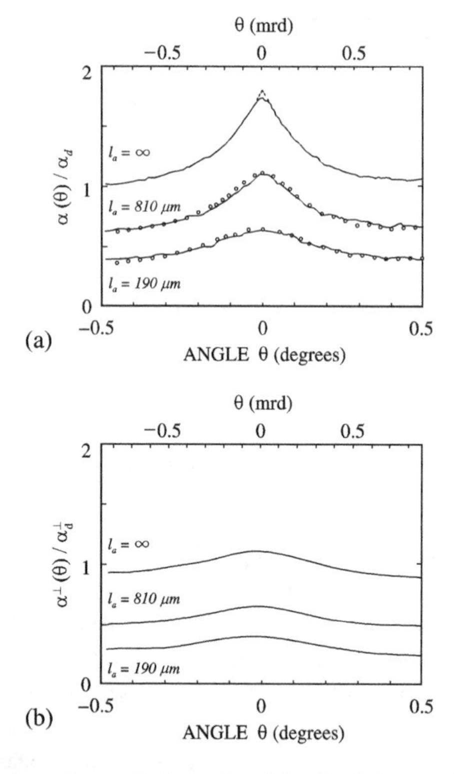

Figure 8.11 Effect of the absorption on the behavior of the albedo in a medium made of a suspension of polystyrene beads immersed in a dye solution. (a) Shows the coherent albedo in the polarization channel parallel to the incident polarization. The points are obtained using the scaling relation (8.49) which relates the coherent albedo in the presence and in the absence of absorption. (b) Same as (a) but in the perpendicular polarization channel [223].

what is observed in the parallel polarization channel [223, 231]. From a closer inspection of Figure 8.11, we also notice the following points relative to the effect of absorption.

- The incoherent contribution to the albedo is equally reduced in each of the two polarization channels, either parallel or perpendicular. This indicates that the contribution of the long trajectories to the incoherent albedo is fully depolarized.
- The height of the coherent albedo peak measured in the parallel polarization channel decreases in the same way as the incoherent background. This is in contrast to the case of the perpendicular polarization channel where the coherent albedo appeared to be much less affected. From this observation, we conclude that, without absorption, all the trajectories contribute to α_c in the parallel channel, whereas only a fraction of these trajectories contribute in the perpendicular channel. This observation is in agreement with the conclusions of section 8.7.

Let us now proceed along the same line of thought but in a more quantitative way. The two expressions (8.49) and (8.50) provide a relation between the coherent albedo curves

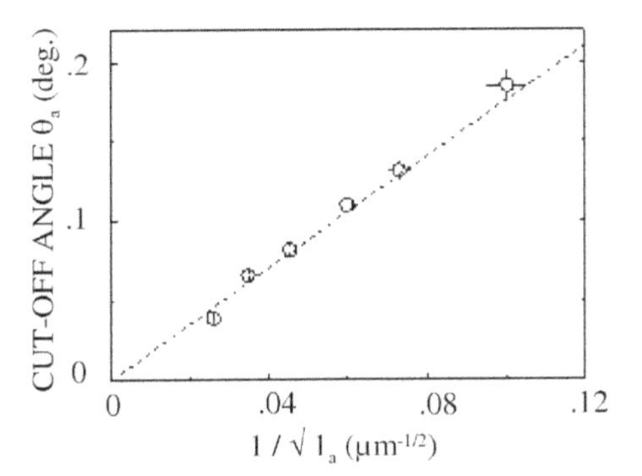

Figure 8.12 Plot of the behavior of the angle θ_a defined by (8.66) as a function of the absorption length. The dashed line corresponds to (8.66) [223].

in the presence and in the absence of absorption, but at different angles. For instance, relation (8.50)

$$\alpha_c(0, l_a) = \alpha_c(k_a, \infty) \tag{8.65}$$

allows us to define for each value of l_a an angle $\theta_a = \lambda k_a / 2\pi$, such that

$$\theta_a = \frac{\lambda}{2\pi}\sqrt{\frac{3}{l_a l^*}}\,. \tag{8.66}$$

Figure 8.12 shows this dependence of θ_a as a function of l_a. From this relation it is possible to deduce an experimental value for the transport mean free path, $l^* = 20 \pm 2\,\mu\text{m}$ which is pretty close to the calculated value (21.5 μm) (see footnote 12, page 343). The validity of the scaling behavior (8.49) which appears clearly in Figure 8.11 justifies once again the use of the diffusion approximation.

From this quantitative agreement, it is possible to infer a value of the parameter β^* given by (8.55). This is an important issue since it allows us to check the predictions presented in section 8.6 concerning the validity of the radiative transfer approach. Within the parallel polarization channel, the coherent and incoherent contributions to the albedo are equally affected by absorption. In this channel, we thus have for the respective behavior of these two quantities, the relation

$$\alpha_d(l_a = \infty) - \alpha_d(l_a) = \alpha_c(k_a = 0) - \alpha_c(k_a) \tag{8.67}$$

which, once we expand $\alpha_c(k_a)$ for small values of k_a using (8.54), gives

$$\alpha_d(l_a) - \alpha_d(\infty) = -\frac{1}{2}\beta^* l^* k_a\,, \tag{8.68}$$

where the factor $1/2$ has been added to account for depolarization effects. The use of this expression is well justified since it describes only the contribution of the long trajectories

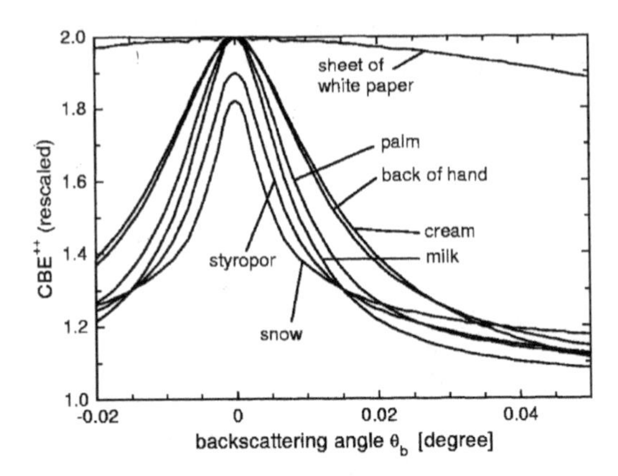

Figure 8.13 Coherent backscattering cones measured using circularly polarized light of wavelength $\lambda = 0.514\,\mu$m in various materials. The scales have been chosen to obtain an enhancement factor of 2 (apart from snow and styropor, for which the true angular variation has been plotted). The enhancement factors actually measured range between 1.6 and 2 [232].

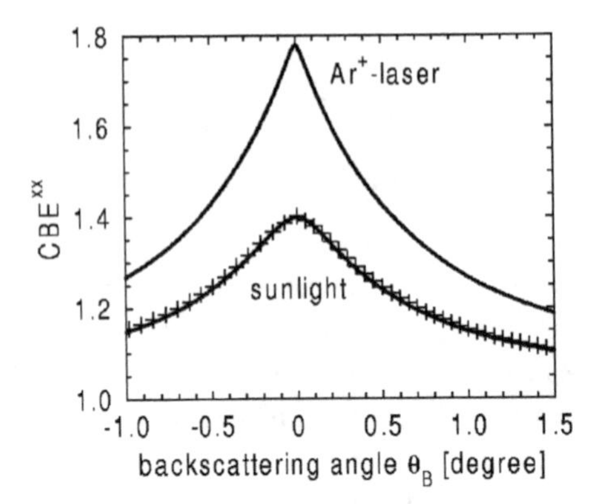

Figure 8.14 Coherent backscattering cone obtained for sunlight on a solid solution of $BaSO_4$ and compared, for the same medium, to a monochromatic light of wavelength $\lambda = 0.514\,\mu$m [232].

Figure 8.15 Glory observed along the Hörnli ridge on mount Cervin (picture G. Montambaux).

(i.e., the limit of small angles). Using $z_0/l^* = 2/3$, we infer from (8.55) the theoretical value $\beta^*/2 = 4.16/4\pi$. From an independent measurement of $\alpha_d(\infty)$ in the absence of absorption and for point-like scatterers [223], we obtain an experimental value $\beta^*/2 = 4.2/4\pi \pm 20\%$ very close to the theoretical value. This is an important result: first it confirms that we cannot fix $z_0 = 0$ for the extrapolation length that enters into the calculation of the albedo of a semi-infinite medium and, moreover, it also justifies using the value $z_0 = 2l^*/3$ obtained within the diffusion approximation. Second, it confirms the assumptions given in section 8.6 which led us to the conclusion that the slope of the albedo at small angles depends solely on l^* and not on l_e.[13]

8.9 Coherent backscattering at large

The interference effect underlying coherent backscattering is shared by many different physical systems and can be observed using a large variety of waves. The very existence of the coherent backscattering effect, the relative ease of measuring it, and the good quantitative understanding we have of it, at least for small angles, has turned it into a tool used relatively frequently in order to display coherent multiple scattering and to obtain in situ precise measurements of the transport mean free path. Moreover, coherent backscattering is a robust effect, which, using light sources, can be measured on a broad range of materials (see Figure 8.13).

The coherent backscattering effect can also be observed using a non-coherent light source such as sunlight. To obtain its expression in this limit, we must perform the convolution of the monochromatic coherent albedo with the spectral correlation function of the light source [232, 233]. For the case of sunlight, this yields a reduction of the height of the cone (Figure 8.14).

8.9.1 Coherent backscattering and the "glory" effect

Other kinds of coherent backscattering effects have been observed for quite a long time. Among them, the most famous is perhaps the *glory* which shows up as a bright halo surrounding the shadow of a plane (or of a mountain hiker) projected onto a sea of clouds (Figure 8.15) [234]. In contrast to the coherent backscattering effect, the glory is a single scattering interference effect.

It results from the interference between equally long light paths inside a spherical drop of water[14] (Figure 8.16). The counting of all paths that contribute to the glory effect constitutes a difficult problem which necessitates use of the Mie theory (section A2.3.2), and which

[13] This result is to be compared with the result obtained in reference [225] which predicts for β^* the expression

$$\beta^* = -3/4\pi(2/3 + l_e/l^*)^2 . \tag{8.69}$$

We would then obtain a much smaller value for β^* because of the large factor $l^*/l_e \simeq 5$ for the case of beads of diameter 0.46 μm.

[14] It might be of interest to note at this point that glory is an effect very different from the rainbow, which apart from supernumerary rainbows inside the main arc, is not an interference effect.

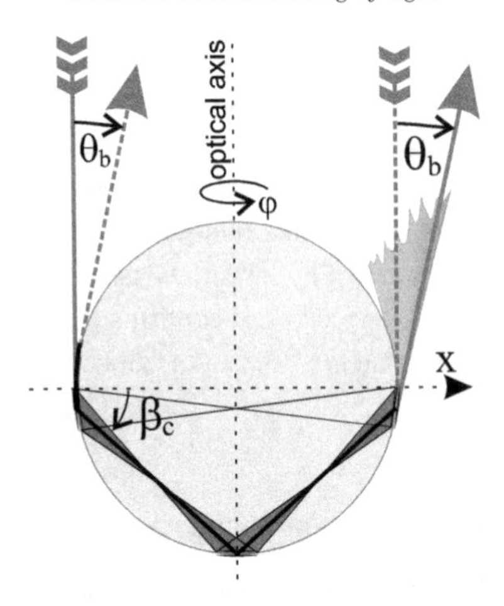

Figure 8.16 Illustration of the typical optical paths that interfere to produce the glory effect. θ_b is the backscattering angle and β_c is the critical angle of total reflection [235].

depends on many parameters such as the wavelength of the incident light, the radius of the drop, and its optical index. The increase in the backscattered intensity occurs inside a cone of aperture of the order of λ/a, where a is the radius of the drop. For water inside clouds, this radius is a few tens of micrometers, i.e., much smaller than the transport mean free path of light in this medium. This explains why the bright halo that is observed is essentially associated with the glory effect and not with coherent backscattering. For big enough scatterers, however, the two effects may coexist, and it could then be possible to cross over continuously from a regime where multiple scattering mainly occurs inside a scatterer to a regime of multiple scattering between scatterers [235].

8.9.2 Coherent backscattering and opposition effect in astrophysics

As early as 1887, it was observed that the sunlight intensity reflected by the rings of Saturn was larger in the backscattering direction [236].[15] This observation was then extended to almost all the planets and their moons when observed in the so-called opposition configuration, i.e. when the Sun, Earth and planet of interest are aligned, thus giving to this increase of the backscattered intensity (Figure 8.17) the name "opposition effect." Various explanations for this effect have been proposed [238], but it is only recently that an explanation based on the coherent backscattering effect has been suggested [239].

It has also been observed that the difference $d_{\parallel} - d_{\perp}$ between the depolarization coefficients defined in (8.57) vanishes in the backscattering direction and becomes negative inside a cone of angular aperture roughly equal to that of the intensity.[16] This effect known

[15] In astronomy, the backscattering angle is often called the phase angle.

[16] The definition (8.57) of the depolarization coefficients is general, but their expression has been obtained in the particular case of Rayleigh scattering.

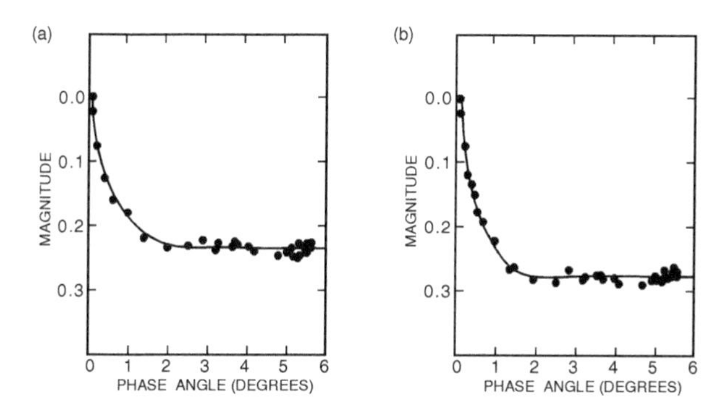

Figure 8.17 Opposition effect observed on the rings A and B of Saturn with natural sunlight (a) and for blue light (b) [237].

as the "polarization opposition effect" was first observed by Lyot in 1929 [240]. It has recently been reexamined in great detail in the framework of polarization effects in multiple scattering. These studies go much beyond the range of validity of the results presented in section 8.7. Actually, the scatterers (for instance small ice crystals) are much larger than the wavelength (Mie regime, section A2.3.2) and of random shape, so that the behavior of the polarization can only be obtained from numerical studies [230, 241]. Interpretation of the opposition effect in terms of coherent backscattering has indeed been fruitful, since it has led to a better understanding of the nature and composition of the measured reflecting surfaces. Nevertheless, we must still use this interpretation cautiously since a number of questions remain unsolved. For instance, the angular aperture of the backscattering cone seems to correspond in certain cases to very short transport mean free paths l^*, a result at odds with the nature of the scatterers.

8.9.3 Coherent backscattering by cold atomic gases

The physics of cold atoms constitutes a very active field of research, especially since the first experimental observation in 1995 of the Bose–Einstein condensation in rubidium atoms. The atomic densities reached in traps are very high, especially near the onset of condensation, and study using the usual spectroscopic methods appears difficult as a result of multiple scattering of photons by atoms. This explains why, although it was first considered a nuisance, multiple scattering is of great interest for study of the properties of cold atoms, as it is for classical scatterers. Moreover, it has also been realized that cold atoms and Bose–Einstein condensates are good candidates for the observation of coherent effects in multiple scattering, not only in the weak disorder regime but also in the Anderson localization regime. Another great advantage of the resonant Rayleigh scattering of photons by atoms resides in the fact that atoms provide an almost perfect realization of point-like scatterers, an assumption that has been underlying most of our previous calculations (section A2.3.3). These various points have triggered the experimental study of the backscattering effect in cold atomic gases (rubidium and strontium) [242]. The behavior of the backscattering

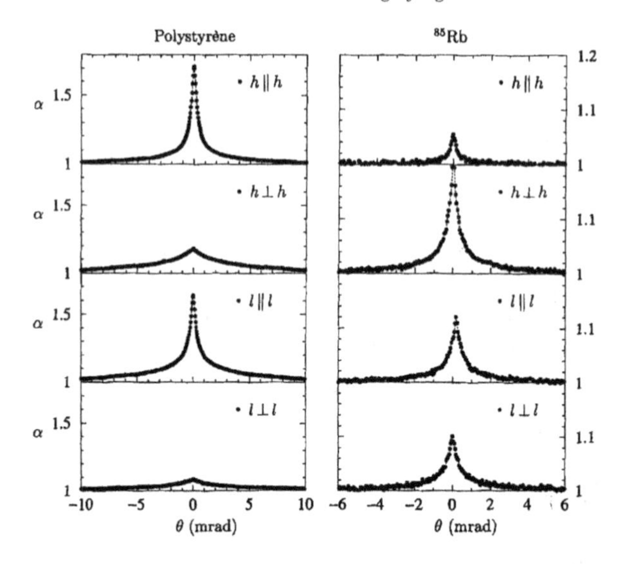

Figure 8.18 Comparison of backscattering cones obtained for classical scatterers (polystyrene) and for cold rubidium atoms. We notice that the enhancement factors are much smaller for atoms. Even more surprising is the behavior as a function of polarization, which differs both qualitatively and quantitatively from the classical case [242].

cone as a function of polarization appears to be qualitatively different from the case of classical scatterers. This shows up clearly in Figure 8.18 where the enhancement factor of the coherent albedo is much smaller in the parallel polarization channel than in the perpendicular one [243]. This is at odds with the case of classical Rayleigh (or Rayleigh–Gans) scatterers where (see sections 6.6.4 and 8.7) a scattered wave analyzed along the incident polarization channel remains unattenuated, whereas it is strongly attenuated in the perpendicular polarization channel. The reason for this unexpected behavior arises from the existence of internal atomic degrees of freedom, i.e., from degenerate Zeeman sublevels which translates into a very different expression for the elementary vertex. As for the classical Rayleigh scattering, we must consider two distinct effects. One is the depolarization of an incident photon that affects equally the Diffuson and the Cooperon, i.e., the incoherent and coherent contributions to the albedo. A second effect is the decrease of the Cooperon contribution in the parallel and perpendicular polarization channels. For Rayleigh scattering by a classical dipole, this decrease occurs only in the perpendicular channel. We shall now see that for atoms with degenerate Zeeman sublevels, it also occurs in the parallel channel (see Appendix A6.5).

Depolarization of the Diffuson

The classical intensity and the incoherent albedo can be measured within either the polarization channel of the incident photons or the perpendicular channel. The characteristic dephasing times $\tau_k^{(d)}$ defined by (6.321) and presented in Table 6.1 account for the loss of polarization of an incident photon. As for Rayleigh scattering, only the scalar mode $k = 0$ survives in the long time limit and contributes equally in all polarization channels. Therefore, for long multiple scattering trajectories of photons, we recover the same depolarization factor $1/2$ obtained in the classical case (see relation 8.57).

Reduction of the enhancement factor of the coherent albedo

As well as depolarization, the Cooperon involves an additional phase shift between multiple scattering amplitudes propagating in opposite directions. This phase shift gives rise to a decrease in the enhancement factor of the coherent albedo relative to the incoherent background.

For classical Rayleigh scattering (section 8.7), the scalar mode $k = 0$ of the Cooperon is a Goldstone mode. Therefore the Cooperon contribution is robust in the parallel polarization channel ($l \parallel l$ or $h \parallel h$) and it is attenuated only in the perpendicular polarization channel ($l \perp l$ or $h \perp h$). However, for atomic gases, the contribution $\Gamma_0^{(c)}(t)$ of this mode decreases exponentially since $\tau_0^{(c)}$ is now finite. Consequently, the component $\Gamma^{(c)}(t)$ of the structure factor is reduced even in the parallel channel. We may also be in a situation opposite to the classical case, i.e., for which the enhancement of the coherent albedo is larger in the perpendicular channel than in the parallel one. The value of the enhancement depends on the nature of the atomic transition and on the total angular momentum of the ground state (J) and of the excited state (J_e). For instance, for rubidium atoms, which correspond to a transition ($J = 3, J_e = 4$), we have $\tau_2^{(c)} > \tau_1^{(c)} > \tau_0^{(c)}$ (see Table 6.1). The exponential decay of the Cooperon is driven by the less rapidly decaying mode, that is, by the largest time $\tau_2^{(c)}$, so that for long enough times, it decays with the same characteristic time $\tau_2^{(c)}$ in all polarization channels. Then, the height of the coherent backscattering cone $\alpha_c^{\parallel}$ in the parallel channel is comparable to that of $\alpha_c^{\perp}$ in the perpendicular channel (Figure 8.18). More quantitatively, the height of the cone in the different channels is given by the coefficient of the $k = 2$ mode which, from Table 8.1 is $2/3, 1/2, 1/6, 1$, respectively for the $l \parallel l$, $l \perp l$, $h \parallel h$ and $h \perp h$ polarization channels. This is in qualitative agreement with the results of Figure 8.18.

Finally, in the case of cold atomic gases, it is important to take into account finite size effects of the cloud that lead to a rounding of the shape of the coherent backscattering cone (Exercise 8.2). Moreover, confinement of the atoms is achieved by means of external lasers, so that the density of the gas is non-homogeneous. This raises the problem of finding the right boundary conditions for the structure factor and for the probability P_d.

Table 8.1 Structure factors $\Gamma^{(d)}$ and $\Gamma^{(c)}$ in the different polarization channels as obtained from equations (6.323, 6.324). See Exercise 2.4 and Table 2.4 for calculation of the scalar product in the different polarization or helicity channels in the backscattering direction. For Rayleigh scattering, $\Gamma_0^{(c)} = \Gamma_0^{(d)}$ are Goldstone modes, while for scattering by rubidium atoms, $\Gamma_0^{(c)}$ is attenuated and decreases faster than $\Gamma_2^{(c)}$.

	$\Gamma^{(d)}$	$\Gamma^{(c)}$
$l \parallel l$	$\dfrac{\Gamma_0^{(d)} + 2\Gamma_2^{(d)}}{3}$	$\dfrac{\Gamma_0^{(c)} + 2\Gamma_2^{(c)}}{3}$
$l \perp l$	$\dfrac{\Gamma_0^{(d)} - \Gamma_2^{(d)}}{3}$	$\dfrac{\Gamma_2^{(c)} - \Gamma_1^{(c)}}{2}$
$h \parallel h$	$\dfrac{\Gamma_0^{(d)}}{3} - \dfrac{\Gamma_1^{(d)}}{2} + \dfrac{\Gamma_2^{(d)}}{6}$	$\dfrac{\Gamma_0^{(c)}}{3} - \dfrac{\Gamma_1^{(c)}}{2} + \dfrac{\Gamma_2^{(c)}}{6}$
$h \perp h$	$\dfrac{\Gamma_0^{(d)}}{3} + \dfrac{\Gamma_1^{(d)}}{2} + \dfrac{\Gamma_2^{(d)}}{6}$	$\Gamma_2^{(c)}$

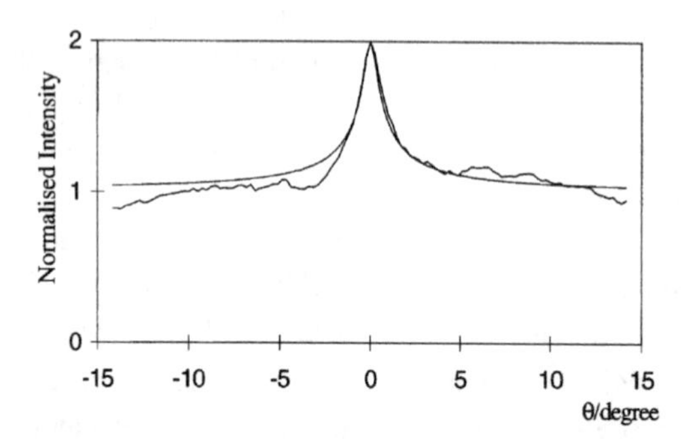

Figure 8.19 Coherent backscattering cone measured with acoustic waves [245].

8.9.4 Coherent backscattering effect in acoustics

In order to close this tour of the various systems in which coherent backscattering has been observed, let us mention the beautiful set of experiments done in acoustics [244,245]. Figure 8.19 shows the backscattering cone for acoustic waves ($\lambda = 0.43$ mm) propagating in a two-dimensional random medium of dimensions 160 mm × 80 mm, composed of 2400 rigid steel rods immersed in a water tank. This medium is characterized by a transport mean free path $l^* \simeq 4$ mm and a diffusion coefficient $D^* \simeq 2.5\ \text{mm}^2/\mu\text{s}$.

The great advantage of the acoustic setup is that the detectors (a network of ultrasonic transducers) can also measure the phase of the detected signal, not only its intensity as in optics. Moreover, because of the lower speed of acoustic waves, it is also possible to

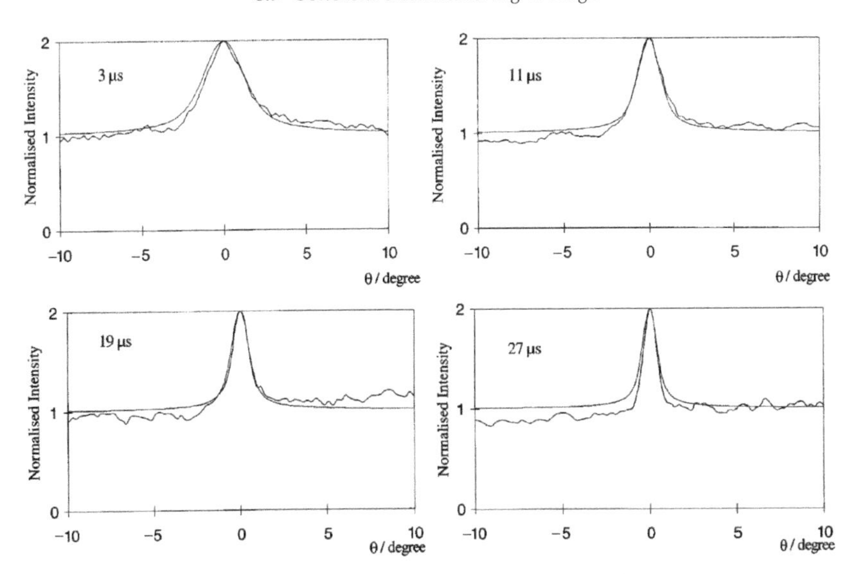

Figure 8.20 Measurement of the time-dependent albedo $\alpha(t)$ with acoustic pulses. The cone has a Gaussian shape and its width varies as $1/\sqrt{Dt}$ [245].

access the time resolved behavior such as the time-dependent albedo $\alpha(\theta, t)$ defined by (8.39, 8.40). This behavior is represented in Figure 8.20. In agreement with relation (8.41), the shape of the coherent cone is Gaussian with a width that varies as $1/\sqrt{Dt}$. A similar experiment in optics would require femtosecond pulses [246].

9

Diffusing wave spectroscopy

9.1 Introduction

In Chapter 6, we studied a variety of dephasing mechanisms whose effect is to modify the Diffuson and the Cooperon by eliminating the contribution of long diffusive trajectories. As a result, we found that the coherent contribution to the integrated return probability $Z(t)$ must be restricted to times smaller than some characteristic time τ_γ. The existence of dephasing thus appears to restrict the observation of coherent effects. However, if its mechanism is well understood, instead of being a nuisance, dephasing can reveal itself as a fruitful contribution to the study of the properties of a diffusive medium. For example, multiple scattering of light can be used as a tool to characterize liquid suspensions of particles or hydrodynamic flows.

The technique usually employed to probe the dynamics of scatterers is quasi-elastic scattering (see [247] for a comprehensive presentation). This technique is routinely used in the single scattering regime, i.e., for extremely dilute suspensions. At higher concentrations, we have to resort to multiple scattering in order to retrieve information on the dynamics of the scatterers. To grasp this point, we come back to the results of section 6.7 on dephasing induced by the motion of scatterers. The time correlation function $g_1(T)$ of a scalar electric field defined by

$$\boxed{g_1(T) = \frac{\langle E(T)E^*(0)\rangle}{\langle |E(0)|^2\rangle}} \tag{9.1}$$

depends on the phase shift between multiple scattering trajectories, and happens to be very useful for the study of short time dynamics and to probe length scales smaller than the wavelength of the incident light. It should be noticed that both the electric field and the correlation function g_1 depend on the observation point $\boldsymbol{r}$ (see section 6.7.1).

In this chapter we show how to obtain the time correlation function $g_1(T)$ of the field in both single and multiple scattering regimes. In an experiment, instead of g_1, the measured quantity is the intensity correlation function defined by

$$\boxed{g_2(T) = \frac{\langle I(T)I(0)\rangle}{\langle I(0)\rangle^2} - 1} \tag{9.2}$$

The intensity $I(T)=|E(T)|^2$ introduced in section 4.7 is measured at a point $\boldsymbol{r}$ for a light source placed at $\boldsymbol{r}_0$ and for a given configuration of the scatterers.[1] The notation $\langle\cdots\rangle$ defined in section 6.7 accounts for an average over all multiple scattering paths and over the dynamics of the scatterers. We will first show the important result that in the multiple scattering regime the two time correlation functions are simply related by $g_2(T)=|g_1(T)|^2$. Then, before moving to the multiple scattering regime, we recall some known results about quasi-elastic scattering. We will next calculate the correlation functions g_1 and g_2 in the multiple scattering regime and for different geometries and different kinds of dynamics of scatterers. For a semi-infinite scattering medium, the expression for the time correlation function $g_1(T)$ measured in reflection can be related to that for the coherent albedo $\alpha_c(\theta)$ studied in Chapter 8. For a slab, by measuring time correlation functions in both transmission and reflection, we can probe the contribution of long multiple scattering trajectories separately.

The origins of fluctuations of an electromagnetic field are numerous. For instance, in an atomic vapor, assuming that atoms collide elastically with each other, the emitted radiation is monochromatic and the total electric field emitted by the ensemble of atoms is a superposition of terms of the form $E(t)=E_0\exp(i\omega t+i\varphi(t))$ where $\varphi(t)$ is a random phase that accounts for the phase shift due to collisions. The correlation functions $g_1(T)$ and $g_2(T)$ are used here to characterize the random phase shift [248]. Although the physical origin of the phase shift in this case is very different from the case of moving scatterers, we observe that a number of results, such as the equality (9.8), are common to these two situations.[2] The generality of this equality results mainly from the fact that statistical properties of light do not depend strongly upon the exact microscopic nature of the phase shift as long as it can be described within the Gaussian approximation. We shall also assume throughout the chapter that the light source is coherent.

9.2 Dynamic correlations of intensity

We now show that the intensity correlation function g_2 is simply related to the correlation function g_1 of amplitudes. We first recall that the intensity $I(T)=|E(T)|^2$ is a product of two amplitudes. The function g_2 is thus built with a product of four amplitudes. To evaluate it, we follow the methodology developed in section 4.4. The scattered electric field issued from a point-like source is given by the Green function (6.181). It takes the form

$$E(T)=\sum_{\mathcal{C}} E_{\mathcal{C}}, \tag{9.3}$$

where $\mathcal{C}$ is a short notation for all the multiple scattering sequences ($\sum_{\mathcal{C}}=\sum_{N=1}^{\infty}\sum_{\boldsymbol{r}_1\cdots\boldsymbol{r}_N}$).

[1] Notice that, in order to simplify notation, the normalization used here for the intensity differs from that of Chapter 4. This choice is of no importance since both g_1 and g_2 appear as a ratio of intensities.

[2] Notice that the function $g_2(T)$ commonly used to characterize the source is defined as in (9.2) but without the -1.

The product of intensities is thus given by

$$I(T)I(0) = \sum_{\mathcal{C}_1,\mathcal{C}_2,\mathcal{C}_3,\mathcal{C}_4} E_{\mathcal{C}_1}(T)E^*_{\mathcal{C}_2}(T)E_{\mathcal{C}_3}(0)E^*_{\mathcal{C}_4}(0). \tag{9.4}$$

Following again the steps in section 4.4, in the preceding summation we keep terms that correspond to either $\mathcal{C}_1 = \mathcal{C}_2$ and $\mathcal{C}_3 = \mathcal{C}_4$, or $\mathcal{C}_1 = \mathcal{C}_4$ and $\mathcal{C}_3 = \mathcal{C}_2$, so that

$$\begin{aligned}\langle I(T)I(0)\rangle &= \sum_{\mathcal{C}_1,\mathcal{C}_2} \langle |E_{\mathcal{C}_1}(T)|^2\rangle\langle |E_{\mathcal{C}_2}(0)|^2\rangle \\ &\quad + \sum_{\mathcal{C}_1,\mathcal{C}_2} \langle E_{\mathcal{C}_1}(T)E^*_{\mathcal{C}_1}(0)\rangle\langle E^*_{\mathcal{C}_2}(T)E_{\mathcal{C}_2}(0)\rangle. \end{aligned} \tag{9.5}$$

The average intensity at the Diffuson approximation is $I_d = \langle I(0)\rangle = \langle I(T)\rangle = \langle \sum_{\mathcal{C}} |E_{\mathcal{C}}|^2\rangle$ (see section 4.7), so that

$$\langle I(T)I(0)\rangle = I_d^2 + \left|\sum_{\mathcal{C}} \langle E_{\mathcal{C}}(T)E^*_{\mathcal{C}}(0)\rangle\right|^2. \tag{9.6}$$

The limit $T = 0$ gives back the Rayleigh law (12.93),[3] namely

$$\langle I^2\rangle = 2\langle I\rangle^2 = 2I_d^2 \tag{9.7}$$

which implies that $g_2(0) = 1$. The last term in (9.6) contains $|g_1(T)|^2$ so that for the correlation function (9.2) we finally obtain the expression

$$\boxed{g_2(T) = |g_1(T)|^2} \tag{9.8}$$

This relation, known as the Siegert relation [249], results here from the Diffuson approximation. A diagrammatic representation of g_2 is shown in Figure 9.1.[4] The calculation of this diagram is straightforward since it decouples into the product of two correlation functions of the field.

The intensity correlation function thus results from that of the field already calculated in section 6.7. To interpret measurements of g_2, we thus need to know only the correlation function of the fields.

[3] Intensity fluctuations at a given point are considered in more detail in Chapter 12, where the conditions of validity of the Rayleigh law (12.76) are discussed.

[4] It is interesting to compare this diagram to those displaying correlation functions of diagonal Green's functions (A4.4) or to those involved in the calculation of conductance fluctuations (Figure 11.3).

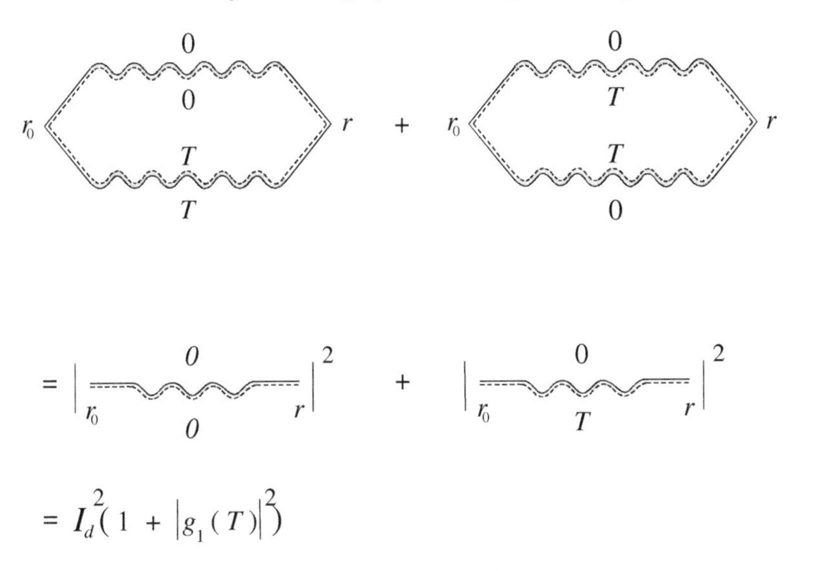

Figure 9.1 Diagrams contributing to the intensity time correlation $\langle I(T)I(0)\rangle$ for a point-like source placed at $\boldsymbol{r}_0$. The equivalence between the first and the second lines relies upon the ergodic assumption, namely, that ensembles of multiple scattering trajectories at different times are identical.

9.3 Single scattering: quasi-elastic light scattering

Before proceeding further with the calculation of the time correlation function in multiple scattering, we first consider some of its main characteristics in single scattering. Quasi-elastic light scattering (QELS) accounts for single scattering in a sufficiently dilute suspension of $\mathcal{N}$ scatterers. For an incident plane wave of wave vector $\boldsymbol{k}_i$, the electric field scattered along a direction $\boldsymbol{k}_f = \boldsymbol{q} + \boldsymbol{k}_i$ depends only on the transfer wave vector $\boldsymbol{q}$ and on the dynamics $\boldsymbol{r}_j(T)$ of the scatterers, namely

$$E(\boldsymbol{q}, T) \propto \sum_{j=1}^{\mathcal{N}} e^{-i\boldsymbol{q}\cdot\boldsymbol{r}_j(T)}. \tag{9.9}$$

The time correlation function $g_1(T)$ is obtained by averaging over all outgoing directions

$$g_1(T) = \frac{1}{\mathcal{N}} \left\langle \sum_{j=1}^{\mathcal{N}} e^{-i\boldsymbol{q}\cdot[\boldsymbol{r}_j(T) - \boldsymbol{r}_j(0)]} \right\rangle_{\boldsymbol{q}, \boldsymbol{r}_j} \tag{9.10}$$

in which the average is taken both over the directions of the transfer wave vector and over the positions of the scatterers. Assuming that the motion of the scatterers is well described by the Brownian motion described in (6.189) and characterized by a diffusion coefficient

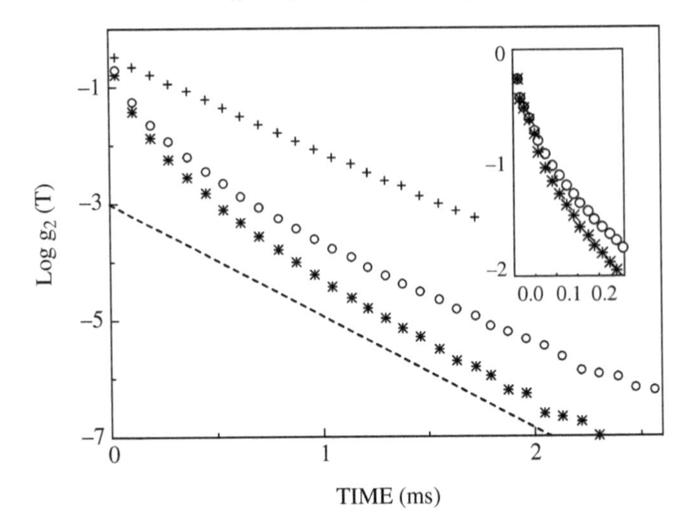

Figure 9.2 Temporal behavior of the intensity correlation function $g_2(T)$ of the light scattered by an aqueous suspension of polystyrene beads. (+): single scattering regime corresponding to a solid fraction of 2×10^{-5}. The expected exponential behavior (9.11) is well observed and corresponds to the dashed line. (∘ and $*$): multiple scattering regime obtained for a solid fraction of 10^{-1}. The two curves shown correspond to different polarizations. The behavior at very short times is represented in the inset [250]. Here it deviates from exponential.

D_b, we obtain for the correlation function the exponential behavior

$$g_1(T) = \left\langle e^{-D_b q^2 T} \right\rangle_{\boldsymbol{q}} \simeq e^{-T/2\tau_b}, \tag{9.11}$$

where $\tau_b = 1/4D_b k^2$ and $k = |\boldsymbol{k}_i| = |\boldsymbol{k}_f|$. The intensity correlation function is again given by (9.8) which applies to single and multiple scattering limits as well. It exhibits an exponential behavior with a relaxation time τ_b, from which it is possible to extract the diffusion coefficient D_b of the Brownian particles in the suspension. This exponential law agrees very well with experimental results at low concentration but it deviates notably at higher concentration where multiple scattering effects become important (see Figure 9.2).

9.4 Multiple scattering: diffusing wave spectroscopy

The time correlation function $g_1(T)$ in the multiple scattering regime was calculated in section 6.7.1 for the specific case of a field created by a point-like source. For Brownian scatterers, $g_1(T)$ is the stretched exponential given by (6.199). It thus differs very much from the exponential behavior (9.11) obtained in single scattering. This difference shows up even more clearly in the short time ($T \ll \tau_b$) behavior of the correlation function measured at larger concentrations.[5] On the other hand, at large times we recover an exponential driven by the time τ_b (Figure 9.2).

In order to have a qualitative understanding of these different behaviors, let us start from expression (6.200) for the time correlation function of the fields in the limit $T \leq \tau_b$,

$$\frac{4\pi}{c}\langle E(\boldsymbol{r},T)E^*(\boldsymbol{r},0)\rangle = \int_0^\infty P(\boldsymbol{r}_0,\boldsymbol{r},t)\, e^{-t/\tau_s}\, dt = P_\gamma(\boldsymbol{r}_0,\boldsymbol{r}) \tag{9.12}$$

where $\gamma = 1/\tau_s$ and $\tau_s = 2\tau_e\tau_b/T$ is the relaxation time of the Diffuson due to the Brownian motion of the scatterers defined in relation (6.197). By comparing this expression with (9.11), we see that it does not involve a single correlation time τ_b but rather a distribution of correlation times $2\tau_b/N$, each related to a given length $N = t/\tau_e$ of a multiple scattering path. This distribution is weighted by the diffusion probability $P(\boldsymbol{r}_0,\boldsymbol{r},t)$ calculated at the approximation of the Diffuson. The largest times T correspond to the shortest paths $N \simeq 1$, and thus to the single scattering limit, that is, the quasi-elastic scattering regime. For longer scattering paths, phase coherence vanishes for much shorter correlation times. This distinction between long and short times T shows up very clearly in Figure 9.2.

9.5 Influence of the geometry on the time correlation function

The expression (9.12) for $g_1(T)$ corresponds to the case of a point-like source inside an infinite medium (see section 6.7.2). However, the correlation function is measured experimentally using a finite slab geometry. Although in single scattering (quasi-elastic scattering) the experimental setup is not expected to play a crucial role, this is no longer true in multiple scattering. There, the geometry plays a role in weighting the contribution of long scattering trajectories, thus affecting the short time expression of $g_1(T)$. More precisely, expression (9.12) involves the Diffuson P_d solution of a diffusion equation. Hence it has a long range contribution whose behavior depends necessarily on the geometry. In the following sections, a thorough study of the correlation function $g_1(T)$ is presented for the case of reflection by a semi-infinite medium and by a slab, and for the case of transmission through a slab.

9.5.1 Reflection by a semi-infinite medium

We consider a semi-infinite geometry such as that described in section 8.2 for the study of the albedo. The interface is illuminated by an incident plane wave incoming from outside the medium. We assume that the outgoing light is detected at a large distance $R \to \infty$ also outside the medium and in the backscattering direction, i.e., along a direction $\hat{s}_e$

[5] As we shall see in the next sections, the exact geometry of the medium plays an important role, but this behavior remains qualitatively correct.

perpendicular to the interface. The time correlation function

$$G_1^r(T) = \frac{R^2}{SI_0} \langle E(T) E^*(0) \rangle \tag{9.13}$$

of the electric field coincides at $T = 0$ with the reflected intensity, i.e., the albedo defined by (8.3) and given by (8.13) within the Diffuson approximation.

For $T \neq 0$, the contribution (9.12) of multiple scattering paths is exponentially reduced at large times t. Using expression (8.38) for the time dependent albedo $\alpha_d(t)$, the time correlation function $G_1^r(T)$ takes the form

$$G_1^r(T) = \int_0^\infty \alpha_d(t) e^{-tT/2\tau_e \tau_b} dt, \tag{9.14}$$

which by means of (8.39) is rewritten as

$$G_1^r(T) = \frac{c}{4\pi l_e^2} \int\int dz\, dz'\, e^{-z/l_e}\, e^{-z'/l_e} \int dt\, P_d(z, z', t)\, e^{-\gamma t}\, dt \tag{9.15}$$

with $\gamma = 1/\tau_s$. The last integral is the Laplace transform $P_\gamma(z, z')$ (see section 5.6) so that

$$\boxed{G_1^r(T) = \frac{c}{4\pi l_e^2} \int\int dz\, dz'\, e^{-z/l_e}\, e^{-z'/l_e} P_\gamma(z, z')} \tag{9.16}$$

For isotropic scattering, $P_\gamma(z, z')$ is given by (5.157). The integrals are straightforward and lead to

$$G_1^r(T) = \frac{3}{8\pi} \frac{1}{(1 + l_e/L_s)^2} \left(\frac{1 - e^{-2z_0/L_s}}{l_e/L_s} + 1 \right) \tag{9.17}$$

where the phase coherence length L_s is deduced from the relaxation time τ_s and reads

$$L_s = \sqrt{D\tau_s} = l_e \sqrt{\frac{2}{3} \frac{\tau_b}{T}} \tag{9.18}$$

The correlation function $g_1^r(T)$, defined by (9.1), is finally given by the dimensionless ratio

$$g_1^r(T) = \frac{G_1^r(T)}{G_1^r(0)} \tag{9.19}$$

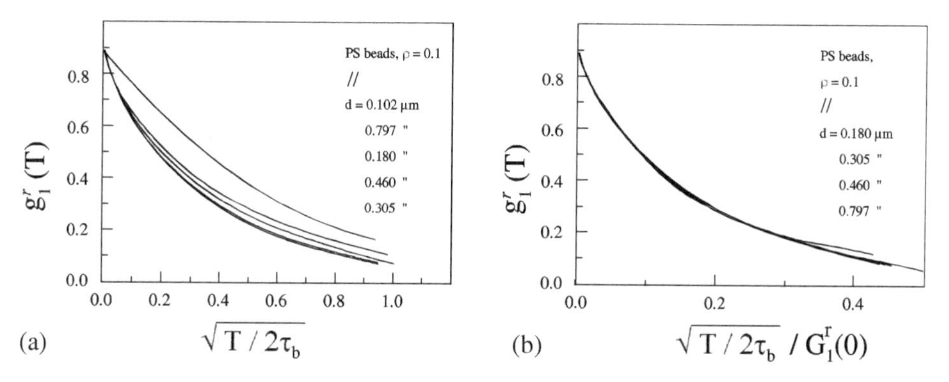

Figure 9.3 Correlation function $g_1^r(T)$ plotted as a function of (a) $\sqrt{T/2\tau_b}$ and (b) $\sqrt{T/2\tau_b}/G_1^r(0)$ for aqueous suspensions of polystyrene beads of different diameters, namely for different values of the ratio l^*/l_e. These results indicate that, unlike the parameter β^*, the quantity $G_1^r(0) = \alpha_c(0)$ depends on the ratio l^*/l_e [251].

9.5.2 Comparison between $G_1^r(T)$ and $\alpha_c(\theta)$

It is quite instructive to compare expression (9.16) for the time correlation $G_1^r(T)$ of the electric field to equation (8.26) giving the coherent albedo $\alpha_c(\theta)$ for $\mu = 1$, within the same approximation:

$$\alpha_c(\theta) = \frac{c}{4\pi l_e^2} \int\int dz\, dz'\, e^{-z/l_e}\, e^{-z'/l_e} P_d(k_\perp, z, z'). \tag{9.20}$$

The expression (5.47) gives a direct relation between these two quantities provided we use the correspondence $\gamma \longleftrightarrow Dk_\perp^2$. Such a relation with the coherent albedo (8.49) in the presence of absorption is also obtained by means of the correspondence

$$Dk_\perp^2 \longleftrightarrow \gamma = \frac{T}{2\tau_e \tau_b} \longleftrightarrow \frac{1}{\tau_a} \tag{9.21}$$

or equivalently

$$k_\perp \longleftrightarrow \frac{1}{L_s} = \frac{1}{l_e}\sqrt{\frac{3T}{2\tau_b}} \longleftrightarrow \sqrt{\frac{3}{l_a l_e}}. \tag{9.22}$$

These correspondences are extremely useful since they provide a relation between distinct physical quantities that depend on a different characteristic length or time, such as l_e, τ_b and l_a.

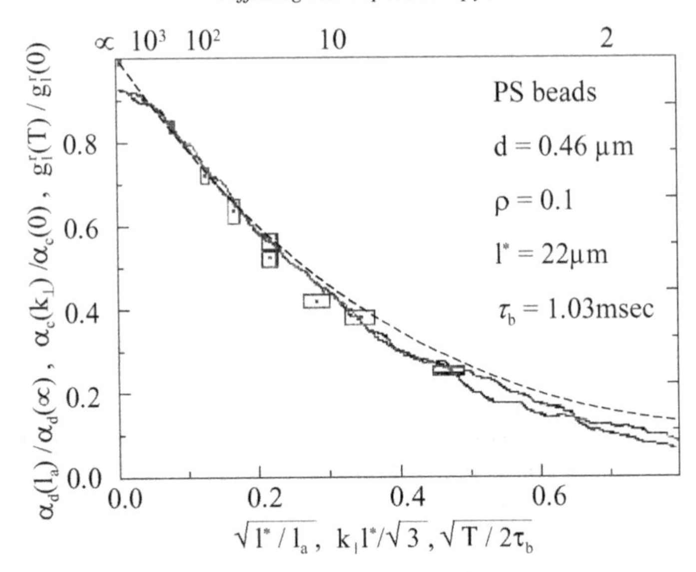

Figure 9.4 Behavior of the incoherent albedo $\alpha_d(l_a)$ as a function of the absorption length (rectangles), of the coherent albedo $\alpha_c(k_\perp)$ as a function of the angle (continuous line), and of the time correlation function $g_1^r(T)$ plotted versus T (dashed line). This behavior is universal when plotted against the scaled parameters $\sqrt{l^*/l_a}, k_\perp l^*/\sqrt{3}$ and $\sqrt{T/2\tau_b}$. This universality results from the correspondence (9.22). The typical size (in units of l^*) of the relevant multiple scattering trajectories is displayed on the upper horizontal axis [251].

For anisotropic collisions ($l^* \neq l_e$), we obtain the following expansions:

$$\alpha_c(k_\perp) \simeq \alpha_c(0) - \beta^* k_\perp l^*$$

$$\alpha_d(l_a) \simeq \alpha_d(l_a = \infty) - \beta^* \sqrt{\frac{3l^*}{l_a}} \tag{9.23}$$

$$G_1^r(T) \simeq G_1^r(0) - \beta^* \sqrt{\frac{3T}{2\tau_b}}$$

where β^* is given by (8.55). The constant term $\alpha_c(0) = \alpha_d(l_a = \infty) = G_1^r(0)$ is not easy to calculate. We have already mentioned in section 8.6 that this term is at first sight a function of l^* and l_e since it depends on multiple scattering trajectories of all lengths. On the other hand, β^* should not depend on the ratio l^*/l_e. These points appear clearly in Figure 9.3, which displays measurements of $G_1^r(T)$ for an aqueous suspension of polystyrene beads. The ratio $g_1^r(T) = G_1^r(T)/G_1^r(0)$ is plotted for various sizes of the beads, i.e., for different values of the ratio l^*/l_e. Figure 9.3(a) displays the non-universal behavior of $g_1^r(T)$ as a function of $\sqrt{T/\tau_b}$. On the other hand, the plot in Figure 9.3(b) of $g_1^r(T)$ as a function of $\sqrt{T/\tau_b}/G_1^r(0)$ indicates that this quantity is indeed independent of the ratio l^*/l_e. Then from (9.23), we are led to the conclusion that, unlike β^*, $G_1^r(0)$ depends on l^*/l_e.

The correspondence (9.22) is displayed in Figure 9.4, which presents measurements of the three different quantities $\alpha_d(l_a)$, $\alpha_c(k_\perp)$ and $G_1^r(T)$, each plotted as a function of a parameter associated to the corresponding dephasing process. The universal behavior shows up in the form of a single plot which results from the correspondence (9.22). From an analysis of these plots it is possible to retrieve the characteristics of the scattering medium, such as the transport mean free path l^*, the characteristic time τ_b or equivalently the diffusion coefficient D_b of the beads [251].

The correspondence (9.22) turns out to be very useful. For instance, the angular dependence of $\alpha_c(\theta)$ together with (9.23) leads immediately to

$$g_1^r(T) \simeq 1 - \frac{z^*}{L_s} \tag{9.24}$$

with

$$z^* = \frac{\beta^* l^*}{G_1^r(0)}, \tag{9.25}$$

where β^* is given by (8.55) and does not depend on the ratio l^*/l_e. Moreover, the characteristic length L_s, in that case, is equal to

$$L_s = l^* \sqrt{\frac{2\tau_b}{3T}}, \tag{9.26}$$

so that we obtain at short times the unusual $\sqrt{T}$ behavior:

$$g_1^r(T) \simeq 1 - \frac{z^*}{l^*} \sqrt{\frac{3T}{2\tau_b}}. \tag{9.27}$$

The behavior of this correlation function, shown in Figure 9.5, has been obtained for an aqueous suspension of polystyrene spheres at room temperature and for high volume concentrations (typically of order 10%) that correspond to the multiple scattering regime. The reported $\sqrt{T}$ behavior is well observed.

9.5.3 *Reflection from a finite slab*

We now extend the results obtained in the previous section to the geometry of a slab of finite width L. The time correlation function $G_1^r(T)$ is still obtained from (9.16), but $P_\gamma(z, z')$ now corresponds to a finite slab and is given by (5.159). In Exercise 8.2 we showed that in order to calculate the albedo of a finite slab, we just need to replace $k_\perp$ by $k_\perp \coth k_\perp L$. This replacement together with the correspondence (9.22) leads to

$$\boxed{g_1^r(T) = 1 - \frac{z^*}{L_s} \coth \frac{L}{L_s}} \tag{9.28}$$

or, making explicit use of the expression for L_s,

$$g_1^r(T) = 1 - \frac{z^*}{l^*} \sqrt{\frac{3T}{2\tau_b}} \coth \frac{L}{l^*} \sqrt{\frac{3T}{2\tau_b}}. \tag{9.29}$$

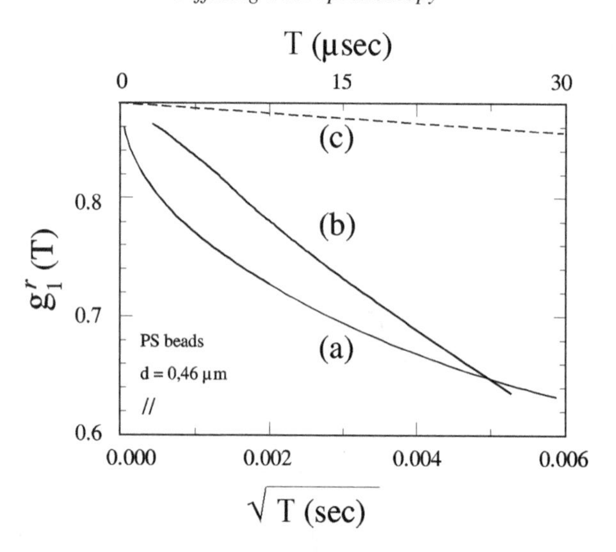

Figure 9.5 Measurement of $g_1^r(T)$ using light reflected from an aqueous suspension of polystyrene spheres. The curves (a) and (b), plotted respectively as a function of T and $\sqrt{T}$, are characteristic of the multiple scattering regime (volume concentration 10^{-1}). The curve (c) accounts for the single scattering limit (volume concentration 10^{-5}) [251].

In this expression, we can identify the characteristic time T_c defined by $L \simeq L_s$, i.e., $T_c = (2\tau_b/3)(l^*/L)^2$. At times smaller than T_c, finite size effects modify the $\sqrt{T}$ behavior previously obtained and instead lead to a linear behavior together with a reduction of the correlation at $T = 0$

$$g_1^r(T) = 1 - \frac{z^*}{L}\sqrt{\frac{T}{T_c}} \coth\sqrt{\frac{T}{T_c}} \underset{T\to 0}{\longrightarrow} 1 - \frac{z^*}{L}\left(1 + \frac{T}{3T_c}\right). \tag{9.30}$$

This modified behavior and the dependence upon varying the width of the slab are indeed clearly observed in Figure 9.6.

9.5.4 Transmission

For a slab of finite width L, it is also possible to consider the time correlation function in transmission. For this purpose, we start again from (9.13) where we insert the corresponding expressions for transmitted electric fields. This leads to an expression formally analogous to (9.16), which takes the form:

$$\boxed{G_1^t(T) = \frac{c}{4\pi l_e^2} \int\int dz\, dz'\, e^{-z/l_e} e^{-|L-z'|/l_e}\, P_\gamma(z, z')} \tag{9.31}$$

with $\gamma = T/2\tau_e\tau_b$. The probability $P_\gamma(z, z')$ is given by (5.159). The net effect of exponential terms in the integral (9.31) is to restrict the values of z and z' around $z \simeq l_e$ and $z' \simeq L - l_e$.

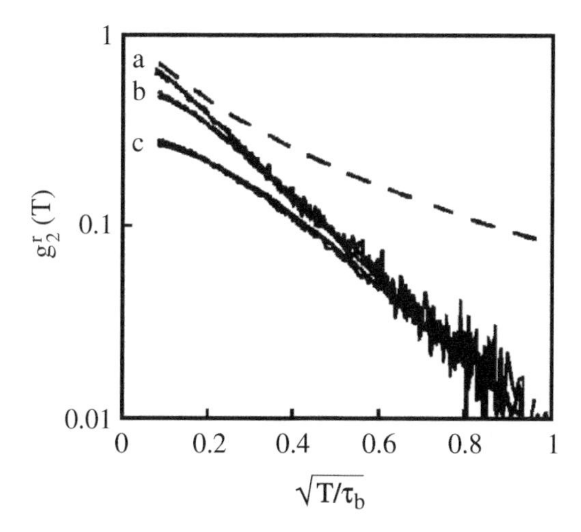

Figure 9.6 Time correlation function $g_2^r(T)$ measured in reflection for slabs of different widths $L=2$ mm (a), $L=1$ mm (b) and $L=0.6$ mm (c). The transport mean free path is $l^* \simeq 143$ μm. This behavior is well accounted for by (9.29) together with $g_2^r(T)=|g_1^r(T)|^2$. For small width, the $\sqrt{T}$ behavior disappears and is replaced by a linear behavior with T for $T \to 0$ [252].

This, together with the limit $l_e \ll L_s$, allows us to rewrite expression (5.159) for $P_\gamma(z,z')$ in the form

$$P_\gamma(z,z') = \frac{1}{DL_s}\frac{(z+z_0)(L+z_0-z')}{\sinh(L+2z_0)/L_s}. \tag{9.32}$$

The integration leads to

$$G_1^t(T) = \frac{3}{4\pi}\frac{(l_e+z_0)^2}{l_e L_s}\frac{1}{\sinh(L+2z_0)/L_s}, \tag{9.33}$$

and for the normalized value $g_1^t = G_1^t(T)/G_1^t(0)$ of the correlation function we obtain (assuming that $z_0 \ll L$):

$$\boxed{g_1^t(T) = \frac{L/L_s}{\sinh L/L_s} = \frac{\frac{L}{l_e}\sqrt{\frac{3T}{2\tau_b}}}{\sinh \frac{L}{l_e}\sqrt{\frac{3T}{2\tau_b}}}} \tag{9.34}$$

The generalization of this expression to anisotropic scattering is straightforward and is obtained simply by replacing l_e by l^*.

Different choices for the geometry of the diffusive medium then lead for the time correlation function $g_1(T)$ to different behaviors that reflect the fact that we are probing different sets of multiple scattering trajectories. When measured in reflection from a

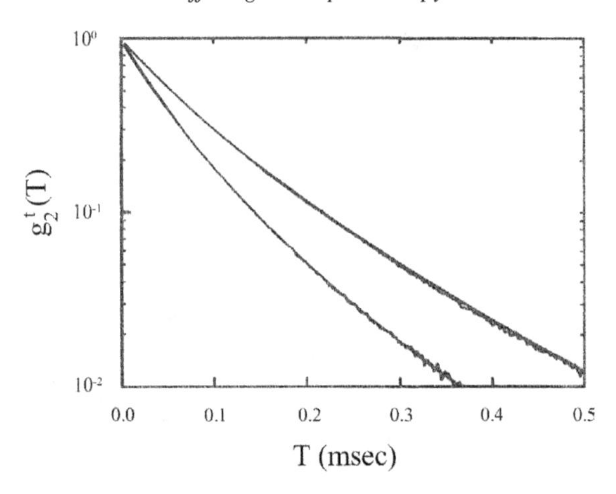

Figure 9.7 Intensity correlation function $g_2^t(T)$ measured in transmission through a cell of width $L=1$ mm and transport mean free path $l^*=166$ μm. The upper curve corresponds to a point-like light source and the lower curve to an incident plane wave [253].

semi-infinite medium, $g_1^r(T)$ behaves as $\sqrt{T}$ at short times, whereas in transmission through a slab, $g_1^t(T)$ behaves linearly in the same limit, namely,

$$g_1^t(T) \simeq 1 - \frac{L^2}{6L_s^2} = 1 - \frac{1}{4}\left(\frac{L}{l^*}\right)^2 \frac{T}{\tau_b}. \tag{9.35}$$

This difference expresses the fact that in transmission only long trajectories traversing the whole slab are probed, while in reflection all trajectories, whatever their length, contribute.

Exercise 9.1 correlation function G_1^t for any light source

The result given in (9.34) corresponds to a uniformly illuminated interface, namely to an incident intensity of the form $I_{inc}(x,y)=I_{inc}$. We now consider a beam described by any smooth function $I_{inc}(x,y)$. Show that the solution (9.33) must now be replaced by the following expression

$$G_1^t(T,\rho) \propto \int \frac{d^2\boldsymbol{q}}{(2\pi)^2} \frac{\tilde{q}}{\sinh(\tilde{q}L)} e^{i\boldsymbol{q}\cdot\boldsymbol{\rho}}\, \tilde{I}_{inc}(\boldsymbol{q}) \tag{9.36}$$

with $\boldsymbol{\rho}=(x,y)$ and $\tilde{q}^2=q^2+1/L_s^2$, and where $\tilde{I}_{inc}(\boldsymbol{q})$ is the Fourier transform of $I_{inc}(x,y)$. For the specific case of a point-like light source, i.e., for $\tilde{I}_{inc}(\boldsymbol{q})=$ constant, show that

$$G_1^t(T,\rho) \propto \int \frac{d^2\boldsymbol{q}}{(2\pi)^2} \frac{\tilde{q}}{\sinh(\tilde{q}L)} e^{i\boldsymbol{q}\cdot\boldsymbol{\rho}} \tag{9.37}$$

and compare the particular value

$$g_1^t(T,\rho=0)=\frac{1}{7\zeta(3)}\int_0^\infty dx\frac{\sqrt{x+(L/L_s)^2}}{\sinh\sqrt{x+(L/L_s)^2}} \tag{9.38}$$

to the corresponding result (9.34) obtained for a plane wave. These different behaviors of $g_2^t(T)=|g_1^t(T)|^2$ between a plane wave and a point-like source appear explicitly in Figure 9.7. In both cases, the short time behavior remains linear, but decreases more rapidly for a plane wave.

Appendix A9.1: Collective motion of scatterers

In the course of Chapter 6 and this chapter, we considered time correlation functions of the field and of the intensity for the case of Brownian scatterers. However, diffusing wave spectroscopy is also a useful tool to probe properties of flows using the dynamics of scatterers spread out in the fluid. Such methods are often used in the single scattering regime, for instance, in the study of turbulent fluids where information concerning the flow is retrieved by measuring the velocity correlation functions of scatterers immersed inside the fluid [254]. Here we limit ourselves to the simple case of a laminar flow in the multiple scattering regime. The velocity profile represented schematically in Figure 9.8 corresponds to velocities oriented along the Ox direction and depending linearly upon the z coordinate, namely $\boldsymbol{v}_x=\Gamma z\hat{\boldsymbol{e}}_x$, where Γ is a velocity gradient. To calculate the correlation function (9.1), we must evaluate the phase shift that appears in the Diffuson, using the method of section 6.7.1. The phase shift $\Delta\phi_N(T)$ that corresponds to a given sequence of $N=t/\tau_e$ collisions is given by the relation (6.186) which is rewritten as

$$\Delta\phi_N(T)=k\sum_{n=1}^{N}\hat{\boldsymbol{e}}_n\cdot[\Delta\boldsymbol{r}_{n+1}(T)-\Delta\boldsymbol{r}_n(T)], \tag{9.39}$$

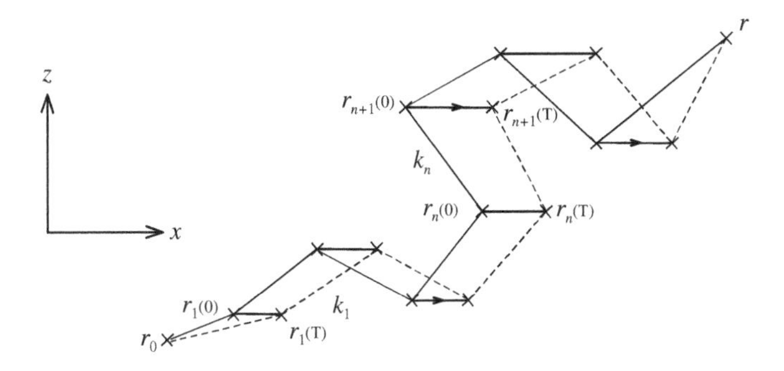

Figure 9.8 Velocity profile of a laminar flow. The flow is along the Ox axis and velocities depend only on the z coordinate normal to the flow.

where we recall that for elastic scattering the modulus $k = |\boldsymbol{k}|$ is conserved. We thus define $\boldsymbol{k}_n = k\hat{\boldsymbol{e}}_n$ where $\hat{\boldsymbol{e}}_n$ is a unit vector and we assume that the velocity of the scatterers is slow enough to neglect safely the change in the direction of wave vectors after a time T. Then, a stationary laminar flow, with the velocity profile defined previously, yields $\Delta\boldsymbol{r}_n(T) = \Gamma\, T(\boldsymbol{r}_0 \cdot \hat{\boldsymbol{e}}_z)\,\hat{\boldsymbol{e}}_x$. Denoting $\Lambda_n = |\boldsymbol{r}_{n+1}(0) - \boldsymbol{r}_n(0)|$, we rewrite the phase shift for a sequence of N collisions in the form

$$\Delta\phi_N(T) = \Gamma T k \sum_{n=1}^{N} \Lambda_n (\hat{\boldsymbol{e}}_n \cdot \hat{\boldsymbol{e}}_x)(\hat{\boldsymbol{e}}_n \cdot \hat{\boldsymbol{e}}_z). \tag{9.40}$$

The last step is to average the quantity $e^{i\Delta\phi_N(T)}$. In contrast to the case of Brownian motion, the average for a laminar flow is taken only over angular variables. Rewriting (9.40) in the form $\Delta\phi_N(T) = \Gamma T k \sum_{n=1}^{N} \Lambda_n \cos\theta_n \sin\theta_n \cos\varphi_n$, and using the angular average $\langle \cos^2\theta \sin^2\theta \cos^2\varphi \rangle = 1/15$, yields[6]

$$\langle \Delta\phi_N^2(T) \rangle = 2N \left(\frac{T}{\tau_l} \right)^2, \tag{9.41}$$

where the characteristic time τ_l is defined by

$$\frac{1}{\tau_l} = \frac{\Gamma k l_e}{\sqrt{30}}. \tag{9.42}$$

Unlike the time τ_b that characterizes a Brownian motion of the scatterers (relation 6.193), the time τ_l depends on the elastic mean free path $l_e = \overline{\Lambda_n}$. From (9.41), we conclude that the motion of the scatterers gives rise to an exponential decrease of the form (6.1), where we have defined the time $t = N\tau_e$:

$$\left\langle e^{i\Delta\phi_N(T)} \right\rangle = e^{-\frac{1}{2}\langle \Delta\phi_N(T)^2 \rangle} = e^{-t/\tau_s} \tag{9.43}$$

and where the dephasing time τ_s is given by

$$\boxed{\tau_s = \tau_e \left(\frac{\tau_l}{T} \right)^2} \tag{9.44}$$

This dephasing time behaves as $1/T^2$ instead of $1/T$ for the Brownian motion case (relation 6.197). For small times T, the correlation function $g_1^r(T)$ measured in reflection thus becomes linear with T. For anisotropic scattering, equation (9.27) needs to be replaced by [255]:

$$g_1^r(T) = 1 - \sqrt{3}\frac{z^*}{l^*}\frac{T}{\tau_l^*}, \tag{9.45}$$

[6] It is straightforward to check that the average value of $\Delta\phi_N(T)$ is proportional to the divergence of the velocity field so that it vanishes for an incompressible fluid [255].

with $1/\tau_l^* = \Gamma k l^*/\sqrt{30}$. In transmission, the time dependence becomes quadratic and instead of (9.35), we have:

$$g_1^t(T) = 1 - \frac{1}{2}\left(\frac{L}{l^*}\right)^2 \left(\frac{T}{\tau_l^*}\right)^2 = 1 - \frac{\Gamma^2 k^2 L^2}{60} T^2. \tag{9.46}$$

This result obtained in transmission is quite remarkable since it turns out to be independent of the elastic mean free path, which is unexpected in the multiple scattering limit.

These behaviors differ qualitatively from those obtained for the case of Brownian motion (9.23 and 9.35). In addition, the scatterers spreading out in the laminar flow also have a Brownian motion. Assuming that it is uncorrelated with the laminar flow, we expect to observe two kinds of behaviors for the time correlation function. In reflection, the short times correlation function probes the Brownian motion of scatterers and thus varies as $\sqrt{T}$. For longer times, it probes the underlying laminar flow and thus behaves linearly with T. A crossover time of order τ_l^2/τ_b separates these two regimes. Hence probing the deterministic motion imposed upon the scatterers by the laminar flow requires that times $T \gg \tau_l^2/\tau_b$ be considered. The same applies in transmission for which the crossover time τ_l^2/τ_b separates linear time behavior in T from quadratic behavior. This crossover has been observed experimentally both in reflection and in transmission [256].

Exercise 9.2: Poiseuille flow

A Poiseuille flow inside a box of finite size L is characterized by a velocity profile

$$\boldsymbol{v}_x(z) = \frac{\Gamma}{L}(Lz - z^2)\hat{\boldsymbol{e}}_x. \tag{9.47}$$

Show that the phase shift corresponding to a multiple scattering sequence of length N is of the form:

$$\Delta\phi_N(T) = \Gamma T k \sum_{n=1}^{N} \Lambda_n (\hat{\boldsymbol{e}}_n \cdot \hat{\boldsymbol{e}}_x)(\hat{\boldsymbol{e}}_n \cdot \hat{\boldsymbol{e}}_z)\left(\frac{z_{n+1} + z_n}{L} - 1\right), \tag{9.48}$$

where Λ_n has been defined in (9.40). Unlike the case of Brownian motion or laminar flow, this phase shift now depends explicitly on the position z_n of the scatterers. Without doing an explicit calculation [255], show that assuming $\langle z_n^2 \rangle = n l_e^2$ leads to the following expression for the fluctuation $\langle \Delta\phi_N^2(T) \rangle$ of the phase shift:

$$\langle \Delta\phi_N^2(T) \rangle = 2\left(\frac{T}{\tau_l}\right)^2 \left[N + N^2 \left(\frac{l_e}{L}\right)^2\right], \tag{9.49}$$

where the time τ_l has been defined in (9.42). Using the results of section 9.5.2, show that the correspondence (9.23) between the time correlation function $G_1^r(T)$ and the coherent albedo no longer holds.

10
Spectral properties of disordered metals

In this chapter we denote by ν_0, the average density of states per spin direction and by $\rho_0 = \nu_0/\Omega$, the density of states per unit volume. The characteristic energy $\Delta = 1/(\rho_0\Omega) = 1/\nu_0$ is the average spacing between energy levels per spin direction. We set $\hbar = 1$.

10.1 Introduction

In the previous chapters, we have considered different signatures of the phase coherence on average values of transport quantities such as the electrical conductivity or the albedo. In order to observe such coherent effects, it is necessary to couple the disordered medium to the outside world, such as electric wires for the conductivity or the free propagation medium for the albedo. However, it is also possible to characterize isolated disordered systems by measuring their spectral properties.

A disordered medium can be viewed as a complex system whose energy or frequency spectrum cannot be described by means of a given deterministic series of numbers. The nature of the spectrum and of its correlations reflects different physical properties. For example, the spectrum of a disordered metal is quite different if it is a good or a bad conductor. The understanding of these properties is essential for the description of thermodynamic properties such as orbital magnetism or persistent current, which we shall study later and which express the sensitivity of the spectrum to an applied magnetic field or to an Aharonov–Bohm flux (Chapter 14).

The purpose of this chapter is to describe the spectral properties of a disordered metal by means of statistical methods. Some of these properties are universal in a sense that they are common to a large class of physical systems (nuclei, atoms or molecules, metals, metallic aggregates, etc.). This universality is well described using the *random matrix theory*, whose main results are presented in section 10.4. In the multiple scattering regime, it is possible to relate some of the spectral properties to the probability of quantum diffusion.

The numerical results presented in this chapter have been obtained using the three-dimensional Anderson model, introduced in section 2.2.3. It is characterized by the transfer energy t from site to site and by the bandwidth W of the distribution of site energies. The weak disorder limit $k_F l_e \gg 1$ is given for this model by $W \ll W_c$, where $W_c = 16.5\,t$ is the critical disorder, corresponding to the metal–insulator transition. Numerically, it is easy to reach the strong disorder regime by increasing W and thus to explore the spectral properties near the metal–insulator transition, i.e., in a limit where the analytical methods developed in this book do not apply anymore. For a thorough study of the metal–insulator transition, we refer the reader to the references [257, 258].

10.1.1 Level repulsion and integrability

Figure 10.1(a) presents two energy spectra. One is the spectrum of a good conductor, in the limit $k_F l_e \gg 1$. The other is a sequence of uncorrelated random numbers, distributed according to a Poisson law. These two spectra are quite different. For the good conductor, we observe a more regular behavior. A first way to account for the difference between these two spectra consists in studying the probability $P(s)$ of two neighboring levels being distant by the energy s.[1] We observe in the figure that for the good conductor, the probability for two levels to be very close to each other vanishes. This is not the case for a spectrum without correlation. This property is called *level repulsion*. In particular, for a disordered metal, degeneracies are lifted.

It turns out that this behavior is quite general. Level repulsion is also observed in the energy (or frequency) spectra of a wide range of physical systems [259–261]. For example, Figure 10.2 shows a histogram of the spacings between highly excited levels of several

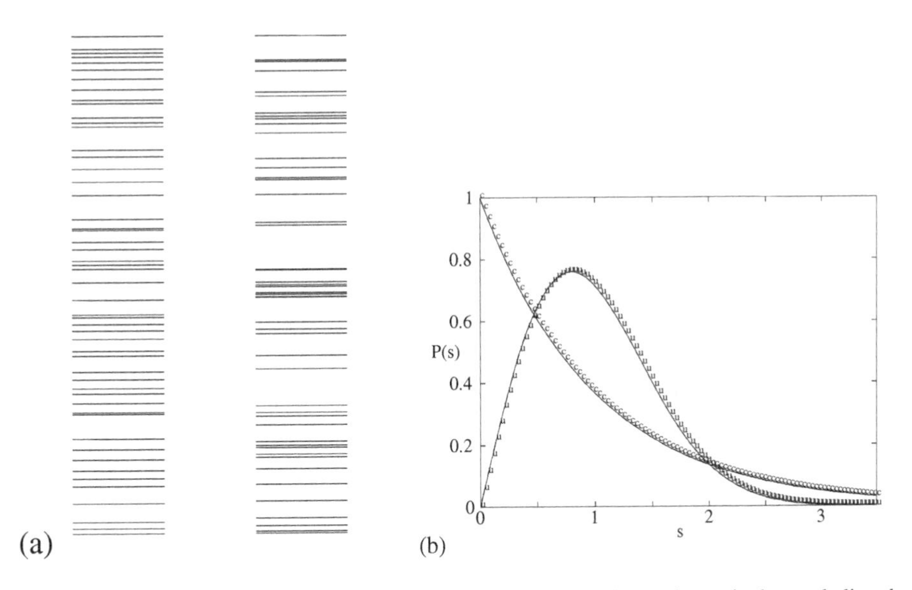

Figure 10.1 (a) Comparison of the energy spectrum of a good metallic conductor in the weak disorder limit (left) with a Poisson spectrum corresponding to a random distribution of uncorrelated levels (right). (b) Wigner–Dyson distribution $P(s)$ for the spectrum of a good metallic conductor (filled circles) and for a Poisson spectrum (open circles).

[1] Here s is a dimensionless energy, normalized to the average spacing Δ between neighboring levels.

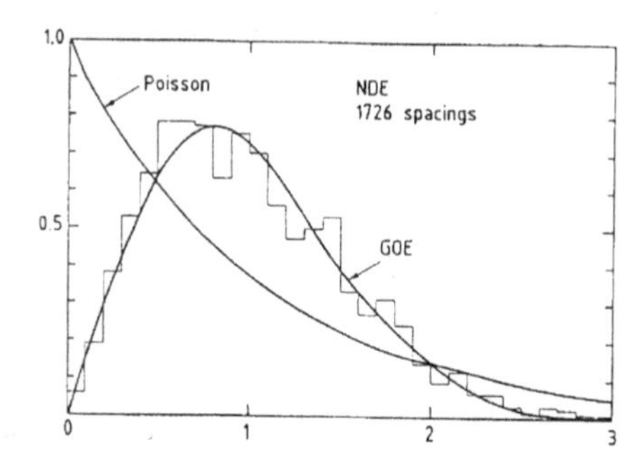

Figure 10.2 Histogram of energy level spacings for several heavy nuclei, obtained by neutron scattering (data are for isotopes of Cd, Sm, Gd, Dy, Er, Yb, W, Th, U) [259].

heavy nuclei. Remarkably enough, this histogram is described by the same distribution $P(s)$ as for the disordered metal. This distribution appears in other complex systems such as heavy nuclei, large molecules, and also in simple systems like the hydrogen atom in a strong magnetic field [262].

These kinds of spectral correlations do not necessarily occur in a physical system with many degrees of freedom. Quite unexpectedly, they also appear in the spectrum of a physical system as simple as a particle in a box (or for the resonance modes of acoustic or electromagnetic cavities [263]). In two dimensions, these boxes, also called billiards, are systems with only two degrees of freedom.

For a rectangular billiard, there are as many integrals of motion (the energy and one component of the momentum) as degrees of freedom, and the classical behavior is perfectly determined. The eigenvalues are labelled by two quantum numbers m and n which correspond to the two components of the momentum and take the form $m^2/a^2 + n^2/b^2$ where a and b are the dimensions of the rectangle. At high energy, the repartition of these eigenvalues reaches a Poisson distribution (see Figure 10.3). This is to be contrasted with the case of a billiard of arbitrary shape for which the energy is the only constant of motion (for example the stadium or the Sinai billiard [261]). In this case, while the momentum is conserved at a reflection on the boundary, this is no longer the case for each of its components separately. The classical behavior is chaotic, in a sense that two, initially close, ballistic trajectories separate exponentially when time increases. The corresponding spectrum exhibits level repulsion, with the same distribution $P(s)$ as for the spectrum of a heavy nucleus or a good conductor!

Quite generally, when it is possible to describe the system by an ensemble of quantum numbers equal to the number of degrees of freedom, the energy levels can be labelled using these quantum numbers, and therefore they are uncorrelated (in the absence of particular symmetries). This is not the case for a billiard of arbitrary shape. The level repulsion thus appears to be characteristic of classically non-integrable systems [264]. A heavy nucleus, a complex molecule, a disordered metal or a chaotic billiard have a common feature,

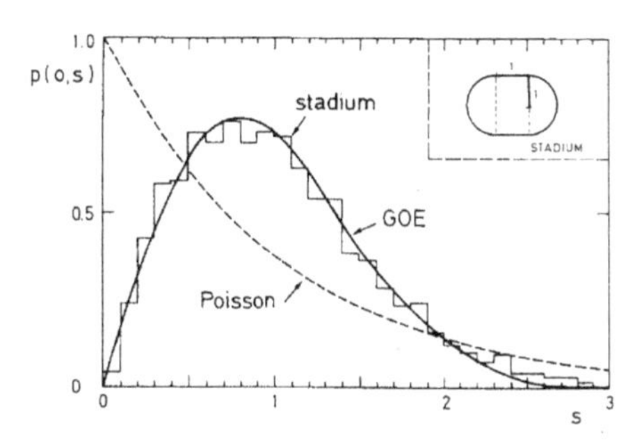

Figure 10.3 Distribution of spacings between first-neighbor energy levels of a rectangular billiard (Poisson distribution) and a stadium billiard [260].

namely their distribution of neighboring energy levels. For these systems, either one-body or many-body, the set of good quantum numbers is smaller than the number of degrees of freedom. Such systems are called *complex systems*.

What are the common features of these different complex systems? How can we characterize the universality of their spectral correlations? To answer these questions, Wigner, Dyson and Mehta [265–267] have proposed a general statistical description of the Hamiltonian of complex systems which is called the *random matrix theory*. It was devised initially to describe the highly excited level of heavy nuclei. This approach consists in replacing the original microscopic Hamiltonian by a matrix whose elements are independent random variables. The advantage of this theory is its generality, and the remarkable result that the energy level distribution depends only on the symmetry of the Hamiltonian. This approach has been used for a large variety of problems in nuclear, atomic and molecular physics. In solid state physics, it was introduced for the first time by Gorkov and Eliashberg [268] in order to explain the polarizability of small metallic aggregates. In that case, the complexity does not result from interactions between particles as in nuclei but is rather due to the collisions of electrons at the irregular boundary of the aggregate which thus constitutes an example of a billiard with a random boundary.

10.1.2 Energy spectrum of a disordered metal

Beyond these common universal features, it is also expected that a disordered metal, a molecule and a heavy nucleus present differences in their spectral properties that are specific to their microscopic underlying nature. The aim of this chapter is to clarify the distinction between universal and non-universal properties [269]. For that purpose we study the spectral properties of a good metal, i.e., of a weakly disordered conductor where electrons are elastically scattered by impurities. Using the approach developed in Chapter 4, we shall recover the universal behavior of random matrix theory and we shall understand the origin of the deviations from this theory. This will give a meaning to the results of random matrix theory in the context of disordered metals.

We shall work within the Diffuson approximation, thus neglecting coherent effects due to crossings of Diffusons (Appendix A4.2.3). We shall thus consider energy scales larger than the level spacing Δ. Conversely, the behavior of spectral properties at energy scales smaller than Δ, that is for time scales larger than the *Heisenberg time* (in units of $\hbar$)

$$\tau_H = \frac{2\pi}{\Delta}, \tag{10.1}$$

involves additional contributions with Diffuson crossings [270]. This long time limit has been studied by means of the supersymmetric non-linear σ model which provides exact results in this limit [271–273].

10.2 Characteristics of spectral correlations

In order to characterize the distribution of eigenenergies, we consider the following quantities.

- The distribution $P(s)$ of spacings (normalized to the mean level spacing Δ) between neighboring levels. This quantity decreases exponentially for large s.
- The probability $P(n, s)$ that the distance between two levels separated by n other levels is equal to s. By definition, we have $P(s) = P(0, s)$.
- The two-point correlation function of the density of states defined by

$$\overline{\rho(\epsilon_1)\rho(\epsilon_2)} - \overline{\rho}(\epsilon_1)\,\overline{\rho}(\epsilon_2), \tag{10.2}$$

where the density of states per unit volume defined from (3.24) is equal to

$$\rho(\epsilon) = \frac{1}{\Omega}\sum_\alpha \delta(\epsilon - \epsilon_\alpha) \tag{10.3}$$

with ϵ_α being the eigenenergies of the microscopic Hamiltonian. The average denoted by $\overline{\cdots}$ is taken here either over a given spectrum by considering different energy intervals or over different realizations of disorder. Numerical simulations show that these two averaging procedures are equivalent (ergodic hypothesis). Notice that for an atomic nucleus or a billiard, there is only one energy spectrum so that only the energy average can be performed [259].

If ϵ_1 and ϵ_2 belong to an energy interval where the average density of states is nearly constant, then the averaged spectrum is translation invariant along the energy axis. The two-point correlation (10.2) depends only on the difference $\omega = \epsilon_1 - \epsilon_2$. We thus introduce the correlation function $K(\omega)$ defined by

$$\boxed{K(\omega) = \frac{\overline{\rho(\epsilon)\rho(\epsilon-\omega)}}{\overline{\rho}^2} - 1} \tag{10.4}$$

The average density of states per unit volume $\overline{\rho}$ is given by the reciprocal of the average energy separation Δ between levels divided by the volume:

$$\boxed{\overline{\rho} = \frac{1}{\Delta\Omega}} \tag{10.5}$$

Remark

$P(s)$ is the probability that two levels are at a distance of s, *with no other level in between*. Therefore $P(s)$ depends on all n-point correlation functions. Inversely, there is a simple relation between the two-point correlation function K and all the $P(n, s)$. Since $P(n, s)$ is the probability that two levels with n other levels intercalated are at a distance of s, we have by definition,

$$K(\omega) = \delta(s) - 1 + \sum_n P(n, s) \tag{10.6}$$

with $s = \omega/\Delta$. The correlation function $K(\omega = s\Delta)$ is the probability that two levels are at a distance of s, whatever the number of levels in between. In particular, when $s \to 0$, $P(n+1, s) \ll P(n, s)$ so that

$$K(\omega) \to \delta(s) - 1 + P(s) \quad \text{if } s \to 0.$$

- The form factor $\tilde{K}(t)$ defined as the Fourier transform of $K(\omega)$

$$\boxed{\tilde{K}(t) = \frac{1}{2\pi}\int_{-\infty}^{\infty} K(\omega)\, e^{-i\omega t}\, d\omega} \tag{10.7}$$

In the diffusive limit, we shall show that $\tilde{K}(t)$ is related to the integrated return probability $Z(t)$ defined by (5.5).

- The fluctuation of the number of levels $N(E)$ contained in an energy interval of width E. This function $N(E)$, called the *counting function*, is nothing but the density of states integrated over an interval of width E:

$$N(E) = \Omega \int_0^E \rho(\epsilon)\, d\epsilon = \sum_\alpha \Theta(E - \epsilon_\alpha). \tag{10.8}$$

It increases by one unit for each additional eigenvalue (Θ is the Heaviside function). The fluctuation of the number of levels is defined by the variance, also called *number variance*

$$\Sigma^2(E) = \overline{N(E)^2} - \overline{N}(E)^2 \tag{10.9}$$

and it can be expressed in a simple way in terms of the two-point correlation function. For that purpose, we use both definitions (10.8) and (10.4) as well as (10.5), to obtain

$$\Sigma^2(E) = \frac{1}{\Delta^2}\int_0^E \int_0^E K(\epsilon_1 - \epsilon_2)\, d\epsilon_1\, d\epsilon_2 \tag{10.10}$$

which can be rewritten, after a change of variables and integration by parts, in the form

$$\Sigma^2(E) = \frac{2}{\Delta^2}\int_0^E (E-\omega)K(\omega)\, d\omega. \tag{10.11}$$

or equivalently,

$$K(\omega) = \frac{\Delta^2}{2}\frac{\partial^2 \Sigma^2(\omega)}{\partial \omega^2}. \tag{10.12}$$

Finally, the variance $\Sigma^2(E)$ may also be expressed directly in terms of the form factor:

$$\boxed{\Sigma^2(E) = \frac{8}{\Delta^2}\int_0^\infty dt \frac{\tilde{K}(t)}{t^2}\sin^2\left(\frac{Et}{2}\right)} \tag{10.13}$$

Remark: unfolding the spectrum
We have assumed that the average density of states $\overline{\rho}$ is independent of energy. This is not usually the case. In order to exhibit universal features of the spectrum of a given system, we must first compare it to a situation where the density of states is constant. This step is called *unfolding* [259]. It goes as follows: we start from the original sequence of eigenvalues $\{\epsilon_\alpha\}$ and fit the counting function by a continuous function $\overline{N}(E)$. We then consider the new sequence defined by $\tilde{\epsilon}_\alpha = \overline{N}(\epsilon_\alpha)$. By definition, the density of states of this new sequence is constant, and the average distance between consecutive levels is 1.

10.3 Poisson distribution

For an uncorrelated sequence of eigenvalues, the probability $P(s)$ is

$$\boxed{P(s) = e^{-s}} \tag{10.14}$$

where s is written in units of the average level spacing.

Exercise 10.1 Establish the expression (10.14) of $P(s)$ for a Poisson sequence. Consider a sequence of N points within an interval $[0, L]$. The probability of a given sequence is constant:

$$P(\epsilon_1, \epsilon_2, \ldots, \epsilon_N) = \left(\frac{1}{L}\right)^N.$$

Consider now two neighboring levels ϵ_1 and ϵ_2, separated by a distance s normalized in units of the average distance $\Delta = L/N$: $s = |\epsilon_1 - \epsilon_2|/\Delta$. The probability $P(s)$ is the product of the probabilities p_j that all other levels ϵ_j are not in the interval $[\epsilon_1, \epsilon_2]$. Since $p_j = (1 - |\epsilon_1 - \epsilon_2|/L)$, we obtain

$$P(s) = \left(1 - \frac{|\epsilon_1 - \epsilon_2|}{L}\right)^{N-2} \simeq \left(1 - \frac{s}{N}\right)^N \underset{N\to\infty}{\longrightarrow} e^{-s}.$$

Using the method of the above exercise, we can show that, for a Poisson sequence, the probability of finding n other levels in between two levels at a distance of s is

$$P(n,s) = \frac{s^n}{n!} e^{-s} \tag{10.15}$$

and that it obeys $\sum_{n=1}^{\infty} P(n,s) = 1$. From (10.6), we obtain that the two-point correlation function $K(\omega)$ vanishes, except for $\omega = 0$, namely

$$K(\omega) = \delta(\omega).$$

The Fourier transform (10.7) thus gives a constant form factor

$$\tilde{K}(t) = \frac{1}{2\pi}.$$

10.4 Universality of spectral correlations: random matrix theory

In this section we introduce the random matrix theory. For more details, the reader may consult references [259, 267].

10.4.1 Level repulsion in 2 × 2 matrices

Before we establish general results about random matrices, we first, after E. P. Wigner [265], consider the case of 2×2 random matrices. This example is the simplest that exhibits *level repulsion*. We start by calculating the distribution $P_o(s)$ for real symmetric matrices

$$\mathcal{H} = \begin{pmatrix} h_{11} & h_{12} \\ h_{12} & h_{22} \end{pmatrix}$$

with a Gaussian normalized distribution of matrix elements $\{h_{ij}\}$:

$$p(h_{ij}) = \sqrt{\frac{\lambda}{\pi}} e^{-\lambda h_{ij}^2}.$$

They are assumed to be independent with a distribution thus given by

$$P(\{h_{ij}\}) = \prod_{ij} p(h_{ij}) = \frac{1}{\mathcal{Z}} e^{-\lambda(h_{11}^2 + h_{22}^2 + 2h_{12}^2)}, \tag{10.16}$$

where $\mathcal{Z}$ is a normalization constant. The distance s between the two eigenvalues is[2]

$$s = |\epsilon_1 - \epsilon_2| = \sqrt{(h_{11} - h_{22})^2 + 4h_{12}^2}.$$

[2] The average distance between levels is set equal to 1.

The probability of having a vanishing value for s requires the two conditions to be satisfied *simultaneously*: $(h_{11} - h_{22}) = 0$ and $h_{12} = 0$. This probability is zero and this corresponds to the so-called "repulsion" of eigenvalues. The probability $P_o(s)$ is given by the integral:

$$P_o(s) = \frac{1}{\mathcal{Z}} \int P(\{h_{ij}\}) \delta\left(s - \sqrt{(h_{11} - h_{22})^2 + 4h_{12}^2}\right) dh_{11}\, dh_{22}\, dh_{12}. \tag{10.17}$$

Integrating over the matrix elements leads to

$$\boxed{P_o(s) = \frac{\pi}{2} s\, e^{-\frac{\pi}{4}s^2}} \tag{10.18}$$

where the probability has been normalized so that $\overline{s} = 1$. The distribution $P_o(s)$ has two remarkable properties that make it very different from the Poisson case: it vanishes *linearly* when s goes to 0 and it decreases more rapidly at large separation ($s \to \infty$). The distribution (10.18) is well known as the *Wigner surmise*.

Exercise 10.2 Establish the expression for $P_o(s)$ starting from relation (10.17).
We define $u = h_{11} - h_{22}$, $v = (h_{11} + h_{22})/2$ and $x = \sqrt{u^2 + 4h_{12}^2}$ so that the integral becomes

$$P_o(s) = \frac{1}{\sqrt{2}} \left(\frac{\pi}{\lambda}\right)^{3/2} \int e^{-(2\lambda v^2 + \lambda s^2/2)} \delta(s - x) \frac{2x}{\sqrt{x^2 - u^2}}\, du\, dv\, dx$$

which, upon integration, leads to

$$P_o(s) = \lambda s e^{-\lambda s^2/2} . \tag{10.19}$$

The normalization ($\overline{s} = 1$) implies that $\lambda = \pi/2$.

This result can be generalized to Hermitian complex matrices. The separation between the two eigenvalues then becomes

$$s = \sqrt{(h_{11} - h_{22})^2 + 4|h_{12}|^2}. \tag{10.20}$$

A vanishing s implies that $(h_{11} - h_{22})$, $\mathrm{Re}(h_{12})$ and $\mathrm{Im}(h_{12})$ simultaneously vanish. There is an additional constraint, compared to real matrices. The probability of having two levels close to each other is therefore smaller and "*level repulsion is stronger*." A calculation similar to the one given in Exercise 10.2 gives the Wigner surmise for the unitary case

$$\boxed{P_u(s) = \frac{32}{\pi^2} s^2\, e^{-\frac{4}{\pi}s^2}} \tag{10.21}$$

For Hermitian matrices, the level repulsion near $s \simeq 0$ is *quadratic*, instead of linear for real matrices. This is a general and remarkable property of the spectra of "complex" systems (Figure 10.4).

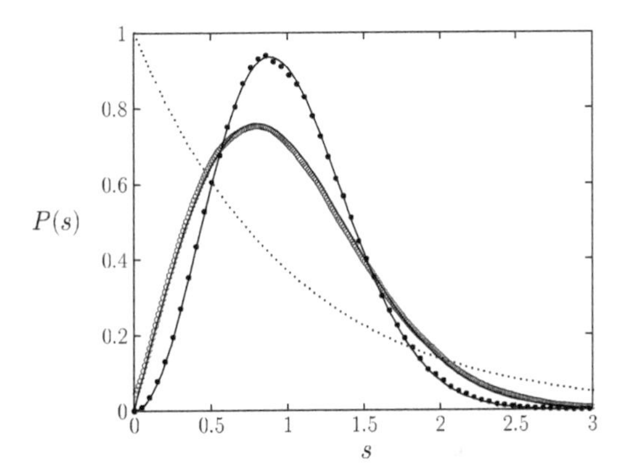

Figure 10.4 Distribution $P(s)$ for a weakly disordered metal, obtained from numerical solution of the Anderson model, with (filled circles) or without (open circles) magnetic field [274]. These distributions are very accurately described by the Wigner results (10.18) and (10.21) for 2×2 random matrices.

Exercise 10.3 Establish the expression for the distribution $P(\epsilon_1, \epsilon_2)$ of the eigenvalues of 2×2 matrices.

The argument of the exponential in the distribution (10.16) is equal to the trace of the square of the matrix, so that the distribution can be rewritten as:

$$P(\{h_{ij}\}) = \frac{1}{\mathcal{Z}} e^{-\lambda \mathrm{Tr} \mathcal{H}^2} = \frac{1}{\mathcal{Z}} e^{-\lambda(\epsilon_1^2 + \epsilon_2^2)} . \tag{10.22}$$

The matrix h_{ij} can be diagonalized by means of an orthogonal transformation (i.e., a rotation of angle θ), so that

$$\begin{pmatrix} h_{11} & h_{12} \\ h_{12} & h_{22} \end{pmatrix} = \begin{pmatrix} \cos\theta & \sin\theta \\ -\sin\theta & \cos\theta \end{pmatrix} \begin{pmatrix} \epsilon_1 & 0 \\ 0 & \epsilon_2 \end{pmatrix} \begin{pmatrix} \cos\theta & -\sin\theta \\ \sin\theta & \cos\theta \end{pmatrix} .$$

The distribution of the eigenvalues ϵ_1 and ϵ_2 satisfies

$$P(\{h_{ij}\})\, dh_{11}\, dh_{22}\, dh_{12} = P(\epsilon_1, \epsilon_2, \theta) \mathcal{J}\, d\epsilon_1\, d\epsilon_2\, d\theta \tag{10.23}$$

where $\mathcal{J}$ is the Jacobian of the transformation. We check that $\mathcal{J} = |\epsilon_1 - \epsilon_2|$. Moreover, $P(\{h_{ij}\})$ has been rewritten is terms of ϵ_1 and ϵ_2. The distribution of eigenvalues is thus given by

$$P(\epsilon_1, \epsilon_2) = \int P(\epsilon_1, \epsilon_2, \theta)\, d\theta$$

so that

$$P(\epsilon_1, \epsilon_2) \propto |\epsilon_1 - \epsilon_2| e^{-\lambda(\epsilon_1^2 + \epsilon_2^2)} . \tag{10.24}$$

We recover the vanishing probability of having two coinciding levels. This results from the structure of the Jacobian of the transformation. The change of variables, $s = |\epsilon_1 - \epsilon_2|$, $\epsilon = (\epsilon_1 + \epsilon_2)/2$, and the integration over ϵ, gives for the probability

$$P(s) \propto s\, e^{-\lambda s^2/2} ,$$

which, after normalization, leads to (10.18).

10.4.2 Distribution of eigenvalues for $N \times N$ matrices

We now generalize the previous results obtained for 2×2 matrices to establish the distribution $P(\epsilon_1, \epsilon_2, \ldots, \epsilon_N)$ of the eigenvalues of $N \times N$ random matrices. Consider an ensemble of real symmetric $N \times N$ random matrices $\mathcal{H}$, with the following distribution of their $N(N+1)/2$ elements h_{ij}:[3]

$$P(\{h_{ij}\}) = \frac{1}{\mathcal{Z}} e^{-\lambda(\sum_i h_{ii}^2 + \sum_{j \neq i} h_{ij}^2)}. \tag{10.25}$$

The argument of the exponential is the trace of the matrix $\mathcal{H}^2$. It is thus independent of the basis and it can be rewritten as the sum of the squares of the eigenvalues ϵ_i:

$$P(\{h_{ij}\}) = \frac{1}{\mathcal{Z}} e^{-\lambda \mathrm{Tr}\mathcal{H}^2} = \frac{1}{\mathcal{Z}} e^{-\lambda \sum_i \epsilon_i^2}. \tag{10.26}$$

We now look for the distribution of eigenvalues of $\mathcal{H}$, using a generalization of the method presented in the previous section. An orthogonal transformation $\mathcal{O}$ diagonalizes this matrix:

$$\mathcal{H} = \mathcal{O}^T \mathcal{D} \mathcal{O},$$

where $\mathcal{D}$ is the diagonal matrix $\mathcal{D}_{ij} = \epsilon_i \delta_{ij}$. Then,

$$h_{ij} = \sum_k \epsilon_k \mathcal{O}_{ik} \mathcal{O}_{jk}. \tag{10.27}$$

The probability is conserved, so that

$$P(\mathcal{D}) d\mathcal{D} = P(\mathcal{H}) d\mathcal{H} = \frac{1}{\mathcal{Z}} e^{-\lambda \mathrm{Tr}\mathcal{H}^2} \, d\mathcal{H} = \frac{1}{\mathcal{Z}} e^{-\lambda \sum_i \epsilon_i^2} \, d\mathcal{H} \tag{10.28}$$

with

$$d\mathcal{H} = dh_{11} \, dh_{22} \, \cdots \, dh_{NN} \, dh_{12} \, dh_{13} \, \cdots \tag{10.29}$$

and

$$d\mathcal{D} = d\epsilon_1 \, d\epsilon_2 \, \cdots \, d\epsilon_N \, d\theta_1 \, d\theta_2 \, \cdots \, d\theta_{N(N-1)/2}. \tag{10.30}$$

Therefore,

$$P(\mathcal{D}) = P(\epsilon_1, \epsilon_2, \ldots, \epsilon_N, \theta_1, \theta_2, \ldots, \theta_{N(N-1)/2}) = \frac{1}{\mathcal{Z}} \mathcal{J} e^{-\lambda \sum_i \epsilon_i^2}$$

[3] We check that $\overline{h_{ij}^2} = 1/(4\lambda)$ for $i \neq j$ and that $\overline{h_{ii}^2} = 1/(2\lambda)$.

where the Jacobian $\mathcal{J}(\{\epsilon_i\}, \{\theta_i\})$ of the orthogonal transformation is a determinant of order $N(N+1)/2$:

$$\mathcal{J} = \begin{vmatrix} \frac{\partial h_{11}}{\partial \epsilon_1} & \frac{\partial h_{11}}{\partial \epsilon_2} & \cdots & \frac{\partial h_{11}}{\partial \epsilon_N} & \frac{\partial h_{11}}{\partial \theta_1} & \frac{\partial h_{11}}{\partial \theta_2} & \cdots & \frac{\partial h_{11}}{\partial \theta_{\frac{N(N-1)}{2}}} \\ \frac{\partial h_{12}}{\partial \epsilon_1} & \frac{\partial h_{12}}{\partial \epsilon_2} & \cdots & \frac{\partial h_{12}}{\partial \epsilon_N} & \frac{\partial h_{12}}{\partial \theta_1} & \frac{\partial h_{12}}{\partial \theta_2} & \cdots & \frac{\partial h_{12}}{\partial \theta_{\frac{N(N-1)}{2}}} \\ \vdots & \vdots & \vdots & \vdots & \vdots & \vdots & \vdots & \vdots \\ \frac{\partial h_{ij}}{\partial \epsilon_1} & \frac{\partial h_{ij}}{\partial \epsilon_2} & \cdots & \frac{\partial h_{ij}}{\partial \epsilon_N} & \frac{\partial h_{ij}}{\partial \theta_1} & \frac{\partial h_{ij}}{\partial \theta_2} & \cdots & \frac{\partial h_{ij}}{\partial \theta_{\frac{N(N-1)}{2}}} \\ \vdots & \vdots & \vdots & \vdots & \vdots & \vdots & \vdots & \vdots \\ \frac{\partial h_{NN}}{\partial \epsilon_1} & \frac{\partial h_{NN}}{\partial \epsilon_2} & \cdots & \frac{\partial h_{NN}}{\partial \epsilon_N} & \frac{\partial h_{NN}}{\partial \theta_1} & \frac{\partial h_{NN}}{\partial \theta_2} & \cdots & \frac{\partial h_{NN}}{\partial \theta_{\frac{N(N-1)}{2}}} \end{vmatrix}. \tag{10.31}$$

Since the transformation (10.27) is linear, the elements in the first N columns of this determinant are independent of the energies ϵ_i. For the same reason, the elements of the next $N(N-1)/2$ columns depend linearly on energies so that the Jacobian $\mathcal{J}(\{\epsilon_i\}, \{\theta_i\})$ is a homogeneous polynomial of degree $N(N-1)/2$ of these energies ϵ_i. Moreover, it vanishes when two energies ϵ_i and ϵ_j coincide. Therefore it can be cast in the form

$$\mathcal{J}(\{\epsilon_i\}, \{\theta_i\}) = \prod_{i<j}^{N} |\epsilon_i - \epsilon_j| f(\{\theta_i\}). \tag{10.32}$$

The integration over the variables θ_i leads to

$$P(\epsilon_1, \epsilon_2, \ldots, \epsilon_N) = \frac{1}{\mathcal{Z}} \prod_{i<j}^{N} |\epsilon_i - \epsilon_j| e^{-\lambda \sum_i \epsilon_i^2}. \tag{10.33}$$

For the case of an ensemble of complex Hermitian matrices, we introduce a unitary transformation to diagonalize the matrix. The number of independent non-diagonal elements is now $N(N-1)$ (notice that each element has a real part and an imaginary part). The Jacobian is therefore a polynomial of degree $N(N-1)$ which vanishes when two eigenvalues are equal. Its dependence on the terms $|\epsilon_i - \epsilon_j|$ is thus quadratic.

To summarize, the eigenvalue distribution of Gaussian random matrices, also known as the Wigner–Dyson distribution, is[4]

$$\boxed{P(\epsilon_1, \epsilon_2, \ldots, \epsilon_N) \propto \prod_{i<j}^{N} |\epsilon_i - \epsilon_j|^{\beta} e^{-\lambda\beta \sum_i \epsilon_i^2}} \tag{10.34}$$

Real and symmetric matrices correspond to systems with time reversal symmetry. The statistical ensemble of these matrices is invariant under orthogonal transformations, and

[4] The presence of the factor β in the exponential implies that the average density of states is independent of β (see footnote 5).

called the Gaussian orthogonal ensemble (GOE). In such a case, the level repulsion is linear and $\beta = 1$. When the time reversal invariance is broken, for example in the presence of a magnetic field, the corresponding matrices are Hermitian and their ensemble is invariant under unitary transformations. This is called the Gaussian unitary ensemble (GUE) and it corresponds to $\beta = 2$. Finally, we note that there is a third ensemble of matrices, whose elements are quaternions, and which corresponds to $\beta = 4$. This last ensemble is called the Gaussian symplectic ensemble (GSE, see references [259, 267] for more details).

10.4.3 Spectral properties of random matrices

Two-point correlation function

Expression (10.34) describes the distribution of eigenvalues of Gaussian random matrices.[5] To calculate the two-point correlation function of the density of states, we must integrate the distribution (10.34) over all variables ϵ_j except for two. This calculation is not easy and a complete derivation is given in [267].

This two-point function is shown in Figure 10.5, in the limit $N \to \infty$, for the GOE and GUE cases. For zero separation, there is a δ-like peak that accounts for the self-correlation of an energy level. When $\omega \to 0$, it tends to -1 since the probability for two levels to be close to each other vanishes so that $\overline{\rho(\epsilon)\rho(\epsilon - \omega)}$ vanishes as well. At large separation the levels become uncorrelated and $K(\omega)$ is also zero. In the GUE case, it is given by

$$K(\omega) = \delta(s) - \left(\frac{\sin(\pi s)}{\pi s}\right)^2 \tag{10.36}$$

with $s = \omega/\Delta$. For the GOE case, we have

$$K(\omega) = \delta(s) - \left(\frac{\sin(\pi s)}{\pi s}\right)^2 - \left[\int_s^\infty \left(\frac{\sin(\pi u)}{\pi u}\right) du\right]\left[\frac{d}{ds}\left(\frac{\sin(\pi s)}{\pi s}\right)\right] \tag{10.37}$$

which, in the limit of large separations, $\omega \gg \Delta$, behaves as

$$K(\omega) = -\frac{\Delta^2}{\beta \pi^2 \omega^2}. \tag{10.38}$$

We recall that, for a Poisson distribution, $K(\omega) = \Delta\delta(\omega)$. The correlation function is zero except at the origin where the δ-like peak accounts for the level self-correlation. For random matrices, the short range correlation is described by a "correlation hole" which is more pronounced for the GUE than for the GOE case, meaning a stronger spectral rigidity.

[5] For these matrices, the average density of states is a semi-circle :

$$\overline{\rho}(\epsilon) = \frac{2\lambda}{\pi}\left(\frac{N}{\lambda} - \epsilon^2\right)^{1/2} \quad \text{for } |\epsilon| < \sqrt{N/\lambda} \tag{10.35}$$

and zero otherwise. The distance between levels behaves as the inverse of $\overline{\rho}(\epsilon)$, and is not constant. For example, in the center of the spectrum, it varies as $N^{-1/2}$.

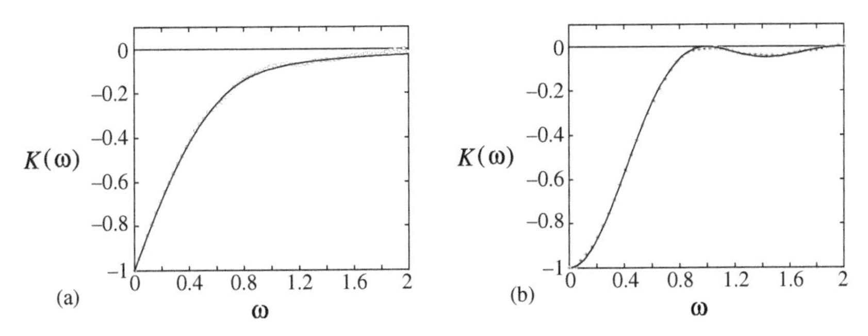

Figure 10.5 Two-point correlation function $K(\omega)$ for the Anderson model in three dimensions and for a weakly disordered metal ($W/t = 4$), (a) without and (b) in the presence of a magnetic field. The continuous line corresponds to random matrix theory ((a): GOE, (b): GUE). The δ function at the origin is not shown [275].

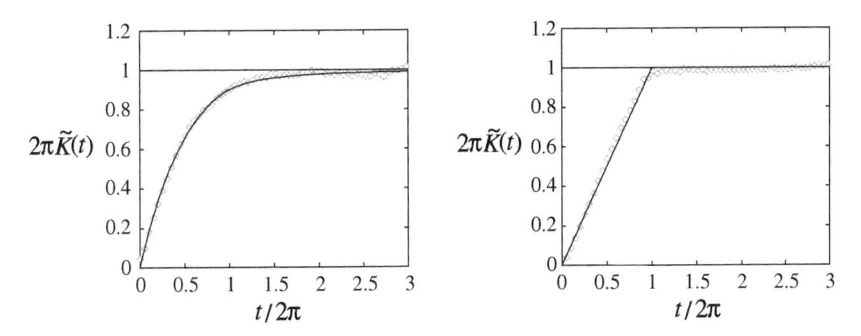

Figure 10.6 Form factor $\tilde{K}(t)$ for a weakly disordered metal ($W/t = 4$), using the Anderson model. The two figures correspond to cases where a magnetic field is applied (a) or not (b). The continuous line is the result of random matrix theory. For a Poisson distribution, $\tilde{K}(t) = 1/2\pi$ [275].

Form factor

Figure 10.6 displays the time dependence of the form factor $\tilde{K}(t)$. For times t much smaller than the Heisenberg time τ_H defined by $\tau_H = 2\pi/\Delta$ (in units of $\hbar$), the form factor behaves linearly with time:

$$\tilde{K}(t) \to \Delta^2 t/(2\pi^2\beta) \quad \text{for } t \ll \tau_H \tag{10.39}$$

and it saturates to a constant value $\Delta/2\pi$ for $t \to \infty$. This constant is nothing but the Fourier transform of the peak $\Delta\delta(\omega)$ that describes the self-correlation of the levels. By introducing $y = t/\tau_H$ and $b(y) = \tau_H \tilde{K}(y\tau_H)$, the form factor is rewritten for $\beta = 2$,

$$\begin{cases} b(y) = y & \text{if } \; y < 1 \\ b(y) = 1 & \text{if } \; y \geq 1, \end{cases} \tag{10.40}$$

whereas for $\beta = 1$ we have,

$$\begin{cases} b(y) = 2y - y\ln(1+2y) & \text{if} \quad y < 1 \\ b(y) = 2 - y\ln\left(\frac{2y+1}{2y-1}\right) & \text{if} \quad y \geq 1. \end{cases} \tag{10.41}$$

Spectral rigidity

Rigidity of the spectrum of random matrices shows up as a very small fluctuation (10.9) of the number of levels in a given energy interval. This fluctuation is obtained from the relations (10.11), (10.36) and (10.37). The exact expression of the variance is rather involved. For $E \to 0$, it reaches Poisson behavior asymptotically, namely, $\Sigma^2(E) \to E/\Delta$. It simplifies in the limit of large energies $E \gg \Delta$, or equivalently at small times $t \ll t_H$. In this case, the two-point correlation function varies as $-1/\omega^2$. The number variance is obtained by integrating this function twice, and therefore it varies logarithmically (Figure 10.7). More precisely,

$$\Sigma^2(E) = \frac{2}{\beta\pi^2}\left[\ln\left(2\pi\frac{E}{\Delta}\right) + c_\beta\right] + \mathcal{O}(\Delta/E) \tag{10.42}$$

with

$$\begin{aligned} c_{\beta=1} &= 1 + e^\gamma - \frac{\pi^2}{8} \\ c_{\beta=2} &= 1 + e^\gamma \\ c_{\beta=4} &= 1 + e^\gamma + \frac{\pi^2}{8} \end{aligned} \tag{10.43}$$

where $\gamma \simeq 0.577$ is the Euler constant. This result is a very remarkable characteristic of the spectral rigidity of the spectra of random matrices. For instance, by considering the spectra

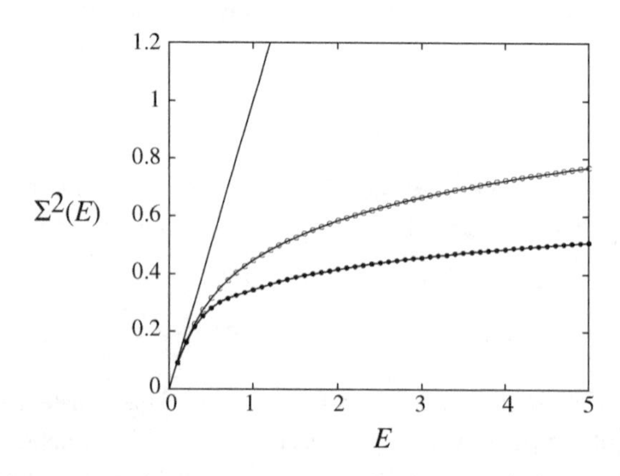

Figure 10.7 Number variance $\Sigma^2(E)$ for a weakly disordered metal with (filled circles) and without (open circles) magnetic field, in the Anderson model with $W/t = 4$. These results are very well described by random matrix theory (continuous lines for $\beta = 1$ and 2). The straight line corresponds to the Poisson distribution [274].

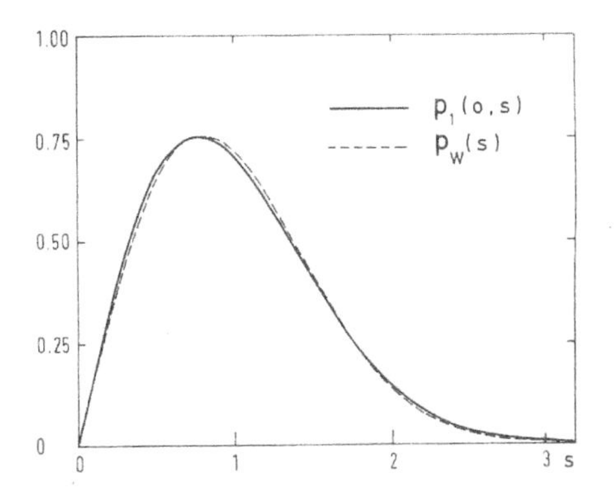

Figure 10.8 Comparison between the distributions $P(s)$ for 2×2 and $N \times N$ (with $N \to \infty$) random matrices [276].

of a set of large random matrices, we obtain that the fluctuation of number of levels in an interval containing 10^6 levels, is only of order 3 or 4. Moreover, it is important to notice that the fluctuation $\Sigma^2(E)$ behaves like $1/\beta$ for $E \gg \Delta$. When time reversal symmetry is broken, the number variance is reduced by a factor 2.

Level spacing distribution

It is rather difficult to obtain the level spacing distribution $P(s)$. Surprisingly enough, in the limit $N \to \infty$, the result (10.18) obtained for 2×2 matrices represents an excellent approximation for $N \times N$ matrices (Figure 10.8). For instance, the slope at small s equals $\pi/2$ for 2×2 matrices, whereas it is $\pi^2/6$ for $N \to \infty$. The striking agreement between these two cases is indeed remarkable. It results from the fact that $P(s)$ probes only short range correlations. The separation between neighboring levels is almost insensitive to the presence of the other levels, which makes the Wigner surmise (10.18, 10.21) quite successful.

10.5 Spectral correlations in the diffusive regime

Random matrix theory accounts for universal spectral properties, which are shared by a large class of physical systems. It does not provide any information about the characteristic energy scales of a given system. The space dimensionality also does not show up although we have seen in many instances (see for example, Chapters 4 and 7) that it plays a primary role in the physics of weakly disordered systems. Moreover, starting from a microscopic description based for instance on the Anderson Hamiltonian (2.43), we notice the difference that exists with a random matrix. The Anderson Hamiltonian matrix is almost empty, only the diagonal elements are random, while the non-diagonal terms are constant. In addition, the Thouless energy (5.35), which plays an important role in disordered metals, does not appear for random matrices. Therefore, we expect the spectral correlations in a disordered metal to deviate from the universal behavior predicted by random matrix theory. It is thus

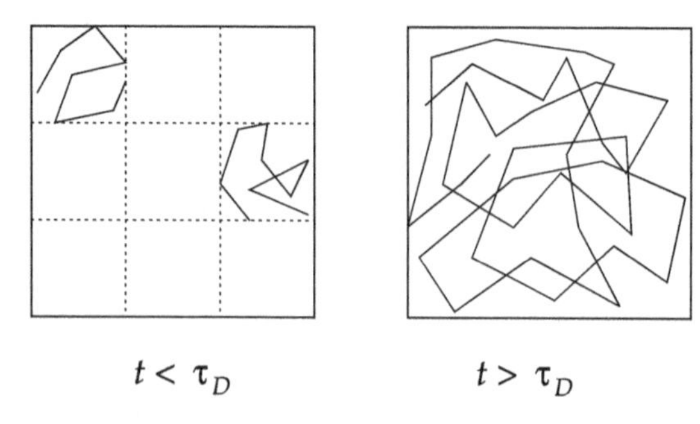

Figure 10.9 Schematic description of the diffusive regime at small times and of the ergodic regime at large times. For small times $t \ll \tau_D$, the system can be split into uncorrelated parts.

important to develop an alternative description of the spectral properties of disordered systems.

Let us first try to understand qualitatively the origin of these deviations. A random Hamiltonian is ergodic, i.e., wave functions are spread out over the entire space, and they have no internal structure. Conversely, consider in a metal the time evolution of a wave packet in the small time limit (or equivalently at large energies). In this limit, a diffusing electron cannot explore the entire volume of the system (Figure 10.9). In other words, for times t smaller than $\tau_D = L^2/D$, spatial correlations do not extend over the entire sample, but only on a scale of order $\sqrt{Dt}$. This corresponds to energies E larger than the Thouless energy, $E \gg E_c$. Spatial correlations extend up to a scale $L_E = \sqrt{D/E} \ll L$. At this energy scale, the system of dimensionality d can be viewed as being composed of $(L/L_E)^d$ independent subsystems (Figure 10.9). Therefore, the variance $\Sigma^2(E)$ is the sum of the contributions of these independent parts, namely

$$\Sigma^2(E) \sim (L/L_E)^d \sim (E/E_c)^{d/2} \, . \tag{10.44}$$

This heuristic argument indicates that spectral correlations in a disordered conductor result from the diffusive motion of electrons [277]. The description of the spectrum using random matrix theory should then be limited to energies smaller than the Thouless energy. This is seen in Figure 10.10 which shows that, beyond some energy scale, the variance $\Sigma^2(E)$ deviates from the logarithmic behavior (10.42) that is expected from random matrix theory.

In the next section, we calculate explicitly the variance for a disordered system, starting from the two-point correlation function of the density of states in the diffusive limit.

10.5.1 Two-point correlation function

From relation (3.30), we obtain for the density of states the expression

$$\rho(\epsilon) = \frac{i}{2\pi\Omega} \int d\boldsymbol{r} \, [G^R(\boldsymbol{r},\boldsymbol{r},\epsilon) - G^A(\boldsymbol{r},\boldsymbol{r},\epsilon)] \, . \tag{10.45}$$

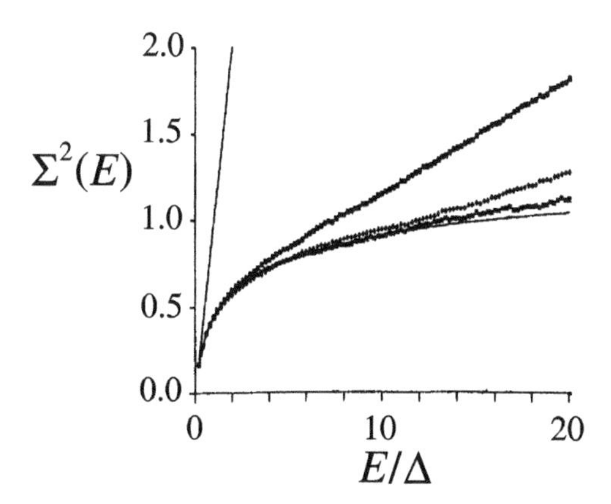

Figure 10.10 Variance of the fluctuation of the number of levels in an interval of energy E for the Anderson model and for different values of the disorder parameter W. The deviation from random matrix behavior (lower continuous line) occurs around the Thouless energy which decreases for increasing W. This underlines the important role played by this energy scale in describing spectral correlations.

The correlation function (10.4) of the density of states can thus be written in terms of the correlator $K(\boldsymbol{r}, \boldsymbol{r}', \omega)$ of the Green functions defined by (4.179) [6]

$$K(\omega) = \frac{\Delta^2}{2\pi^2} \int d\boldsymbol{r} d\boldsymbol{r}' \mathrm{Re} K(\boldsymbol{r}, \boldsymbol{r}', \omega). \tag{10.46}$$

The function $K(\boldsymbol{r}, \boldsymbol{r}', \omega)$ is long range.[7] It is the sum of two terms, one containing Diffusons (4.188) and the other Cooperons (4.190) (see Figure 4.27(d)). One obtains

$$\boxed{K(\boldsymbol{r}, \boldsymbol{r}', \omega) = P_d(\boldsymbol{r}, \boldsymbol{r}', \omega) P_d(\boldsymbol{r}', \boldsymbol{r}, \omega) + P_c(\boldsymbol{r}, \boldsymbol{r}', \omega) P_c(\boldsymbol{r}', \boldsymbol{r}, \omega)} \tag{10.47}$$

We recall that $P_d(\boldsymbol{r}, \boldsymbol{r}', \omega)$ and $P_c(\boldsymbol{r}, \boldsymbol{r}', \omega)$ take the form (5.3). Since the eigenfunctions $\psi_n(\boldsymbol{r})$ are normalized, the integration of $K(\boldsymbol{r}, \boldsymbol{r}', \omega)$ over $\boldsymbol{r}$ and $\boldsymbol{r}'$ leads to two contributions of the form [277]:

$$\frac{\Delta^2}{2\pi^2} \mathrm{Re} \sum_n \frac{1}{(-i\omega + E_n^{(d,c)})^2}, \tag{10.48}$$

where the energies $E_n^{(d,c)}$ are eigenvalues of the Diffuson and of the Cooperon. It is interesting to compare this expression with the integrated return probability $Z(t)$ that contains two terms Z_d and Z_c related to the Diffuson and to the Cooperon, given by (5.5).

[6] The contribution of the terms $G^R G^R$ and $G^A G^A$ is negligible as seen in Appendix A4.5.1 and relation (4.208). From the definition (4.179) of the correlation function $K(\omega)$, we see that P_0 does not contribute.

[7] There is another contribution to the correlation function $K(\omega)$ that arises from the function $K^{(1)}(\boldsymbol{r}, \boldsymbol{r}', \omega)$ given in Appendix A4.4. It is short range and its contribution to $K(\omega)$ is negligible (Exercise 10.5).

Their Fourier transform[8] $Z(\omega) = \int_0^\infty Z(t)e^{i\omega t}dt$ can be expressed as

$$Z_{d,c}(\omega) = \sum_n \frac{1}{-i\omega + E_n^{(d,c)}}, \tag{10.49}$$

so that $K(\omega)$ written as a function of $Z(\omega) = Z_d(\omega) + Z_c(\omega)$ becomes

$$K(\omega) = \frac{\Delta^2}{2\pi^2}\mathrm{Im}\frac{\partial}{\partial\omega}Z(\omega). \tag{10.50}$$

Fourier transforming this relation, we deduce the form factor [278]

$$\boxed{\tilde{K}(t) = \frac{\Delta^2}{4\pi^2}|\,t\,|Z(\,|t|\,)} \tag{10.51}$$

From equation (10.13), we obtain for the variance $\Sigma^2(E)$ the expression

$$\Sigma^2(E) = \frac{2}{\pi^2}\int_0^\infty dt\frac{Z(t)}{t}\sin^2\left(\frac{Et}{2}\right) \tag{10.52}$$

with $Z(t) = \int_\Omega P(\boldsymbol{r},\boldsymbol{r},t)d\boldsymbol{r}$ and $P = P_d + P_c$. The form factor $\tilde{K}(t)$ is proportional to the integrated return probability $Z(|t|)$. The origin of the multiplicative factor $|t|$ can be qualitatively understood as follows. The two trajectories in Figure 4.27(d), whose pairing contributes to the form factor, involve distinct starting points but they follow the same sequence of scattering events and the starting points can be chosen arbitrarily along this sequence. Therefore, integration over the relative position $\boldsymbol{r} - \boldsymbol{r}'$ provides a volume term proportional to the length $v_F|t|$ of the sequence, whence the multiplicative factor $|t|$ in equation (10.51). This result exhibits an interesting connection between spectral properties and the Diffuson. It has been obtained here for the Schrödinger equation but it holds for the Helmholtz scalar equation as well. A similar relation has also been established in the context of the quantum behavior of classically chaotic systems [279].

If phase coherence is not preserved [277], we must introduce a cutoff time $\tau_\gamma = 1/\gamma$ and replace $Z(t)$ by $Z(t)e^{-\gamma t}$ (see Chapter 6) so that the correlation function $K(\omega)$ is now given by the sum of the two terms:

$$\frac{\Delta^2}{2\pi^2}\mathrm{Re}\sum_n \frac{1}{(-i\omega + \gamma + E_n^{(d,c)})^2}. \tag{10.53}$$

The validity of this calculation is limited to energy scales ω larger than the level spacing Δ [280]. The relation (10.53) is thus meaningful only for $\omega \gg \Delta$ or $\gamma \gg \Delta$.

[8] We recall that $Z(t) = 0$ for $t < 0$.

Exercise 10.4: calculation of the correlation function in momentum space
In momentum space representation, the correlation function has the structure shown in Figure 10.11. Neglecting the $\boldsymbol{q}$ dependence in the Green functions, the two boxes become equal and the correlation function is

$$K(\omega) = \frac{\Delta^2}{2\pi^2\gamma_e^2}|f^{2,1}|^2 \,\mathrm{Re}\sum_{\boldsymbol{q}} \Gamma_\omega^2(\boldsymbol{q}),$$

where $f^{2,1}$, given by equations (3.107, 3.108), is the contribution of the three averaged Green functions that appear on each side of the diagrams. By replacing $f^{2,1}$ by its value (Table 3.2) and the structure factor by the probability P_d (relation 4.37), we obtain

$$K(\omega) = \frac{\Delta^2}{2\pi^2}\mathrm{Re}\sum_{\boldsymbol{q}} P_d^2(\boldsymbol{q},\omega).$$

The contribution of the Cooperon must be added to this expression. We recover equation (10.48) for the particular case where the eigenvalues are labelled by the wave vector $\boldsymbol{q}$.

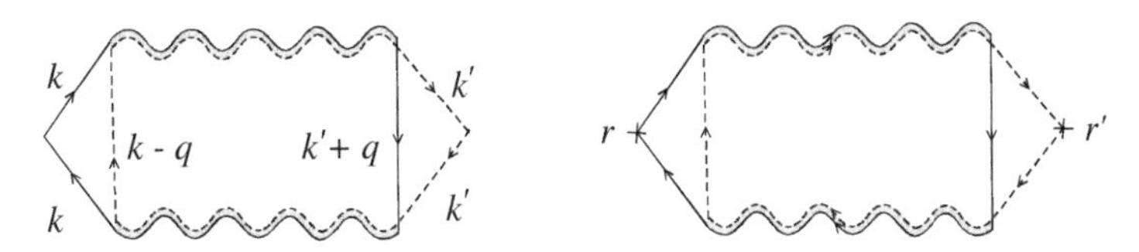

Figure 10.11 Diagrammatic representation of the correlation function K in momentum and real spaces.

Exercise 10.5 Show that, in the diffusive regime, the diagram of Figure 10.12 is negligible.
This diagram is the one-Diffuson contribution (diagram $K^{(1)}$ of Figure 4.27 (c)) to the correlation function (4.191). Using equations (10.46) and (4.191), we obtain

$$\begin{aligned} K^{(1)}(\omega) &= \frac{\Delta^2}{2\pi^2}\int d\boldsymbol{r}\, d\boldsymbol{r}'\, \mathrm{Re}\, K^{(1)}(\boldsymbol{r},\boldsymbol{r}',\omega) \\ &= \frac{\Delta^2}{\pi}\rho_0\Omega\int d\boldsymbol{R}\, g^2(\boldsymbol{R})\, \mathrm{Re}\,[P_d(\boldsymbol{r},\boldsymbol{r},\omega)+P_c(\boldsymbol{r},\boldsymbol{r},\omega)] \end{aligned} \tag{10.54}$$

with $\boldsymbol{R} = \boldsymbol{r}-\boldsymbol{r}'$. Using $\int d\boldsymbol{R}\, g^2(\boldsymbol{R}) = \tau_e/(\pi\rho_0)$ leads to

$$\begin{aligned} K^{(1)}(\omega) &= \frac{\Delta^2}{\pi^2}\tau_e\Omega\, \mathrm{Re}\,[P_d(\boldsymbol{r},\boldsymbol{r},\omega)+P_c(\boldsymbol{r},\boldsymbol{r},\omega)] \\ &= \frac{\Delta^2}{\pi^2}\tau_e\, \mathrm{Re}Z(\omega) \end{aligned} \tag{10.55}$$

where $Z(\omega) = Z_d(\omega) + Z_c(\omega)$. Then, adding the main contribution (10.50) yields

$$K(\omega) = \frac{\Delta^2}{\pi^2}\left[\frac{1}{2}\mathrm{Im}\frac{\partial Z(\omega)}{\partial\omega} + \tau_e \mathrm{Re}Z(\omega)\right]. \tag{10.56}$$

The second term is smaller that the first one in the ratio $\omega\tau_e \ll 1$. It is negligible in the diffusive limit.

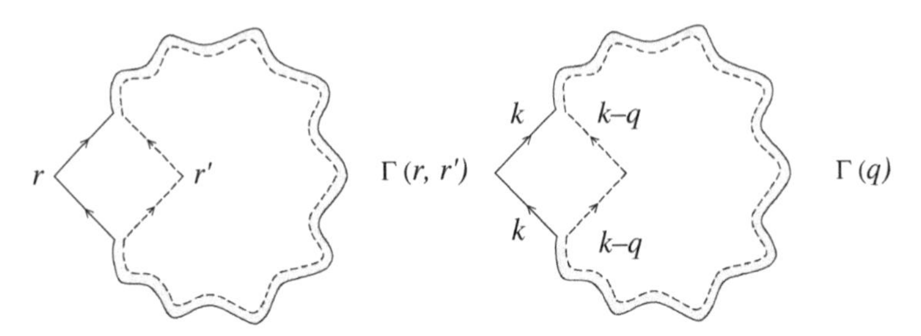

Figure 10.12 Contribution $K^{(1)}(\omega)$ to the correlation function $K(\omega)$ of the diagram with one Diffuson.

10.5.2 The ergodic limit

In the long time diffusive limit, i.e., for time scales much larger than the Thouless time $\tau_D = L^2/D$, a wave that propagates inside an isolated volume L^d explores uniformly all the available space. In this so-called ergodic limit, the return probability is independent of the starting point (section 5.5.3) and since it depends only on the zero mode contribution, it is also time independent. In the presence of a strong enough magnetic field, the Cooperon contribution vanishes and the Diffuson alone remains, so that $Z(t) = Z_d(t) = 1$. In the opposite case, i.e., in the absence of a magnetic field, the probability is doubled due to the zero mode of the Cooperon. Finally, the probability $Z(t)$ can be written in the form:

$$Z(t) = \frac{2}{\beta}, \tag{10.57}$$

where $\beta = 1$ if there is time reversal symmetry (GOE), while $\beta = 2$ in the case of broken time reversal symmetry (GUE). Therefore, using equation (10.51), for the form factor in the ergodic limit we obtain the expression

$$\tilde{K}(t) = \frac{\Delta^2}{2\pi^2\beta}\,|t|\,. \tag{10.58}$$

At intermediate times $\tau_D \ll t \ll 2\pi/\Delta$, we recover the form factor given by random matrix theory. The reduction by a factor 2 between the GOE and GUE results from the vanishing of the Cooperon when time reversal symmetry is broken. Notice that the perturbative calculation does not apply at energy scales such that $\omega \lesssim \Delta$, or equivalently times $t \gtrsim \tau_H$. It is meaningful only in the presence of a cutoff time $\tau_\gamma = 1/\gamma$ smaller than τ_H, that may result from dephasing processes. In this case, $Z(t)$ is replaced by $Z(t)e^{-\gamma t}$ so that equation (10.53) is now rewritten as

$$K(\omega) = \frac{\Delta^2}{\beta\pi^2} \mathrm{Re} \frac{1}{(-i\omega + \gamma)^2}. \tag{10.59}$$

Using equation (10.13), the variance $\Sigma^2(E)$ takes the form

$$\Sigma^2(E) = \frac{1}{\beta\pi^2} \ln\left(1 + \frac{E^2}{\gamma^2}\right). \tag{10.60}$$

Without dephasing ($\gamma = 0$), the only energy scale is the mean level spacing Δ. Expression (10.60) coincides with the result (10.42) of random matrix theory, provided we choose $\gamma \propto \Delta$.

10.5.3 Free diffusion limit

For energies larger than E_c, or equivalently for times smaller than τ_D, a wave packet experiences free diffusion and boundaries play no role.[9] The corresponding integrated return probability (5.23) depends on the space dimensionality d and $Z(t) = Z_d + Z_c = \frac{2}{\beta} Z_d$ is thus given by (5.24)

$$Z(t) = \frac{2}{\beta} \frac{\Omega}{(4\pi D t)^{d/2}}. \tag{10.61}$$

From equation (10.51), we deduce the form factor

$$\tilde{K}(t) = \frac{\Delta^2}{2\pi^2\beta} |t| \frac{\Omega}{(4\pi D|t|)^{d/2}} \propto |t|^{1-d/2} \tag{10.62}$$

represented in Figure 10.13.

Likewise, equation (10.13) together with (15.86) gives the expression for the number variance:

$$\Sigma^2(E) = \frac{c_d}{\beta} \left(\frac{E}{E_c}\right)^{d/2} \tag{10.63}$$

with $c_d^{-1} = d 2^{d-1} \pi^{1+d/2} \Gamma(d/2) \sin \pi d/4$. In particular, $c_1 = \sqrt{2}/\pi^2$, $c_2 = 1/4\pi^2$ and $c_3 = \sqrt{2}/6\pi^3$. The correlation function $K(\omega)$ is obtained from the energy derivative of the previous expression and with the help of (10.12):

$$K(\omega) = \frac{d}{4}\left(\frac{d}{2} - 1\right) \frac{\Delta^2}{\beta} c_d \frac{1}{\omega^2} \left(\frac{\omega}{E_c}\right)^{d/2}. \tag{10.64}$$

[9] We assume that $t \gg \tau_e$ which corresponds to the diffusive limit. The ballistic regime $t < \tau_e$ is described in reference [281].

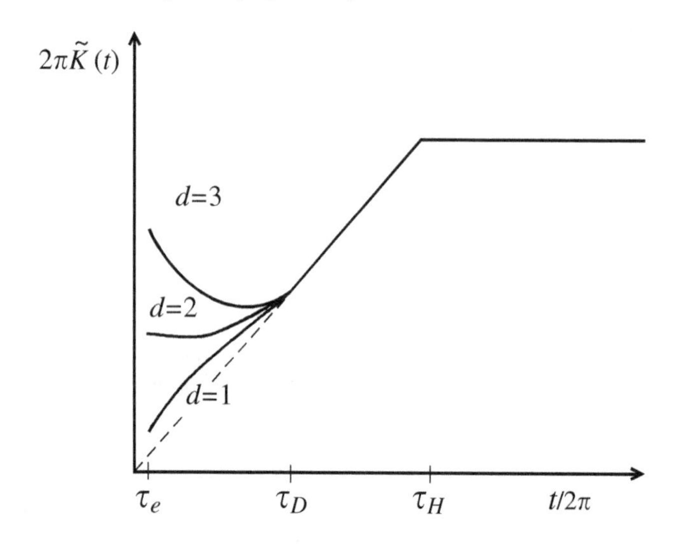

Figure 10.13 Time dependence of the form factor $\tilde{K}(t)$. Four regimes can be identified. For $t \gg \tau_H$, the form factor tends to a value close to 1 (see also Figure 10.6). For $\tau_D < t < \tau_H$, we observe a universal behavior which corresponds here to the GUE case. For $\tau_e < t < \tau_D$, the time dependence is non-universal and well described by the diffusion approximation [277]. In this limit, $\tilde{K}(t)$ behaves as $t^{1-d/2}$. The ballistic regime $t < \tau_e$ is not represented.

We notice that the sign of this correlation function depends on d and that it vanishes for $d = 2$. Moreover, in the diffusive limit the variance is much larger than its value in the ergodic limit. This loss of rigidity, which takes place beyond the Thouless energy, confirms the simple argument leading to equation (10.44) presented in the introduction of section 10.5.

The crossover between the universal ergodic (10.60) and the diffusive regimes (10.63) is well described by a discrete summation over diffusion modes. From (10.53) and (10.11), we obtain for the variance the expression

$$\Sigma^2(E) = \frac{1}{\beta\pi^2} \sum_{\boldsymbol{q}} \ln\left[1 + \frac{E^2}{(\gamma + Dq^2)^2}\right]. \tag{10.65}$$

In this relation, the Thouless energy $E_c = D/L^2$ naturally shows up from the quantization of the diffusion modes $q_i = 2\pi n_i/L$ (obtained here using periodic boundary conditions). Figure 10.14 displays the crossover between ergodic and diffusive regimes.

In summary, the various behaviors that we have identified for spectral correlations reflect the underlying classical dynamics of the electronic motion, as represented in Figure 10.15. For intermediate times $\tau_e \lesssim t \lesssim \tau_D$, or equivalently for energies $E_c \lesssim E \lesssim 1/\tau_e$, an electronic wave packet experiences a free diffusion characterized by a return probability that depends on the space dimensionality d. This leads to a power law behavior of the spectral

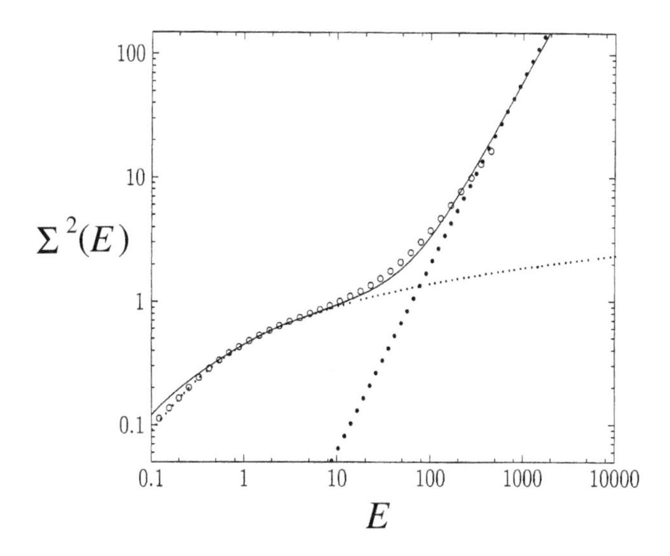

Figure 10.14 Number variance $\Sigma^2(E)$ for a weakly disordered conductor described using the Anderson model with $20 \times 20 \times 20$ sites and $E_c \simeq 2.5\Delta$. The figure displays the crossover between the universal ergodic regime described by random matrix theory (dotted line) and the diffusive behavior described by an $E^{3/2}$ power law (filled circles). Deviation from the universal regime occurs for $E \simeq E_c$. The crossover is well described using the discrete sum (10.48) over diffusion modes (continuous line).

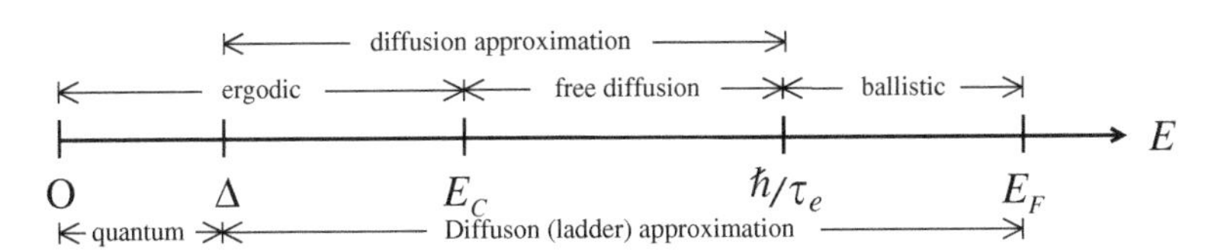

Figure 10.15 Characteristic energy scales corresponding to the different behaviors of spectral correlations.

rigidity. For long times $t \gtrsim \tau_D$ (i.e., for energies $E \lesssim E_c$), we are in the ergodic regime for which a wave packet explores all the available space uniformly. This limit corresponds to the universal behavior of the random matrix theory.

This classical behavior relies upon the Diffuson approximation and it does not account for quantum corrections. It is therefore limited to times smaller than the Heisenberg time, that is energies larger than the mean level spacing Δ. To go beyond this limit, we must resort either to random matrix theory or to the supersymmetric non-linear σ-model (see references [271–273]).

The small time limit $t \lesssim \tau_e$ corresponds to a ballistic motion of electronic wave packets for which the Diffuson approximation still works, but not the diffusion approximation [281].

The return probability becomes very small and so do the spectral fluctuations at energies $E \gtrsim 1/\tau_e$.

Appendix A10.1: The GOE-GUE transition

In this chapter, we have studied two limiting cases corresponding to the GOE ($\beta = 1$) and GUE ($\beta = 2$) symmetries. Random matrix theory also describes the transition between these two symmetries. To this purpose, we follow reference [282] and we define the following set of matrices

$$\mathcal{H} = \mathcal{H}(S) + i\alpha\mathcal{H}(A), \tag{10.66}$$

where $\mathcal{H}(S)$ and $\mathcal{H}(A)$ are symmetric and antisymmetric random matrices of size N with a Gaussian distribution of their matrix elements of variance v^2 so that $\overline{h_{ij}^2} = v^2$ for $i \neq j$ and $\overline{h_{ii}^2} = 2v^2$.[10] The value $\alpha = 0$ corresponds to the orthogonal case, whereas $\alpha = 1$ describes the unitary case. On increasing α, spectral correlations cross over between the GOE and GUE ensembles. This crossover is driven by the single parameter $\Lambda = v^2\alpha^2/\Delta^2$ where Δ is the average level spacing. For Gaussian random matrices, Δ depends on the position in the spectrum since the density of states is a semi-circle (see footnote 5, page 382). For instance, at the center of the spectrum, $\Delta = \pi v/\sqrt{N}$, so that the parameter Λ which drives the crossover is $\Lambda = N\alpha^2/\pi$. Therefore the larger the matrices, the faster the crossover. We do not provide here a closed expression of the form factor [282], but we rather mention that in the limit $t \ll \tau_H$, it is simply given by the expression

$$\tilde{K}(t) = \frac{\Delta^2}{4\pi^2}[1 + e^{-4\pi\Lambda\Delta t}] \tag{10.67}$$

which interpolates between the GOE and GUE ensembles.

A physical realization of this crossover is provided by a metallic ring pierced by an Aharonov–Bohm flux. We have shown in section 6.4.1 that the integrated return probability $Z(t)$ is given in that case by (6.61)

$$Z(t, \phi) = 1 + e^{-16\pi^2 E_c \varphi^2 t} \tag{10.68}$$

when $\tau_D \ll t \ll \tau_H$ and in the small flux limit. From equation (10.51), we infer the following expression for the form factor

$$\tilde{K}(t) = \frac{\Delta^2}{4\pi^2}[1 + e^{-16\pi^2 E_c \varphi^2 t}]. \tag{10.69}$$

[10] The variance v^2 is related to the parameter λ defined in section 10.4.2 through $v^2 = 1/(4\lambda)$.

Therefore, the effect of a magnetic flux on a conductor in the ergodic regime is well accounted for by the GOE-GUE transition of random matrices, provided we make the replacement

$$\frac{N\alpha^2}{\pi} \rightarrow 4\pi \frac{E_c \varphi^2}{\Delta}. \tag{10.70}$$

This shows that the Thouless energy which measures the correlation between energy levels is related to the size N of random matrices and that the Aharonov–Bohm flux is the time reversal symmetry breaking parameter α. The different spectral correlations are universal functions of the dimensionless parameter $(E_c/\Delta)\varphi^2$ [274, 283].

11

Universal conductance fluctuations

In this chapter, $\hbar$ is restored and we denote by s the spin degeneracy. We set the Boltzmann constant as $k_B = 1$.

11.1 Introduction

In the multiple scattering regime, the coherent effects associated with the Cooperon modify the average values of the electrical conductivity (Chapter 7) and of the albedo of a diffusive medium (Chapter 8). Similarly, correlation functions (of the density of states, of the intensity) are affected by these coherent effects. We now want to study the higher moments of the distribution of these physical quantities, starting with the conductance fluctuations in a metal.

In the elastic multiple scattering regime, the phase coherence is preserved for each disorder configuration. In optics, it leads to a speckle pattern specific to each configuration. Analysis of these speckles shows the existence of fluctuations with new and unexpected behavior due to phase coherence. They will be studied in Chapter 12. Similarly, in an electronic system, the conductance depends on the disorder configuration and constitutes a unique signature, a "*fingerprint*," of this configuration. From one sample to another, or simply by varying a physical parameter such as the magnetic field, the conductance presents fluctuations which are characteristic of this phase coherence and which are studied in this chapter.

In order to introduce the problem, let us consider a statistical ensemble of conductors of size $L \gg l_e$ and of conductance G. We denote by $\overline{G}$ the average conductance and we define the variance $\overline{\delta G^2} = \overline{G^2} - \overline{G}^2$. Assuming that the conductance is determined by the microscopic configuration of impurities at the scale of the elastic mean free path l_e, each conductor can be considered as the juxtaposition of $N = (L/l_e)^d$ independent subsystems. Therefore, one could expect that the relative fluctuations of the conductance $\sqrt{\overline{\delta G^2}}/\overline{G}$ would be of order $1/\sqrt{N}$ and vary as

$$\frac{\sqrt{\overline{\delta G^2}}}{\overline{G}} \propto \left(\frac{l_e}{L}\right)^{d/2}. \tag{11.1}$$

The average conductance $\overline{G}$ being given by Ohm's law $\overline{G} = \sigma_0 L^{d-2}$ (see 7.138), one should get for the variance

$$\overline{\delta G^2} \propto L^{d-4}. \tag{11.2}$$

The amplitude of these fluctuations should depend on the strength of disorder, related to l_e and, for $d < 4$, it should vanish in the macroscopic limit $L \to \infty$.

However, a totally different behavior is observed experimentally. Remarkably, in a conductor where the phase coherence is preserved, that is in the *mesoscopic regime* where $L < L_\phi$, the fluctuations around the average value do not depend on disorder – they are independent of the mean free path l_e – and they depend only on the geometry of the conductor: *the conductance fluctuations are said to be universal* [284–289].

This behavior is seen in Figure 11.1 which presents measurements of the conductance G as a function of the magnetic field for three samples of quite different nature whose average conductance differs by several orders of magnitude. The amplitude of the fluctuations remains the same, is independent of the disorder strength, and is of order:

$$\overline{\delta G^2} = \overline{G^2} - \overline{G}^2 \sim \left(\frac{e^2}{h}\right)^2. \tag{11.3}$$

Moreover, the aperiodic variations observed when the magnetic field is varied, as in Figure 11.1, are *reproducible*. They are the signature of the modification by the magnetic field of the interference pattern between multiple scattering trajectories. Other parameters, such as the Fermi energy, can also modify the interference pattern (section 11.4).

In order to describe these results, one needs to calculate not only the average conductance but also the distribution and more specifically the variance of this distribution $\overline{\delta G^2}$, where the

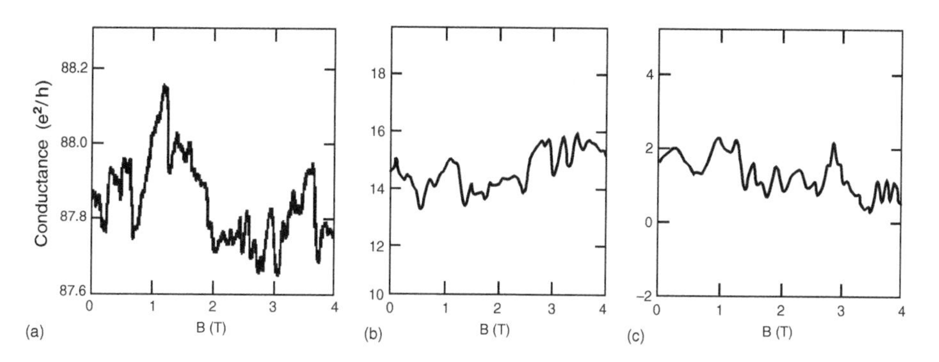

Figure 11.1 Aperiodic variations of the magnetoconductance for three different systems. (a) A gold ring, (b) a Si-MOSFET sample, and (c) the result of numerical simulations using the Anderson model. The conductance varies by several orders of magnitude from one system to another but the fluctuations stay of order e^2/h [289].

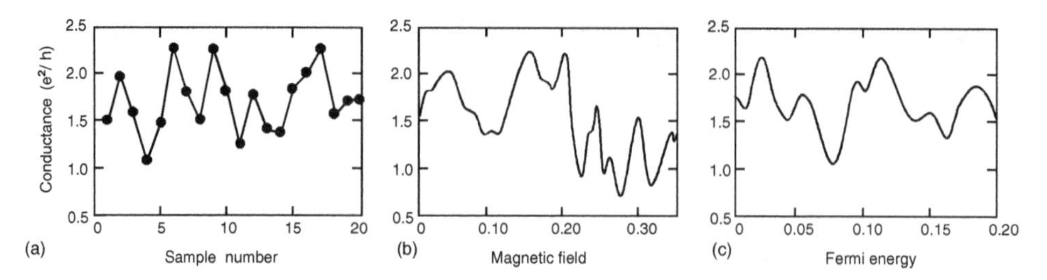

Figure 11.2 Dependence of the conductance fluctuations versus different parameters: (a) disorder configurations, (b) applied magnetic field, (c) Fermi energy. These results support the ergodic hypothesis by which a variation of the magnetic field or of the Fermi energy is equivalent to a change in the disorder configuration [289].

average is performed on the disorder configurations. The ergodic hypothesis (Figure 11.2) states that the average on disorder configurations is the same as the average obtained by varying a physical parameter like the Fermi energy or the magnetic field.

The goal of this chapter is to explain the origin of the universality of these fluctuations. This universality can be explained heuristically with the property of spectral rigidity discussed in Chapter 10 [288]. It consists in expressing the dimensionless average conductance $g = \overline{G}/(e^2/h)$ of a finite system as the ratio of the Thouless energy E_c and of the average energy level spacing Δ (relation 7.25). This ratio is the number of energy levels in a window of width E_c:

$$g \propto \frac{E_c}{\Delta} = \overline{N(E_c)}. \tag{11.4}$$

The average $\overline{\cdots}$ on the energy spectrum was defined in section 10.2. This expression concerns the average conductance. Suppose, however, that it can be generalized to the conductance as a random variable. Then its fluctuation $\overline{\delta g^2}$ is directly related to the fluctuation $\Sigma^2(E_c)$ of the number of energy levels in a window of width E_c, given by (10.9). Moreover we have seen that the phenomenon of spectral rigidity corresponds to an extremely small fluctuation (10.42) of the number of levels in a window of given energy. The variance of the conductance therefore has to be of order unity:

$$\overline{\delta g^2} \propto \Sigma^2(E_c) \propto \ln\left(\frac{E_c}{\Delta}\right) \simeq 1. \tag{11.5}$$

This argument is very qualitative and should be used with extreme caution. In fact, we shall see that the fluctuations are universal if the sample is *perfectly connected* to external leads (remark on page 412), whereas the fluctuation $\Sigma^2(E)$ of the number of levels is defined for an *isolated* system.

An alternative, and to our opinion more rigorous, way to understand the origin of the universality of the conductance fluctuations is proposed in Chapter 1, page 22.

The correlation function of two conductances involves two Diffusons with two quantum crossings, each of them giving a weight $1/g$. Therefore the correlation function involves the product of two average conductances g divided by g^2 so that $\delta g^2 \simeq 1$.

To conclude this introduction, let us notice that, in order to calculate the strength of the conductance fluctuations from the Kubo formula (7.2) for the conductivity, we shall have to evaluate terms of the form

$$\overline{G^R(\boldsymbol{r},\boldsymbol{r}')G^A(\boldsymbol{r}',\boldsymbol{r})G^R(\boldsymbol{r}'',\boldsymbol{r}''')G^A(\boldsymbol{r}''',\boldsymbol{r}'')} \tag{11.6}$$

with gradient operators which are not written explicitly here. As for the calculation of the probability (4.4), upon disorder averaging, we shall select in this product only multiple scattering processes in which the corresponding trajectories are paired. The reason for this is that unpaired trajectories have lengths differing by more than l_e and in the weak disorder limit $kl_e \gg 1$, the relative dephasing is large so their contribution cancels after averaging.

Whereas for the probability we had to pair products of two trajectories, here we have to pair off four trajectories. Such pairings lead to products of two Diffusons (or two Cooperons) connected to the points $\boldsymbol{r},\boldsymbol{r}',\boldsymbol{r}'',\boldsymbol{r}'''$ by average Green's functions. Thus one obtains diagrams made of long range contributions, the Diffusons or the Cooperons, connected to the different points $\boldsymbol{r}$ by short range functions,[1] of range l_e. Two contractions are then possible (Figure 11.3):

- $\boldsymbol{r} \simeq \boldsymbol{r}', \boldsymbol{r}'' \simeq \boldsymbol{r}'''$ (diagram 11.3(a)); this diagram has the same structure as that describing density of states fluctuations (Figure 10.11);
- $\boldsymbol{r} = \boldsymbol{r}'', \boldsymbol{r}' = \boldsymbol{r}'''$ (diagram 11.3(b)).

We shall see that the unusual behavior of the fluctuations originates in the long range behavior of the Diffusons (or of the Cooperons).

11.2 Conductivity fluctuations

Calculation of the variance of the conductivity distribution from the Kubo formula follows the same lines as for the weak localization correction or for spectral fluctuations (Chapters 7 and 10). We consider the most general case of fluctuations of the components $\sigma_{\alpha\beta}$, which are not necessarily diagonal, of the conductivity. We shall limit ourselves to the case of isotropic collisions, meaning that the elastic and transport mean free paths are equal. The case of anisotropic collisions is discussed in Appendix A11.1. We denote by $\sigma_{\alpha\beta}(\epsilon)$ the static ($\omega = 0$) real part of the conductivity at the Fermi energy ϵ. We consider the

[1] This is true only for isotropic collisions. As for the average conductivity, the absence of Diffusons directly attached to the points $\boldsymbol{r},\boldsymbol{r}',\boldsymbol{r}'',\boldsymbol{r}'''$ is due to the vertex currents. For anisotropic collisions, one must follow a procedure similar to that used for calculation of the average conductivity (see Appendices A7.4 and A11.1 and section 7.2.3), namely one must "dress" the current vertices by Diffusons. One then gets more complex diagrams with Diffusons directly attached to the points $\boldsymbol{r},\boldsymbol{r}',\boldsymbol{r}'',\boldsymbol{r}'''$, as shown in Figure 11.3(c). These diagrams do not contribute for isotropic collisions because of the current vertices.

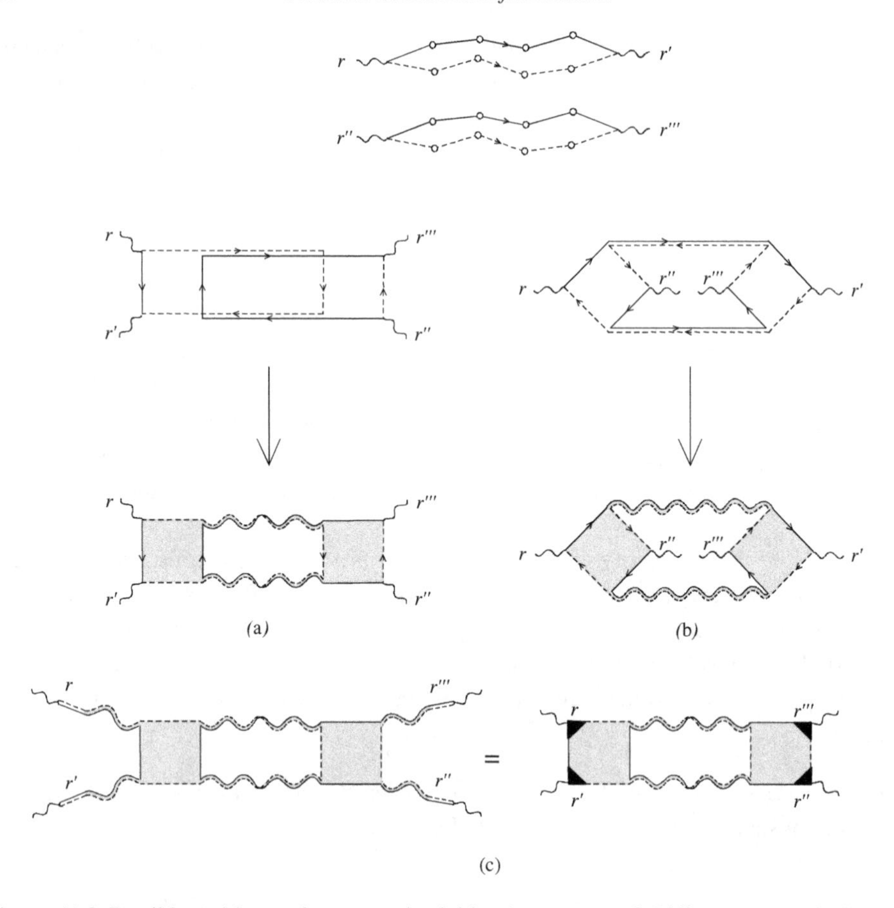

Figure 11.3 Possible pairings of two conductivities by means of Diffusons. We obtain two contributions quite different in nature. Each of them is made of two Diffusons (or Cooperons) connected by "boxes" built with short range functions. These "boxes" may have to be dressed with additional impurity lines. The symbols $\sim$ indicate the current vertices for the conductivity. Diagram (c), in which Diffusons are directly attached to the current vertices, vanishes for isotropic collisions.

correlation function $\overline{\delta\sigma_{\alpha\beta}(\epsilon)\delta\sigma_{\gamma\delta}(\epsilon')}$, where the two conductivities correspond to different Fermi energies ϵ and ϵ'.[2]

Use of the Einstein relation

The Einstein relation (7.14) relates the conductivity σ_0 to the diffusion constant D. We assume that this relation can be generalized to the random variable $\sigma_{\alpha\beta}(\epsilon)$, so that we can write

$$\sigma_{\alpha\beta}(\epsilon) = se^2 D_{\alpha\beta}\, \rho(\epsilon). \tag{11.7}$$

[2] Let us insist on the fact that the argument of the conductivity is not the frequency measurement as in definition (7.1), but the Fermi energy.

Then the product $\overline{\delta\sigma_{\alpha\beta}(\epsilon)\delta\sigma_{\gamma\delta}(\epsilon')}$, where $\delta\sigma = \sigma - \overline{\sigma}$, contains a priori two contributions related on the one hand to the density of states fluctuations and on the other hand to the fluctuations of the diffusion coefficient $D_{\alpha\beta}$, namely

$$\overline{\delta\sigma_{\alpha\beta}\delta\sigma_{\gamma\delta}} = s^2 e^4 \left(\overline{D}_{\alpha\beta}\, \overline{D}_{\gamma\delta}\, \overline{\delta\rho(\epsilon)\delta\rho(\epsilon')} + \overline{\delta D_{\alpha\beta}(\epsilon)\delta D_{\gamma\delta}(\epsilon')}\, \rho_0^2 \right) \tag{11.8}$$

with $\overline{D}_{\alpha\beta} = D\delta_{\alpha\beta}$, where D is the average diffusion constant (supposed to be isotropic) defined by relation (4.35). $\rho_0 = \overline{\rho}$ is the average density of states. We have[3]

$$\overline{\delta\sigma_{\alpha\beta}\delta\sigma_{\gamma\delta}} = \sigma_0^2 \left(\delta_{\alpha\beta}\delta_{\gamma\delta}\, \frac{\overline{\delta\rho(\epsilon)\delta\rho(\epsilon')}}{\rho_0^2} + \frac{\overline{\delta D_{\alpha\beta}(\epsilon)\delta D_{\gamma\delta}(\epsilon')}}{D^2} \right). \tag{11.9}$$

We can interpret the conductivity fluctuations as being due partly to the fluctuations of the density of states, and partly to the fluctuations of the diffusion constant. The first contribution exists only for the diagonal components of the conductivity ($\alpha = \beta, \gamma = \delta$). This separation of the conductivity fluctuations into two contributions will show up in the explicit calculation.

Calculation from the Kubo formula

Before disorder averaging, the zero frequency conductivity $\sigma_{\alpha\beta}(\epsilon)$ at a given energy ϵ is given by

$$\begin{aligned}
\sigma_{\alpha\beta}(\epsilon) &= \frac{se^2\hbar^3}{4\pi m^2\Omega} \sum_{\boldsymbol{k},\boldsymbol{k}'} k_\alpha k'_\beta \left[G^R_\epsilon(\boldsymbol{k},\boldsymbol{k}') G^A_\epsilon(\boldsymbol{k}',\boldsymbol{k}) + G^A_\epsilon(\boldsymbol{k},\boldsymbol{k}') G^R_\epsilon(\boldsymbol{k}',\boldsymbol{k}) \right] \\
&= \frac{se^2\hbar^3}{4\pi m^2\Omega} \sum_{\boldsymbol{k},\boldsymbol{k}'} (k_\alpha k'_\beta + k_\beta k'_\alpha) G^R_\epsilon(\boldsymbol{k},\boldsymbol{k}') G^A_\epsilon(\boldsymbol{k}',\boldsymbol{k}).
\end{aligned} \tag{11.10}$$

This expression generalizes equation 7.2 for the diagonal conductivity. As a result, we obtain for the correlation function $\overline{\delta\sigma_{\alpha\beta}\delta\sigma_{\gamma\delta}} = \overline{\sigma_{\alpha\beta}\sigma_{\gamma\delta}} - \overline{\sigma}_{\alpha\beta}\, \overline{\sigma}_{\gamma\delta}$:

$$\begin{aligned}
\overline{\delta\sigma_{\alpha\beta}(\epsilon)\delta\sigma_{\gamma\delta}(\epsilon')} = \left(\frac{se^2\hbar^3}{4\pi m^2\Omega} \right)^2 &\sum_{\boldsymbol{k},\boldsymbol{k}',\boldsymbol{k}'',\boldsymbol{k}'''} (k_\alpha k'_\beta + k_\beta k'_\alpha)(k''_\gamma k'''_\delta + k''_\delta k'''_\gamma) \\
&\times \overline{G^R_\epsilon(\boldsymbol{k},\boldsymbol{k}') G^A_\epsilon(\boldsymbol{k}',\boldsymbol{k}) G^R_{\epsilon'}(\boldsymbol{k}'',\boldsymbol{k}''') G^A_{\epsilon'}(\boldsymbol{k}''',\boldsymbol{k}'')}^{\,c}
\end{aligned} \tag{11.11}$$

where the average $\overline{\cdots}^{\,c}$ involves only non-factorizable diagrams (in which the product of the average values has been subtracted). These diagrams are represented in Figure 11.4 and correspond to those of Figure 11.3, but in reciprocal space. Diagram 11.4(a) describes the fluctuations of the density of states whereas 11.4(b) describes the fluctuations of the diffusion coefficient [288].

[3] We assume here that fluctuations of the diffusion coefficient and of the density of states are uncorrelated. This is shown in Exercise 11.3.

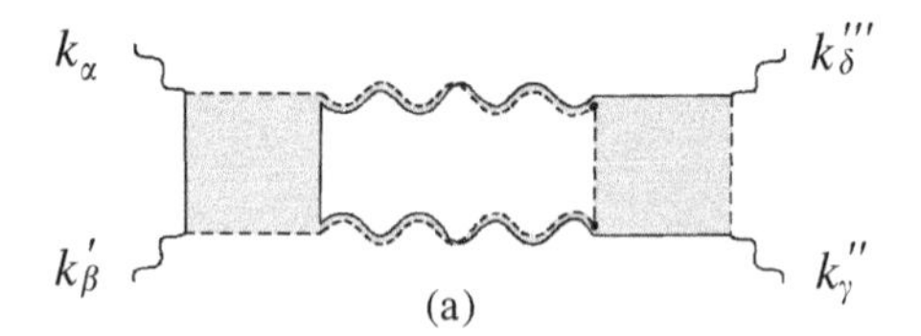

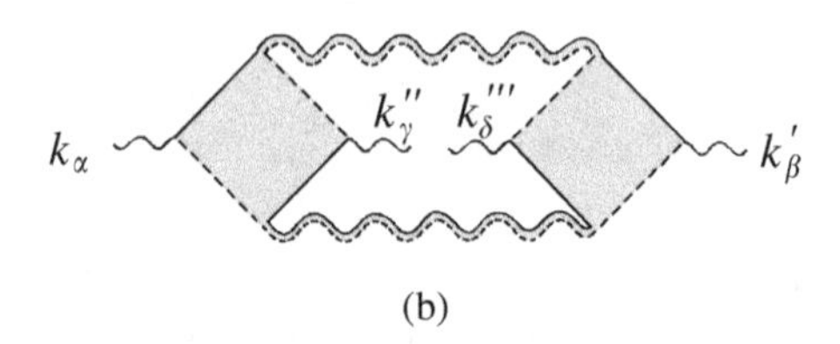

Figure 11.4 Contributions to the conductivity fluctuation $\overline{\delta\sigma_{\alpha\beta}\delta\sigma_{\gamma\delta}}$. The symbols $\sim$ represent current vertices carrying a momentum $\boldsymbol{k}_i$. There is another diagram equivalent to (a) in which the Green functions $\overline{G}^R$ (full lines) and $\overline{G}^A$ (dashed lines) are permuted. These diagrams involve two Diffusons. By exchanging k'' and k''', we construct similar diagrams with two Cooperons.

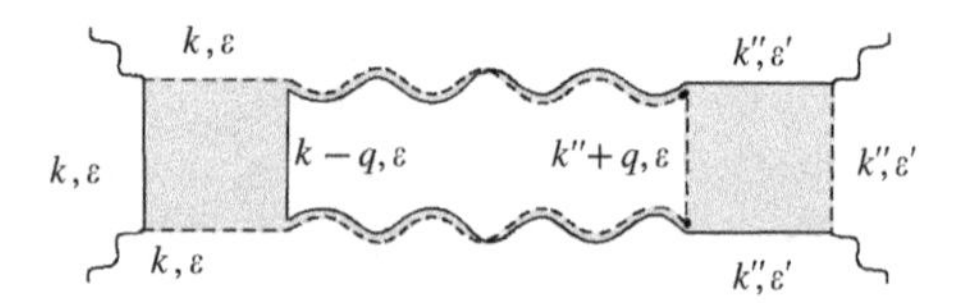

Figure 11.5 Diagram for the conductivity fluctuations, where we have represented the momentum and the energy for each Green's function.

11.2.1 Fluctuations of the density of states

Consider first the diagram 11.4(a). In the two boxes, there are Green's functions of the form $\overline{G_\epsilon(\boldsymbol{k},\boldsymbol{k}')} = \overline{G}_\epsilon(\boldsymbol{k})\delta_{\boldsymbol{k}\boldsymbol{k}'}$. Therefore, omitting for the moment the current vertices, the average in (11.11) takes the form

$$\sum_{\boldsymbol{k},\boldsymbol{k}'',\boldsymbol{q}} \overline{G}^R_\epsilon(\boldsymbol{k})\overline{G}^R_{\epsilon'}(\boldsymbol{k}-\boldsymbol{q})\overline{G}^{A\,2}_\epsilon(\boldsymbol{k})\,\Gamma^2_{\epsilon-\epsilon'}(\boldsymbol{q})\,\overline{G}^{R\,2}_{\epsilon'}(\boldsymbol{k}'')\overline{G}^A_\epsilon(\boldsymbol{k}''+\boldsymbol{q})\overline{G}^A_{\epsilon'}(\boldsymbol{k}'') \tag{11.12}$$

which corresponds to the diagram of Figure 11.5 in which we have represented the energy and the momentum for each Green's function. $\Gamma_{\epsilon-\epsilon'}$ is the structure factor of the Diffuson. In the diffusion approximation (section 4.5), the dependences in $\boldsymbol{q}$ and in the energy difference $\epsilon-\epsilon'$ can be neglected in the average Green functions.

Then expression (11.12) factorizes as a product of a long range contribution of two Diffusons, and two identical short range contributions corresponding to the boxes, so

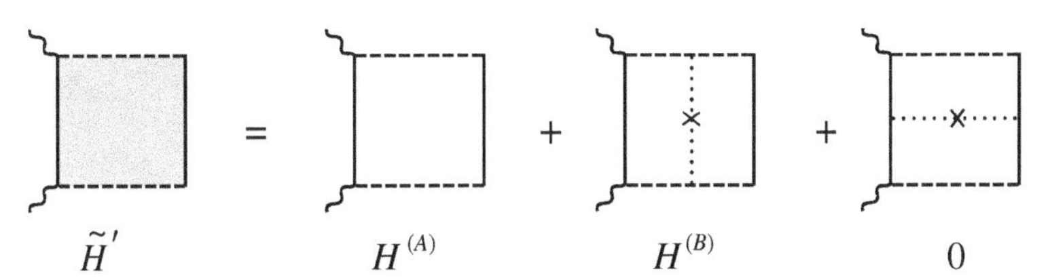

Figure 11.6 In the diagram (a) of Figure 11.4, the boxes can be dressed with one or more additional impurity lines. The third diagram cancels because the impurity line lies in between two current vertices (see Exercise 11.1).

we have

$$\sum_{q} \Gamma^2_{\epsilon-\epsilon'}(q) \left(\sum_{k} \left[\overline{G}^R_\epsilon(k) \overline{G}^A_\epsilon(k) \right]^2 \right)^2 . \tag{11.13}$$

This factorization is valid only in the diffusive regime, for which the rapid spatial variation of the average Green functions is decoupled from the slow variation of the Diffusons. Moreover, by dressing each box with an additional impurity line, we obtain a priori two more contributions of the same order (section A4.2.1) shown in Figure 11.6.

It remains to count the diagrams of the form 11.4(a). Let us first remember that the angular integration over wave vectors amounts to replacing terms of the form $k_\alpha k_\beta$ by the average $(k_F^2/d)\delta_{\alpha\beta}$ (see Chapter 7). The non-vanishing terms are those for which $\alpha = \beta$ and $\gamma = \delta$. In expression (11.11), the product of wave vectors is thus equal to $4(k_F^2/d)^2\, \delta_{\alpha\beta}\delta_{\gamma\delta}$. Moreover, the sum of the diagram 11.4(a) and of its complex conjugate gives twice its real part. Therefore the contribution to the conductivity fluctuation is

$$\overline{\delta\sigma_{\alpha\beta}(\epsilon)\delta\sigma_{\gamma\delta}(\epsilon')}^{(a)} = 8\delta_{\alpha\beta}\delta_{\gamma\delta} \left(\frac{se^2\hbar v_F^2}{4\pi d\Omega} \right)^2 \tilde{H}'^2 \sum_{q} \mathrm{Re}\left[\Gamma^2_{\epsilon-\epsilon'}(q) \right]. \tag{11.14}$$

The quantity $\tilde{H}'$ is the sum of the terms of Figure 11.6, in which only $H^{(A)}$ and $H^{(B)}$ contribute and are given in Table 3.2 and Figure 3.6:[4]

$$H^{(A)} = \frac{1}{\Omega} \sum_{k} \left[\overline{G}^R_\epsilon(k) \overline{G}^A_\epsilon(k) \right]^2 = \frac{1}{\gamma_e} f^{2,2} = \frac{4\pi \rho_0 \tau_e^3}{\hbar^3} \tag{11.15}$$

and

$$\begin{aligned} H^{(B)} &= \frac{\gamma_e}{\Omega^2} \sum_{k} \overline{G}^R_\epsilon(k) \overline{G}^{A^2}_\epsilon(k) \sum_{k'} \overline{G}^{A^2}_\epsilon(k') \overline{G}^R_\epsilon(k') \\ &= \frac{1}{\gamma_e} \left(f^{1,2} \right)^2 = -\frac{2\pi \rho_0 \tau_e^3}{\hbar^3} = -\frac{H^{(A)}}{2}. \end{aligned} \tag{11.16}$$

[4] In the calculation of these diagrams, the dependence in ϵ and ϵ' can be neglected in the limit $|\epsilon - \epsilon'|\tau_e \ll 1$. We use expressions given in Table 3.1, restoring $\hbar$.

Therefore

$$\tilde{H}' = H^{(A)} + H^{(B)} = \frac{H^{(A)}}{2} = \frac{2\pi \rho_0 \tau_e^3}{\hbar^3}. \tag{11.17}$$

Finally, by inserting (11.17) in (11.14), we obtain

$$\overline{\delta\sigma_{\alpha\beta}\delta\sigma_{\gamma\delta}}^{(a)} = 8\left(\frac{se^2 v_F^2 \rho_0 \tau_e^3}{2d\hbar^2 \Omega}\right)^2 \delta_{\alpha\beta}\delta_{\gamma\delta} \sum_{\boldsymbol{q}} \mathrm{Re}\big[\Gamma^2_{\epsilon-\epsilon'}(\boldsymbol{q})\,\big]. \tag{11.18}$$

The relation (4.87) between $\Gamma_\omega(\boldsymbol{q})$ and $P_d(\boldsymbol{q}, \omega)$,

$$\Gamma_\omega(\boldsymbol{q}) = \frac{\hbar}{2\pi \rho_0 \tau_e^2} P_d(\boldsymbol{q}, \omega), \tag{11.19}$$

leads to

$$\overline{\delta\sigma_{\alpha\beta}(\epsilon)\delta\sigma_{\gamma\delta}(\epsilon')}^{(a)} = 2\delta_{\alpha\beta}\delta_{\gamma\delta}\left(\frac{se^2 D}{h\Omega}\right)^2 \sum_{\boldsymbol{q}} \mathrm{Re}[P_d^2(\boldsymbol{q}, \omega)], \tag{11.20}$$

where $\omega = \epsilon - \epsilon'$ is the difference of the Fermi energies, and *not* the frequency at which the conductivity is measured; this frequency is zero since we consider static conductivity. This relation can be expressed simply in terms of the correlation function $K(\epsilon - \epsilon')$ of the density of states (10.46). By using the Einstein relation (7.14), $\sigma_0 = se^2 D\rho_0$, and the relation $\Delta\rho_0\Omega = 1$ (10.5), we obtain

$$\overline{\delta\sigma_{\alpha\beta}(\epsilon)\delta\sigma_{\gamma\delta}(\epsilon')}^{(a)} = \sigma_0^2\, \delta_{\alpha\beta}\delta_{\gamma\delta}\, K(\epsilon - \epsilon'). \tag{11.21}$$

This result corresponds to the first term of the relation (11.9) and justifies it a posteriori.

The diagram 11.4(a) involves two Diffusons. By permuting the variables $\boldsymbol{k}''$ and $\boldsymbol{k}'''$, one constructs a similar diagram where the Diffusons are replaced by Cooperons, with the structure factor Γ' (relation 4.43). By replacing the probability P_d by P_c,[5] we obtain

$$\boxed{\overline{\delta\sigma_{\alpha\beta}(\epsilon)\delta\sigma_{\gamma\delta}(\epsilon')}^{(a)} = 2\delta_{\alpha\beta}\delta_{\gamma\delta}\left(\frac{se^2 D}{h\Omega}\right)^2 \sum_{\boldsymbol{q}} \mathrm{Re}\Big[P_d^2(\boldsymbol{q}, \omega) + P_c^2(\boldsymbol{q}, \omega)\Big]} \tag{11.22}$$

with[6] $\omega = \epsilon - \epsilon'$.

[5] See remark on page 119 about the relation $P_c(\boldsymbol{q}, \omega) = (2\pi\rho_0\tau_e^2/\hbar)\Gamma'_\omega(\boldsymbol{q})$.

[6] To lighten the notation in the sequel of this chapter, we shall consider ω indifferently as a frequency or as an energy which amounts to taking $\hbar = 1$ in front of it. We keep $\hbar$ everywhere else.

Exercise 11.1 Show that the third diagram of Figure 11.6 cancels out.
This diagram is given by

$$\frac{1}{2\pi\rho_0\tau_e\Omega^2}\sum_{kk'} k_\alpha k'_\alpha \overline{G}^R_\epsilon(\boldsymbol{k})\overline{G}^{A^2}_\epsilon(\boldsymbol{k})\overline{G}^{A^2}_\epsilon(\boldsymbol{k}')\overline{G}^R_\epsilon(\boldsymbol{k}'). \tag{11.23}$$

By summing over wave vectors, we obtain the angular average of the product $k_\alpha k'_\alpha$ which cancels out.

11.2.2 Fluctuations of the diffusion coefficient

We consider now the contribution of diagram 11.4(b). As for diagram (a), the angular integration amounts to replacing terms of the form $k_\alpha k_\gamma$ by $(k_F^2/d)\delta_{\alpha\gamma}$. Thus, one obtains a finite contribution when $\alpha=\gamma$, $\beta=\delta$ or $\alpha=\delta$, $\beta=\gamma$. The product of wave vectors is equal to $2(k_F^2/d)^2(\delta_{\alpha\gamma}\delta_{\beta\delta}+\delta_{\alpha\delta}\delta_{\beta\gamma})$, so that the contribution of the Diffusons corresponding to 11.4(b) can be written as:

$$\overline{\delta\sigma_{\alpha\beta}(\epsilon)\delta\sigma_{\gamma\delta}(\epsilon')}^{(b)} = 2\left(\frac{se^2\hbar v_F^2}{4\pi d\Omega}\right)^2(\delta_{\alpha\gamma}\delta_{\beta\delta}+\delta_{\alpha\delta}\delta_{\beta\gamma})\tilde{H}^2\sum_{\boldsymbol{q}}|\Gamma_\omega(\boldsymbol{q})|^2. \tag{11.24}$$

The box $\tilde{H}$ is represented in Figure 11.7. The contributions associated with additional impurity lines cancel out because of the current vertices, so that $\tilde{H}=H^{(A)}$ (Exercise 11.1).

We can check that, in contrast to diagram 11.4(a), the two Diffusons are taken at opposite frequencies $\epsilon-\epsilon'$ and $\epsilon'-\epsilon$. They are thus complex conjugates, which explains the modulus $|\Gamma_\omega(\boldsymbol{q})|^2$ in relation (11.24). Therefore, contrary to (11.20), this contribution cannot be written in terms of the correlation function of the density of states. By adding the contribution of the Cooperons, we finally obtain

$$\boxed{\overline{\delta\sigma_{\alpha\beta}(\epsilon)\delta\sigma_{\gamma\delta}(\epsilon')}^{(b)} = 2(\delta_{\alpha\gamma}\delta_{\beta\delta}+\delta_{\alpha\delta}\delta_{\beta\gamma})\left(\frac{se^2D}{h\Omega}\right)^2\sum_{\boldsymbol{q}}\left(|P_d(\boldsymbol{q},\omega)|^2+|P_c(\boldsymbol{q},\omega)|^2\right)} \tag{11.25}$$

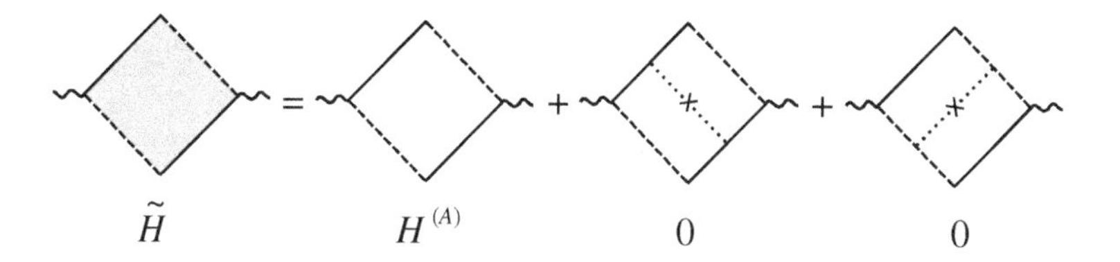

Figure 11.7 Among the three diagrams constituting the Hikami box of diagram 11.4(b), the two diagrams with an additional impurity line cancel out, because of the position of the current vertices. Because of the impurity line, the two incoming momenta are uncorrelated and their average value is zero (see Exercise 11.1).

11.3 Universal conductance fluctuations

The total conductivity fluctuation is given by the sum of expressions (11.22) and (11.25):

$$\overline{\delta\sigma_{\alpha\beta}(\epsilon)\delta\sigma_{\gamma\delta}(\epsilon')} = 2\left(\frac{se^2D}{h\Omega}\right)^2 \sum_{\boldsymbol{q}}\Big(\delta_{\alpha\beta}\delta_{\gamma\delta}\mathrm{Re}[P_d^2(\boldsymbol{q},\omega) + P_c^2(\boldsymbol{q},\omega)] + (\delta_{\alpha\gamma}\delta_{\beta\delta} + \delta_{\alpha\delta}\delta_{\beta\gamma})(|P_d(\boldsymbol{q},\omega)|^2 + |P_c(\boldsymbol{q},\omega)|^2)\Big) \tag{11.26}$$

which can be rewritten in the form

$$\overline{\delta\sigma_{\alpha\beta}(\epsilon)\delta\sigma_{\gamma\delta}(\epsilon')} = \sigma_0^2\big[\delta_{\alpha\beta}\delta_{\gamma\delta}K_\rho(\epsilon-\epsilon') + (\delta_{\alpha\gamma}\delta_{\beta\delta} + \delta_{\alpha\delta}\delta_{\beta\gamma})K_D(\epsilon-\epsilon')\big] \tag{11.27}$$

where K_ρ and K_D are respectively the relative fluctuations of the density of states (denoted K in Chapter 10) and of the diffusion coefficient:

$$\begin{aligned} K_\rho(\omega) &= \frac{\overline{\delta\rho(\epsilon)\delta\rho(\epsilon')}}{\rho_0^2} = \frac{\Delta^2}{2\pi^2}\sum_{\boldsymbol{q}} \mathrm{Re}\big[P_d(\boldsymbol{q},\omega)^2 + P_c(\boldsymbol{q},\omega)^2\big] \\ K_D(\omega) &= \frac{\overline{\delta D(\epsilon)\delta D(\epsilon')}}{D^2} = \frac{\Delta^2}{2\pi^2}\sum_{\boldsymbol{q}}\big(|P_d(\boldsymbol{q},\omega)|^2 + |P_c(\boldsymbol{q},\omega)|^2\big) \end{aligned} \tag{11.28}$$

with $\omega = \epsilon - \epsilon'$.

In order to calculate the conductance of a cubic conductor of size L in d dimensions, we use Ohm's law (7.21), $G = \sigma L^{d-2}$ which relates the conductance G to the conductivity σ. We now consider the two following situations: a system with time reversal symmetry (both Diffuson and Cooperon contributions exist) and a conductor where this symmetry is broken (for example in a magnetic field). In the latter case, the Cooperon contribution vanishes (see section 6.3). It is possible to account for both cases by writing the variance of the conductance in the form

$$\overline{\delta G_{\alpha\beta}\delta G_{\gamma\delta}} = G_\rho^2\delta_{\alpha\beta}\delta_{\gamma\delta} + G_D^2(\delta_{\alpha\gamma}\delta_{\beta\delta} + \delta_{\alpha\delta}\delta_{\beta\gamma}) \tag{11.29}$$

with

$$\begin{aligned} G_\rho^2 &= \frac{4}{\beta}\left(\frac{se^2D}{hL^2}\right)^2 \sum_{\boldsymbol{q}} \mathrm{Re}P_d^2(\boldsymbol{q}) \\ G_D^2 &= \frac{4}{\beta}\left(\frac{se^2D}{hL^2}\right)^2 \sum_{\boldsymbol{q}} |P_d(\boldsymbol{q})|^2 \end{aligned} \tag{11.30}$$

where $\beta = 1$ if the system is time reversal invariant and $\beta = 2$ if it is not. Moreover, at zero temperature, both conductances G_D and G_ρ are taken at the same Fermi energy, $\epsilon = \epsilon' = \epsilon_F$, so that $\omega = 0$ in the expressions for P_d and P_c. Taking $\alpha = \beta = \gamma = \delta$, we obtain

$$\overline{\delta G^2} = \frac{12 s^2}{\beta} \left(\frac{e^2}{h}\right)^2 \sum_{\boldsymbol{q}} \left(\frac{1}{q^2 L^2}\right)^2 . \tag{11.31}$$

This relation shows that $\overline{\delta G^2}$ does not depend on l_e, i.e., on the strength of disorder: the fluctuations are said to be *universal*. They depend only on the sample geometry. It is interesting to point out that the origin of this universality lies in the cancellation of the diffusion constants appearing in the numerator and in the denominator.[7]

Suppose now that the sample is connected to leads along the Ox direction and that it is isolated in the other directions. The leads play the role of a reservoir, and the boundary conditions in that direction correspond to absorbing walls (section 5.5.2) so that the probability $P_d(\boldsymbol{r}, \boldsymbol{r}')$ vanishes at the boundary. The diffusion modes in this direction are thus quantized as $q_i = n_i \pi / L$ where $n_i = 1, 2, 3, \ldots$. In the other directions, the boundary conditions are those of hard walls (the probability current vanishes) and the contribution of the mode $n = 0$ is added (see section 5.5.3). Therefore,

$$\overline{\delta G^2} = \frac{12 s^2}{\beta \pi^4} \left(\frac{e^2}{h}\right)^2 \sum_{n_x \neq 0, n_y, n_z} \frac{1}{(n_x^2 + n_y^2 + n_z^2)^2} . \tag{11.32}$$

For a quasi-one-dimensional sample (meaning for a one-dimensional diffusive motion), there is only one sum in expression (11.32), which equals $\pi^4/90$, so that [288]

$$\boxed{\overline{\delta G^2} = \frac{2}{15} \frac{s^2}{\beta} \left(\frac{e^2}{h}\right)^2} \tag{11.33}$$

More generally, for the geometry of a cube of size L, we obtain in d dimensions

$$\overline{\delta G_{\alpha\beta} \delta G_{\gamma\delta}} = \frac{s^2}{\beta} \left(\frac{e^2}{h}\right)^2 \frac{4}{\pi^4} b_d \left(\delta_{\alpha\beta}\delta_{\gamma\delta} + \delta_{\alpha\gamma}\delta_{\beta\delta} + \delta_{\alpha\delta}\delta_{\beta\gamma}\right) \tag{11.34}$$

where $b_1 = \pi^4/90$, $b_2 \simeq 1.51$ and $b_3 \simeq 2.52$.

One might wonder how to generalize this result to anisotropic collisions. To that purpose, we need to replace the diffusion coefficient D by D^* in the expression of the probability P_d (see Appendix A4.3). The contributions associated with the Hikami boxes remain unchanged since they correspond to short range contributions. We would therefore expect an expression

[7] The weak localization correction to the conductance appears also as the ratio of a diffusion constant divided by a diffusion pole. The diffusion constant also disappears and the weak localization correction has a priori a universal character if $L \ll L_\phi$. However, because of the divergence in the diffusion pole for $d \leq 2$, a cutoff has to be introduced which depends on l_e (see relations 7.56 and Exercise 11.9).

for $\overline{\delta G^2}$ that depends on the ratio D/D^* and is no longer universal. Obviously this cannot be correct since in relations (11.7, 11.9) only the coefficient D^* should appear if universality holds. In order to recover this universality, we need to consider additional diagrams (an example of which is shown in Figure 11.3(c), whose contribution vanishes in the isotropic case. This point is discussed further in Appendix A11.1.

In this chapter, we considered only the variance of the conductance distribution. In section 12.7, we shall discuss why this distribution is Gaussian in the limit $g \gg 1$.

Exercise 11.2 Show that, in the diffusion approximation, the fluctuations of the density of states and of the diffusion coefficient are not correlated [288], namely

$$\overline{\delta D_{\alpha\beta}\delta\rho} = 0.$$

Prove first that

$$\overline{\delta\sigma_{\alpha\beta}\delta\rho} = \frac{\overline{\sigma}_{\alpha\beta}}{\rho_0}\,\overline{\delta\rho^2},$$

and deduce the relation (11.9).

Exercise 11.3 Show that the Fourier transforms of the correlation functions $K_\rho(\omega)$ and $K_D(\omega)$ introduced in (11.28) can be written as

$$\tilde{K}_D(t) = \frac{\Delta^2}{2\beta\pi^2\hbar^2}\int_{|t|}^{\infty} Z(t')\,dt'$$

and

$$\tilde{K}_\rho(t) = \frac{\Delta^2}{2\beta\pi^2\hbar^2}\,|t|Z(|t|),$$

where $Z(t) = \int P(\boldsymbol{r},\boldsymbol{r},t)d\boldsymbol{r}$ is the integrated return probability (5.5).
Hint: notice that $\partial P(\omega)/\partial\omega = iP^2(\omega)$ (see also 10.51).
Deduce that the conductivity fluctuation (11.27) for the case $\alpha = \beta = \delta = \gamma$ can be written in the form

$$\overline{\delta\sigma^2}(\omega) = \sigma_0^2\int_{-\infty}^{\infty}[\tilde{K}_\rho(t) + 2\tilde{K}_D(t)]\,e^{i\omega t}\,dt \tag{11.35}$$

with $\omega = \epsilon - \epsilon'$. Show that

$$\overline{\delta\sigma^2}(\omega) = \frac{\sigma_0^2\Delta^2}{\beta\pi^2\hbar^2}\int_0^{\infty} t\,Z(t)\left(\cos\omega t + 2\frac{\sin\omega t}{\omega t}\right)dt, \tag{11.36}$$

so that when $\epsilon = \epsilon'$,

$$\overline{\delta\sigma^2} = \frac{3}{\beta}\frac{\sigma_0^2\Delta^2}{\pi^2\hbar^2}\int_0^{\infty} t\,Z(t)\,dt. \tag{11.37}$$

11.4 Effect of external parameters

For a given realization of disorder, the conductance depends on the relative phases of all multiple scattering trajectories. These phases can be modified by tuning an external parameter X, such as the Fermi energy or an applied magnetic field. In the following sections, we study this dependence by means of the correlation function $\overline{\delta G(X)\delta G(X')}$.

11.4.1 Energy dependence

The relative phase of two identical multiple scattering trajectories taken at different energies ϵ and ϵ' depends on the difference $\omega = \epsilon - \epsilon'$. After a time t, the phase shift between two such trajectories is $\delta\varphi = \omega t$. Since the characteristic time scale for diffusive trajectories through a sample of size L is the Thouless time $\tau_D = L^2/D$, the typical phase shift between these trajectories is of order $\delta\varphi \simeq \omega/E_c$. Correlations are thus expected to vanish when this phase shift becomes larger than unity, that is, for $\omega \simeq E_c$.

The energy dependence of the conductivity correlation $\overline{\delta\sigma(\epsilon)\delta\sigma(\epsilon-\omega)}$ is given by (11.27). The variance of the conductance, $\overline{\delta G^2}(\omega) = \overline{\delta G(\epsilon)\delta G(\epsilon-\omega)}$ is thus obtained from (11.29, 11.30) shifting the diffusion pole to finite frequency $Dq^2 \longrightarrow -i\omega + Dq^2$, so that

$$\overline{\delta G^2}(\omega) = \frac{4s^2}{\beta}\left(\frac{e^2}{h}\right)^2 \sum_q \left[\mathrm{Re}\frac{1}{(q^2L^2 - i\frac{\omega}{E_c})^2} + \frac{2}{|q^2L^2 - i\frac{\omega}{E_c}|^2}\right]. \qquad (11.38)$$

The energy dependence of this correlation function of the conductance is indeed driven by the Thouless energy E_c. At large energy ω, the sum (11.38) can be replaced by an integral, which leads to the algebraic dependence:

$$\overline{\delta G^2}(\omega) \propto \left(\frac{E_c}{\omega}\right)^{(4-d)/2}. \qquad (11.39)$$

In two-dimensional structures such as MOSFETs, it is possible to monitor the Fermi energy by applying a gate voltage. An applied voltage eV larger than E_c decorrelates the "fingerprints" corresponding to energies ϵ_F and $\epsilon_F + eV$.

11.4.2 Temperature dependence

The sensitivity of conductance fluctuations to a variation of the Fermi energy must show up in the temperature dependence of the correlations. The temperature dependence of the conductance is obtained from its energy dependence through the integral (relation 7.125):

$$G(T) = -\int \frac{\partial f}{\partial \epsilon} G(\epsilon)\, d\epsilon. \qquad (11.40)$$

The average conductance does not depend on energy and therefore it does not explicitly depend on temperature.[8] However, the variance $\overline{\delta G^2(T)}$ depends on temperature, and it is obtained from the correlation function $\overline{\delta G(\epsilon)\delta G(\epsilon')}$. It can be written in the form

$$\overline{\delta G^2}(T) = \int\int \frac{\partial f}{\partial \epsilon}\frac{\partial f}{\partial \epsilon'}\,\overline{\delta G^2}(\epsilon - \epsilon')\, d\epsilon\, d\epsilon'.$$

From (15.113), we deduce

$$\overline{\delta G^2}(T) = \int_{-\infty}^{\infty} d\left(\frac{\omega}{2T}\right) F\left(\frac{\omega}{2T}\right)\overline{\delta G^2}(\omega),$$

where $F(x) = (x \coth x - 1)/\sinh^2 x$. The two contributions to the conductance fluctuations have different temperature dependences:

$$\begin{aligned} G^2_\rho(T) &= \frac{4s^2}{\beta}\left(\frac{e^2 E_c}{h}\right)^2 \int_{-\infty}^{\infty} \frac{d\omega}{2T} F\left(\frac{\omega}{2T}\right) \sum_{\boldsymbol{q}} \mathrm{Re} P_d^2(\boldsymbol{q}, \omega) \\ G^2_D(T) &= \frac{4s^2}{\beta}\left(\frac{e^2 E_c}{h}\right)^2 \int_{-\infty}^{\infty} \frac{d\omega}{2T} F\left(\frac{\omega}{2T}\right) \sum_{\boldsymbol{q}} |P_d(\boldsymbol{q}, \omega)|^2. \end{aligned} \tag{11.41}$$

From (11.38), we see that the temperature dependence is also driven by the Thouless energy E_c. At high temperature, $T \gg E_c$, we obtain an algebraic decrease which follows directly from the energy dependence (11.39) obtained for $\omega \gg E_c$. Nevertheless, care must be paid to the fact that the thermal function $F(\omega/2T)$ introduces an upper cutoff for energies of order T. Then from relations (11.41) we have

$$\overline{\delta G^2}(T) \propto \int_{E_c}^{T} \frac{d\omega}{T}\left(\frac{E_c}{\omega}\right)^{(4-d)/2}. \tag{11.42}$$

Thus, for $d \leq 3$,

$$\overline{\delta G^2}(T) \propto \left(\frac{E_c}{T}\right)^{(4-d)/2} \tag{11.43}$$

with a logarithmic correction for $d = 2$. On the other hand, for $d = 1$, the integral depends on the lower cutoff so that

$$\overline{\delta G^2}(T) \propto \frac{E_c}{T}. \tag{11.44}$$

The significant dependence of the conductance fluctuations on space dimensionality is a consequence of the diffusive nature of the electronic transport.

[8] The average conductivity has an indirect dependence on temperature that arises from the phase coherence time $\tau_\phi(T)$ (see section 7.4.3). But this dependence is not related to thermal broadening of the energy levels.

Exercise 11.4 Show that the expression for conductivity fluctuations can be written as

$$\overline{\delta\sigma^2}(T) = \frac{2}{\beta}\sigma_0^2 \int_0^\infty R^2(t)[\tilde{K}_\rho(t) + 2\tilde{K}_D(t)]\, dt, \tag{11.45}$$

where $R(t) = \pi T t / \sinh(\pi T t)$. Hint: use relation (15.113).

11.4.3 Phase coherence and the mesoscopic regime

Universal conductance fluctuations are very sensitive to phase coherence and they diminish when coherence is reduced. This is because both Diffusons and Cooperons which contribute to $\overline{\delta G^2}$ are affected by a finite phase coherence time τ_ϕ, as for the weak localization correction (section 7.4). The sensitivity of the Cooperon to phase coherence is expected because of the relative dephasing of time-reversed trajectories. A dependence of the Diffuson on phase coherence is at first sight more surprising. Indeed, we have seen in Chapter 4 that the Diffuson is built out of pairs of trajectories with identical phases. It is thus a classical, phase independent, object. Moreover, the Diffuson is the Goldstone mode that expresses conservation of particle number (it can be obtained from Fick's law and the equation of conservation of particles). It cannot be sensitive to dephasing.

The Diffuson which appears in conductivity correlations is of a different nature: it is built by pairing multiple scattering trajectories which may correspond to *different* configurations of the external degrees of freedom. In particular, the Coulomb interaction between electrons can be described as the interaction of an electron with a fluctuating electromagnetic field (section 13.7). Since a conductance experiment requires a finite lapse, paired multiple scattering trajectories necessarily probe different configurations of the electrostatic potential (see also footnote 12, page 420). Therefore the pairing of trajectories is affected by dephasing and Coulomb interaction leads to a finite phase coherence time and a reduction of the conductance fluctuations, for both the Diffuson and the Cooperon contributions.[9]

With a finite coherence time τ_ϕ, the Cooperon *and* the Diffuson now have a shifted diffusion pole

$$P_d(\boldsymbol{q}, \omega) = P_c(\boldsymbol{q}, \omega) = \frac{1}{-i\omega + \gamma + Dq^2} \tag{11.46}$$

with $\gamma = 1/\tau_\phi$. The pole now exhibits two characteristic energy scales: the Thouless energy E_c and the phase coherence rate $1/\tau_\phi$, or correspondingly two length scales, L and $L_\phi = \sqrt{D\tau_\phi}$. The fluctuations are universal for $1/\tau_\phi \ll E_c$, that is $L_\phi \gg L$. This is the so-called *mesoscopic* regime. In the opposite limit ($L \gg L_\phi$), we deduce from (11.39) that

[9] More technically, it is possible to show that, in the presence of Coulomb interaction, the diagram of Figure 4.4(d) which corresponds to the Diffuson P_d must be modified by insertion of interaction lines. These lines connect either the Green functions G^R and G^A, or each Green's function to itself (self-energy correction, see section 13.4). Conservation of the number of particles is ensured by an identity which states that the two types of correction exactly cancel out. For Diffusons which appear in the fluctuations diagrams, the Green functions G^R and G^A cannot be connected by interaction lines *since they correspond to two copies of the sample*. Therefore the above cancellation no longer exists. This leads to a finite time τ_ϕ [289, 290].

the fluctuations decrease as

$$\overline{\delta G^2}(L, L_\phi) \propto \left(\frac{E_c}{\gamma}\right)^{(4-d)/2} \propto \left(\frac{L_\phi}{L}\right)^{4-d}. \tag{11.47}$$

In this limit, we recover the L^{d-4} dependence characteristic of classical conductors (relation 11.2).

Remarks:

Characteristic lengths

The above discussion can be used to recover simply the energy (ω) or the temperature (T) dependence of the correlations (except for $d = 1$, see Exercise 11.6). To that purpose, we introduce the characteristic lengths L_T and L_ω:

$$\begin{aligned} L_T^2 &= \frac{\hbar D}{T} \\ L_\omega^2 &= \frac{D}{\omega}. \end{aligned} \tag{11.48}$$

The relations (11.39, 11.43, 11.47) can be written in a single form

$$\overline{\delta G^2}(L, L_c) \propto \left(\frac{L_c}{L}\right)^{4-d} \tag{11.49}$$

where $L_c = L_\phi, L_\omega$ or L_T is such that $L \gg L_c$. The situation is slightly more complicated when there are several characteristic lengths. This is clearly the case at finite temperature which induces both a thermal broadening of energy levels described by the thermal length L_T and a reduction of phase coherence described by the length $L_\phi(T)$. Then, we are led to compare the lengths L_T and $L_\phi(T)$. The phase coherence length L_ϕ is limited by several processes. The phase coherence length associated with electron–electron interaction is larger than the thermal length $L_\phi^{ee} > L_T$ (see section 13.7). However, other mechanisms (electron–phonon interaction, magnetic impurities) may limit phase coherence and lead to a phase coherence length smaller than the thermal length. Several situations are described in reference [289].

Weakly coupled system

It is important to notice that the universality of fluctuations in the mesoscopic limit $L \ll L_\phi$ holds only when the sample is perfectly connected to leads which play the role of incoherent reservoirs. The corresponding boundary conditions (Dirichlet) exclude the zero mode (see section 5.5.3) in the sum (11.31). For a weakly connected sample (the contact with each reservoir being for example a tunnel barrier), the boundary conditions are of Neumann type (section 5.5.3). In the sum (11.32), we must add the non-universal contribution of the zero mode, that is, a term proportional to $(E_c/\gamma)^2 = (L_\phi/L)^4$. In this case, the variance $\overline{\delta G^2}$ is no longer universal [291].

In the following exercises, we recall that the dimensionless conductance g is defined as $g = G/(e^2/h)$. Several results obtained in these exercises are summarized in Table 11.1.

Exercise 11.5: universality of conductance fluctuations and of the weak localization

For a sample of volume L^d, write the weak localization correction Δg and the variance $\overline{\delta g^2}$ in the form

$$\begin{aligned} \Delta g &= -2s \int_0^{\tau_D} Z(t) \frac{dt}{\tau_D} \\ \overline{\delta g^2} &= 12 \frac{s^2}{\beta} \int_0^{\tau_D} Z(t) \frac{t\, dt}{\tau_D^2}. \end{aligned} \tag{11.50}$$

In the limit $L \ll L_\phi$, the upper cutoff is the time spent by a diffusing particle in the sample, that is, the Thouless time τ_D. For times smaller than τ_D, the integrated return probability is

$$Z(t) = \left(\frac{\tau_D}{4\pi t}\right)^{d/2}. \tag{11.51}$$

Deduce from this result that conductance fluctuations are universal for $d < 4$ and that the weak localization correction is universal for $d < 2$.

Exercise 11.6: conductance fluctuations in a quasi-one-dimensional wire

The result (11.44) shows that, for $L_T \ll L \ll L_\phi$, the temperature dependence of the variance $\overline{\delta g^2}$ has the form [289]

$$\overline{\delta g^2} \propto \left(\frac{L_T}{L}\right)^2, \tag{11.52}$$

where $L_T^2 = \hbar D/T$.

Similarly, (11.47) shows that for $L_\phi \ll L, L_T$, $\overline{\delta g^2}$ varies as

$$\overline{\delta g^2} \propto \left(\frac{L_\phi}{L}\right)^3. \tag{11.53}$$

From (11.38) and (11.42), show that for $L_T \ll L_\phi \ll L$, we have

$$\overline{\delta g^2} \propto \frac{L_\phi L_T^2}{L^3}. \tag{11.54}$$

Exercise 11.7: variance $\overline{\delta g^2}(L_\phi, L_T)$ for a diffusive wire

- Show that for a quasi-one-dimensional diffusive wire with finite L_ϕ, the variance (11.32) of the conductance distribution at low temperature ($L_T \gg L$) is given by the series

$$\overline{\delta g^2}(L_\phi) = \frac{12 s^2}{\pi^4 \beta} \sum_{n=1}^{\infty} \frac{1}{\left[n^2 + (L/\pi L_\phi)^2\right]^2}, \tag{11.55}$$

which can be rewritten as

$$\overline{\delta g^2}(L_\phi) = \frac{s^2}{\beta} F_3(L/L_\phi), \tag{11.56}$$

where $F_3(x) = 3(2+2x^2-2\cosh 2x + x\sinh 2x)/(2x^4\sinh^2 x)$. By expanding this function, verify the following limiting behaviors

$$\overline{\delta g^2}(L_\phi) \to \frac{2s^2}{15\beta} \qquad \text{if} \quad L \ll L_\phi \ll L_T$$

$$\overline{\delta g^2}(L_\phi) \to \frac{3s^2}{\beta}\frac{L_\phi^3}{L^3} \qquad \text{if} \quad L_\phi \ll L \ll L_T.$$

The reader should try to understand the relation between this result and the temporal dependence (12.69) of the correlation function $C^{(3)}(t)$ for speckle fluctuations.

- Show that, at high temperature ($L_T \ll L$), the variance $\overline{\delta g^2}$ depends both on L_T and on L_ϕ, and that it is written as

$$\overline{\delta g^2}(L_\phi, L_T) = \frac{4s^2}{3\beta\pi}\frac{L_T^2}{L^2}\sum_{n=1}^{\infty}\frac{1}{n^2+(L/\pi L_\phi)^2} = \frac{2\pi s^2}{3\beta}\frac{L_T^2}{L^2}\left(\frac{1}{x}\coth x - \frac{1}{x^2}\right) \tag{11.57}$$

with $x = L/L_\phi$. Hint: starting from relations (11.41), notice that the thermal function $F(\omega/2T)$ saturates to $1/3$ for $T \to \infty$ and integrate over ω. Show that the contribution related to fluctuations of density of states becomes negligible. Verify the asymptotic behaviors (11.52, 11.54):

$$\overline{\delta g^2}(L_\phi, L_T) \to \frac{2\pi s^2}{9\beta}\frac{L_T^2}{L^2} \qquad \text{if} \quad L_T \ll L \ll L_\phi$$

$$\overline{\delta g^2}(L_\phi, L_T) \to \frac{2\pi s^2}{3\beta}\frac{L_T^2 L_\phi}{L^3} \qquad \text{if} \quad L_T \ll L_\phi \ll L.$$

Exercise 11.8: conductance fluctuations in d dimensions

In the limit $L_\phi \ll L, L_T$, show that the variance of the conductance distribution is given by

$$\overline{\delta g^2} = \frac{12 s^2}{\beta}\frac{\Gamma(2-d/2)}{2^d\pi^{d/2}}\left(\frac{L_\phi}{L}\right)^{4-d}. \tag{11.58}$$

Hint: in this limit, after the substitution $q^2 \to q^2 + 1/L_\phi^2$, the sum (11.31) can be replaced by an integral. Use (15.67, 15.2) and the properties of the Γ functions (see the formulary).

Exercise 11.9: conductance fluctuations and weak localization

From (11.55) and expression (7.60) for the weak localization correction, establish the relation [292]

$$\overline{\delta g^2}(L_\phi) = -\frac{3s}{2\beta}\frac{L_\phi^3}{L^2}\frac{\partial}{\partial L_\phi}\Delta g(L_\phi) \qquad \text{if} \qquad L \ll L_T \tag{11.59}$$

where $\Delta g(L_\phi) = \Delta\sigma(L_\phi)L^{d-2}/(e^2/h)$ is the weak localization correction to the dimensionless conductance. Compare relations (7.61) and (11.56).
Show that the relation between conductance fluctuations and the weak localization correction [290]:

$$\overline{\delta g^2}(L_\phi, L_T) = -\frac{2\pi s}{3\beta}\frac{L_T^2}{L^2}\Delta g(L_\phi, L_T) \qquad \text{if} \qquad L_T \ll L, \tag{11.60}$$

holds at high temperature. $\Delta g(L_\phi, L_T)$ is the expression (7.63) for the weak localization correction, where the short distance cutoff l_e has been replaced by L_T (if $L \ll L_\phi$, replace L_ϕ by L in equation 11.60).

Table 11.1 Summary of the results obtained for the variance $\overline{\delta g^2}$ of the dimensionless conductance for a conductor of volume L^d, and dimensions $d = 1, 2, 3$. The multiplicative factor s^2/β shared by all these relations, has been set to unity here.

	$L \ll L_T, L_\phi$	$L_\phi \ll L, L_T$	$L_T \ll L \ll L_\phi$	$L_T \ll L_\phi \ll L$
1d	$\frac{2}{15}$	$3\left(\frac{L_\phi}{L}\right)^3$	$\frac{2\pi}{9}\frac{L_T^2}{L^2}$	$\frac{2\pi}{3}\frac{L_T^2 L_\phi}{L^3}$
2d	0.186...	$\frac{3}{\pi}\left(\frac{L_\phi}{L}\right)^2$	$\propto \frac{L_T^2}{L^2}\ln\frac{L}{L_T}$	$\propto \frac{L_T^2}{L^2}\ln\frac{L_\phi}{L_T}$
3d	0.310...	$\frac{3}{2\pi}\frac{L_\phi}{L}$	$\propto \frac{L_T}{L}$	$\propto \frac{L_T}{L}$

11.4.4 Magnetic field dependence

In the presence of a magnetic field, we observe aperiodic and reproducible variations of the conductance $G(B)$ such as those displayed in Figure 11.1(b). They are called "*magnetofingerprints*." Each print represents a unique signature of a given disorder realization. Figure 11.8(a) shows the prints associated with 46 different disorder configurations. In order to describe these variations of conductance, we consider the correlation function $\overline{\delta G(B)\delta G(B')}$.[10]

[10] One must be careful to avoid possible confusion in the writing of the correlation functions. We have seen that $\overline{\delta G(\epsilon)\delta G(\epsilon - \omega)}$ does not depend on the energy ϵ and it has been denoted by $\overline{\delta G^2}(\omega)$ where ω is the difference of arguments. But the correlation function $\overline{\delta G(B)\delta G(B')}$ depends on both fields B and B' and not only on their difference. We shall keep the notation $\overline{\delta G^2}(B)$ for the correlation function $\overline{\delta G(B)\delta G(B)}$.

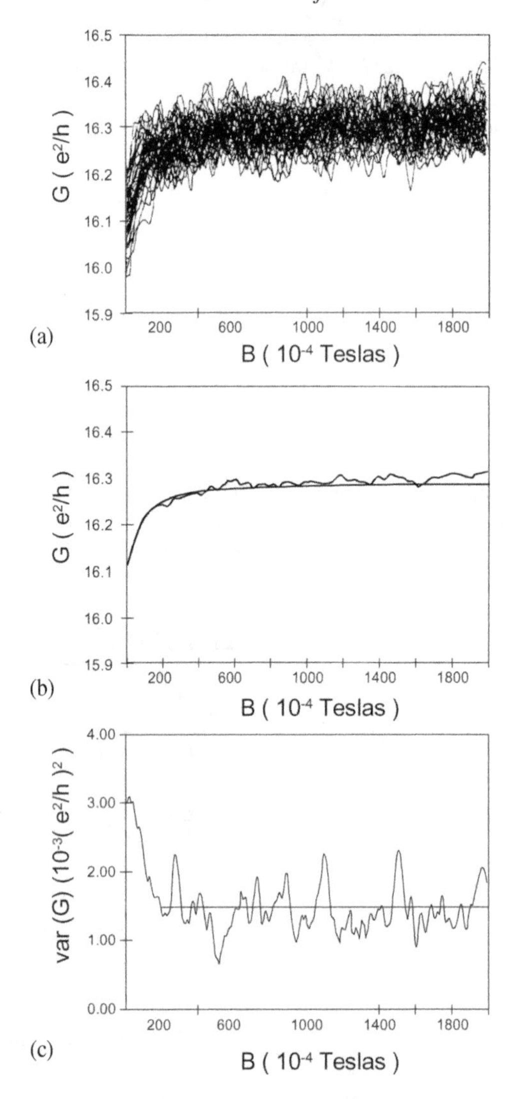

Figure 11.8 Reproducible fluctuations of the magnetoconductance in units of e^2/h, at $T = 45\,\text{mK}$ for a silicon doped GaAs sample, annealed 46 times. (a) Displays the 46 field dependences, each of which is associated with a disorder configuration ("magnetofingerprint"). (b) Represents the average conductance as a function of magnetic field. The weak localization correction vanishes above a characteristic field. Above this same field, the variance of conductance fluctuations is reduced by a factor 2. This corresponds to the vanishing of the Cooperon contribution [293].

First we aim to determine the field dependence of the variance $\overline{\delta G^2}(B) = \overline{\delta G(B)\delta G(B)}$. We have seen that the effect of a magnetic field is to reduce the contribution of the Cooperon (section 6.2). For a sufficiently large field B, we thus expect a reduction of the variance by a factor 2, since the Diffuson contribution remains unchanged:

$$\overline{\delta G^2}(B) = \frac{1}{2}\overline{\delta G^2}(B = 0). \tag{11.61}$$

More precisely, the variation $\overline{\delta G^2}(B)$ is obtained from (11.29, 11.30) and by replacing the eigenvalues Dq^2 by those of the diffusion equation for the Cooperon in a magnetic field, $E_n(B)$ (section 6.2). The sum $\sum_q$ is transformed according to

$$\sum_q \frac{1}{(\gamma + Dq^2)^2} \longrightarrow \sum_n g_n(B) \frac{1}{[\gamma + E_n(B)]^2}, \tag{11.62}$$

where $g_n(B)$ is the degeneracy associated with the eigenvalue $E_n(B)$. For instance in two dimensions, for an infinite plane, the eigenvalues are given by (6.39), namely

$$E_n = (n + 1/2)4eDB/\hbar \tag{11.63}$$

and their degeneracy is $g_n(B) = 2eB/h$. By summing over n, we obtain

$$\overline{\delta G^2}(B) = \frac{1}{2}\overline{\delta G^2}(0)\left[1 + \frac{B_\phi}{B}\Psi'\left(\frac{1}{2} + \frac{B_\phi}{B}\right)\right], \tag{11.64}$$

where the digamma function Ψ is defined by (15.43). The characteristic field $B_\phi = \hbar/(4eD\tau_\phi)$ corresponds to a flux quantum through an area $8\pi L_\phi^2$. The variance of conductance fluctuations is indeed divided by a factor 2 above a field of order B_ϕ. This same characteristic field corresponds also to the vanishing of the weak localization correction (7.80). This shows up in Figures 11.8(b,c) which display the vanishing of the weak localization correction and the reduction of the variance by a factor 2 for about the same magnetic field.[11] For a finite mesoscopic system, such that $L < L_\phi$, the Cooperon vanishes for a field which corresponds to a flux quantum through the system.

Exercise 11.10: From the relation (11.59) between weak localization correction and conductance fluctuations, recover expression (11.64) using (7.71) for the magnetoconductivity.

Exercise 11.11: Show that, for a quasi-one-dimensional wire in a perpendicular magnetic field B, the field dependence of the variance $\overline{\delta g^2}(B)$ is given by

$$\overline{\delta g^2}(B) = \frac{s^2}{2}\left(F_3\left[\frac{L}{L_\phi(0)}\right] + F_3\left[\frac{L}{L_\phi(B)}\right]\right) \tag{11.65}$$

where the function F_3 is defined in Exercise 11.7, and where the length $L_\phi(B)$ is given by (7.75).

---

[11] In fact, the experiment shown here is performed on quasi-one-dimensional wires, for which expressions (7.71) and (11.64) do not apply. For this specific geometry, see Exercise 11.11.

The magnetic field modifies the relative phase of multiple scattering trajectories. This modification is characterized by the correlation function of conductance measured at *different* values B and B' of the field. The Diffuson and the Cooperon, solutions of (6.38), are thus both affected. To obtain their behavior in a uniform field, we use the substitution

$$\frac{1}{\gamma + Dq^2} \longrightarrow \frac{1}{\gamma + (n+\frac{1}{2})\frac{2eD}{\hbar}(B \pm B')}. \tag{11.66}$$

Consequently, the correlation function $\overline{\delta G(B)\delta G(B')}$ appears as a sum of two terms

$$\overline{\delta G(B)\delta G(B')} = f\left(\frac{B-B'}{2B_\phi}\right) + f\left(\frac{B'+B}{2B_\phi}\right), \tag{11.67}$$

respectively associated with the Diffuson and the Cooperon. We have defined

$$f(x) = \overline{\delta G^2}(0)\frac{1}{2x}\Psi'\left(\frac{1}{2}+\frac{1}{x}\right). \tag{11.68}$$

The above correlation function thus depends on B and on the difference $\Delta B = B' - B$:

$$\overline{\delta G(B)\delta G(B')} = f\left(\frac{B}{B_\phi} + \frac{\Delta B}{2B_\phi}\right) + f\left(\frac{\Delta B}{2B_\phi}\right). \tag{11.69}$$

Beyond the characteristic field B_ϕ and in the limit $B \gg B_\phi$, the correlation function depends only on the difference $B - B'$ and becomes

$$\overline{\delta G(B)\delta G(B')} = f\left(\frac{B-B'}{2B_\phi}\right). \tag{11.70}$$

Therefore the characteristic correlation field is twice the field B_ϕ characterizing the decrease of the Cooperon.

11.4.5 Motion of scatterers

It may happen that between two conductance measurements, the position of scatterers has changed (section 6.7). As in diffusing wave spectroscopy (Chapter 9), measurements of the autocorrelation of the conductance at different times 0 and T may provide information on the motion of impurities. We therefore want to calculate $\langle \delta G(0)\delta G(T)\rangle$. The pairs of trajectories which constitute the Cooperons or the Diffusons involved in this function correspond now to different disorder configurations. Consequently, this function has the same structure as (11.20), with a shifted diffusion pole. From (6.14) and (6.196), we find that the diffusion pole is given by the substitution

$$\frac{1}{Dq^2} \to \frac{1}{Dq^2 + 1/\tau_\gamma}, \tag{11.71}$$

where

$$\tau_\gamma = \tau_e \frac{b(T)}{1 - b(T)} \tag{11.72}$$

and $b(T) = \langle V(0)V(T)\rangle / \langle V^2 \rangle$ is the time correlation function of the disorder potential [294]. By measuring τ_γ and thus $b(T)$, and by using (6.194, 6.193), it is possible to obtain the diffusion coefficient D_b of the impurities. The related but different problem of the effect of the motion of a single impurity has been considered in reference [295].

11.4.6 Spin-orbit coupling and magnetic impurities

Spin-orbit coupling and scattering by magnetic impurities lead to a dephasing which modifies the weak localization correction (section 7.5.2). For a strong spin-orbit coupling, the triplet contribution cancels out in relation (6.138) so that only the singlet contribution remains (Figure 6.8).

For conductance fluctuations, we have to pair four multiple scattering sequences (Figure 11.3) so as to construct Diffusons or Cooperons. The paired trajectories may have different spin states and therefore both Cooperons *and* Diffusons are modified by spin-orbit coupling or by magnetic impurities (because the paired trajectories do not belong to the same configuration of magnetic impurities, see section 6.5.3). The structure factors of the Diffuson and of the Cooperon are thus modified [296, 297] and their expression is given by (6.118) and (6.124).

Figure 11.9 shows how to modify the conductivity fluctuation diagrams in order to account for spin degrees of freedom. Each diagram becomes proportional to $\sum_{\alpha\beta,\gamma\delta} \Gamma_{\alpha\beta,\gamma\delta}\Gamma_{\gamma\delta,\alpha\beta}$. In the absence of spin-orbit coupling and magnetic impurities, the structure factor Γ is spin independent (scalar impurities) and the sum is simply equal to $4\Gamma^2$, the factor 4 being the square of the spin degeneracy $s = 2$ (relation 11.26). In the presence of these couplings and with the help of relations (6.118) and (6.124), we find that this sum becomes

$$\sum_{\alpha\beta\gamma\delta} \Gamma^2_{\alpha\beta,\gamma\delta} = 3\Gamma_T^2 + \Gamma_S^2 \tag{11.73}$$

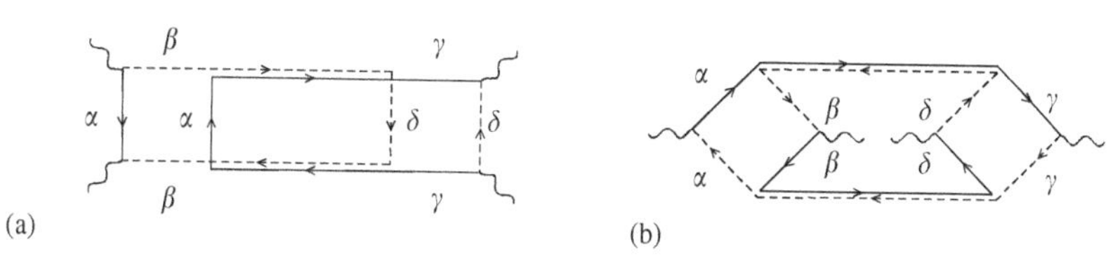

Figure 11.9 Examples of diagrams for conductance fluctuations. Spin degrees of freedom are displayed. (a) Contribution of the Diffuson to density of states fluctuations. (b) Contribution of the Cooperon to fluctuations of the diffusion coefficient. These diagrams involve the product of either two Diffusons or two Cooperons, of the form $\sum_{\alpha\beta,\gamma\delta} \Gamma_{\alpha\beta,\gamma\delta}\Gamma_{\gamma\delta,\alpha\beta}$.

either for the Cooperon or for the Diffuson. This expression displays the contribution to conductance fluctuations from each subspace, singlet (Γ_S) or triplet (Γ_T).

In the expressions for Γ_T and Γ_S, a characteristic time describes the relaxation of the magnetic impurities and depends on their nature. Therefore conductance fluctuations depend on the dynamics of their relaxation. More precisely, the correlation function $\langle \delta G(0) \delta G(T) \rangle$ of conductances measured at different times 0 and T depends on the relaxation of magnetic impurities. This relaxation is described by the function $f(T) = \langle S(0)S(T)\rangle / \langle S^2 \rangle$, which is of the form $f(T) = e^{-T/\tau_K}$, with a characteristic Korringa relaxation time τ_K [298]. By inserting in (11.73), the eigenmodes Γ_J given by (6.117) and (6.128), we obtain for the conductivity fluctuations an expression similar to (11.26), with a contribution of the Diffusons and of the Cooperons of the form [299]

$$P_d^2(\boldsymbol{q}) \longrightarrow \frac{3}{4} \frac{1}{\left(Dq^2 + \frac{4}{3\tau_{so}} + \frac{[1+f(T)/3]}{\tau_m}\right)^2} + \frac{1}{4} \frac{1}{\left(Dq^2 + \frac{[1-f(T)]}{\tau_m}\right)^2} \tag{11.74}$$

$$P_c^2(\boldsymbol{q}) \longrightarrow \frac{3}{4} \frac{1}{\left(Dq^2 + \frac{4}{3\tau_{so}} + \frac{[1-f(T)/3]}{\tau_m}\right)^2} + \frac{1}{4} \frac{1}{\left(Dq^2 + \frac{[1+f(T)]}{\tau_m}\right)^2}. \tag{11.75}$$

Two limits are of special interest.[12]

- *Frozen impurities*. The time T between the two measurements is shorter than the relaxation time τ_K of the magnetic impurities. The two measurements correspond to the same configuration of impurities. In this case,

$$T \ll \tau_K: \quad \begin{aligned} P_d^2(\boldsymbol{q}) &\longrightarrow \frac{3}{4} \frac{1}{\left(Dq^2 + \frac{4}{3\tau_{so}} + \frac{4}{3\tau_m}\right)^2} + \frac{1}{4} \frac{1}{(Dq^2)^2} \\ P_c^2(\boldsymbol{q}) &\longrightarrow \frac{3}{4} \frac{1}{\left(Dq^2 + \frac{4}{3\tau_{so}} + \frac{2}{3\tau_m}\right)^2} + \frac{1}{4} \frac{1}{\left(Dq^2 + \frac{2}{\tau_m}\right)^2}. \end{aligned} \tag{11.78}$$

[12] It must be stressed that a conductance experiment necessarily requires a finite duration T, so that the experimentalist measures

$$G = \frac{1}{T} \int_0^T G(t)\, dt. \tag{11.76}$$

Consequently, the measured static conductance fluctuation involves the time correlation function of the conductance, namely,

$$\overline{\delta G^2} = \frac{1}{T^2} \int_0^T \int_0^T \overline{\delta G(t) \delta G(t')}\, dt\, dt'. \tag{11.77}$$

The time scale T of the experiment has to be compared with the characteristic relaxation times of each external degree of freedom, and the variance $\overline{\delta G^2}$ of the *measured* conductance corresponds, in the limiting cases, either to $\overline{\delta G^2(0)}$ or to $\lim_{T \to \infty} \overline{\delta G(0) \delta G(T)}$.

- *Free impurities*. During the time T between two measurements, magnetic impurities have relaxed and they are no longer correlated. In this case, $f(\infty) = 0$ so that

$$T \gg \tau_K : \qquad P^2_{d,c}(\boldsymbol{q}) \longrightarrow \frac{3}{4}\frac{1}{\left(Dq^2 + \frac{4}{3\tau_{so}} + \frac{1}{\tau_m}\right)^2} + \frac{1}{4}\frac{1}{\left(Dq^2 + \frac{1}{\tau_m}\right)^2}. \tag{11.79}$$

In this latter case, there is no correlation between impurity spins corresponding to times 0 and T. There is no magnetic impurity line connecting Green's functions (section 6.5.3).

To summarize, *mesoscopic conductance fluctuations are only reduced by frozen impurities whereas they are suppressed by free impurities*. In the first case, the singlet channel in the Diffuson is affected neither by magnetic impurities nor by spin-orbit coupling. In the second case, the diffusion pole in this channel is removed. Depending on the relative values of the different parameters which characterize the strength of magnetic impurities and spin-orbit coupling, the variance of the conductance distribution is reduced by simple ratios (we can also consider the effect of a magnetic field which, by breaking time reversal symmetry, suppresses the contribution of the Cooperon). For example, a strong magnetic field reduces the variance by a factor 2. In the absence of magnetic impurities, spin-orbit coupling[13] reduces the variance by a factor 4. Table 11.2 summarizes different possible situations. For a more general discussion of these results, and in particular for their interpretation in terms of symmetry classes, one may consult Table I of [288].

Table 11.2 Dependence of the amplitude of the conductance fluctuations on the strength of the spin-orbit coupling and of the coupling to magnetic impurities. For each case, the two figures of the first line are the respective amplitudes of the Diffuson and of the Cooperon, taking the (GOE) case as unity. The second line gives the reduction of δG^2, taking the (GOE) case as unity. For a mesoscopic sample, the external field B and the characteristic fields B_{so} defined by (7.80) and B_m defined by (7.83) must be compared to the field B_c corresponding to a flux quantum through the sample.

	$B, B_m \ll B_c$	$B_m \ll B_c \ll B$	$B_c \ll B_m, B$	
			$T \ll \tau_K$	$T \gg \tau_K$
$B_{so} \ll B_c$	1 1	1 0	1/4 0	0 0
	1	**1/2**	**1/8**	**0**
$B_{so} \gg B_c$	1/4 1/4	1/4 0	1/4 0	0 0
	1/4	**1/8**	**1/8**	**0**

[13] For the same reason, spectral fluctuations (see for instance relation 10.53) are also divided by a factor 4. In the framework of random matrix theory, this corresponds to the symplectic ensemble (GSE) described by the parameter $\beta = 4$.

Exercise 11.12: conductance fluctuations and Zeeman effect

In the presence of a magnetic field, not only is the contribution of the Cooperon suppressed, but the Zeeman effect also breaks the degeneracy of triplet states. In strong field it leads to the vanishing of the contributions of two among the three triplet states. Calculate in this case the amplitude of conductance fluctuations [288].

Appendix A11.1: Universal conductance fluctuations and anisotropic collisions

We have discussed the universality of conductance fluctuations for the case of anisotropic collisions (page 407). Here we consider this point in greater detail in order to show how universality still holds in this case (see also Appendix A7.4 devoted to the weak localization correction for anisotropic collisions).

First we notice that, for anisotropic collisions, we must use the structure factor Γ^* given by (4.169) rather than Γ in the diagrams of Figure 11.4, so we must perform the substitution

$$\Gamma \propto \frac{1}{-i\omega + Dq^2} \quad \longrightarrow \quad \Gamma^* \propto \frac{1}{-i\omega + D^* q^2}. \tag{11.80}$$

Another dependence on D originates from Hikami boxes which enter in these diagrams (Figures 11.6 and 11.7). What changes for anisotropic collisions?

For fluctuations related to the diffusion constant, the contribution of each box (Figure 11.7) becomes $\tilde{H} = \tilde{H}^{(A)} + \tilde{H}^{(B)} + \tilde{H}^{(C)}$ with (Figure 11.10(a))

$$\begin{aligned} \tilde{H}^{(A)} &= \left(\frac{\tau^*}{\tau_e}\right)^2 H^{(A)} \\ \tilde{H}^{(B)} = \tilde{H}^{(C)} &= -\frac{1}{2}\frac{\gamma_1}{\gamma_e}\left(\frac{\tau^*}{\tau_e}\right)^2 H^{(A)}, \end{aligned} \tag{11.81}$$

where γ_1 is given by (4.157) and τ^* by (4.167). These terms were calculated in Appendix A7.4 and their sum is

$$\tilde{H} = \frac{\tau^*}{\tau_e} H^{(A)} \tag{11.82}$$

instead of $H^{(A)}$ for the isotropic case. Therefore, the transport time τ^* also appears in the contribution of Hikami boxes.

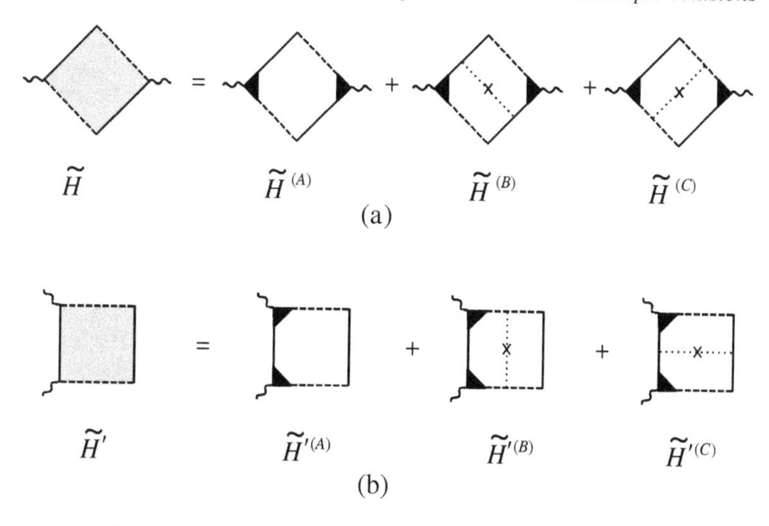

Figure 11.10 Hikami boxes appearing in the calculation of conductance fluctuations, for anisotropic collisions. (a) Fluctuations of the diffusion constant. (b) Fluctuations of the density of states.

For density of states fluctuations, the contribution of each box (Figure 11.6) becomes $\tilde{H}' = \tilde{H}'^{(A)} + \tilde{H}'^{(B)} + \tilde{H}'^{(C)}$ with (Figure 11.10(b)):

$$\begin{aligned}
\tilde{H}'^{(A)} &= \left(\frac{\tau^*}{\tau_e}\right)^2 H^{(A)} \\
\tilde{H}'^{(B)} &= -\frac{1}{2}\left(\frac{\tau^*}{\tau_e}\right)^2 H^{(A)} \\
\tilde{H}'^{(C)} &= -\frac{1}{2}\frac{\gamma_1}{\gamma_e}\left(\frac{\tau^*}{\tau_e}\right)^2 H^{(A)}.
\end{aligned} \tag{11.83}$$

The sum of these terms is

$$\tilde{H}' = \frac{\tau^*}{\tau_e}\frac{H^{(A)}}{2} \tag{11.84}$$

instead of $H^{(A)}/2$ for the isotropic case (Figure 11.11).

We conclude that the contribution of each box is multiplied by $\tau^*/\tau_e = D^*/D$, so that the variance $\delta\sigma^2$ of conductivity fluctuations becomes

$$\delta\sigma^2 \propto D^{*2}\sum_{q}\frac{1}{(D^*q^2)^2}. \tag{11.85}$$

Therefore the fluctuations stay universal, namely disorder independent.

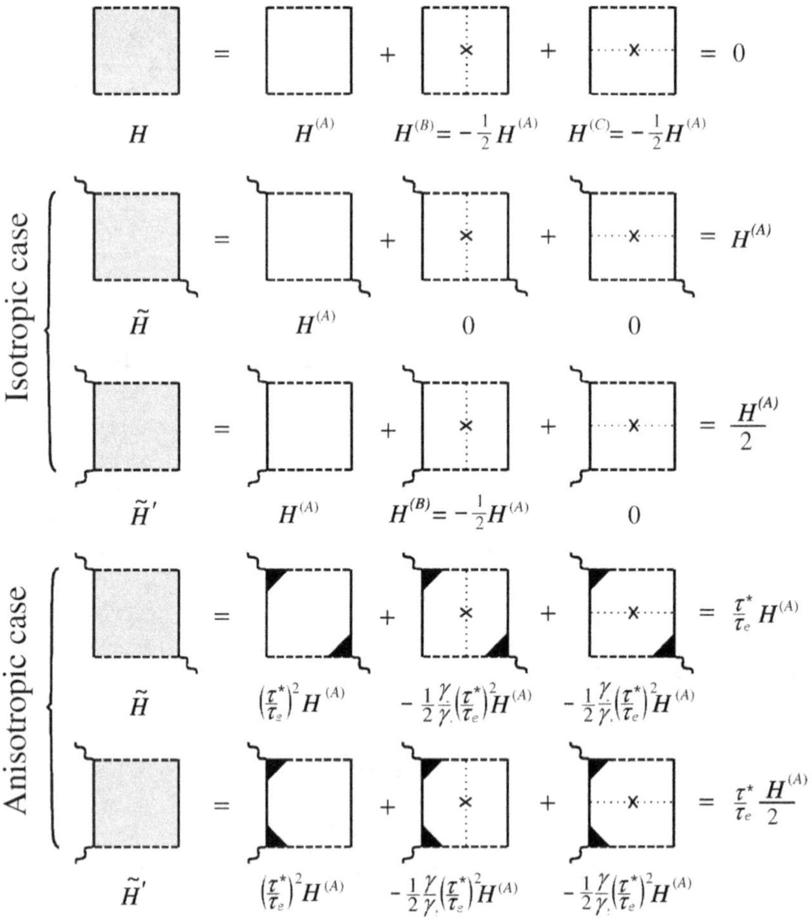

Figure 11.11 This figure summarizes the different combinations of Hikami boxes calculated in this chapter and in Chapter 7.

Appendix A11.2: Conductance fluctuations in the Landauer formalism

It is possible to calculate the weak localization correction using the Landauer approach instead of the Kubo formalism (section A7.2.5). Similarly, for a quasi-one-dimensional wire, one can recover the universal conductance fluctuations (11.33) using the Landauer formalism. The derivation is simple for this geometry and use of the Landauer formalism becomes essential for more complex geometries.

This approach is described in great detail in Chapter 12 for the study of speckle fluctuations. It is interesting that in optics it is possible to measure directly the correlations between different transverse channels. In this appendix, we simply sketch the Landauer approach for the calculation of conductance fluctuations.

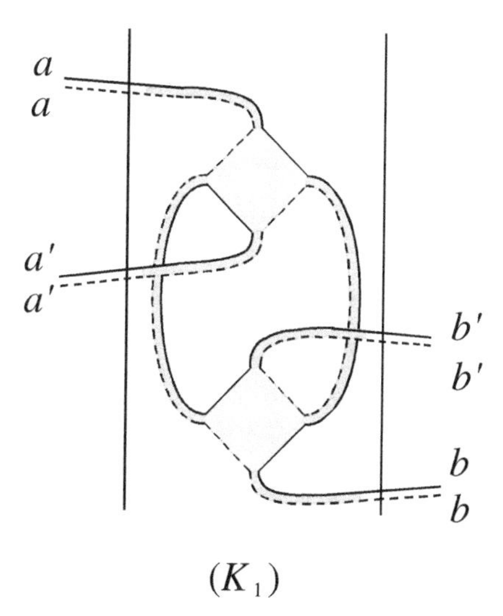

Figure 11.12 The correlation function of the transmission coefficient $\overline{\delta T_{ab}\delta T_{a'b'}}^{(3)}$ involves the correlation of two Diffusons as the result of their crossings. In Chapter 12 we show that diagrams with a single crossing give a negligible contribution upon summation over transverse modes.

By means of the Landauer formula (7.153), we relate the variance of conductance fluctuations to the correlation functions of transmission coefficients T_{ab}:

$$\overline{\delta G^2} = s^2 \frac{e^4}{\hbar^2} \sum_{aa'bb'} \overline{\delta T_{ab}\delta T_{a'b'}}. \tag{11.86}$$

Each correlation $\overline{\delta T_{ab}\delta T_{a'b'}}$ involves crossings of two Diffusons. A systematic expansion of $\overline{\delta T_{ab}\delta T_{a'b'}}$ in terms of the number of crossings based on the ideas of section A4.2.3 is presented in Chapter 12. Here we do not detail the calculation; we simply mention that the no-crossing and one-crossing terms vanish upon summation over the modes (they are actually negligible, see section 12.4.4), so that the remaining contribution contains two crossings, as represented in Figure 11.12. This term has the same structure as the diagram 11.4(b), which corresponds to conductivity fluctuations in the Kubo formalism.

A calculation analogous to the one leading to expression (7.185) for the weak localization correction to the average transmission coefficient $\overline{T}_{ab}$ leads us to write diagram 11.12 in the form

$$\overline{\delta T_{ab}\delta T_{a'b'}} = 4\frac{\tau_e^4}{S^4} h_4^2 \int \partial_x \Gamma(l_e\mu_a, x)\, \partial_x \Gamma(l_e\mu_{a'}, x)\, \Gamma(x, x')^2$$
$$\times\, \partial_{x'} \Gamma(x', L - l_e\mu_b)\, \partial_{x'} \Gamma(x', L - l_e\mu_{b'})\, dx\, dx' \tag{11.87}$$

where h_4 is given by (4.127). In the geometry of a quasi-one-dimensional conductor, the structure factor $\Gamma(\boldsymbol{\rho}, x, x')$ depends on the difference $\boldsymbol{\rho}$ of transverse coordinates and on positions x and x' along the conductor. We denote by $\Gamma(x,x')$ the Fourier transform $\Gamma(\boldsymbol{q}_\perp,x,x')$ for $\boldsymbol{q}_\perp = 0$. The projections μ_i are defined by (7.160). Using relation (4.37) for Γ and the probability P_d, we obtain

$$\overline{\delta T_{ab}\delta T_{a'b'}} = \frac{4D^2}{(2\pi\rho_0 S)^4}\int \partial_x P_d(l_e\mu_a,x)\partial_x P_d(l_e\mu_{a'},x)P_d(x,x')^2 \times \partial_x P_d(x',L-l_e\mu_b)\partial_x P_d(x',L-l_e\mu_{b'})\,dx\,dx'. \tag{11.88}$$

The derivatives yield constant terms given in (7.187). Introducing the number M of channels given by (7.173), $M = \pi\rho_0 v_F S/2$, we deduce

$$\overline{\delta T_{ab}\delta T_{a'b'}} = \frac{27D^2}{64M^2L^4}\prod_i\left(\mu_i+\frac{2}{3}\right)\int P_d(x,x')^2\,dx\,dx'. \tag{11.89}$$

After summation over all channels and using (7.167, 7.173), we obtain

$$\sum_{q_a}\left(\mu_a+\frac{2}{3}\right) = \frac{2\pi\rho_0 v_F S}{3} = \frac{4M}{3} \tag{11.90}$$

and

$$\overline{\delta T^2} = 4\frac{D^2}{L^4}\int P_d(x,x')^2\,dx\,dx' \tag{11.91}$$

so that, from (5.58)

$$\overline{\delta T^2} = \frac{4}{L^4}\int_0^L x_m^2\left(1-\frac{x_M}{L}\right)^2 dx\,dx' = \frac{2}{45} \tag{11.92}$$

where $x_m = \min(x,x')$ and $x_M = \max(x,x')$. By adding the other contributions represented in Figure 12.11, we obtain $\overline{\delta T^2} = 2/15$, which is the result (11.33) obtained in the Kubo formalism. For completeness, we add here the contributions of the no-crossing and one-crossing diagrams that have been obtained in section 12.4.4:

$$\overline{\delta T^2} = \frac{g^2}{M^2}+\frac{4g}{3M}+\frac{2}{15}. \tag{11.93}$$

Here g is the spinless conductance. These contributions are negligible in the diffusive regime since $g/M \propto l_e/L \ll 1$.

12
Correlations of speckle patterns

12.1 What is a speckle pattern?

In the course of Chapters 8 and 9, we studied the albedo and the time correlation function of the intensity either reflected or transmitted by a diffusive medium. We then limited ourselves to the study of the average intensity. For moving scatterers, this average is easy to obtain in practice by time averaging. We could imagine another set of experiments in which we have either an incident pulse which is very narrow in time or immobile scatterers. In such situations, we obtain a snapshot of the scattering medium or a so-called *speckle pattern* (Figure 12.1) [300]. The main characteristic of this pattern is its "granularity," i.e., a random distribution of dark and bright spots. The appearance of dark regions means that the intensity vanishes at these points implying relative intensity fluctuations of order unity. This observation constitutes the well-known *Rayleigh law*, which expresses the fact that intensity fluctuations are comparable to the average intensity.

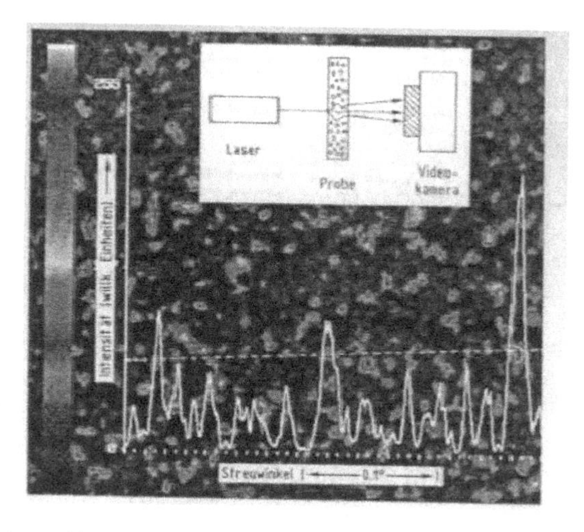

Figure 12.1 A typical speckle pattern. The white and noisy curve represents the angular dependence of the light intensity along the cut represented by the dashed line. The relative fluctuations are of order unity (courtesy of G. Maret).

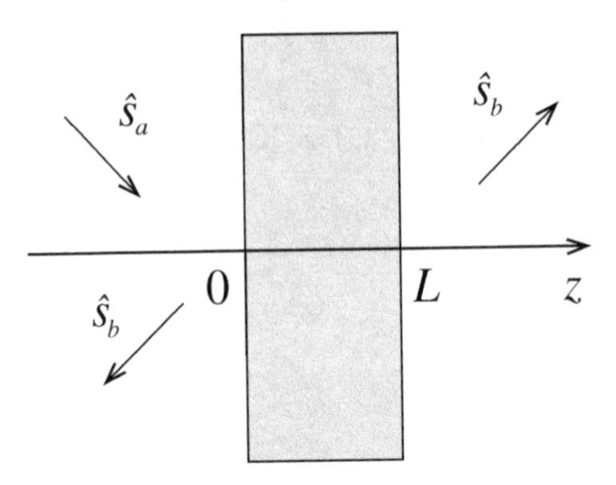

Figure 12.2 Geometric setup used for the definition of reflection and transmission coefficients through a slab of width L.

In contrast to electronic systems (Chapter 11), the setup for electromagnetic waves permits a broader range of measurable physical quantities such as the angular correlation function of the transmission (or the reflection) coefficient between two distinct directions of the transmitted (or reflected) wave. Such measurements cannot be achieved with electronic conductance, which is related to the total transmission through the conductor (see Appendix A7.2), i.e., a sum over all transmitted modes (relation 7.153). Thus a conductance experiment cannot analyze angular correlations between transmission channels.

We shall see that a phenomenon equivalent to the universal conductance fluctuations also exists in optics, although it is not very easy to observe. It is revealed in the transmission coefficient when the angular average over all the incident and emergent channels is considered. The fluctuations of this angular average coefficient are very small relative to the very large fluctuations of the light intensity in speckle patterns. The relationship between speckle and conductance fluctuations is discussed in section 12.4.4.

12.2 How to analyze a speckle pattern

Coherent effects that show up in speckle patterns are best studied using the correlation functions of the transmission and reflection coefficients for the geometry of a slab. For this purpose consider the setup of Figure 12.2. A light source placed at infinity outside the scattering medium emits a plane wave of incident direction $\hat{\boldsymbol{s}}_a$ onto the interface plane $z = 0$. After this wave propagates through the slab, it emerges in either reflection or transmission as a spherical wave[1] which is analyzed by a detector placed at infinity, i.e., at a position defined by the direction $\hat{\boldsymbol{s}}_b$. In both cases (reflection or transmission), the measured quantity is the outgoing energy flux per unit time and per unit solid angle, so that the reflection

[1] Practically speaking, this means that the detector is placed at a distance much larger than any of the characteristic dimensions of the slab.

coefficient is nothing but the albedo defined in (8.3) and is proportional to the reflected intensity along the direction $\hat{s}_b$. The transmission coefficient, denoted by $\mathcal{T}_{ab}$, is defined as the intensity transmitted along the direction $\hat{s}_b$ for an incident wave along $\hat{s}_a$. The definition is the same as for the albedo, namely,

$$\boxed{\mathcal{T}_{ab} = \frac{R^2}{S}\frac{I(R\hat{s}_b, \hat{s}_a)}{I_0}} \tag{12.1}$$

The outgoing intensity is equal to

$$I(R\hat{s}_b, \hat{s}_a) = \frac{4\pi}{c}|\psi_{ab}|^2. \tag{12.2}$$

The amplitude ψ_{ab} of the outgoing wave takes the form (8.4),[2] namely,

$$\psi_{ab} \propto \int d\mathbf{r}\, d\mathbf{r}'\, e^{ik(\hat{s}_a\cdot\mathbf{r}-\hat{s}_b\cdot\mathbf{r}')}\, G(\mathbf{r},\mathbf{r}'), \tag{12.3}$$

where $G(\mathbf{r},\mathbf{r}')$ is the Green function that describes the propagation in the scattering medium.

Our purpose in this chapter is to provide a complete description of the fluctuations of the transmission coefficient around its average value $\overline{\mathcal{T}}_{ab}$. Section 12.3 is devoted to the calculation of this average value using the methods already encountered in Chapters 8 and 9. Fluctuations of $\mathcal{T}_{ab}$ are characterized by the behavior of the angular correlation function $\overline{\delta\mathcal{T}_{ab}\delta\mathcal{T}_{a'b'}}$, or by the ratio

$$C_{aba'b'} = \frac{\overline{\delta\mathcal{T}_{ab}\delta\mathcal{T}_{a'b'}}}{\overline{\mathcal{T}}_{ab}\overline{\mathcal{T}}_{a'b'}}, \tag{12.4}$$

where $\delta\mathcal{T}_{ab} = \mathcal{T}_{ab} - \overline{\mathcal{T}}_{ab}$. We shall consider more specifically the function $C_{abab'}$, which describes the correlations of a speckle pattern which is the image of a single incident beam along the direction $\hat{s}_a$. The function (12.4) provides additional information about the behavior of the speckle pattern while changing the incident direction ($a \to a'$).

We also define the transmission coefficients obtained by integrating over all the emergent directions b as well as over all the incident directions a:[3]

$$\begin{aligned} \mathcal{T}_a &= \int \mathcal{T}_{ab}\, db \\ \mathcal{T} &= \int \mathcal{T}_a\, da = \int \mathcal{T}_{ab}\, da\, db. \end{aligned} \tag{12.6}$$

[2] The wave frequency ω_0 will no longer be explicitly mentioned.

[3] Here a and b define angular directions and are thus continuous variables. In a wave guide geometry, these directions must be understood as propagation "modes" or channels quantized by the transverse directions of the scattering medium (see section A7.2.2). The transmission coefficients T_{ab} depend on these discrete variables and the integrals in (12.6) are replaced by discrete sums:

$$\mathcal{T}_a = \sum_a \mathcal{T}_{ab} \quad \text{and} \quad \mathcal{T} = \sum_{ab} \mathcal{T}_{ab}. \tag{12.5}$$

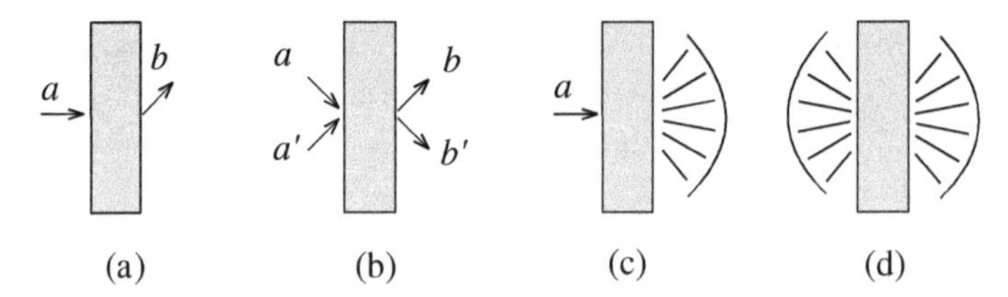

Figure 12.3 Schematic representation of the various setups designed for measurement of the correlations of the transmission coefficients: (a) transmission coefficient $\mathcal{T}_{ab}$, (b) correlation function $C_{aba'b'}$, (c) transmission coefficient $\mathcal{T}_a$ for an incident plane wave $\hat{s}_a$ obtained by integration over all emergent directions, (d) transmission coefficient $\mathcal{T}$ obtained by integration over all incident and emergent directions.

These various quantities are represented schematically in Figure 12.3. As we shall see, this is the total transmission coefficient $\mathcal{T}$ that will play a role analogous to the electrical conductance. This is the physical content of the Landauer formula (7.141).

Just like electrical conductance fluctuations, the correlation function $\overline{\delta\mathcal{T}_{ab}\delta\mathcal{T}_{a'b'}}$ involves the disorder average of a product of four amplitudes. In order to exhibit the various contributions to this correlation function, we follow the strategy developed in section 9.2. The complex scattering amplitude (12.3) takes the form

$$\psi_{ab} \propto \int d\boldsymbol{r}\, d\boldsymbol{r}'\, e^{ik(\hat{s}_a\cdot\boldsymbol{r}-\hat{s}_b\cdot\boldsymbol{r}')} \sum_{\mathcal{C}} E_{\mathcal{C}}, \tag{12.7}$$

where the amplitude $E_{\mathcal{C}}$ corresponds to a given multiple scattering trajectory $\mathcal{C}$. We thus obtain for the correlation function the average product

$$\sum_{\mathcal{C}_1,\mathcal{C}_2,\mathcal{C}_3,\mathcal{C}_4} \overline{E^{a,b}_{\mathcal{C}_1} E^{*a,b}_{\mathcal{C}_2} E^{a',b'}_{\mathcal{C}_3} E^{*a',b'}_{\mathcal{C}_4}}, \tag{12.8}$$

where the notation used is defined in Figure 12.4. The non-vanishing contributions to this average value of four amplitudes involve two Diffusons. We thus retain in the summation (12.8) the two terms such that $\mathcal{C}_1 = \mathcal{C}_2$ and $\mathcal{C}_3 = \mathcal{C}_4$, or $\mathcal{C}_1 = \mathcal{C}_4$ and $\mathcal{C}_3 = \mathcal{C}_2$. These two combinations of amplitudes are depicted schematically in Figure 12.5. The first is nothing but the product of the average values of the transmission coefficient. The second combination gives

$$\overline{\delta\mathcal{T}_{ab}\delta\mathcal{T}_{a'b'}}^{(1)} = \left(\frac{4\pi R^2}{cSI_0}\right)^2 \left|\overline{\psi_{ab}\psi^*_{a'b'}}\right|^2 \tag{12.9}$$

and the corresponding term in the correlation function $C_{aba'b'}$ is denoted by $C^{(1)}_{aba'b'}$. For the specific combination $a = a'$ and $b = b'$, which corresponds to a speckle pattern generated

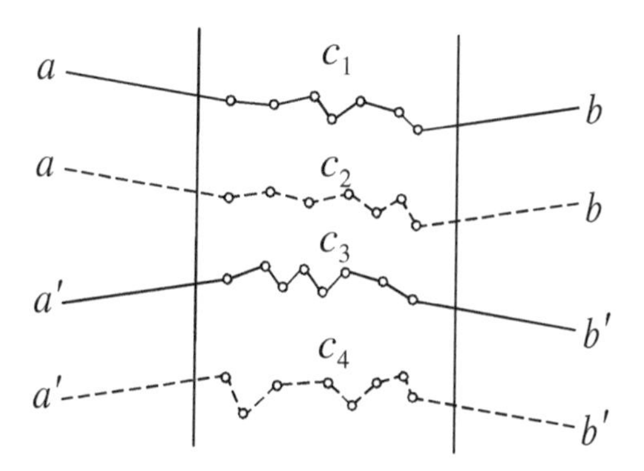

Figure 12.4 Schematic representation of the product of four amplitudes ψ that correspond to four waves incident along the directions $\hat{s}_a$ and $\hat{s}_{a'}$ and emergent along $\hat{s}_b$ and $\hat{s}_{b'}$. Non-vanishing contributions are obtained by pairing these amplitudes into Diffusons that may possibly cross each other.

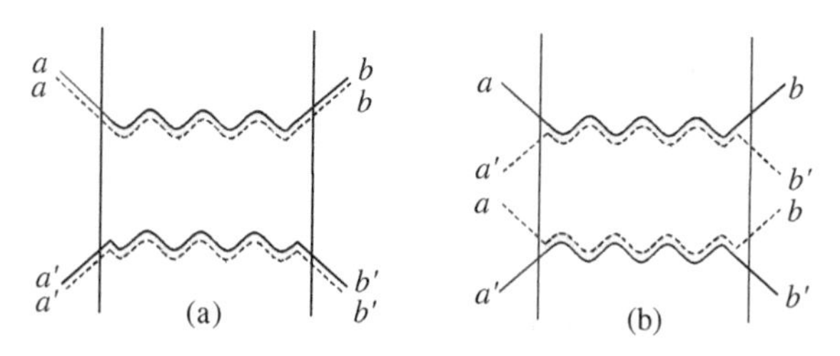

Figure 12.5 Schematic representation of the two contributions to the product $\overline{\mathcal{T}_{ab}\mathcal{T}_{a'b'}}$ that correspond respectively to the contractions $\mathcal{C}_1 = \mathcal{C}_2$, $\mathcal{C}_3 = \mathcal{C}_4$ and $\mathcal{C}_1 = \mathcal{C}_4$, $\mathcal{C}_2 = \mathcal{C}_3$. The first (a) corresponds to the product $\overline{\mathcal{T}}_{ab}\overline{\mathcal{T}}_{a'b'}$. The second (b) gives a contribution to the correlation function that we denote by $C^{(1)}_{aba'b'}$.

by a single beam with fixed incident and emergent directions, we obtain

$$\boxed{\overline{\delta\mathcal{T}^2_{ab}} = \overline{\mathcal{T}}^2_{ab}} \tag{12.10}$$

or equivalently $\overline{\mathcal{T}^2_{ab}} = 2\overline{\mathcal{T}}^2_{ab}$. This result constitutes the well-known Rayleigh law[4] which accounts for the characteristic granular aspect of a speckle pattern: the relative fluctuations are of order unity. This is the pairing depicted in Figure 12.5(b) that provides the main contribution to the correlation function and is responsible for the Rayleigh behavior. There are however other additional terms whose contribution is much more difficult to observe.

To proceed further, we recall that in multiple scattering, the Diffuson is long range which may give rise to new and interesting behaviors. By seeking in (12.8) all the possible pairings of two trajectories among four, we can construct diagrams with a crossing of two Diffusons. Such a crossing is described by a Hikami box (4.22), namely by a new pairing of

[4] In section 9.2 we obtained a similar expression. See also section 12.7.1.

four amplitudes that constitute these two Diffusons. In order to evaluate the contribution of terms involving quantum crossings, we must estimate the probability of their occurrence. To achieve this, we follow the physical argument presented in section A4.2.3.

For a sample with the geometry of a slab of width L and section S, the time spent by a diffusive trajectory in crossing over the sample is the Thouless time $\tau_D = L^2/D$. The *total length* of this trajectory is $\mathcal{L} = c\tau_D = 3L^2/l_e$. The volume corresponding to the crossing of the two diffusive trajectories, namely the volume of a Hikami box, is $\lambda^2 l_e$ (Figure 4.11). We can thus ascribe to a Diffuson a characteristic length $\mathcal{L}$, a cross section λ^2, and a volume $\mathcal{L}\lambda^2$. The crossing probability of two Diffusons is then proportional to the ratio between the volume of a Diffuson and the total volume $\Omega = SL$ of the scattering medium, namely $\lambda^2 \mathcal{L}/\Omega = \lambda^2 L/l_e S \propto 1/g$ where we have defined the dimensionless number

$$\boxed{g = \frac{k^2 l_e S}{3\pi L}} \tag{12.11}$$

with $k = 2\pi/\lambda$. This number happens to be the dimensionless, disordered average conductance (apart from the spin factor) of a conducting wire of length L and section S (relation 7.23). This number is very large in the weak disorder limit defined by $kl_e \gg 1$, and is of the order of 10^2 in the liquid suspensions usually studied in optics [301]. In this limit, we can safely assume that the quantum crossings are uncorrelated so that the probability of having n crossings of two Diffusons is simply proportional to $1/g^n$. This remark allows us to generate systematically all the contributions to the intensity correlation function from the number of crossings involved. We thus obtain the various contributions depicted in Figure 12.6, the first one being the Rayleigh contribution of Figure 12.5(b). In

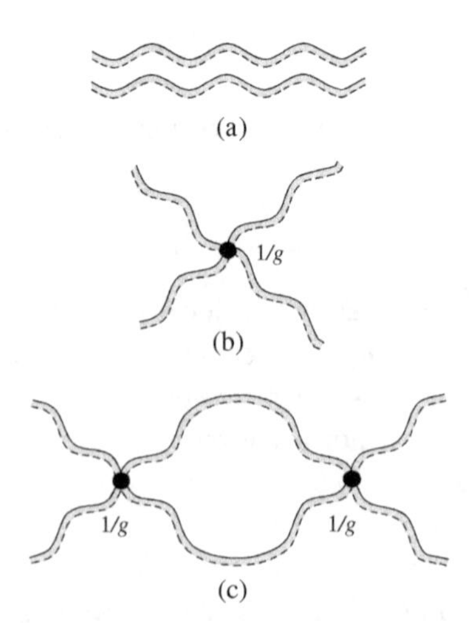

Figure 12.6 Classification of the various contributions to the correlation function $C_{aba'b'}$ given in terms of the number of crossings of two Diffusons. At each crossing, the corresponding contribution is multiplied by a factor $1/g \ll 1$. The three contributions represented are denoted by $C^{(1)}$, $C^{(2)}$, and $C^{(3)}$.

the following sections, we evaluate these various contributions, corresponding respectively to zero, one and two crossings.

12.3 Average transmission coefficient

In order to calculate the average of the transmission and reflection coefficients, we use the diffusion approximation to calculate the product of amplitudes depicted in Figure 12.7. In reflection, the average outgoing intensity is given by (8.6), and the reflection coefficient is given by the albedo (8.13) (see Exercise 8.2 for the slab geometry). For the average transmission coefficient, we obtain an analogous relation,[5] namely

$$\boxed{\overline{\mathcal{T}}_{ab} = \frac{4\pi R^2}{cI_0 S}\int d\boldsymbol{r}_1\, d\boldsymbol{r}_2\, |\overline{\psi}_a(\boldsymbol{r}_1)|^2 \Gamma(\boldsymbol{r}_1,\boldsymbol{r}_2)|\overline{G}^R(\boldsymbol{r}_2,\boldsymbol{r})|^2} \tag{12.12}$$

where $\Gamma(\boldsymbol{r}_1,\boldsymbol{r}_2)$ is the structure factor taken at zero frequency, and $\overline{\psi}_a(\boldsymbol{r}_1)$ accounts for an incident plane wave attenuated over the elastic mean free path while entering into the scattering medium (relation 8.7):

$$\overline{\psi}_a(\boldsymbol{r}_1) = \sqrt{\frac{cI_0}{4\pi}}\, e^{-|\boldsymbol{r}_1-\boldsymbol{r}|/2l_e}\, e^{ik\hat{\boldsymbol{s}}_a\cdot\boldsymbol{r}_1}. \tag{12.13}$$

$\boldsymbol{r}$ is a point on the interface (Figure 12.7(b)). The average Green function $\overline{G}^R(\boldsymbol{r}_2,\boldsymbol{R})$ accounts for the propagation of the wave between the last scattering event up to a point $\boldsymbol{R}$ far away from the scattering medium. In this so-called Fraunhoffer limit (relation 8.9), we have

$$\overline{G}^R(\boldsymbol{r}_2,\boldsymbol{R}) = e^{-|\boldsymbol{r}'-\boldsymbol{r}_2|/2l_e}\, e^{-ik\hat{\boldsymbol{s}}_b\cdot\boldsymbol{r}_2}\, \frac{e^{ikR}}{4\pi R}, \tag{12.14}$$

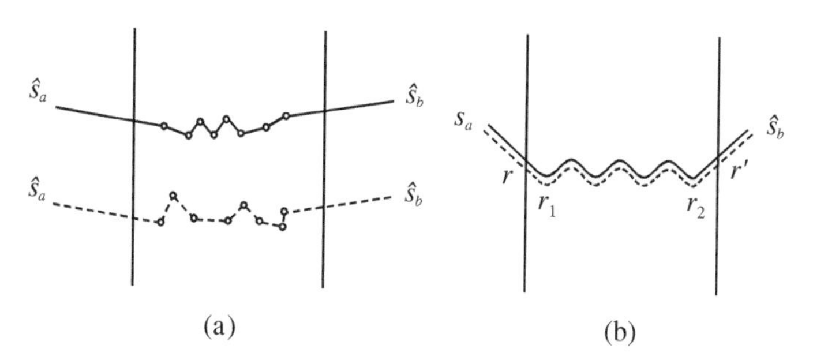

Figure 12.7 Schematic representation (a) of the product of two amplitudes ψ corresponding to two incident plane waves along the directions $\hat{\boldsymbol{s}}_a$ emergent along $\hat{\boldsymbol{s}}_b$, (b) of the Diffuson that results from the pairing of these two amplitudes and that represents the main contribution to the average transmission coefficient $\overline{\mathcal{T}}_{ab}$. $\boldsymbol{r}$ and $\boldsymbol{r}'$ are points on the interface plane whereas $(\boldsymbol{r}_1,\boldsymbol{r}_2)$ are the extremities of a multiple scattering trajectory.

[5] We assume that the difference in refraction index between the two media is negligible.

where $\boldsymbol{r}'$ is a point on the interface. From these expressions, we deduce

$$\overline{T}_{ab} = \frac{1}{(4\pi)^2 S} \int d\boldsymbol{r}_1\, d\boldsymbol{r}_2\, e^{-z_1/\mu_a l_e}\, e^{-|L-z_2|/\mu_b l_e}\, \Gamma(\boldsymbol{r}_1, \boldsymbol{r}_2) \tag{12.15}$$

with μ_a (μ_b) being the projection of the vector $\hat{\boldsymbol{s}}_a$ ($\hat{\boldsymbol{s}}_b$) along the z axis. This last expression should be compared with (7.161). For a slab geometry, the structure factor $\Gamma(\boldsymbol{r}_1, \boldsymbol{r}_2)$ depends only on the variables z_1, z_2 and $\boldsymbol{\rho} = (\boldsymbol{r}_2 - \boldsymbol{r}_1)_\perp$. Performing the integrals over z_1 and z_2 and making use of (4.63), we obtain[6]

$$\overline{T}_{ab} = \frac{c}{4\pi}\mu_a\mu_b \int_S d^2\boldsymbol{\rho}\, P_d(\boldsymbol{\rho}, l_e\mu_a, L - l_e\mu_b) = \frac{c}{4\pi}\mu_a\mu_b P_d(k_\perp = 0, l_e\mu_a, L - l_e\mu_b) \tag{12.16}$$

where $P_d(k_\perp, z, z')$ is the two-dimensional Fourier transform

$$P_d(\boldsymbol{k}_\perp, z, z') = \int d^2\boldsymbol{\rho} P_d(\boldsymbol{\rho}, z, z')\, e^{-i\boldsymbol{k}_\perp \cdot \boldsymbol{\rho}}. \tag{12.17}$$

For $k_\perp = 0$, and using the right choice of boundary conditions, this function is given by the relation (5.158):

$$P_d(k_\perp = 0, z, z') = P_d(z, z') = \frac{z_m + z_0}{D} \frac{L + z_0 - z_M}{L + 2z_0} \tag{12.18}$$

with $z_m = \min(z, z')$, $z_M = \max(z, z')$ and $z_0 = 2\, l_e/3$. Inserting this expression into (12.16) yields

$$\overline{T}_{ab} = \frac{3}{4\pi} \frac{l_e}{L} \mu_a\mu_b \left(\mu_a + \frac{z_0}{l_e}\right)\left(\mu_b + \frac{z_0}{l_e}\right). \tag{12.19}$$

This expression for the average transmission coefficient is weakly dependent on the angles, just like the average reflection coefficient (see 8.16 and Figure 8.7 for the average albedo).

[6] Notice that all the integrals over exponential terms that appear in transmission are such that

$$\int_0^\infty dz\, e^{-z/l_e} f(z) = l_e f(l_e)$$

since the function $f(z)$ we consider behaves linearly near the interface.

This weak dependence on μ_a and μ_b arises from the projections of unit vectors that appear in the exponential terms. At small angles, it is negligible and we shall not consider it anymore. Thus, by taking $\mu_a = \mu_b = 1$:[7]

$$\boxed{\overline{T}_{ab} = \frac{3}{4\pi}\frac{(l_e + z_0)^2}{l_e L} = \frac{25}{12\pi}\frac{l_e}{L}} \tag{12.20}$$

To conclude this section, it is important to notice that the various effects described in this chapter do not depend strongly on the precise nature of the conversion of an incident ballistic wave into a diffusive one. All the correlation functions are normalized by the average transmission coefficient and therefore they do not depend on the extrapolation length z_0. In what follows, and in order to simplify our results, we shall take $z_0 = 0$. It is easy to modify the obtained expressions by restoring z_0.

12.4 Angular correlations of the transmitted light

The purpose of this section is the study of each contribution to $C_{aba'b'}$ that corresponds to the different sets of multiple scattering trajectories represented in Figure 12.6.

12.4.1 Short range $C^{(1)}$ correlations

This type of correlation is given by Figure 12.5(b). Calculation of this correlation is very similar to that of the average transmission coefficient $\overline{T}_{ab}$, except that the distinct directions a and a' as well as b and b' give rise to additional phase factors. Just as for (12.15) we obtain

$$\overline{\psi_{ab}\psi^*_{a'b'}} = \frac{cI_0}{(4\pi)^3 R^2}\int d\boldsymbol{r}_1\, d\boldsymbol{r}_2\, e^{ik[\Delta\hat{\boldsymbol{s}}_a\cdot\boldsymbol{r}_1 - \Delta\hat{\boldsymbol{s}}_b\cdot\boldsymbol{r}_2]}\, e^{-z_1/l_e}\, e^{-|L-z_2|/l_e}\,\Gamma(\boldsymbol{r}_1,\boldsymbol{r}_2) \tag{12.21}$$

where we have defined $\Delta\hat{\boldsymbol{s}}_a = \hat{\boldsymbol{s}}_a - \hat{\boldsymbol{s}}_{a'}$ and $\Delta\hat{\boldsymbol{s}}_b = \hat{\boldsymbol{s}}_b - \hat{\boldsymbol{s}}_{b'}$. We consider a limit where $\Delta\hat{\boldsymbol{s}}_a$ and $\Delta\hat{\boldsymbol{s}}_b$ are small enough for their projection along the z axis to be negligible. Making use of (12.1) and (12.2) yields

$$\overline{\delta T_{ab}\delta T_{a'b'}} = \left(\frac{1}{(4\pi)^2 S}\int d\boldsymbol{r}_1\, d\boldsymbol{r}_2\, e^{ik[\Delta\hat{\boldsymbol{s}}_a\cdot\boldsymbol{r}_1 - \Delta\hat{\boldsymbol{s}}_b\cdot\boldsymbol{r}_2]}\, e^{-z_1/l_e}\, e^{-|L-z_2|/l_e}\,\Gamma(\boldsymbol{r}_1,\boldsymbol{r}_2)\right)^2. \tag{12.22}$$

[7] This result can be cast in the form $\beta\, l_e/L$, where β is the same coefficient that enters the angular dependence of the coherent backscattering cone (equation 8.32). The reader may try to understand why this is so.

For the geometry of a slab, the structure factor $\Gamma(\boldsymbol{r}_1,\boldsymbol{r}_2)$ depends only on the variables z_1, z_2 and $\boldsymbol{\rho} = (\boldsymbol{r}_2 - \boldsymbol{r}_1)_\perp$. Performing the integrals over z_1 and z_2 (see footnote 6, page 434) and making use of (4.63), we obtain

$$\overline{\delta \mathcal{T}_{ab} \delta \mathcal{T}_{a'b'}} = \left(\frac{c}{4\pi} \delta_{\Delta \hat{s}_a, \Delta \hat{s}_b} \int_S d^2 \boldsymbol{\rho} \, e^{ik \boldsymbol{\rho} \cdot \Delta \hat{s}_a} \, P_d(\boldsymbol{\rho}, l_e, L - l_e) \right)^2 , \tag{12.23}$$

so that

$$\overline{\delta \mathcal{T}_{ab} \delta \mathcal{T}_{a'b'}} = \left(\frac{c}{4\pi} \delta_{\Delta \hat{s}_a, \Delta \hat{s}_b} P_d(q_a, l_e, L - l_e) \right)^2 \tag{12.24}$$

where $q_a = k|\Delta \hat{s}_a|$. The Fourier transform $P(q_a, l_e, L - l_e)$ is given by (5.55), namely,

$$P_d(q_a, z, z') = \frac{1}{D} \frac{\sinh q_a z_m \sinh q_a (L - z_M)}{q_a \sinh q_a L} \tag{12.25}$$

so that in the limit $q_a l_e \ll 1$, we have

$$P_d(q_a, l_e, L - l_e) = \frac{1}{D} \frac{\sinh^2 q_a l_e}{q_a \sinh q_a L} \simeq \frac{3}{c} \frac{q_a l_e}{\sinh q_a L} . \tag{12.26}$$

Putting these expressions together leads to

$$\overline{\delta \mathcal{T}_{ab} \delta \mathcal{T}_{a'b'}} = \overline{\mathcal{T}}_{ab} \, \overline{\mathcal{T}}_{a'b'} \delta_{\Delta \hat{s}_a, \Delta \hat{s}_b} F_1(q_a L) \tag{12.27}$$

with

$$F_1(x) = \left(\frac{x}{\sinh x} \right)^2 . \tag{12.28}$$

The correlation function $C^{(1)}_{aba'b'}$ is thus given by [302]

$$\boxed{C^{(1)}_{aba'b'} = \delta_{\Delta \hat{s}_a, \Delta \hat{s}_b} F_1(q_a L) = \delta_{\Delta \hat{s}_a, \Delta \hat{s}_b} \left(\frac{q_a L}{\sinh q_a L} \right)^2} \tag{12.29}$$

with $q_a = k|\Delta \hat{s}_a|$. This contribution to the correlation function accounts for the main characteristics of speckle patterns. For $a' = a$, $b' = b$, we recover the expression of the magnitude of the speckle fluctuations, namely, $\overline{\delta \mathcal{T}^2_{ab}} = \overline{\mathcal{T}}^2_{ab}$. This is the Rayleigh law (12.10) which states that relative fluctuations of the intensity are of order unity. This granularity is the most prominent characteristic of a speckle pattern as shown in Figure 12.1.

Remark: speckle spots and transmission channels

Expression (12.27) was obtained by assuming an infinitely extended plane interface illuminated by a plane wave. For an incident beam of finite section S, the Kronecker symbol $\delta_{\Delta\hat{s}_a,\Delta\hat{s}_b}$ needs to be replaced by a function $f(\Delta\hat{s}_a - \Delta\hat{s}_b)$ which is related to the Fourier transform of the incident beam and whose exact expression is easily obtained from (12.22)

$$f(\Delta\hat{s}_a - \Delta\hat{s}_b) = \left(\frac{1}{S}\int_S d^2\boldsymbol{\rho}\, e^{ik\boldsymbol{\rho}\cdot(\Delta\hat{s}_a - \Delta\hat{s}_b)}\right)^2. \tag{12.30}$$

For instance, for a beam of cylindrical shape and of width W, the function takes the form,

$$f(\Delta\hat{s}_a - \Delta\hat{s}_b) = \left(\frac{J_1(q_{ab}W)}{q_{ab}W/2}\right)^2 \tag{12.31}$$

with $q_{ab} = k|\Delta\hat{s}_a - \Delta\hat{s}_b|$. A beam of Gaussian shape $e^{-\rho^2/W^2}$ (the effective width is then given by $S = \pi W^2$), leads to

$$f(\Delta\hat{s}_a - \Delta\hat{s}_b) = e^{-q_{ab}^2 W^2/4}. \tag{12.32}$$

For a given incident beam, the far-field intensity on a screen, i.e., the angular dependence of the signal, is given by the correlation function taken for $a = a'$

$$\overline{\delta T_{ab}\delta T_{ab'}} = \overline{T}_{ab}^2 f(\Delta\hat{s}_b) \tag{12.33}$$

where the angular width of f is $1/kW$. The angular width of a given speckle spot is therefore of the order of $1/kW$ and its spatial extension on a screen placed at a distance R away is of the order of $\lambda R/W$. Furthermore, each spot can be viewed as a *mode* or a *transverse propagation channel* for the incident beam. The number of such modes is then of order k^2W^2. Such modes are analogous to the propagation modes in a wave guide. They play a very important role in the Landauer description of transport (Appendix A7.2).

The correlation function (12.29) predicts another interesting and unexpected feature. From expression (12.30) for f, we infer that a small shift $\Delta\hat{s}_a$ of the direction of the incident beam leads to a similar shift $\Delta\hat{s}_b \simeq \Delta\hat{s}_a$ of the outgoing beam. Therefore, by changing the direction of the incident beam, we observe a shift of the whole speckle pattern. This feature, also called the *memory effect*, is indeed remarkable since it demonstrates that beyond the multiple scattering process, the light beam retains some information about the incident beam [303]. This effect disappears for $\Delta\hat{s}_a \simeq 1/kL$, namely for vanishing values of F_1 (Figure 12.8(b,c)).

There are additional, although smaller, contributions to $C_{aba'b'}$ which have different angular behavior. $C^{(1)}$ is built upon paired amplitudes that correspond to distinct angular directions for both incident and emergent light. This specific pairing displayed in Figure 12.5(b) can be represented schematically

$$\boxed{C^{(1)} : (aa')(aa') \longrightarrow (bb')(bb')} \tag{12.34}$$

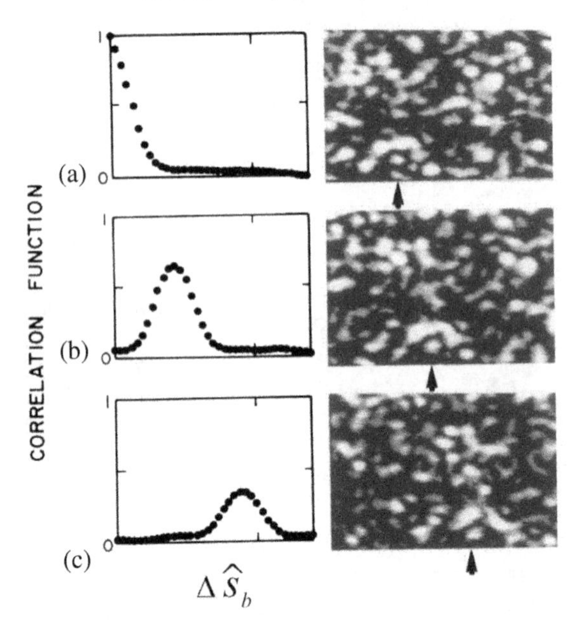

Figure 12.8 Right: speckle patterns measured in transmission for different directions of the incident beam. The arrow is a guide to show the evolution of a particular feature. The first picture is the reference pattern, corresponding to a given direction of the incident beam. By slowly tilting the incident beam (10 mdeg, then 20 mdeg), we notice a corresponding shift of the speckle pattern which retains some "memory" of the reference pattern. If the tilt angle becomes too large, the speckle is distorted. Left: corresponding behavior of the correlation $C_{aba'b'}$ as a function of the angle $\Delta\hat{s}_b$. The upper curve (a) is the autocorrelation (i.e., for $a = a'$) of the reference speckle pattern. The width is of order $1/kW$. The two other figures (b, c) display the correlation of the second and of the third speckle patterns with the reference pattern (i.e., $a \neq a'$). The correlation is maximum for $\Delta\hat{s}_b = \Delta\hat{s}_a$, and this maximum is $F_1(kL|\Delta\hat{s}_a|)$. It vanishes for angles larger than $|\Delta\hat{s}_a| \simeq 1/kL$ [303].

Exercise 12.1: correlation of a speckle pattern measured in reflection

We denote by $\mathcal{R}_{ab}$ the reflection coefficient (i.e., the albedo defined in Chapter 8). Using a modification of (12.22), show that the angular correlation of $\mathcal{R}_{ab}$ is

$$\overline{\delta\mathcal{R}_{ab}\delta\mathcal{R}_{a'b'}} = \left(\frac{1}{(4\pi)^2 S}\int d\boldsymbol{r}_1\, d\boldsymbol{r}_2\, e^{ik[\Delta\hat{s}_a\cdot\boldsymbol{r}_1 - \Delta\hat{s}_b\cdot\boldsymbol{r}_2]}\, e^{-z_1/l_e}\, e^{-z_2/l_e}\, \Gamma(\boldsymbol{r}_1,\boldsymbol{r}_2)\right)^2 \tag{12.35}$$

or equivalently,

$$\overline{\delta\mathcal{R}_{ab}\delta\mathcal{R}_{a'b'}} = \delta_{\Delta\hat{s}_a,\Delta\hat{s}_b}\left(\frac{c}{4\pi l_e^2}\int_0^\infty dz_1\, dz_2\, e^{-z_1/l_e}\, e^{-z_2/l_e}\, P_d(q_a, z_1, z_2)\right)^2 \tag{12.36}$$

with $q_a = k|\Delta\hat{s}_a|$.

Show that this expression is simply related to the square of the coherent albedo $\alpha_c(q_a)$ given by (8.26)

$$\frac{\overline{\delta\mathcal{R}_{ab}\delta\mathcal{R}_{a'b'}}}{\overline{\mathcal{R}}_{ab}\overline{\mathcal{R}}_{a'b'}} = \delta_{\Delta\hat{s}_a,\Delta\hat{s}_b}\left(\frac{\alpha_c(q_a)}{\alpha_c(0)}\right)^2 . \tag{12.37}$$

This implies that, in reflection, speckle patterns are made of spots of angular width $1/kW$ and that they exhibit a memory effect just as in transmission. But the angular range for observation of the memory effect is broader, of order $1/kl_e$ instead of $1/kL$ in transmission. As for the average albedo (section 8.8), it is important to notice that a complete description of the correlations is more difficult in reflection than in transmission, since we need to include properly the contribution of short trajectories which depends on the nature of the scatterers and on the properties of the interface. We expect, nevertheless, that our description faithfully describes the contribution of the long trajectories, i.e., of small angles ($q_a l_e < 1$). Furthermore, in a measurement of speckle correlations near the backscattering direction, phase coherence described by the Cooperon leads to additional effects described in [304].

12.4.2 Long range correlations $C^{(2)}$

As stated previously, there are additional contributions to the correlation function which involve crossings of Diffusons (Figure 12.4). These contributions are smaller but they have a much larger angular range.

We consider first the term with a single crossing. It is described by means of a Hikami box and Figure 12.9 illustrates the only two possible pairings. We notice that the two incoming Diffusons necessarily show up along a diagonal of the box. As a result of the Diffuson crossing, we expect a contribution smaller by a factor $1/g$. Furthermore, the crossing leads to an exchange of the channels a and a' and therefore to an angular behavior which is different from that obtained in (12.34) for $C^{(1)}$ and which can be represented by

$$\boxed{C^{(2)}: \quad \begin{matrix} (aa)(a'a') \longrightarrow (bb')(bb') \\ \text{and} \\ (aa')(aa') \longrightarrow (bb)(b'b') \end{matrix}} \tag{12.38}$$

The corresponding correlation function is

$$\overline{\delta\mathcal{T}_{ab}\delta\mathcal{T}_{a'b'}}^{(2)} = \frac{1}{(4\pi)^4 S^2}\int\prod_{i=1}^{4} d\boldsymbol{r}_i[e^{ik\Delta\hat{s}_b\cdot(\boldsymbol{r}_2-\boldsymbol{r}_4)} + e^{ik\Delta\hat{s}_a\cdot(\boldsymbol{r}_1-\boldsymbol{r}_3)}]E(z_i)$$
$$\times\int\prod_{i=1}^{4} d\mathbf{R}_i\, H(\mathbf{R}_i)\Gamma(\boldsymbol{r}_1,\mathbf{R}_1)\Gamma(\boldsymbol{r}_3,\mathbf{R}_3)\Gamma(\mathbf{R}_2,\boldsymbol{r}_2)\Gamma(\mathbf{R}_4,\boldsymbol{r}_4), \tag{12.39}$$

where we have defined

$$E(z_i) = e^{-(z_1+z_3)/l_e}\, e^{-|L-z_2|/l_e}\, e^{-|L-z_4|/l_e}. \tag{12.40}$$

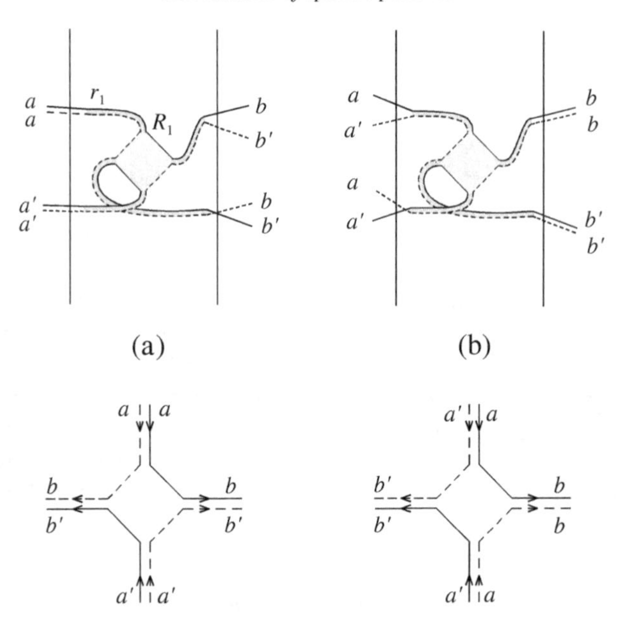

Figure 12.9 Contributions to $\overline{\delta\mathcal{T}_{ab}\delta\mathcal{T}_{a'b'}}$ that involve one Diffuson crossing. The two cases correspond to distinct configurations of incident plane waves incoming along $\hat{\boldsymbol{s}}_a$ and $\hat{\boldsymbol{s}}_{a'}$ and outgoing along $\hat{\boldsymbol{s}}_b$ and $\hat{\boldsymbol{s}}_{b'}$. (a) Shows a contribution to the correlation function that depends on $\Delta\hat{\boldsymbol{s}}_b$ but not on $\Delta\hat{\boldsymbol{s}}_a$, whereas the opposite situation is depicted in (b).

Let us start with the first term in (12.39), which corresponds to Figure 12.9(a). For the Hikami box we choose expression (4.150) in which the gradient terms apply to the incoming Diffusons. Making use of (4.63), we obtain

$$\overline{\delta\mathcal{T}_{ab}\delta\mathcal{T}_{a'b'}}^{(2)} = \frac{2h_4}{(4\pi)^4 S^2}\left(\frac{4\pi c}{l_e^2}\right)^4 \int \prod_{i=1}^{4} d\boldsymbol{r}_i \, e^{ik\Delta\hat{\boldsymbol{s}}_b\cdot(\boldsymbol{r}_2-\boldsymbol{r}_4)} \, E(z_i)$$
$$\times \int d\boldsymbol{r}\, [\boldsymbol{\nabla}_{\boldsymbol{r}} P_d(\boldsymbol{r}_1,\boldsymbol{r})\cdot \boldsymbol{\nabla}_{\boldsymbol{r}} P_d(\boldsymbol{r}_3,\boldsymbol{r})]\, P_d(\boldsymbol{r},\boldsymbol{r}_2) P_d(\boldsymbol{r},\boldsymbol{r}_4) \tag{12.41}$$

with $h_4 = l_e^5/(48\pi k^2)$. Performing integrations over z_i amounts to considering $z_1 = z_3 = l_e$ and $z_2 = z_4 = L - l_e$.[8] Making use of the Fourier transform (12.17) leads to

$$\overline{\delta\mathcal{T}_{ab}\delta\mathcal{T}_{a'b'}}^{(2)} = \frac{l_e c^4}{24\pi k^2 S}\int_0^L dz\, [\partial_z P_d(0,l_e,z)]^2\, P_d(q_b,z,L-l_e)^2, \tag{12.42}$$

where $q_b = k|\Delta\hat{\boldsymbol{s}}_b|$. Furthermore, considering the limit $q_b l_e \ll 1$ for which

$$\partial_z P_d(0,l_e,z) = -\frac{l_e}{DL} \qquad\qquad P_d(q_b,z,L-l_e) = \frac{l_e}{D}\frac{\sinh q_b z}{\sinh q_b L} \tag{12.43}$$

[8] See footnote 6, page 434.

yields[9]

$$\overline{\delta \mathcal{T}_{ab} \delta \mathcal{T}_{a'b'}}^{(2)} = \frac{81}{48\pi} \frac{l_e}{k^2 L S} F_2(q_b L). \tag{12.44}$$

We have introduced the function

$$F_2(x) = \frac{1}{\sinh^2 x} \left(\frac{\sinh 2x}{2x} - 1 \right) \tag{12.45}$$

whose limiting behavior is

$$\begin{array}{ll} x \to 0 & F_2(x) \to 2/3 \\ x \to \infty & F_2(x) \to 1/x. \end{array}$$

Using (12.20) for the average transmission coefficient $\overline{\mathcal{T}}_{ab}$ and adding the other contribution of diagram 12.9(b) obtained with a similar calculation, we obtain for the corresponding correlation function the final expression [305–307]

$$\boxed{C^{(2)}_{aba'b'} = \frac{\overline{\delta \mathcal{T}_{ab} \delta \mathcal{T}_{a,b'}}}{\overline{\mathcal{T}}^2_{ab}} = \frac{1}{g} \left[F_2(q_a L) + F_2(q_b L) \right]} \tag{12.46}$$

where $q_a = k|\Delta \hat{s}_a|$ and $q_b = k|\Delta \hat{s}_b|$. As expected, $C^{(2)}$ is proportional to the inverse of the dimensionless "conductance" g given by (12.11) and is thus smaller than $C^{(1)}$ by a factor $1/g$. This latter factor also appears in the weak localization correction to conductivity (section 7.4).

Unlike $C^{(1)}_{aba'b'}$, which decreases exponentially, the contribution $C^{(2)}$ to the correlation decreases algebraically and vanishes only if *both* $\Delta \hat{s}_a$ *and* $\Delta \hat{s}_b$ are large. For a speckle pattern generated from a single incident beam $a = a'$, we then expect to observe long range angular correlations of weak amplitude,

$$C^{(2)}_{abab'} = \frac{1}{g} \left(\frac{2}{3} + F_2(q_b L) \right) \xrightarrow{b \neq b'} \frac{2}{3g}. \tag{12.47}$$

This term is small but, unlike $C^{(1)}$, it is long range and it is therefore possible to observe it experimentally. In sections 12.5 and 12.6 we show how to single out the $C^{(2)}$ contribution experimentally in either the time or frequency domain.

12.4.3 Two-crossing contribution and $C^{(3)}$ correlation

The next contribution $C^{(3)}_{aba'b'}$ to the correlation function involves two crossings of Diffusons and hence two Hikami boxes. Figure 12.10 shows the various possible combinations, with

[9] The reader may check that the exact result obtained by restoring the extrapolation length z_0 consists in replacing l_e by $(l_e+z_0)^4/l_e^3$. As has been stated previously, this dependence in z_0 will disappear upon dividing this expression by $\overline{\mathcal{T}}^2_{ab}$ in order to obtain $C^{(2)}$.

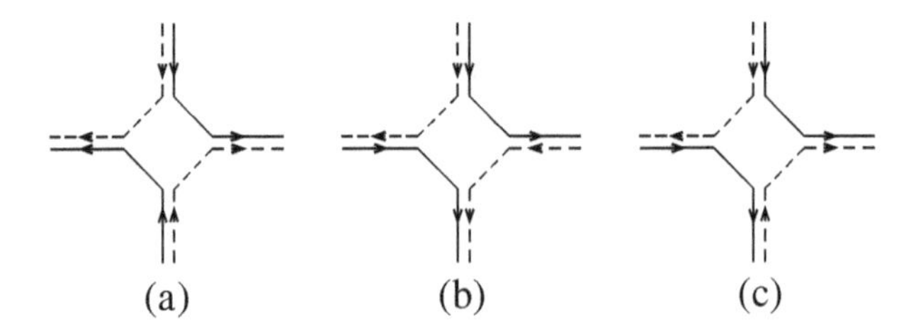

Figure 12.10 Possible combinations of Hikami boxes with one incoming Diffuson and one outgoing Diffuson.

the constraint of having both an incoming and an outgoing Diffuson. It is easy to be convinced that the only possible combinations of multiple scattering trajectories are those of Figure 12.11 (see also Figure 12.12). The two boxes that appear in the diagram K_1^d are of the type represented in Figure 12.10(a), and they are linked together by Diffusons. On the other hand, for the diagram K_2^c, the boxes are type 12.10(b) and are thus linked by two Cooperons. There are two more diagrams K_3 where the boxes are either of type 12.10(a) or 12.10(c), and thus are linked either by Diffusons or by Cooperons. To calculate these four diagrams, we use the same method as for $C^{(2)}$. Furthermore, we notice that the diagrams of Figure 12.11 lead to the absence of angular correlation and that they correspond to the combination

$$\boxed{C^{(3)} : (aa)(a'a') \longrightarrow (bb)(b'b')} \tag{12.48}$$

To proceed further, we start by evaluating K_1^d. For the two Hikami boxes, we apply the gradients on the external (either incoming or outgoing) Diffusons, which yields

$$\overline{\delta T_{ab} \delta T_{a'b'}}^{(3)}|_{K_1} = (2h_4)^2 \frac{l_e^4}{(4\pi)^4 S^2} \left(\frac{4\pi c}{l_e^2}\right)^6$$
$$\times \int_0^L \int_0^L dz\, dz' \, [\partial_z P_d(0, l_e, z)]^2 \, P_d(0, z, z')^2 \, [\partial_{z'} P_d(0, z', L - l_e)]^2. \tag{12.49}$$

The probability $P_d(z, z')$ as given by (5.58), behaves linearly with z and z' so that gradients give constant terms and

$$C^{(3)}_{aba'b'}|_{K_1} = \frac{4}{g^2} \frac{D^2}{L^4} \int_0^L \int_0^L dz\, dz' \, P_d(z, z')^2. \tag{12.50}$$

Upon integrating, this term is found equal to $2/(45g^2)$. The calculation of K_2^c is identical to the calculation giving K_1^d, except for replacement of the internal Diffusons by Cooperons.

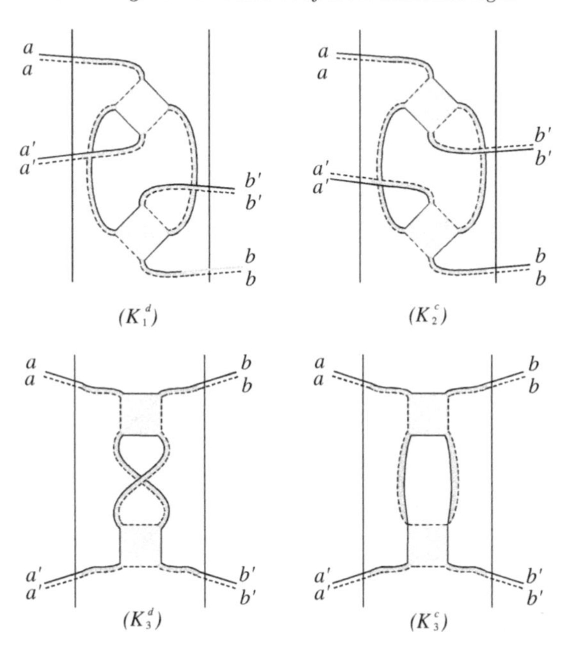

Figure 12.11 Representation of the contributions to $\overline{\delta\mathcal{T}_{ab}\delta\mathcal{T}_{a'b'}}^{(3)}$ that involve two crossings of Diffusons. The rules of Figure 12.10 imply that K_1^d involves necessarily two internal Diffusons, whereas K_2^c involves two internal Cooperons. This simply results from the orientation of the trajectories as shown in Figure 12.12. There are two K_3 diagrams: the first, K_3^d, involves two internal Diffusons, the second, K_3^c, contains two internal Cooperons. Diagrams K_3^d and K_1^d have the same structure as those used in Figures 11.3(a and b) to calculate the conductance fluctuations. Diagrams K_3^c and K_2^c are their counterparts for Cooperons.

The evaluation of K_3 is slightly more involved since the products of derivative terms that appear in Hikami boxes apply on both internal and external Diffusons. This difficulty being understood, a further integration by parts leads to the result that the contribution of K_3^d and K_3^c turns out to be half of the contribution of K_1^d and K_2^c. We obtain finally [305, 308, 309]

$$\boxed{C^{(3)}_{aba'b'} = \frac{\overline{\delta\mathcal{T}_{ab}\delta\mathcal{T}_{a'b'}}^{(3)}}{\overline{\mathcal{T}}^2_{ab}} = \frac{2}{15}\frac{1}{g^2}} \tag{12.51}$$

or using (12.20),

$$\overline{\delta\mathcal{T}_{ab}\delta\mathcal{T}_{a'b'}}^{(3)} = \frac{27}{40k^4S^2}\left(\frac{l_e+z_0}{l_e}\right)^4 = \frac{125}{24k^4S^2}. \tag{12.52}$$

This last contribution is *universal*, meaning that it is independent of l_e, i.e., independent of the strength of disorder. It is analogous to the universal conductance fluctuations, a point

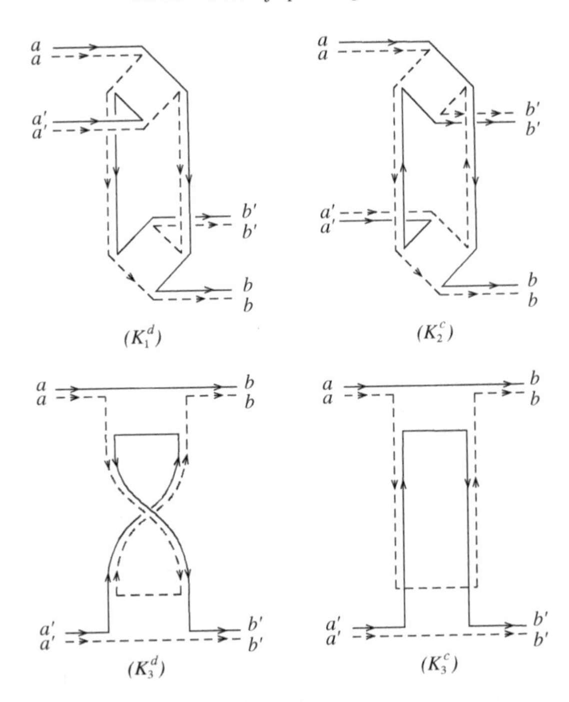

Figure 12.12 Diagrams of Figure 12.11 in which the pairing between trajectories has been represented. These diagrams do not lead to any angular behavior of the correlation function. On the other hand, the combination in which channels a and a' are paired leads to diagrams that give a contribution with the angular structure of $C^{(1)}$. We shall not consider them.

that shows up clearly if we notice the similarity between the structure of the diagrams of Figures 12.11 and 11.4.

It is straightforward to check that other contributions to the correlation function exist which involve two crossings of Diffusons and display an angular dependence identical to that of $C^{(1)}$ or $C^{(2)}$.[10] For instance, we obtain diagrams with the angular structure of $C^{(1)}$ (combination 12.34) by considering the diagrams of Figure 12.11 and by coupling the incoming amplitudes a and a' instead of a and a. The outgoing coupled amplitudes turn out to be b and b'. The interested reader may check that to obtain a diagram having the $C^{(2)}$ angular structure (combination 12.38) the two crossings must be different, one being a "square" and the second being "diamond shaped." We shall not dwell any longer on the structure of these contributions since they represent corrections much smaller (of order $1/g^2$) than $C^{(1)}$ and $C^{(2)}$ previously calculated [307, 308].

[10] It is usual [304, 306, 310] to classify the contributions $C^{(1)}$, $C^{(2)}$ and $C^{(3)}$ according to their angular dependence and not according to the number of Diffuson crossings. But if for each contribution we consider only the dominant term (of order 1 for $C^{(1)}$, $1/g$ for $C^{(2)}$, and $1/g^2$ for $C^{(3)}$), then these two classifications coincide.

12.4.4 Relation with universal conductance fluctuations

Summarizing the results obtained so far for the angular correlation function of the transmission coefficient, we arrive at the following expression which involves three terms of different strength and angular dependence:

$$\boxed{C_{aba'b'} = \delta_{\Delta\hat{s}_a,\Delta\hat{s}_b} F_1(q_a L) + \frac{1}{g}\big[F_2(q_a L) + F_2(q_b L)\big] + \frac{2}{15}\frac{1}{g^2}} \tag{12.53}$$

with $q_a = k|\Delta\hat{s}_a|$ and $q_b = k|\Delta\hat{s}_b|$. Experiments measuring the angular dependence give access mostly to the first term since the two others are much too small to be detected. But upon integrating over all the transmitted channels, the contribution of this first term is negligible, and we end up with the following expression for the fluctuations of the coefficient $\mathcal{T}_a$:

$$\frac{\overline{\delta\mathcal{T}_a \delta\mathcal{T}_{a'}}}{\overline{\mathcal{T}}_a \overline{\mathcal{T}}_{a'}} = \frac{1}{g} F_2(kL\Delta\hat{s}_a) + \frac{2}{15}\frac{1}{g^2}. \tag{12.54}$$

Furthermore, upon integrating over the incident channels as well, the contribution of F_2 also becomes negligible, and we deduce the relative fluctuations of the total transmission coefficient $\mathcal{T}$,

$$\boxed{\frac{\overline{\delta\mathcal{T}^2}}{\overline{\mathcal{T}}^2} = \frac{2}{15}\frac{1}{g^2}} \tag{12.55}$$

Given that the average total transmission coefficient $\overline{\mathcal{T}}$ is itself proportional to the mean free path and thus to g (equations 7.199 and 7.196), the variance $\overline{\delta\mathcal{T}^2}$ is universal in the sense that it is independent of disorder:

$$\overline{\mathcal{T}} = \frac{4\pi}{3}\frac{l_e}{L} \qquad \overline{\delta\mathcal{T}^2} = \frac{32\pi^4}{15k^4 S^2}. \tag{12.56}$$

For a wave guide geometry, more appropriate to the description of electric conductance, the angular variables are quantized into transverse modes or channels. In section A7.2.2, we defined $\mathcal{T}_{ab}$ as the transmission coefficient between channels a and b. The angular functions in (12.53) must be replaced by discrete Kronecker symbols. Since $F_1(0) = 0$ and $F_2(0) = 2/3$, equation (12.53) becomes [311]

$$\boxed{\frac{\overline{\delta\mathcal{T}_{ab}\,\delta\mathcal{T}_{a'b'}}}{\overline{\mathcal{T}}_{ab}\,\overline{\mathcal{T}}_{a'b'}} = \delta_{aa'}\delta_{bb'} + \frac{2}{3g}(\delta_{aa'} + \delta_{bb'}) + \frac{2}{15}\frac{1}{g^2}} \tag{12.57}$$

The first term expresses the Rayleigh law and corresponds to uncorrelated channels, while the next terms describe correlations between channels. The last term corresponds to a uniform correlation between channels. Summing over the outgoing channels b, we find for the correlations of the transmission coefficient $T_a = \sum_b T_{ab}$:

$$\frac{\overline{\delta T_a \, \delta T_{a'}}}{\overline{T}_a \, \overline{T}_{a'}} = \frac{\delta_{aa'}}{M} + \frac{2}{3g}\left(\delta_{aa'} + \frac{1}{M}\right) + \frac{2}{15g^2}. \tag{12.58}$$

Finally, the summation over the incoming channels a leads, for the correlations of the total transmission coefficient $T = \sum_{ab} T_{ab}$, to:

$$\frac{\overline{\delta T^2}}{\overline{T}^2} = \frac{1}{M^2} + \frac{4}{3Mg} + \frac{2}{15g^2} \tag{12.59}$$

or, since $\overline{T} = g$,

$$\boxed{\overline{\delta T^2} = \frac{g^2}{M^2} + \frac{4g}{3M} + \frac{2}{15}} \tag{12.60}$$

This expression allows us to identify clearly the different contributions to the conductance fluctuations. The last term which describes universal conductance fluctuations results from correlations between channels induced by two quantum crossings. Since $g \propto Ml_e/L$, the first two terms which correspond respectively to the Rayleigh contribution (no correlation between channels) and one quantum crossing contribution are smaller respectively by factors $(l_e/L)^2$ and l_e/L.

As for the transmission coefficients, correlations between channels are essential to describe the correlations between reflection coefficients R_{ab}. It is again possible to classify the various contributions in powers of $1/g$ arising from quantum crossings and to calculate $\overline{\delta R_{ab}\delta R_{a'b'}}$, $\overline{\delta R_a \delta R_{a'}}$ and $\overline{\delta R^2}$. The universality of $\overline{\delta R^2}$ results from the channel correlations induced by the two-crossing contribution. This delicate point has been thoroughly discussed in reference [311].

12.5 Speckle correlations in the time domain

As has already been stated, it is difficult to resolve experimentally the different contributions to the angular correlation function (12.53). This function is mostly driven by the first term (memory effect). The two other terms are much smaller, being respectively proportional to $1/g$ and $1/g^2$ with typically $g \simeq 10^2$ [301]. Furthermore, $C^{(3)}$ is constant, meaning that fluctuations affect all the speckle spots in the same way: in other words, if one spot is brighter, all the other spots are equally brighter. An independent measurement of each of the three contributions $C^{(i)}$ requires finding in each of them distinct behavior as a function of an external parameter. Such a parameter might be, for instance, the time associated with the motion of scatterers. We are thus led to measure a time correlation function of the light intensity, as was done for diffusing wave spectroscopy presented in Chapters 6 and 9.

We thus consider the correlation function[11]

$$C_{aba'b'}(t) = \frac{\overline{\delta\mathcal{T}_{ab}(0)\delta\mathcal{T}_{a'b'}(t)}}{\overline{\mathcal{T}}_{ab}^2} \tag{12.61}$$

and we now show that each of the contributions $C^{(1)}$, $C^{(2)}$ and $C^{(3)}$ has its own particular time dependence that results from different pairings of two trajectories taken respectively at time 0 and t. The dominant contribution $C^{(1)}$ decreases exponentially with t whereas the two remaining contributions have an algebraic decrease. These very different behaviors allow us to measure each of them separately.

12.5.1 Time dependent correlations $C^{(1)}(t)$ and $C^{(2)}(t)$

We reconsider the derivation of $C^{(1)}_{aba'b'}$ but now the two amplitudes involved are taken at different times (Figure 12.13(a)). The motion of the scatterers is assumed to be Brownian so that the probability $P_d(\boldsymbol{r},\boldsymbol{r}')$ needs to be replaced by $P_\gamma(\boldsymbol{r},\boldsymbol{r}')$. The parameter γ and the time τ_s defined in (6.197) are such that $\gamma = 1/\tau_s = t/(2\tau_e\tau_b)$ where the time τ_b characterizes the Brownian motion of scatterers (section 9.5).

The determination of $C^{(1)}_{aba'b'}(t)$ is straightforward if we formally replace the probability $P_d(q_a,z,z')$ by $P_\gamma(q_a,z,z')$ in the derivation leading to (12.27). Using the correspondence (5.48), this simply amounts to keeping track of expressions obtained for $t=0$, and replacing q_a by $\sqrt{q_a^2+1/L_s^2}$ where $L_s = \sqrt{D/\gamma} \propto 1/\sqrt{t}$. In place of (12.27), we thus obtain

$$C^{(1)}_{aba'b'}(t) = \delta_{\Delta\hat{s}_a,\Delta\hat{s}_b} F_1(L/\mathcal{L}_a), \tag{12.62}$$

where the length $\mathcal{L}_a$ is defined by

$$\frac{1}{\mathcal{L}_a} = \sqrt{q_a^2 + \frac{1}{L_s^2}}\,. \tag{12.63}$$

In particular, for fixed incoming and outgoing directions, the time dependent correlation function is rewritten as

$$\boxed{C^{(1)}(t) = C^{(1)}_{abab}(t) = F_1(L/L_s) = \left(\frac{L/L_s}{\sinh L/L_s}\right)^2} \tag{12.64}$$

We recover the result (9.34), while noticing that $g_2(t) = |g_1(t)|^2$. In the long time limit, this function decreases exponentially as displayed in Figure 12.14.

The time dependence of $C^{(2)}_{aba'b'}(t)$ is derived using the same procedure. As already stated, in diagram 12.13(b), only the two Diffusons placed *after* the crossing are affected by the phase shift. There is another diagram, similar to 12.9(b), where the phase shift modifies the

[11] This function is nothing but the intensity correlation $g_2(T)$ defined by (9.2). In the definition (12.61), we change the notation and use t instead of T to describe the time associated with motion of the scatterers.

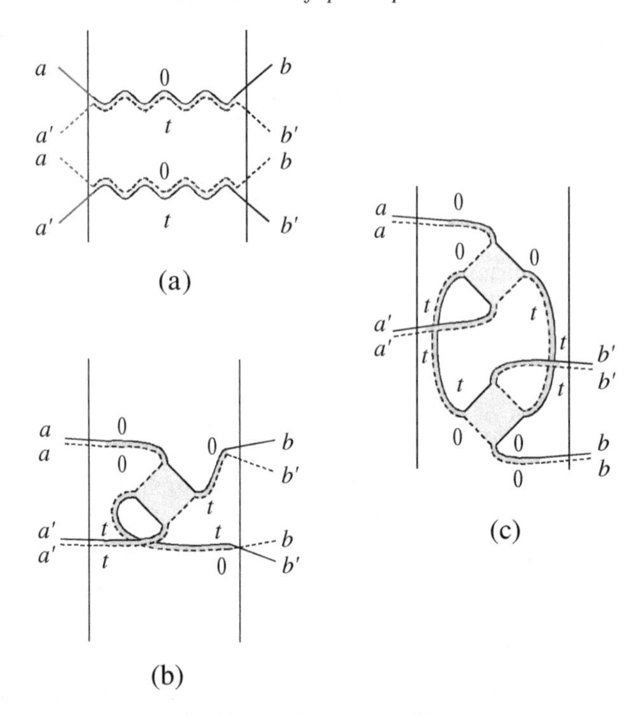

Figure 12.13 Time dependences of $C^{(1)}(t)$, $C^{(2)}(t)$ and $C^{(3)}(t)$. (*a*) For $C^{(1)}(t)$, the time t induces a phase shift between the two Diffusons. (*b*) For $C^{(2)}(t)$, only the two Diffusons placed *after* the crossing are affected by the phase shift. For $C^{(3)}(t)$, only the two internal Diffusons which form the closed loop between the two crossings are modified by the time dependent phase shift.

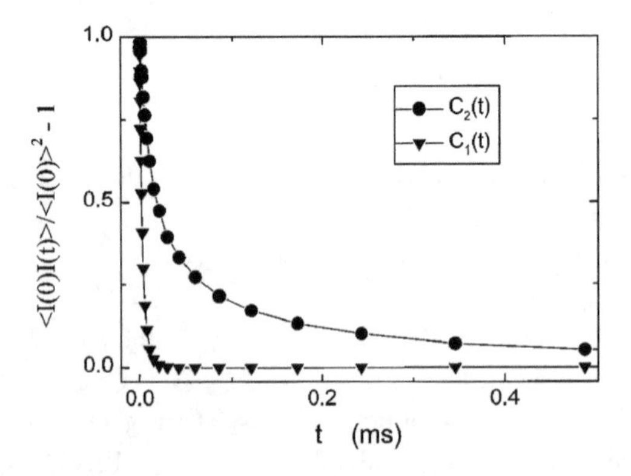

Figure 12.14 Time dependence of $C^{(1)}(t)$ and $C^{(2)}(t)$. The former decreases exponentially at long times, while the latter decreases only as a power law [301].

Diffusons placed *before* the crossing instead. This emphasizes again the role of the crossing as an exchange of pairing of the amplitudes. To account for the phase shift, we need to replace in (12.42) the probability $P_d(q_b, z, L - l_e)$ by $P_\gamma(q_b, z, L - l_e)$ as given by (5.55) with $L_\gamma \to L_s$, taking care not to change $P_d(0, l_e, z)$. For the case $a = a'$, $b = b'$, we obtain

$$\boxed{C^{(2)}(t) = C^{(2)}_{abab}(t) = \frac{2}{g} F_2(L/L_s)} \tag{12.65}$$

where $F_2(x)$ was defined in (12.45).[12] We notice in this expression that, unlike $C^{(1)}(t)$, the correlation function $C^{(2)}(t)$ behaves at large times as $L_s/L \propto 1/\sqrt{t}$. This long time decrease is much slower than for $C^{(1)}(t)$, as shown in Figure 12.14.

The measurement of $C^{(1)}(t)$ was discussed Chapter 9: for a fixed incident direction a, the outgoing light signal is measured along a direction b. To measure $C^{(2)}(t)$, the contribution of $C^{(1)}$ is removed upon integrating the outgoing signal over all the angular channels, or equivalently upon averaging over a large number of speckle spots, the incident channel a being fixed [301].

Exercise 12.2: correlation function $C^{(2)}(t)$ in the presence of absorption

• Show that in the presence of finite absorption characterized by a length l_a, but for motionless scatterers, $C^{(2)}(t)$ is obtained through the replacement in (12.42) of $P_d(q_b, z, z')$ by $P_\gamma(q_b, z, z')$ and of $P_d(0, z, z')$ by $P_\gamma(0, z, z')$, together with $\gamma = c/l_a$.

Notice that, in the presence of finite absorption, the correspondence (9.22) obtained in Chapter 9 for the diffusing wave spectroscopy cannot be readily extended to the calculation of $C^{(2)}(t)$ since absorption affects *all* Diffusons equally. This correspondence applies only to the calculation of $C^{(1)}(t)$, i.e., in the absence of crossing of Diffusons (see Figure 9.1).

• Show that, in the presence of finite absorption and for moving scatterers, we need to introduce the two quantities

$$\gamma_1 = \frac{c}{l_a} \tag{12.66}$$

and

$$\gamma_2 = \frac{c}{l_a} + \frac{D}{L_s^2} \tag{12.67}$$

in order to describe the behavior of Diffusons before and after crossing.

---

[12] Including the angular dependences leads to

$$C^{(2)}_{aba'b'}(t) = \frac{1}{g}\left[F_2(L/\mathcal{L}_a) + F_2(L/\mathcal{L}_b)\right]$$

with

$$\frac{1}{\mathcal{L}_a} = \sqrt{q_a^2 + \frac{1}{L_s^2}} \qquad \frac{1}{\mathcal{L}_b} = \sqrt{q_b^2 + \frac{1}{L_s^2}}.$$

12.5.2 Time dependent correlation $C^{(3)}(t)$

The calculation of $C^{(3)}_{aba'b'}(t)$ is more difficult. In the steady state limit ($t = 0$), the corresponding contribution (12.51) has no angular dependence so we cannot simply replace the wave vector q_a by the inverse of the length $\mathcal{L}_a$ as for the calculation of $C^{(1)}(t)$ and $C^{(2)}(t)$.

As shown in Figure 12.13(c), only Diffusons that contribute to the internal loop are affected by the motion of scatterers. The correlation $C^{(3)}(t)$ is obtained through the replacement of $P_d(q=0,z,z')$ by $P_\gamma(q=0,z,z')$ in (12.50). This yields[13]

$$C^{(3)}(t) = \frac{12\, D^2}{g^2\, L^4} \int_0^L \int_0^L dz\, dz'\; P_\gamma^2(q=0,z,z') \tag{12.68}$$

where $P_\gamma(q=0,z,z')$ is given by (5.55) with $L_\gamma \to L_s$. Upon integrating, we obtain

$$\boxed{C^{(3)}(t) = \frac{1}{g^2} F_3(L/L_s)} \tag{12.69}$$

where the function $F_3(x)$ is defined by

$$F_3(x) = \frac{3}{2} \frac{2 + 2x^2 - 2\cosh 2x + x \sinh 2x}{x^4 \sinh^2 x} \tag{12.70}$$

and behaves asymptotically as

$$\begin{aligned} x \to 0 \qquad & F_3(x) \to 2/15 \\ x \to \infty \qquad & F_3(x) \to 3/x^3. \end{aligned}$$

Notice that we recover (12.51) for $C^{(3)}(0)$ while taking the limit $L_s \to \infty$.

Given that $C^{(2)}(t)$ decreases faster than $C^{(3)}(t)$ and that its amplitude is larger by a factor g, observation of $C^{(3)}(t)$ is not easy. It has nevertheless been measured [312] and is represented in Figure 12.15. The measurement requires eliminating the contributions of $C^{(1)}(t)$ and $C^{(2)}(t)$. The first disappears upon integrating over the transmitted channels. The key to splitting the two contributions $C^{(2)}$ and $C^{(3)}$ is to enhance geometrically the crossing probability $1/g \propto L/W^2$ by means of an additional aperture of width W such as that represented in Figure 12.16. As a result, the main contribution to $C^{(3)}$, which originates from closed loops located in the area of length L, is enhanced. Moreover, by decreasing the width L_1 of the left part of the slab, we also decrease the contribution to $C^{(2)}$ which originates from multiple scattering paths located in this region.

It is also worthwhile to notice that the behavior of $C^{(3)}(t)$ in the presence of absorption cannot simply be obtained through the substitution of $P_d(z,z')$ by $P_\gamma(z,z')$, namely using the correspondence (9.22). The reason why is that only some of the Diffusons have a time dependence (Figure 12.13) whereas they all depend on absorption.

[13] An additional factor 3 accounts for the other contributions shown in Figure 12.12.

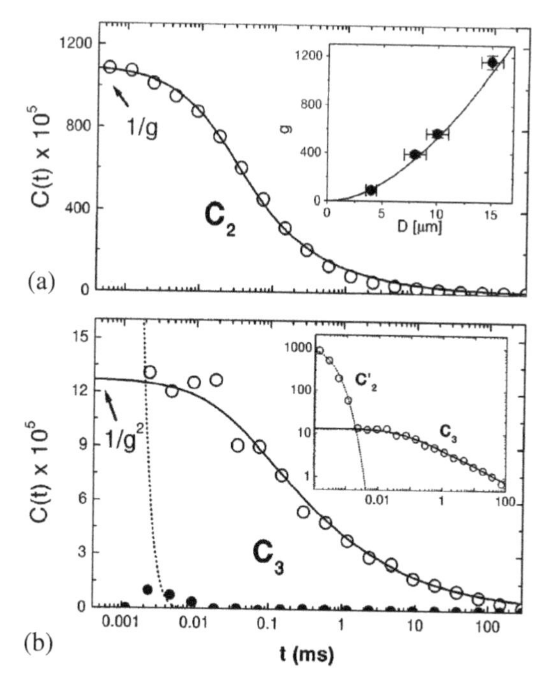

Figure 12.15 Time dependences of $C^{(2)}(t)$ and $C^{(3)}(t)$ measured by multiple scattering of light in an aqueous colloidal suspension of TiO_2 beads. Notice the long time power law behavior of both functions [312]. The theoretical expression (12.69) corresponds to a faster decrease than is observed experimentally.

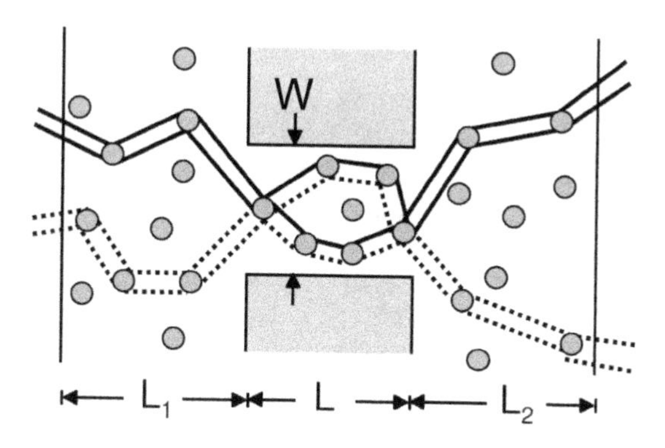

Figure 12.16 Schematic experimental setup used to measure $C^{(3)}(t)$. Inclusion of the middle area of length L leads to an amplification of the probability of closed loops and then to an enhancement of $C^{(3)}$ [312].

12.6 Spectral correlations of speckle patterns

For the sake of completeness, we now turn to another important parametric dependence of the transmission coefficient, namely the frequency dependence of the correlations $C^{(1)}$ and $C^{(2)}$. This is obtained by correlating two speckle patterns respectively generated by two waves taken at frequencies ω_0 and $\omega_0 + \Delta\omega$. These frequencies play the role of external parameters just like the times 0 and t for the time dependent correlations (see Figure 12.13). As for the previous cases, the calculation of $C^{(1)}(\Delta\omega)$ is straightforward and results from the correspondence (5.46), i.e., by the replacement in (12.62) of the length $\mathcal{L}_a$ by

$$\frac{1}{\mathcal{L}_a(\Delta\omega)} = \sqrt{q_a^2 + \frac{1}{L_s^2} - i\frac{\Delta\omega}{D}} \tag{12.71}$$

so that for $q_a = 0$ and $L_s \to \infty$, we obtain[14]

$$C^{(1)}(\Delta\omega) = \frac{\left(L/\sqrt{2}L_\omega\right)^2}{\cosh\left(L/\sqrt{2}L_\omega\right) - \cos\left(L/\sqrt{2}L_\omega\right)}, \tag{12.72}$$

where $L_\omega = \sqrt{D/\Delta\omega}$. The fit of experimental data [313] with this function is shown in Figure 12.17. In the limit $\Delta\omega \gg E_c$, it behaves as $e^{-\sqrt{\Delta\omega/E_c}}$.

The calculation of $C^{(2)}(\Delta\omega)$ is similar to that of $C^{(2)}(t)$, provided that in (12.42) we replace $P_d(q_b, z, z')$ by $P_{\gamma=-i\Delta\omega}(z, z')$. The latter quantity is complex valued, a result that can be traced back to the diagram 12.13(b) which involves two Diffusons, one taken at

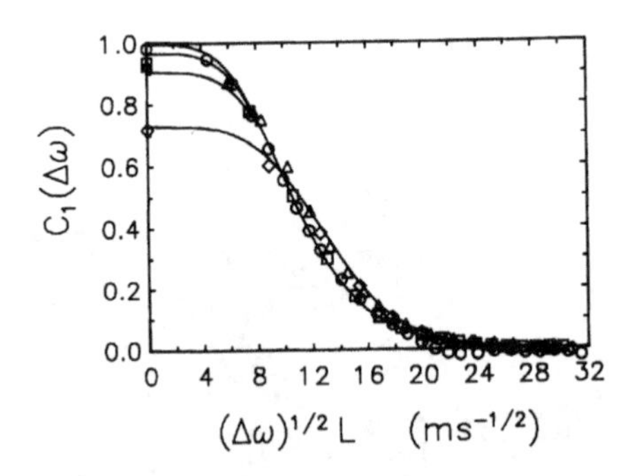

Figure 12.17 Frequency dependence of the correlation function $C^{(1)}(\Delta\omega)$. The various curves represented correspond to different values of the width L of the slab, ranging from 13 μm to 45 μm. The measured diffusion coefficient is $D = 12\,\mathrm{m}^2/\mathrm{s}$. The discrepancy with the predicted scaling behavior versus $L/L_\omega \propto L\sqrt{\Delta\omega}$ observed at low frequency results from the finite spectral width of the laser. This explains why this discrepancy decreases for increasing frequencies. At high frequency, namely for $\Delta\omega \gg E_c$ where $E_c = D/L^2$ is the Thouless frequency (5.35), we have $C^{(1)}(\Delta\omega) \propto e^{-\sqrt{\Delta\omega/E_c}}$ [313].

[14] Since $\mathcal{L}_a$ happens to be a complex valued function, care must be taken to replace F_1 by $|x/\sinh x|^2$.

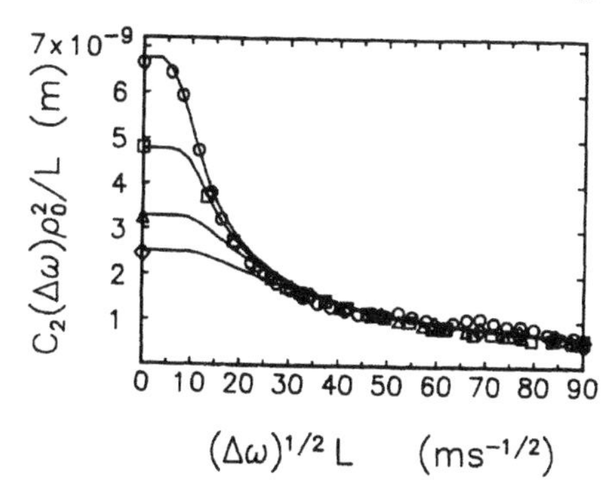

Figure 12.18 Frequency dependence of $C^{(2)}$. The various curves correspond to different values of the width L of the slab, ranging from 13 μm to 78 μm. The incident beam has a Gaussian shape of diameter $W = 26$ μm (denoted ρ_0 in the figure). The disagreement observed at low frequency between the data and the predicted scaling behavior is discussed in the text. The power law dependence $\propto 1/\sqrt{\Delta\omega}$ is well reproduced at high frequency, i.e., for $\Delta\omega \gg E_c$ where E_c is the Thouless frequency (5.35) [313].

frequency $\Delta\omega$ and the second at $-\Delta\omega$. For $q_a = q_b = 0$, we finally obtain the expression

$$C^{(2)}(\Delta\omega) = \frac{4}{g}\left(\frac{L_\omega}{\sqrt{2}L}\right)\frac{\sinh(\sqrt{2}L/L_\omega) - \sin(\sqrt{2}L/L_\omega)}{\cosh(\sqrt{2}L/L_\omega) - \cos(\sqrt{2}L/L_\omega)}. \tag{12.73}$$

This function is plotted in Figure 12.18 and decreases as $1/\sqrt{\Delta\omega}$. The experimental points represented in the figure were obtained for a Gaussian beam of width W, and not for a plane wave. Furthermore, the plotted function is, in our notation, the product $g\, C^{(2)}(\Delta\omega)$ which, according to (12.73), is a function of the scaled variable L/L_ω only. This is in disagreement with the results displayed in Figure 12.18. The reason for this spurious disagreement is the same convolution effect mentioned for $C^{(1)}(\Delta\omega)$. Moreover, for this experiment, the configuration average is not taken over the random potential but instead over a finite frequency interval. This choice underestimates the contribution to the correlation function in the low frequency range of the order of this finite frequency interval. This appears to be of less importance for $C^{(1)}(\Delta\omega)$ than for $C^{(2)}(\Delta\omega)$.[15]

12.7 Distribution function of the transmission coefficients

Up to now, we have characterized fluctuations of either the electrical conductance or speckle patterns by means of their second moment. Our next purpose is to obtain the entire distribution function, i.e., higher order moments. We shall limit ourselves to the case $g \gg 1$. For more details, the reader may consult references [309, 314].

[15] For more details concerning this experiment, see section 5 of reference [313].

12.7.1 Rayleigh distribution law

We consider first the distribution of the transmission coefficient $\mathcal{T}_{ab}$. The relation

$$\overline{\mathcal{T}_{ab}^2} = 2\,\overline{\mathcal{T}}_{ab}^2 \tag{12.74}$$

was obtained in section 12.2. The extension of this result to $\overline{\mathcal{T}_{ab}^n}$ is straightforward. This quantity is proportional to $\overline{\psi_{ab}^*\psi_{ab}\cdots\psi_{ab}^*\psi_{ab}}$, i.e., to a product of n amplitudes ψ_{ab} and n conjugate amplitudes ψ_{ab}^*. To lowest order in $1/g$, we discard contributions of diagrams that involve crossings of Diffusons. We are thus led to replace the previous disorder average by a product of n terms such as $\overline{\psi_{ab}^*\psi_{ab}}$. Since there are $n!$ distinct pairings of ψ_{ab}^* and ψ_{ab}, we end up with

$$\overline{\mathcal{T}_{ab}^n} = n!\,\overline{\mathcal{T}}_{ab}^n. \tag{12.75}$$

From the expression for these moments, it is straightforward to obtain the corresponding distribution function, namely

$$\boxed{P(\mathcal{T}_{ab}) = \frac{1}{\overline{\mathcal{T}}_{ab}}\; e^{-\mathcal{T}_{ab}/\overline{\mathcal{T}}_{ab}}} \tag{12.76}$$

which constitutes the well-known Rayleigh distribution law, and is also the distribution law $P(\mathcal{R}_{ab})$ of the reflection coefficient. It reproduces the measurements quite faithfully, in both single scattering and multiple scattering regimes (Figure 12.19), and corresponds to the limit $g \to \infty$.

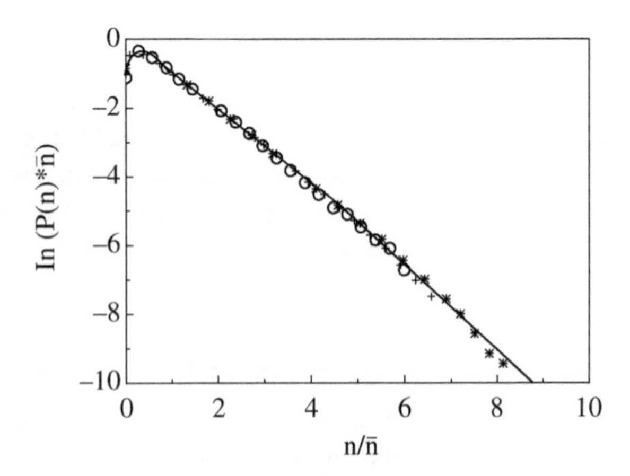

Figure 12.19 Probability distribution of the intensity reflected in the multiple scattering regime, by a rotating solid sample of $BaSO_4$ parallel to the incident polarization channel. The intensity is proportional to the number n of photons, and corresponds to the reflection coefficient $\mathcal{R}_{ab}$. The distribution $P(\mathcal{R}_{ab})$ is in perfect agreement with the Rayleigh distribution (12.76) [315].

Remark: Rayleigh distribution and central limit theorem

The Rayleigh distribution simply expresses the fact that a speckle pattern results from the coherent superposition of N uncorrelated random complex valued amplitudes, $A = \sum_j A_j$. Given that these amplitudes are independently distributed, their moduli and phases are uncorrelated random variables as well. Hence the total amplitude A turns out to be the sum of a large number of independent random variables whose distribution in the limit of large N is Gaussian. The distribution functions of the real A_r and imaginary A_i parts of A are also Gaussian of zero average and of variance σ^2. The joint law $P(A_r, A_i)$ is thus given by

$$P(A_r, A_i) = \frac{1}{\pi\sigma^2} e^{-(A_r^2 + A_i^2)/\sigma^2}. \tag{12.77}$$

The resulting distribution function of the intensity $I = |A|^2$ then appears to be the Rayleigh law (12.76). In the single scattering regime, the Rayleigh distribution results from the assumption that scatterers are randomly distributed over length scales comparable to the wavelength of the light so that the corresponding amplitudes are random. In multiple scattering we obtain the Rayleigh distribution by simply assuming uncorrelated outgoing amplitudes.

Higher order corrections in the parameter $1/g$ account for crossings of Diffusons. They lead to correlations between outgoing amplitudes and as such to deviations from the Rayleigh distribution.

12.7.2 Gaussian distribution of the transmission coefficient $\mathcal{T}_a$

The probability distribution of $\mathcal{T}_a$ is obtained from the expression of its cumulants $\overline{\mathcal{T}_a^n}^c$. Making use of (12.47) yields for the second cumulant:

$$\overline{\mathcal{T}_a^2}^c = \overline{\delta\mathcal{T}_a^2} = \frac{2}{3g}\overline{\mathcal{T}_a}^2. \tag{12.78}$$

Cumulants of order n correspond to an ensemble of n Diffusons crossing each other. The smallest number of crossings is $n-1$ (Figure 12.20), each of which gives a reduction by a factor $1/g$, thus leading to

$$\overline{\mathcal{T}_a^n}^c \propto \frac{\overline{\mathcal{T}_a}^n}{g^{n-1}}. \tag{12.79}$$

For $g \to \infty$, higher order cumulants are negligible, and keeping only the contribution of the second cumulant leads to the Gaussian distribution:

$$P(\mathcal{T}_a) = \frac{1}{\sqrt{2\pi\overline{\delta\mathcal{T}_a^2}}} e^{-(\mathcal{T}_a - \overline{\mathcal{T}_a})^2/2\overline{\delta\mathcal{T}_a^2}}. \tag{12.80}$$

This distribution has been measured and is represented in Figure 12.21.

Deviations from the Gaussian behavior can be systematically accounted for by including corrections to the cumulants that result from crossings of Diffusons [309, 316–319]. These

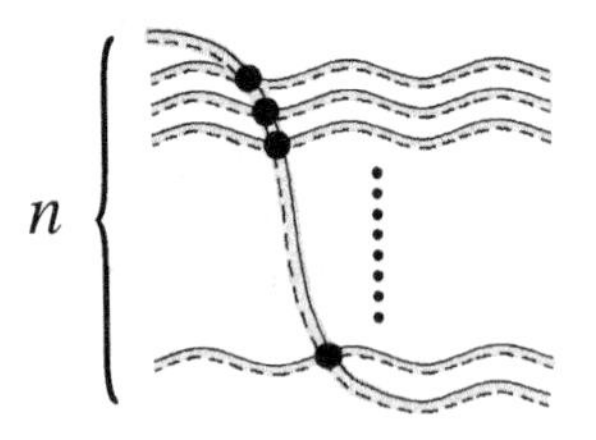

Figure 12.20 The smallest number of crossings of n Diffusons to lowest order in the parameter $1/g$ is $n-1$.

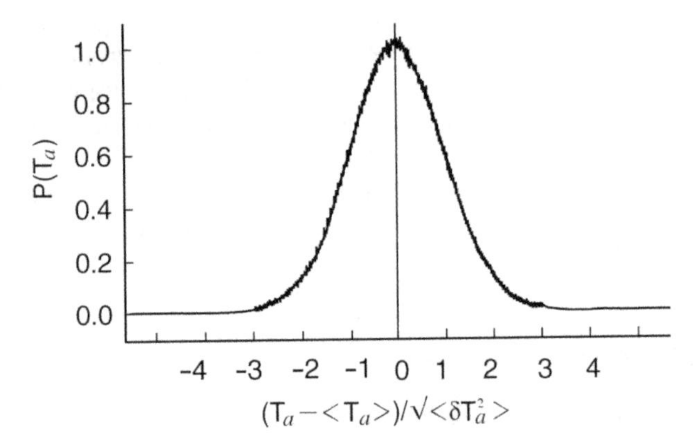

Figure 12.21 Measurement of the probability distribution $P(\mathcal{T}_a)$ of the transmission coefficient $\mathcal{T}_a$ for an ensemble of TiO_2 particles with $l_e \simeq 0.8\,\mu$m and $g \geq 10^3$. The vertical scale is chosen so that $P(0) = 1$. The Gaussian behavior (12.80) is well reproduced [316].

deviations have been measured for values of the disorder parameter g ranging between 1 and 10, and they are represented in Figure 12.22.

12.7.3 Gaussian distribution of the electrical conductance

The probability distribution of the total transmission coefficient $\mathcal{T}$ or of the electrical conductance (Appendix A7.2) can also be obtained using the approach of the previous section. The variance $\overline{\delta\mathcal{T}^2}$ is obtained upon integrating the correlation function $\overline{\delta\mathcal{T}_{ab}\delta\mathcal{T}_{a'b'}}$ given by (12.53) over all angular channels. Given that only diagrams with two Diffusons crossing contribute to that function, the corresponding variance (12.51) takes the form

$$\overline{\delta\mathcal{T}^2} \propto \frac{\overline{\mathcal{T}}^2}{g^2}. \tag{12.81}$$

To proceed further, we consider the cumulant $\overline{\mathcal{T}^n}^c$, which is built out of a product of n Diffusons crossing in pairs. Given that each Diffuson must have at least two crossings in

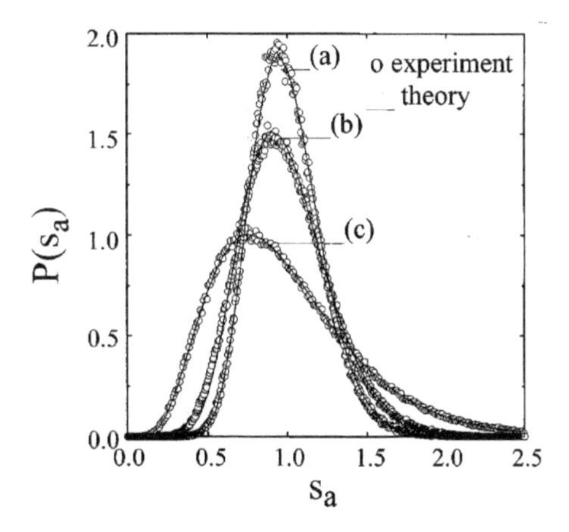

Figure 12.22 Measurement of the probability distribution of the transmission coefficient $\mathcal{T}_a$ for microwaves propagating in a diffusive medium composed of polystyrene beads. We have defined $s_a = \mathcal{T}_a/\overline{\mathcal{T}_a}$. The different curves correspond to the following values of the parameter g: (a) $g = 15$, (b) $g = 9$ and (c) $g = 2.25$. The Gaussian behavior is well reproduced for a very weak disorder (a) and the discrepancy becomes more and more pronounced for stronger disorders, i.e., for smaller values of g. These deviations are well accounted for by including corrections due to crossing of Diffusons (solid lines) in higher cumulants [309, 317, 320].

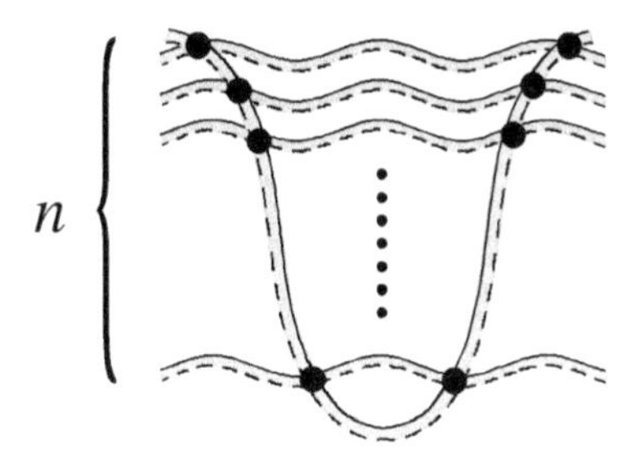

Figure 12.23 The smallest number of crossings of n Diffusons to lowest order in $1/g$ is $2n - 2$.

order to get rid of any further angular dependence, the smallest number of crossings is equal to $2n - 2$ (Figure 12.23). Since each crossing provides an additional reduction factor $1/g$, we have

$$\overline{\mathcal{T}^n}^c \propto \frac{\overline{\mathcal{T}}^n}{g^{2n-2}}. \tag{12.82}$$

This result can be transposed to the electrical conductance G (7.153), namely

$$\overline{G^n}^c \propto \frac{1}{g^{n-2}}. \tag{12.83}$$

For $g \to \infty$, moments of order $n > 2$ do not contribute so that the resulting distribution function is Gaussian just like $P(\mathcal{T}_a)$. As shown before, it is also possible to incorporate corrections that arise at finite g and lead to deviations from the Gaussian behavior [314].

Appendix A12.1: Spatial correlations of light intensity

In this chapter, we studied angular correlation functions of the transmission coefficient through a diffusive medium. Here we are interested instead in spatial correlations of intensity measured inside the bulk of the medium and issued from a monochromatic source of wave vector $k = \epsilon/c$. The methodology used to tackle this problem turns out to be very similar to that used for angular correlations (section 12.4). We thus expect the correlation functions to exhibit comparable behavior.

Consider a point-like source placed in a point $\boldsymbol{r}_0$ in the bulk of the diffusive medium. The corresponding intensity is defined by (4.54) so that its correlation function is given by

$$\overline{I(\boldsymbol{r})I(\boldsymbol{r}')} = \left(\frac{4\pi}{c}\right)^2 \overline{G^R_\epsilon(\boldsymbol{r}_0,\boldsymbol{r})G^A_\epsilon(\boldsymbol{r},\boldsymbol{r}_0)G^R_\epsilon(\boldsymbol{r}_0,\boldsymbol{r}')G^A_\epsilon(\boldsymbol{r}',\boldsymbol{r}_0)}\,. \tag{12.84}$$

To calculate this quantity we proceed as in section 12.2. A first contribution is obtained through replacement of the average product by a product of averages, namely

$$\overline{I(\boldsymbol{r})I(\boldsymbol{r}')} = \overline{I}(\boldsymbol{r})\overline{I}(\boldsymbol{r}')\,. \tag{12.85}$$

This term is represented by the diagram of Figure 12.24, i.e., by a product of two independent Diffusons.

Spatial correlations of the intensity are described by means of the correlation function[16]

$$\overline{\delta I(\boldsymbol{r})\delta I(\boldsymbol{r}')} = \overline{I(\boldsymbol{r})I(\boldsymbol{r}')} - \overline{I}(\boldsymbol{r})\overline{I}(\boldsymbol{r}') \tag{12.86}$$

and they are generated by connected diagrams. We define as well the normalized correlation function

$$C(\boldsymbol{r},\boldsymbol{r}') = \frac{\overline{\delta I(\boldsymbol{r})\delta I(\boldsymbol{r}')}}{I_d(\boldsymbol{r})I_d(\boldsymbol{r}')}. \tag{12.87}$$

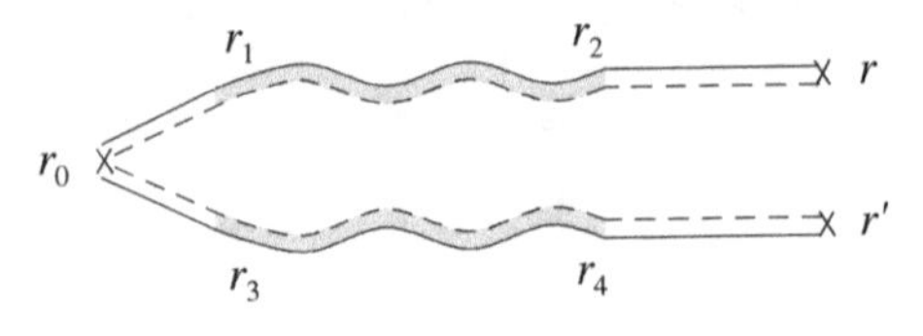

Figure 12.24 Diagrammatic representation of the correlation function of intensity evaluated at the Drude–Boltzmann approximation for which $\overline{I(\boldsymbol{r})I(\boldsymbol{r}')} = \overline{I}(\boldsymbol{r})\,\overline{I}(\boldsymbol{r}')$.

16 The average intensity $\overline{I}(\boldsymbol{r})$ at the Diffuson approximation is equal to $I_d(\boldsymbol{r})$ (see section 4.7). From now on, we shall keep the latter notation.

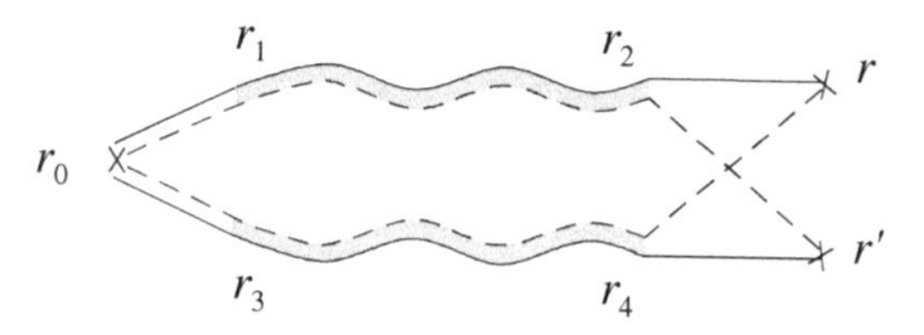

Figure 12.25 Diagrammatic representation of the intensity correlation function $\overline{\delta I(\boldsymbol{r})\delta I(\boldsymbol{r}')}^{(1)}$. The crossing of the two amplitudes between points $\boldsymbol{r}_2, \boldsymbol{r}_4, \boldsymbol{r}$ and $\boldsymbol{r}'$ can be rewritten as the square of the short range function $g(\boldsymbol{r}-\boldsymbol{r}')$ defined in (3.97).

To calculate this function we proceed as before and generate each term according to its number of Diffuson crossings, i.e., we perform a perturbation expansion with the small parameter $1/g$.

A12.1.1 Short range correlations

The main contribution to the intensity correlation is given by the diagram 12.25, so its expression is

$$\overline{\delta I(\boldsymbol{r})\delta I(\boldsymbol{r}')}^{(1)} = \left(\frac{4\pi}{c}\right)^2 \int d\boldsymbol{r}_1\, d\boldsymbol{r}_2\, d\boldsymbol{r}_3\, d\boldsymbol{r}_4\, |\overline{G}^R(\boldsymbol{r}_0,\boldsymbol{r}_1)|^2 |\overline{G}^R(\boldsymbol{r}_0,\boldsymbol{r}_3)|^2$$
$$\times\, \Gamma(\boldsymbol{r}_1,\boldsymbol{r}_2)\Gamma(\boldsymbol{r}_3,\boldsymbol{r}_4)\, \overline{G}^R(\boldsymbol{r}_2,\boldsymbol{r})\overline{G}^A(\boldsymbol{r}',\boldsymbol{r}_2)\overline{G}^R(\boldsymbol{r}_4,\boldsymbol{r}')\overline{G}^A(\boldsymbol{r},\boldsymbol{r}_4). \quad (12.88)$$

All average single-particle Green functions are taken at the same frequency. Working within the approximation of slow spatial variations allows us to take the product $\Gamma(\boldsymbol{r}_1,\boldsymbol{r}_2)\Gamma(\boldsymbol{r}_3,\boldsymbol{r}_4)$ outside the integral and approximate it by $\Gamma^2(\boldsymbol{r}_0,\boldsymbol{r})$, thus leaving four independent integrals over the variables $\boldsymbol{r}_1, \boldsymbol{r}_2, \boldsymbol{r}_3$ and $\boldsymbol{r}_4$. We recall that

$$\int d\boldsymbol{r}_2\, \overline{G}^R(\boldsymbol{r}_2,\boldsymbol{r})\overline{G}^A(\boldsymbol{r}',\boldsymbol{r}_2) = \frac{l_e}{4\pi} g(\boldsymbol{r}-\boldsymbol{r}'), \quad (12.89)$$

where $g(\boldsymbol{r}-\boldsymbol{r}')$ was defined in (3.97). The structure factor $\Gamma(\boldsymbol{r}_0,\boldsymbol{r})$ is given by (4.63)

$$\Gamma(\boldsymbol{r}_0,\boldsymbol{r}) = \frac{4\pi c}{l_e^2} I_d(r) = \frac{4\pi c}{l_e^2} \frac{1}{4\pi D r} \quad (12.90)$$

with $r = |\boldsymbol{r}-\boldsymbol{r}_0|$, so that[17]

$$\overline{\delta I(\boldsymbol{r})\delta I(\boldsymbol{r}')}^{(1)} = I_d^2(\boldsymbol{r}) g(\boldsymbol{r}-\boldsymbol{r}')^2. \quad (12.91)$$

[17] Keep in mind that positions are all measured relative to the position $\boldsymbol{r}_0$ of the point-like source.

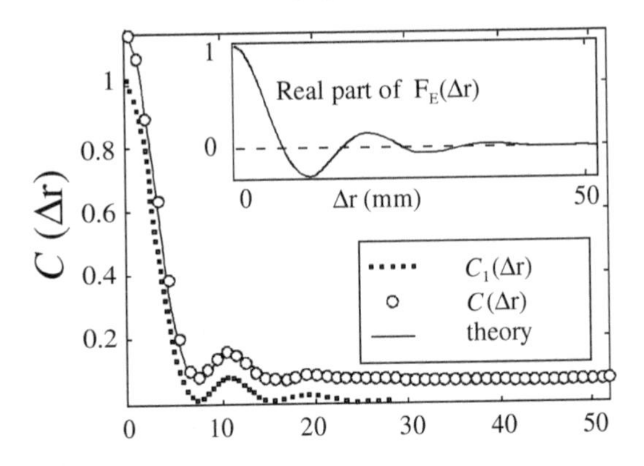

Figure 12.26 Measurement of the local intensity correlation $C(\Delta \boldsymbol{r})$ by varying the position of the detector on the outgoing surface located at $z = L$. $C_1(\Delta \boldsymbol{r})$ is obtained upon squaring the correlation function of the electric field displayed in the inset. The discrepancy $C - C_1$ accounts for the contribution of long range correlations studied in the next section. These results were obtained with microwaves that propagate in a random ensemble of polystyrene beads, and the average is taken over an ensemble of distinct configurations of the positions of the beads [323].

Finally, the normalized correlation function (12.87) acquires the simple form [302]

$$\boxed{C_1(\Delta \boldsymbol{r}) = \frac{\overline{\delta I(\boldsymbol{r})\delta I(\boldsymbol{r}')}^{(1)}}{I_d^2(\boldsymbol{r})} = \left(\frac{\sin k\Delta r}{k\Delta r}\right)^2 e^{-\Delta r/l_e}} \tag{12.92}$$

where $\Delta r = |\boldsymbol{r} - \boldsymbol{r}'|$. For $\boldsymbol{r} = \boldsymbol{r}'$, we recover the Rayleigh result (12.10), namely

$$\overline{I^2}(\boldsymbol{r}) = 2\ I_d^2(\boldsymbol{r}). \tag{12.93}$$

The correlation function $C_1(\boldsymbol{r}, \boldsymbol{r}')$ has been measured experimentally [321, 322]. Figure 12.26 shows this spatial dependence for the geometry of a long cylinder. The point source is located on the interface section of the cylinder whereas the points $\boldsymbol{r}$ and $\boldsymbol{r}'$ are located on the outgoing surface.

The extension of this result to a situation where the point source located in $\boldsymbol{r}_0$ is also allowed to move, leading to the correlation between $I(\boldsymbol{r}_0, \boldsymbol{r})$ and $I(\boldsymbol{r}'_0, \boldsymbol{r}')$, is straightforward. The corresponding expression for the normalized correlation function $C_1(\Delta \boldsymbol{r}_0, \Delta \boldsymbol{r})$ follows from Figure 12.27 and it generalizes (12.92) which now is rewritten as [323]

$$C_1(\Delta \boldsymbol{r}_0, \Delta \boldsymbol{r}) = g^2(\Delta \boldsymbol{r}_0) g^2(\Delta \boldsymbol{r}), \tag{12.94}$$

where $\Delta \boldsymbol{r}_0 = \boldsymbol{r}_0 - \boldsymbol{r}'_0$ and $\Delta \boldsymbol{r} = \boldsymbol{r} - \boldsymbol{r}'$ account respectively for the displacement of the point source and of the detector. The measured behaviors of C while varying either $\Delta \boldsymbol{r}$ or $\Delta \boldsymbol{r}_0$ are identical [323].

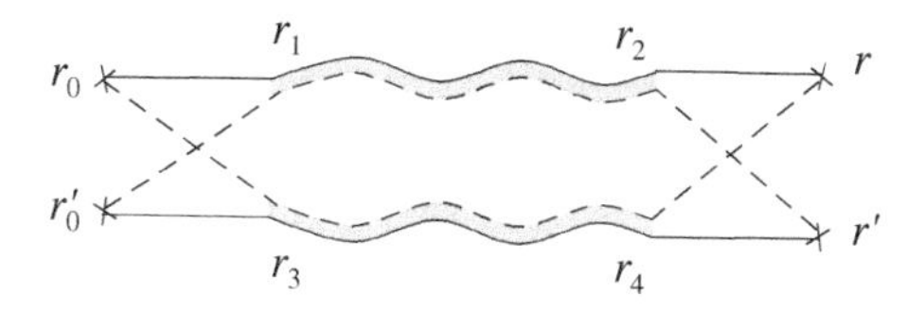

Figure 12.27 Schematic representation of the local correlation function of intensity obtained by varying the position of both the source $\boldsymbol{r}_0$ and of the measurement point $\boldsymbol{r}$. The crossings at each extremity of this diagram can be represented by means of the Hikami box $H^{(A)}$ of Figure 4.16.

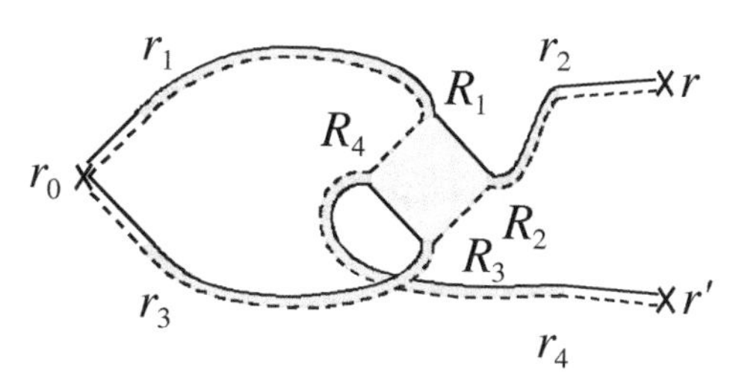

Figure 12.28 Diagram with a long range contribution to the intensity correlation. The interference resulting from the crossing of the Diffusons is described by a Hikami box.

A12.1.2 Long range correlations

The correlation function $\overline{\delta I(\boldsymbol{r})\delta I(\boldsymbol{r}')}$ includes additional contributions. As for angular correlations, we identify for instance the term denoted by $\overline{\delta I(\boldsymbol{r})\delta I(\boldsymbol{r}')}^{(2)}$ and given by the diagram of Figure 12.28. This contribution involves a single crossing of two Diffusons and gives rise to long range spatial correlations.

The crossing of two Diffusons is described by the same Hikami box that already showed up in the calculation of transmission correlations $C^{(2)}$ (Figure 12.9). The diagram 12.28 is similar to 12.9(a) and it leads to

$$\overline{\delta I(\boldsymbol{r})\delta I(\boldsymbol{r}')}^{(2)} = \left(\frac{4\pi}{c}\right)^2 \int \prod_{i=1}^{4} d\boldsymbol{R}_i \prod_{j=1}^{4} d\boldsymbol{r}_j |\overline{G}^R(\boldsymbol{r}_0,\boldsymbol{r}_1)|^2 |\overline{G}^R(\boldsymbol{r}_0,\boldsymbol{r}_3)|^2 H(\boldsymbol{R}_i)$$
$$\times\, \Gamma(\boldsymbol{r}_1,\boldsymbol{R}_1)\Gamma(\boldsymbol{r}_3,\boldsymbol{R}_3)\Gamma(\boldsymbol{R}_2,\boldsymbol{r}_2)\Gamma(\boldsymbol{R}_4,\boldsymbol{r}_4)|\overline{G}^R(\boldsymbol{r}_2,\boldsymbol{r})|^2 |\overline{G}^R(\boldsymbol{r}_4,\boldsymbol{r}')|^2, \tag{12.95}$$

where $H(\boldsymbol{R}_i)$ accounts for the crossing (Appendix A4.2.1). In the diffusive limit, integrals of single-particle Green's functions become decoupled and are all equal to $1/\gamma_e = l_e/4\pi$. Using (4.63) which relates $\Gamma(\boldsymbol{r}_0,\boldsymbol{r})$ to $I_d(\boldsymbol{r})$, we obtain

$$\overline{\delta I(\boldsymbol{r})\delta I(\boldsymbol{r}')}^{(2)} = \int \prod_{i=1}^{4} d\boldsymbol{R}_i \; I_d(\boldsymbol{R}_1) I_d(\boldsymbol{R}_3)\, H(\boldsymbol{R}_i)\, \Gamma(\boldsymbol{R}_2,\boldsymbol{r})\Gamma(\boldsymbol{R}_4,\boldsymbol{r}') \tag{12.96}$$

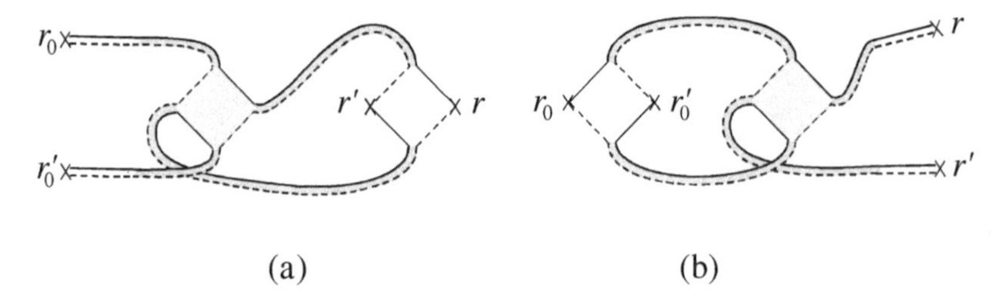

Figure 12.29 Diagrams representing the correlation $C_2(\Delta \boldsymbol{r}_0, \Delta \boldsymbol{r})$.

which, with the help of (4.150), yields[18]

$$\boxed{\overline{\delta I(\boldsymbol{r})\delta I(\boldsymbol{r}')}^{(2)} = \frac{2\pi l_e c^2}{3k^2} \int d\boldsymbol{R}\, I_d^2(\boldsymbol{R}) \left[\nabla_{\boldsymbol{R}} P_d(\boldsymbol{R},\boldsymbol{r})\right] \cdot \left[\nabla_{\boldsymbol{R}} P_d(\boldsymbol{R},\boldsymbol{r}')\right]} \tag{12.98}$$

This expression corresponds to long range spatial correlations of the intensity [324]. Because of the Diffuson crossing, it scales as the reciprocal of the dimensionless conductance (see Exercise 12.4):

$$C_2(\Delta \boldsymbol{r}) \propto \frac{1}{g}\, G(\Delta \boldsymbol{r}), \tag{12.99}$$

where $G(\Delta \boldsymbol{r})$ is a long range function since $\boldsymbol{r}$ and $\boldsymbol{r}'$ are linked by Diffusons. It is helpful to discuss further the structure of diagram 12.28 in order to gain more understanding of the origin of this long range behavior. The diagram can be split into two distinct parts. The first one includes the point source $\boldsymbol{r}_0$ and extends to the crossing point $\boldsymbol{R}$ of the two Diffusons. Its contribution in (12.98) is described by the term $I_d^2(\boldsymbol{R})$. The second part includes the two Diffusons that propagate towards the end points $\boldsymbol{r}$ and $\boldsymbol{r}'$. The Hikami box contribution is expressed in the form of the two gradients.

As for C_1, it is possible to account for the motion of the point source $\boldsymbol{r}_0$. The corresponding correlation $C_2(\Delta \boldsymbol{r}_0, \Delta \boldsymbol{r})$ represented by the diagram 12.29(b) is driven by the short range interference term (3.98) $g(\Delta \boldsymbol{r}_0)$ and it is such that

$$C_2(\Delta \boldsymbol{r}_0, \Delta \boldsymbol{r}) \propto \frac{1}{g}\, g(\Delta \boldsymbol{r}_0) G(\Delta \boldsymbol{r}), \tag{12.100}$$

where we are careful not to confuse the dimensionless conductance g and the short range function $g(\Delta r_0)$. Together with the contribution of 12.29(a), we obtain

$$C_2(\Delta \boldsymbol{r}_0, \Delta \boldsymbol{r}) \propto \frac{1}{g}\left[g(\Delta \boldsymbol{r}_0) G(\Delta \boldsymbol{r}) + g(\Delta \boldsymbol{r}) G(\Delta \boldsymbol{r}_0)\right]. \tag{12.101}$$

[18] Notice that applying the gradient terms to the intensity I_d leads to the equivalent expression:

$$\overline{\delta I(\boldsymbol{r})\delta I(\boldsymbol{r}')}^{(2)} = \frac{2\pi l_e c^2}{3k^2} \int d\boldsymbol{R} \left[\nabla_{\boldsymbol{R}} I_d(\boldsymbol{R})\right]^2 P_d(\boldsymbol{R},\boldsymbol{r}) P_d(\boldsymbol{R},\boldsymbol{r}'). \tag{12.97}$$

The spatial correlation function $C_2(\Delta \boldsymbol{r}_0, \Delta \boldsymbol{r})$ has been measured experimentally using microwaves propagating in solid suspensions of polystyrene beads [321–323]. Furthermore, previous expressions have also been extended to include effects of a finite absorption [325].

Exercise 12.3 Draw diagrams describing the spatial intensity correlation function $C_3(\Delta \boldsymbol{r}_0, \Delta \boldsymbol{r})$ that correspond to two crossings of Diffusons.
Show that diagrams of Figure 12.11 give a contribution that can be written as

$$C_3(\Delta \boldsymbol{r}_0, \Delta \boldsymbol{r}) \propto \frac{1}{g^2} \left[1 + g(\Delta \boldsymbol{r}_0) g(\Delta \boldsymbol{r})\right]. \tag{12.102}$$

Find another diagram which, at this order in perturbation, gives a contribution of the form $(1/g^2)\left[g(\Delta \boldsymbol{r}_0) + g(\Delta \boldsymbol{r})\right]$.

Exercise 12.4: correlation function $C_2(\Delta \boldsymbol{r})$ for an incident plane wave [324]
The correlation expressed by (12.98) corresponds to a point-like light source. We would like to express it for the geometry of a slab of finite length L illuminated by an incident plane wave as sketched in Figure 12.30.

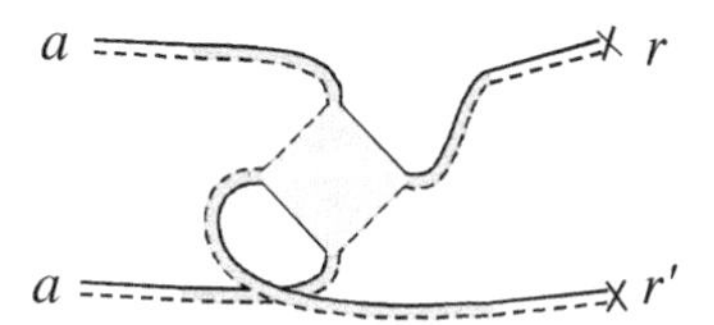

Figure 12.30 Diagrammatic representation of the spatial correlation $\overline{\delta I(\boldsymbol{r})\delta I(\boldsymbol{r}')}^{(2)}$ that corresponds to a measure of intensity on the outgoing interface plane and an incident plane wave along direction $\hat{s}_a$. This diagram describes the correlation calculated in Exercise 12.4.

Show that the correlation function $\overline{\delta I(\boldsymbol{r})\delta I(\boldsymbol{r}')}^{(2)}$ taken at $z = z' = L - l_e$ is a function of the projection $\boldsymbol{\rho} = (\boldsymbol{r} - \boldsymbol{r}')_\perp$ onto the xOy plane only

$$\overline{\delta I(\boldsymbol{r})\delta I(\boldsymbol{r}')}^{(2)} \simeq \int d^2\boldsymbol{q}\, e^{i\boldsymbol{q}\cdot\boldsymbol{\rho}} F_2(qL), \tag{12.103}$$

where $F_2(x)$ was calculated in (12.45). This correlation thus appears to be the Fourier transform of the angular correlation derived in section 12.4.2. Show that this Fourier transform leads for the normalized correlation function to the expression

$$C_2(\Delta \boldsymbol{r}) = \frac{9}{2k^2 l_e^2} \left(\frac{l_e}{L}\right)^3 G(\Delta r / L) \tag{12.104}$$

where $G(\Delta r/L)$ is such that

$$G(\Delta r/L) = \frac{L}{\Delta r} + \int_0^\infty dx J_0(x\Delta r/L)\left(\frac{\sinh 2x - 2x}{2\sinh^2 x} - 1\right) \tag{12.105}$$

where J_0 is a Bessel function.
Infer from this expression that the correlation $C_2(\Delta \boldsymbol{r})$ at this order in perturbation is long range and that it can be written as

$$C_2(\Delta \boldsymbol{r}) = \frac{3}{2\pi}\frac{1}{g}\left(\frac{S}{L^2}\right) G(\Delta r/L). \tag{12.106}$$

Notice that it is proportional to $1/g$, so it is much smaller than C_1.

13
Interactions and diffusion

In this chapter ν_0 is the average density of states per spin direction and $\rho_0 = \nu_0/\Omega$ is the density of states per unit volume. The energy $\Delta = 1/(\rho_0\Omega) = 1/\nu_0$ is the average level spacing per spin direction. Most results will be presented in the CGS system, in the form most commonly found in the literature. Unless specified, we take $\hbar = 1$.

13.1 Introduction

Up to this point, electron–electron interaction has been neglected in the description of spectral properties and electronic transport. Although electrons interact through the Coulomb interaction, the free electron model is a very good approximation for the description of many physical properties. This is due to the screening of the Coulomb interaction which occurs on a length of the order of the average distance between electrons. However, the electron–electron interaction has important physical consequences which can be classified in two categories.

- Each electron is sensitive not only to the disorder potential but also to the electronic density fluctuations induced by other electrons. As a result, the energy levels are shifted and the thermodynamic and transport properties are modified, particularly the density of states and the conductivity. The change in density of states is maximum around the Fermi level, thus constituting a direct signature of the interaction. Moreover, this change is important since it affects the orbital magnetism of the electron gas and the persistent current (Chapter 14). The change in conductivity is of the same order of magnitude as the weak localization correction, but its nature is quite different. In particular, it does not depend on the magnetic field, making it more difficult to observe.
- The interaction between electrons is an inelastic process (the total energy is conserved but the energy of each electron is modified). Each electron stays in an eigenstate of the single particle Schrödinger equation only during a finite time, and thus phase sensitive processes are affected. The electron–electron interaction, like the interaction with other degrees of freedom (e.g. phonons), destroys the phase coherence after a characteristic time τ_ϕ^{ee}. We shall see that this phase coherence time can also be understood as resulting from a fluctuating electromagnetic field that dephases the trajectories paired in either the

Diffuson or the Cooperon, in a way qualitatively similar to the dephasing induced by the motion of scatterers (Chapter 6). The equivalence between the effects of the Coulomb interaction and of a fluctuating electromagnetic field is far from being obvious. It relies on the fluctuation–dissipation theorem.

The diffusive nature of the electronic motion plays an essential role because it *strengthens* the interaction effect. This can be understood in the following qualitative way. As a result of the diffusive motion, the probability that two electrons interact is enhanced since an electron moves less rapidly than if its motion were ballistic. The effective interaction between two electrons is thus enhanced since each of them has an increased probability of staying in the interaction region. The modification of physical quantities due to the interaction must be proportional to the time spent in this region. More precisely, for a physical quantity $X(E)$ that depends on some energy scale E, we expect a modification proportional to the probability of return (5.5) into the interaction region during the time $\hbar/E$,

$$\frac{\delta X(E)}{X} \propto \frac{1}{\rho_0 \hbar \Omega} \int_0^{\hbar/E} Z(t)\, dt. \tag{13.1}$$

This chapter is devoted to the study of the interplay between disorder and interaction. The latter is treated as a perturbation to the model of independent electrons in a random potential.

13.2 Screened Coulomb interaction

The Coulomb interaction between two electrons at a distance R is described by the potential[1] $U_0(\boldsymbol{R}) = e^2/R$. In $d = 3$, the Fourier transform is $U_0(\boldsymbol{q}) = 4\pi e^2/q^2$. The case of other space dimensions is discussed in Appendix A13.1. In a metal, the Coulomb interaction is screened owing to the presence of the other electrons. In the Thomas–Fermi approximation, it becomes [326]

$$U(\boldsymbol{q}) = \frac{4\pi e^2}{q^2 + \kappa^2} \tag{13.2}$$

or

$$U(\boldsymbol{R}) = \frac{e^2}{R} e^{-\kappa R}. \tag{13.3}$$

The Thomas–Fermi wave vector κ, reciprocal of the screening length, is given by [326]

$$\boxed{\kappa^2 = 8\pi e^2 \rho_0 = 8\pi e^2 \frac{\nu_0}{\Omega}} \tag{13.4}$$

[1] In CGS units.

where ρ_0 is the density of states per unit volume and per spin direction. The $\boldsymbol{q}=0$ value of the screened interaction is simply related to the density of states by

$$U = U(\boldsymbol{q}=0) = \int U(\boldsymbol{R})\, d\boldsymbol{R} = \frac{1}{2\rho_0}. \tag{13.5}$$

For a metal, $\kappa \simeq k_F$, the screening is quite efficient and acts over a length of order λ_F (see remark on page 468). In the weak disorder limit, the screening length κ^{-1} is much smaller than the elastic mean free path l_e.

In a diffusive system, the screening is not instantaneous and it is important to describe its dynamics properly. The effective interaction between electrons is a function $U(\boldsymbol{q},\omega)$ of wave vector *and* of frequency. The charge reorganization is described by means of the dielectric function $\epsilon(\boldsymbol{q},\omega)$ which is related to the density-density response function $\overline{\chi}_0(\boldsymbol{q},\omega)$ by the relation $\epsilon(\boldsymbol{q},\omega) = 1 + U_0(\boldsymbol{q})\overline{\chi}_0(\boldsymbol{q},\omega)$. We have [326]

$$U(\boldsymbol{q},\omega) = \frac{U_0(\boldsymbol{q})}{\epsilon(\boldsymbol{q},\omega)} = \frac{U_0(\boldsymbol{q})}{1 + U_0(\boldsymbol{q})\overline{\chi}_0(\boldsymbol{q},\omega)}. \tag{13.6}$$

In section A7.1.2, we showed that, in the diffusion approximation,

$$\overline{\chi}_0(\boldsymbol{q},\omega) = 2\rho_0 \frac{Dq^2}{-i\omega + Dq^2}. \tag{13.7}$$

Therefore the effective interaction also depends on frequency: the interaction is said to be *dynamically screened*, a consequence of the diffusive nature of the electronic motion. Using (13.6), we obtain

$$U(\boldsymbol{q},\omega) = \frac{4\pi e^2}{q^2} \frac{-i\omega + Dq^2}{-i\omega + Dq^2 + D\kappa^2}. \tag{13.8}$$

The dielectric function $\epsilon(\boldsymbol{q},\omega)$ deduced from (13.7) is

$$\epsilon(\boldsymbol{q},\omega) = 1 + 2\rho_0 \frac{4\pi e^2}{q^2} \frac{Dq^2}{-i\omega + Dq^2}, \tag{13.9}$$

or

$$\boxed{\epsilon(\boldsymbol{q},\omega) = 1 + 4\pi\sigma_0 P_d(\boldsymbol{q},\omega)} \tag{13.10}$$

where $\sigma_0 = 2e^2 D\rho_0$ is the Drude conductivity given by the Einstein relation (7.14) and P_d is the Diffuson (4.88) written in the diffusion approximation, $ql_e \ll 1$ so that a fortiori $q \ll \kappa$. Finally the dielectric function becomes

$$\epsilon(\boldsymbol{q},\omega) \simeq \frac{4\pi\sigma_0}{-i\omega + Dq^2} \tag{13.11}$$

and the dynamical effective interaction is simply related to the density-density response function $\overline{\chi}_0(\boldsymbol{q},\omega)$ by

$$U(\boldsymbol{q},\omega) \simeq \frac{1}{\overline{\chi}_0(\boldsymbol{q},\omega)} = \frac{1}{2\rho_0}\frac{-i\omega + Dq^2}{Dq^2}. \tag{13.12}$$

The static limit, $\overline{\chi}_0 = 2\rho_0$, is obtained by taking $\omega = 0$.

Remark: the parameter r_s

A measure of the strength of the electronic correlations is provided by the dimensionless parameter r_s, which is the ratio between Coulomb potential energy and kinetic energy. The potential energy is of order e^2/a where $a = n^{-1/d}$ is the average distance between electrons in a gas of density n in d dimensions, while the kinetic energy is the Fermi energy $p_F^2/2m \propto \hbar^2/(2ma^2)$. The parameter r_s is thus defined as the ratio

$$r_s = \frac{a}{a_0} \propto \frac{\text{potential energy}}{\text{kinetic energy}}, \tag{13.13}$$

where $a_0 = \hbar^2/(m_0 e^2)$ is the Bohr radius and m_0 is the free electron mass (see section 2.1.1). The parameter r_s is thus proportional to the average distance between electrons. In three dimensions,

$$\frac{4\pi}{3} r_s^3 a_0^3 = \frac{1}{n}. \tag{13.14}$$

Since $k_F^3 = 3\pi^2 n$,

$$r_s = \left(\frac{9\pi}{4}\right)^{1/3} \frac{1}{k_F a_0}. \tag{13.15}$$

The screening vector κ is given by $\kappa^2 = 8\pi e^2 \rho_0$, where $\rho_0 = mk_F/(2\pi^2\hbar^2)$ is the density of states at the Fermi level, per unit volume and per spin direction. The ratio κ/k_F is thus given by:

$$\frac{\kappa}{k_F} = \left(\frac{16}{3\pi^2}\right)^{1/3} \left(\frac{m}{m_0}\right)^{1/2} \sqrt{r_s} \simeq 0.81 \left(\frac{m}{m_0}\right)^{1/2} \sqrt{r_s}. \tag{13.16}$$

For a metal such as copper, we have $r_s = 2.67$ and $m/m_0 = 1.3$, so that $\kappa/k_F \simeq 1.51$. The screening is thus very efficient in this metal. For other metals, see Table 1.1 in reference [326], taking note of the different definitions.

13.3 Hartree–Fock approximation

In order to describe the effects of the Coulomb interaction, we now use the Hartree–Fock approximation, the main points of which are recalled here. For more details, see [326]. First, the Hartree approximation consists in finding the solutions (ϵ_i, ϕ_i) of the non-linear equation

$$\epsilon_i \phi_i(\boldsymbol{r}) = -\frac{1}{2m}\Delta\phi_i(\boldsymbol{r}) + V_{ion}(\boldsymbol{r})\phi_i(\boldsymbol{r}) + \int U(\boldsymbol{r}-\boldsymbol{r}')n(\boldsymbol{r}')\phi_i(\boldsymbol{r})\,d\boldsymbol{r}', \tag{13.17}$$

where $V_{ion}(\boldsymbol{r})$ is the one-body potential describing the interaction of the electrons with the lattice and the impurities, and $U(\boldsymbol{r}-\boldsymbol{r}')$ is the two-body screened Coulomb interaction between electrons. The electronic density is

$$n(\mathbf{r}) = 2\sum_j f(\epsilon_j)|\phi_j(\mathbf{r})|^2 \tag{13.18}$$

and $f(\epsilon)$ is the Fermi factor. Electrical neutrality implies

$$\overline{V}_{ion}(\mathbf{r}) + \int U(\mathbf{r}-\mathbf{r}')\overline{n(\mathbf{r}')}\, d\mathbf{r}' = 0, \tag{13.19}$$

where we have replaced the spatial average by the disorder average. The Hartree equation is then rewritten as

$$\epsilon_i\phi_i(\mathbf{r}) = -\frac{1}{2m}\Delta\phi_i(\mathbf{r}) + V(\mathbf{r})\phi_i(\mathbf{r}) + \int U(\mathbf{r}-\mathbf{r}')(n(\mathbf{r}') - \overline{n})\,\phi_i(\mathbf{r})\, d\mathbf{r}', \tag{13.20}$$

in which $V(\mathbf{r}) = V_{ion}(\mathbf{r}) - \overline{V}_{ion}$ is the disorder potential defined in Chapter 2, and $\overline{n} = \overline{n(\mathbf{r})}$. In the Hartree approximation, the non-linear equation (13.20) is an effective Schrödinger equation in which the potential seen by one electron depends on the electronic density, that is on the wave functions of the other electrons. In this approximation, the total wave function is the product of single particle wave functions and does not satisfy the Pauli principle.

In order to take into account the antisymmetry of the total wave function, we have to add the Fock term, which describes *exchange* between particles of the same spin. Equation (13.20) then becomes [326, 327]:

$$\begin{aligned}\epsilon_i\phi_i(\mathbf{r}) &= -\frac{1}{2m}\Delta\phi_i(\mathbf{r}) + V(\mathbf{r})\phi_i(\mathbf{r}) + \int U(\mathbf{r}-\mathbf{r}')(n(\mathbf{r}') - \overline{n})\,\phi_i(\mathbf{r})\, d\mathbf{r}' \\ &\quad - \sum_j f(\epsilon_j)\int U(\mathbf{r}-\mathbf{r}')\,\phi_j^*(\mathbf{r}')\phi_j(\mathbf{r})\phi_i(\mathbf{r}')\, d\mathbf{r}'.\end{aligned} \tag{13.21}$$

In principle, the non-linear equation (13.21) must be solved self-consistently, which is a difficult problem with no analytical solution in the presence of disorder. Here we shall consider the interaction U as a perturbation and we shall limit ourselves to lowest order. The unperturbed states are the eigenstates $\{\epsilon_i, \phi_i(\mathbf{r})\}$ of the Hamiltonian (2.1) in the presence of disorder.[2] We calculate the effect of the Coulomb interaction in the framework of the Hartree–Fock approximation, following the method developed in Chapter 3. In principle, we have to evaluate first the diagonal Green function in the state ϕ_i by writing a Dyson equation analogous to (3.67). A self-energy $\Sigma_i = \delta\epsilon_i + i\Gamma_i$ is then obtained, whose real part $\delta\epsilon_i$ measures the displacement of the energy level ϵ_i and whose imaginary part Γ_i gives the width of this level, that is its inverse lifetime.[3] These two components of the self-energy give the correction to the one-particle density of states and the electronic lifetime. Here we present a simplified version of this formalism.

[2] We could have followed the same line as in Chapter 3 and treated on the same footing the disorder potential and the interaction as a perturbation on the basis of plane waves. This choice would not be convenient since a Fermi golden rule argument such as in (3.1) would introduce a width $\hbar/\tau_e$ resulting from the disorder effect. As we shall see, this width is much larger than the width due to interactions. It is thus more convenient to treat the disorder in a non-perturbative way.

[3] For more details on the N-body problem, see [328] and more specifically [329] for the Hartree–Fock approximation in the presence of disorder.

13.4 Density of states anomaly

13.4.1 Static interaction

We first evaluate the one-particle density of states in the presence of electron–electron interaction by calculating the shift $\delta\epsilon_i$ of the energy levels for a static screened interaction $U(\boldsymbol{r}-\boldsymbol{r}')$ given by (13.3) in $d=3$. The shift $\delta\epsilon_i$ is obtained from (13.21) and takes the form $\delta\epsilon_i = \delta\epsilon_i^H + \delta\epsilon_i^F$ where the Hartree contribution $\delta\epsilon_i^H$ is given, in the lowest order, by

$$\delta\epsilon_i^H = \int U(\boldsymbol{r}-\boldsymbol{r}')|\phi_i(\boldsymbol{r})|^2\big(n(\boldsymbol{r}')-\overline{n}\big)\,d\boldsymbol{r}\,d\boldsymbol{r}'. \tag{13.22}$$

Similarly, the exchange, or Fock, term $\delta\epsilon_i^F$ is written

$$\delta\epsilon_i^F = -\sum_{j,\sigma} f(\epsilon_j)\int U(\boldsymbol{r}-\boldsymbol{r}')\phi_j^*(\boldsymbol{r}')\phi_j(\boldsymbol{r})\phi_i^*(\boldsymbol{r})\phi_i(\boldsymbol{r}')\,d\boldsymbol{r}\,d\boldsymbol{r}'. \tag{13.23}$$

The total energy E_T is

$$\begin{aligned} E_T = E_T^0 &+ \frac{1}{2}\int U(\boldsymbol{r}-\boldsymbol{r}')n(\boldsymbol{r})\delta n(\boldsymbol{r}')\,d\boldsymbol{r}\,d\boldsymbol{r}' \\ &- \sum_{i,j} f(\epsilon_i)f(\epsilon_j)\int U(\boldsymbol{r}-\boldsymbol{r}')\phi_j^*(\boldsymbol{r}')\phi_j(\boldsymbol{r})\phi_i^*(\boldsymbol{r})\phi_i(\boldsymbol{r}')\,d\boldsymbol{r}\,d\boldsymbol{r}' \end{aligned} \tag{13.24}$$

where E_T^0 is the total energy in the absence of interaction and $\delta n(\boldsymbol{r}') = n(\boldsymbol{r}') - \overline{n}$. The factor $1/2$ avoids double counting of the interaction in the total energy. The mean shift of an energy level ϵ is defined by

$$\Delta_\epsilon = \frac{1}{\nu_0}\overline{\sum_i \delta(\epsilon-\epsilon_i)\delta\epsilon_i}. \tag{13.25}$$

Since, on average, each energy level ϵ is changed into $\epsilon + \Delta_\epsilon$, the distance between two levels ϵ_1 and ϵ_2 becomes $(\epsilon_2-\epsilon_1)[1+\partial\Delta_\epsilon/\partial\epsilon]$. This shift leads to a relative change in the density of states

$$\frac{\delta\nu}{\nu_0} = -\frac{\partial\Delta_\epsilon}{\partial\epsilon}. \tag{13.26}$$

Consider first the *exchange term* (13.23). By inserting the relation $f(\epsilon_j) = \int d\epsilon'\delta(\epsilon'-\epsilon_j)\, f(\epsilon')$ and using (3.26) to transform the product of four wave functions into a product of two non-local densities of states, this term can be rewritten as:[4]

$$\Delta_\epsilon^F = -\frac{1}{\nu_0}\int_{-\infty}^{\infty} f(\epsilon-\omega)\,d\omega\int U(\boldsymbol{r}-\boldsymbol{r}')\overline{\rho_\epsilon(\boldsymbol{r},\boldsymbol{r}')\rho_{\epsilon-\omega}(\boldsymbol{r}',\boldsymbol{r})}\,d\boldsymbol{r}\,d\boldsymbol{r}'. \tag{13.27}$$

[4] The zero of the energies is taken at the Fermi level.

Expressing the correlation function of the non-local density of states $\rho_\epsilon(\boldsymbol{r},\boldsymbol{r}')$ with the help of (4.206) and using (13.26), we obtain[5]

$$\delta\nu^F(\epsilon) = \frac{\rho_0}{\pi}\int_{-\infty}^{\infty} f'(\epsilon-\omega)\,d\omega \int U(\boldsymbol{r}-\boldsymbol{r}')\,\mathrm{Re}P_d(\boldsymbol{r},\boldsymbol{r}',\omega)\,d\boldsymbol{r}\,d\boldsymbol{r}'. \tag{13.28}$$

The interaction (13.3) being short range (much shorter than l_e), the integral factorizes

$$\delta\nu^F(\epsilon) \simeq \frac{U\rho_0}{\pi}\int_{-\infty}^{\infty} f'(\epsilon-\omega)\,d\omega \int \mathrm{Re}P_d(\boldsymbol{r},\boldsymbol{r},\omega)\,d\boldsymbol{r}, \tag{13.29}$$

where the parameter U is defined by (13.5). Introducing the temporal Fourier transform of relation (5.5), $\int \mathrm{Re}P_d(\boldsymbol{r},\boldsymbol{r},\omega)d\boldsymbol{r} = \int_0^\infty Z(t)\cos\omega t dt$, the density of states correction due to the interaction can be written as a function of the return probability $Z(t)$. Using (15.109) and $U\rho_0 = 1/2$, we obtain[6]

$$\delta\nu^F(\epsilon) = -\frac{1}{2\pi}\int_0^{\infty} \frac{\pi T t}{\sinh \pi T t} Z(t)\cos\epsilon t\,dt. \tag{13.30}$$

The contribution of the *Hartree term* to the density of states correction is obtained from the relations (13.22) and (13.25)

$$\Delta_\epsilon^H = \frac{2}{\nu_0}\int_{-\infty}^{\infty} f(\epsilon-\omega)\,d\omega \int U(\boldsymbol{r}-\boldsymbol{r}')\,\overline{\rho_\epsilon(\boldsymbol{r},\boldsymbol{r})\rho_{\epsilon-\omega}(\boldsymbol{r}',\boldsymbol{r}')}^c\,d\boldsymbol{r}\,d\boldsymbol{r}'. \tag{13.31}$$

The product of the local densities that appears in this expression is given by (4.209). Due to the short range potential, the main contribution has only one Diffuson. The relation (13.26) for the density of states gives

$$\delta\nu^H(\epsilon) = -2\frac{\rho_0}{\pi}\int_{-\infty}^{\infty} f'(\epsilon-\omega)\,d\omega \int g^2(\boldsymbol{R})\,U(\boldsymbol{R})\mathrm{Re}P_d(\boldsymbol{r},\boldsymbol{r},\omega)\,d\boldsymbol{r}\,d\boldsymbol{r}' \tag{13.32}$$

where the function $g(\boldsymbol{R})$ is defined by (3.98). The short range term $g^2(\boldsymbol{R})U(\boldsymbol{R})$ can be integrated separately, and the Hartree contribution is written as

$$\delta\nu^H(\epsilon) = -\frac{F}{\pi}\int_{-\infty}^{\infty} f'(\epsilon-\omega)\,d\omega \int \mathrm{Re}P_d(\boldsymbol{r},\boldsymbol{r},\omega)\,d\boldsymbol{r} \tag{13.33}$$

where we have introduced the parameter (Figure 13.1)

$$\boxed{F = \frac{\int g^2(\boldsymbol{R})U(\boldsymbol{R})d\boldsymbol{R}}{\int U(\boldsymbol{R})d\boldsymbol{R}} = \frac{1}{U}\int g^2(\boldsymbol{R})\,U(\boldsymbol{R})\,d\boldsymbol{R}} \tag{13.34}$$

with $U = 1/2\rho_0$ (relation 13.5). The parameter F varies between 0 for strong screening ($\kappa \to \infty$) and 1 for weak screening ($\kappa \to 0$). For more details, see Exercises 13.3 and 13.4.

[5] We keep only the contribution of the Diffuson, and we also check that the product of average values gives a negligible contribution.
[6] Notice that the result does not depend on the coupling constant e^2. This is because the screening length κ^{-1} is much smaller than l_e, and thus the relation (13.29) contains the coefficient $U = U(q=0) = 4\pi e^2/\kappa^2$. This ratio no longer depends on e^2.

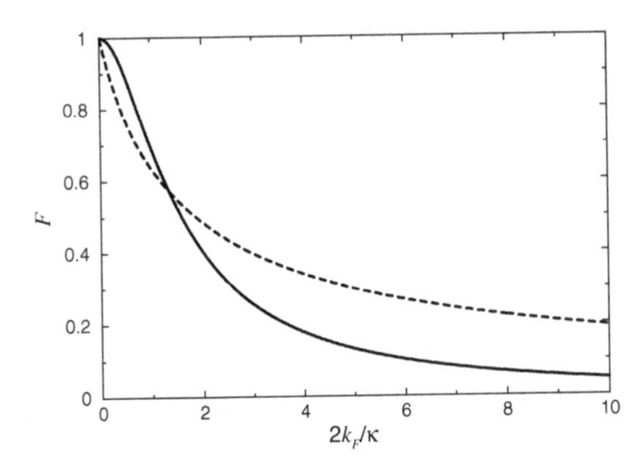

Figure 13.1 Variation of the parameter F as a function of the ratio $2k_F/\kappa$, in dimensions $d=2$ (dashed line) and $d=3$ (full line).

The expression of $\delta\nu^H(\epsilon)$ is proportional to (13.29) so the total correction to the density of states per spin direction is

$$\boxed{\delta\nu(\epsilon) = -\frac{\lambda_\nu}{2\pi}\int_0^\infty \frac{\pi T t}{\sinh \pi T t} Z(t)\cos\epsilon t\, dt} \tag{13.35}$$

where the interaction is described by the parameter λ_ν which here takes the value $\lambda_\nu = 1-2F$. This value corresponds to a static interaction and is different from the value obtained for a dynamically screened interaction $U(\boldsymbol{q},\omega)$ (page 480). Moreover, λ_ν depends on the range of the interaction through the parameter F. Since the latter varies between 0 and 1, the sign of the density of states correction seems to depend on the nature of the screened interaction. If the exchange term is larger than the Hartree term ($F \ll 1$), that is for a potential whose range is larger than the Fermi wavelength, the density of states correction is negative as observed experimentally. However, it seems that $\delta\nu(\epsilon)$ might become positive for a very short range interaction. For example, in copper where $F \simeq 0.6$, a positive correction might be expected. We shall see in section 13.4.3 that this result is an *artefact* and that taking into account the dynamical character of the screened interaction modifies the amplitude of the correction which always stays negative.

The density of states correction (13.35) reflects obviously the diffusive motion of the electrons and thus depends on space dimensionality[7] through the probability $Z(t)$ which, in free space, is given by (5.24). For example, in a quasi-one-dimensional system of volume Ω at zero temperature, we have $Z(t) = \Omega/\sqrt{4\pi D t}$. Thus, from (13.35) and (15.82),

$$\delta\rho(\epsilon) = -\frac{\lambda_\nu}{4\pi\sqrt{2}}\frac{1}{\sqrt{D\epsilon}} \tag{13.36}$$

[7] Recall that the diffusive motion depends on the dimensionality d, but the Coulomb interaction is always three dimensional if all the dimensions of the sample are larger than κ^{-1}. In a semiconductor where κ^{-1} is large, the nature of the interaction may change in the presence of gates, due to the existence of image charges [327].

is obtained *per spin direction and unit volume*. Similarly, in two dimensions and in the limit $\epsilon\tau_e \ll 1$,

$$\delta\rho(\epsilon) = \frac{\lambda_\nu}{8\pi^2 D} \ln \epsilon\tau_e, \tag{13.37}$$

is obtained from (15.83), and in three dimensions (15.84) gives

$$\delta\rho(\epsilon) = \frac{\lambda_\nu}{8\pi^2\sqrt{2D}} \left(\sqrt{\frac{\epsilon}{D}} - C\right) \tag{13.38}$$

where C is a constant independent of ϵ.

The dependence of the density of states correction on the return probability exhibits the same functional dependence as the weak localization correction $\Delta\sigma$ (see 7.56). Formally we have

$$\frac{\delta\rho}{\rho_0} \propto \frac{\Delta\sigma}{\sigma_0}(L_\phi = L_\epsilon), \tag{13.39}$$

where $L_\epsilon = \sqrt{D/\epsilon}$. Here the long time cutoff is determined by the energy ϵ instead of the phase coherence time $1/\tau_\phi$. We summarize the above relations for the density of states correction, also called *density of states anomaly* by

$$\boxed{\delta\rho(\epsilon) \propto -\frac{1}{D} \begin{cases} L_\epsilon - l_e & d=1 \\ \ln \dfrac{L_\epsilon}{l_e} & d=2 \\ \dfrac{1}{l_e} - \dfrac{1}{L_\epsilon} & d=3 \end{cases}} \tag{13.40}$$

Notice that the amplitude of the relative correction is of order $1/g$, where g is the dimensionless conductance.

Exercise 13.1 Show that

$$\mathrm{Re}P(\boldsymbol{r},\boldsymbol{r}',\epsilon) = \mathrm{Im}\int_\epsilon^\infty d\omega \int P(\boldsymbol{r},\boldsymbol{r}'',\omega)P(\boldsymbol{r}'',\boldsymbol{r}',\omega)\, d\boldsymbol{r}''$$

and more generally that

$$\mathrm{Re}\int_{-\infty}^{\infty} f'(\epsilon-\omega)P(\boldsymbol{r},\boldsymbol{r}',\omega)\,d\omega = -\,\mathrm{Im}\int_{-\infty}^{\infty} f(\epsilon-\omega)\,d\omega \int P(\boldsymbol{r},\boldsymbol{r}'',\omega)P(\boldsymbol{r}'',\boldsymbol{r}',\omega)\,d\boldsymbol{r}''$$

or

$$\int_{-\infty}^{\infty} f'(\epsilon-\omega)\mathrm{Re}\,P(\boldsymbol{q},\omega)\,d\omega = -\int_{-\infty}^{\infty} f(\epsilon-\omega)\mathrm{Im}\,P^2(\boldsymbol{q},\omega)\,d\omega \tag{13.41}$$

where P is either the Diffuson P_d, or the Cooperon P_c (see also 15.8).

Exercise 13.2 Show that the density of states correction (13.35) can also be written in the equivalent forms

$$\delta\nu(\epsilon) = -\frac{1-2F}{2\pi}\int_{-\infty}^{\infty} d\omega f(\epsilon-\omega)\sum_q \mathrm{Im}\,P_d^2(\boldsymbol{q},\omega) \tag{13.42}$$

or

$$\delta\nu(\epsilon) = -\frac{1-2F}{4\pi}\int_0^{\infty} d\omega\left[\tanh\frac{\epsilon+\omega}{2T} + \tanh\frac{\omega-\epsilon}{2T}\right]\sum_q \mathrm{Im}\,P_d^2(\boldsymbol{q},\omega). \tag{13.43}$$

Exercise 13.3 The parameter F represents the ratio of the Hartree and Fock (exchange) contributions. By calculating the Hartree contribution (13.32) in reciprocal space, show that F can also be written in the form [327],

$$F = \frac{\langle U(\boldsymbol{p}-\boldsymbol{p}')\rangle}{U}, \tag{13.44}$$

where $U(\boldsymbol{p}-\boldsymbol{p}')$ is the Fourier transform of the interaction $U(\boldsymbol{r})$ and the average is made over the momenta $\boldsymbol{p}$ and $\boldsymbol{p}'$ taken on the Fermi surface. Check directly that

$$\langle U(\boldsymbol{p}-\boldsymbol{p}')\rangle = \int a(\boldsymbol{q})\,U(\boldsymbol{q})\,d\boldsymbol{q} = \int g^2(\boldsymbol{R})\,U(\boldsymbol{R})\,d\boldsymbol{R} \tag{13.45}$$

where $a(\boldsymbol{q})$ is the Fourier transform of $g^2(\boldsymbol{R})$ defined in (3.98).

Exercise 13.4: calculation of F in the Thomas–Fermi approximation

In three dimensions, using the expression $U(\boldsymbol{R}) = (e^2/R)e^{-\kappa R}$ and the relation (3.98) for $g(\boldsymbol{R})$, show that in the limit $\kappa l_e \gg 1$,

$$F = \frac{\kappa^2}{4k_F^2}\ln\left(1+\frac{4k_F^2}{\kappa^2}\right). \tag{13.46}$$

F reaches the value 1 for a perfectly screened interaction (see Figure 13.1). For copper where $\kappa/k_F \simeq 1.51$, F is of the order of 0.6.

In two dimensions, show that in the limit $\kappa l_e \gg 1$,

$$F = \frac{2}{\pi}\frac{\kappa}{\sqrt{\kappa^2-4k_F^2}}\arctan\frac{\sqrt{\kappa^2-4k_F^2}}{2k_F} \tag{13.47}$$

or

$$F = \int \frac{d\theta}{2\pi} \frac{1}{1 + (2k_F/\kappa) \sin\theta/2}. \tag{13.48}$$

Show first that in two dimensions $U(\boldsymbol{q}) = 2\pi e^2/(q+\kappa)$ and use the corresponding expression (3.102) for $a(\boldsymbol{q})$.

13.4.2 Tunnel conductance and density of states anomaly

The change in the density of states in the vicinity of the Fermi level can be observed experimentally by tunnel conductance measurements. The experiment consists of connecting the metal we want to study to another metal whose density of states is known. One of the two metals is oxidized before growing a layer of the second metal, thus creating an oxide barrier between the two conductors. The width of the oxide layer can be controlled and constitutes a tunnel barrier. The tunnel current is proportional to the density of states of the two metals, and its measurement gives access to the density of states anomaly. The dip in the tunnel conductance is not specific to the weak disorder regime considered here, but rather is a general characteristic of the Coulomb interaction which subsists even for strong disorder (near the metal–insulator transition) or for semiconductors.

Let us recall the measurement principle. The tunnel current $I(V)$, for a voltage $V > 0$ applied between two metals a and b, depends on the tunnel probability of transferring electrons between the two metals. The tunnel rate between an initial state i of metal a and final state f of metal b is given by the Fermi golden rule

$$\Gamma_{i \to f}(V) = \frac{2\pi}{\hbar} |t_{if}|^2 \delta(E_i - E_f + eV), \tag{13.49}$$

where t_{if} is a matrix element which describes the coupling between the two states and which depends on the geometry of the junction. The tunnel rate between metal a and metal b depends on the occupation numbers of the initial and final states. It is given by

$$\Gamma_{ab}(V) = \frac{2\pi}{\hbar} \sum_{if} |t_{if}|^2 f(E_i)[1 - f(E_f)] \delta(E_i - E_f + eV) \tag{13.50}$$

where $f(E)$ is the Fermi distribution. At finite temperature, there is also a finite transition probability $\Gamma_{ba}(V)$ from b to a, so the tunnel current between a and b is

$$I = e(\Gamma_{ab} - \Gamma_{ba}) = 2\pi \frac{e}{\hbar} \sum_{if} |t_{if}|^2 [f(E_i) - f(E_f)] \delta(E_i - E_f + eV). \tag{13.51}$$

Assuming that the tunnel matrix element depends only weakly on the energy and that the voltage V and the temperature T are small compared to the Fermi energy and the height of the tunnel barrier, the sums can be replaced by integrals, and we can introduce the respective densities of states $\rho_a(\epsilon)$ and $\rho(\epsilon)$ of the reference electrode and of the conductor being studied. We then get

$$I(V) = 2\pi \frac{e}{\hbar} |t|^2 \int_{-\infty}^{\infty} \rho_a(\epsilon)\rho(\epsilon + eV) [f(\epsilon) - f(\epsilon + eV)] \, d\epsilon. \tag{13.52}$$

If the densities of states vary only weakly near the Fermi level (they are denoted ρ_a and ρ_0), the integral takes a very simple form, and for a small voltage gives a tunnel current $I(V)$ proportional to V, and thus a linear characteristic which defines the tunnel conductance G_t

$$G_t = 2\pi \frac{e^2}{\hbar} |t|^2 \rho_a \rho_0. \tag{13.53}$$

Assuming that the density of states of the reference electrode is energy independent, a variation $\delta\rho(\epsilon)$ of the density of states of the metal being studied leads to a variation $\delta I(V)$ of the current and thus to a variation $\delta G_t(V)$ of the tunnel conductance given by

$$\delta G_t(V) = \frac{d\delta I}{dV} = -2\pi \frac{e^2}{\hbar} \rho_a |t|^2 \int_{-\infty}^{\infty} d\epsilon \, \delta\rho(\epsilon) f'(\epsilon - eV) \tag{13.54}$$

so that, at zero temperature

$$\frac{\delta G_t(V)}{G_t} = \frac{\delta\rho(eV)}{\rho_0}. \tag{13.55}$$

The reduction of the tunnel conductance is thus a direct measurement of the variation of the density of states due to Coulomb interaction.

The first experimental evidence for the variations of the tunnel conductance as given by (13.40) is shown in Figure 13.2. It displays the tunnel conductance G_t as a function of the voltage V for a tunnel contact InO_x–insulator–lead with indium oxide films of different thickness. When the film thickness increases, we see a crossover from a two-dimensional behavior (logarithmic) of the density of states anomaly towards a three-dimensional behavior proportional to $\sqrt{V}$ [330].

The one-dimensional behavior has also been observed in aluminum wires [331]. The behavior (13.36) in $1/\sqrt{V}$ is visible[8] in Figure 13.3.

[8] A quantitative comparison with the theoretical prediction (13.36) is more complicated [331]. It takes into account other effects neglected in our calculation, such as the influence of the electromagnetic environment on the conductance as well as finite geometry effects. However, it is interesting that these additional effects leave the dependence of the tunnel conductance as the function of the voltage unchanged, and simply give a renormalization of the diffusion constant which appears in the function $Z(t)$ (see Exercise 13.9).

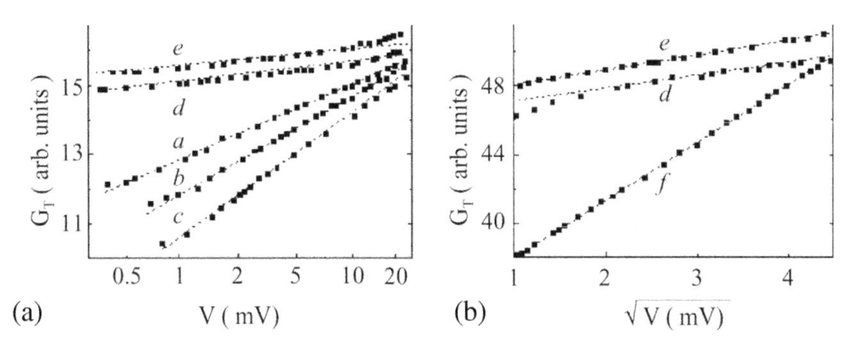

Figure 13.2 Tunnel conductance versus lnV (a) and $\sqrt{V}$ (b), for a junction InO_x–insulator–lead. The different curves are obtained by varying the thickness of the indium oxide film: (a) $a = 160$Å; (b) $a = 190$Å; (c) $a = 210$Å; (d) $a = 310$Å; (e) $a = 460$Å; (f) $a = 2600$Å [330].

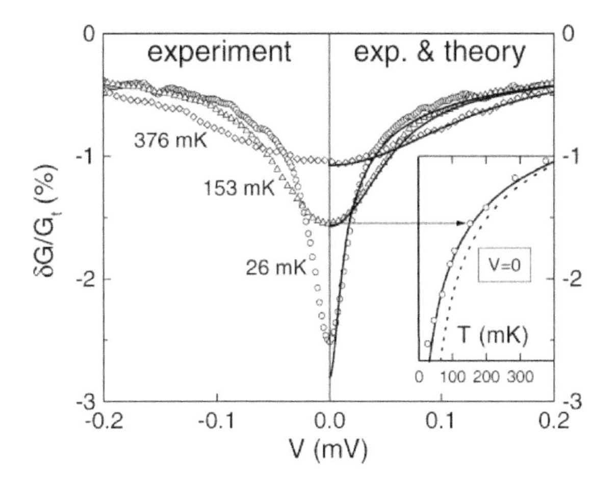

Figure 13.3 Measurement of the tunnel conductance in aluminum wires as a function of the voltage and for different temperatures. The continuous lines present the theoretical prediction (13.36), with a diffusion coefficient D^* renormalized by the geometry (Exercise 13.9) [331].

Exercise 13.5: tunnel conductance at finite temperature

Using (13.35) and (13.54), show that, at finite temperature, the relative correction to the tunnel conductance is written (exchange term):

$$\frac{\delta G_t(V,T)}{G_t} = -\frac{1}{2\pi \nu_0} \int_0^\infty Z(t) R^2(t) \cos eVt \, dt \tag{13.56}$$

where the function $R(t)$ is given by $R(t) = \pi T t / \sinh(\pi T t)$.

13.4.3 Dynamically screened interaction

Since it does not account for the dynamical screening of the interaction, that is, its frequency dependence, the above derivation of the density of states is an approximation. The systematic calculation [328] of the density of states correction in the Hartree–Fock approximation using expression (13.6) for the effective potential is obtained by evaluating the two diagrams of Figure 13.4 (see also Figure 13.5). The two upper diagrams are the usual representations of the Hartree and exchange (Fock) corrections in the many-body formalism [332]. The two lower diagrams propose a topologically equivalent representation which shows more explicitly the role of the structure factor Γ describing the diffusive nature of the electronic motion. The exchange (Fock) term contributing to the variation δG^F of the Green function due to interactions [332] is obtained by a separation of slow and rapid spatial variations and, using Table 3.2, we find

$$\delta G^F(\boldsymbol{r}_0,\boldsymbol{r}_0) = -\int_{-\infty}^{\infty}\frac{d\omega}{2i\pi}f(\epsilon-\omega)\frac{f^{2,1}}{\gamma_e}\int \Gamma_\omega(\boldsymbol{r}_0,\boldsymbol{r})\frac{1}{\gamma_e^2}U_\omega(\boldsymbol{r},\boldsymbol{r}')\Gamma_\omega(\boldsymbol{r}',\boldsymbol{r}_0)\,d\boldsymbol{r}\,d\boldsymbol{r}'. \tag{13.57}$$

Using (4.37) and the expression for $f^{2,1}$ given in Table 3.2, we deduce

$$\delta G^F(\boldsymbol{r}_0,\boldsymbol{r}_0) = \rho_0\int_{-\infty}^{\infty} f(\epsilon-\omega)\,d\omega\int P_d(\boldsymbol{r}_0,\boldsymbol{r},\omega)U_\omega(\boldsymbol{r},\boldsymbol{r}')P_d(\boldsymbol{r}',\boldsymbol{r}_0,\omega)\,d\boldsymbol{r}\,d\boldsymbol{r}'. \tag{13.58}$$

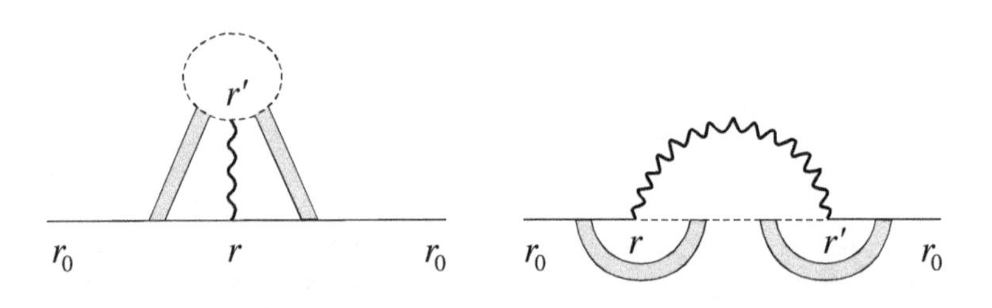

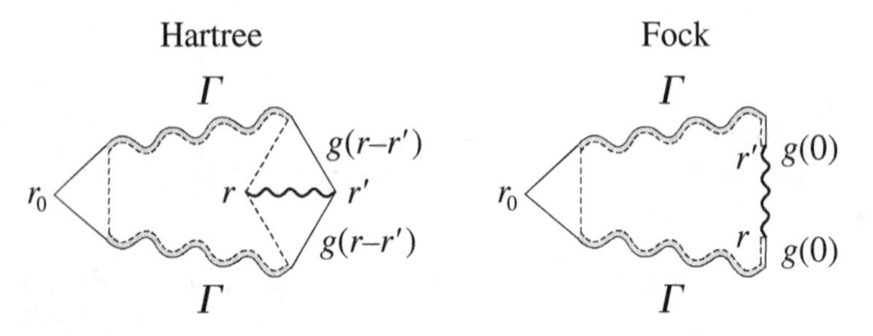

Figure 13.4 Hartree and exchange (Fock) diagrams for the local density of states anomaly. The Hartree diagram contains the function $g^2(\boldsymbol{r}-\boldsymbol{r}')$ and the Fock diagram contains the factor $g^2(0) = 1$.

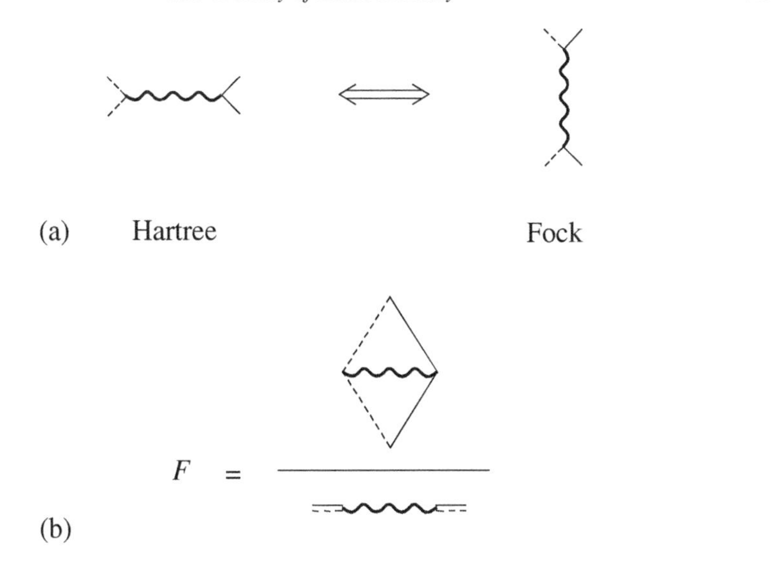

Figure 13.5 The two diagrams, Hartree and Fock, are identical given the correspondence (a). (b) Representation of the parameter F (13.34) as the ratio of two diagrams.

$U_\omega(\boldsymbol{r})$ is the Fourier transform of $U(\boldsymbol{q},\omega)$ given by (13.12). The corresponding correction to the local density of states is related by (3.25) to the imaginary part of $\delta G^F(\boldsymbol{r}_0,\boldsymbol{r}_0)$, so that

$$\delta\rho^F(\boldsymbol{r}_0,\boldsymbol{r}_0) = -\frac{\rho_0}{\pi}\mathrm{Im}\int_{-\infty}^{\infty} f(\epsilon-\omega)\,d\omega \int P_d(\boldsymbol{r}_0,\boldsymbol{r},\omega)U_\omega(\boldsymbol{r},\boldsymbol{r}')P_d(\boldsymbol{r}',\boldsymbol{r}_0,\omega)\,d\boldsymbol{r}\,d\boldsymbol{r}'. \tag{13.59}$$

The correction to the total density of states is obtained by integrating over r_0, and is written as[9]

$$\delta\nu^F = -\frac{\rho_0}{\pi}\mathrm{Im}\int_{-\infty}^{\infty} f(\epsilon-\omega)\,d\omega \sum_{\boldsymbol{q}} U(\boldsymbol{q},\omega)P_d^2(\boldsymbol{q},\omega). \tag{13.61}$$

In the limit of a static interaction $U(\boldsymbol{q})$, the result (13.29) is recovered (see Exercise 13.2). On the other hand, for a dynamically screened interaction and in the limit $q \ll \kappa$, we have

$$U(\boldsymbol{q},\omega) \simeq U\frac{-i\omega + Dq^2}{Dq^2}. \tag{13.62}$$

In three dimensions, the result (13.28) obtained for a static interaction is recovered, up to a factor 2 (see Exercise 13.6).

[9] A tunnel conductance measurement allows access to the *local* density of states $\delta\rho(\boldsymbol{r}_0,\boldsymbol{r}_0)$. In a non-translation-invariant system, the density of states anomaly depends on the measurement point and is proportional to the return probability to $\boldsymbol{r}_0$

$$\delta\rho(\epsilon,\boldsymbol{r}_0) \propto \int_0^{\infty} \frac{\pi T t}{\sinh \pi T t} P(\boldsymbol{r}_0,\boldsymbol{r}_0,t)\cos\epsilon t\,dt. \tag{13.60}$$

We now consider the Hartree term, which is obtained from the corresponding diagram of Figure 13.4 and gives a contribution δG^H to the local Green function of the form[10]

$$\delta G^H(\boldsymbol{r}_0,\boldsymbol{r}_0) = -2\rho_0 \int_{-\infty}^{\infty} f(\epsilon-\omega)\, d\omega \int P_d(\boldsymbol{r}_0,\boldsymbol{r},\omega) g^2(\boldsymbol{r}-\boldsymbol{r}') U(\boldsymbol{r},\boldsymbol{r}') P_d(\boldsymbol{r},\boldsymbol{r}_0,\omega)\, d\boldsymbol{r}\, d\boldsymbol{r}'. \tag{13.63}$$

For this term, the energy exchange during the interaction is zero, and the interaction $U(\boldsymbol{r},\boldsymbol{r}')$ remains static. Therefore the local density of states correction is

$$\delta\rho^H(\boldsymbol{r}_0,\boldsymbol{r}_0) = -2\frac{\rho_0}{\pi}\mathrm{Im}\int_{-\infty}^{\infty} f(\epsilon-\omega)\, d\omega \int P_d(\boldsymbol{r}_0,\boldsymbol{r},\omega) g^2(\boldsymbol{r}-\boldsymbol{r}') U(\boldsymbol{r},\boldsymbol{r}') P_d(\boldsymbol{r},\boldsymbol{r}_0,\omega)\, d\boldsymbol{r}\, d\boldsymbol{r}'. \tag{13.64}$$

Integrating over $\boldsymbol{r}_0$ and using (13.45), we obtain

$$\delta\nu^H = 2F\frac{U\rho_0}{\pi}\int_{-\infty}^{\infty} f(\epsilon-\omega)\, d\omega \sum_{\boldsymbol{q}} \mathrm{Im} P_d^2(\boldsymbol{q},\omega). \tag{13.65}$$

For the Hartree term, the interaction is *not* dynamically screened and the static result (13.33) is recovered. Finally, for a dynamically screened interaction, the correction to the density of states is obtained by adding the two contributions, so that

$$\boxed{\delta\nu = -\frac{\rho_0}{\pi}\int_{-\infty}^{\infty} f(\epsilon-\omega)\, d\omega \sum_{\boldsymbol{q}} \mathrm{Im}[(U(\boldsymbol{q},\omega) - 2FU)P_d^2(\boldsymbol{q},\omega)]} \tag{13.66}$$

In three dimensions, Exercise 13.6 shows that, taking into account the dynamically screened interaction, the exchange term is multiplied by a factor 2. The parameter λ_ν, which gives the strength of the density of states anomaly (13.35), thus becomes $\lambda_\nu = 2-2F$. Moreover, the perturbative calculation which leads to (13.35) and (13.66) is valid only for $F \ll 1$. It can be shown that the prefactor of the exchange term is 2 only to first order in perturbation. To the next order there is a term linear in F (equal to $F/2$) so that the prefactor λ_ν of the Hartree–Fock correction is indeed [327, 333]:

$$\lambda_\nu = 2 - \frac{3F}{2}, \quad d = 3. \tag{13.67}$$

By definition (13.34), $0 < F < 1$, and the correction is thus always *negative*. For the case of other dimensionalities, see reference [327].[11]

[10] The Hartree (13.63) and Fock (13.58) contributions differ in sign and by a factor 2 which originate respectively from the exchange and the spin.

[11] The results presented here are obtained in the perturbative limit, that is when $\delta\nu/\nu \ll 1$. A non-perturbative expression has been obtained in reference [334].

Remark

Another contribution to the density of states anomaly is obtained by replacing the Diffusons by Cooperons in the diagrams of Figure 13.4. In this case it can be shown that truncating the perturbation series to lowest order is meaningless, and we must take into account the infinite series (called *Cooper channel renormalization*) of diagrams built with interaction lines and Cooperons. This gives an additional contribution to the density of states anomaly proportional to the Diffuson contribution up to a small multiplicative factor $1/\ln(T_c/\epsilon)$ (or $\ln(\ln T_c\tau_e/\ln(T_c/\epsilon))$ in two dimensions). For the Coulomb interaction, T_c is a characteristic energy of the order of the bandwidth. For an attractive interaction, T_c is the superconducting temperature [327].

Exercise 13.6 Show that, in three dimensions, the contribution of the exchange term is multiplied by a factor 2 when the dynamical character of the screening is taken into account.
For a static interaction $U(\boldsymbol{q},\omega)=U$, relation (13.61) shows that the exchange term contains the integral

$$U\,\mathrm{Im}\int_0^\infty \frac{1}{(-i\omega+Dq^2)^2}q^2\,dq.$$

If the interaction is dynamically screened, that is for $U(\boldsymbol{q},\omega)=U(-i\omega+Dq^2)/Dq^2$, the integral becomes

$$\frac{U}{D}\mathrm{Im}\int_0^\infty \frac{1}{-i\omega+Dq^2}dq.$$

An integration by parts shows that the dynamical screening of the interaction multiplies the result of the static case by a factor 2. For similar calculations in other dimensionalities, see reference [327].

Exercise 13.7 From relations (13.61) and (15.114), show that the exchange term contribution to the density of states anomaly can also be written in the form

$$\delta\rho^F=-\frac{\rho_0}{2\pi\Omega}\int_0^\infty d\omega\left[\tanh\frac{\omega+\epsilon}{2T}+\tanh\frac{\omega-\epsilon}{2T}\right]\sum_{\boldsymbol{q}}\mathrm{Im}[U(\boldsymbol{q},\omega)P_d^2(\boldsymbol{q},\omega)]. \tag{13.68}$$

Exercise 13.8 Show that the relations (13.61) and (13.29) are equivalent for a static interaction $U(\boldsymbol{r}-\boldsymbol{r}')$.

13.4.4 Capacitive effects

In this section we show how the geometry of the system can play an important role in determining the density of states anomaly. Consider for example a quasi-one-dimensional wire of length L and of square section W^2, placed on a metallic electrode and separated from this electrode by a tunnel junction of thickness a. In this case, the Coulomb interaction depends on the capacitance of the junction, and the uniform component of the interaction has the form

$$U(\boldsymbol{q}=0,\omega)=\frac{e^2}{C}, \tag{13.69}$$

where C is the capacitance per unit length of the junction. At short distance, that is for large wave vectors (but still in the diffusive limit), the dynamically screened interaction keeps the form (13.12) provided we replace ρ_0 by the one-dimensional density of states $\rho_{1d} = \rho_0 W^2$:

$$U(\boldsymbol{q},\omega) = \frac{1}{2\rho_{1d}} \frac{-i\omega + Dq^2}{Dq^2}. \tag{13.70}$$

In order to account for both behaviors, it is shown in Appendix A13.1 that the interaction has to be written in the form [331, 335, 336]

$$U(\boldsymbol{q},\omega) = \frac{-i\omega + Dq^2}{2\rho_{1d} Dq^2 - i\omega C/e^2} \tag{13.71}$$

or equivalently

$$U(\boldsymbol{q},\omega) = \frac{1}{2\rho_{1d}} \frac{D^*}{D} \frac{-i\omega + Dq^2}{-i\omega + D^* q^2}, \tag{13.72}$$

where the coefficient $D^* = 2\rho_0 W^2 e^2 D/C \gg D$ can be interpreted as an effective diffusion coefficient which describes the propagation of the electromagnetic field in the junction.[12] Since the conductance of the wire is $\sigma = 2e^2 D\rho_0$, this diffusion coefficient can be rewritten as $D^* = 1/(RC)$ where $R = 1/(\sigma W^2)$ is the resistance of the wire per unit length and C is the capacitance of the junction per unit length (equation 13.209).

Exercise 13.9 Calculate the density of states anomaly for a wire placed in the vicinity of a metallic electrode [331, 335].
By inserting interaction (13.72) in the expression (13.66) for the density of states correction and after integrating over $\boldsymbol{q}$ and ω, we obtain:

$$\delta\rho(\epsilon) = -\frac{1}{2\pi\sqrt{2}} \frac{1}{\sqrt{D\epsilon}} \frac{D^*/D}{1+\sqrt{D^*/D}} \tag{13.73}$$

for the exchange term at zero temperature. Show that for a wire whose thickness is larger than the screening length, the ratio $D^*/D \propto (\kappa W)^2$ is large. Thus the contribution of the exchange term to the density of states correction (13.36) must be multiplied by the factor $\sqrt{D^*/D}$. Show that the Hartree term is not modified.

---

[12] This effective diffusion coefficient of the electromagnetic field should not be confused with the diffusion coefficient $D^* = v_F l^*/d$ introduced in (4.170) to describe anisotropic collisions.

Remark: dynamical Coulomb blockade

Another way to describe the tunnel conductance is to relate the density of states anomaly to the impedance of the environment of the system being studied. The relevant quantity is then the probability that an electron crossing the tunnel barrier transfers a given energy to its environment [335,337,338]. For a geometry where this environment is the conductor itself, this formulation is equivalent to the approach presented in this chapter.

13.5 Correction to the conductivity

Taking into account the interaction between electrons also leads to a reduction in the conductivity [339, 340]. Without going into the details of the calculation, we can argue that the reduction is a consequence of the correction to the density of states. Both effects result from the scattering of an electron by the charge fluctuations induced by disorder. The temperature dependence of the conductivity $\sigma(T)$ is related to its energy dependence at $T=0\text{K}$ by relation (7.125). Since the conductivity is proportional to the density of states (Einstein relation), we expect the density of states anomaly to lead to a correction of the conductivity given by

$$\frac{\delta\sigma(T)}{\sigma_0} = \int d\epsilon \left(-\frac{\partial f}{\partial \epsilon}\right) \frac{\delta\nu(\epsilon)}{\nu_0}, \tag{13.74}$$

where σ_0 is the Drude conductivity (7.14). For a static interaction, the density of states correction is given by (13.35), and using (15.109) we have[13]

$$\boxed{\delta\sigma(T) = -\lambda_\sigma \left(\frac{e^2 D}{\pi\Omega}\right) \int_0^\infty \left(\frac{\pi T t}{\sinh \pi T t}\right)^2 Z(t)\, dt} \tag{13.75}$$

with $\lambda_\sigma = 1-2F$. As for the density of states correction, this value corresponds to a static interaction. For a quasi-one-dimensional system, $Z(t)=\Omega/\sqrt{4\pi D t}$, so that with the help of (15.89), we obtain

$$\delta\sigma(T) = -\lambda_\sigma \frac{e^2}{\pi^2}\frac{3}{8}\sqrt{\frac{\pi}{2}}\,\zeta\left(\frac{3}{2}\right)\left(\frac{D}{T}\right)^{1/2}. \tag{13.76}$$

In two dimensions $Z(t)=\Omega/4\pi D t$, and in the limit $T\tau_e \ll 1$, we obtain, using (15.90),

$$\delta\sigma(T) = -\lambda_\sigma \frac{e^2}{4\pi^2} \ln \frac{e^\gamma}{2\pi T \tau_e} \tag{13.77}$$

[13] In order to compare this relation and the following ones with the weak localization in Chapter 7, we have to reintroduce $\hbar$. Result (13.75) can also be obtained using another method called "quasi-classical" and is presented in this form in reference [340].

where $\gamma \simeq 0.577$ is the Euler constant. In three dimensions, $Z(t) = \Omega/(4\pi Dt)^{3/2}$ and (15.93) lead to

$$\delta\sigma(T) = -\lambda_\sigma \frac{e^2}{\pi^2} \frac{\sqrt{\pi}}{8\sqrt{2}} \zeta\left(\frac{1}{2}\right)\left(\frac{T}{D}\right)^{1/2} \tag{13.78}$$

up to a subtractive constant.

As for the density of states anomaly, the results differ slightly when the dynamic character of the interaction is taken into account. A full-fledged treatment shows that the results (13.76, 13.77, 13.78) are still valid provided λ_σ is given by (see Exercise 13.11):

$$\lambda_\sigma = \frac{4}{d} - \frac{3F}{2}. \tag{13.79}$$

Remark

As for the density of states anomaly, another contribution to $\delta\sigma(T)$ is obtained by replacing Diffusons by Cooperons (see remark on page 505). This contribution is proportional to the Diffuson contribution times a reduction term $1/\ln(T_c/T)$ (or $\ln(\ln T_c\tau_e/\ln(T_c/T))$ in two dimensions). For the Coulomb interaction, T_c is a characteristic energy of the order of the bandwidth so that $T \ll T_c$. The correction is thus negative. For an attractive interaction, T_c is the superconducting temperature, which leads, even for $T \gg T_c$, to an increase in conductivity. There are two other classes of diagrams. The so-called Maki–Thomson diagrams describe the diffusion of electrons by superconducting fluctuations and give a correction proportional to the weak localization correction [339, 340]. The so-called Aslamasov–Larkin correction is related to the Cooper pair fluctuations and is important only in the vicinity of T_c. For a review of theoretical and experimental results on this subject, see [327, 342].

It is interesting to compare the correction (13.75) to the conductivity with the weak localization correction (7.53). Both are of the same order. The physical mechanisms at the origin of these corrections are different, but both are related to the integrated return probability. The temperature dependences, however, are different. In the case of the weak localization correction, the temperature enters only through the phase coherent time $\tau_\phi(T) \propto T^{-p}$ (section 7.4.3), leading to different temperature dependences in $d = 1$ and in $d = 3$, whereas they are both logarithmic in two dimensions (compare (7.66) with (13.76, 13.77) and (13.78)). The usual way to extract experimentally the correction due to electron–electron interactions is to apply a magnetic field in order to suppress the weak localization correction.

Exercise 13.10 Show that

$$\int_{-\infty}^{\infty} [f(\epsilon - \omega) - f(\epsilon + \omega)] \left(-\frac{\partial f}{\partial \epsilon}\right) d\epsilon = \frac{\partial}{\partial \omega}\left(\omega \coth \frac{\beta\omega}{2}\right). \tag{13.80}$$

Exercise 13.11 Show that the correction to the conductivity for a static interaction $U = 1/(2\rho_0)$ can be written in the form

$$\frac{\delta\sigma}{\sigma_0} = -\frac{1-2F}{4\pi\nu_0}\int_{-\infty}^{\infty} d\omega \frac{\partial}{\partial\omega}\left(\omega\coth\frac{\beta\omega}{2}\right)\sum_{\boldsymbol{q}} \mathrm{Im} P_d^2(\boldsymbol{q},\omega). \tag{13.81}$$

To that purpose, start from the relation (13.42) for the density of states correction. For an interaction $U(\boldsymbol{q},\omega)$, it is shown in reference [343] that the correction to the exchange term is written (see the next exercise):

$$\frac{\delta\sigma}{\sigma_0} = -\frac{2}{\pi d\Omega}\int_{-\infty}^{\infty} d\omega \frac{\partial}{\partial\omega}\left(\omega\coth\frac{\beta\omega}{2}\right)\sum_{\boldsymbol{q}} Dq^2 \mathrm{Im}[U(\boldsymbol{q},\omega)P_d^3(\boldsymbol{q},\omega)]. \tag{13.82}$$

Show that for the dynamically screened interaction $U(\boldsymbol{q},\omega) = U(-i\omega + Dq^2)/Dq^2$, the correction due to exchange is:

$$\frac{\delta\sigma}{\sigma_0} = -\frac{1}{\pi\nu_0 d}\int_{-\infty}^{\infty} d\omega \frac{\partial}{\partial\omega}\left(\omega\coth\frac{\beta\omega}{2}\right)\sum_{\boldsymbol{q}} \mathrm{Im} P_d^2(\boldsymbol{q},\omega) \tag{13.83}$$

and differs from the exchange term in (13.81) only by a factor $4/d$.

Exercise 13.12: correction to the conductivity

Reference [343] presents the calculation of the interaction contribution to the conductivity. For the exchange term, this calculation involves the three diagrams of Figure 13.6. Show that the sum of these three diagrams is zero.

To prove this, it is useful to redraw the diagrams in a different way so that the long range and short range parts are more explicitly separated (Figure 13.6). This representation involves Hikami boxes whose structure is similar to the boxes used in diagrams for conduction fluctuations. It can be shown that their sum is zero, by using the results of section 11.2.1. From (11.17), the sum of diagrams (a) + (b) is proportional to:

$$2\frac{k_F^2}{d}(H^{(A)} + H^{(B)}) = 2\frac{k_F^2}{d}\tilde{H}' = 2\frac{k_F^2}{d}2\pi\rho_0\tau_e^3. \tag{13.84}$$

The factor 2 accounts for diagrams similar to (a) and (b) but where the retarded and advanced parts have been exchanged. Diagram (c) is proportional to

$$-\frac{k_F^2}{d}H^{(A)} = -\frac{k_F^2}{d}4\pi\rho_0\tau_e^3. \tag{13.85}$$

The minus sign comes from the average of incoming momenta which are opposite. The sum of the three diagrams is thus zero.

The non-vanishing diagrams contributing to the conductivity correction are represented in Figure 13.7. They are built of two retarded (or two advanced) Green's functions. From relation (7.1), we know that they are smaller than the above diagrams by a factor $1/k_F l_e$. However, they are divergent. Using the expansion $\overline{G}(\boldsymbol{k}-\boldsymbol{q}) = \overline{G}(\boldsymbol{k}) - \boldsymbol{v}\cdot\boldsymbol{q}\,\overline{G}(\boldsymbol{k})^2$, show that each "triangle" is proportional to the wave vector and that both diagrams are thus proportional to:

$$\sum_{\boldsymbol{q}} q_x^2 \operatorname{Im} P_d^3(\boldsymbol{q}, \omega) U(\boldsymbol{q}, \omega). \tag{13.86}$$

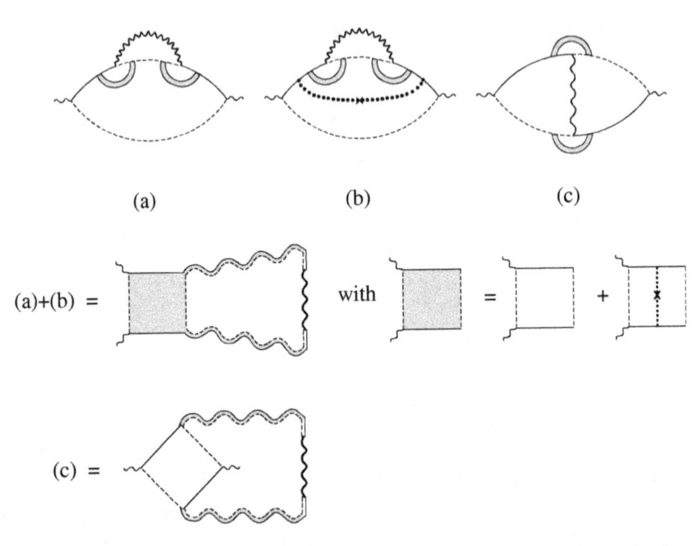

Figure 13.6 Diagrams for the exchange contribution to the conductivity. The topologically equivalent representation drawn below makes it clear that their sum is zero. The Hartree terms, whose sum is also zero, are obtained by the transformation displayed in Figure 13.5.

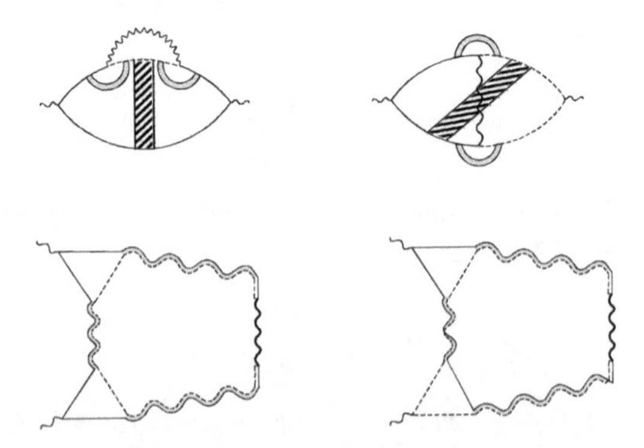

Figure 13.7 Conductivity diagrams (exchange). The two upper diagrams are drawn in the usual representation, whereas the two lower diagrams provide a topologically equivalent representation.

13.6 Lifetime of a quasiparticle

13.6.1 Introduction: Landau theory and disorder

The Coulomb interaction between two electrons is strongly screened by the presence of the other electrons (section 13.2). Each electron, "dressed" by the screening cloud, is called a *quasiparticle*. The Landau theory of "Fermi liquids" formalizes this concept of quasiparticle and shows that their properties are essentially the same as those of non-interacting electrons, given a renormalization of physical parameters such as the electron mass [344].

In fact, the quasiparticles interact weakly via the screened interaction, also called the residual interaction, and, because of this interaction, a quasiparticle acquires a finite lifetime. The Landau theory relies on the fact that this quasiparticle lifetime diverges near the Fermi level.

In this section, we study the quasiparticle lifetime and show that the disorder plays an essential role. Determination of this lifetime is crucial to determining whether the low energy properties, i.e., close to the Fermi level, can still be described within the framework of the Fermi liquid theory, that is with non-interacting quasiparticles. This is also important in order to understand what limits phase coherence in the interacting electron gas, a central question in mesoscopic physics.

In the absence of residual interaction, a quasiparticle has an infinite lifetime. With the interaction, the probability $\mathcal{P}(t)$ that a quasiparticle stays in its initial state has the form [344]:

$$\mathcal{P}(t) = e^{-t/\tau_{ee}}, \tag{13.87}$$

where $\tau_{ee}(\epsilon, T)$ defines the lifetime of the quasiparticle. This lifetime depends on the energy ϵ measured with respect to the Fermi level, as well as on the temperature T. In the absence of disorder, Landau has shown that, in three dimensions, the lifetime τ_{ee} of a quasiparticle is given by

$$\frac{1}{\tau_{ee}(\epsilon, T)} \simeq \max\left(\frac{\epsilon^2}{\epsilon_F}, \frac{T^2}{\epsilon_F}\right). \tag{13.88}$$

Near the Fermi level, the quasiparticle is well defined since the width $1/\tau_{ee}$ of a state goes to zero more rapidly than its energy ϵ when approaching the Fermi level (see Appendix A13.2).

13.6.2 Lifetime at zero temperature

In this section, we show that in the presence of disorder, multiple scattering increases the probability that two quasiparticles interact and thus reduces the electronic lifetime. More precisely, in the diffusion approximation and at zero temperature, the lifetime can be written [327][14]

$$\frac{1}{\tau_{ee}(\epsilon)} \simeq \Delta \left(\frac{\epsilon}{E_c}\right)^{d/2} \qquad \epsilon \gg E_c. \tag{13.89}$$

The dependence of this power law on space dimensionality d is the signature of the diffusive regime. E_c is the Thouless energy and $\Delta = 1/\nu_0$ is the average level spacing at the Fermi energy in the absence of interaction. This expression is limited to the case where the excitation energy ϵ is larger than E_c. In the opposite limit, the lifetime varies as [345]

$$\frac{1}{\tau_{ee}(\epsilon)} \simeq \Delta \left(\frac{\epsilon}{E_c}\right)^{2} \qquad \epsilon \ll E_c. \tag{13.90}$$

In this section, we derive these two limiting behaviors. In order to evaluate the electronic lifetime, we consider an eigenstate $|\alpha\rangle$ of the non-interacting disordered Hamiltonian, whose energy ϵ_α is above the Fermi level.[15] This state interacts with a filled state $|\gamma\rangle$ of energy ϵ_γ (Figure 13.8). The lifetime of the state $|\alpha\rangle$ is given by the Fermi golden rule

$$\frac{1}{\tau_\alpha} = 4\pi \sum_{\beta\gamma\delta} |\langle \alpha\gamma|U|\beta\delta\rangle|^2 \delta(\epsilon_\alpha + \epsilon_\gamma - \epsilon_\beta - \epsilon_\delta). \tag{13.91}$$

A factor 2 accounts for the spin degeneracy of the state $|\gamma\rangle$. The matrix element $\langle \alpha\gamma|U|\beta\delta\rangle$ describes the interaction between the two states $|\alpha\rangle$ and $|\gamma\rangle$ which evolve into the two final states $|\beta\rangle$ and $|\delta\rangle$. Let us notice that $\epsilon_\gamma < 0$ and that the final states must be empty so that their energies obey the constraints $\epsilon_\beta > 0$ and $\epsilon_\delta > 0$.

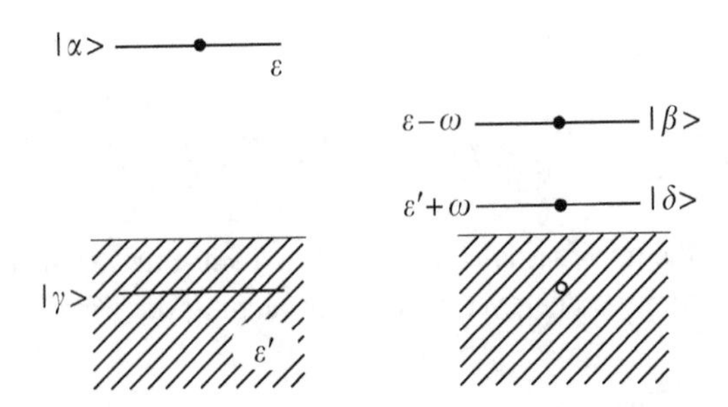

Figure 13.8 A quasiparticle in a state $|\alpha\rangle$ of energy $\epsilon_\alpha = \epsilon$ interacts with another quasiparticle $|\gamma\rangle$ of energy $\epsilon_\gamma = \epsilon'$ in the Fermi sea. The final state is made of two quasiparticles above the Fermi sea and one hole.

[14] One might expect that the temperature dependence of $1/\tau_{ee}(T)$ could be obtained by replacing ϵ by T as in (13.88) for the ballistic case, thus leading to $1/\tau_{ee}(T) \propto T^{d/2}$. In section 13.6.3, we show that this is not correct for $d \leq 2$.

[15] More precisely, we consider the non-interacting quasiparticle states, whose spectrum is assumed to have the same statistical properties as the non-interacting electrons.

Exercise 13.13 Check that, for a constant matrix element, the lifetime $\tau_{ee}(\epsilon)$ given by (13.91) varies as $1/\epsilon^2$.
The lifetime has the form

$$\frac{1}{\tau_{ee}(\epsilon)} = 2\pi U^2 \nu_f(\epsilon),$$

where $\nu_f(\epsilon) = 2\sum'_{\beta\gamma\delta} \delta(\epsilon + \epsilon_\gamma - \epsilon_\beta - \epsilon_\delta)$ is the density of final states. The sum $\sum'$ is limited to states such that $\epsilon_\gamma < 0$, $\epsilon_\beta > 0$ and $\epsilon_\delta > 0$. Replacing this sum by integrals and using the density of states ν_0, we obtain

$$\nu_f(\epsilon) = 2\nu_0^3 \int_{-\infty}^{0} d\epsilon_\gamma \int_0^\infty d\epsilon_\beta \int_0^\infty d\epsilon_\delta \, \delta(\epsilon + \epsilon_\gamma - \epsilon_\beta - \epsilon_\delta) = \nu_0^3 \epsilon^2.$$

If the matrix elements do not depend on the energy, the lifetime varies as $1/\epsilon^2$. This energy dependence is simply related to the density of final states, whence the dependence (13.88) obtained by Landau [344] (see also Appendix A13.2).

In order not to single out a given state, we must average the lifetime over all states having the same energy ϵ. Thus, we calculate

$$\frac{1}{\tau_{ee}(\epsilon)} = \frac{4\pi}{\nu_0} \sum_{\alpha\beta\gamma\delta} |\langle\alpha\gamma|U|\beta\delta\rangle|^2 \delta(\epsilon_\alpha + \epsilon_\gamma - \epsilon_\beta - \epsilon_\delta)\delta(\epsilon - \epsilon_\alpha). \tag{13.92}$$

Denoting by ϵ' the energy of the states $|\gamma\rangle$, energy conservation implies that the final states $|\beta\rangle$ and $|\delta\rangle$ have energies $\epsilon - \omega$ and $\epsilon' + \omega$, where ω is the energy transfer due to the interaction (Figure 13.8). The inverse lifetime becomes

$$\begin{aligned}\frac{1}{\tau_{ee}(\epsilon)} = \frac{4\pi}{\nu_0} \int_0^\epsilon d\omega \int_{-\omega}^0 d\epsilon' \sum_{\alpha\beta\gamma\delta} |\langle\alpha\gamma|U|\beta\delta\rangle|^2 \\ \times \delta(\epsilon - \epsilon_\alpha)\delta(\epsilon' - \epsilon_\gamma)\delta(\epsilon - \omega - \epsilon_\beta)\delta(\epsilon' + \omega - \epsilon_\delta).\end{aligned} \tag{13.93}$$

Upon averaging over disorder, we obtain for the lifetime

$$\boxed{\frac{1}{\tau_{ee}(\epsilon)} = 4\pi \nu_0^3 \int_0^\epsilon \omega W^2(\omega) d\omega} \tag{13.94}$$

with

$$W^2(\omega) = \frac{1}{\nu_0^4} \overline{\sum_{\alpha\beta\gamma\delta} |\langle\alpha\gamma|U|\beta\delta\rangle|^2 \delta(\epsilon - \epsilon_\alpha)\delta(\epsilon' - \epsilon_\gamma)\delta(\epsilon - \omega - \epsilon_\beta)\delta(\epsilon' + \omega - \epsilon_\delta)}. \tag{13.95}$$

We will see that the characteristic matrix element $W(\omega)$ depends only on the energy transfer ω.[16] In the literature, the interaction "kernel", defined by $K(\omega)=4\pi\nu_0^3 W^2(\omega)$, is frequently used, for instance to rewrite the inverse lifetime as $1/\tau_{ee}(\epsilon)=\int_0^\epsilon \omega K(\omega)d\omega$.

We now calculate $1/\tau_{ee}$ at the diffusion approximation. The matrix element $\langle\alpha\gamma|U|\beta\delta\rangle$, calculated in the basis of the eigenfunctions $\phi_i(\boldsymbol{r})$ of the Hamiltonian (2.1), is

$$\langle\alpha\gamma|U|\beta\delta\rangle = \int d\boldsymbol{r}_1\, d\boldsymbol{r}_2\, \phi_\alpha^*(\boldsymbol{r}_1)\phi_\gamma^*(\boldsymbol{r}_2)\phi_\beta(\boldsymbol{r}_1)\phi_\delta(\boldsymbol{r}_2)U_\omega(\boldsymbol{r}_1-\boldsymbol{r}_2), \tag{13.96}$$

where $U_\omega(\boldsymbol{r})$ is the dynamically screened potential. Making use of (3.26) which relates the wave functions to the non-local density of states $\rho_\epsilon(\boldsymbol{r},\boldsymbol{r}')$, $W^2(\omega)$ can be rewritten as

$$W^2(\omega)=\frac{1}{\nu_0^4}\int d\boldsymbol{r}_1\, d\boldsymbol{r}_2\, d\boldsymbol{r}_1'\, d\boldsymbol{r}_2' U_\omega(\boldsymbol{r}_1-\boldsymbol{r}_2)U_\omega(\boldsymbol{r}_1'-\boldsymbol{r}_2')$$
$$\times\, \overline{\rho_\epsilon(\boldsymbol{r}_1,\boldsymbol{r}_1')\rho_{\epsilon-\omega}(\boldsymbol{r}_1',\boldsymbol{r}_1)}\;\; \overline{\rho_{\epsilon'}(\boldsymbol{r}_2,\boldsymbol{r}_2')\rho_{\epsilon'+\omega}(\boldsymbol{r}_2',\boldsymbol{r}_2)}, \tag{13.97}$$

where the disorder average of the product of four Green's functions has been decoupled into a product of two averages. To evaluate $W^2(\omega)$, we use (4.206) and (3.99) so that[17]

$$\overline{\rho_\epsilon(\boldsymbol{r},\boldsymbol{r}')\rho_{\epsilon-\omega}(\boldsymbol{r}',\boldsymbol{r})} = \frac{\rho_0}{\pi}\mathrm{Re}P_d(\boldsymbol{r},\boldsymbol{r}',\omega)+\rho_0^2 g^2(\boldsymbol{R}), \tag{13.98}$$

whose Fourier transform is given by

$$\frac{\rho_0}{\pi}\left(\mathrm{Re}P_d(\boldsymbol{q},\omega)+\frac{\pi}{2|q|v_F}\theta(2k_f-|q|)\right). \tag{13.99}$$

The second term is independent of disorder in the limit $kl_e \gg 1$ (see equation 4.108) and it gives exactly the Landau contribution (13.216) for large values of q. For small q ($ql_e \ll 1$), the contribution of the Diffuson dominates, so that (13.97) becomes

$$W^2(\omega)=\frac{1}{\pi^2\nu_0^2\Omega^2}\int d\boldsymbol{r}_1\, d\boldsymbol{r}_2 d\boldsymbol{r}_1'\, d\boldsymbol{r}_2' U_\omega(\boldsymbol{r}_1-\boldsymbol{r}_2)U_\omega(\boldsymbol{r}_1'-\boldsymbol{r}_2')$$
$$\times\, \mathrm{Re}P_d(\boldsymbol{r}_1,\boldsymbol{r}_1',\omega)\mathrm{Re}P_d(\boldsymbol{r}_2,\boldsymbol{r}_2',-\omega), \tag{13.100}$$

or, after a Fourier transform

$$W^2(\omega)=\frac{1}{\pi^2\nu_0^2\Omega^2}\sum_{\boldsymbol{q}\neq 0}|U(\boldsymbol{q},\omega)|^2[\mathrm{Re}P_d(\boldsymbol{q},\omega)]^2. \tag{13.101}$$

[16] For that reason, the integral over ϵ' in (13.93) provides simply a factor ω.
[17] The Cooperon gives a negligible contribution.

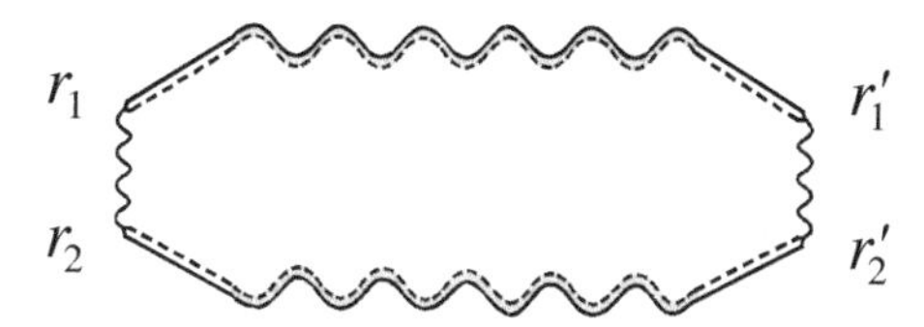

Figure 13.9 Diagrammatic representation of $W^2(\omega)$ as given by relation (13.100). Notice that this structure is quite similar to that of the density of states correlation function $K_\rho(\omega)$ (Figure 10.11).

A diagrammatic representation of this quantity is shown in Figure 13.9. In the diffusion approximation, the dynamically screened potential is given by (13.12) and therefore[18]

$$W^2(\omega) = \frac{1}{4\pi^2 \nu_0^4} \sum_{q \neq 0} \frac{1}{\omega^2 + D^2 q^4}, \tag{13.104}$$

which depends solely on ω and can be expressed in terms of the integrated return probability $Z(t)$

$$W^2(\omega) = \frac{1}{4\pi^2 \nu_0^4} \frac{1}{\omega} \int_0^\infty Z(t) \sin \omega t \, dt. \tag{13.105}$$

Finally, the electronic lifetime (13.94) is written as

$$\boxed{\frac{1}{\tau_{ee}(\epsilon)} = \frac{2}{\pi \nu_0} \int_0^\infty \frac{Z(t)}{t} \sin^2 \frac{\epsilon t}{2} dt} \tag{13.106}$$

For a metal of volume Ω, we can identify two different regimes: $t \ll \tau_D$ and $t \gg \tau_D$.

- For $t \ll \tau_D$, where τ_D is the Thouless time (5.34), an electron described as a diffusive wave packet is insensitive to the boundaries and behaves as in an infinite medium where, according to (5.24), $Z(t) = \Omega/(4\pi D t)^{d/2}$. From (15.86), we obtain for the integral (13.105)[19]

$$W^2(\omega) = \frac{d c_d}{16} \frac{1}{\nu_0^4 \omega^2} \left(\frac{\omega}{E_c}\right)^{d/2} \tag{13.107}$$

[18] An equivalent expression for $W^2(\omega)$ is

$$W^2(\omega) = \frac{1}{4\pi^2 \nu_0^4} \sum_{q \neq 0} \frac{1}{Dq^2} \mathrm{Re} P_d(\boldsymbol{q}, \omega) = \frac{1}{4\pi^2 \nu_0^4} \sum_{q \neq 0} \frac{1}{\omega} \mathrm{Im} P_d(\boldsymbol{q}, \omega), \tag{13.102}$$

which yields for the lifetime:

$$\frac{1}{\tau_{ee}(\epsilon)} = \frac{1}{\pi \nu_0} \int_0^\epsilon \omega \, d\omega \sum_{q \neq 0} \frac{1}{Dq^2} \mathrm{Re} P_d(\boldsymbol{q}, \omega) = \frac{1}{\pi \nu_0} \int_0^\epsilon d\omega \sum_{q \neq 0} \mathrm{Im} P_d(\boldsymbol{q}, \omega). \tag{13.103}$$

[19] The small time regime corresponds to the case where the excitation energy ϵ is much larger than the Thouless energy, $\epsilon \gg E_c$. In this case, the sum (13.104) on wave vectors can be replaced by an integral and we recover (13.107).

so that the electronic lifetime is equal to

$$\boxed{\frac{1}{\tau_{ee}(\epsilon)} = \frac{\pi}{2} c_d \Delta \left(\frac{\epsilon}{E_c}\right)^{d/2} \qquad \epsilon \gg E_c} \tag{13.108}$$

where c_d is a constant defined in (10.63), with $c_1 = \sqrt{2}/\pi^2$, $c_2 = 1/4\pi^2$, $c_3 = \sqrt{2}/6\pi^3$. Such behavior has indeed been observed in silver wires ($d = 1$) for which $W^2(\omega) \propto \omega^{-3/2}$ and $1/\tau_{ee}(\epsilon) \propto \epsilon^{1/2}$, although the measured prefactor was larger than the value predicted here [335, 346]. Other behavior in disagreement with these results has been observed in gold and copper wires and has been attributed to other relaxation mechanisms such as the coupling to two-level systems [346] or magnetic impurities [347].

Remarks

Screening effect

The behavior of the lifetime depends only weakly on the exact nature of the screened potential. Indeed, assuming a static potential, that is $U(\boldsymbol{q}, \omega = 0) = \Omega/2\nu_0$, instead of a dynamically screened potential, the sum (13.104) becomes

$$W^2(\omega) = \frac{1}{4\pi^2 \nu_0^4} \sum_{\boldsymbol{q} \neq 0} \frac{D^2 q^4}{(\omega^2 + D^2 q^4)^2}. \tag{13.109}$$

A high frequencies $\omega \gg E_c$, the sum can be replaced by an integral and we recover a power law similar to (13.107):

$$W^2(\omega) \propto \frac{1}{\nu_0^4 \omega^2} \left(\frac{\omega}{E_c}\right)^{d/2}, \tag{13.110}$$

where only the prefactor has been modified.

Lifetime and spectral rigidity

It is interesting to compare the expressions for $1/\tau_{ee}(\epsilon)$ and the variance $\Sigma^2(E)$ of the distribution of energy levels in the diffusive regime (relation 10.52). We find

$$\frac{1}{\tau_{ee}(\epsilon)} = \frac{\pi \Delta}{2} [\Sigma^2(\epsilon) - \Sigma_0^2(\epsilon)] \tag{13.111}$$

where the contribution $\Sigma_0^2(\epsilon)$ of the zero mode has been subtracted. This relation can also be understood from the similarity of the diagrams for the correlation function $K(\omega)$ and for $W^2(\omega)$ shown respectively in Figures 10.11 and 13.9.

• The limit $t \gg \tau_D$ corresponds to the ergodic regime in which the diffusive electronic wave packet explores all the accessible volume Ω. Thus we would expect $Z(t)$ to be driven only by the zero mode. This is not so, because in expression (13.104) this mode has been removed in order to ensure electronic neutrality. The excitation energy ϵ is smaller than E_c and it is not possible to replace the sum (13.104) by an integral. In this limit, we obtain

$$W^2(\omega) = \frac{a_d}{4\pi^6} \frac{\Delta^4}{E_c^2} \propto \frac{\Delta^2}{g^2}, \tag{13.112}$$

where the coefficient a_d is defined by the series

$$a_d = \sum_{n_x, n_y, n_z} \frac{1}{(n_x^2 + n_y^2 + n_z^2)^2}. \tag{13.113}$$

The ratio E_c/Δ is the dimensionless conductance g defined by (7.25). For $\omega \ll E_c$, the characteristic matrix element of the interaction is thus energy independent and of order Δ/g. The inverse lifetime in this case [345] is

$$\boxed{\frac{1}{\tau_{ee}(\epsilon)} = \frac{a_d}{2\pi^5} \Delta \left(\frac{\epsilon}{E_c}\right)^2 \qquad \epsilon \ll E_c} \tag{13.114}$$

It is noteworthy that this lifetime depends on boundary conditions through the coefficient a_d. For example, if the sample is connected to reservoirs only in the Ox direction, the boundary conditions $n_x \in \mathbb{N}^*, n_y \in \mathbb{N}, n_z \in \mathbb{N}$ are the same as in the sum (11.32) for conductance fluctuations, i.e., $a_d = b_d$. For an isolated sample, only the mode $n_x = n_y = n_z = 0$ is excluded, so that $a_1 = b_1 = \pi^4/90$, $a_2 = b_1 + b_2 = 2.59$ and $a_3 = b_1 + b_2 + b_3 = 5.11$.

Exercise 13.14 Show that there is an additional contribution to (13.97) which involves the product $\overline{\rho_\epsilon(\boldsymbol{r}_1, \boldsymbol{r}_1')\rho_{\epsilon'+\omega}(\boldsymbol{r}_2', \boldsymbol{r}_2)}\ \overline{\rho_{\epsilon'}(\boldsymbol{r}_2, \boldsymbol{r}_2')\rho_{\epsilon-\omega}(\boldsymbol{r}_1', \boldsymbol{r}_1)}$. Draw the corresponding diagram with the help of Figures 13.9 and 13.5. Show that this contribution is of order F^2 [327]. Note that this term does not depend only on ω as in (13.97), but also on $\epsilon - \epsilon' - \omega$.

Remark: lifetime and dielectric function

The previous expressions for the electronic lifetime can be reformulated to introduce the dielectric function $\epsilon(\boldsymbol{q},\omega)$. To do this, we show from (13.10) that the following identity holds:

$$\frac{1}{2\rho_0}\mathrm{Im}\left[\frac{-1}{\epsilon(\boldsymbol{q},\omega)}\right]=\omega\frac{4\pi e^2}{q^2}\frac{\mathrm{Re}P_d(\boldsymbol{q},\omega)}{|\epsilon(\boldsymbol{q},\omega)|^2}, \tag{13.115}$$

so that the combination $|U(\boldsymbol{q},\omega)|^2\,[\mathrm{Re}P_d(\boldsymbol{q},\omega)]^2$ which enters expression (13.101) for $W^2(\omega)$ fulfills

$$\omega\,|U(\boldsymbol{q},\omega)|^2[\mathrm{Re}P_d(\boldsymbol{q},\omega)]^2=\frac{4\pi e^2}{2\rho_0 q^2}\mathrm{Im}\left[-\frac{1}{\epsilon(\boldsymbol{q},\omega)}\right]\mathrm{Re}P_d(\boldsymbol{q},\omega). \tag{13.116}$$

For $W^2(\omega)$, this leads to

$$W^2(\omega)=\frac{1}{2\pi^2 v_0^3\Omega}\sum_{\boldsymbol{q}}\frac{4\pi e^2}{q^2\omega}\mathrm{Im}\left[\frac{-1}{\epsilon(\boldsymbol{q},\omega)}\right]\mathrm{Re}P_d(\boldsymbol{q},\omega), \tag{13.117}$$

and for the lifetime (13.94):

$$\frac{1}{\tau_{ee}(\epsilon)}=\frac{2}{\pi\Omega}\int_0^{\epsilon}d\omega\sum_{\boldsymbol{q}}\frac{4\pi e^2}{q^2}\mathrm{Im}\left[\frac{-1}{\epsilon(\boldsymbol{q},\omega)}\right]\mathrm{Re}P_d(\boldsymbol{q},\omega). \tag{13.118}$$

13.6.3 Quasiparticle lifetime at finite temperature

The time $\tau_{ee}(\epsilon)$ represents the lifetime of a quasiparticle injected above the Fermi sea at $T=0$ K. At finite temperature, the probability for a quasiparticle to stay in its initial state is assumed to keep the form [344]

$$\mathcal{P}(t,\epsilon,T)=e^{-t/\tau_{ee}(\epsilon,T)}, \tag{13.119}$$

where $\tau_{ee}(\epsilon,T)$ is the quasiparticle lifetime at finite energy and finite temperature. To calculate this lifetime, we just need to include the Fermi factors in (13.94) [327],[20]

$$\frac{1}{\tau_{ee}(\epsilon,T)}=4\pi v_0^3\int_{-\infty}^{\infty}d\omega\int_{-\infty}^{\infty}d\epsilon' F(\epsilon,\epsilon',\omega)W^2(\omega), \tag{13.120}$$

where $F(\epsilon,\epsilon',\omega)$ is a combination of Fermi factors $f_\epsilon=1/(e^{\beta\epsilon}+1)$:

$$F(\epsilon,\epsilon',\omega)=f_{\epsilon'}(1-f_{\epsilon-\omega})(1-f_{\epsilon'+\omega})+(1-f_{\epsilon'})f_{\epsilon-\omega}f_{\epsilon'+\omega}. \tag{13.121}$$

The first term in this expression is larger when $\epsilon>0$. It describes the decay of an electron-like state above the Fermi level. The second term dominates when $\epsilon<0$ and describes the

[20] Notice that we consider here the temperature effect coming from the Fermi statistics and not from the coupling to other degrees of freedom such as phonons.

decay of a hole-like state into the Fermi sea. For $\epsilon = 0$, both terms are equal. Integrating on ϵ' (relation 15.115), we obtain

$$\frac{1}{\tau_{ee}(\epsilon, T)} = 4\pi \nu_0^3 \int_{-\infty}^{\infty} d\omega\, \omega W^2(\omega) f_{\epsilon-\omega} \frac{e^{\beta\epsilon}+1}{e^{\beta\omega}-1}. \tag{13.122}$$

This lifetime can also be obtained from the imaginary part of the self-energy of a quasiparticle in the presence of a screened interaction [329].[21] At zero temperature, we recover the result (13.108).

Remark: relaxation towards equilibrium

The time $\tau_{ee}(\epsilon, T)$ can be also interpreted as the relaxation time towards Fermi equilibrium distribution. This relaxation is defined from the Boltzmann equation [327, 348]

$$\begin{aligned} \frac{\partial n_\epsilon}{\partial t} = & -4\pi \nu_0^3 \int_{-\infty}^{\infty} d\omega\, W^2(\omega) \\ & \times \int_{-\infty}^{\infty} d\epsilon' \, [n_\epsilon n_{\epsilon'}(1-n_{\epsilon-\omega})(1-n_{\epsilon'+\omega}) - n_{\epsilon-\omega} n_{\epsilon'+\omega}(1-n_\epsilon)(1-n_{\epsilon'})]. \end{aligned} \tag{13.123}$$

The relaxation term contains two contributions which describe respectively the quasiparticles leaving a given quantum state ("out" contribution) and reaching this state ("in" contribution). At equilibrium, n_ϵ is equal to the Fermi factor $f_\epsilon = 1/(e^{\beta\epsilon}+1)$ and the term in brackets is zero. By linearizing around the equilibrium distribution $n_\epsilon = f_\epsilon + \delta n_\epsilon$, we obtain the equation

$$\begin{aligned} \frac{\partial \delta n_\epsilon}{\partial t} = & -4\delta n_\epsilon \pi \nu_0^3 \int_{-\infty}^{\infty} d\omega\, W^2(\omega) \\ & \times \int_{-\infty}^{\infty} d\epsilon' \, [f_{\epsilon'}(1-f_{\epsilon-\omega})(1-f_{\epsilon'+\omega}) + (1-f_{\epsilon'}) f_{\epsilon-\omega} f_{\epsilon'+\omega}]. \end{aligned} \tag{13.124}$$

The right hand side term is of the form $-\delta n_\epsilon / \tau_{ee}(\epsilon, T)$. Thus $\tau_{ee}(\epsilon, T)$ can be interpreted as the relaxation time towards equilibrium distribution.

13.6.4 *Quasiparticle lifetime at the Fermi level*

We now consider more specifically the lifetime of a quasiparticle *near the Fermi level* ($\epsilon = 0$) *and at finite temperature*. Physical properties such as conductance are expressed in terms of *single-particle* states at the Fermi level. It is thus essential to understand the range of validity of the description in terms of independent quasiparticles. In the following, we denote by

$$\boxed{\tau_{in}(T) = \tau_{ee}(\epsilon = 0, T)} \tag{13.125}$$

[21] Using (15.117), the relation (13.122) can also be written in the form [329]:

$$\frac{1}{\tau_{ee}(\epsilon, T)} = 2\pi \nu_0^3 \int_{-\infty}^{\infty} d\omega\, W^2(\omega)\, \omega \left(\coth \frac{\beta\omega}{2} + \tanh \frac{\beta}{2}(\epsilon - \omega) \right).$$

the relaxation time of a quasiparticle at the Fermi level.[22] From relation (13.122), we have [349][23]

$$\frac{1}{\tau_{in}(T)} = \frac{1}{\tau_{ee}(0,T)} = 8\pi\nu_0^3 \int_0^\infty d\omega W^2(\omega) \frac{\omega}{\sinh\beta\omega}. \tag{13.127}$$

From relations (13.127) and (13.105), it is also possible to obtain an expression for the lifetime as a function of the return probability $Z(t)$:

$$\frac{1}{\tau_{in}(T)} = \frac{2}{\pi\nu_0} \int_0^\infty \frac{d\omega}{\sinh\beta\omega} \int_0^\infty dt\, Z(t) \sin\omega t \tag{13.128}$$

or, using (15.116),

$$\frac{1}{\tau_{in}(T)} = \frac{T}{2\nu_0} \int_0^\infty Z(t) \tanh\frac{\pi T t}{2} dt. \tag{13.129}$$

For the diffusion in free space, $Z(t)$ is given by (5.24), so the time integral (13.128) is proportional to $\omega^{d/2-1}$ (relation 15.28)

$$\frac{1}{\tau_{in}(T)} = \frac{\pi d c_d}{2\nu_0} \int_0^\infty \frac{d\omega}{\omega \sinh\beta\omega} \left(\frac{\omega}{E_c}\right)^{d/2}. \tag{13.130}$$

Therefore, in three dimensions, we have

$$\frac{1}{\tau_{in}(T)} = \frac{\sqrt{2}}{4\pi^2\nu_0} \int_0^\infty \frac{d\omega}{\omega\sinh\beta\omega} \left(\frac{\omega}{E_c}\right)^{3/2} \simeq \frac{T}{\nu_0} \int_0^T \frac{d\omega}{\omega^2} \left(\frac{\omega}{E_c}\right)^{3/2} \tag{13.131}$$

so that

$$\boxed{\frac{1}{\tau_{in}(T)} \simeq \Delta \left(\frac{T}{E_c}\right)^{3/2} \qquad d=3} \tag{13.132}$$

up to a numerical factor. Note that the exponent of the power law is the same as the exponent for the energy dependence of the lifetime at zero temperature (13.108). This result follows at once if we notice that relevant processes in the quasiparticle relaxation described by $\omega W^2(\omega)$ are those for which the energy transfer ω is of order T.

[22] We should not confuse the lifetime of a quasiparticle at the Fermi level and at finite temperature, $\tau_{in}(T) = \tau_{ee}(\epsilon = 0, T)$, with the time $\tau_{ee}(\epsilon = T, T = 0)$ sometimes introduced in the literature by means of the substitution $\epsilon \longrightarrow T$ in the expression for the zero temperature relaxation time. This second time has no physical significance.

[23] From relations (13.127) and (13.117), we can express the quasiparticle relaxation time in terms of the dielectric function

$$\frac{1}{\tau_{in}(T)} = \frac{4}{\pi\Omega} \int_0^\infty \frac{d\omega}{\sinh\beta\omega} \sum_{\boldsymbol{q}} \frac{4\pi e^2}{q^2} \mathrm{Im}\left(\frac{-1}{\epsilon(\boldsymbol{q},\omega)}\right) \mathrm{Re}P_d(\boldsymbol{q},\omega). \tag{13.126}$$

It would be tempting to generalize this result to any dimension and to conclude that $1/\tau_{in}(T) \propto T^{d/2}$. This is not correct for $d \leq 2$. In this case, the contribution of electron–electron processes with *low energy transfer* $\omega \simeq 0$ dominates and leads to a divergence in the integral (13.130). In order to cure this divergence, it is worth noticing that $\tau_{in}(T)$ represents precisely the lifetime of an eigenstate, so the energy transfer ω cannot be defined with an accuracy better than $1/\tau_{in}$. Consequently, there is no energy transfer smaller than $1/\tau_{in}(T)$, so the integral (13.130) needs to be cut off self-consistently for ω smaller than $1/\tau_{in}(T)$. For $d \leq 2$, we thus obtain a self-consistent relation for τ_{in}:

$$\frac{1}{\tau_{in}(T)} \simeq \frac{1}{\nu_0} \int_{1/\tau_{in}}^{\infty} \frac{d\omega}{\omega \sinh \beta\omega} \left(\frac{\omega}{E_c}\right)^{d/2} \simeq \frac{T}{\nu_0} \int_{1/\tau_{in}}^{T} \frac{d\omega}{\omega^2} \left(\frac{\omega}{E_c}\right)^{d/2} \tag{13.133}$$

where the thermal factor has been replaced by a cutoff at $\omega \sim T$. In two dimensions, $1/\tau_{in}(T)$ is proportional to the temperature (within logarithmic corrections):

$$\frac{1}{\tau_{in}(T)} \simeq \Delta \frac{T}{E_c} \ln \frac{E_c}{\Delta} \qquad d = 2. \tag{13.134}$$

In one dimension, and since $T\tau_{in} \gg 1$, the integral becomes proportional to $\sqrt{\tau_{in}}$ so that the self-consistent relation leads to

$$\boxed{\frac{1}{\tau_{in}(T)} \simeq \Delta \left(\frac{E_c}{\Delta}\right)^{1/3} \left(\frac{T}{E_c}\right)^{2/3} \qquad d = 1} \tag{13.135}$$

Remarks

Non-exponential relaxation of quasiparticles in dimension $d \leq 2$

The introduction of the low energy cutoff may appear as a handwaving and artificial way to handle the low energy divergence in (13.130). The profound reason for this divergence is that, for $d \leq 2$, *the relaxation of quasiparticles is not exponential* [350]. The relaxation rate $-d \ln \mathcal{P}/dt$ is no longer constant as was assumed in (13.119). Indeed, we know from the Fermi golden rule that, after a time t, energy must be conserved within $1/t$. Thus the energy transfer ω cannot be defined with a precision better than $1/t$ and we have to cut off contributions of energies smaller than $1/t$ [350]. Thus (13.126) becomes

$$\ln \mathcal{P} = -\frac{4t}{\pi\Omega} \int_{1/t}^{\infty} \frac{d\omega}{\sinh \beta\omega} \sum_{\boldsymbol{q}} \frac{4\pi e^2}{q^2} \mathrm{Im}\left(\frac{-1}{\epsilon(\boldsymbol{q}, \omega)}\right) \mathrm{Re}P_d(\boldsymbol{q}, \omega), \tag{13.136}$$

that is

$$\ln \mathcal{P} = -\frac{\pi d c_d}{2\nu_0} t \int_{1/t}^{\infty} \frac{d\omega}{\omega \sinh \beta\omega} \left(\frac{\omega}{E_c}\right)^{d/2}. \tag{13.137}$$

The lower cutoff does not affect the relaxation in dimension $d = 3$, since the integral converges at low frequency. For $d \leq 2$, however, the low frequency behavior drives the relaxation. Consider the case $d=1$. We obtain, for times $t \gg 1/T$:

$$\ln \mathcal{P} \simeq -\frac{\sqrt{2}}{2\pi \nu_0 \sqrt{E_c}} T t \int_{1/t}^{T} \frac{d\omega}{\omega^{3/2}} \simeq -\frac{\sqrt{2}T}{\pi \nu_0 \sqrt{E_c}} t^{3/2} \qquad (13.138)$$

which leads to non-exponential behavior for the quasiparticle relaxation:

$$\mathcal{P}(t,T) \sim e^{-[t/\tau_{in}(T)]^{3/2}} \quad d=1 \qquad (13.139)$$

with

$$\frac{1}{\tau_{in}(T)} \sim \left(\frac{\Delta T}{E_c^{1/2}}\right)^{2/3}. \qquad (13.140)$$

This argument shows that the low frequency divergence is indeed the signature of a non-exponential behavior. Moreover, we recover the characteristic time obtained in (13.135). In dimension $d=2$, we have

$$\mathcal{P}(t,T) \sim e^{-t/(\tau_{in} \ln Tt)} \quad d=2. \qquad (13.141)$$

Validity of the Fermi liquid description

The relaxation rate $1/\tau_{in}(T)$ stays smaller than the temperature T. With the help of (13.132) and (13.134), we check that this is always the case for $d \geq 2$. In one dimension, $1/\tau_{in}$ decreases more slowly than temperature. We might wonder whether quasiparticles are still well defined at low temperature and question the validity of the Fermi liquid description. However, $1/\tau_{in}$ becomes of order T at extremely low temperature of order Δ/g, with $g \sim E_c/\Delta \gg 1$, which so far is not accessible and which is zero in the thermodynamic limit.

13.7 Phase coherence

13.7.1 Introduction

In the preceding section, we studied the relaxation of a quasiparticle at the Fermi level and at finite temperature. This relaxation is characterized by the time $\tau_{in}(T)$, and in dimension $d \leq 2$ it is no longer exponential.

In this section, we wish to study the nature of the processes which limit the phase coherence and therefore the observation of interference effects such as weak localization. We shall denote by τ_ϕ^{ee} the characteristic time associated with the loss of phase coherence.

A first simple approach is to consider that phase coherence is limited by the lifetime of quasiparticles. Since the multiple scattering trajectories that are paired in the Cooperon are defined for a given energy state, they cannot interfere for times larger than $\tau_{in}(T)$. This results in an irreversible dephasing between the trajectories and thus a loss of phase coherence. It is therefore appealing to assume that

$$\tau_\phi^{ee}(T) = \tau_{in}(T) = \tau_{ee}(\epsilon = 0, T). \qquad (13.142)$$

We have also shown that quasiparticle relaxation is not exponential for $d \leq 2$ and we might ask whether phase relaxation is also non-exponential.

A second approach consists in calculating *directly* the dephasing $\langle e^{i\Phi(t)} \rangle$ resulting from electron–electron interaction and accumulated between time-reversed conjugated multiple scattering sequences. To that purpose, we replace the interaction between electrons by an effective interaction which describes the coupling of a single electron to the electromagnetic field created by the other electrons [351–353]. This electric noise is called Nyquist noise.

In developing this second approach, we shall see not only that these two characteristic times τ_{in} and τ_ϕ^{ee} are equal, but also that the two processes, *quasiparticle relaxation* and *phase relaxation*, are very similar so that the loss of phase coherence, described by the average $\langle e^{i\Phi(t)} \rangle$, behaves like the probability $\mathcal{P}(t, \epsilon = 0, T)$.

Remark: definition of the phase coherence time

This definition is not unique. It depends on the physical quantity we consider. For states which contribute to electronic transport and which are close to the Fermi level, the definition (13.142) is quite natural. However, at finite temperature T, states which contribute to transport are located in an energy interval of width T around the Fermi level and the dephasing time $\tau_\phi^{ee}(\epsilon, T)$ depends in principle on the energy. Thus it would be equally natural to consider an average of $\tau_\phi^{ee}(\epsilon, T)$ over this energy range. Consider for example the weak localization correction given by

$$\delta\sigma(T) = \int \delta\sigma(\epsilon) \left(\frac{-\partial f}{\partial \epsilon} \right) d\epsilon, \tag{13.143}$$

where $\delta\sigma(\epsilon)$ is the correction for a given energy ϵ. As an example, in dimension $d = 2$, we have

$$\delta\sigma(\epsilon) = -\frac{e^2}{\pi h} \ln \frac{\tau_\phi^{ee}(\epsilon, T)}{\tau_e}, \tag{13.144}$$

so that the weak localization correction $\delta\sigma(T)$ can be written in the form

$$\delta\sigma(T) = -\frac{e^2}{\pi h} \ln \frac{\tau_\phi^{ee}(T)}{\tau_e} \tag{13.145}$$

and the phase coherence time $\tau_\phi^{ee}(T)$ is then defined by the average [354]

$$\ln \tau_\phi^{ee}(T) = \int \left(\frac{-\partial f}{\partial \epsilon} \right) \ln \tau_\phi^{ee}(\epsilon, T)\, d\epsilon. \tag{13.146}$$

We may check that $\tau_\phi^{ee}(T)$ does not depend significantly on the method used to determine the energy average, so we shall keep the definition (13.142) in the following.

13.7.2 Phase coherence in a fluctuating electric field

We now want to determine how the electron–electron interaction leads to a dephasing between time-reversed trajectories. To that purpose, we assume that the total electric field acting on a given electron and resulting from all other electrons can be replaced by an effective fluctuating electric field whose characteristics are imposed by the fluctuation–dissipation theorem [351].

To proceed further, we consider the contribution of the Cooperon to the return probability $P_c(\boldsymbol{r},\boldsymbol{r},t)$ in a time-dependent electric potential $V(\boldsymbol{r},t)$. In Appendix A6.3, we showed that this contribution can be written in the form (6.247),

$$P_c(\boldsymbol{r},\boldsymbol{r},t) = P_c^{(0)}(\boldsymbol{r},\boldsymbol{r},t)\langle e^{i\Phi}\rangle_{\mathcal{C}}, \tag{13.147}$$

where $P_c^{(0)}$ is the probability in the absence of the fluctuating potential, and Φ is the relative phase accumulated along a pair of time-reversed trajectories after a time t (relations 6.245 and 6.246):

$$\Phi = \frac{e}{\hbar}\int_0^t [V(\boldsymbol{r}(\tau),\tau) - V(\boldsymbol{r}(\tau),\overline{\tau})]\, d\tau \tag{13.148}$$

where $\overline{\tau} = t - \tau$ and $\langle\cdots\rangle_{\mathcal{C}}$ is the average taken over the distribution of diffusion paths.

We also have to average the thermal fluctuations of the electric potential and we denote this average by $\langle\cdots\rangle_T$. These fluctuations being Gaussian,[24] the average $\langle e^{i\Phi}\rangle_T$ is given by

$$\langle e^{i\Phi}\rangle_T = e^{-\frac{1}{2}\langle\Phi^2\rangle_T}. \tag{13.149}$$

We now need to determine the average $\langle e^{i\Phi}\rangle_{T,\mathcal{C}}$ on *both* diffusion paths and thermal fluctuations. We will start with the calculation of $\langle\Phi^2\rangle_T$. From (13.148), we have:

$$\langle\Phi^2\rangle_T = \frac{e^2}{\hbar^2}\int_0^t\int_0^t \langle [V(\tau_1) - V(\overline{\tau}_1)][V(\tau_2) - V(\overline{\tau}_2)]\rangle_T\, d\tau_1\, d\tau_2 \tag{13.150}$$

where $V(\tau) = V(\boldsymbol{r}(\tau),\tau)$ and $V(\overline{\tau}) = V(\boldsymbol{r}(\tau),\overline{\tau})$. We define the correlator $\langle VV\rangle_T(\boldsymbol{q},\omega)$ by the Fourier transform

$$\langle V(\boldsymbol{r},\tau)V(\boldsymbol{r}',\tau')\rangle_T = \int \frac{d\boldsymbol{q}}{(2\pi)^d}\frac{d\omega}{2\pi}\langle VV\rangle_T(\boldsymbol{q},\omega)\, e^{i[\boldsymbol{q}\cdot(\boldsymbol{r}-\boldsymbol{r}')-\omega(\tau-\tau')]}. \tag{13.151}$$

Its thermal average is related by the fluctuation–dissipation theorem to the dielectric function [344, 355, 356]:[25]

$$e^2\langle VV\rangle_T(\boldsymbol{q},\omega) = \frac{4\pi e^2}{q^2}\mathrm{Im}\left[\frac{-1}{\epsilon(\boldsymbol{q},\omega)}\right]\frac{2}{1-e^{-\beta\omega}} \tag{13.152}$$

where, according to (13.11), $\mathrm{Im}(-1/\epsilon(\boldsymbol{q},\omega)) = \omega/4\pi\sigma_0$. The processes that contribute to the dephasing have an energy $|\omega|$ lower than temperature. Indeed, because of the Pauli principle, an electron at the Fermi level cannot exchange an energy larger than T with its

[24] The modes of the electromagnetic field are quadratic and their fluctuations are thus Gaussian.

[25] Here we consider only the longitudinal fluctuations of the electromagnetic field. The transverse fluctuations are screened by the skin effect [351–353] and can be neglected, except in confined geometries.

environment. Then for $|\omega| < T$, we replace the thermal function by its high temperature limit:

$$\boxed{e^2\langle VV\rangle_T(\boldsymbol{q},\omega) = \frac{2e^2T}{\sigma_0 q^2}} \tag{13.153}$$

The integrand in (13.151) no longer depends on frequency, so integrating on ω gives[26]

$$\langle V(\boldsymbol{r},\tau)V(\boldsymbol{r}',\tau')\rangle_T = \frac{\delta(\tau-\tau')}{(2\pi)^d}\frac{2T}{\sigma_0}\int \frac{d\boldsymbol{q}}{q^2}e^{i\boldsymbol{q}\cdot(\boldsymbol{r}-\boldsymbol{r}')} \tag{13.154}$$

and by inserting this expression in (13.150):

$$\boxed{\left\langle \Phi^2\right\rangle_T = \frac{4e^2T}{\sigma_0\hbar^2}\int_0^t d\tau \int \frac{d\boldsymbol{q}}{(2\pi)^d}\frac{1}{q^2}[1-\cos \boldsymbol{q}\cdot(\boldsymbol{r}(\tau)-\boldsymbol{r}(\overline{\tau}))]} \tag{13.155}$$

This quantity depends on paths $\boldsymbol{r}(\tau)$. We now have to calculate the average

$$\left\langle e^{-\frac{1}{2}\langle\Phi^2\rangle_T}\right\rangle_C \tag{13.156}$$

over the distribution of closed diffusive paths which contribute to the Cooperon $P_c(\boldsymbol{r},\boldsymbol{r},t)$. This average is obtained following the functional integral approach developed in reference [351]. The Cooperon can be written as

$$\begin{aligned} P_c(\boldsymbol{r},\boldsymbol{r},t) &= \int_{\boldsymbol{r}(0)=\boldsymbol{r}}^{\boldsymbol{r}(t)=\boldsymbol{r}} \mathcal{D}\{\boldsymbol{r}\}\exp\left(-\int_0^t \frac{\dot{\boldsymbol{r}}^2(\tau)}{4D}d\tau - \frac{1}{2}\langle\Phi^2\rangle_T\right) \\ &= P_c^{(0)}(\boldsymbol{r},\boldsymbol{r},t)\left\langle e^{-\frac{1}{2}\langle\Phi^2\rangle_T}\right\rangle_C, \end{aligned} \tag{13.157}$$

where $P_c^{(0)}(\boldsymbol{r},\boldsymbol{r},t)$ is the return probability in the absence of a fluctuating electric field. In order to decouple the paths $\boldsymbol{r}(\tau)$ and $\boldsymbol{r}(\overline{\tau})$ which enter in $\langle\Phi^2\rangle_T$, we first use the semi-group relation (5.7) in the form

$$P_c(\boldsymbol{r},\boldsymbol{r},0,t) = \int d\boldsymbol{R}\; P_c(\boldsymbol{r},\boldsymbol{R},0,t/2)P_c(\boldsymbol{R},\boldsymbol{r},t/2,t). \tag{13.158}$$

We have denoted explicitly the initial and final times. Then, we perform the change of variables $\boldsymbol{R}(\tau) = [\boldsymbol{r}(\tau)+\boldsymbol{r}(\overline{\tau})]/\sqrt{2}$ and $\boldsymbol{\rho}(\tau) = [\boldsymbol{r}(\tau)-\boldsymbol{r}(\overline{\tau})]/\sqrt{2}$. After integration on $\boldsymbol{R}$, $P_c(\boldsymbol{r},\boldsymbol{r},t) = C(\boldsymbol{\rho}=0,\boldsymbol{\rho}=0,t/2)$ with [351]:

$$C\left(\boldsymbol{\rho}=0,\boldsymbol{\rho}=0,\frac{t}{2}\right) = \frac{1}{\sqrt{2}}\int_{\boldsymbol{\rho}(0)=0}^{\boldsymbol{\rho}(t/2)=0}\mathcal{D}\{\boldsymbol{\rho}\}\exp\left(-\int_0^{t/2}\left[\frac{\dot{\boldsymbol{\rho}}^2(\tau)}{4D}+U(\boldsymbol{\rho})\right]d\tau\right) \tag{13.159}$$

[26] As a result of the cutoff T on energy transfer ω, the δ function that shows up in (13.154) is in fact a strongly peaked function of width $1/T$. But since relevant times, including the quasiparticle relaxation time, are much larger than $1/T$ (remark on page 497) this function can be safely replaced by a δ function [353, 357].

where we have introduced the effective potential $U(\boldsymbol{\rho})$ defined by

$$U(\boldsymbol{\rho}) = \frac{4e^2T}{\sigma_0\hbar^2}\int\frac{d\boldsymbol{q}}{(2\pi)^d q^2}\left[1-\cos\boldsymbol{q}\cdot\boldsymbol{\rho}\sqrt{2}\right]. \tag{13.160}$$

The integral $C(0,0,t)$ obeys the differential equation:[27]

$$\left[\frac{\partial}{\partial t} - D\Delta_\rho + U(\boldsymbol{\rho})\right]C(0,\boldsymbol{\rho},t) = \frac{1}{\sqrt{2}}\delta(\boldsymbol{\rho})\delta(t). \tag{13.163}$$

We now derive an expression for the dephasing from the solution of this differential equation.

Exercise 13.15: Nyquist noise

Using the fluctuation–dissipation theorem written in the form (13.153), recover the Nyquist expression for the voltage noise in a conductor (restoring the Boltzmann constant k_B) [358, 359]:

$$\langle V^2\rangle_T(\omega) = 2k_BT\,R \tag{13.164}$$

for ω both positive *and* negative. $V = V(L) - V(0)$ is the voltage drop at the edge of a wire of resistance R, of length L and of section S.
Hint: from (13.151) and (13.153), calculate the thermal fluctuations of the voltage for the wire geometry (using Ohm's law $R = L/(S\sigma_0)$):

$$\langle V(r)V(r')\rangle_T(\omega) = \int\frac{dq}{2\pi}\frac{2Rk_BT}{q^2L}e^{iq(r-r')}. \tag{13.165}$$

By fixing $r=0$ and $r'=L$ at the edges of the samples and using the integral (15.97), relation (13.164) is obtained.

13.7.3 Phase coherence time in dimension $d = 1$

We consider now the case $d = 1$, for which we have seen that the quasiparticle relaxation presents peculiar characteristics due to the diverging contribution of energy exchanges with small energy transfer (section 13.6.3). Analogous features are thus expected for the phase relaxation. We consider a quasi-one-dimensional wire of section S. Using (15.97), the phase

[27] We should remember that the functional integral

$$F(\boldsymbol{r},\boldsymbol{r}',t) = \int_{\boldsymbol{r}(0)=\boldsymbol{r}}^{\boldsymbol{r}(t)=\boldsymbol{r}'}\mathcal{D}\{\boldsymbol{r}\}\exp\left(-\int_0^t\left[\frac{\dot{\boldsymbol{r}}^2(\tau)}{4D}+U(\boldsymbol{r})\right]d\tau\right) \tag{13.161}$$

obeys the differential equation:

$$\left[\frac{\partial}{\partial t} - D\Delta_{\boldsymbol{r}'} + U(\boldsymbol{r}')\right]F(\boldsymbol{r},\boldsymbol{r}',t) = \delta(\boldsymbol{r}-\boldsymbol{r}')\delta(t). \tag{13.162}$$

fluctuation (13.155) can be written:

$$\langle \Phi^2 \rangle_T = \frac{2e^2 T}{\hbar^2 \sigma_0 S} \int_0^t |r(\tau) - r(\overline{\tau})| d\tau, \tag{13.166}$$

so that

$$U(\rho) = \frac{2\sqrt{2}e^2 T}{\hbar^2 \sigma_0 S} |\rho|. \tag{13.167}$$

The differential equation (13.163) for C becomes

$$\left[\frac{\partial}{\partial t} - D\frac{\partial^2}{\partial \rho^2} + \frac{2\sqrt{2}e^2 T}{\hbar^2 \sigma_0 S}|\rho|\right] C(0, \rho, t) = \frac{1}{\sqrt{2}}\delta(\rho)\delta(t) \tag{13.168}$$

and its solution gives $C(0,0,t)$ and therefore $P_c(r,r,t)$, from which we obtain the time dependence of the function $\langle e^{-\frac{1}{2}\langle \Phi^2 \rangle_T} \rangle_C$. This function is calculated in Exercise 13.16, and we consider instead the weak localization correction given by relation (7.53):

$$\Delta\sigma = -\frac{se^2 D}{\pi\hbar} \int_0^\infty dt\, P_c(r,r,t)\, e^{-\gamma t}, \tag{13.169}$$

with $\gamma = 1/\tau_\gamma$, where τ_γ accounts for any dephasing time due to processes other than electron–electron interaction and is assumed to induce an exponential relaxation of the phase. This weak localization correction is nothing but the Laplace transform of the probability $P_c(r,r,t)$:

$$\begin{aligned} P_\gamma(r,r) &= \int_0^\infty P_c(r,r,t)\, e^{-\gamma t}\, dt \\ &= \int_0^\infty C(0,0,t/2)\, e^{-\gamma t}\, dt \\ &= 2C_{2\gamma}(0,0), \end{aligned} \tag{13.170}$$

where $C_\gamma(\rho, \rho')$, the Laplace transform of $C(\rho, \rho', t)$, enters directly expression (13.169),

$$\Delta\sigma = -\frac{2se^2 D}{\pi\hbar} C_{2\gamma}(0,0). \tag{13.171}$$

It obeys the differential equation:

$$\left[-D\frac{\partial^2}{\partial \rho^2} + \frac{2\sqrt{2}e^2 T}{\hbar^2 \sigma_0 S}|\rho| + 2\gamma\right] C_{2\gamma}(0,\rho) = \frac{1}{\sqrt{2}}\delta(\rho). \tag{13.172}$$

Introducing the characteristic time known as the Nyquist time in the literature[28] [358],

$$\boxed{\tau_N = \left(\frac{\hbar^2 \sigma_0 S}{e^2 T \sqrt{D}}\right)^{2/3}} \tag{13.173}$$

we obtain the dimensionless differential equation for $x = \rho\sqrt{2/D\tau_N}$:

$$\left[-\frac{\partial^2}{\partial x^2} + |x| + \frac{\tau_N}{\tau_\gamma}\right] C_{2\gamma}(0,x) = \frac{1}{2}\sqrt{\frac{\tau_N}{D}}\,\delta(x). \tag{13.174}$$

With the help of (15.101, 15.102), we obtain:

$$P_\gamma(r,r) = 2C_{2\gamma}(0,0) = -\frac{1}{2}\sqrt{\frac{\tau_N}{D}}\frac{\mathrm{Ai}(\tau_N/\tau_\gamma)}{\mathrm{Ai}'(\tau_N/\tau_\gamma)}, \tag{13.175}$$

where Ai and Ai′ are respectively the Airy function and its derivative [360]. From relation (13.171), the weak localization correction becomes

$$\boxed{\Delta\sigma = s\frac{e^2}{hS}\sqrt{D\tau_N}\,\frac{\mathrm{Ai}(\tau_N/\tau_\gamma)}{\mathrm{Ai}'(\tau_N/\tau_\gamma)}} \tag{13.176}$$

instead of

$$\Delta\sigma = -s\frac{e^2}{hS}\sqrt{D\tau_\gamma} \tag{13.177}$$

in the absence of electron–electron interaction (see 7.56). In the limit $\tau_\gamma \ll \tau_N$, we obtain

$$\Delta\sigma = -s\frac{e^2}{hS}\sqrt{D\tau_\gamma}\left[1 - \frac{1}{4}\left(\frac{\tau_\gamma}{\tau_N}\right)^{3/2}\right]. \tag{13.178}$$

Inversely, for $\tau_N \ll \tau_\gamma$,

$$\Delta\sigma = -1.372 s\frac{e^2}{hS}\sqrt{D\tau_N}, \tag{13.179}$$

where we have used the asymptotic forms (15.103, 15.104). Finally, using the approximation (15.105) for the ratio Ai/Ai′, we obtain a very good approximation for the weak localization correction as a function of the times τ_γ and τ_N [361]:

$$\Delta\sigma \simeq -s\frac{e^2}{hS}\sqrt{D}\left(\frac{1}{2\tau_N} + \frac{1}{\tau_\gamma}\right)^{-1/2}. \tag{13.180}$$

This approximation amounts to assuming an exponential relaxation of the phase, $\langle e^{i\Phi}\rangle_{T,\mathcal{C}} \simeq e^{-t/2\tau_N}$. In Exercise 13.16 and in Figure 13.10 we give the exact form of this relaxation which differs from an exponential. However, at this approximation, it is found that, in order to estimate the phase relaxation time in the presence of other dephasing processes, it is sufficient to add their inverse dephasing times. The phase coherence time τ_ϕ^{ee}, or phase relaxation time, in the presence of electron–electron interaction is thus

[28] This time is proportional to the quasiparticle relaxation time τ_{in} given by (13.135). We return to this remark later.

given by:

$$\boxed{\tau_\phi^{ee} = 2\tau_N = 2\left(\frac{\hbar^2\sigma_0 S}{e^2 T\sqrt{D}}\right)^{2/3}} \tag{13.181}$$

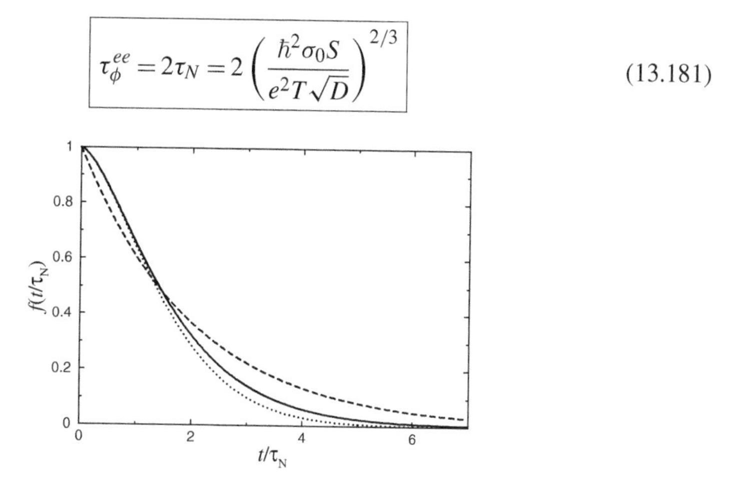

Figure 13.10 Relaxation of the phase. The continuous line represents the function $\langle e^{i\Phi}\rangle_{T,\mathcal{C}} = f(t/\tau_N)$ obtained from equation (13.182). The dotted line corresponds to the approximation $e^{-(\sqrt{\pi}/4)(t/\tau_N)^{3/2}}$ obtained from the small time expansion (13.184). The phase relaxation is clearly non-exponential. The approximation $e^{-t/2\tau_N}$ (dashed line) allows us to identify the characteristic time $\tau_\phi^{ee} = 2\tau_N$.

Exercise 13.16: relaxation of the phase in a quasi-one-dimensional conductor
Find the time dependence of the phase factor $\langle e^{i\Phi}\rangle_{T,\mathcal{C}} = f(t/\tau_N)$. Using the Laplace transform of $P_c(r,r,t)$, show that relation (13.175) can be written as

$$\int_0^\infty \frac{1}{\sqrt{4\pi Dt}} f\left(\frac{t}{\tau_N}\right) e^{-t/\tau_\gamma}\, dt = -\frac{1}{2}\sqrt{\frac{\tau_N}{D}}\frac{\mathrm{Ai}(\tau_N/\tau_\gamma)}{\mathrm{Ai}'(\tau_N/\tau_\gamma)}. \tag{13.182}$$

Noticing that the zeros of the Airy function and of its derivative are distributed on the negative real axis, calculate the integral in the complex plane using $\mathrm{Ai}''(x) = x\mathrm{Ai}(x)$ and show that [350]

$$\langle e^{i\Phi}\rangle_{T,\mathcal{C}} = \sqrt{\frac{\pi t}{\tau_N}}\sum_{n=1}^{\infty}\frac{e^{-|u_n|t/\tau_N}}{|u_n|}, \tag{13.183}$$

where the u_n' are the zeros of the function Ai$'$. For large n, they are well approximated by $|u_n| = \left(\frac{3\pi}{2}(n-\frac{3}{4})\right)^{2/3}$ [360]. Show that for $t < \tau_N$:

$$\langle e^{i\Phi}\rangle_{T,\mathcal{C}} \simeq e^{-\sqrt{\pi}/4(t/\tau_N)^{3/2}}. \tag{13.184}$$

We check that the short time behavior of the phase relaxation is identical to the relaxation of a quasi particle given by (13.139). As displayed in Figure 13.10, the function $f(t/\tau_N)$ is not an exponential. Calculate explicitly the average $\langle\Phi^2\rangle_{T,\mathcal{C}}$ over closed Brownian trajectories of duration t. Show first that $\langle|\boldsymbol{r}(\tau)|\rangle_{\mathcal{C}} = \sqrt{\frac{4D}{\pi t}\tau(t-\tau)}$, and then use this result to obtain an approximation for $\langle e^{i\Phi}\rangle_{T,\mathcal{C}}$:

$$\langle e^{i\Phi}\rangle_{T,\mathcal{C}} = \left\langle e^{-\frac{1}{2}\langle\Phi^2\rangle_T}\right\rangle_{\mathcal{C}} \simeq e^{-\frac{1}{2}\langle\Phi^2\rangle_{T,\mathcal{C}}} = e^{-\frac{\sqrt{\pi}}{4}(t/\tau_N)^{3/2}}. \tag{13.185}$$

Remark: dephasing and geometry, qualitative arguments

The main results of this section devoted to phase relaxation in quasi-one-dimensional wires can be recovered with a simple argument. From (13.148), it is easy to find that for a δ-correlated potential of the form (13.154), the time dependence of the dephasing is given by

$$\frac{d\langle\Phi^2(t)\rangle}{dt} = \frac{e^2}{2\hbar^2}\langle V_t^2\rangle, \tag{13.186}$$

where $\langle V_t^2\rangle$ represents the voltage fluctuation on a time scale t. These fluctuations are related through the Nyquist theorem (13.164), $\langle V_t^2\rangle = 2k_BTR_t$, to the resistance $R_t = \sigma_0 r_t/S$ of a wire of typical length r_t which is a typical distance reached after a time t by diffusive trajectories. Therefore,

$$\boxed{\frac{d\langle\Phi^2(t)\rangle}{dt} = \frac{e^2k_BT}{\hbar^2\sigma_0S}r_t} \tag{13.187}$$

In an infinite wire, the typical length varies as $r_t \sim \sqrt{Dt}$, so the variance of the phase, $\langle\Phi^2(t)\rangle$, varies non-linearly with time:

$$\langle\Phi^2(t)\rangle \sim \left(\frac{t}{\tau_N}\right)^{3/2}$$

with the characteristic Nyquist time τ_N given by (13.173).

It is also interesting to consider the case of a *finite* wire of length L. For times t larger than the Thouless time $\tau_D = L^2/D$, an electron has explored the system completely. This is the ergodic regime, defined in section 5.5.3. In this regime the typical distance r_t is no longer time dependent and it is set by the length L of the system. Consequently, the time dependence of the dephasing is linear:

$$\langle\Phi^2(t)\rangle \sim \frac{t}{\tau_c}$$

where τ_c is size dependent and is given by

$$\frac{\hbar}{\tau_c} = \frac{e^2}{\hbar}\frac{k_BTL}{\sigma_0S} = \frac{e^2}{\hbar}k_BTR, \tag{13.188}$$

where R is the resistance of the wire. The two times τ_N and τ_c are related by $\tau_N^3 = \tau_c^2\tau_D$. They have a different temperature dependence. This difference has been stressed recently and should show up in Aharonov–Bohm experiments on rings [362, 363].

13.7.4 Phase coherence and quasiparticle relaxation

It is very interesting to compare the results obtained for the *phase relaxation* with those obtained in section 13.6.4 to describe the *quasiparticle relaxation*. We notice first that the phase coherence time $\tau_\phi^{ee}(T)$ obtained in dimension $d = 1$ is parametrically identical to the quasiparticle relaxation time $\tau_{in}(T)$ given by (13.135).

In order to understand this similarity, we return in more detail to the structure of the expressions which describe the quasiparticle and the phase relaxations. The quasiparticle relaxation is given by (13.136):[29]

$$-\ln \mathcal{P}(t)=\frac{4}{\pi\hbar^2}t\int_{1/t}^{\infty}\frac{d\omega}{\sinh\beta\omega}\int\frac{d\boldsymbol{q}}{(2\pi)^d}\frac{4\pi e^2}{q^2}\mathrm{Im}\left(\frac{-1}{\epsilon(\boldsymbol{q},\omega)}\right)\mathrm{Re}P_d(\boldsymbol{q},\omega). \quad (13.190)$$

On the other hand, we have directly calculated the dephasing induced by fluctuations of the electric field on the Cooperon. This dephasing is characterized by the average values (13.150) and (13.151). Using the relation (5.22), $\langle e^{i\boldsymbol{q}\cdot\boldsymbol{r}(\tau)}\rangle_C=e^{-Dq^2\tau}$, we obtain the phase fluctuation in the form[30]

$$\frac{1}{2}\langle\Phi^2\rangle_{T,C}=\frac{1}{\pi\hbar^2}\int_0^t\int_0^t d\tau_1\,d\tau_2\int\frac{d\boldsymbol{q}}{(2\pi)^d}\int_{|\omega|<T}d\omega\frac{4\pi e^2}{q^2}\mathrm{Im}\left(\frac{-1}{\epsilon(\boldsymbol{q},\omega)}\right)\frac{1}{1-e^{-\beta\omega}}$$
$$\times -e^{-Dq^2|\tau_1-\tau_2|}\mathrm{Re}\big(e^{-i\omega(\tau_1-\tau_2)}-e^{i\omega(\tau_1-\overline{\tau}_2)}\big), \quad (13.191)$$

where $\overline{\tau}=t-\tau$. The integration over ω is, up to T, a consequence of the Pauli principle discussed on page 501. In order to show the equivalence of the approaches leading to relations (13.190) and (13.191), consider this latter expression.

Instead of integrating on the frequency as was done to obtain (13.154), let us first integrate on time. After a tedious calculation, we obtain:[31]

$$\frac{1}{2}\langle\Phi^2\rangle_{T,C}\simeq\frac{2}{\pi\hbar^2}t\int_{|\omega|<T}\frac{d\omega}{1-e^{-\beta\omega}}\int\frac{d\boldsymbol{q}}{(2\pi)^d}\frac{4\pi e^2}{q^2}$$
$$\times\,\mathrm{Im}\left(\frac{-1}{\epsilon(\boldsymbol{q},\omega)}\right)\mathrm{Re}P_d(\boldsymbol{q},\omega)\left(1-\frac{\sin\omega t}{\omega t}\right). \quad (13.193)$$

[29] Expressions (13.190) and (13.193) can be greatly simplified by noticing that

$$\frac{4\pi e^2}{q^2}\mathrm{Im}\left(\frac{-1}{\epsilon}\right)\mathrm{Re}P_d(\boldsymbol{q},\omega)=\frac{1}{2\rho_0}\mathrm{Im}P_d(\boldsymbol{q},\omega). \quad (13.189)$$

We have chosen to keep the expressions where the dielectric function appears explicitly.

[30] More precisely, it can be shown that $\langle e^{i\boldsymbol{q}\cdot\boldsymbol{r}(\tau)}\rangle_C=e^{-Dq^2\tau(t-\tau)/t}$. But the approximation used here is sufficient in the large time limit [352].

[31] We notice that, for an odd function $F(\omega)$,

$$\int_{-\infty}^{\infty}\frac{F(\omega)d\omega}{1-e^{-\beta\omega}}=\int_0^{\infty}F(\omega)\coth\frac{\beta\omega}{2}d\omega.$$

We can then rewrite (13.193) in the equivalent form

$$\frac{1}{2}\langle\Phi^2\rangle_{T,C}\simeq\frac{t}{\pi\hbar^2}\int_0^T d\omega\coth\frac{\beta\omega}{2}\int\frac{d\boldsymbol{q}}{(2\pi)^d}\frac{4\pi e^2}{q^2}$$
$$\times\,\mathrm{Im}\left(\frac{-1}{\epsilon(\boldsymbol{q},\omega)}\right)\mathrm{Re}P_d(\boldsymbol{q},\omega)(1-\frac{\sin\omega t}{\omega t}). \quad (13.192)$$

Quite remarkably, we notice that the compensation between the two correlators $\langle V(\tau_1)V(\tau_2)\rangle_T$ and $\langle V(\tau_1)V(\overline{\tau_2})\rangle_T$ introduced in (13.150) provides naturally, through the term $(1 - \sin \omega t/\omega t)$, a low frequency cutoff of order $1/t$, and leads to a time dependence $\langle \Phi^2 \rangle_{T,\mathcal{C}}$ which varies as $t^{3/2}$.[32]

Relations (13.190) and (13.193) show that the relaxations of a quasiparticle state and of the phase are identical, provided that the following remarks hold true.

- Although it is natural to expect that phase coherence is limited by the lifetime of quasiparticles, it is far from being obvious a priori that the two relaxation processes are identical. In particular, the frequency transfer ω has a different meaning in each case. For quasiparticle relaxation, it means an energy transfer between quantum states, while for the phase relaxation, it means the frequency of the fluctuating modes of the electric field. The correspondence between the two mechanisms is provided by the fluctuation–dissipation theorem.
- In (13.193), the lower cutoff on energies ω results quite naturally from the difference between the two correlators and thus from the structure of the Cooperon which couples two time-reversed trajectories. But in (13.190), it results from the Fermi golden rule. We emphasize the essential role played by the correlator $\langle V(\tau_1)V(\overline{\tau_2})\rangle_T$ which describes the potential correlation between time-reversed trajectories.
- The thermal function $\omega/\sinh\beta\omega$ which shows up in the Fermi golden rule derivation of (13.190) can be rewritten in the form

$$\frac{\omega}{\sinh\beta\omega} = \frac{2\omega}{1-e^{-\beta\omega}}[1-f(-\omega)]. \tag{13.194}$$

We recognize the thermal factor $\omega/(1-e^{-\beta\omega})$ that characterizes thermal fluctuations of the electromagnetic field. It is multiplied by a Fermi factor $1-f(-\omega)=f(\omega)$ which cuts the contribution of energy exchanges up to a value of order T. This term expresses the constraint due to the Pauli principle, which is explicitly taken into account by the Fermi factors of relation (13.121). By contrast, in the calculation of the phase relaxation leading to equation (13.193), only the first term appears, $\omega/(1-e^{-\beta\omega})$, because it describes the interaction of a single electron with a fluctuating electromagnetic field.

However, an additional constraint follows from the existence of the Fermi sea, namely that a quasiparticle of energy ϵ can relax only to an available empty state $\epsilon-\omega$, whence the Fermi factor $1-f(\epsilon-\omega)$. For quasiparticles at the Fermi level, this leads to the result (13.194). This multiplicative factor leads to the cutoff $\omega < T$ in the integral (13.193). The energy exchanged with the fluctuating field due to the other electrons cannot be larger than T. With this precaution, it appears that the relaxation of quasiparticles and of the phase are

[32] In the calculation leading to the relation (13.193), we have supposed that $Dq^2t \gg 1$. This hypothesis, also considered in reference [352], amounts to neglecting other terms which have the same time dependence.

equivalent mechanisms [357].[33]

> Quasiparticle and phase relaxations are driven by the same time scale
>
> $$\tau_\phi^{ee}(T) = \tau_{in}(T)$$

13.7.5 Phase coherence time in dimensions $d = 2$ and $d = 3$

In the quasi-one-dimensional case, the relaxation of the phase is mainly driven by the low frequency fluctuations. It is instructive to reconsider the results obtained in section 13.7.2, for dimensionalities 2 and 3. As for $d = 1$, the potential $U(\boldsymbol{\rho})$ could be determined, and we could solve the differential equation (13.162). Here we limit ourselves to more qualitative considerations by evaluating the fluctuations $\langle \Phi^2 \rangle_T$ of the phase.

For $d = 2$, the integral (13.155) on the wave vector diverges and must be cut off at $Dq^2 \simeq T$ since there is no energy exchange larger than T with the fluctuations of the electric field. By using (15.98), we obtain

$$\frac{1}{2}\langle \Phi^2 \rangle_T = \frac{e^2 T}{\pi \sigma_0 \hbar^2 a} \int_0^t d\tau \ln \frac{2}{q_c |\boldsymbol{r}(\tau) - \boldsymbol{r}(\overline{\tau})|} \sim \frac{e^2 T}{\pi \sigma_0 \hbar^2 a} t \ln \frac{1}{Tt} \tag{13.195}$$

where $q_c \simeq \sqrt{T/D}$ and a is the film thickness. Therefore, in dimension $d = 2$, the relaxation of the phase is exponential with a logarithmic correction. The phase coherence rate, defined by $\frac{1}{2}\left\langle \Phi^2 \right\rangle_T \simeq 1$, varies linearly with the temperature [351]:

$$\frac{1}{\tau_N} \simeq \frac{e^2 T}{2\pi \sigma_0 \hbar^2 a} \ln \frac{2\pi \sigma_0 \hbar a}{e^2} \tag{13.196}$$

and we notice that this time is the same as the quasiparticle relaxation time (13.134).

In dimension $d = 3$, the integral (13.155) diverges for the large wave vectors. As for $d = 2$, it must be cut off for $q_c \sim \sqrt{T/D}$. We thus obtain an exponential relaxation with the characteristic time τ_N given by

$$\frac{1}{\tau_N} = \frac{e^2 T^{3/2}}{\pi^2 \sigma_0 \hbar^2 \sqrt{D}} \tag{13.197}$$

and we recover a result similar to (13.132).

We conclude by noticing that, in dimensions 2 and 3, the phase relaxation and the quasiparticle relaxation are driven by energy transfers of the order of temperature T.

[33] This subtlety was at the origin of a debate [364,365] triggered by experimental results [366] which seemed to exhibit a saturation of the phase coherence time at low temperature. The assumption that the energy an electron can exchange with its electric environment is not limited by the temperature, but by the elastic collision rate, leads to a saturation of the phase coherence time at low temperature (within this assumption, the divergence which appears at high energy is not cut off by temperature but by the inverse collision rate).

13.7.6 Measurements of the phase coherence time τ_ϕ^{ee}

The usual method of determining experimentally the phase coherence time consists in measuring the magnetoresistance and deducing the weak localization correction (section 7.5). The latter is sensitive to phase coherence and is suppressed by different dephasing mechanisms, which result from a magnetic field, spin-orbit coupling (τ_{so}), coupling to magnetic impurities (τ_m), electron–phonon interaction (τ_{ph}) or from the electron–electron interaction (τ_ϕ^{ee}). Let us also recall that the weak localization correction does not involve any intrinsic temperature dependence, in contrast for example to the conductance fluctuations: it depends solely on L_ϕ but not on L_T (see 11.48). Some of these mechanisms are well controlled: the magnetic field is an external parameter and by measuring the magnetoresistance, it is possible to determine the various characteristic times. The strength of the spin-orbit coupling can be tuned by a change in the concentration of heavy substitution atoms. The electron–phonon coupling decreases rapidly at low temperature $1/\tau_{ph} \propto T^3$ and can then be neglected. What remains at low temperature is the dephasing due to magnetic impurities and to electron–electron interaction.

In practice, fitting the experimental curves by the theoretical expression for the magnetoconductance allows us to extract the different characteristic times. For example, in $d = 2$, the magnetoconductance can be fitted with expression (7.82). In $d = 1$, expression (7.56) gives the magnetoresistance of a wire of length L, in a perpendicular magnetic field (Exercise 7.3 and equation 7.84):

$$\frac{\Delta R}{R} = s\frac{e^2}{hRL}\left(\frac{3}{2}L_{trip} - \frac{1}{2}L_{sing}\right). \tag{13.198}$$

The triplet and singlet contributions are respectively given by (equation 7.75)

$$\frac{1}{L_{trip}^2} = \frac{1}{D}\left(\frac{1}{\tau_\phi^{ee}} + \frac{4}{3\tau_{so}} + \frac{2}{3\tau_m}\right) + \frac{W^2}{12L_B^4} \tag{13.199}$$

and

$$\frac{1}{L_{sing}^2} = \frac{1}{D}\left(\frac{1}{\tau_\phi^{ee}} + \frac{2}{\tau_m}\right) + \frac{W^2}{12L_B^4}, \tag{13.200}$$

where $L_B = \sqrt{\hbar/2eB}$ is the magnetic length.

Figure 13.11 presents the temperature dependence of the phase coherence time measured in a gold wire. The phase coherence time is found to vary as $T^{-2/3}$ in good agreement with (13.181). Figure 13.12 shows results obtained on metallic wires made of gold, silver and copper. The $T^{-2/3}$ dependence is also observed on gold and silver wires. At higher temperature, the time τ_ϕ deviates from the behavior predicted by relation (13.181). This indicates the existence of an additional dephasing mechanism due to the coupling to other degrees of freedom, such as the electron–phonon coupling [352], the latter giving a contribution proportional to T^{-3} which is not calculated here. The experimental results obtained in gold and silver wires are well described by the

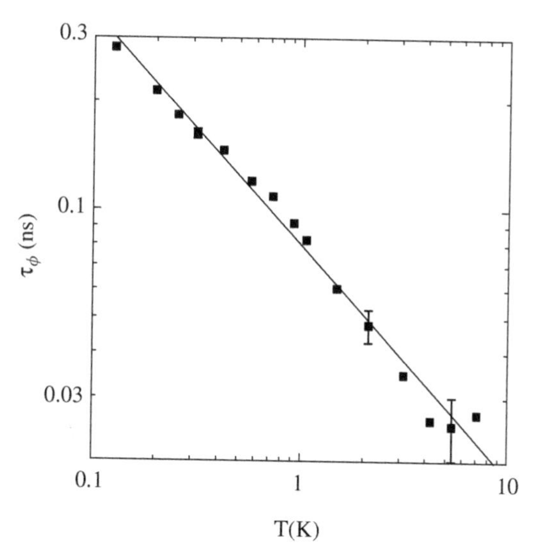

Figure 13.11 Phase coherence time τ_ϕ as a function of temperature in a gold wire. The straight line corresponds to $\tau_\phi \propto T^{-0.64}$ [367].

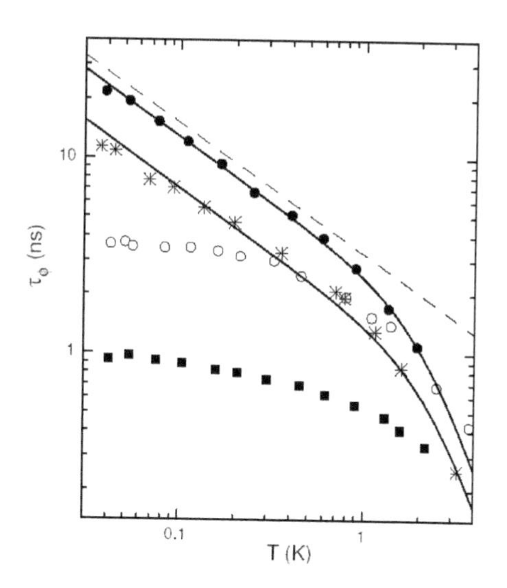

Figure 13.12 Temperature dependence of the phase coherence time τ_ϕ of four metallic wires (silver (• and ∘), gold (∗) and copper (■)) obtained by magnetoresistance measurements [361]. For the purest samples (silver (•) and gold (∗)), a dependence $\tau_\phi^{-1}(T) = AT^{2/3} + BT^3$ is observed (solid curves). The dashed line represents the contribution $AT^{2/3}$ for the silver samples. For less pure silver (∘) and for copper (■), this power law dependence is no longer observed. The low temperature saturation has been attributed to other dephasing mechanisms.

relation

$$\frac{1}{\tau_\phi} = AT^{2/3} + BT^3 \tag{13.201}$$

where A and B are fitting parameters.

We may conclude by saying that the low temperature saturation of the phase coherence time results from the coupling to degrees of freedom such as magnetic impurities [347,361, 368] or two-level systems [361] and not from electron–electron interaction. This conclusion is reinforced by Figure 13.12 which shows that the saturation in the case of silver wires depends on the purity of the sample.[34]

Appendix A13.1: Screened Coulomb potential in confined geometry

In section 13.2, we described the screened potential for a three-dimensional sample. More generally, the form of the interaction depends on space dimensionality and on the environment of the sample. For an isolated sample, the Fourier transform of the Coulomb potential e^2/R depends on the dimensionality d as

$$U_0(\boldsymbol{q}) = \begin{cases} \dfrac{4\pi e^2}{q^2} & (d=3) \\ \dfrac{2\pi e^2}{q} & (d=2) \\ 2e^2 \ln \dfrac{1}{qW} & (d=1). \end{cases} \tag{13.202}$$

The last expression corresponds to a quasi-one-dimensional wire of section W^2 and is valid for distances larger than W, that is, for $qW \ll 1$.[35] From (13.6), the static screened potential ($\omega = 0$) is given by the relation $U(\boldsymbol{q}) = U_0(\boldsymbol{q})/[1 + 2\rho_0 U_0(\boldsymbol{q})]$ where ρ_0 is the density of states. We thus obtain:

$$U(\boldsymbol{q}) = \begin{cases} \dfrac{4\pi e^2}{q^2 + \kappa_3^2} & (d=3) \\ \dfrac{2\pi e^2}{q + \kappa_2} & (d=2) \\ \dfrac{2e^2}{\ln^{-1} \dfrac{1}{qW} + 4e^2 \rho_{1d}} & (d=1). \end{cases} \tag{13.203}$$

[34] Recent experimental studies confirm that the low temperature dependence of τ_ϕ is driven by magnetic impurities and may be non-monotonic: it first saturates and then it further increases at lower temperature [361,368]. This behavior has been attributed to the Kondo effect which tends to screen the magnetic impurities at low T (see remark on page 213). At lower temperature the RKKY coupling between impurities may also be relevant [369].

[35] For the case $d = 1$, the cutoff in the logarithm results from the finite thickness W of the wire and it is given up to a numerical factor.

For $d=3$, the screening vector κ_3 is given by $\kappa_3^2=\kappa^2=8\pi e^2\rho_{3d}$ where ρ_{3d} is the density of states per unit volume and per spin direction in $d=3$. In $d=2$, the screening vector is given by $\kappa_2=4\pi e^2\rho_{2d}$ where ρ_{2d} is the density of states in $d=2$. For example in a quasi-two-dimensional sample of thickness W, $\rho_{2d}=\rho_{3d}W$, so that $\kappa_{2d}=\kappa_{3d}^2 W/2$. The dynamically screened interaction is given by (13.6), that is,

$$U(\boldsymbol{q},\omega)=\begin{cases}\dfrac{4\pi e^2}{q^2+\kappa_3^2\frac{Dq^2}{-i\omega+Dq^2}} & (d=3)\\[2ex] \dfrac{2\pi e^2}{q+\kappa_2\frac{Dq^2}{-i\omega+Dq^2}} & (d=2)\\[2ex] \dfrac{2e^2}{\ln^{-1}\dfrac{1}{qW}+4e^2\rho_{1d}\frac{Dq^2}{-i\omega+Dq^2}} & (d=1).\end{cases} \tag{13.204}$$

In any dimension and in the diffusive limit $ql_e \ll 1$, the screened Coulomb interaction retains the unique form

$$U(\boldsymbol{q},\omega)\underset{q\to 0}{\longrightarrow}\frac{1}{2\rho_d}\frac{-i\omega+Dq^2}{Dq^2}. \tag{13.205}$$

However, in $d=1$, this expression for the screened potential may lead to divergences (e.g. for the calculation of the density of states correction in equation (13.66)) and the full form of the interaction must be retained.

In a confined geometry, the expression for the screened interaction depends on the nature of the environment (see section 13.4.4 and reference [327]). For instance, if we consider a two-dimensional sample placed at a distance a from a metallic gate, the image charges induced by the gate modify the static interaction which becomes

$$U_0(\boldsymbol{q})=\frac{2\pi e^2}{q}(1-e^{-2qa}), \tag{13.206}$$

so that at a distance smaller than a, that is, for $qa \gg 1$, we recover the two-dimensional static Coulomb interaction. At large distance, that is for $qa \ll 1$, the interaction becomes

$$U_0(q\to 0)=4\pi e^2 a, \tag{13.207}$$

and it can be cast in the form $U_0(\boldsymbol{q}\to 0)=e^2/C$, where the capacitance per unit surface is defined as $C=1/(4\pi a)$.

Similarly, for a wire of section $W\times W$ at a distance a from a gate, the static interaction is

$$U_0(\boldsymbol{q})=2e^2\left[\ln\frac{e^{-\gamma}}{qW}-K_0(2qa)\right], \tag{13.208}$$

where $\gamma \simeq 0.577$ is the Euler constant and K_0 is a modified Bessel function [360] (15.76). In the limit $qa \gg 1$, we recover the one-dimensional result.[36] At a large distance, that is for $qa \ll 1$, we obtain

$$U_0(q \to 0) = 2e^2 \ln \frac{a}{W} \tag{13.209}$$

which is again of the form $U_0(q \to 0) = e^2/C$, where we define the capacitance per unit length $C = 1/(2 \ln a/W)$.

Once the static interaction $U_0(\boldsymbol{q})$ is obtained for these specific geometries, we deduce the dynamically screened interaction from the relation (13.6):

$$U(\boldsymbol{q}, \omega) = \frac{1}{U_0^{-1}(q) + 2\rho_d \frac{Dq^2}{-i\omega + Dq^2}} \tag{13.210}$$

and in the diffusive limit where $U_0(q \to 0) = e^2/C$,

$$U(q \to 0, \omega) \to \frac{-i\omega + Dq^2}{2\rho_d Dq^2 - i\omega C/e^2} \tag{13.211}$$

which is expression (13.71).

Lastly, it is important to notice that the form of the screened interaction depends on the length scale under consideration. For a wire and a length scale smaller than its width W, the interaction retains its three-dimensional form, whereas it is of a one-dimensional type for larger values of the length.

The expression for the screened interaction depends on space dimensionality and the energy scale ϵ can be used to monitor the crossover between dimensionalities $d = 1$ and $d = 3$ by comparison of the energy dependent length $L_\epsilon = \sqrt{D/\epsilon}$ and the thickness W.

Appendix A13.2: Lifetime in the absence of disorder

This appendix recalls the main steps of the calculation of the quasiparticle lifetime for an interacting electron gas in the absence of disorder. The result is at the basis of the Landau theory of Fermi liquids.

In the absence of disorder, the eigenstates are plane waves indexed by their momentum, that is $|\alpha\rangle = |\boldsymbol{p}\rangle$, $|\gamma\rangle = |\boldsymbol{p}'\rangle$, $|\beta\rangle = |\boldsymbol{p} - \boldsymbol{q}\rangle$ and $|\delta\rangle = |\boldsymbol{p}' + \boldsymbol{q}\rangle$. Therefore, relation (13.92) becomes:

$$\frac{1}{\tau_{ee}(\epsilon)} = \frac{4\pi}{\nu_0} \int_0^\epsilon d\omega \int_{-\omega}^0 d\epsilon' \sum_{\boldsymbol{p}\boldsymbol{p}'\boldsymbol{q}} \frac{|U_{\boldsymbol{q}}|^2}{\Omega^2}$$
$$\times\, \delta(\tilde{\epsilon} - \epsilon_{\boldsymbol{p}})\delta(\tilde{\epsilon}' - \epsilon_{\boldsymbol{p}'})\delta(\tilde{\epsilon} - \omega - \epsilon_{\boldsymbol{p}-\boldsymbol{q}})\delta(\tilde{\epsilon}' + \omega - \epsilon_{\boldsymbol{p}'+\boldsymbol{q}}) \tag{13.212}$$

where $\tilde{\epsilon} = \epsilon + \epsilon_F$ and $\tilde{\epsilon}' = \epsilon' + \epsilon_F$. We denote by $U_{\boldsymbol{q}}$ the Fourier transform of the screened interaction potential, as given by relation (13.2), in the Thomas–Fermi approximation. The

[36] See footnote 35.

quadratic dispersion relation for electrons implies that $\epsilon_{p-q} = \epsilon_p - v_F \cdot q + q^2/2m$ and $\sum_p = \nu_0 \int d\epsilon_p d\varpi$. Integration over the momenta p and p' leads to

$$\frac{1}{\tau_{ee}(\epsilon)} = 2\pi \nu_0 \int_0^\epsilon \omega\, d\omega \sum_q \frac{|U_q|^2}{\Omega^2} \int \delta\left(\omega - v \cdot q + \frac{q^2}{2m}\right) \delta\left(\omega - v' \cdot q - \frac{q^2}{2m}\right) d\varpi\, d\varpi' \tag{13.213}$$

where ϖ (respectively ϖ') is the solid angle (v, q) (respectively (v', q)). Upon angular integration and in the limit $\omega \ll \epsilon_F$, we obtain

$$\frac{1}{\tau_{ee}(\epsilon)} = 2\pi \nu_0^3 \epsilon^2 \langle |U|^2 \rangle, \tag{13.214}$$

where the interaction parameter is

$$\langle |U|^2 \rangle = \frac{1}{4\nu_0^2 \Omega^2} \sum_{|q|<2k_F} \frac{|U_q|^2}{(v_F q)^2}. \tag{13.215}$$

In dimension $d = 3$, from expression (13.2) for the screened potential and in the limit $\kappa \ll k_F$ [344], we deduce

$$\boxed{\frac{1}{\tau_{ee}} = \frac{\pi^2}{64} \frac{\kappa}{k_F} \frac{\epsilon^2}{\epsilon_F}} \tag{13.216}$$

where κ is the inverse screening length (13.4). We notice that the strength of the interaction is of order

$$\langle |U|^2 \rangle \propto \frac{\Delta^3}{\epsilon_F}. \tag{13.217}$$

For any κ and k_F, we obtain

$$\begin{aligned} \frac{1}{\tau_{ee}} &= \frac{\pi \epsilon^2}{16} \frac{\kappa^4}{k_F \epsilon_F} \int_0^{2k_F} \frac{dq}{(q^2 + \kappa^2)^2} \\ &= \frac{\kappa}{k_F} \frac{\pi \epsilon^2}{16 \epsilon_F} \left(\frac{k_F \kappa}{\kappa^2 + 4k_F^2} + \frac{1}{2} \arctan \frac{2k_F}{\kappa} \right), \end{aligned} \tag{13.218}$$

which indeed reproduces (13.216) in the limit $\kappa \ll k_F$, whereas for $\kappa \gg k_F$ it becomes

$$\frac{1}{\tau_{ee}} = \frac{\pi}{8} \frac{\epsilon^2}{\epsilon_F}. \tag{13.219}$$

14

Orbital magnetism and persistent currents

In this chapter, we restore $\hbar$ and we set the Boltzmann constant $k_B = 1$ unless specified otherwise.

14.1 Introduction

We have seen how phase coherence affects transport properties of electronic systems, for example the conductance and its fluctuations. The aim of the present chapter is to study the influence of these coherent effects on thermodynamic properties, such as magnetization of disordered conductors. In particular, we shall consider the problem of persistent current in a non-superconducting disordered metallic mesoscopic ring. We shall show that magnetization and persistent current are identical phenomena but correspond to different sample geometries. We shall first recall general characteristics of the magnetization, more specifically in the geometry of an infinite plane placed in a uniform and perpendicular magnetic field. Then we shall consider the persistent current flowing in a ring threaded by a magnetic Aharonov–Bohm flux.

The problem of persistent current may appear simpler at first glance than the study of conductance, since it deals with an equilibrium rather than a transport property. However, it happens that the calculation raises a number of non-trivial issues, such as the role of electron–electron interaction for the average current.

A conductor placed in a uniform magnetic field B acquires a magnetic moment $\mathcal{M}$ given by the derivative of the thermodynamic grand potential $\Phi(T, \mu, B) = -k_B T \ln \mathcal{Z}_g$ with respect to the applied field:

$$\mathcal{M}(B) = -\left(\frac{\partial \Phi}{\partial B}\right)_{\mu,T}, \tag{14.1}$$

where $\mathcal{Z}_g$ is the grand canonical partition function. We define the magnetization M as the magnetic moment per unit volume and the magnetic susceptibility χ as the linear response to a weak field:

$$M = \frac{\mathcal{M}}{\Omega} \qquad \chi = \lim_{B \to 0} \frac{dM}{dB}. \tag{14.2}$$

For the specific case of a ring, the magnetic moment $\mathcal{M}$ is that of a current-carrying loop of area $S = \pi R^2$, where R is the radius. The current $I = \mathcal{M}/S$ is best known as a *persistent current*, and it can be written in the form

$$I(\phi) = -\left(\frac{\partial \Phi}{\partial \phi}\right)_{\mu,T}, \tag{14.3}$$

where $\phi = BS$ is the flux through the ring.

We intend in this chapter to characterize the distribution of the magnetic moment $\mathcal{M}$. For a weakly disordered metal, we will show that the average magnetization $\overline{\mathcal{M}}$ and its fluctuation $\overline{\delta\mathcal{M}^2}^{1/2}$ are expressed in terms of the integrated return probability $Z(t)$, very much like the average conductance or its fluctuations studied in Chapters 7 and 11. Since magnetization and persistent current are two manifestations of the same physical phenomenon, their dependence on $Z(t)$ is identical. The persistent current will be simply obtained using the formal replacement of the magnetic moment $\mathcal{M}$ by the current I and of the magnetic field B by the flux ϕ.

We start by studying the case of a non-interacting electron gas. The corresponding grand potential Φ is a function of the single-particle energies ϵ_n, which are eigenvalues of the Hamiltonian (2.2) with appropriate boundary conditions:[1]

$$\Phi = -2k_B T \sum_n \ln(1 + e^{\beta(\epsilon_F - \epsilon_n)}). \tag{14.4}$$

The factor 2 accounts for spin degeneracy. This potential can be rewritten as an integral

$$\Phi(T, B) = -2k_B T \int_0^\infty \nu(\epsilon, B) \ln\left(1 + e^{\beta(\epsilon_F - \epsilon)}\right) d\epsilon, \tag{14.5}$$

where the magnetic field dependence of the energy spectrum shows up in the density of states $\nu(\epsilon, B) = \sum_n \delta(\epsilon - \epsilon_n(B))$. For the magnetic moment, we obtain the relation

$$\boxed{\mathcal{M}(T, B) = 2k_B T \int_0^\infty \frac{\partial \nu(\epsilon, B)}{\partial B} \ln\left(1 + e^{\beta(\epsilon_F - \epsilon)}\right) d\epsilon} \tag{14.6}$$

By integrating the grand potential Φ twice by parts, we obtain the following expressions

$$\Phi(B) = -2\int_0^\infty N(\epsilon, B) f(\epsilon)\, d\epsilon \tag{14.7}$$

and

$$\Phi(B) = 2\int_0^\infty \mathcal{N}(\epsilon, B) \frac{\partial f}{\partial \epsilon}\, d\epsilon, \tag{14.8}$$

[1] We consider a degenerate electron gas, such that $T \ll T_F$. In this limit, the chemical potential is temperature independent and is equal to its value at $T = 0$, namely $\epsilon_F(B) = \epsilon_F$.

where $N(\epsilon, B)$ is the integrated density of states (the counting function) per spin direction, namely $N(\epsilon, B) = \int_0^\epsilon \nu(\epsilon', B) d\epsilon'$. We denote by $\mathcal{N}(\epsilon, B)$ the doubly integrated density of states and by $f(\epsilon)$ the Fermi factor. At zero temperature, $\Phi(B)$ takes the simple form

$$\Phi(B) = -2\mathcal{N}(\epsilon_F, B). \tag{14.9}$$

Using equation (14.7), we obtain for the magnetic moment

$$\mathcal{M}(B) = 2\int_0^\infty \frac{\partial N(\epsilon, B)}{\partial B} f(\epsilon)\, d\epsilon = -2\sum_n f(\epsilon_n)\frac{\partial \epsilon_n}{\partial B} \tag{14.10}$$

i.e., a sum of contributions of each occupied level. At zero temperature, this simplifies to

$$\mathcal{M}(T = 0, B) = -2\int_0^{\epsilon_F} \frac{\partial \nu(\epsilon, B)}{\partial B}(\epsilon - \epsilon_F)\, d\epsilon \tag{14.11}$$

or

$$\mathcal{M}(T = 0, B) = -2\sum_{\epsilon_n < \epsilon_F} \frac{\partial \epsilon_n}{\partial B}. \tag{14.12}$$

14.2 Free electron gas in a uniform field

14.2.1 A reminder: the case of no disorder

Landau susceptibility and de Haas–van Alphen effect

Consider a two-dimensional gas. The eigenstates of the Schrödinger equation are Landau levels, of energy $\epsilon_n = (n + 1/2)\hbar\omega_c$, $\omega_c = eB/m$ being the cyclotron frequency, and with degeneracy eB/h per unit area and spin direction. The density of states ν per spin direction is thus given by

$$\nu(\epsilon, B) = \frac{eB}{h} S \sum_{n>0} \delta\left(\epsilon - (n + 1/2)\,\hbar\omega_c\right), \tag{14.13}$$

where S is the area of the system. It is convenient to symmetrize the sum over n using a $\theta(\epsilon)$ Heaviside function:

$$\nu(\epsilon, B) = \frac{eB}{h} S \sum_n \delta\left(\epsilon - (n + 1/2)\,\hbar\omega_c\right)\theta(\epsilon). \tag{14.14}$$

The Poisson transformation (15.106) leads to the Fourier decomposition of the density of states,

$$\nu(\epsilon, B) = \frac{mS}{2\pi\hbar^2}\left[1 + 2\sum_{p=1}^{\infty}(-1)^p \cos\frac{2\pi p\epsilon}{\hbar\omega_c}\right]\theta(\epsilon), \tag{14.15}$$

and the following expression for both the integrated density of states,

$$N(\epsilon, B) = \frac{mS}{2\pi\hbar^2}\left[\epsilon + 2\sum_{p=1}^{\infty}(-1)^p \frac{\hbar\omega_c}{2\pi p}\sin\frac{2\pi p\epsilon}{\hbar\omega_c}\right]\theta(\epsilon), \tag{14.16}$$

and the doubly integrated density of states,

$$\mathcal{N}(\epsilon) = \frac{mS}{2\pi\hbar^2}\left[\frac{1}{2}\epsilon^2 + 2\sum_{p=1}^{\infty}(-1)^p\left(\frac{\hbar\omega_c}{2\pi p}\right)^2\left(1-\cos\frac{2\pi p\epsilon}{\hbar\omega_c}\right)\right]\theta(\epsilon). \tag{14.17}$$

From equation (14.9), we obtain that at zero temperature the variation with the field of the grand potential $\Phi(B)$ is

$$\delta\Phi(B) = \Phi(B) - \Phi(0) = -\frac{mS\omega_c^2}{2\pi^3}\sum_{p=1}^{\infty}\frac{(-1)^p}{p^2}\left(1-\cos\frac{2\pi p\epsilon_F}{\hbar\omega_c}\right). \tag{14.18}$$

For a fixed Fermi energy ϵ_F, the thermodynamic potential and the magnetic moment $\mathcal{M}$ given by equation (14.1) oscillate strongly with the magnetic field.[2] These oscillations constitute the *de Haas–van Alphen effect*. They vanish when the temperature increases so that only the non-oscillating term remains

$$\delta\Phi(B) = -\frac{e^2S}{2\pi^3 m}B^2\sum_{p=1}^{\infty}\frac{(-1)^p}{p^2} = \frac{e^2S}{24\pi m}B^2, \tag{14.19}$$

which allows us to define the two-dimensional Landau susceptibility, per unit area

$$\chi_L = -\frac{e^2}{12\pi m} = -\frac{2}{3}\mu_B^2\rho_0(\epsilon_F), \tag{14.20}$$

where $\rho_0(\epsilon_F) = \nu_0/\Omega$ is the density of states at the Fermi level per spin direction and $\mu_B = e\hbar/2m$ is the Bohr magneton.

Effect of temperature

At finite temperature, the grand potential is given by equation (14.8). The non-oscillating term in expression (14.18) remains unchanged since $\int_0^\infty -\partial f/\partial\epsilon d\epsilon = 1$. The oscillating term involves the integral

$$\int_0^\infty\left(-\frac{\partial f}{\partial\epsilon}\right)\cos\frac{2\pi p\epsilon}{\hbar\omega_c}d\epsilon = \frac{2\pi^2 p/\beta\hbar\omega_c}{\sinh 2\pi^2 p/\beta\hbar\omega_c}\cos\frac{2\pi p\epsilon_F}{\hbar\omega_c}.$$

[2] We assume that the chemical potential ϵ_F does not vary with the magnetic field. It is indeed known that for an electron gas with a fixed electronic density, the chemical potential oscillates with the field. These oscillations are strong in two dimensions. For more details on the de Haas–van Alphen effect, see references [370, 371].

To obtain the right hand side, we have replaced the lower bound of the integral by $-\infty$ and we have used relations (15.108) and (15.62). Then, the grand potential at finite temperature is

$$\delta\Phi(B) = -\frac{S}{2}\chi_L B^2 \left[1 + \frac{12}{\pi^2}\sum_{p=1}^{\infty}\frac{(-1)^p}{p^2} R\left(\frac{2\pi^2 p}{\beta\hbar\omega_c}\right)\cos\frac{2\pi p\epsilon_F}{\hbar\omega_c}\right], \tag{14.21}$$

with $R(x) = x/\sinh x$. The oscillations of the grand potential $\delta\Phi(B)$ and of the magnetic moment are damped at finite temperature and vanish for $k_B T$ larger than $\hbar\omega_c$.

Exercise 14.1: Landau susceptibility and de Haas–van Alphen oscillations in a three-dimensional electron gas

In $d = 3$, the energy spectrum becomes

$$\epsilon = (n + 1/2)\,\hbar\omega_c + \frac{\hbar^2 k_z^2}{2m}.$$

The corresponding doubly integrated density of states $\mathcal{N}^{3d}$ appears to be simply related to its two-dimensional counterpart $\mathcal{N}^{2d}$ (14.17) by

$$\mathcal{N}^{3d}(\epsilon) = L_z \int_0^{\sqrt{2m\epsilon}/\hbar} \frac{dk_z}{\pi}\mathcal{N}^{2d}\left(\epsilon - \frac{\hbar^2 k_z^2}{2m}\right),$$

where L_z is the sample size along the direction parallel to the field. The non-oscillating term of the potential $\delta\Phi(B)$ is thus multiplied by $k_F L_z/\pi$ so that

$$\chi_L^{3d} = \frac{k_F}{\pi}\chi_L^{2d} = -\frac{e^2 k_F}{12\pi^2 m} = -\frac{1}{3}\mu_B^2 \rho_0(\epsilon_F), \tag{14.22}$$

where $\rho_0(\epsilon_F)$ is now the three-dimensional density of states per unit volume and per spin direction. The oscillating term involves the integral

$$I = \int_0^{k_F} \cos\left(\frac{2\pi p}{\hbar\omega_c}\left(\epsilon - \frac{\hbar^2 k_z^2}{2m}\right)\right)\frac{dk_z}{\pi}.$$

In the limit $\epsilon_F \gg \hbar\omega_c$, the upper bound of the integral can be safely taken to $+\infty$. Using the Fresnel relation (15.64), this integral is rewritten as

$$I = \frac{1}{\pi\sqrt{2}}\left(\frac{m\omega_c}{2\hbar p}\right)^{1/2}\cos\left(\frac{2\pi p\epsilon}{\hbar\omega_c} - \frac{\pi}{4}\right), \tag{14.23}$$

so that we finally obtain

$$\delta\Phi(B) = -\frac{\Omega}{2}\chi_L^{3d} B^2\left[1 + \frac{6}{\pi^2}\sum_{p=1}^{\infty}\frac{(-1)^p}{p^2}\left(\frac{\hbar\omega_c}{2p\epsilon_F}\right)^{1/2}\cos\left(\frac{2\pi p\epsilon_F}{\hbar\omega_c} - \frac{\pi}{4}\right)\right].$$

14.2.2 Average magnetization

We have seen in section 3.3 that in the presence of a disordered potential, characterized solely by its elastic mean free time τ_e, the energy levels are broadened, so that the average density of states (3.93) is now given by the convolution:

$$\rho(\epsilon) = \int_0^\infty g(\epsilon - \epsilon')\rho_0(\epsilon')\, d\epsilon' \tag{14.24}$$

with

$$g(\epsilon) = \frac{\hbar/(2\pi\tau_e)}{\epsilon^2 + (\hbar/2\tau_e)^2}. \tag{14.25}$$

As a result of this broadening, the average (14.8) of the grand potential at zero temperature becomes

$$\Phi(B) = 2\int_0^\infty \mathcal{N}(\epsilon, B) g(\epsilon - \epsilon_F)\, d\epsilon. \tag{14.26}$$

The oscillating term in expression (14.18) now acquires the form

$$\int_{-\infty}^\infty g(\epsilon - \epsilon_F)\cos\frac{2\pi p\epsilon}{\hbar\omega_c} d\epsilon = e^{-\pi p/\omega_c\tau_e}\cos\frac{2\pi p\epsilon_F}{\hbar\omega_c}. \tag{14.27}$$

Since the function $g(\epsilon - \epsilon_F)$ peaks around $\epsilon = \epsilon_F$, we have replaced the lower bound of the integral by $-\infty$ and we have used (15.63). This leads to an exponential decrease of the harmonics of the grand potential when the disorder increases:

$$\delta\Phi(B) = -\frac{1}{2}\chi_L B^2\left[1 + \frac{12}{\pi^2}\sum_{p=1}^\infty \frac{(-1)^p}{p^2} e^{-\pi p/\omega_c\tau_e}\cos\frac{2\pi p\epsilon_F}{\hbar\omega_c}\right]. \tag{14.28}$$

Measurement of the damping of these oscillations is the usual way to extract the elastic collision time τ_e . The other physical quantities described in this book actually depend on τ^*.[3] The average magnetic moment $\overline{\mathcal{M}}$ is given by (14.1). If $1/\tau_e$ is no longer negligible compared to ϵ_F, the non-oscillating term vanishes as well, since the integral $\int_0^\infty g(\epsilon)d\epsilon$ is no longer equal to 1. The average Landau susceptibility $\chi_L(\tau_e)$ then becomes

$$\chi_L(\tau_e) = \chi_L(\tau_e = \infty)\int_0^\infty \frac{\rho_0(\epsilon)}{\rho_0(\epsilon_F)} g(\epsilon - \epsilon_F)\, d\epsilon. \tag{14.29}$$

[3] In most cases, we have considered isotropic collisions so that $\tau_e = \tau^*$. The reader may check that for anisotropic collisions, τ_e needs to be replaced by τ^*, as may be seen in the example of the diffusion coefficient or the conductivity (Appendices A4.3 and A7.4). Here the damping factor depends only on τ_e, because the grand potential is related to the average density of states, that is to the single-particle Green function whose broadening depends only on τ_e.

The disorder leads to corrections of order $1/\epsilon_F\tau_e$ in dimension $d = 2$ and of order $1/(\epsilon_F\tau_e)^2$ for $d = 3$. They result from the correction to the average density of states (section 3.3) [372]. This is the only disorder effect left in the average susceptibility.

We now study the fluctuations of the magnetization. Calculation of these fluctuations involves the disorder average of a product of two densities of states so they are expected to be related to the integrated return probability $Z(t)$. We thus expect to deal with physical quantities that are sensitive to phase coherence just as those studied in Chapter 10 and related to spectral correlations.

14.2.3 Fluctuations of the magnetization

The average magnetization is only weakly affected by disorder. This is not the case for fluctuations. Changing the origin of energies, the magnetic moment (14.11) becomes[4]

$$\mathcal{M}(T=0,B) = -2\int_{-\epsilon_F}^{0} \frac{\partial \nu(\epsilon,B)}{\partial B}\,\epsilon\, d\epsilon. \tag{14.30}$$

The variance $\overline{\delta\mathcal{M}^2}$ defined by $\overline{\delta\mathcal{M}^2} = \overline{\mathcal{M}^2} - \overline{\mathcal{M}}^2$ is

$$\overline{\delta\mathcal{M}^2} = \frac{4}{\Delta^2}\partial_B\partial_{B'}\int_{-\epsilon_F}^{0}\int_{-\epsilon_F}^{0} \epsilon\,\epsilon' K(\epsilon-\epsilon',B,B')\, d\epsilon\, d\epsilon'\bigg|_{B'=B}, \tag{14.31}$$

where ∂_B stands for $\partial/\partial B$ and $K(\omega,B,B')$ is the normalized correlation function of the density of states, defined by (10.4) and calculated here for two different values of the magnetic field. The density of states $\nu(\epsilon,B)$ has been written as $\nu(\epsilon,B) = \Omega\rho(\epsilon,B) = \rho(\epsilon,B)/(\Delta\rho_0)$, where Δ is the average spacing between energy levels for a given spin. At the diffusion approximation, the correlation function K appears as the sum of two contributions K_d and K_c (section A4.4):[5]

$$K(\omega,B,B') = K_d\left(\omega,\frac{B-B'}{2}\right) + K_c\left(\omega,\frac{B+B'}{2}\right) \tag{14.32}$$

so that

$$\overline{\delta\mathcal{M}^2} = \frac{\hbar^2}{\Delta^2}\int_{-\epsilon_F}^{0}\int_{-\epsilon_F}^{0} \epsilon\epsilon'\left[K_c''(\epsilon-\epsilon',0) - K_d''(\epsilon-\epsilon',B)\right] d\epsilon\, d\epsilon', \tag{14.33}$$

where $''$ stands for the second derivative $\partial^2/\partial B^2$ with respect to the field. Using both the form factor $\tilde{K}(t)$, the Fourier transform of the correlation function $K(\omega)$ and the integral [6]

[4] We recall that $\nu = \rho\Omega$.

[5] See footnote 4, page 201, and the remark on page 203.

[6] Because of its lower bound, the integral $\int_{-\epsilon_F}^{0} \epsilon\, e^{i\epsilon t} d\epsilon$ contains a contribution which oscillates with the Fermi energy ϵ_F. One can check that this contribution is negligible when the Fermi energy is much larger than the Thouless energy, $\epsilon_F \gg E_c$. Moreover, the energy levels at the bottom of the band do not contribute to the persistent current [373]. This justifies using the integral $\int_{-\infty}^{0} \epsilon\, e^{i\epsilon t} d\epsilon = 1/t^2$.

$\int_{-\infty}^{0} \epsilon\, e^{i\epsilon t} d\epsilon = 1/t^2$, we obtain the relation between the variance and the field dependence of the form factor

$$\overline{\delta \mathcal{M}^2} = \frac{2\hbar^2}{\Delta^2} \int_0^\infty \frac{\tilde{K}_c''(t,B) - \tilde{K}_d''(t,0)}{t^4} dt. \tag{14.34}$$

Finally, using (10.51) between the form factor and the integrated return probability $Z(t)$ leads to

$$\boxed{\overline{\delta \mathcal{M}^2} = \frac{\hbar^2}{2\pi^2} \int_0^\infty \frac{Z_c''(t,B) - Z_d''(t,0)}{t^3} dt} \tag{14.35}$$

Depending on the physical situation we are considering, the integral may not converge, either for short times or for long times. We thus need to introduce the corresponding cutoff τ_e or τ_ϕ in the form of a multiplicative factor $(e^{-t/\tau_\phi} - e^{-t/\tau_e})$ in the integrand. This cutoff procedure has already been implemented for the calculation of the weak localization correction (7.54).

Equation (14.35) depends on the geometry of the system through the expression of $Z(t)$. As an example, for the diffusion in a plane, the return probability is given by (6.41). For $BDt \ll \phi_0$, it can be expanded as[7]

$$\begin{aligned} Z(t,B) &= \frac{BS/\phi_0}{\sinh 4\pi BDt/\phi_0} \\ &= S\left(\frac{1}{4\pi Dt} - \frac{2\pi}{3}Dt\frac{B^2}{\phi_0^2} + \frac{56\pi^3}{45}D^3t^3\frac{B^4}{\phi_0^4} + \cdots\right). \end{aligned} \tag{14.36}$$

The integral (14.35) is thus proportional to $B^4\, \tau_\phi$, so that

$$\overline{\delta \mathcal{M}^2} = \frac{112\pi}{15}\hbar^2 D^3 \tau_\phi \frac{B^2}{\phi_0^4} S. \tag{14.37}$$

This leads, for the relative fluctuation of the susceptibility [374, 375], to

$$\frac{(\delta\chi^2)^{1/2}}{|\chi_L|} = \sqrt{\frac{84}{5\pi}\frac{L_\phi}{\sqrt{S}}}\, k_F l_e \qquad (d=2). \tag{14.38}$$

In contrast to the average magnetization, the amplitude of the fluctuations depends on the phase coherence length L_ϕ. These fluctuations, obtained here for a macroscopic system of size $L > L_\phi$, can be understood as resulting from the superposition of $N = (L/L_\phi)^d$ incoherent subsystems. Therefore the relative fluctuation of the magnetization varies as $1/\sqrt{N} \propto (L_\phi/L)^{d/2}$. Within each subsystem of size L_ϕ, phase coherence is fully preserved and the fluctuations are larger than the average, in the ratio $k_F l_e \gg 1$.[8]

[7] The expression (6.41) for $Z(t,B)$ has been obtained at the eikonal approximation (section 6.3), that is for weak fields corresponding to $\omega_c \tau_e \ll 1$.

[8] The reader can verify that, for anisotropic collisions, the factor $k_F l_e$ should be replaced by $k_F l^*$, where l^* is the transport mean free path.

At finite temperature, (14.31) becomes

$$\overline{\delta\mathcal{M}^2} = \frac{\hbar^2}{\beta^2\Delta^2}\partial_B\partial_{B'}\int\int \ln(1+e^{-\beta\epsilon})\ln(1+e^{-\beta\epsilon'})K(\epsilon-\epsilon',B,B')\,d\epsilon\,d\epsilon'\bigg|_{B'=B}. \tag{14.39}$$

Relation (10.51) between the form factor and the return probability together with equation (15.112) lead to

$$\boxed{\overline{\delta\mathcal{M}^2} = \frac{\hbar^2}{2\pi^2}\int_0^\infty \frac{Z_c''(t,B)-Z_d''(t,0)}{t^3}\left(\frac{\pi Tt}{\sinh \pi Tt}\right)^2 dt} \tag{14.40}$$

The effect of temperature is to cut off the long time contribution so as to limit the fluctuations, even for $L_\phi \to \infty$. Since

$$\overline{\delta\mathcal{M}^2} \propto \int_0^\infty \left(\frac{\pi Tt}{\sinh \pi Tt}\right)^2 dt = \frac{\pi}{6T} \tag{14.41}$$

we obtain[9] [376]

$$\frac{(\delta\chi^2)^{1/2}}{|\chi_L|} = \sqrt{\frac{14}{5}\frac{L_T}{\sqrt{S}}}\,k_F l_e \qquad (d=2), \tag{14.42}$$

where the thermal length L_T is defined by $L_T^2 = \hbar D/T$.

Exercise 14.2: magnetization fluctuations in $d = 3$
In this case, the return probability $Z(t)$ is multiplied by the factor $1/\sqrt{4\pi Dt}$ which accounts for the diffusion along the direction of the magnetic field. Show that

$$\frac{(\delta\chi^2)^{1/2}}{|\chi_L|} = \sqrt{\frac{21\pi}{5}\frac{l_e}{L_\phi}}\left(\frac{L_\phi}{L}\right)^{3/2} \qquad \text{if } L_\phi \ll L_T, \tag{14.43}$$

$$\frac{(\delta\chi^2)^{1/2}}{|\chi_L|} = \sqrt{\frac{42I}{5}\frac{l_e}{L_T}}\left(\frac{L_T}{L}\right)^{3/2} \qquad \text{if } L_T \ll L_\phi, \tag{14.44}$$

where, with the help of (15.89), we obtain the prefactor $I = 3/4\sqrt{\pi/2}\zeta(3/2) \simeq 2.456$. This result has also been obtained using another method [376].

14.3 Effect of Coulomb interaction

Up to now we have considered the magnetic response of a free electron gas. Taking into account interaction between electrons may lead to a large additional contribution to the average magnetization that results from phase coherence.

[9] See footnote 8.

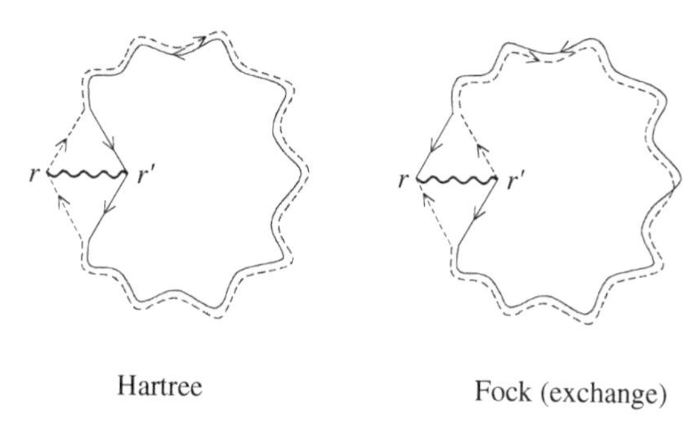

Figure 14.1 Hartree–Fock contribution to the magnetization.

14.3.1 Hartree–Fock approximation

Within the Hartree–Fock approximation (Figure 14.1, section 13.3) [377], we consider first the zero temperature limit. Using equation (13.24), the contribution $\delta\overline{E}_T$ of the interaction to the total average energy appears as a function of the local density $n(\boldsymbol{r}) = 2\sum_j^{occ}|\phi_j(\boldsymbol{r})|^2 = 2\int_{-\epsilon_F}^{0}\rho_\epsilon(\boldsymbol{r})d\epsilon$:

$$\delta\overline{E}_T = \frac{1}{2}\int U(\boldsymbol{r}-\boldsymbol{r}')\,\overline{\delta n(\boldsymbol{r})\delta n(\boldsymbol{r}')}\,d\boldsymbol{r}\,d\boldsymbol{r}' - \int U(\boldsymbol{r}-\boldsymbol{r}')\overline{\sum_{ij}^{occ}\phi_j^*(\boldsymbol{r}')\phi_j(\boldsymbol{r})\phi_i^*(\boldsymbol{r})\phi_i(\boldsymbol{r}')}\,d\boldsymbol{r}\,d\boldsymbol{r}'. \tag{14.45}$$

The sums $\sum^{occ}$ are taken over occupied states. This expression involves products of wave functions that were studied in Appendix A4.5. Here, we retain only the contribution of the Cooperon, since it depends on the magnetic field. From (4.210), we obtain

$$\overline{\delta n(\boldsymbol{r})\delta n(\boldsymbol{r}')} = \frac{4}{\pi}\rho_0\int_{-\epsilon_F}^{0}\int_{-\epsilon_F}^{0} g^2(\boldsymbol{R})P_c(\boldsymbol{r},\boldsymbol{r},\epsilon_1-\epsilon_2)\,d\epsilon_1\,d\epsilon_2, \tag{14.46}$$

where $\boldsymbol{R} = \boldsymbol{r}-\boldsymbol{r}'$. The factor 4 accounts for spin degeneracy. The second term in (4.210) is long range. Once it is multiplied by the short range interaction $U(\boldsymbol{r}-\boldsymbol{r}')$ given by (13.3) , it becomes negligible upon spatial integration. The exchange term can be written as a function of $\overline{\mathrm{Im}G_{\epsilon_1}(\boldsymbol{r},\boldsymbol{r}')\mathrm{Im}G_{\epsilon_2}(\boldsymbol{r}',\boldsymbol{r})}$, and using (4.206), we have

$$\overline{\sum_{ij}^{occ}\phi_j^*(\boldsymbol{r}')\phi_j(\boldsymbol{r})\phi_i^*(\boldsymbol{r})\phi_i(\boldsymbol{r}')} = \frac{2}{\pi}\rho_0\int_{-\epsilon_F}^{0}\int_{-\epsilon_F}^{0} g^2(\boldsymbol{R})P_c(\boldsymbol{r},\boldsymbol{r},\epsilon_1-\epsilon_2)\,d\epsilon_1\,d\epsilon_2. \tag{14.47}$$

Only wave functions with identical spin contribute to the exchange term, which leads to the factor 2. By inserting (14.46) and (14.47) into (14.45), the spatial integration leads to the factor F defined by (13.34) which accounts for the strength of the electronic interaction. Notice that the contributions of the Cooperon to the Hartree and Fock terms are identical except for a factor 2, due to spin degeneracy. Both are proportional to F. Let us remind ourselves that this is in contrast to the case of the Hartree–Fock correction leading to the density of states anomaly (section 13.4.1), where only the Hartree term is proportional to F. Finally, the contribution $\mathcal{M}_{ee} = -\partial\delta\overline{E}_T/\partial B$ of the Coulomb interaction to the average magnetization, after using (13.5), can be written as

$$\mathcal{M}_{ee} = -\frac{F\hbar}{2\pi}\partial_B \int_{-\epsilon_F}^{0}\int_{-\epsilon_F}^{0} \mathrm{Re} Z_c(\epsilon_1 - \epsilon_2, B)\, d\epsilon_1\, d\epsilon_2 \tag{14.48}$$

or

$$\mathcal{M}_{ee} = -\frac{F\hbar}{2\pi}\partial_B \int_0^{\infty} \omega\, \mathrm{Re} Z_c(\omega, B)\, d\omega. \tag{14.49}$$

A Fourier transform and use of the relation[10] $\int_{-\infty}^{0} e^{-i\epsilon t} d\epsilon = 1/t$ lead to the following expression for the magnetization:[11]

$$\boxed{\mathcal{M}_{ee} = -\frac{F\hbar}{2\pi}\int_0^{\infty} \frac{Z_c'(t,B)}{t^2}\, dt} \tag{14.50}$$

As an example, we consider the magnetization of an infinite plane. Expansion (14.36) shows that the term linear in the field in $Z'(t, B)$ also varies linearly with t. Thus the integral diverges logarithmically and must be cut off at short times by τ_e and at large times by τ_ϕ:

$$\mathcal{M}_{ee} = -\frac{F\hbar}{2\pi}\int Z_c'(t,B)\frac{dt}{t^2}(e^{-t/\tau_\phi} - e^{-t/\tau_e}). \tag{14.51}$$

In weak fields, we obtain[12]

$$\mathcal{M}_{ee} = \frac{2F}{3}\frac{\hbar D S}{\phi_0^2} B \ln\frac{\tau_\phi}{\tau_e}. \tag{14.52}$$

The Coulomb interaction thus contributes to the average magnetic response of the electron gas. This response, proportional to $k_F l_e$, is large compared to the Landau susceptibility χ_L (14.20) and it depends logarithmically on the phase coherence time τ_ϕ:[13]

$$\chi_{ee} = \chi_L \frac{F}{\pi} k_F l_e \ln\frac{\tau_\phi}{\tau_e}. \tag{14.53}$$

14.3.2 Cooper renormalization

The calculation that leads to expressions (14.48–14.53) is not complete. This is a perturbative calculation where interaction is taken to the lowest order. To obtain higher order corrections,

10 See footnote 6, page 522.
11 In the literature, we also find the notation $\lambda_0 = FU\rho_0 = F/2$.
12 One can also obtain the result (14.52) from (14.49) and (15.31) by limiting the integral between $1/\tau_\phi$ and $1/\tau_e$.
13 See footnote 8, page 523.

it is necessary to sum a perturbation series to all orders in the interaction parameter [378, 379]. This sum leads to a result similar to (14.49),

$$\mathcal{M}_{ee} = -\frac{\hbar}{\pi} \partial_B \int_0^\infty \omega \, \mathrm{Re}\lambda(\omega) \, Z_c(\omega, B) \, d\omega, \tag{14.54}$$

but with a renormalized interaction $\lambda(\omega)$ which depends on energy $\hbar\omega$:[14]

$$\lambda(\omega) = \frac{\lambda_0}{1 + \lambda_0 \Pi_c(\omega)} \tag{14.55}$$

where $\lambda_0 = F/2$ and $\Pi_c(\omega) = \ln \omega_{max}/\omega$. The expression for this effective interaction results from the sum of a geometric series which describes successive orders of the perturbation expansion. This structure is very similar to the sum of diagrams describing the superconducting instability in a metal. For this reason, this renormalization of the interaction is called "Cooper channel renormalization." The form (14.55) is general and does not depend on the nature of the bare interaction λ_0. However, the characteristic energy $\hbar\omega_{max}$ depends on the nature of the interaction. For the Coulomb interaction, the parameter $\lambda_0 = F/2$ is positive and the characteristic energy $\hbar\omega_{max}$ is the Fermi energy ϵ_F. Introducing the temperature $T_0 = \epsilon_F e^{1/\lambda_0}$, the effective interaction may be rewritten in the form

$$\lambda(\omega) = \frac{1}{\ln T_0/\hbar\omega}. \tag{14.56}$$

Coming back to the magnetization of the infinite plane and replacing $\lambda_0 = F/2$ by the effective interaction $1/\ln(T_0/\omega)$, for the susceptibility (14.54 and 15.31), we obtain the expression[15]

$$\chi_{ee} = \frac{4}{3} \frac{\hbar D S}{\phi_0^2} \int_{1/\tau_\phi}^{1/\tau_e} \frac{d\omega}{\omega \ln T_0/\hbar\omega}. \tag{14.57}$$

The integral (15.68) leads to[16]

$$\chi_{ee} = \frac{4\hbar D}{3\phi_0^2} S \ln \frac{\ln T_0 \tau_\phi/\hbar}{\ln T_0 \tau_e/\hbar} = \frac{2}{\pi} \chi_L \, k_F l_e \, \ln \frac{\ln T_0 \tau_\phi/\hbar}{\ln T_0 \tau_e/\hbar} \tag{14.58}$$

instead of equation (14.53). This contribution to the susceptibility thus appears to be much larger than the Landau susceptibility, by a factor $k_F l_e$ [378].

[14] Expression (14.54) is an approximation. The renormalized interaction also depends on the eigenmodes E_n of the diffusion equation. More details are can be found in [380].
[15] see footnote 12.
[16] See footnote 8, page 523.

14.3.3 Finite temperature

We discuss here the effect of temperature to first order in the interaction. Expression (14.48) for the magnetic moment becomes

$$\mathcal{M}_{ee} = -\frac{F\hbar}{2\pi}\partial_B \int_{-\infty}^{\infty}\int_{-\infty}^{\infty} \mathrm{Re}Z_c(\epsilon_1 - \epsilon_2, B) f(\epsilon_1) f(\epsilon_2)\, d\epsilon_1\, d\epsilon_2 \tag{14.59}$$

where $f(\epsilon)$ is the Fermi factor. After a Fourier transform, one obtains (see 15.111)

$$\boxed{\mathcal{M}_{ee} = -\frac{F\hbar}{2\pi}\int_0^{\infty} Z_c'(t, B)\frac{\pi^2 T^2}{\sinh^2 \pi T t}dt} \tag{14.60}$$

For the infinite plane and for $L_\phi \to \infty$, the upper bound of the integral is given by the inverse temperature, while its lower bound is still τ_e, so that (see 15.90)

$$\mathcal{M}_{ee} = \frac{2F}{3}\frac{\hbar DS}{\phi_0^2}B \ln \frac{\hbar e^{\gamma}}{2\pi T \tau_e} \tag{14.61}$$

where $\gamma \simeq 0.577$ is the Euler constant. We recover a result similar to (14.52), where $\hbar/\tau_\phi$ has been replaced by T. Taking into account the renormalized interaction leads to a result for the susceptibility χ_{ee} similar to (14.58), provided we replace $\hbar/\tau_\phi$ by T [378].

Exercise 14.3 Show that (14.59) for the average magnetic moment $\mathcal{M}_{ee}$ is equivalent to the form [381, 382]

$$\mathcal{M}_{ee} = -\frac{F\hbar}{2\pi}\partial_B \int_0^{\infty} \omega \coth\frac{\beta\omega}{2} Z_c(\omega)\, d\omega. \tag{14.62}$$

To this purpose, use the relation (15.111).

14.4 Persistent current in a ring

Until now, we have considered simply connected geometries such as the infinite plane. We now study the case of a ring pierced by a magnetic flux ϕ. This flux might result from an Aharonov–Bohm flux line or simply from a uniform magnetic field, provided that the ring is thin enough that no flux penetrates into the metal. We have already considered this geometry. In section 1.2, we saw that the sample specific conductance of a ring connected to wires is a periodic function of the Aharonov–Bohm flux with a period equal to the flux quantum ϕ_0. Then in section 7.6.1, we showed that the average conductance exhibits oscillations of period $\phi_0/2$, which originate from weak localization effects.

In this section, we consider first a clean one-dimensional ring. Then, we shall study the case of a quasi-one-dimensional ring, with electronic motion in transverse directions, i.e., with transverse channels.

The existence of a persistent current at thermal equilibrium in a metallic ring was first proposed by Hund [383] in the 1930s. It was then calculated by Bloch and by Kulik for a clean one-dimensional ring [384, 385]. The proposition by Büttiker, Imry and Landauer in 1983 that such a current could exist even in the presence of elastic disorder has largely motivated research activity in this topic [386]. The experimental difficulty arises from the small value of the current and of the induced magnetic moment. At present, experimental results have not yet been described in a proper quantitative way by the theory developed here and based on the Hartree–Fock approximation in the diffusive regime. This problem has given rise to a large number of theoretical works based on different premises. We have chosen to develop the Hartree–Fock approach which is, so far, the only quantitative one.

The persistent current defined by (14.3) is given, in the absence of interaction, by the expression

$$I = 2\frac{\partial}{\partial \phi}\int_0^\infty \mathcal{N}(\epsilon,\phi)\left(-\frac{\partial f}{\partial \epsilon}\right) d\epsilon \tag{14.63}$$

and at zero temperature, it reduces to

$$I = 2\frac{\partial \mathcal{N}(\epsilon_F,\phi)}{\partial \phi}. \tag{14.64}$$

14.4.1 Clean one-dimensional ring: periodicity and parity effects

The case of a strictly one-dimensional ring may be regarded as rather academic. However, its study allows us to highlight a few characteristics of the persistent current, such as its periodicity with flux. It is also useful for gaining some insight about the averaging procedure over an ensemble of rings. As for the Landau magnetization, we do not consider the spin of the electron which simply leads to a degeneracy factor 2.

The geometry is that of a one-dimensional ring, with perimeter L and without disorder, threaded by an Aharonov–Bohm flux (Figure 14.2). Written in the symmetric gauge,

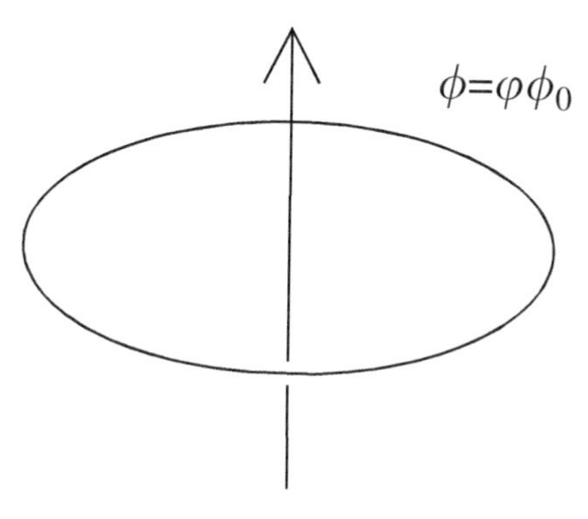

Figure 14.2 One-dimensional ring threaded by an Aharonov–Bohm flux $\phi = \varphi\phi_0$.

the vector potential $\boldsymbol{A}$ has only one constant azimuthal component, $\boldsymbol{A} = \boldsymbol{e}_\theta \phi / L$. The corresponding Schrödinger equation is

$$-\frac{\hbar^2}{2m}\left(\frac{\partial}{\partial x} + 2i\pi\frac{\varphi}{L}\right)^2 \phi_n(x) = E_n \phi_n(x), \tag{14.65}$$

where the wave function satisfies the continuity equation

$$\phi_n(x+L) = \phi_n(x) \tag{14.66}$$

and where $\varphi = \phi/\phi_0$ is the reduced flux written in units of $\phi_0 = h/e$.

Remark: Aharonov–Bohm flux and boundary conditions
By performing a gauge transformation, the new wave function becomes

$$\phi_n'(x) = \phi_n(x)\, e^{-i\frac{e}{\hbar}\int_0^x A dl} \tag{14.67}$$

namely,

$$\phi_n'(x) = \phi_n(x)\, e^{-2i\pi\varphi\frac{x}{L}}. \tag{14.68}$$

$\phi_n'(x)$ is a solution of the Schrödinger equation without flux, i.e.,

$$-\frac{\hbar^2}{2m}\frac{d^2}{dx^2}\phi_n'(x) = \epsilon_n \phi_n'(x), \tag{14.69}$$

but the flux dependence is now contained in the new boundary condition

$$\phi_n'(x+L) = \phi_n'(x)\, e^{-2i\pi\varphi}. \tag{14.70}$$

As a result of this gauge transformation, the vector potential has been removed from the Hamiltonian. However, the new wave function fulfills a new boundary condition which depends on the magnetic flux. We thus see that an Aharonov–Bohm flux is equivalent to a change in the boundary conditions.

The energy levels, shown in Figure 14.3, are given by[17]

$$\epsilon_n = \frac{\hbar^2}{2m}\left(\frac{2\pi}{L}\right)^2 (n-\varphi)^2 \tag{14.71}$$

where $n \in \mathbb{Z}$ and $\varphi = \phi/\phi_0$. Using the Poisson summation (15.107), the density of states defined as

$$\nu(\epsilon, \phi) = \sum_{n=-\infty}^{\infty} \delta(\epsilon - \epsilon_n) \tag{14.72}$$

17 A similar spectrum has also been obtained for the diffusion equation associated with the Cooperon for the geometry of a disordered ring (section 6.4.1) provided the reduced flux φ is replaced by 2φ.

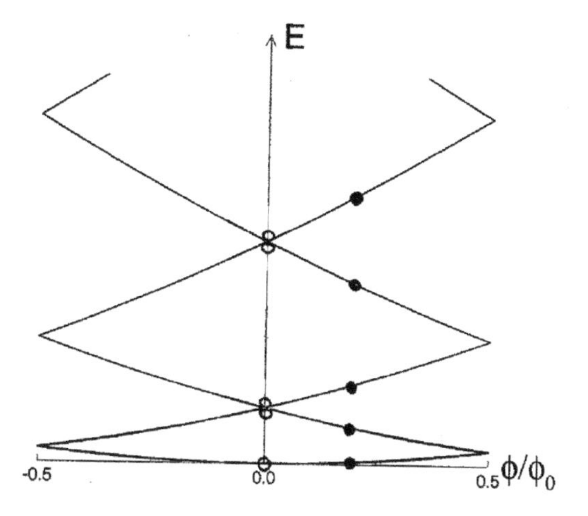

Figure 14.3 Spectrum of a one-dimensional ring in a reduced zone representation.

can be cast in the form

$$\nu(\epsilon, \phi) = \frac{L}{2\pi} \sum_{p=-\infty}^{\infty} \int_{-\infty}^{\infty} \delta\left(\epsilon - \epsilon(k)\right) e^{ipkL}\, dk\ \cos 2\pi p\varphi \tag{14.73}$$

with $\epsilon(k) = \hbar^2 k^2/2m$. Therefore, the integrated density of states

$$N(\epsilon, \phi) = \int_0^{\epsilon} \nu(\epsilon', \phi)\, d\epsilon' \tag{14.74}$$

admits the Fourier expansion:

$$N(\epsilon, \phi) = N_0(\epsilon) + \frac{2}{\pi} \sum_{p=1}^{\infty} \frac{\sin pk(\epsilon)L}{p} \cos 2\pi p\varphi, \tag{14.75}$$

where $k(\epsilon) = \frac{1}{\hbar}\sqrt{2m\epsilon}$ and where $N_0(\epsilon) = k(\epsilon)L/\pi$ is the integrated average density of states. Upon integrating the density of states twice, we obtain the following expression for the resulting function taken at the Fermi energy:

$$\mathcal{N}(\epsilon_F, \phi) = \int_0^{\epsilon_F} N(\epsilon, \phi)\, d\epsilon = \int_0^{k_F} N(\epsilon, \phi) \frac{\partial \epsilon}{\partial k}\, dk \tag{14.76}$$

or equivalently,

$$\mathcal{N}(\epsilon_F, \phi) = \mathcal{N}_0(\epsilon_F) - \frac{2\hbar v_F}{\pi L} \sum_{p=1}^{\infty} \frac{1}{p^2} \left(\cos pk_F L - \frac{\sin pk_F L}{pk_F L} \right) \cos 2\pi p\varphi, \tag{14.77}$$

$v_F = \hbar k_F/m$ being the Fermi velocity. Equation (14.64) leads to the following expression for the current

$$I(\phi) = \frac{2}{\pi} I_0 \sum_{p=1}^{\infty} \frac{1}{p} \left(\cos pk_F L - \frac{\sin pk_F L}{pk_F L} \right) \sin 2\pi p\varphi \tag{14.78}$$

where $I_0 = ev_F/L$ is the persistent current carried by a single level at the Fermi energy. For a large number of electrons, that is for $k_F L \gg 1$, we obtain [387]

$$\boxed{I(\phi) = \frac{2}{\pi} I_0 \sum_{p=1}^{\infty} \frac{\cos pk_F L}{p} \sin 2\pi p\varphi} \tag{14.79}$$

This expression for the persistent current applies to a ring whose chemical potential (i.e., its wave vector k_F) is fixed and flux independent. We may also consider the situation where *the number N of electrons is fixed in a ring*. In this case, the Fermi wave vector is equal to $k_F = N\pi/L$. The current thus becomes

$$I(N,\phi) = \frac{2}{\pi} I_0 \sum_{p=1}^{\infty} \frac{(-1)^{pN}}{p} \sin 2\pi p\varphi. \tag{14.80}$$

We see in this expression that the current depends on the *parity of the number N of electrons* in the ring. This is illustrated in Figure 14.4 where we notice in particular that adding one electron changes the sign of the current.

Exercise 14.4 Show that, in the interval $[-\pi, \pi]$, the current (14.80) can be rewritten as

$$I = I_0 (1 - 2\varphi) \qquad \varphi > 0$$
$$I = I_0 (-1 - 2\varphi) \qquad \varphi < 0$$

if N is an even integer, while

$$I = -2I_0\varphi$$

if N is an odd integer. These expressions can be obtained directly by summing up all the single level currents $i_n(\phi)$ given by

$$i_n(\phi) = -\frac{\partial \epsilon_n}{\partial \phi} = -\frac{1}{\phi_0}\frac{\partial \epsilon_n}{\partial \varphi} = -\frac{4\pi^2\hbar^2}{mL^2}(n - \varphi).$$

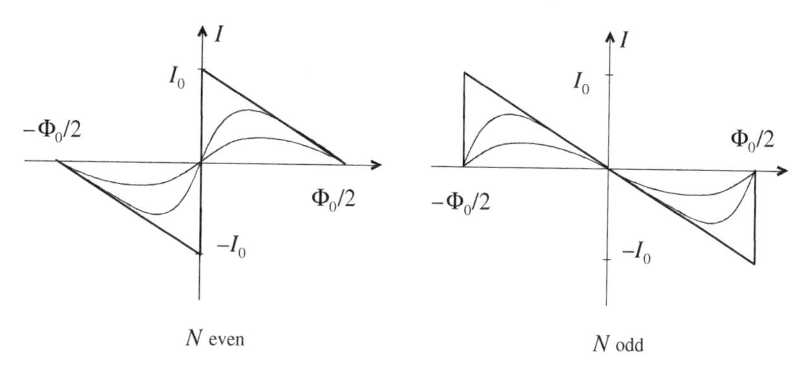

Figure 14.4 Persistent current of a non-disordered one-dimensional ring. The variation $I(\phi)$ presents a discontinuity which disappears at finite temperature. Note that the sign of the current depends on the parity of the number of electrons.

Temperature effect

At finite temperature, the harmonics of the current can be obtained, as for the magnetization, from relation (14.63). By linearizing the dispersion relation around the Fermi level $\epsilon = \epsilon_F + \hbar v_F(k - k_F)$, we obtain

$$\int_{-\infty}^{\infty} \left(-\frac{\partial f}{\partial \epsilon}\right) \cos pkL \, d\epsilon = \frac{\pi pL}{\beta \hbar v_F} \frac{1}{\sinh \frac{\pi pL}{\beta \hbar v_F}} \cos pk_F L. \tag{14.81}$$

Each harmonic of the current $I(\phi)$ is multiplied by the following function characterizing temperature effects in a Fermi gas (13.35, 14.40 and 14.60)

$$R(T/T_p) = \frac{T/T_p}{\sinh T/T_p} \tag{14.82}$$

so that current oscillations are exponentially damped, with a characteristic temperature $T_p = \Delta/p\pi^2$ for each harmonic p. The energy $\Delta = \pi \hbar v_F/L$ is the average level spacing. The total current vanishes exponentially, as soon as the temperature becomes of the order of the mean level spacing.

Disorder effect

A finite elastic collision time leads to a broadening of the Fermi level, described by the Lorentzian function $g(\epsilon)$ (14.25),

$$\int_{-\infty}^{\infty} \frac{\cos pkL}{p} g(\epsilon - \epsilon_F) \, d\epsilon = e^{-pL/2l_e} \frac{\cos pk_F L}{p}. \tag{14.83}$$

This broadening of the Fermi level $\epsilon_F \to \epsilon_F + i\hbar/2\tau_e$ was discussed in section 3.3 and it can be described by an imaginary part for the wave vector $k_F \to k_F + i/2l_e$. For a given parity of the electron number, the average current becomes

$$\overline{I}(\phi) = \frac{2}{\pi} I_0 \sum_{p=1}^{\infty} \frac{(-1)^{pN}}{p} e^{-pL/2l_e} \sin 2\pi p\varphi. \tag{14.84}$$

Each harmonics decreases exponentially. For an odd number, the behavior of the current as $\phi \to 0$ is such that [388]

$$\overline{I}(\phi) \simeq -2\frac{I_0}{\phi_0}\phi\left(1 - \tanh\frac{L}{4l_e}\right). \tag{14.85}$$

Then, for an elastic mean free path l_e that becomes smaller than the perimeter L of the ring, the average persistent current vanishes exponentially.

Remark: persistent current and de Haas–van Alphen oscillations
It is useful at this point to compare the structures of the persistent current (14.79) and of the de Haas–van Alphen oscillations considered in section 14.2.1. In both cases, and for a fixed field, the thermodynamic potential oscillates with the Fermi energy. For the de Haas–van Alphen oscillations, the phase of the oscillations is equal to $2\pi p\epsilon_F/\hbar\omega_c$ whereas for the persistent current, it is $pk_FL = p\pi\epsilon_F/2\Delta$. One can thus define a characteristic energy scale ϵ^*, of the order of $\hbar\omega_c$ in the first case and of the order of Δ in the second case. This characteristic energy measures the "rigidity" of the oscillations, since a variation of the chemical potential of order ϵ^* affects the phase of these oscillations. It is also possible to characterize this rigidity by the number of electrons $n^* = \epsilon^*/\Delta$ to be added to the system to change the phase of the oscillations.
It becomes apparent that the de Haas–van Alphen oscillations are quite rigid since $n^* \to \infty$ in the thermodynamic limit, whereas for the persistent current adding only one electron ($n^* = 1$) affects the phase of the oscillations.
In the limit $\omega\tau_e \gg 1$ and $k_BT < \hbar\omega_c$, the de Haas–van Alphen oscillations still exist whereas the average persistent current vanishes as soon as the temperature or the scattering rate $\hbar/\tau_e$ are of the order of Δ.

14.4.2 Average current

Consider now an ensemble of isolated rings whose total magnetization is measured. By dividing this magnetization by the total number of rings, we define the average current circulating in a single ring. The average is performed both on the number of electrons and on disorder configurations. We have seen in the previous section that, in one dimension, the sign of the current depends on the parity of the number of electrons. Therefore the average current is a priori expected to vanish, so that an experiment performed on a large number of isolated rings should measure a vanishing total magnetic moment. However, Figure 14.4 shows that the current is very anharmonic. By averaging the expression (14.80) over N, that is over a large number of rings with different numbers of electrons, one obtains a *finite* value of the average current *which oscillates with the period* $\phi_0/2$:

$$\overline{I}(\phi) = \frac{1}{\pi} I_0 \sum_{p=1}^{\infty} \frac{1}{p} e^{-pL/l_e} \sin 4\pi p\varphi. \tag{14.86}$$

This current becomes exponentially small when the perimeter L exceeds the elastic mean free path l_e.

In order to describe a situation which is more realistic than the strictly one-dimensional ring, we must first take into account the electronic dispersion in transverse directions. Consider for instance a cylinder of perimeter L, of height L_z and of negligible thickness. The Schrödinger equation takes the form

$$\left(-\frac{\hbar^2}{2m} \left[\left(\frac{\partial}{\partial x} + 2i\pi\varphi \right)^2 + \frac{\partial^2}{\partial z^2} \right] + V(\boldsymbol{r}) \right) \phi_n(\boldsymbol{r}) = \epsilon_n \phi_n(\boldsymbol{r}). \tag{14.87}$$

The continuity of the wave function around the cylinder and the vanishing of the wave function on the edges quantize the energy levels:

$$\epsilon_n = \frac{\hbar^2}{2m} \left(\frac{2\pi}{L} \right)^2 (n+\varphi)^2 + \frac{\hbar^2 k_z^2}{2m} \tag{14.88}$$

for which the transverse wave vector is $k_z = n_z\pi/L_z$ with $n_z \in \mathbb{N}$. The number M of transverse channels[18] is equal to $M = k_F W/\pi$. Since the energy appears as a sum of the decoupled longitudinal and transverse contributions, the total current obtained from (14.79) is the sum of the independent contributions of the transverse channels [389]:

$$I(\phi) = \frac{2}{\pi} I_0 \sum_{m=1}^{M} \sum_{p=1}^{\infty} \frac{k_x}{k_F} \frac{\cos pk_x L}{p} \sin 2\pi p\varphi \tag{14.89}$$

with $k_x^2 = k_F^2 - k_z^2$. For the case of a quasi-one-dimensional ring such that $L_z \ll L$, the cosine is a rapidly varying function of k_z so that the average vanishes. To get an estimation of the typical current, we assume that the arguments of the cosine that correspond to different values of k_z are completely uncorrelated so that the typical current varies as $\sqrt{M}$ [389].

[18] The transverse channels have been introduced in section A7.2.1 (relations 7.142 and 7.147).

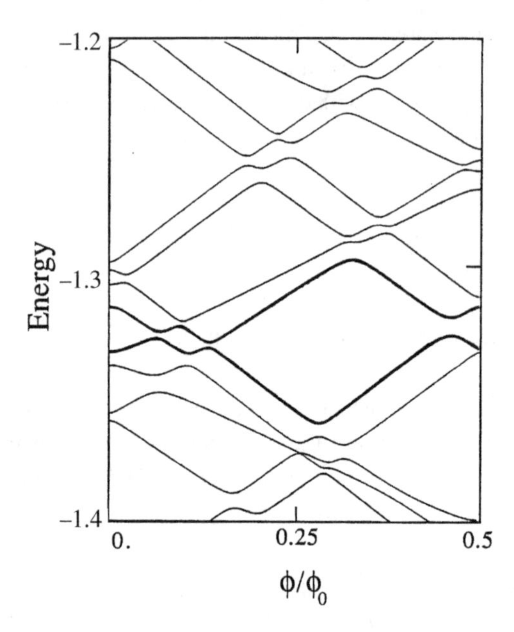

Figure 14.5 Spectrum of a multichannel quasi-one-dimensional ring threaded by an Aharonov–Bohm flux ϕ. Disorder, even if it is very weak, lifts the degeneracies. This spectrum was obtained using the Anderson model for a sample of size $L_x = 50a$, $L_z = 10a$ (a is the lattice spacing) folded along the x-direction. The disorder parameter is $W = 0.2t$ [390].

This description is actually not very physical since we have not taken into account disorder effects which remove the energy level degeneracies, and lead to avoided crossings in the energy spectrum (Figures 14.5 and 14.6).

14.5 Diffusive limit and persistent current

The persistent current is, by definition, a function of the density of states and therefore it can be expressed in terms of a Green's function. For a weak disorder, the average Green function is given by (3.90) and the average current is thus vanishing exponentially for $L > l_e$. In this section, we first go beyond the average current and show that in the diffusive regime, the variance of the distribution of the persistent current remains finite. Then, we shall include interaction effects and show that they lead to a finite value for the average persistent current.

To this purpose, we consider an ensemble of quasi-one-dimensional rings, i.e., with a finite section. We assume that the electronic motion is three dimensional (see Figure 14.6) and we solve the corresponding Schrödinger equation, also accounting for spin degeneracy. As a result of the very anisotropic geometry, we evaluate averages over the disorder by considering a one-dimensional diffusive motion (section 5.5.4).

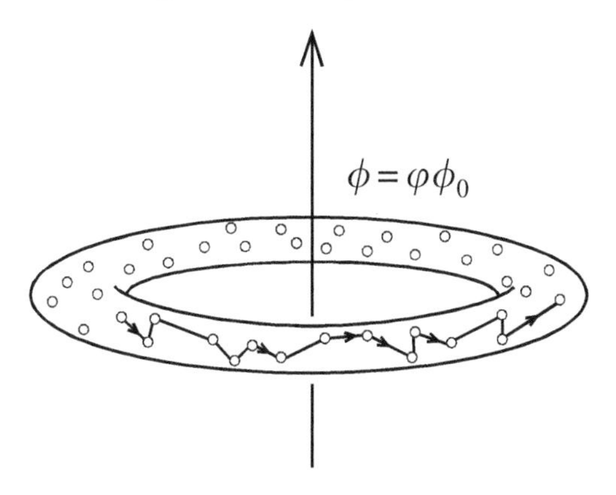

Figure 14.6 Spectrum of a multichannel quasi-one-dimensional ring pierced by an Aharonov–Bohm flux $\phi = \varphi\phi_0$, in the diffusive regime. This spectrum was obtained using the Anderson model for a sample of size $L_x = 20a$, $L_z = 10a$ folded along the x-direction and pierced by the flux. $W = t$ and a is the lattice spacing [390].

14.5.1 Typical current of a disordered ring

The variance $\overline{\delta I^2}$ of the distribution of the persistent current is defined by $\overline{\delta I^2} = \overline{I^2} - \overline{I}^2$, and the calculation is very similar to that presented in section 14.2.3 for the variance $\overline{\delta \mathcal{M}^2}$ of the magnetic moment. We only need to replace the magnetic moment $\mathcal{M}$ by the current I and the magnetic field by the flux ϕ. In the literature the quantity $\overline{\delta I^2}^{1/2}$ is often called the *typical current* and is denoted by I_{typ}. Equation (14.35) leads to the expression

$$\boxed{I_{typ}^2 = \frac{\hbar^2}{2\pi^2} \int_0^\infty \frac{Z_c''(t,\phi) - Z_d''(t,0)}{t^3}\, dt} \tag{14.90}$$

where we have defined $Z_c'' = \partial^2 Z/\partial\phi^2$. Using the winding number expansion (6.54) of the return probability,

$$Z_c(t,\phi) = \frac{1}{\sqrt{4\pi E_c t}} \sum_{p=-\infty}^{\infty} e^{-p^2 L^2/4Dt} \cos 4\pi p\varphi, \tag{14.91}$$

we obtain the Fourier expansion of the typical current

$$I_{typ}^2 = \sum_{p=1}^{\infty} I_p^2 \sin^2(2\pi p\varphi), \tag{14.92}$$

with the harmonics

$$I_p^2 = \frac{32p^2}{\phi_0^2} \int_0^\infty \frac{e^{-p^2/4E_c t}}{\sqrt{4\pi E_c}\, t^{7/2}}\, dt. \tag{14.93}$$

The integral (15.73) leads to

$$I_p^2 = \frac{384}{p^3} \left(\frac{E_c}{\phi_0}\right)^2. \tag{14.94}$$

It is important to note that, in contrast to the typical magnetization in a plane (14.35), the integral converges at small times. This convergence results from the fact that the weight of diffusive loops which involve at least one full turn around the ring is exponentially small at small time.

In the presence of dephasing, there is a multiplicative factor e^{-t/τ_ϕ} in the integrand, which results from the finite phase coherence time τ_ϕ (Chapter 6). The harmonics of the current are therefore reduced by a factor e^{-pL/L_ϕ}. More precisely, using relation (15.71), the harmonics (14.93) now take the form

$$I_p^2 = \frac{384}{p^3} \left(\frac{E_c}{\phi_0}\right)^2 \left[1 + p\frac{L}{L_\phi} + \frac{p^2}{3}\left(\frac{L}{L_\phi}\right)^2\right] e^{-pL/L_\phi}. \tag{14.95}$$

For $L \ll L_\phi$, the first term inside the brackets is the largest one, so the typical current is

$$\boxed{I_{typ} = 8\sqrt{6}\frac{E_c}{\phi_0} \sin 2\pi\varphi} \tag{14.96}$$

Since $E_c = \hbar/\tau_D$, this current can be interpreted as the current created by a single electron which encircles the ring diffusively in a time τ_D [391], namely

$$I_{typ} \simeq \frac{e}{\tau_D}. \tag{14.97}$$

Comparison of this value with that obtained in the absence of disorder, $I_0 = ev_F/L$ (equation 14.78), shows that disorder reduces the current in the ratio $I_{typ}/I_0 \simeq l_e/L$.

Exercise 14.5: typical current at finite temperature
At finite temperature, the integrand in (14.90) needs to be multiplied by the thermal factor $(\pi Tt/\sinh \pi Tt)^2$, just as the magnetization in (14.40). Establish that the temperature dependence of the harmonics of the typical current is of the form

$$I_p^2(T) = I_p^2(0) f^2(T), \tag{14.98}$$

with

$$f^2(T) = \frac{4}{3\sqrt{\pi}} \int_0^\infty e^{-1/x} R^2\left(\frac{\pi T p^2 x}{4E_c}\right) \frac{dx}{x^{7/2}} \tag{14.99}$$

and $R(x) = \sinh x/x$.

Exercise 14.6: typical current in a cylinder
Consider a cylinder of perimeter L and height L_z threaded by a magnetic flux parallel to its axis. In the limit $L_\phi \to \infty$, show that, for $L_z \gg pL$, the harmonics of the current are given by

$$I_p^2 = \frac{1024}{\pi p^3}\left(\frac{E_c}{\phi_0}\right)^2 \frac{L_z}{pL}.$$

Comparison with the current in a ring as given by (14.95) shows that the harmonic p corresponds to the incoherent addition of L_z/pL rings.
Show that for finite L_ϕ, the current becomes

$$I_p^2 = \frac{128}{\pi p^3}\left(\frac{E_c}{\phi_0}\right)^2 \frac{p^2 L^2 L_z}{L_\phi^3} K_3\left(p\frac{L}{L_\phi}\right)$$

where K_3 is a modified Bessel function [392].

14.5.2 *Effect of the Coulomb interaction on the average current*

In the absence of electron–electron interaction, the average current vanishes (section 14.4.2). Does this conclusion remain in the presence of Coulomb interaction? Its net effect treated within the Hartree–Fock approximation is to add a further contribution to the average magnetization as given by (14.50). For the geometry of a ring, this additional contribution corresponds to a finite average current:

$$\boxed{I_{ee} = -\frac{F\hbar}{2\pi}\int_0^\infty \frac{Z_c'(t,\phi)}{t^2}\,dt} \tag{14.100}$$

The expression (14.91) for $Z(t,\phi)$ yields the following Fourier expansion of the current:

$$I_{ee}(\phi) = \sum_{p=1}^{\infty} I_{ee,p} \sin(4\pi p\varphi), \tag{14.101}$$

whose harmonics are

$$I_{ee,p} = 2p\frac{F}{\phi_0}\int_0^\infty \frac{e^{-p^2/4E_c t - t/\tau_\phi}}{\sqrt{4\pi E_c}\, t^{5/2}}\, dt \tag{14.102}$$

and, using equation (15.71), this becomes

$$I_{ee,p} = 8\frac{F}{\phi_0}\frac{E_c}{p^2}\left(1 + p\frac{L}{L_\phi}\right) e^{-pL/L\phi}. \tag{14.103}$$

Thus, retaining only the first harmonic ($p = 1$), and considering the mesoscopic limit $L \ll L_\phi$, the average persistent current behaves as

$$I_{ee}(\phi) = 8F\frac{E_c}{\phi_0}\sin 4\pi\varphi. \tag{14.104}$$

As for the magnetization of the infinite plane (section 14.3), this expression results from a perturbation calculation up to first order in the interaction, and higher order terms must be included. This is the Cooper channel renormalization described in section 14.3, which amounts to replacing the interaction $\lambda_0 = F/2$ by an effective interaction $\lambda(\omega) = 1/\ln(T_0/\hbar\omega)$ with $T_0 = \epsilon_F e^{1/\lambda_0}$. Let us rewrite (14.49) in the form

$$I_{ee}(\phi) = -\frac{\hbar}{\pi}\partial_\phi \int_0^\infty \frac{\omega}{\ln T_0/\hbar\omega}\mathrm{Re}Z_c(\omega,\phi)\, d\omega. \tag{14.105}$$

By inserting the function $Z_c(\omega,\phi)$, the Fourier transform of $Z_c(t,\phi)$ given by (6.54),

$$Z(\omega,\phi) = \sum_{p=-\infty}^{\infty}\sqrt{\frac{i}{4\hbar\omega E_c}}\, e^{-p\sqrt{-i\hbar\omega/E_c}+4ip\pi\varphi}, \tag{14.106}$$

we recover the Fourier decomposition

$$I_{ee} = \sum_{p=1}^{\infty} I_{ee,p}\sin(4\pi p\varphi), \tag{14.107}$$

with the harmonics

$$I_{ee,p} = -\frac{4p}{\phi_0}\int_0^\infty \frac{1}{\ln T_0/\hbar\omega}\mathrm{Re}\sqrt{\frac{i\hbar\omega}{4E_c}}\, e^{-p\sqrt{-i\hbar\omega/E_c}}\, d\omega. \tag{14.108}$$

The logarithm function varies slowly so that the integral behaves roughly as

$$I_{ee,p} = 16\frac{\lambda_{eff}(p)}{\phi_0}\frac{E_c}{p^2}, \tag{14.109}$$

with

$$\lambda_{eff}(p) \simeq \frac{1}{\ln\left(p^2\frac{T_0}{E_c}\right)}. \tag{14.110}$$

We notice that the Cooper channel renormalization does not affect all harmonics in the same way. The energy scale associated with the harmonic p is E_c/p^2, and it corresponds to a time p^2L^2/D which is the time needed to encircle the ring p times in diffusive motion. An approximate expression for the average persistent current is obtained by neglecting the harmonic content [380], namely,

$$\boxed{I_{ee} \simeq 16\lambda_{eff}\frac{E_c}{\phi_0}\sin 2\pi\varphi} \tag{14.111}$$

with

$$\lambda_{eff} \sim \frac{1}{\ln T_0/E_c}. \tag{14.112}$$

Exercise 14.7: average current at finite temperature
At finite temperature, the integrand that appears in (14.100) should be multiplied by the thermal factor $(\pi Tt/\sinh \pi Tt)^2$. Show that the temperature dependence of the contribution to the average persistent current due to Coulomb interaction is

$$I_{ee}(T) = I_{ee}(0)g(T) \tag{14.113}$$

with

$$g(T) = \frac{2}{\sqrt{\pi}}\int_0^\infty e^{-1/x}R^2\left(\frac{\pi Tp^2x}{4E_c}\right)\frac{dx}{x^{5/2}} \tag{14.114}$$

and $R(x) = x/\sinh x$. This function $g(T)$ is pretty well approximated by the function [381]

$$g(T) = e^{-T/3E_c}. \tag{14.115}$$

Exercise 14.8: average current in a cylinder
Consider a cylinder of perimeter L and height L_z, threaded by a magnetic flux parallel to its axis. Show that, for $L_\phi \to \infty$, the harmonics of the contribution to the average persistent current due to Coulomb interaction are given by:

$$I_{ee,p} = \frac{8F}{\pi p^2}\frac{E_c}{\phi_0}\left(\frac{L_z}{pL}\right).$$

If we compare this expression for the current with expression (14.103) obtained for the geometry of a ring, we notice that it corresponds to the coherent addition of L_z/pL such rings.
For a finite value of L_ϕ, show that

$$I_{ee,p} = \frac{4F}{\pi p}\frac{E_c}{\phi_0}\frac{LL_z}{L_\phi^2}K_2\left(p\frac{L}{L_\phi}\right),$$

where K_2 is a modified Bessel function [392].

14.5.3 Persistent current and spin-orbit coupling

Spin-orbit coupling may affect the coherent transport substantially and may also change the sign of the magnetoresistance (section 7.5.2). It modifies the relative phase between two paired trajectories that contribute to the Cooperon. Quantitatively, it leads to an exponential decrease of the Cooperon contribution to the return probability $Z(t)$ (6.262).

Since the persistent current can also be written in terms of the integrated return probability $Z(t)$, it is worth addressing the question of the effect of spin-orbit coupling in that context. We show here that it does not affect the average persistent current, but it does reduce the typical current (Exercise 14.11). The reason for this has to be found in the form of the Hartree–Fock correction (14.45) to the total energy, which is independent of the spin-orbit coupling. To see this, we recall that the Hartree term involves the following combination of wave functions

$$\text{Hartree} \propto \sum_{\alpha\beta} \phi^*_{i,\alpha}(\boldsymbol{r})\phi_{i,\alpha}(\boldsymbol{r})\phi^*_{j,\beta}(\boldsymbol{r}')\phi_{j,\beta}(\boldsymbol{r}'), \tag{14.116}$$

whereas for the Fock term

$$\text{Fock} \propto -\sum_{\alpha\beta} \phi^*_{i,\alpha}(\boldsymbol{r})\phi_{i,\beta}(\boldsymbol{r}')\phi^*_{j,\beta}(\boldsymbol{r}')\phi_{j,\alpha}(\boldsymbol{r})\delta_{\alpha\beta}, \tag{14.117}$$

where α and β denote spins. The spatial dependences of these two contributions are identical (section 13.3). So let us concentrate only on the spin part, namely

$$\begin{aligned} \text{Hartree} &\propto \sum_{\alpha\beta} \langle \alpha\beta|\alpha\beta\rangle = 4 \\ \text{Fock} &\propto -\sum_{\alpha\beta} \langle \alpha\beta|\beta\alpha\rangle = -2, \end{aligned} \tag{14.118}$$

so that Hartree + Fock $\propto \sum_{\alpha\beta} (\langle \alpha\beta|\alpha\beta\rangle - \langle \alpha\beta|\beta\alpha\rangle) = 2$. The insertion of the closure relation $|S\rangle\langle S| + \sum_i |T_i\rangle\langle T_i| = 1$ together with the identities $\langle T_i|\alpha\beta\rangle = \langle T_i|\beta\alpha\rangle$ and $\langle S|\alpha\beta\rangle = -\langle S|\beta\alpha\rangle$ yields

$$\text{Hartree} + \text{Fock} \propto 2\sum_{\alpha\beta} \langle \alpha\beta|S\rangle\langle S|\alpha\beta\rangle. \tag{14.119}$$

There is no triplet contribution in the Hartree–Fock Hamiltonian. But these triplet terms are the only ones that are affected by the spin-orbit coupling (section 6.5). Therefore, both the average magnetization and the persistent current are unaffected by the existence of spin-orbit coupling.

Exercise 14.9 Establish the following identities:

$$\sum_{\alpha\beta} \langle \beta\alpha | S \rangle \langle S | \alpha\beta \rangle = -1$$

$$\sum_{\alpha\beta} \langle \alpha\beta | S \rangle \langle S | \alpha\beta \rangle = 1$$

$$\langle \alpha\alpha | S \rangle \langle S | \alpha\alpha \rangle = 0, \tag{14.120}$$

and use them to show that for strong spin-orbit coupling, we have

$$\text{Fock} \propto \sum_{\alpha\beta} \langle \beta\alpha | S \rangle \langle S | \alpha\beta \rangle = -1$$

$$\text{Hartree} \propto \sum_{\alpha\beta} \langle \alpha\beta | S \rangle \langle S | \alpha\beta \rangle = 1. \tag{14.121}$$

Comparing with (14.118), we notice that the Fock term is thus multiplied by a factor $-1/2$. This is the well-known factor that appears in the magnetoresistance in the presence of spin-orbit coupling (section A6.4.2). The Hartree term is multiplied by a factor $1/4$, so that the total contribution (Hartree + Fock) both remains unchanged and equal to 2 [380].

Exercise 14.10 Show that in the presence of a strong spin-orbit coupling, the typical persistent current is reduced by a factor 2. For this purpose, notice that

$$I_{typ}^2 \propto \sum_{\alpha\beta} \langle \alpha\beta | \alpha\beta \rangle \tag{14.122}$$

is reduced to

$$I_{typ}^2 \propto \sum_{\alpha\beta} \langle \alpha\beta | S \rangle \langle S | \alpha\beta \rangle. \tag{14.123}$$

Exercise 14.11 Show that the coupling to magnetic impurities reduces the persistent current. Using the results of section A6.4.2, show that the average current is still given by (14.100), but where the return probability $Z(t, B)$ is reduced by the exponential factor e^{-2t/τ_m}, τ_m being the magnetic scattering time [380].

14.5.4 A brief overview of experiments

In the preceding sections we have presented a theoretical description of persistent current which has not yet been confirmed by experiment. We now briefly describe the present status of the experimental work. The measurement of a persistent current consists actually in measuring a magnetization. This can be achieved by means of a SQUID. The magnetization signal is extremely weak so this kind of experiment is indeed very delicate. Let us first estimate the order of magnitude of the current. Both typical and average currents have been shown to be of the order of E_c/ϕ_0 , namely e/τ_D or $esabv_F l_e/L^2$. This corresponds to a magnetic moment $\mathcal{M}$ of the order of $ev_F l_e$ which, in units of the Bohr magneton $\mu_B = e\hbar/2m$, yields

$$\mathcal{M} \simeq \mu_B k_F l_e.$$

In a metal, $k_F l_e$ varies between 10 and 100, so that the magnetization per ring is small and difficult to measure. With this aim, two kinds of experiments have been designed.

Experiments involving a large number of isolated rings

In this case, the magnetization of a large number of isolated rings is measured. This is the method that was originally used to detect a persistent current [393]. The ensemble involved 10^7 copper rings, of perimeter $L = 2\mu\text{m}$ and with a mean free path estimated to be $l_e \simeq 20$ nm. In this experiment the measured average current oscillates with the applied magnetic flux, with a period $\phi_0/2$. From the estimate $E_c = \hbar D/L^2 \simeq 20$ mK, we obtain a net average current of the order of 0.4 nA which can be written as

$$I_{exp} \simeq 6\,\frac{E_c}{\phi_0}$$

where we have used the value $v_F = 1.6 \times 10^6$ m/s. By comparing this result with the expression (14.111) we obtain an estimate of the effective strength of the interaction $\lambda_{eff} \simeq 0.4$, which is larger by roughly one order of magnitude than the theoretical estimate $\lambda_{eff} = 1/\ln(T_0/E_c)$ given by (14.112). However, the relation (14.115) accounts properly for the temperature dependence.

Two other experiments, one with 10^4 semiconducting rings [394], and the second with 10^5 silver rings [395], have confirmed these results. The average current oscillates with a period $\phi_0/2$. Its amplitude is larger than that predicted by the model presented in section 14.5.2. Moreover, it is diamagnetic whereas from (14.111) we expect the current to carry the sign of the effective interaction, namely a paramagnetic current. Nevertheless, we have assumed the dominant interaction to be the Coulomb one. If this is not the case, for example if the dominant interaction is attractive which might result for instance from the electron–phonon coupling, the metallic ring may undergo a superconducting transition. The characteristic frequency ω_{max} introduced in (14.55) is then the Debye frequency ω_D and the critical temperature is $T_c = \hbar\omega_D e^{-1/|\lambda_0|}$. If the temperature T_c is lower than the Thouless energy, it can be shown that the effective interaction λ_{eff} becomes [380]

$$\lambda_{eff} \simeq -\frac{1}{\ln E_c/T_c}. \tag{14.124}$$

This mechanism is quite interesting since it shows that a metal with an extremely low superconducting temperature (which could be the case for copper or silver) may host diamagnetic persistent current well above the critical temperature.

Single ring experiments

Other experiments have been carried out on a single ring or on a small number of rings, where one measures a current which oscillates with the period ϕ_0. The first experiment [396] performed on a gold ring seemed to exhibit a current whose strength was of the order of $\sim 0.3 - 2.0\, ev_F/L$, namely one or two orders of magnitude higher than the theoretical prediction (14.96) for the typical current. In a more recent experiment [397], the strength of the measured current appeared to be compatible with (14.96), but with a sign corresponding to a diamagnetic current.

Another set of experiments has been performed on rings made of a semiconducting heterojunction [398]. There, the disorder is relatively weak and the mean free path $l_e \sim 11\,\mu\text{m}$ is comparable to the perimeter of the ring, $l_e/L \simeq 1.3$. The system is in between the diffusive and ballistic regimes.

In conclusion, the physics of the persistent current is not yet understood, either its amplitude or its sign. The aim of this chapter was to show the similarity that exists between persistent current in a ring and the magnetization of a film. These two quantities are of the same nature. The only difference resides in the kind of diffusive trajectories involved. It would be interesting to measure the susceptibility of metallic films at low temperature to check whether the orbital response is indeed enhanced compared to the Landau contribution, and whether it is diamagnetic as for the persistent current.

Appendix A14.1: Average persistent current in the canonical ensemble

Throughout Chapter 14, we have assumed the Fermi energy to be independent of the magnetic flux. However, experimentally this is the number of particles N which is fixed in each ring. This constraint, $N = \int_0^{\epsilon_F} \rho(\epsilon, \phi) d\epsilon$, leads to a dependence of the Fermi level on the flux, $\epsilon_F(\phi)$, so that the density of states itself depends on the flux, thus leading by means of (14.8) to the following expression for the persistent current at zero temperature[19]

$$I_N = \frac{\partial \mathcal{N}(\epsilon_F(\phi))}{\partial \phi}. \tag{14.125}$$

The disorder average of this current may contribute significantly. This current, called the canonical current,[20] can be calculated by performing an expansion around the Fermi level. By writing the average Fermi level $\epsilon_F = \overline{\mu(\phi)}$, we obtain

$$I_N = - \left.\frac{\partial \Phi}{\partial \phi}\right|_{\epsilon_F(\phi)} = - \left.\frac{\partial \Phi}{\partial \phi}\right|_{\epsilon_F} - \left.\frac{\partial^2 \Phi}{\partial \mu \partial \phi}\right|_{\epsilon_F} (\epsilon_F(\phi) - \epsilon_F). \tag{14.126}$$

By definition, the first term in this expression is the "grand-canonical" current studied in section 14.4.2. Upon averaging, it vanishes for $L \gg l_e$. The derivative $-\partial \Phi / \partial \mu$ is the

[19] If the spin is ignored.

[20] The word "canonical" may lead to some confusion. The calculation is still done in the grand canonical ensemble, but with the additional constraint that the average number of electrons is fixed and flux independent.

number N of electrons so that,

$$I_N = I_{\epsilon_F} + \left.\frac{\partial N}{\partial \phi}\right|_{\epsilon_F} (\epsilon_F(\phi) - \epsilon_F). \tag{14.127}$$

The thermodynamic relation $\delta\mu|_N = -\Delta\delta N|_\mu$ relates a variation of the chemical potential at fixed number of particles to a variation of the number of particles at fixed chemical potential. Upon disorder averaging and neglecting the first contribution, $\overline{I}_{\epsilon_F}$, we obtain [399]

$$\overline{I}_N = -\frac{\Delta}{2}\frac{\partial}{\partial \phi}\Sigma^2(\epsilon_F, \phi) = -\frac{\Delta}{2}\frac{\partial}{\partial \phi}\int_0^{\epsilon_F}\int_0^{\epsilon_F} K(\epsilon, \epsilon', \phi)\, d\epsilon\, d\epsilon', \tag{14.128}$$

where $\Sigma(E) = \overline{N(E)^2} - \overline{N(E)}^2$ is the fluctuation (10.9) of the number of levels. Then, for a fixed number of electrons in each ring, there is a finite average current which results from the fluctuation of the number of energy levels from one sample to another [382, 399–401]. With the help of relations (10.13) and (10.51), we obtain[21]

$$\overline{I}_N = -\Delta\frac{\partial}{\partial \phi}\int_0^\infty \frac{\tilde{K}(t, \varphi)}{t^2}\, dt = -\frac{\Delta}{4\pi^2}\frac{\partial}{\partial \phi}\int_0^\infty \frac{Z(t, \varphi)}{t}\, dt. \tag{14.129}$$

The Fourier expansion of this current as obtained from (14.91) is

$$\overline{I}_N = \frac{2}{\pi}\frac{\Delta}{\phi_0}\sum_{p=1}^\infty e^{-pL/L_\phi}\sin(4\pi p\varphi). \tag{14.130}$$

The sum can be performed so that we are led to [402]

$$\overline{I}_N = \frac{\Delta}{\pi\phi_0}\frac{\sin 4\pi\varphi}{\cosh L/L_\phi - \cos 4\pi\varphi}. \tag{14.131}$$

This expression for the current must be multiplied by a factor 2 to account for the spin degeneracy. The current oscillates with period $\phi_0/2$ and is *paramagnetic* at small flux. This behavior can be easily understood starting from relation (14.128). The average current measures the change in the spectral rigidity when time reversal invariance is broken by the magnetic flux. The variance $\Sigma^2(\epsilon_F, \phi)$ decreases when ϕ is finite, thus leading to a paramagnetic current.

This current is very tiny, of the order of Δ/ϕ_0, i.e., comparable to the current carried by a single energy level. This contribution to the average current is thus much smaller than the contribution resulting from the electron–electron interaction as obtained in section 14.3.

Finally, it is of interest to notice the analogy between the functional form of the canonical current and expression (7.90), which gives the weak localization correction to the conductivity of a ring threaded by an Aharonov–Bohm flux.

[21] See footnote 6, page 522.

15
Formulary

15.1 Density of states and conductance

Density of states

For a spectrum $\epsilon = \hbar^2 k^2/2m$

$$\rho_0(\epsilon) = \frac{dA_d}{(2\pi)^d}\frac{mk^{d-2}}{\hbar^2} \tag{15.1}$$

where A_d is the volume of the unit sphere:

$$A_d = \frac{\pi^{d/2}}{\Gamma(1+d/2)} \qquad A_3 = \frac{4\pi}{3} \qquad A_2 = \pi \qquad A_1 = 2 \tag{15.2}$$

In particular

$$\begin{aligned} \rho_0(\epsilon) &= \frac{mk}{2\pi^2\hbar^2} \quad (d=3) \\ \rho_0(\epsilon) &= \frac{m}{2\pi\hbar^2} \quad (d=2) \\ \rho_0(\epsilon) &= \frac{m}{\pi k\hbar^2} \quad (d=1) \end{aligned} \tag{15.3}$$

Classical conductivity and conductance

$$\sigma_0 = se^2 D\rho_0 = s\frac{e^2}{\hbar}\frac{A_d}{(2\pi)^d}k_F^{d-1}l_e \tag{15.4}$$

$$G = \sigma_0 L^{d-2} = s\frac{e^2}{\hbar}\frac{A_d}{(2\pi)^d}(k_F L)^{d-2}k_F l_e \tag{15.5}$$

In dimension $d=3$:

$$G = s\frac{e^2}{h}\frac{k_F^2 l_e}{3\pi}\frac{A}{L} = s\frac{e^2}{h}\left(\frac{l_e}{3L}\right)\left(\frac{k_F^2 A}{\pi}\right) \tag{15.6}$$

In dimension $d = 2$:

$$G = s\frac{e^2}{h}\left(\frac{l_e}{2L}\right)k_F L \tag{15.7}$$

In dimension $d = 1$:

$$G = s\frac{e^2}{h}\frac{2l_e}{L} \tag{15.8}$$

15.2 Fourier transforms: definitions

$$\tilde{F}(\boldsymbol{k}) = \int F(\boldsymbol{r})\, e^{-i\boldsymbol{k}\cdot\boldsymbol{r}}\, d\boldsymbol{r} \tag{15.9}$$

$$\tilde{f}(\omega) = \int f(t)\, e^{i\omega t}\, dt \tag{15.10}$$

$$F(\boldsymbol{r}) = \frac{1}{\Omega}\sum_{\boldsymbol{k}} \tilde{F}(\boldsymbol{k})\, e^{i\boldsymbol{k}\cdot\boldsymbol{r}} = \frac{1}{(2\pi)^d}\int \tilde{F}(\boldsymbol{k})\, e^{i\boldsymbol{k}\cdot\boldsymbol{r}}\, d\boldsymbol{k} \tag{15.11}$$

$$f(t) = \frac{1}{2\pi}\int \tilde{f}(\omega) e^{-i\omega t}\, d\omega \tag{15.12}$$

$$\frac{1}{\Omega}\sum_{\boldsymbol{k}} e^{i\boldsymbol{k}\cdot\boldsymbol{r}} = \frac{1}{(2\pi)^d}\int e^{i\boldsymbol{k}\cdot\boldsymbol{r}} d\boldsymbol{k} = \delta(\boldsymbol{r}) \tag{15.13}$$

$$\int e^{-i\boldsymbol{k}\cdot\boldsymbol{r}}\, d\boldsymbol{r} = \Omega\delta_{\boldsymbol{k},0} = (2\pi)^d\delta(\boldsymbol{k}) \tag{15.14}$$

$\delta(\boldsymbol{k})$ is the Dirac function and $\delta_{\boldsymbol{k},0}$ is the Kronecker symbol.

$$\int_{-\infty}^{\infty} e^{-i\omega t}\, dt = 2\pi\delta(\omega) \qquad \int_{-\infty}^{\infty} e^{i\omega t}\, d\omega = 2\pi\delta(t) \tag{15.15}$$

15.3 Collisionless probability $P_0(\boldsymbol{r},\boldsymbol{r}',t)$

The Fourier transform of P_0 is defined in section A4.1.1. See Exercise 4.4, page 115 and Table 15.1.

15.4 Probability $P(\boldsymbol{r},\boldsymbol{r}',t)$

The probability $P(\boldsymbol{r},\boldsymbol{r}',t)$ verifies:

$$\begin{aligned} P(\boldsymbol{r},\boldsymbol{r}',t) &= P^*(\boldsymbol{r}',\boldsymbol{r},t) \\ P(\boldsymbol{r},\boldsymbol{r}',\omega) &= P^*(\boldsymbol{r}',\boldsymbol{r},-\omega) \end{aligned} \tag{15.16}$$

Table 15.1 Probability P_0 for a particle to have no collision after a time t. $R = |\boldsymbol{R}| = |\boldsymbol{r}-\boldsymbol{r}'|$.

	$d=3$	$d=2$	$d=1$
$P_0(\boldsymbol{q},\omega)$	$\frac{1}{qv}\arctan\frac{ql_e}{1-i\omega\tau_e}$	$\frac{\tau_e}{\sqrt{(1-i\omega\tau_e)^2+q^2l_e^2}}$	$\frac{\tau_e(1-i\omega\tau_e)}{(1-i\omega\tau_e)^2+q^2l_e^2}$
$P_0(\boldsymbol{q},t)$	$e^{-t/\tau_e}\frac{\sin qvt}{qvt}$	$e^{-t/\tau_e}J_0(qvt)$	$e^{-t/\tau_e}\cos qvt$
$P_0(\boldsymbol{r},\omega)$	$\frac{e^{i\omega R/v-R/l_e}}{4\pi R^2 v}$	$\frac{e^{i\omega R/v-R/l_e}}{2\pi Rv}$	$\frac{e^{i\omega R/v-R/l_e}}{2v}$
$P_0(\boldsymbol{r},t)$	$\frac{\delta(R-vt)}{4\pi R^2}e^{-t/\tau_e}$	$\frac{\delta(R-vt)}{2\pi R}e^{-t/\tau_e}$	$\frac{\delta(R-vt)}{2}e^{-t/\tau_e}$

If there is time reversal symmetry, it obeys:

$$
\begin{aligned}
P(\boldsymbol{r},\boldsymbol{r}',t) &= P(\boldsymbol{r}',\boldsymbol{r},t) \\
P(\boldsymbol{r},\boldsymbol{r}',\omega) &= P(\boldsymbol{r}',\boldsymbol{r},\omega)
\end{aligned}
\tag{15.17}
$$

The Fourier transform of $P(\boldsymbol{r},\boldsymbol{r}',t)$ is $P(\boldsymbol{q},\omega) = 1/(-i\omega + Dq^2)$.
The Fourier transform of $P(\boldsymbol{r},\boldsymbol{r}',|t|)$ is $2\mathrm{Re}\,P(\boldsymbol{q},\omega)$.

$$
P(\boldsymbol{q},t_1)P(\boldsymbol{q},t_2) = P(\boldsymbol{q},t_1+t_2)
$$

$$
\int d\boldsymbol{r}'\,P(\boldsymbol{r},\boldsymbol{r}',t_1)P(\boldsymbol{r}',\boldsymbol{r}'',t_2) = P(\boldsymbol{r},\boldsymbol{r}'',t_1+t_2)
$$

$$
\mathrm{Re}P(\boldsymbol{q},\epsilon) = \mathrm{Im}\int_{\epsilon}^{\infty} d\omega\,P^2(\boldsymbol{q},\omega) \tag{15.18}
$$

Free space

$P(\boldsymbol{R},\omega)$ is the Fourier transform of $P(\boldsymbol{R},t)$ defined by (5.20) where $R = |\boldsymbol{r}-\boldsymbol{r}'|$. For $d \geq 2$, the integral diverges for short times and thus must be cut off by the elastic collision time τ_e. Therefore we define:

$$
P(\boldsymbol{r},\omega) = \frac{1}{(4\pi D)^{d/2}}\int_0^{\infty}\frac{1}{t^{d/2}}\,e^{-R^2/4Dt}(1-e^{-t/\tau_e})\,e^{i\omega t}\,dt \tag{15.19}
$$

$$
P^{(1d)}(\boldsymbol{r},\omega) = \frac{1}{\sqrt{-4i\omega D}}\,e^{-R\sqrt{-i\omega/D}} \tag{15.20}
$$

$$
P^{(2d)}(\boldsymbol{r},\omega) = \frac{1}{2\pi D}\left[K_0\left(R\sqrt{-i\omega/D}\right) - K_0\left(R/\sqrt{D\tau_e}\right)\right] \tag{15.21}
$$

$$
P^{(3d)}(\boldsymbol{r},\omega) = \frac{1}{4\pi DR}\left[e^{-R\sqrt{-i\omega/D}} - e^{-R/\sqrt{D\tau_e}}\right] \tag{15.22}
$$

The integrated probability of return to the origin is written, for a system of volume L^d (for $t \ll L^2/D$),

$$Z(t) = \frac{L^d}{(4\pi Dt)^{d/2}} = \left(\frac{\tau_D}{4\pi t}\right)^{d/2} \tag{15.23}$$

where $\tau_D = L^2/D$ is the Thouless time. The Fourier transform $Z(\omega)$ is obtained using (15.81). For $d \geq 2$, the integral diverges for short times and must be cut off by the collision time τ_e. We define:

$$Z(\omega) = \int_0^\infty Z(t)(1 - e^{-t/\tau_e})\, e^{i\omega t}\, dt \tag{15.24}$$

From (15.82–15.84), we obtain for $\omega\tau_e \ll 1$:

$$Z^{(1d)}(\omega) = \sqrt{\frac{i\tau_D}{4\omega}} \tag{15.25}$$

$$Z^{(2d)}(\omega) = \frac{\tau_D}{4\pi} \ln i/\omega\tau_e \tag{15.26}$$

$$Z^{(3d)}(\omega) = \frac{\tau_D^{3/2}}{4\pi}(\sqrt{1/\tau_e} - \sqrt{-i\omega}) \tag{15.27}$$

For $d < 4$, we have:

$$\mathrm{Im}Z(\omega) = \frac{\pi^2 d c_d}{4} \frac{1}{\omega} \left(\frac{\omega}{E_c}\right)^{d/2} \tag{15.28}$$

where c_d is given by

$$c_d = \frac{1}{d 2^{d-1} \pi^{\frac{d}{2}+1} \sin(\pi d/4)\, \Gamma(d/2)} \qquad c_1 = \frac{\sqrt{2}}{\pi^2} \qquad c_2 = \frac{1}{4\pi^2} \qquad c_3 = \frac{\sqrt{2}}{6\pi^3} \tag{15.29}$$

Infinite plane ($d = 2$) *in a magnetic field, for* $\omega\tau_e \ll 1$

$$\begin{aligned} Z_c(\omega, B) &= \int_0^\infty \frac{BS/\phi_0}{\sinh 4\pi BDt/\phi_0} e^{i\omega t}(1 - e^{-t/\tau_e})\, dt \\ &= \frac{S}{4\pi D}\left[\psi\left(\frac{1}{2} + \frac{\hbar}{4eDB\tau_e}\right) - \psi\left(\frac{1}{2} - i\frac{\hbar\omega}{4eDB}\right)\right] \end{aligned} \tag{15.30}$$

If $B \to 0$:

$$Z(\omega, B) = \frac{S}{4\pi D}\left[\ln i/\omega\tau_e + \frac{2e^2B^2D^2}{3\hbar^2\omega^2}\right] \tag{15.31}$$

One-dimensional ring and Aharonov–Bohm flux

The Fourier transform of (6.54) is ($\varphi = \phi/\phi_0$):

$$Z(\omega,\phi) = \sum_{m=-\infty}^{\infty} \sqrt{\frac{1}{-4i\omega E_c}}\, e^{-m\sqrt{-i\omega/E_c}+4im\pi\varphi} \tag{15.32}$$

15.5 Wigner–Eckart theorem and 3j-symbols

For general references, see for example [403, 404].

• A matrix element of a component $T_q^{(k)}$ of a tensor operator $T^{(k)}$ of integer order k evaluated between eigenstates of the angular momentum is given by

$$\langle J_e m_e|T_q^{(k)}|Jm\rangle = \frac{\langle J_e||T^{(k)}||J\rangle}{\sqrt{2J_e+1}}\langle Jkmq|J_e m_e\rangle \tag{15.33}$$

$$\langle J_e m_e|T_q^{(k)}|Jm\rangle = (-1)^{J_e-m_e}\langle J_e||T^{(k)}||J\rangle \begin{pmatrix} J_e & k & J \\ -m_e & q & m \end{pmatrix} \tag{15.34}$$

where $\langle J_e||T^{(k)}||J\rangle$ is a coefficient independent of m, m_e and q, which only depends on J, J_e and on the amplitude of the tensor operator $T^{(k)}$.[1] The second term $\langle Jkmq|J_e m_e\rangle$ is a Clebsch–Gordan coefficient. In general, for the problem of addition of angular momenta, $\boldsymbol{J} = \boldsymbol{j}_1 + \boldsymbol{j}_2$, these coefficients are used to decompose the vectors $|JM\rangle$, as functions of the states $|j_1 j_2 m_1 m_2\rangle$, that is $\langle j_1 j_2 m_1 m_2|JM\rangle$. The Clebsch–Gordan coefficients are real and non-zero for $m_e - m = q$ and $|J_e - J| \leq k \leq J_e + J$, which determine the selection rules of the corresponding transition.

• Clebsch–Gordan coefficients and 3j-symbols are related by:

$$\begin{pmatrix} j_1 & j_2 & J \\ m_1 & m_2 & -M \end{pmatrix} = \frac{(-1)^{j_1-j_2+M}}{\sqrt{2J+1}}\langle j_1 j_2 m_1 m_2|JM\rangle \tag{15.35}$$

with the selection rules $M = m_1 + m_2$ and $|j_1 - j_2| \leq J \leq j_1 + j_2$.

$$\begin{pmatrix} j_1 & j_2 & j_3 \\ m_1 & m_1 & m_3 \end{pmatrix} \tag{15.36}$$

is invariant in a circular permutation of columns, multiplied by $(-1)^{j_1+j_2+j_3}$ in a permutation of two columns or if the signs of m_1, m_2 and m_3 are changed.

[1] The term $\sqrt{2J_e+1}$ is purely conventional.

For an integer value of J:

$$\sum_{m_e}(-1)^{J_e-m_e}\begin{pmatrix} J_e & 1 & J \\ -m_e & -\alpha & m_i \end{pmatrix}\begin{pmatrix} J_e & 1 & J \\ -m_e & -\gamma & m_f \end{pmatrix}$$

$$= \sum_{kq}(2k+1)(-1)^q \begin{Bmatrix} 1 & 1 & k \\ J & J & J_e \end{Bmatrix}\begin{pmatrix} J & J & k \\ m_i & -m_f & -q \end{pmatrix}\begin{pmatrix} 1 & 1 & k \\ -\alpha & \gamma & q \end{pmatrix} \quad (15.37)$$

where $\{\cdots\}$ is a 6j-symbol. The 3j-symbols are non-zero for integers (k, q) such that $0 \leq k \leq 2$ and $|q| \leq k$. Due to selection rules, J_e is also integer and $|J - 1| \leq J_e \leq J + 1$. The first sum contains only one term $m_e = \alpha + m_i = \gamma + m_f$ and the sum over q contains also one term $q = \gamma - \alpha = m_i - m_f$.

$$\sum_{m_i m_f}\begin{pmatrix} J & J & k \\ m_i & -m_f & -q \end{pmatrix}\begin{pmatrix} J & J & k' \\ m_i & -m_f & -q' \end{pmatrix} = \frac{1}{2k+1}\delta_{kk'}\delta_{qq'} \quad (15.38)$$

$$(-1)^{\gamma+\delta}(2k+1)\sum_q \begin{pmatrix} 1 & 1 & k \\ -\alpha & \gamma & q \end{pmatrix}\begin{pmatrix} 1 & 1 & k \\ -\beta & \delta & q \end{pmatrix} = \mathcal{T}^{(k)}_{\alpha\gamma,\beta\delta} \quad (15.39)$$

where the tensors $\mathcal{T}^{(k)}$ are defined in (6.311).

15.6 Miscellaneous

Pauli matrices

$$\sigma_x = \begin{pmatrix} 0 & 1 \\ 1 & 0 \end{pmatrix} \qquad \sigma_y = \begin{pmatrix} 0 & -i \\ i & 0 \end{pmatrix} \qquad \sigma_z = \begin{pmatrix} 1 & 0 \\ 0 & -1 \end{pmatrix} \quad (15.40)$$

Γ *function*

$$\Gamma(s) = \int_0^\infty t^{s-1}\, e^{-t}\, dt \quad (15.41)$$

$$\Gamma(1) = 1$$

$$\Gamma(\tfrac{1}{2}) = \sqrt{\pi}$$

$$\Gamma(x+1) = x\Gamma(x)$$

$$\Gamma(n) = (n-1)!$$

$$\Gamma(n+\tfrac{1}{2}) = \frac{\sqrt{\pi}}{2^n}(2n-1)!!$$

$$\Gamma(\tfrac{1}{2}-n) = (-1)^n \frac{2^n\sqrt{\pi}}{(2n-1)!!}$$

$$\Gamma(d/2) = \frac{\sqrt{\pi}}{2^{(d-2)}}(d-2)!! \tag{15.42}$$

$$\Gamma(1-x)\Gamma(x) = \frac{\pi}{\sin \pi x}$$

Digamma function $\Psi(x)$

$$\Psi(x) = \frac{d \ln \Gamma(x)}{dx} \tag{15.43}$$

$$\Psi(x) = \int_0^\infty \left(\frac{e^{-t}}{t} - \frac{e^{-xt}}{1-e^{-t}} \right) dt \tag{15.44}$$

$$\Psi(\tfrac{1}{2}+x) \simeq \ln x + \frac{1}{24x^2} + \cdots \qquad \text{for} \quad x \to \infty \tag{15.45}$$

$$\Psi(\tfrac{1}{2}+x) \simeq \Psi(\tfrac{1}{2}) + x\Psi'(\tfrac{1}{2}) + \cdots \qquad \text{for} \quad x \to 0 \tag{15.46}$$

$$\Psi(\tfrac{1}{2}) = -\gamma - 2\ln 2 \tag{15.47}$$

where $\gamma \simeq 0.577$ is the Euler constant.

$$\sum_{n=0}^{N} \frac{1}{n+x} = \ln N - \Psi(x) \qquad \text{for} \quad N \to \infty \tag{15.48}$$

$$\frac{1}{2}\int_0^\infty \frac{e^{-yt} - e^{-zt}}{\sinh(t/2)} dt = \Psi(z+\tfrac{1}{2}) - \Psi(y+\tfrac{1}{2}) \tag{15.49}$$

$$\int_0^\infty e^{-yx} \left(\frac{1}{x} - \frac{1}{\sinh x} \right) dx = \Psi(y+\tfrac{1}{2}) - \ln y/2$$

Riemann zeta function $\zeta(x)$

$$\zeta(\tfrac{1}{2}) \simeq -1.460$$

$$\zeta(\tfrac{3}{2}) \simeq 2.612$$

$$\zeta(2) = \sum_1^\infty \frac{1}{n^2} = \frac{\pi^2}{6}$$

$$\zeta(4) = \sum_1^\infty \frac{1}{n^4} = \frac{\pi^4}{90}$$

$$\sum_1^\infty \frac{(-1)^{n+1}}{n^2} = \frac{\pi^2}{12} \tag{15.50}$$

Function $E_n(z)$

$$E_n(z) = \int_1^\infty dt\, \frac{e^{-zt}}{t^n} \tag{15.51}$$

$$\int_0^\infty E_n(z)\, dz = \frac{1}{n} \qquad \int_0^\infty zE_n(z)\, dz = \frac{1}{n+1} \tag{15.52}$$

$$\int_0^\infty dz'\, E_1(|z-z'|) = 2 - E_2(z) \tag{15.53}$$

$$\int_0^\infty dz'\, z'E_1(|z-z'|) = 2z + E_3(z) \tag{15.54}$$

$$\int_0^\infty dz'\, z'E_1(z+z') = E_3(z) \tag{15.55}$$

In the limit $z \to \infty$,

$$\begin{aligned} I_1 &= \int_0^\infty dz'\, E_1(|z-z'|) \to 2 \\ I_2 &= \int_0^\infty dz'\, (z-z')E_1(|z-z'|) \to 0 \\ I_3 &= \int_0^\infty dz'\, (z-z')^2 E_1(|z-z'|) \to \frac{4}{3} \end{aligned} \tag{15.56}$$

$$\begin{aligned} \int_0^\infty E_2(z)E_3(z)\, dz &= \frac{1}{8} \\ \int_0^\infty E_2(z)^2\, dz &= \frac{2}{3}(1 - \ln 2) \\ \int_0^\infty E_3(z)^2\, dz &= \frac{1}{5}(2\ln 2 - 1) \\ \int_0^\infty E_2(z)\, dz \int_0^\infty E_1(|z-z'|)\, dz' &= \frac{1}{3}(1 + 2\ln 2) \\ \int_0^\infty E_3(z)\, dz \int_0^\infty E_1(|z-z'|)\, dz' &= \frac{13}{24} \\ \int_0^\infty dz\, E_2(z) \int_0^\infty dz' E_2(z')E_1(|z-z'|) &= \frac{1}{3} + \frac{\pi^2}{12} - \frac{4}{3}\ln 2 \\ \int_0^\infty dz\, E_3(z) \int_0^\infty dz' E_2(z')E_1(|z-z'|) &= \frac{1}{3}\ln 2 - \frac{1}{12} \\ \int_0^\infty dz\, E_3(z) \int_0^\infty dz' E_3(z')E_1(|z-z'|) &= \frac{56}{45}\ln 2 - \frac{79}{360} - \frac{\pi^2}{18} \end{aligned} \tag{15.57}$$

Useful relations

Defining

$$\delta_\gamma(x) = \frac{\gamma/\pi}{x^2+\gamma^2} \tag{15.58}$$

we have

$$\int_{-\infty}^{\infty} dz\, \delta_{\gamma/2}(x-z)\delta_{\gamma/2}(y-z) = \delta_\gamma(x-y). \tag{15.59}$$

$$\int_0^\infty x J_0^2(ax) J_0(bx)\, dx = \frac{2}{\pi b}\frac{1}{\sqrt{4a^2-b^2}} \quad \text{if } b < 2a \tag{15.60}$$

$$\int d^2\boldsymbol{\rho}\, J_0(\lambda\rho)\, e^{i\boldsymbol{k}_\perp\cdot\boldsymbol{\rho}} = \frac{2\pi}{k_\perp}\delta(\lambda - k_\perp) \tag{15.61}$$

$$\int_{-\infty}^{\infty} \frac{\cos ax}{\cosh^2 bx}\, dx = \frac{a\pi}{b^2 \sinh \frac{a\pi}{2b}} \tag{15.62}$$

$$\int_{-\infty}^{\infty} \frac{b}{(x-a)^2+b^2} \cos cx\, dx = \pi\, e^{-bc} \cos ac \tag{15.63}$$

$$\int_0^\infty \cos \tfrac{1}{2}\pi x^2\, dx = \int_0^\infty \sin \tfrac{1}{2}\pi x^2\, dx = \tfrac{1}{2} \tag{15.64}$$

$$\sum_{n>0} \frac{\cos nx}{n^2+a^2} = \frac{\pi}{2a}\frac{\cosh a(\pi - |x|)}{\sinh \pi a} - \frac{1}{2a^2} \tag{15.65}$$

$$\int_0^\infty \frac{x^{1/2}}{(x-a)^2+b^2}\, dx = \pi (a^2+b^2)^{1/4} \cos\left(\frac{1}{2}\arctan\frac{b}{a}\right) \tag{15.66}$$

$$\int_0^\infty \frac{x^{d-1}\, dx}{(x^2+a^2)^2} = \frac{\pi}{2}(1-d/2)\frac{a^{d-4}}{\sin \pi d/2} \tag{15.67}$$

$$\int_a^b \frac{dx}{x \ln x} = \ln \frac{b}{a} \tag{15.68}$$

$$e^{z\cos\theta} = \sum_{n=-\infty}^{\infty} I_n(z)\, e^{in\theta} \tag{15.69}$$

$$\int_0^\infty \frac{e^{-\beta/x-\gamma x}}{x^\alpha}\, dx = 2\left(\frac{\gamma}{\beta}\right)^{(\alpha-1)/2} K_{\alpha-1}(2\sqrt{\beta\gamma}) \tag{15.70}$$

For half-integer α, the modified Bessel functions K are related to finite polynomials:

$$\begin{aligned}\int_0^\infty \frac{e^{-\beta/x-\gamma x}}{x^{n+1/2}}\, dx &= 2\left(\frac{\gamma}{\beta}\right)^{(2n-1)/4} K_{n-1/2}(2\sqrt{\beta\gamma}) \\ &= \sqrt{\pi}\frac{\gamma^{(n-1)/2}}{\beta^{n/2}}\, e^{-2\sqrt{\beta\gamma}} \sum_{k=0}^{n-1} \frac{(n+k-1)!}{k!(n-k-1)!(4\sqrt{\beta\gamma})^k}.\end{aligned} \tag{15.71}$$

In particular,

$$\int_0^\infty \frac{e^{-\beta/x-\gamma x}}{x^{1/2}}\,dx = \sqrt{\frac{\pi}{\gamma}}\,e^{-2\sqrt{\beta\gamma}} \tag{15.72}$$

and in the limit $\gamma \to 0$, one has

$$\int_0^\infty \frac{e^{-\beta/x}}{x^{n+1/2}}\,dx = \frac{\sqrt{\pi}}{\beta^{n-1/2}}\frac{(2n-2)!}{(n-1)!4^{n-1}} = \frac{\sqrt{\pi}}{\beta^{n-1/2}}\frac{(2n-3)!!}{2^{n-1}} \tag{15.73}$$

$$\int_0^\infty \frac{e^{-\beta/x}}{x^{n}}\,dx = \frac{(n-2)!}{\beta^{n-1}}. \tag{15.74}$$

$$K_0(x) \simeq \sqrt{\frac{\pi}{2x}}\,e^{-x} \qquad \text{for} \quad x \to \infty \tag{15.75}$$

$$K_0(x) \sim -\gamma - \ln x/2 \qquad \text{for} \quad x \to 0 \tag{15.76}$$

where $\gamma \simeq 0.577$ is the Euler constant.

$$z^n K_n(z) \to (n-1)!2^{n-1} \qquad \text{for} \quad z \to 0 \quad n > 0 \tag{15.77}$$

$$z^{n-1/2}K_{n-1/2}(z) \to \sqrt{\frac{\pi}{2}}\frac{(2n-2)!}{(n-1)!2^{n-1}} = \sqrt{\frac{\pi}{2}}(2n-3)!! \qquad z \to 0 \tag{15.78}$$

$$(2n)!! = n!\,2^n \tag{15.79}$$

$$(2n+1)!! = \frac{(2n+1)!}{(2n)!!} = \frac{(2n+1)!}{n!2^n} \tag{15.80}$$

$$\int_0^\infty \frac{1}{t^{d/2}}\,e^{-i\omega t}\,dt = \Gamma(1-d/2)\,(i\omega)^{d/2-1} \qquad \text{for} \quad d < 2 \tag{15.81}$$

$$\int_0^\infty \frac{1}{\sqrt{t}}\cos\epsilon t\,dt = \sqrt{\frac{\pi}{2\epsilon}} \tag{15.82}$$

$$\int_0^\infty \frac{1}{t}(1-e^{-t/\tau_e})\cos\epsilon t\,dt = \frac{1}{2}\ln\left(1+\frac{1}{\epsilon^2\tau_e^2}\right) \underset{\epsilon\tau_e\to\infty}{\longrightarrow} -\ln\epsilon\tau_e \tag{15.83}$$

$$\int_0^\infty \frac{1}{t^{3/2}}(1-e^{-t/\tau_e})\cos\epsilon t\,dt = \sqrt{\pi}\left[-\sqrt{2\epsilon}+2\left(\frac{1}{\tau_e^2}+\epsilon^2\right)^{1/4}\cos\frac{\arctan\epsilon\tau_e}{2}\right]$$
$$\underset{\epsilon\tau_e\to\infty}{\longrightarrow} 2\left(\frac{1}{\sqrt{\tau_e}}-\sqrt{\epsilon}\right) \tag{15.84}$$

$$\int_0^\infty \frac{1}{t^{d/2}}(e^{-\gamma t}-e^{-\delta t})\,dt = \Gamma(1-d/2)\,[\gamma^{d/2-1}-\delta^{d/2-1}] \quad d < 4 \tag{15.85}$$

$$\int_0^\infty \sin \omega t / t^{d/2}\, dt = \frac{\pi \omega^{d/2-1}}{2 \sin(\pi d/4)\Gamma(d/2)} \quad d < 4 \tag{15.86}$$

$$\int_0^\infty \frac{\sin^2 \epsilon t/2}{t^{d/2}}\, dt = -\frac{\pi \epsilon^{d/2-1}}{4 \cos(\pi d/4)\Gamma(d/2)} \quad d > 2 \tag{15.87}$$

$$\int_0^\infty x^\mu\, e^{-\gamma x}\, dx = \frac{1}{\gamma^{\mu+1}}\Gamma(\mu) \quad \mu > -1$$

$$\int_0^\infty \sin^2(ax)\, e^{-bx} \frac{dx}{x} = \frac{1}{4} \ln\left(1 + \frac{4a^2}{b^2}\right) \tag{15.88}$$

$$\int_0^\infty \frac{x^{\mu-1}}{\sinh^2 ax}\, dx = \frac{4}{(2a)^\mu}\Gamma(\mu)\zeta(\mu-1) \quad \mu > 2 \tag{15.89}$$

$$\int_0^\infty x \frac{1 - e^{-ax}}{\sinh^2 x}\, dx \simeq \ln \frac{ae^\gamma}{2} \quad a \gg 1 \tag{15.90}$$

where $\gamma \simeq 0.577$ is the Euler constant.

$$\int_0^\infty \frac{1}{x^{3/2}}(1 - e^{-ax})\, dx = 2\sqrt{\pi a} \tag{15.91}$$

$$\int_0^\infty \sqrt{x}\left(\frac{1}{\sinh^2 x} - \frac{1}{x^2}\right) dx = -\sqrt{\frac{\pi}{2}}\zeta(1/2) \tag{15.92}$$

$$\int_0^\infty \frac{\sqrt{x}}{\sinh^2 x}(1 - e^{-ax})\, dx \simeq 2\sqrt{\pi a} - \sqrt{\frac{\pi}{2}}\zeta(1/2) \quad a \gg 1 \tag{15.93}$$

$$\int_0^\infty \frac{\tanh x}{x^{3/2}}\, dx = \sqrt{\frac{2}{\pi}}(2\sqrt{2} - 1)\zeta(3/2) \tag{15.94}$$

$$\int_0^\infty \frac{\tanh x}{x}\, e^{-\beta x}\, dx \simeq \ln \frac{4}{\pi \beta} \quad \text{for} \quad \beta \to 0 \tag{15.95}$$

$$\int_0^\infty \frac{\tanh x}{\sqrt{x}}\, e^{-\beta x}\, dx \simeq \sqrt{\frac{\pi}{\beta}} \quad \text{for} \quad \beta \to 0 \tag{15.96}$$

$$\int_{-\infty}^\infty \frac{dx}{x^2}(1 - \cos px) = \pi |p| \tag{15.97}$$

$$\int_0^\infty \frac{dx}{x}[J_0(x) - e^{-px}] = \ln 2p \tag{15.98}$$

$$\int_0^\infty e^{-px} \sin qx \frac{dx}{x} = \arctan \frac{q}{p} \tag{15.99}$$

$$\int_0^\infty x^a\, e^{-bx^2}\, dx = \frac{1}{2b^{(a+1)/2}}\Gamma\left(\frac{a+1}{2}\right) \tag{15.100}$$

The differential equation

$$\left(-\frac{\partial^2}{\partial x^2}+x_0+|x|\right)C(x)=A\delta(x) \tag{15.101}$$

admits the solution

$$C(x)=-\frac{A}{2}\frac{\mathrm{Ai}(x_0+|x|)}{\mathrm{Ai}'(x_0)}. \tag{15.102}$$

$$\frac{\mathrm{Ai}(x)}{\mathrm{Ai}'(x)}\simeq-\frac{1}{\sqrt{x}}\left(1-\frac{1}{4x^{3/2}}\right)\quad\text{for}\quad x\to\infty \tag{15.103}$$

$$\frac{\mathrm{Ai}(x)}{\mathrm{Ai}'(x)}\simeq-\frac{\Gamma(1/3)}{3^{1/3}\Gamma(2/3)}+x\simeq-1.372+x\quad\text{for}\quad x\to 0 \tag{15.104}$$

$$\frac{\mathrm{Ai}(x)}{\mathrm{Ai}'(x)}\simeq-\frac{1}{\sqrt{1/2+x}}\quad\forall x>0\quad\text{within }4\% \tag{15.105}$$

15.7 Poisson formula

$$\sum_{n=-\infty}^{\infty}f(n)=\sum_{m\in\mathbb{Z}}\int_{-\infty}^{\infty}f(y)\,e^{2i\pi my}\,dy \tag{15.106}$$

$$\sum_{n=-\infty}^{\infty}f(n+\varphi)=\sum_{m\in\mathbb{Z}}e^{2i\pi m\varphi}\int_{-\infty}^{\infty}f(y)\,e^{2i\pi my}\,dy$$

If f is an even function:

$$\sum_{n=-\infty}^{\infty}f(n-\varphi)=\sum_{m\in\mathbb{Z}}\cos 2\pi m\varphi\int_{-\infty}^{\infty}f(y)\,e^{2i\pi my}\,dy \tag{15.107}$$

15.8 Temperature dependences

$f(\epsilon)=1/(e^{\beta\epsilon}+1)$ is the Fermi factor.

$$f'(\epsilon)=-\frac{\beta}{4\cosh^2\frac{\beta\epsilon}{2}} \tag{15.108}$$

$$f'(\omega)=-\frac{1}{2\pi}\int_{-\infty}^{\infty}\frac{\pi Tt}{\sinh\pi Tt}\cos\omega t\,dt \tag{15.109}$$

$$f(\omega)=-\frac{1}{2\pi}\int_{-\infty}^{\infty}\frac{\pi Tt}{\sinh\pi Tt}\frac{\sin\omega t}{t}\,dt \tag{15.110}$$

$$\int d\epsilon \int d\epsilon' M(\epsilon - \epsilon') f(\epsilon) f(\epsilon') = \int \frac{\pi^2 T^2}{(\sinh \pi T t)^2} \tilde{M}(t)\, dt$$
$$= -\int_0^\infty d\omega\, M(\omega)\, \omega \coth \frac{\beta\omega}{2} \tag{15.111}$$

where $\tilde{M}(t)$ is the Fourier transform of $M(\epsilon)$:

$$\tilde{M}(t) = \frac{1}{2\pi} \int M(\epsilon)\, e^{-i\epsilon t}\, d\epsilon$$

$F(\epsilon) = -\frac{1}{\beta} \ln(1 + e^{-\beta\epsilon})$ is the integral of the Fermi factor.

$$\int d\epsilon \int d\epsilon' M(\epsilon - \epsilon') F(\epsilon) F(\epsilon') = \int \frac{\pi^2 T^2}{(t \sinh \pi T t)^2} \tilde{M}(t)\, dt \tag{15.112}$$

$$\int d\epsilon \int d\epsilon' M(\epsilon - \epsilon') \frac{\partial f}{\partial \epsilon} \frac{\partial f}{\partial \epsilon'} = \int \frac{\pi^2 T^2 t^2}{(\sinh \pi T t)^2} \tilde{M}(t)\, dt$$
$$= \frac{1}{2} \int_{-\infty}^{\infty} \left(\frac{\beta\omega}{2} \coth \frac{\beta\omega}{2} - 1 \right) \frac{\beta}{\sinh^2 \beta\omega/2} M(\omega) d\omega \tag{15.113}$$

For an odd function $I(\omega)$:

$$\int_{-\infty}^{\infty} f(\epsilon - \omega) I(\omega)\, d\omega = \frac{1}{2} \int_0^\infty d\omega \left[\tanh \frac{\epsilon + \omega}{2T} + \tanh \frac{\omega - \epsilon}{2T} \right] I(\omega)\, d\omega \tag{15.114}$$

$$\int_{-\infty}^{\infty} f(\epsilon)[1 - f(\epsilon + \omega)]\, d\epsilon = \frac{\omega}{1 - e^{-\beta\omega}} \tag{15.115}$$

$$\int_0^\infty \frac{\sin \omega t}{\sinh \beta\omega} d\omega = \frac{\pi}{2\beta} \tanh \frac{\pi t}{2\beta} \tag{15.116}$$

$$\frac{e^{\beta\epsilon} + 1}{e^{\beta\omega} - 1} f(\epsilon - \omega) = \frac{1}{2} \left(\coth \frac{\beta\omega}{2} - \tanh \frac{\beta}{2}(\omega - \epsilon) \right) \tag{15.117}$$

For an even function $P(\omega)$:

$$\int_{-\infty}^{\infty} \frac{P(\omega)}{1 - e^{-\beta\omega}} d\omega = \int_0^\infty P(\omega) \coth \frac{\beta\omega}{2} d\omega \tag{15.118}$$

15.9 Characteristic times introduced in this book

This table summarizes the various characteristic times introduced in the book, and it recalls their exact location in the text.

τ_H	Heisenberg time	(10.1)
τ_D	Thouless time	(5.34)

Single-particle times

τ_e	elastic collision time	(3.73)
τ^*	transport time	(4.168)
τ_m	magnetic scattering time	(6.82)
τ_{so}	spin-orbit collision time	(6.82)
τ_{ee}	quasiparticle lifetime	section 13.6
τ_{in}	quasiparticle lifetime at the Fermi level	(13.125)
τ_{tot}	Matthiessen time	(3.75)
τ_{pol}	elastic collision time for a polarized wave	(6.150)
τ_{at}	elastic collision time for photon–atom scattering	(6.276)

Dephasing times

τ_γ	generic cutoff time or dephasing time	section 5.2.2, (6.15)
τ_B	magnetic time	(6.43)
τ_k	dephasing time due to polarization	(6.166)
τ_s	relaxation time of the Diffuson due to laminar flow	(9.44)

Phase coherence times[2]

τ_ϕ	generic phase coherence time	section 6.8
τ_J	relaxation time of the eigenmode J of the structure factor	(6.19)
$\tau_S^{(d,c)}$	relaxation times of the Diffuson (d) and of the Cooperon (c) in the singlet channel	(6.114, 6.125)
$\tau_T^{(d,c)}$	relaxation times of the Diffuson (d) and of the Cooperon (c) in the triplet channel	(6.114, 6.125)
$\tau_k^{(d,c)}$	relaxation times of the Diffuson (d) and of the Cooperon (c) in mode k	(6.321)
τ_s	relaxation time of the Diffuson due to Brownian motion of the scatterers	(6.197)
τ_N	Nyquist time	(13.173)
τ_ϕ^{ee}	phase coherence time due to electron–electron interactions	section 13.7

Relaxation of scatterers

τ_b	Brownian scatterer time	(6.193)
τ_l	laminar flow time	(9.42)
τ_K	Korringa time	section 6.5.3

Miscellaneous

τ_R	recurrence time of a classical random walk	(5.16)
τ_a	absorption time	(8.48)

[2] Notice that the times $\tau_{S,T}^{(d,c)}$ include both spin-orbit scattering which introduces dephasing only and magnetic impurities which lead to decoherence.

References

Chapter 1

[1] M. Kaveh, M. Rosenbluh and I. Freund, Speckle patterns permit direct observation of phase breaking, *Nature*, **326**, 778 (1987).

[2] R. A. Webb and S. Washburn, Quantum interference fluctuations in disordered metals, *Physics Today*, 2 (December 1988).

[3] Y. Aharonov and D. Bohm, Significance of electromagnetic potentials in quantum theory, *Phys. Rev.*, **115**, 485 (1959).

[4] Y. Imry and R. A. Webb, Quantum interference and the Aharonov–Bohm effect, *Sci. Am.*, **260**, 36 (1989).

[5] J. J. Sakurai, Comments on quantum-mechanical interference due to the Earth's rotation, *Phys. Rev. D*, **21**, 2993 (1980); G. Rizzi and M. L. Ruggiero, The relativistic Sagnac effect: two derivations, in *Relativity in Rotating Frames*, eds. G. Rizzi and M. L. Ruggiero, Dordrecht: Kluwer Academic Publishers, 2003, also accessible on arxiv.org/abs/gr-qc/0305084.

[6] M. Peshkin and A. Tonomura, *The Aharonov–Bohm Effect*, Lecture Notes in Physics, Vol. 340, Heidelberg: Springer-Verlag, 1989.

[7] R. A. Webb, S. Washburn, C. P. Umbach and R. P. Laibowitz, Observation of h/e Aharonov–Bohm oscillations in normal-metal rings, *Phys. Rev. Lett.*, **54**, 2696 (1985).

[8] D. Yu. Sharvin and Yu. V. Sharvin, Magnetic-flux quantization in a cylindrical film of a normal metal, *JETP Lett.*, **34**, 272 (1981).

[9] A. Schuster, Radiation through a foggy atmosphere, *Astrophys. J.*, **21**, 1 (1905); H. C. van de Hulst, *Multiple Light Scattering* Vols. 1 and 2, New York: Academic Press, 1980; S. Chandrasekhar, *Radiative Transfer*, New York: Dover, 1960.

[10] Y. Imry, The physics of mesoscopic systems, in *Directions in Condensed Matter Physics*, eds. G. Grinstein and G. Mazenko, 2nd World Scientific, 1986.

[11] Y. Imry, *Introduction to Mesoscopic Physics*, 2nd edition, Oxford: Oxford University Press, 2002.

[12] P. A. Lee, A.D. Stone and H. Fukuyama, Universal conductance fluctuations in metals: effects of finite temperature, interactions, and magnetic field, *Phys. Rev. B*, **35**, 1039 (1987).

[13] The denomination "Hikami box" is the most commonly used: S. Hikami, Anderson localization in a nonlinear-σ-model representation, *Phys. Rev. B*, **24**, 2671 (1981); this idea has also been introduced by L. P. Gorkov, A. Larkin and D. E. Khmelnitskii, Particle conductivity in a two-dimensional random potential, *JETP Lett.*, **30**, 228 (1979).

[14] F. Scheffold and G. Maret, Universal conductance of light, *Phys. Rev. Lett.*, **81**, 5800 (1998).
[15] E. Akkermans and G. Montambaux, Mesoscopic physics of photons, *J. Opt. Soc. Am. B*, **21**, 101 (2004).
[16] M. Born and E. Wolf, *Principles of optics*, 7th edition, Cambridge: Cambridge University Press, 1999, Chapter X.

Chapter 2

[17] D. Pines and P. Nozières, *The Theory of Quantum Liquids*, Vol. 1, New York: Addison-Wesley, 1989.
[18] U. Frisch, Wave propagation in random media, in *Probabilistic Methods in Applied Mathematics*, ed. A.T. Barucha-Reid, Vol. 1, New York: Academic Press, 1968, pp. 76–198; one may also consult, A. Ishimaru, *Wave Propagation and Scattering in Random Media*, New York: Academic Press, 1978.
[19] B. Douçot and R. Rammal, On Anderson localization in nonlinear random media, *Europhys. Lett.*, **3**, 969 (1987); P. Devillard and B. Souillard, Polynomially decaying transmission for the nonlinear Schrödinger equation in a random medium, *J. Stat. Phys.*, **43**, 423 (1986); B. Spivak and A. Zyuzin, Propagation of nonlinear waves in disordered media, *J. Opt. Soc. Am. B*, **21**, 177 (2004); E. Akkermans, S. Ghosh and Z. Muslimani, Numerical study of one dimensional interacting Bose–Einstein condensates in a random potential, *Phys. Rev. A*, in press, arXiv: cond mat/0610579.
[20] L. Landau and E. Lifchitz, *Fluid Mechanics*, Pergamon, 1989.
[21] M. Belzons, E. Guazzelli and O. Parodi, Gravity waves on a rough bottom: experimental evidence of one-dimensional localization, *J. Fluid Mech.*, **186**, 539 (1988); M. Belzons, P. Devillard, F. Dunlop, E. Guazzelli, O. Parodi and B. Souillard, Localization of surface waves on a rough bottom: theories and experiments, *Europhys. Lett.*, **4**, 909 (1987).
[22] M. V. Berry, R. G. Chambers, M. D. Large, C. Upstill and J. C. Walmsley, Wavefront dislocations in the Aharonov–Bohm effect and its water wave analogue, *Eur. J. Phys.*, **1**, 154 (1980).
[23] A. Tourin, A. Derode, P. Roux, B. A. van Tiggelen and M. Fink, Time-dependent coherent backscattering of acoustic waves, *Phys. Rev. Lett.*, **79**, 3637 (1997); A. Tourin, M. Fink and A. Derode, Multiple scattering of sound, *Waves Random Media*, **10**, 31 (2000).
[24] D. Sornette and B. Souillard, Strong localization of waves by internal resonances, *Europhys. Lett.*, **7**, 269 (1988).
[25] L. Landau and E. Lifchitz, *Theory of Elasticity*, Pergamon, 1959.
[26] R. L. Weaver, Wave chaos in elastodynamics, in *Waves and Imaging through Complex Media*, ed. P. Sebbah, Dordrecht: Kluwer, 2001; B. A. van Tiggelen, L. Margerin and M. Campillo, Coherent backscattering of elastic waves: Specific role of source, polarization, and near field, *J. Acoust. Soc. Am.*, **110**, 1291 (2001).
[27] S. M. Cohen and J. Machta, Localization of third sound by a disordered substrate, *Phys. Rev. Lett.*, **54**, 2242 (1985).
[28] C. Dépollier, J. Kergomard and F. Laloë, Anderson localisation of waves in acoustical random one-dimensional lattices, *Ann. Phys. (France)*, **11**, 457 (1986).
[29] E. Akkermans and R. Maynard, Weak localization and anharmonicity of phonons, *Phys. Rev. B*, **32**, 7850 (1985).

[30] K. Arya, Z. B. Su and J. Birman, Anderson localization of electromagnetic waves in a dielectric medium of randomly distributed metal particles, *Phys. Rev. Lett.*, **57**, 2725 (1986).
[31] M. Campillo and A. Paul, Long range correlations in the diffuse seismic coda, *Science*, **229**, 547 (2003).
[32] C. W. J. Beenakker and H. van Houten, Quantum transport in semiconductor nanostructures, *Solid State Physics*, eds. H. Ehrenreich and D. Turnbull, Vol. 44, New York: Academic Press, 1991, pp. 1–228.
[33] C. M. Marcus, R. M. Westervelt, P. F. Hopkins and A. C. Gossard, Conductance fluctuations and quantum chaotic scattering in semiconductor microstructures, *Chaos*, **3**, 643 (1993); H. Baranger, R. A. Jalabert and A. D. Stone, Quantum-chaotic scattering effects in semiconductor microstructures, *Chaos*, **3**, 665 (1993); K. Richter, D. Ullmo and R. A. Jalabert, Orbital magnetism in the ballistic regime: geometrical effects, *Phys. Rep.*, **276**, 1 (1996).
[34] R. Prange and S. Girvin eds., *The quantum Hall Effect*, Springer, 1990.
[35] I. M. Lifschits, S. A. Gredeskul and L. A. Pastur, *Introduction to the Theory of Disordered Systems*, New York: Wiley, 1988.
[36] J. M. Luck, *Systèmes Désordonnés Unidimensionnels*, Aléa Saclay, 1992.
[37] S. F. Edwards, A new method for the evaluation of electric conductivity in metals, *Philos. Mag.*, **3**, 1020 (1958).
[38] P. W. Anderson, Absence of diffusion in certain random lattices, *Phys. Rev.*, **109**, 1492 (1958).
[39] B. Kramer and A. MacKinnon, Localization: theory and experiment, *Rep. Prog. Phys.*, **56**, 1469 (1993).
[40] For a discussion on Peierls substitution, one may consult: J. Callaway, *Quantum Theory of Solids*, New York: Academic Press, 1974, Chapter 6.
[41] R. Peierls, Zur theorie des diamagnetismus von leitungselektronen, *Z. Phys.*, **80**, 763 (1933); W. Kohn, Theory of Bloch electrons in a magnetic field: the effective Hamiltonian, *Phys. Rev.*, **115**, 1460 (1959); G. H. Wannier, Dynamic of band electrons in electric and magnetic fields, *Rev. Mod. Phys.*, **34**, 645 (1962).
[42] M. L. Goldberger and K. M. Watson, *Collision Theory*, New York: Wiley, 1964.
[43] R. G. Newton, *Scattering Theory of Waves and Particles*, Springer-Verlag, 1982.
[44] M. Born and E. Wolf, *Principles of Optics*, 7th edition, Cambridge: Cambridge University Press, 1999, chapter X.
[45] K. Gottfried, *Quantum Mechanics*, New York: Addison-Wesley, 1989.
[46] R. Loudon, *The Quantum Theory of Light*, Oxford: Clarendon Press, 1986.
[47] K. Huang, *Statistical Mechanics*, New York: Wiley, 1987.
[48] L. D. Landau and E. M. Lifshitz, *Quantum Mechanics (Non-relativistic Theory)*, Pergamon, 1977, sections 140, 144; J. J. Sakurai, *Modern Quantum Mechanics*, Addison-Wesley, 1994, section 7.10.
[49] D. S. Saxon, Tensor scattering matrix for the electromagnetic field, *Phys. Rev.*, **100**, 1771 (1955).
[50] B. A. van Tiggelen and R. Maynard, Reciprocity and coherent backscattering of light, in *Wave Propagation in Complex Media*, ed. G. Papanicolaou, IMA, Springer, 1997.
[51] M. Kerker, *The Scattering of Light and Other Electromagnetic Radiations*, New York: Academic Press, 1969; C. F. Bohren and D. R. Huffman, *Absorption and Scattering of Light by Small Particles*, New York: John Wiley, 1983.

[52] H. C. van de Hulst, *Multiple Light Scattering*, Vols. 1 and 2, New York: Academic Press, 1980.
[53] J. D. Jackson, *Classical Electrodynamics*, New York: Wiley, 1975.
[54] N. Ashcroft and D. Mermin, *Solid State Physics*, Philadelphia, PA: Saunders College, 1976.
[55] M. Abramowitz and I. A. Stegun, *Handbook of Mathematical Functions*, New York: Dover, 1972.
[56] D. S. Wiersma, Light in strongly scattering and amplifying random media, *Ph.D. Thesis*, Amsterdam, 1995.
[57] A. Lagendijk and B. A. van Tiggelen, Resonant multiple scattering of light, *Phys. Rep.*, **270**, 143 (1996).
[58] This topic is generally covered in quantum mechanics textbooks. For example, one may consult: G. Baym, *Lectures on Quantum Mechanics*, New York: W. A. Benjamin, 1973.
[59] C. Cohen-Tannoudji, J. Dupont-Roc and G. Grynberg, *Atom–Photon Interactions: Basic Processes and Applications*, New York: Wiley, 1998.
[60] A. Messiah, *Quantum Mechanics*, New York: Dover, 1999.
[61] C. A. Mueller, T. Jonckheere, C. Miniatura and D. Delande, Weak localization of light by cold atoms: the impact of quantum internal structure, *Phys. Rev. A*, **64**, 053804 (2001).
[62] C. A. Müller and C. Miniatura, Multiple scattering of light by atoms with internal degeneracy, *J. Phys. A*, **35**, 10163 (2002)
[63] L. D. Landau and E. M. Lifshitz, *Quantum Electrodynamics*, Pergamon, 1982, section 60.

Chapter 3

[64] M. L. Goldberger and K. M. Watson, *Collision Theory*, New York: Wiley, 1964.
[65] E. N. Economou, *Green's Functions in Quantum Physics*, Springer Verlag, 1979.
[66] S. Doniach and E. H. Sondheimer, *Green's Functions for Solid State Physicists*, Frontiers in Physics, New York: W. A. Benjamin, 1974.
[67] A. A. Abrikosov, L. P. Gorkov and I. Y. Dzyaloshinskii, *Quantum Field Theoretical Methods in Statistical Physics*, Pergamon, 1965.
[68] M. Abramowitz and I. A. Stegun, *Handbook of Mathematical Functions*, New York: Dover, 1972.
[69] N. Ashcroft and D. Mermin, *Solid State Physics*, Philadelphia, PA: Saunders College, 1976.
[70] V. L. Berezinskii, Kinetics of a quantum particle in a one-dimensional random potential, *JETP*, **38**, 620 (1974); A. A. Abrikosov and I.A. Rizhkin, Conductivity of quasi-one-dimensional metal systems, *Adv. Phys.*, **27**, 147 (1978).
[71] S. Hershfield, Current conservation and resistance fluctuations in a four probe geometry, *Ann. Phys.*, **196**, 12 (1989).
[72] R. Berkovits and S. Feng, Correlations in coherent multiple scattering, *Phys. Rep.*, **238**, 135 (1994).

Chapter 4

[73] S. Chakravarty and A. Schmid, Weak-localization: the quasiclassical theory of electrons in a random potential, *Phys. Rep.*, **140**, 193 (1986).

[74] B. A. van Tiggelen and R. Maynard, Reciprocity and coherent backscattering of light, in *Wave Propagation in Complex Media*, ed. G. Papanicolaou, IMA, Springer, 1997.
[75] V. N. Prigodin, B. L. Altshuler, K. B. Efetov and S. Iida, Mesoscopic dynamical echo in quantum dots, *Phys. Rev. Lett.*, **72**, 546 (1994).
[76] S. Chandrasekhar, *Radiative Transfer*, New York: Dover, 1960.
[77] P. Sheng, *Introduction to Wave Scattering, Localization, and Mesoscopic Phenomena*, New York: Academic Press, 1995.
[78] L. Brillouin, *Wave Propagation and Group Velocity*, New York: Academic Press, 1960; R. Loudon, The propagation of electromagnetic energy through an absorbing dielectric, *J. Phys. A*, **3**, 233 (1970).
[79] M. P. van Albada, B. A. van Tiggelen, A. Lagendijk and A. Tip, Speed of propagation of classical waves in strongly scattering media, *Phys. Rev. Lett.*, **66**, 3132 (1991); B. A. van Tiggelen, A. Lagendijk, M. P. van Albada and A. Tip, Speed of light in random media, *Phys. Rev. B*, **45**, 12233 (1992); A. Lagendijk and B. A. van Tiggelen, Resonant multiple scattering of light, *Phys. Rep.*, **270**, 3 (1996).
[80] A. Gero and E. Akkermans, Effect of superradiance on transport of diffusing photons in cold atomic gases, *Phys. Rev. Lett.*, **96**, 093601 (2006).
[81] J. S. Langer and T. Neal, Breakdown of the concentration expansion for the impurity potential resistivity of metals, *Phys. Rev. Lett.*, **16**, 984 (1966); T. Neal, Breakdown of the concentration expansion for the zero-temperature impurity resistivity, *Phys. Rev.*, **169**, 508 (1968).
[82] Y. Pomeau and P. Résibois, Time dependent correlation functions and mode-mode coupling theories, *Phys. Rep.*, **19C**, 63 (1975); T. R. Kirkpatrick and J. R. Dorfman, Divergences and long-time tails in two- and three-dimensional quantum Lorentz gases, *Phys. Rev. A*, **28**, 1022 (1983).
[83] L. P. Gorkov, A. Larkin and D. E. Khmelnitskii, Particle conductivity in a two-dimensional random potential, *JETP Lett.*, **30**, 228 (1979).
[84] S. Hikami, Anderson localization in a nonlinear σ-model representation, *Phys. Rev. B*, **24**, 2671 (1981).
[85] M. C. W. van Rossum and T. M. Nieuwenhuizen, Multiple scattering of classical waves: microscopy, mesoscopy and diffusion, *Rev. Mod. Phys.*, **71**, 313 (1999).
[86] D. Vollhardt and P. Wölfle, Diagrammatic, self-consistent treatment of the Anderson localization problem in $d \leq 2$ dimensions, *Phys. Rev. B*, **22**, 4666 (1980).
[87] More elaborate and subsequent calculations for the scaling of D assume that it is a function of both $\mathbf{Q}$ and ω. See for example: T. Brandes, B. Huckestein and L. Schweitzer, Critical dynamics and multifractal exponents at the Anderson transition in $3d$ disordered systems, *Ann. Phys.*, **5**, 633 (1996).
[88] C. L. Kane, R. A. Serota and P. A. Lee, Long-range correlations in disordered metals, *Phys. Rev. B*, **37**, 6701 (1988).
[89] M. B. Hastings, A. D. Stone and H. U. Baranger, Inequivalence of weak-localization and coherent backscattering, *Phys. Rev. B*, **50**, 8230 (1994).
[90] G. D. Mahan, *Many-Particle Physics*, 2nd edition, New York: Plenum, 1990.
[91] M. V. Berry, Regular and irregular semiclassical wavefunctions, *J. Phys. A*, **10**, 2083 (1977).
[92] S. McDonald and A. Kaufman, Wave chaos in the stadium: statistical properties of short-wave solutions of the Helmholtz equation, *Phys. Rev. A*, **37**, 3067 (1988).

[93] Ya. M. Blanter and A. D. Mirlin, Correlations of eigenfunctions in disordered systems, *Phys. Rev. E*, **55**, 6514 (1997).

Chapter 5

[94] W. Feller, *An introduction to Probability Theory and its Applications*, New York: Wiley, 1990; S. Chandrasekar, in *Selected Papers on Noise and Stochastic Processes*, ed. A. Wax, New York: Dover, 1994.
[95] H. S. Carlslaw and J. C. Jaeger, *Operational Methods in Applied Mathematics*, New York: Dover, 1963.
[96] M. Berger, P. Gauduchon and E. Mazet, *Le Spectre d'une Variété Riemannienne*, Lecture Notes in Mathematics 194, Springer Verlag, 1971.
[97] A. Ishimaru, *Wave Propagation and Scattering in Random Media*, Vol. 1, New York: Academic Press, 1978, Chapers 7 and 9.
[98] A. Schuster, Radiation through a foggy atmosphere, *Astrophys. J.*, **21**, 1 (1905).
[99] S. Chandrasekhar, *Radiative Transfer*, New York: Dover, 1960.
[100] U. Frisch, in *Probabilistic Methods in Applied Mathematics*, ed. A. T. Bharucha-Reid, Vol. 1, New York: Academic Press, 1968.
[101] J. X. Xhu, D. J. Pine and D. A. Weitz, Internal reflection of diffusive light in random media, *Phys. Rev. A*, **44**, 3948 (1991).
[102] A. Ishimaru, Y. Kuga, R. Cheung and K. Shimizu, Scattering and diffusion of a beam wave in randomly distributed scatterers, *J. Opt. Soc. Am.*, **73**, 131 (1983).
[103] E. Akkermans, P. E. Wolf and R. Maynard, Coherent backscattering of light by disordered media: analysis of the peak line shape, *Phys. Rev. Lett.*, **56**, 1471 (1986).
[104] P. M. Morse and H. Feshbach, *Methods of Theoretical Physics*, New York: McGraw Hill, 1953.
[105] M. Kac, Can you hear the shape of a drum?, *Am. Math. Month.* **73**S, 1 (1966).
[106] H. P. Baltes and E. R. Hilf, *Spectra of Finite Systems*, Vienna: B. I. Wissenschaftsverlag, 1976.
[107] S. Rosenberg, *The Laplacian on a Riemannian Manifold*, Cambridge: Cambridge University Press, 1997.
[108] K. Stewartson and R. T. Waechter, On hearing the shape of a drum: further results, *Proc. Cambridge Philos. Soc.*, **69**, 353 (1971).
[109] V. E. Kravtsov and V. I. Yudson, Topological spectral correlations in 2D disordered systems, *Phys. Rev. Lett.*, **82**, 157 (1999).
[110] S. Alexander, Superconductivity of networks, *Phys. Rev. B*, **27**, 1541 (1983).
[111] B. Douçot and R. Rammal, Interference effects and magnetoresistance oscillations in normal-metal networks: weak localization approach, *J. Phys. (Paris)*, **47**, 973 (1986).
[112] M. Pascaud and G. Montambaux, Persistent currents on networks, *Phys. Rev. Lett.*, **82**, 4512 (1999).
[113] E. Akkermans, A. Comtet, J. Desbois, G. Montambaux and C. Texier, Spectral determinant on quantum graphs, *Ann. Phys.*, **284**, 10 (2000).
[114] M. Pascaud, Magnétisme orbital de conducteurs mésoscopiques désordonnés, *Thèse*, Université Paris-Sud, 1998.
[115] D. Hofstadter, Energy levels and wave functions of Bloch electrons in rational and irrational magnetic fields, *Phys. Rev. B*, **14**, 2239 (1976).

[116] C. Texier and G. Montambaux, Weak localization in multiterminal networks of diffusive wires, *Phys. Rev. Lett.*, **92**, 186801 (2004).

Chapter 6

[117] M. Robnik and M. V. Berry, False time reversal violation and energy level statistics: the role of anti-unitary symmetry, *J. Phys. A*, **19**, 669 (1986).
[118] Y. Aharonov and D. Bohm, Significance of electromagnetic potentials in quantum theory, *Phys. Rev.*, **115**, 485 (1959).
[119] The eikonal approximation is also used to describe the effect of a magnetic field on a type II superconductor, for which effects related to Landau quantization are usually neglected: L. P. Gorkov, Microscopic derivation of the Ginzburg–Landau equations in the theory of superconductivity, *Sov. Phys. JETP*, **9**, 1364 (1959). An excellent discussion of this approximation is given in A. L. Fetter and J. D. Walecka, *Quantum Theory of Many-Particle Systems*, New York: MacGraw-Hill, 1971, p. 468. Deviations from this approximation and the effect of Landau quantization are studied in M. Rasolt and Z. Tesanovic, Theoretical aspects of superconductivity in very high magnetic fields, *Rev. Mod. Phys.*, **64**, 709 (1992).
[120] P. A. Lee and M. G. Payne, Pair propagator approach to fluctuation-induced diamagnetism in superconductors – effects of impurities, *Phys. Rev. B*, **5**, 923 (1972).
[121] P. G. de Gennes, *Superconductivity of Metals and Alloys*, New York: Addison Wesley, 1989.
[122] P. Levy, *Processus Stochastiques et Mouvement Brownien*, 2nd edition, Paris: Gauthier-Villard, 1965, Chapter 7, section 55.
[123] J. D. Jackson, *Classical Electrodynamics*, New York: Wiley 1975.
[124] G. Baym, *Lectures on Quantum Mechanics*, New York: W. A. Benjamin, 1973.
[125] S. Hikami, A. Larkin and Y. Nagaoka, Spin-orbit interaction and magnetoresistance in the two-dimensional random systems, *Prog. Theor. Phys.*, **63**, 707 (1980); A. A. Golubentsev, Direct calculation of the conductivity of films with magnetic impurities, *JETP Lett.*, **41**, 642 (1985).
[126] S. Chakravarty and A. Schmid, Weak localization: the quasiclassical theory of electrons in a random potential, *Phys. Rep.*, **140**, 193 (1986).
[127] L. D. Landau and E. M. Lifshitz, *Quantum Mechanics (Non-relativistic Theory)*, Pergamon, 1977, sections 140, 144.
[128] G. Grüner and F. Zawadowski, Magnetic impurities in non-magnetic metals, *Rep. Prog. Phys.*, **37**, 1497 (1974).
[129] S. Doniach and E. Sondheimer, *Green's Functions for Solid State Physicists*, New York: W. A. Benjamin, 1975.
[130] M. B. Maple, *Magnetism*, ed. H. Suhl, Vol. 5, New York: Academic Press, 1973; C. Van Haesendonck, J. Vranken and Y. Bruynseraede, *Phys. Rev. Lett.*, **58**, 1968 (1987).
[131] F. Pierre, A. B. Gougam, A. Anthore, M. Pothier, D. Esteve and N. Birge, Dephasing of electrons in mesoscopic metal wires, *Phys. Rev. B*, **68**, 85413 (2003).
[132] T. Micklitz, A. Altland, T. A. Costi and A. Rosch, Universal dephasing rate due to diluted Kondo impurities, *Phys. Rev. Lett.*, **96**, 226601 (2006).
[133] W. Götze and P. Schlottmann, On the longitudinal static and dynamic susceptibility of spin 1/2 Kondo systems, *J. Low Temp. Phys.*, **16**, 87 (1974).

[134] M. G. Vavilov and L. I. Glazman, Conductance of mesoscopic systems with magnetic impurities, *Phys. Rev. B*, **67**, 115310 (2003).
[135] P. W. Anderson, Lectures on amorphous systems and spin glasses, J. Joffrin, Disordered systems, an experimental viewpoint, J. Souletie, About spin glasses, in *Les Houches, Ill-condensed matter, session XXXI*, eds. R. Balian, R. Maynard and G. Toulouse, Amsterdam: North-Holland, 1978.
[136] V. I. Falko, Effect of impurity spin dynamics on weak localization, *Sov. Phys. JETP Lett.*, **53**, 340 (1991).
[137] A. A. Bobkov, V. I. Falko and D. E. Khmelnitskii, Mesoscopics in metals with magnetic impurities, *Sov. Phys. JETP*, **71**, 393 (1990).
[138] S. Hershfield, Resistance fluctuations with magnetic impurities in a four-terminal geometry, *Phys. Rev. B*, **44**, 3320 (1991).
[139] A. Messiah, *Quantum Mechanics*, New York: Dover, 1999.
[140] G. Bergmann, Weak-localization in thin films, *Phys. Rep.*, **107**, 1 (1984).
[141] E. Akkermans, P. E. Wolf, R. Maynard and G. Maret, Theoretical study of the coherent backscattering of light by disordered media, *J. Phys. (France)*, **49**, 77 (1988).
[142] P. E. Wolf, G. Maret, E. Akkermans and R. Maynard, Optical coherent backscattering by random media: an experimental study, *J. Phys. (France)*, **49**, 63 (1988).
[143] A. A. Golubentsev, Suppression of interference effects in multiple scattering of light, *Sov. Phys. JETP*, **59**, 26 (1984).
[144] M. J. Stephen and G. Cwillich, Rayleigh scattering and weak localization: Effects of polarization, *Phys. Rev. B*, **34**, 7564 (1986).
[145] G. Maret and P. E. Wolf, Multiple light scattering from disordered media. The effect of brownian motion of scatterers, *Z. Phys. B*, **65**, 409 (1987).
[146] M. J. Stephen, Temporal fluctuations in wave propagation in random media, *Phys. Rev. B*, **37**, 1 (1988).
[147] D. J. Pine, D. A. Weitz, P. M. Chaikin and E. Herbolzheimer, Diffusing wave spectroscopy, *Phys. Rev. Lett.*, **60**, 1134 (1988).
[148] Y. Imry, *Introduction to Mesoscopic Physics*, 2nd edition, Oxford: Oxford University Press, 2002; A. Stern, Y. Aharonov and Y. Imry, Phase uncertainty and loss of interference: a general picture, *Phys. Rev. A*, **41**, 3436 (1990).
[149] P. B. Gilkey, *Invariance Theory, the Heat Equation and the Atiyah–Singer Index Theorem*, Boca Raton, FL: CRC Press, 1995.
[150] R. P. Feynman and A. R. Hibbs, *Quantum Mechanics and Path Integrals*, New York: McGraw-Hill, 1965; G. Roepstorff, *Path Integral Approach to Quantum Physics: an Introduction*, New York: Springer Verlag, 1994.
[151] F. W. Wiegel, *Introduction to Path-Integral Methods in Physics and Polymer Science*, World Scientific, 1986.
[152] A. Comtet, J. Desbois and S. Ouvry, Winding of planar brownian curves, *J. Phys A*, **23**, 3562 (1990); B. Duplantier, Areas of planar Brownian curves, *J. Phys A*, **22**, 3033 (1989).
[153] S. F. Edwards, Statistical mechanics with topological constraints: I, *Proc. Phys. Soc.*, **91**, 513 (1967).
[154] B. L. Altshuler, A. G. Aronov and D. E. Khmelnitskii, Effects of electron–electron collisions with small energy transfers on quantum localisation, *J. Phys. C*, **15**, 7367 (1982).
[155] G. Bergmann, Weak anti-localization: an experimental proof for the destructive interference of rotated spin 1/2, *Solid State Commun.*, **42**, 815 (1982).

[156] E. Akkermans, Ch. Miniatura and C. A. Müller, *Phase Coherence Times in the Multiple Scattering of Photons by Cold Atoms*, arXiv:cond-mat/0206298, 2002.
[157] C. A. Mueller, C. Miniatura, E. Akkermans and G. Montambaux, Weak localization of light by cold atoms: the impact of quantum internal structure, *Phys. Rev. A*, **64**, 053804 (2001).
[158] C. A. Müller and C. Miniatura, Multiple scattering of light by atoms with internal degeneracy, *J. Phys. A*, **35**, 10163 (2002).
[159] O. Assaf and E. Akkermans, Intensity correlations and mesoscopic fluctuations of diffusing photons in cold atoms, *Phys. Rev. Lett.*, **98**, 83601 (2007).
[160] C.A. Müller, C. Miniatura, E. Akkermans and G. Montambaux, Mesoscopic scattering of spin *s* particles, *J. Phys. A*, **38**, 7807 (2005).

Chapter 7

[161] C. L. Kane, R. A. Serota and P. A. Lee, Long-range correlations in disordered metals, *Phys. Rev. B*, **37**, 6701 (1988).
[162] S. Datta, *Electronic Transport in Mesoscopic Systems*, Cambridge studies in Semiconductor Physics and Micoelectronic Engineering, Cambridge: Cambridge University Press, 1995.
[163] S. Chakravarty and A. Schmid, Weak-localization: the quasiclassical theory of electrons in a random potential, *Phys. Rep.*, **140**, 193 (1986).
[164] J. Rammer, *Quantum Transport Theory*, Frontiers in Physics, Perseus, 1998.
[165] N. Ashcroft and D. Mermin, *Solid State Physics*, Philadelphia, PA: Saunders College, 1976.
[166] D. J. Thouless, Electrons in disordered systems and the theory of localization, *Phys. Rep.*, **13**, 93 (1974); J. T. Edwards and D. J. Thouless, Numerical studies of localization in disordered systems, *J. Phys. C*, **5**, 807 (1972); E. Akkermans and G. Montambaux, Conductance and statistical properties of metallic spectra, *Phys. Rev. Lett.*, **68**, 642 (1992).
[167] E. Abrahams, P. W. Anderson, D. C. Licciardello and T. V. Ramakrishnan, Scaling theory of localization: absence of quantum diffusion in two dimensions, *Phys. Rev. Lett.*, **42**, 673 (1979).
[168] A. A. Abrikosov, L. P. Gorkov and I. Y. Dzyaloshinskii, *Quantum Field Theoretical Methods in Statistical Physics*, Pergamon, 1965.
[169] S. Doniach and E. H. Sondheimer, *Green's Functions for Solid State Physicists*, Frontiers in Physics, New York: W. A. Benjamin, 1974.
[170] P. Wölfle and R. N. Bhatt, Electron localization in anisotropic systems, *Phys. Rev. B*, **30**, 3452 (1984).
[171] The short time cutoff τ_e has been introduced by hand. For a detailed discussion on this point, see: A. Cassam-Chenai and B. Shapiro, Two dimensional weak localization beyond the diffusion approximation, *J. Phys. I (France)*, **4**, 1527 (1994).
[172] H. Bouchiat, Experimental signatures of phase coherent transport, in *Les Houches Summer School*, eds. E. Akkermans, G. Montambaux, J.-L. Pichard and J. Zinn-Justin, Amsterdam: Elsevier, 1995, Session LXI, Mesoscopic Quantum Physics.
[173] B. Kramer and A. MacKinnon, Localization: theory and experiment, *Rep. Prog. Phys.*, **56**, 1469 (1993).

[174] P. A. Lee and T. V. Ramakrishnan, Disordered electronic systems, *Rev. Mod. Phys.*, **57**, 287 (1985).
[175] M. Janssen, *Fluctuations and Localization in Mesoscopic Electron Systems*, Lecture Notes in Physics, Vol. 44, Singapore: World Scientific, 2001.
[176] G. Dolan and D. Osheroff, Nonmetallic conduction in thin metal films at low temperatures, *Phys. Rev. Lett.*, **43**, 721 (1979).
[177] B. L. Altshuler, A. G. Aronov, M. E. Gershenson and Yu. V. Sharvin, Quantum effects in disordered metal films, *Phys. Rev., Sov. Sci. Rev.*, **9**, 225 (1987).
[178] Y. Imry, *Introduction to Mesoscopic Physics*, Oxford: Oxford University Press, 2002.
[179] G. Bergmann, Weak localization in thin films, *Phys. Rep.*, **107**, 1 (1984).
[180] J. C. Licini, G. J. Dolan, and D. J. Bishop, Weakly localized behavior in quasi-one-dimensional Li films, *Phys. Rev. Lett.*, **54**, 1585 (1985).
[181] B. L. Altshuler and A. G. Aronov, Magnetoresistance of thin films and of wires in a longitudinal magnetic field, *JETP Lett.*, **33**, 499 (1981).
[182] S. Hikami, A. Larkin and Y. Nagaoka, Spin-orbit interaction and magnetoresistance in the two-dimensional random systems, *Prog. Theor. Phys.*, **63**, 707 (1980).
[183] B. Pannetier, J. Chaussy, R. Rammal and P. Gandit, Magnetic flux quantization in the weak-localizalization regime of a nonsuperconducting metal, *Phys. Rev. Lett.*, **53**, 718 (1984); G. J. Dolan, J. C. Licini and D. J. Bishop, Quantum interference effects in lithium ring arrays, *Phys. Rev. Lett.*, **56**, 1493 (1986).
[184] I. S. Gradshsteyn and I. M Ryzhik, *Table of Integrals, Series and Products*, New York: Academic Press, 1980.
[185] B. L. Altshuler, A. G. Aronov and B. Z. Spivak, The Aharonov–Bohm effect in disordered conductors, *JETP Lett.*, **33**, 94 (1981).
[186] D. Yu. Sharvin and Yu. V. Sharvin, Magnetic flux quantization in a cylindrical film of a normal metal, *JETP Lett.*, **34**, 272 (1981).
[187] B. L. Altshuler, A. G. Aronov, B. Z. Spivak, D. Yu. Sharvin and Yu. V. Sharvin, Observation of the Aharonov-Bohm effect in hollows metal cylinders, *JETP Lett.*, **35**, 588 (1982).
[188] R. A. Webb, S. Washburn, C. P. Umbach and R. P. Laibowitz, Observation of h/e Aharonov–Bohm oscillations in normal-metal rings, *Phys. Rev. Lett.*, **54**, 2696 (1985).
[189] M. Murat, Y. Gefen and Y. Imry, Ensemble and temperature averaging of quantum oscillations in normal-metal rings, *Phys. Rev. B*, **34**, 657 (1986).
[190] M. Büttiker, Symmetry of electrical conduction, *IBM J. Res. Develop.*, **32**, 317 (1988).
[191] S. Washburn and R. A. Webb, Aharonov–Bohm effect in normal metal, quantum coherence and transport, *Adv. Phys.*, **35**, 412 (1986).
[192] A. G. Aronov and Yu. V. Sharvin, Magnetic flux effects in disordered conductors, *Rev. Mod. Phys.*, **59**, 755 (1987).
[193] B. Pannetier, J. Chaussy and R. Rammal, First observation of Altshuler–Aronov–Spivak effect in gold and copper, *Phys. Rev. B*, **31**, 3209 (1985).
[194] B. Pannetier, J. Chaussy and R. Rammal, Quantum interferences in superconducting and normal metal arrays, *Phys. Scripta.*, **13**, 245 (1986).
[195] B. Douçot and R. Rammal, Quantum oscillations in normal-metal networks, *Phys. Rev. Lett.*, **55**, 1148 (1985); Interference effects and magnetoresistance oscillations in normal metal networks: 1–weak localization approach, *J. Phys. (France)*, **47**, 973 (1986).

[196] S. J. Bending, K. von Klitzing and K. Ploog, Weak localization in a distribution of magnetic flux tubes, *Phys. Rev. Lett.*, **65**, 1060 (1990).
[197] J. Rammer and A. L. Shelankov, Weak localization in inhomogeneous magnetic fields, *Phys. Rev. B*, **36**, 3135 (1987).
[198] R. P. Feynman, *The Feynman Lectures on Physics, Electromagnetism*, New York: Addison-Wesley, 1970, section 22.6.
[199] D. Vollhardt and P. Wölfle, Self-consistent theory of anderson localization, *Proc. Fourth Taniguchi International Symposium on the Theory of Condensed Matter*, Springer, 1982, p. 26; Diagrammatic self-consistent treatment of the anderson localization problem in $d \leq 2$, *Phys. Rev. B*, **22**, 4666 (1980).
[200] D. Forster, *Hydrodynamic Fluctuations, Broken Symmetry, and Correlation Functions*, New York: Addison-Wesley, 1983.
[201] R. Landauer, Spatial variation of currents and fields due to localized scatterers in metallic conduction, *IBM J. Res. Develop.*, **1**, 233 (1957); Electrical resistance of disordered one-dimensional lattices, *Philos. Mag.*, **21**, 863 (1970).
[202] For more details, the reader may consult: S. Datta, *Electronic Transport in Mesoscopic Systems*, Cambridge: Cambridge University Press, 1995; Y. Imry, *Introduction to Mesoscopic Physics*, Oxford: Oxford University Press, 2002; *IBM J. Res. Develop.*, **32**, 304 (1988).
[203] D. S. Fisher and P. A. Lee, Relation between the conductivity and the transmission matrix, *Phys. Rev. B*, **23**, 6851 (1981).
[204] C. W. J. Beenakker, Random-matrix theory of quantum transport, *Rev. Mod. Phys.*, **69**, 731 (1997).
[205] M. B. Hastings, A. D. Stone and H. U. Baranger, Inequivalence of weak-localization and coherent backscattering, *Phys. Rev. B*, **50**, 8230 (1994).

Chapter 8

[206] A. Schuster, Radiation through a foggy atmosphere, *Astrophys. J.*, **21**, 1 (1905).
[207] K. M. Watson, Multiple scattering of electromagnetic waves in an underdense plasma, *J. Math. Phys.*, **10**, 688 (1969).
[208] D. A. de Wolf, Electromagnetic reflection from an extended turbulent medium: cumulative forward-scatter single-backscatter approximation, *IEEE Trans. Antennas Propag.*, **19**, 254 (1971).
[209] Yu. N. Barabanenkov, Wave corrections to the transport equation for backscattering, *Izv. Vyssh. Uchebn. zaved. Radiofiz.*, **16**, 88 (1973).
[210] A. A. Golubentsev, Suppression of interference effects in multiple scattering of light, *Sov. Phys. JETP*, **59**, 26 (1984).
[211] E. Akkermans and R. Maynard, Weak localization of waves, *J. Phys. Lett. (France)*, **46**, L1045 (1985).
[212] Y. Kuga and A. Ishimaru, Retroreflectance from a dense distribution of spherical particles, *J. Opt. Soc. Am. A*, **8**, 831 (1984); L. Tsang and A. Ishimaru, Backscattering enhancement of random discrete scatterers, *J. Opt. Soc. Am. A*, **1**, 836 (1984); L. Tsang and A. Ishimaru, Theory of backscattering enhancement of random discrete isotropic scatterers based on the summation of all ladder and cyclical diagrams, *J. Opt. Soc. Am. A*, **2**, 1331 (1985); L. Tsang and A. Ishimaru, Radiative wave and cyclical transfer equation for dense nontenuous media, *J. Opt. Soc. Am. A*, **2**, 2187 (1985).

[213] P. E. Wolf and G. Maret, Weak localization and coherent backscattering of photons in disordered media, *Phys. Rev. Lett.*, **55**, 2696 (1985).

[214] M. P. van Albada and A. Lagendijk, Observation of weak localization of light in a random medium, *Phys. Rev. Lett.*, **55**, 2692 (1985).

[215] For a reference list up to 1991, one may consult: Yu. N. Barabenenkov, Yu. A. Kravtsov, V. D. Ozrin and A. I. Saichev, *Enhanced Backscattering in Optics*, Progress in Optics, Vol. XXIX, Amsterdam: North Holland, 1991.

[216] E. Akkermans, P. E. Wolf, and R. Maynard, Coherent backscattering of light by disordered media: analysis of the peak line shape, *Phys. Rev. Lett.*, **56**, 1471 (1986).

[217] E. Akkermans, P. E. Wolf, R. Maynard and G. Maret, Theoretical study of the coherent backscattering of light by disordered media, *J. Phys. (France)*, **49**, 77 (1988).

[218] M. B. van der Mark, M. P. van Albada and A. Lagendijk, Light scattering in strongly scattering media: multiple scattering and weak localization, *Phys. Rev. B*, **37**, 3575 (1988).

[219] H. C. van de Hulst, *Multiple Light Scattering* Vols. 1 and 2, New York: Academic Press, 1980.

[220] A. Ishimaru, *Wave Propagation and Scattering in Random Media*. Vols. 1 and 2, New York: Academic Press, 1978.

[221] D. S. Wiersma, Light in strongly scattering and amplifying random media, *Ph.D. Thesis*, Amsterdam, 1995.

[222] For an exact solution of the coherent backscattering problem for scalar waves and isotropic collisions, see: E. E. Gorodnichev, S. L. Dudarev and D. B. Rogozkin, Coherent wave backscattering by random medium. Exact solution of the albedo problem, *Phys. Lett. A*, **144**, 48 (1990).

[223] P. E. Wolf, G. Maret, E. Akkermans and R. Maynard, Optical coherent backscattering by random media: an experimental study, *J. Phys. (France)*, **49**, 63 (1988).

[224] A. Ishimaru, Y. Kuga, R. L. T. Cheung and K. Shimizu, Scattering and diffusion of a beam wave in randomly distributed scatterers, *J. Opt. Soc. Am.*, **73**, 131 (1983).

[225] A. Ishimaru and L. Tsang, Backscattering enhancement of random discrete scatterers of moderate sizes, *J. Opt. Soc. Am. A*, **5**, 228 (1988).

[226] M. Stephen and G. Cwillich, Rayleigh scattering and weak localization: effects of polarization, *Phys. Rev. B*, **34**, 7564 (1986).

[227] M. Kaveh, M. Rosenbluh, I. Edrei and I. Freund, Weak localization and light scattering from disordered solids, *Phys. Rev. Lett.*, **57**, 2049 (1986).

[228] S. Etemad, R. Thomson and M. J. Andrejco, Weak localization of photons: universal fluctuations and ensemble averaging, *Phys. Rev. Lett.* **57**, 575 (1986).

[229] D. S. Wiersma, M. P. van Albada, B. A. van Tiggelen and A. Lagendijk, Experimental evidence for recurrent multiple scattering events of light in disordered media, *Phys. Rev. Lett.*, **74**, 4193 (1995); B. A. van Tiggelen, D. S. Wiersma and A. Lagendijk, Self-consistent theory for the enhancement factor in coherent backscattering, *Europhys. Lett.*, **30**, 1 (1995).

[230] M. I. Mishchenko, Enhanced backscattering of polarized light from discrete random media: calculations in exactly the backscattering direction, *J. Opt. Soc. Am. A*, **9**, 4578 (1992); M. I. Mishchenko, Diffuse and coherent backscattering by discrete random media I. Radar reflectivity, polarization ratios, and enhancement factors for a half-space of polydisperse, nonabsorbing and absorbing spherical particles, *J. Quant. Spectrosc. Radiat. Transfer*, **56**, 673 (1996).

[231] M. P. van Albada, M. B. van der Mark and A. Lagendijk, Observation of weak localization of light in a finite slab: anisotropy effects and light path classification, *Phys. Rev. Lett.*, **58**, 361 (1987).
[232] R. Lenke and G. Maret, Multiple scattering of light: coherent backscattering and transmission, *Prog. Colloid Polym. Sci.*, **104**, 126 (1997).
[233] T. Okamoto and T. Asakura, Enhanced backscattering of partially coherent light, *Opt. Lett.*, **21**, 369 (1996).
[234] H. C. Bryant and J. Nelson, The glory, *Sci. Am.*, **60** (July 1974).
[235] R. Lenke, U. Mack and G. Maret, Comparison of the "glory" with coherent backscattering of light in turbid media, *J. Opt. A: Pure Appl. Opt.*, **4**, 309 (2002); R. Lenke, R. Tweer and G. Maret, Coherent backscattering of turbid samples containing large Mie spheres, *J. Opt. A: Pure Appl. Opt.*, **4**, 293 (2002).
[236] G. Muller, Seeliger analyzed observations, *Publ. Obs. Potsdam*, **8**, 193 (1893).
[237] F. A. Franklin and A. F. Cook, Optical properties of Saturn rings, *Astrophys. J.*, **70**, 704 (1965).
[238] W. W. Montgomery and R. H. Kohl, Opposition effect experimentation, *Optics Lett.*, **5**, 546 (1980).
[239] B. Hapke and D. Blewett, Coherent backscatter model for the unusual radar reflectivity of icy satellites, *Nature*, **352**, 46 (1991); B. Hapke, R. M. Nelson and W. D. Smythe, The opposition effect of the moon: the contribution of coherent backscatter, *Science*, **260**, 509 (1993).
[240] B. Lyot, Recherche sur la polarisation de la lumière des planètes et de quelques substances terrestres, *Ann. Obs. Paris*, **8**, 89 (1929).
[241] M. I. Mishchenko, On the nature of the polarization opposition effect exhibited by Saturn's rings, *Astrophys. J.*, **411**, 351 (1993).
[242] G. Labeyrie, F. de Tomasi, J.-C. Bernard, C. A. Müller, C. Miniatura and R. Kaiser, Coherent backscattering of light by cold atoms, *Phys. Rev. Lett.*, **83**, 5266 (1999); G. Labeyrie, C. A. Müller, D. S. Wiersma, C. Miniatura and R. Kaiser, Observation of coherent backscattering of light by cold atoms, *J. Opt. B*, **2**, 672 (2000).
[243] C. A. Müller, T. Jonckheere, C. Miniatura and D. Delande, Weak localization of light by cold atoms: the impact of quantum internal structure, *Phys. Rev. A*, **64**, 053804 (2001); T. Jonckheere, C. A. Müller, R. Kaiser, C. Miniatura and D. Delande, Multiple scattering of light by atoms in the weak localization regime, *Phys. Rev. Lett.*, **85**, 4269 (2000); C. A. Müller and C. Miniatura, Weak localisation of light by atoms with quantum internal structure, in *Wave Scattering in Complex Media, from Theory to Applications*, eds. S. E. Skipetrov and B. A. van Tiggelen, NATO Series, Dordrecht: Kluwer, 2003.
[244] M. Fink, Time reversed acoustics, *Physics Today*, 34 (March 1997).
[245] A. Tourin, A. Derode, P. Roux, B. A. van Tiggelen and M. Fink, Time-dependent coherent backscattering of acoustic waves, *Phys. Rev. Lett.*, **79**, 3637 (1997).
[246] R. Vreeker, M. P. van Albada, R. Sprik and A. Lagendijk, Femtosecond time-resolved measurements of weak localization of light, *Phys. Lett. A*, **132**, 51 (1988).

Chapter 9

[247] B. J. Berne and R. Pecora, *Dynamic Light Scattering with Applications to Chemistry, Biology and Physics*, New York: John Wiley, 1976.
[248] R. Loudon, *The Quantum Theory of Light*, Oxford: Clarendon Press, 1986.

[249] M. C. Teich and B. Saleh, Photon bunching and antibunching, *Prog Opt.*, **26**, 1 (1988).
[250] G. Maret and P. E. Wolf, Multiple light scattering from disordered media. The effect of brownian motion of scatterers, *Z. Phys. B*, **65**, 409 (1987).
[251] P. E. Wolf and G. Maret, Dynamics of brownian particles from strongly multiple light scattering, in *Scattering in Volumes and Surfaces*, eds. M. Nieto-Vesperinas and J. C. Dainty, Amsterdam: North-Holland, 1990, p. 37.
[252] D. J. Pine, D. A. Weitz, P. M. Chaikin and E. Herbolzheimer, Diffusing wave spectroscopy, *Phys. Rev. Lett.*, **60**, 1134 (1988).
[253] D. J. Pine, D. A. Weitz, J. X. Zhu and E. Herbolzheimer, Diffusing-wave spectroscopy: dynamic light scattering in the multiple scattering limit, *J. Phys. (France)*, **51**, 2101 (1990).
[254] E. Guyon, J. P. Hulin and L. Petit, Hydrodynamique physique, InterEditions, Editions du CNRS, 1991; P. Tong, W. I. Goldburg, C. K. Chan and A. Sirivat, Turbulent transition by photon-correlation spectroscopy, *Phys. Rev. A*, **37**, 2125 (1988).
[255] D. Bicout, E. Akkermans and R. Maynard, Dynamical correlations for multiple light scattering in laminar flow, *J. Phys. I*, **1**, 471 (1991).
[256] X. L. Wu, D. J. Pine, J. S. Huang, P. M. Chaikin and D. A. Weitz, Diffusing-wave spectroscopy in a shear flow, *J. Opt. Soc. Am. B*, **7**, 15 (1990).

Chapter 10

[257] B. Kramer and A. MacKinnon, Localization: theory and experiment, *Rep. Prog. Phys.*, **56**, 1469 (1993).
[258] T. Ohtsuki, K. Slevin and T. Kawarabayashi, Review of recent progress on numerical studies of the Anderson transition, *Ann. Phys.*, **8**, 655 (1999).
[259] O. Bohigas, Random matrix theories and chaotic dynamics, in *Chaos and Quantum Physics, Proceedings of Les Houches Summer School*, eds. M. J. Giannoni, A. Voros and J. Zinn-Justin, Amsterdam: North-Holland, 1991, Session LII, p. 91.
[260] O. Bohigas, M. J. Giannoni and C. Schmit, in *Quantum Chaos and Statistical Nuclear Physics*, eds. T. H. Seligman and N. Nishioka, Lecture Notes in Physics, Vol. 263, Berlin: Springer, 1986, p. 18.
[261] O. Bohigas, M. J. Giannoni and C. Schmit, Characterization of chaotic quantum spectra and universality of level fluctuation laws, *Phys. Rev. Lett.*, **52**, 1 (1984).
[262] D. Delande, Chaos in atomic and molecular physics, in *Chaos and Quantum Physics, Proceedings of Les Houches Summer School*, eds. M. J. Giannoni, A. Voros and J. Zinn-Justin, Amsterdam: North-Holland, 1991, Session LII, p. 665.
[263] H.-J. Stöckmann, *Quantum Chaos: an Introduction*, Cambridge: Cambridge University Press, 1999.
[264] The literature on the topic of so-called "quantum chaos" is huge! This subject is beyond the scope of this book and we propose here only a few general references: A. Ozorio de Almeida, *Hamiltonian Systems: Chaos and Quantization*, Cambridge: Cambridge University Press, 1988; M. Tabor, *Chaos and Integrability in Nonlinear Dynamics. An Introduction*, New York: Wiley, 1989; F. Haake, *Quantum Signatures of Chaos*, Berlin: Springer, 1992; B. Eckardt, Quantum mechanics of classically non-integrable systems, *Phys. Rep.*, **163**, 205 (1988).

[265] E. P. Wigner, *Statistical Properties of Real Symmetric Matrices with Many Dimensions*, Can. Math. Congr. Proc., Toronto: University of Toronto Press, 1957, p. 174.
[266] E. P. Wigner, On the distribution of the roots of certain symmetric matrices, *Ann. Math.*, **67**, 325 (1958); F. J. Dyson, Statistical theory of energy levels of complex systems I–III, *J. Math. Phys.*, **3**, 140 (1962); F. J. Dyson and M. L. Mehta, Statistical theory of energy levels of complex systems IV–V, *J. Math. Phys.*, **4**, 701 (1963).
[267] M. L. Mehta, *Random Matrices*, New York: Academic Press, 1991.
[268] L. P. Gor'kov and G. M. Eliashberg, Minute metallic particles in an electromagnetic field, *Sov. Phys.* JETP, **21**, 940 (1965).
[269] B. L. Altshuler and B. Simons, *Universalities: from Anderson localization to quantum chaos*, in *Les Houches Summer School Mesoscopic Quantum Physics*, eds. E. Akkermans, G. Montambaux, J.-L. Pichard and J. Zinn-Justin, Amsterdam: Elsevier, 1995, Session LXI.
[270] R. S. Whitney, I. V. Lerner and R. A. Smith, Can the trace formula describe weak localization?, *Waves Random Media*, **9**, 179 (1999).
[271] K. B. Efetov, Supersymmetry and theory of disordered metals, *Adv. Phys.*, **32**, 53 (1983); K. B. Efetov, *Supersymmetry in Disorder and Chaos*, Cambridge: Cambridge University Press, 1997.
[272] T. Guhr, A. Müller-Groeling and H.A. Weidenmüller, Random-matrix theories in quantum physics: common concepts, *Phys. Rep.*, **299**, 189 (1998).
[273] A. D. Mirlin, Statistics of energy levels and eigenfunctions in disordered and chaotic systems: supersymmetry approach, *Proceedings of the International School of Physics Enrico Fermi*, eds. G. Casati, I. Guarneri and U. Smilansky, Amsterdam: IOS Press, 2000, Course CXLIII.
[274] N. Dupuis and G. Montambaux, Aharonov–Bohm flux and statistics of energy levels in metals, *Phys. Rev. B*, **43**, 14390 (1991).
[275] D. Braun and G. Montambaux, Spectral correlations from the metal to the mobility edge, *Phys. Rev. B* **52**, 13903 (1995).
[276] M. Gaudin, Sur la loi limite de l'espacement de valeurs propres d'une matrice aléatoire, *Nucl. Phys.*, **25**, 447 (1961).
[277] B. L. Al'tshuler and B. Shklovskiĭ, Repulsion of energy levels and conductivity in small metal samples, *Sov. Phys. JETP*, **64**, 127 (1986).
[278] N. Argaman, Y. Imry and U. Smilansky, Semiclassical analysis of spectral correlations in mesoscopic systems, *Phys. Rev. B*, **47**, 4440 (1993).
[279] J. H. Hannay and A. M. Ozorio de Almeida, Periodic orbits and a correlation function for the semiclassical density of states, *J. Phys. A*, **17**, 3429 (1983).
[280] V. N. Prigodin, B. L. Al'tshuler, K. B. Efetov and S. Iida, Mesoscopic dynamical echo in quantum dots, *Phys. Rev. Lett.*, **72**, 546 (1994).
[281] A. Altland and Y. Gefen, Spectral statistics of nondiffusive disordered electron systems: a comprehensive approach, *Phys. Rev. B*, **51**, 10671 (1995).
[282] A. Pandey and M. L. Mehta, Gaussian ensembles of random hermitian matrices intermediate between orthogonal and unitary ones, *Commum. Math. Phys.*, **87**, 449 (1983).
[283] A. Altland, S. Iida and K. B. Efetov, The cross-over between orthogonal and unitary symmetry in small disordered systems: a supersymmetry approach, *J. Phys. A*, **26**, 3545 (1993).

Chapter 11

[284] S. Washburn, Fluctuations in the extrinsic conductivity of disordered metals, *IBM J. Res. Develop*, **32**, 335 (1988).
[285] A. D. Stone, Magnetoresistance fluctuations in mesoscopic wires and rings, *Phys. Rev. Lett.*, **54**, 2692 (1985).
[286] P. A. Lee and A. D. Stone, Universal conductance fluctuations in metals, *Phys. Rev. Lett.*, **55**, 1622 (1985).
[287] B.L. Al'tshuler, Fluctuations in the extrinsic conductivity of disordered conductors, *Sov. Phys. JETP Lett.*, **41**, 648 (1985).
[288] B. L. Al'tshuler and B. Shklovskiĭ, Repulsion of energy levels and conductivity of small metal samples, *Sov. Phys. JETP*, **64**, 127 (1986).
[289] P. A. Lee, A. D. Stone and H. Fukuyama, Universal conductance fluctuations in metals: effects of finite temperature, interactions, and magnetic field, *Phys. Rev. B*, **35**, 1039 (1987).
[290] I. Aleiner and Ya. Blanter, Inelastic scattering time for conductance fluctuations, *Phys. Rev. B*, **65**, 115317 (2002).
[291] R. A. Serota, S. Feng, C. Kane and P. A. Lee, Conductance fluctuations in small disordered conductors: thin-lead and isolated geometries, *Phys. Rev. B*, **36**, 5031 (1987).
[292] M. Pascaud and G. Montambaux, Interference effects in mesoscopic disordered rings and wires, *Phys. Uspek.*, **41**, 182 (1998).
[293] D. Mailly and M. Sanquer, Sensitivity of quantum conductance fluctuations and of 1/f noise to time reversal symmetry, *J. Phys. I (France)*, **2**, 357 (1992).
[294] This problem has been considered in : B. L. Altshuler and B. Z. Spivak, Variation of the random potential and the conductivity of samples of small dimensions, *JETP Lett.*, **42**, 447 (1986). Notice that their result (2) is at odds with our relation (11.72). It amounts to replacing $b(T)$ by 1 in the numerator of (11.72).
[295] S. Feng, P. A. Lee and A. D. Stone, Sensitivity of the conductance of a disordered metal to the motion of a single atom: implications for $1/f$ noise, *Phys. Rev. Lett.*, **56**, 1960 (1986).
[296] V. Chandrasekhar, P. Santhanam and D. E Prober, Effect of spin-orbit and spin-flip scattering on conductance fluctuations, *Phys. Rev. B*, **42**, 6823 (1990).
[297] S. Feng, Mesoscopic conductance fluctuations in the presence of spin-orbit coupling and Zeeman splitting, *Phys. Rev. B*, **39**, 8722 (1989).
[298] M. G. Vavilov and L. I. Glazman, Conductance of mesoscopic systems with magnetic impurities, *Phys. Rev. B*, **67**, 115310 (2003).
[299] A. A. Bobkov, V. I. Falko and D. E. Khmelnitskii, Mesoscopics in metals with magnetic impurities, *Sov. Phys. JETP*, **71**, 393 (1990).

Chapter 12

[300] J. W. Goodman, *Statistical Optics*, New York: Wiley, 1985.
[301] F. Scheffold, W. Hartl, G. Maret and E. Matijevic, Observation of long-range correlations in temporal intensity fluctuations of light, *Phys. Rev. B*, **56**, 10942 (1997).
[302] B. Shapiro, Large intensity fluctuations for wave propagation in random media, *Phys. Rev. Lett.*, **57**, 2168 (1986).

[303] I. Freund, M. Rosenbluh and S. Feng, Memory effects in propagation of optical waves through disordered media, *Phys. Rev. Lett.*, **61**, 2328 (1988).

[304] R. Berkovits and S. Feng, Correlations in coherent multiple scattering, *Phys. Rep.*, **238**, 135 (1994).

[305] S. Feng and P. A. Lee, Mesoscopic conductors and correlations in laser speckle patterns, *Science*, **251**, 633 (1991).

[306] S. Feng, C. Kane, P. A. Lee and A. D. Stone, Correlations and fluctuations of coherent wave transmission through disordered media, *Phys. Rev. Lett.*, **61**, 834 (1988).

[307] E. Akkermans and G. Montambaux, Mesoscopic physics of photons, *J. Opt. Soc. Am. B*, **21**, 101 (2004).

[308] M. C. V. van Rossum and Th. M. Nieuwenhuizen, Multiple scattering of classical waves: microscopy, mesoscopy and diffusion, *Rev. Mod. Phys.*, **71**, 313 (1999).

[309] R. Pnini, Correlation of speckle in random media, in *Waves and Imaging Through Complex Media*, ed. P. Sebbah, Dordrecht: Kluwer, 2001, p. 391.

[310] M. J. Stephen, Interference, fluctuations and correlations in the diffusive scattering from a disordered medium, in *Mesoscopic Phenomena in Solids*, eds. B. L. Altshuler, P. A. Lee and R. A. Webb, Amsterdam: North-Holland, 1991.

[311] P. A. Mello, E. A. Akkermans and B. Shapiro, Macroscopic approach to correlations in the electronic transmission and reflection from disordered conductors, *Phys. Rev. Lett.*, **61**, 459 (1988); E. Akkermans, *Universal fluctuations and long-range correlations for wave propagation in random media, Physica A*, **157**, 101 (1989).

[312] F. Scheffold and G. Maret, Universal conductance of light, *Phys. Rev. Lett.*, **81**, 5800 (1998).

[313] J. F. de Boer, M. P. van Albada and A. Lagendijk, Transmission and intensity correlations in wave propagation through random media, *Phys. Rev. B*, **45**, 658 (1992).

[314] B. L. Altshuler, V. E. Kravtsov and I. V. Lerner, Distribution of mesoscopic fluctuations and relaxation processes in disordered conductors, in *Mesoscopic Phenomena in Solids*, eds. B. L. Altshuler, P. A. Lee and R. A. Webb, Amsterdam: North-Holland, 1991; I. V. Lerner, Rigorous perturbation results in the theory of mesoscopic fluctuations: distribution functions and time-dependent phenomena, in *Quantum coherence in mesoscopic systems*, ed. B. Kramer, New York: Plenum, 1991.

[315] P. E. Wolf, G. Maret, E. Akkermans and R. Maynard, Optical coherent backscattering by random media: an experimental study, *J. Phys. (France)*, **49**, 63 (1988).

[316] J. F. de Boer, M. C. W. van Rossum, M. P. van Albada, T. M. Nieuwenhuizen and A. Lagendijk, Probability distribution of multiple scattered light measured in total transmission, *Phys. Rev. Lett.*, **73**, 2567 (1994).

[317] T. M. Nieuwenhuizen and M. C. W. van Rossum, Intensity distributions of waves transmitted through a multiple scattering medium, *Phys. Rev. Lett.*, **74**, 2674 (1995).

[318] E. Kogan and M. Kaveh, Random matrix theory approach to the intensity distributions of waves propagating in a random medium, *Phys. Rev. B*, **52**, R3813 (1995).

[319] S. A. van Langen, P. W. Brouwer and C. W. J. Beenakker, Non perturbative calculation of the probability distribution of plane wave transmission through a disordered waveguide, *Phys. Rev. B*, **53**, R1344 (1996).

[320] M. Stoytchev and A. Z. Genack, Measurement of the probability distribution of total transmission in random waveguides, *Phys. Rev. Lett.*, **79**, 309 (1997).
[321] A. Z. Genack, N. Garcia and W. Polkosnik, Long-range intensity correlation in random media, *Phys. Rev. Lett.*, **65**, 2129 (1990).
[322] P. Sebbah, R. Pnini and A. Z. Genack, Field and intensity correlation in random media, *Phys. Rev. E*, **62**, 7348 (2000).
[323] P. Sebbah, B. Hu, A. Z. Genack, R. Pnini and B. Shapiro, Spatial-field correlation: the building block of mesoscopic fluctuations, *Phys. Rev. Lett.*, **88**, 123901 (2002).
[324] M. J. Stephen and G. Cwilich, Intensity correlations and fluctuations of light scattered from a random medium, *Phys. Rev. Lett.*, **59**, 285 (1987).
[325] R. Pnini and B. Shapiro, Intensity correlation in absorbing random media, *Phys. Lett. A*, **157**, 265 (1991).

Chapter 13

[326] N. Ashcroft and D. Mermin, *Solid State Physics*, Philadelphia, PA: Saunders College, 1976.
[327] B. L. Altshuler and A. G. Aronov, Electron–electron interaction in disordered conductors, in *Electron–electron Interactions in Disordered Systems*, eds. A. L. Efros and M. Pollak, Amsterdam: Elsevier, 1985, p. 1.
[328] A. L. Fetter and J. D. Walecka, *Quantum Theory of Many-particle Systems*, New York: McGraw-Hill, 1971; S. Doniach and E. H. Sondheimer, *Green's Functions for Solid State Physicists*, New york: W. A. Benjamin, 1974.
[329] E. Abrahams, P. W. Anderson, P. A. Lee and T. V. Ramakrishnan, Quasiparticle lifetime in disordered two-dimensional metals, *Phys. Rev. B*, **24**, 6783 (1981).
[330] Y. Imry and Z. Ovadyahu, Density of states anomalies in a disordered conductor: a tunneling study, *Phys. Rev. Lett.*, **49**, 841 (1982).
[331] F. Pierre, H. Pothier, P. Joyez, N. O. Birge, D. Esteve and M. Devoret, Electrodynamic dip in the local density of states of a metallic wire, *Phys. Rev. Lett.*, **86**, 1590 (2001).
[332] A. A. Abrikosov, L. P. Gorkov and I. Y. Dzyaloshinskii, *Quantum Field Theoretical Methods in Statistical Physics*, Pergamon, 1965.
[333] A. M. Finkelshtein, Influence of Coulomb interaction on the properties of disordered metals, *Sov. Phys. JETP*, **57**, 97 (1983).
[334] A. Kamenev and A. Andreev, Electron–electron interactions in disordered metals: Keldysh formalism, *Phys. Rev. B*, **60**, 2218 (1999).
[335] F. Pierre, Interactions électron-électron dans les fils mésoscopiques, *Ann. Phys.*, **26** (4) (2001).
[336] B. L. Altshuler, A. G. Aronov and A. Yu. Zyuzin, Size effects in disorder conductors, *JETP*, **59**, 415 (1984).
[337] G. L. Ingold and Yu. Nararov, Charge tunneling rates in ultrasmall junctions, in *Single Charge Tunneling*, eds. H. Grabert and M. Devoret, New York: Plenum, 1982.
[338] Yu. Nazarov, Anomalous current-voltage characteristics of tunnel junctions, *JETP*, **68**, 561 (1990).
[339] B. L. Altshuler, A. G. Aronov and P. A. Lee, Interaction effects in disordered Fermi systems in two dimensions, *Phys. Rev. Lett.*, **44**, 1288 (1980).
[340] P. Schwab and R. Raimondi, Quasiclassical theory of charge transport in disordered interacting electron systems, *Ann. Phys.*, **12**, 471 (2003).

[341] A. I. Larkin, Reluctance of two-dimensional systems, *JETP Lett.*, **31**, 219 (1980).
[342] M. Gijs, C. Van Haesendonck and Y. Bruynseraede, Quantum oscillations in the superconducting fluctuation regime of cylindrical Al films, *Phys. Rev. B*, **30**, 2964 (1984).
[343] A. G. Aronov and Yu. V. Sharvin, Magnetic flux effects in disordered conductors, *Rev. Mod. Phys.*, **59**, 755 (1987).
[344] D. Pines and P. Nozières, *The Theory of Quantum Liquids*, Vol.1, New York: Addison-Wesley, 1989.
[345] U. Sivan, Y. Imry and A. G. Aronov, Quasi-particle lifetime in a quantum dot, *Europhys. Lett.*, **28**, 115 (1994).
[346] F. Pierre, H. Pothier, D. Esteve and M. H. Devoret, Energy redistribution between quasiparticles in mesoscopic silver wires, *J. Low Temp. Phys.*, **118**, 437 (2000).
[347] A. Kaminski and L. I. Glazman, Electron energy relaxation in the presence of magnetic impurities, *Phys. Rev. Lett.*, **86**, 2400 (2001); M. G. Vavilov, A. Kaminski and L. I. Glazman, *Electron energy and phase relaxation on magnetic impurities, Proceedings of LT23 Conference*, 2002.
[348] A. Schmid, On the dynamics of electrons in an impure metal, *Z. Phys.*, **271**, 251 (1974).
[349] W. Eiler, Electron–electron interaction and weak-localization, *J. Low Temp. Phys.*, **56**, 481 (1984).
[350] G. Montambaux and E. Akkermans, Non exponential quasiparticle decay and phase relaxation in low dimensional conductors, *Phys. Rev. Lett.*, **95**, 016403 (2005).
[351] B. L. Altshuler, A. G. Aronov and D. E. Khmelnitskii, Effects of electron–electron collisions with small energy transfers on quantum localization, *J. Phys. C*, **15**, 7367 (1982).
[352] J. J. Lin and J. P. Bird, Recent experimental studies of electron dephasing in metal and semiconductor structures, *J. Phys.: Condens. Matter*, **14**, R501, (2002).
[353] A. Stern, Y. Aharonov and Y. Imry, Phase uncertainty and loss of interference: a general picture, *Phys. Rev. A*, **41**, 3436 (1990).
[354] B. N. Narozhny, G. Zala and I. L. Aleiner, Interaction corrections at intermediate temperatures: dephasing time, *Phys. Rev. B*, **65**, 180202 (2002).
[355] R. Kubo, The fluctuation–dissipation theorem, *Rep. Prog. Phys.*, **29**, 255 (1966).
[356] E. M. Lifshitz and L. P. Pitaevskii, *Statistical Physics*, Vol. 2, Pergamon, 1980.
[357] Y. Imry, *Introduction to Mesoscopic Physics*, 2nd edition, Oxford: Oxford University Press, 2002.
[358] H. Nyquist, Thermal agitation of electric charge in conductors, *Phys. Rev.*, **32**, 110 (1928).
[359] H. R. Callen and T. A. Welton, Irreversibility and generalized noise, *Phys. Rev. B*, **83**, 34 (1951).
[360] M. Abramowitz and I. Stegun eds., *Handbook of Mathematical Functions*, New York: Dover, 1970.
[361] F. Pierre, A. B. Gougam, A. Anthore, H. Pothier, D. Esteve and N. Birge, Dephasing of electrons in mesoscopic metal wires, *Phys. Rev. B*, **68**, 85413 (2003).
[362] T. Ludwig and A. Mirlin, Interaction-induced dephasing of Aharonov–Bohm oscillations, *Phys. Rev. B*, **69**, 193306 (2004).
[363] C. Texier and G. Montambaux, Dephasing due to electron–electron interaction in a diffusive ring, *Phys. Rev. B*, **72**, 115327 (2005).

[364] D. S. Golubev and A. D. Zaikin, Quantum decoherence in disordered mesoscopic systems, *Phys. Rev. Lett.*, **81**, 1074 (1998).
[365] I. Aleiner, B. L. Altshuler and M. E. Gershenson, Interaction effects and phase relaxation in disordered systems, *Waves Random Media*, **9**, 201 (1999).
[366] P. Mohanty, E. M. Q. Jariwala and R. A. Webb, Intrinsic decoherence in mesoscopic systems, *Phys. Rev. Lett.*, **78**, 3366 (1997).
[367] P. M. Echternach, M. E. Gershenson, H. M. Bozler, A. L. Bogdanov and B. Nilsson, Nyquist phase relaxation in one-dimensional metal films, *Phys. Rev. B* **48**, 11516 (1993).
[368] B. Huard, A. Anthore, N. O. Birge, H. Pothier and D. Esteve, Effect of magnetic impurities on energy exchange between electrons, *Phys. Rev. Lett.*, **95**, 036802 (2005).
[369] F. Schopfer, C. Bäuerle, W. Rabaud and L. Saminadayar, Anomalous temperature dependence of the dephasing time in mesoscopic Kondo wires, *Phys. Rev. Lett.*, **90**, 56801 (2003).

Chapter 14

[370] D. Shoenberg, *Magnetic Oscillations in Metals*, Cambridge: Cambridge University Press, 1984.
[371] T. Champel, Chemical potential oscillations and de Haas–van Alphen effect, *Phys. Rev. B*, **64**, 54407 (2001).
[372] R. B. Dingle, Some magnetic properties of metals, *Proc. R. Soc. London, ser. A*, **211**, 517 (1952).
[373] A. Szafer and B. L. Altshuler, Universal correlation in the spectra of disordered systems with an Aharonov–Bohm flux, *Phys. Rev. Lett.*, **70**, 587 (1993).
[374] H. Fukuyama, Fluctuations of the Landau diamagnetism in mesoscopic systems, *J. Proc. Soc. Jpn.*, **58**, 47 (1989).
[375] E. Akkermans and B. Shapiro, Fluctuations in the diamagnetic response of disordered metals, *Europhys. Lett.*, **11**, 467 (1990).
[376] A. Raveh and B. Shapiro, Fluctuations in the orbital magnetic response of mesoscopic conductors, *Europhys. Lett.*, **19**, 109 (1992).
[377] N. Ashcroft and D. Mermin, *Solid State Physics*, Philadelphia, PA: Saunders College, 1976.
[378] B. L. Altshuler, A. G. Aronov and A. Y. Zyuzin, Thermodynamic properties of disordered conductors, *Sov. Phys. JETP*, **57**, 889 (1983).
[379] A. Altshuler and A. Aronov, in *Electron–Electron Interactions in Disordered Systems*, ed. A. Efros and M. Pollack, Amsterdam: North-Holland, 1985.
[380] U. Eckern, Coherence and destruction of coherence in mesoscopic rings, *Z. Phys. B*, **82**, 393 (1991).
[381] V. Ambegaokar and U. Eckern, Coherence and persistent current in mesoscopic rings, *Phys. Rev. Lett.*, **65**, 381 (1990); **67**, 3192 (1991).
[382] A. Schmid, Persistent currents in mesoscopic rings by suppression of charge fluctuations, *Phys. Rev. Lett.*, **66**, 80 (1991).
[383] F. Hund, Rechnungen über das magnetische verhalten von kleinen metallstücken bei tiefen temperaturen, *Ann. Phys. (Leipzig)*, **32**, 102 (1938).
[384] F. Bloch, Off-diagonal long-range order and persistent currents in a hollow cylinder, *Phys. Rev. A*, **137**, 787 (1965).

[385] I. O. Kulik, Magnetic flux quantization in the normal state, *Sov. Phys. JETP*, **31**, 1172 (1970).
[386] M. Büttiker, Y. Imry and R. Landauer, Josephson behavior in small normal one-dimensional rings, *Phys. Lett. A*, **96**, 365 (1983).
[387] H. F. Cheung, Y. Gefen, E. K. Riedel and W.-H. Shih, Persistent currents in small one-dimensional metallic rings, *Phys. Rev. B*, **37**, 6050 (1988).
[388] O. Entin-Wohlman and Y. Gefen, Persistent currents in two-dimensional metallic cylinders, *Europhys. Lett.*, **8**, 477 (1989).
[389] H. F. Cheung, Y. Gefen and E. K. Riedel, Isolated rings of mesoscopic dimensions. Quantum coherence and persistent currents, *IBM J. Res. Develop.*, **32**, 359 (1988).
[390] H. Bouchiat and G. Montambaux, Persistent currents in mesoscopic rings: ensemble averages and half-flux-quantum periodicity, *J. Phys. (Paris)*, **50**, 2695 (1989).
[391] H. F. Cheung, E. K. Riedel and Y. Gefen, Persistent currents in mesoscopic rings and cylinders, *Phys. Rev. Lett.*, **62**, 587 (1989).
[392] I. Gradshteyn and I. Ryzhik, *Table of Integrals, Series and Products*, London: Academic Press, 1980.
[393] L. P. Lévy, G. Dolan, J. Dunsmuir and H. Bouchiat, Persistent currents in mesoscopic copper rings, *Phys. Rev. Lett.*, **64**, 2074 (1990).
[394] B. Reulet, M. Ramin, H. Bouchiat and D. Mailly, Dynamic response of isolated Aharonov–Bohm rings coupled to an electromagnetic resonator, *Phys. Rev. Lett.*, **75**, 124 (1995).
[395] R. Deblock, R. Bel, B. Reulet, H. Bouchiat and D. Mailly, Diamagnetic orbital response of mesoscopic silver rings, *Phys. Rev. Lett.*, **89**, 206803 (2002).
[396] V. Chandrasekhar, R. A. Webb, M. J. Brady, M. B. Ketchen, W.J. Gallaghem and A. Kleinsasser, Magnetic response of a single, isolated gold loop, *Phys. Rev. Lett.*, **67**, 3578 (1991).
[397] E. M. Q. Jariwala, P. Mohanty, M. B. Ketchen and R. A. Webb, Diamagnetic persistent current in diffusive normal-metal rings, *Phys. Rev. Lett.*, **86**, 1594 (2001).
[398] D. Mailly, C. Chapelier and A. Benoit, Experimental observation of persistent current in GaAs–AlGaAs single loop, *Phys. Rev. Lett.*, **70**, 2020 (1993).
[399] B. L. Altshuler, Y. Gefen and Y. Imry, Persistent differences between canonical and grand canonical averages in mesoscopic ensembles: large paramagnetic orbital susceptibilities, *Phys. Rev. Lett.*, **66**, 88 (1991).
[400] F. von Oppen and E. Riedel, Average persistent current in a mesoscopic ring, *Phys. Rev. Lett.*, **66**, 84 (1991).
[401] E. Akkermans, Scattering phase shift analysis of persistent currents in mesoscopic Aharonov–Bohm geometries, *Europhys. Lett.*, **15**, 709 (1991).
[402] S. Oh, A. Yu. Zyuzin and R. A. Serota, Orbital magnetism of mesoscopic systems, *Phys. Rev. B*, **44**, 8858 (1991).

Chapter 15

[403] A. Messiah, *Quantum Mechanics*, New York: Dover, 1999.
[404] A. R. Edmonds, *Angular Momentum in Quantum Mechanics*, Princeton, NJ: Princeton University Press, 1960.

Index

CPSIA information can be obtained
at www.ICGtesting.com
Printed in the USA
BVHW011948120220
572196BV00007B/84

9 780521 349475